Eco Polymeric Materials and Natural Polymer

Eco Polymeric Materials and Natural Polymer

Editors

Jingpeng Li
Yun Lu
Huiqing Wang

Basel • Beijing • Wuhan • Barcelona • Belgrade • Novi Sad • Cluj • Manchester

Editors

Jingpeng Li
Laboratory of Efficient
Utilization of Bamboo and
Wood
China National Bamboo
Research Center
Hangzhou
China

Yun Lu
Research Institute of Wood
Industry
Chinese Academy of Forestry
Beijing
China

Huiqing Wang
Department of Polymer
Science and Engineering
Hefei University of
Technology
Hefei
China

Editorial Office
MDPI
St. Alban-Anlage 66
4052 Basel, Switzerland

This is a reprint of articles from the Special Issue published online in the open access journal *Polymers* (ISSN 2073-4360) (available at: www.mdpi.com/journal/polymers/special_issues/Eco_Polymeric_Materials_Natural_Polymer).

For citation purposes, cite each article independently as indicated on the article page online and as indicated below:

Lastname, A.A.; Lastname, B.B. Article Title. *Journal Name* **Year**, *Volume Number*, Page Range.

ISBN 978-3-0365-9149-0 (Hbk)
ISBN 978-3-0365-9148-3 (PDF)
doi.org/10.3390/books978-3-0365-9148-3

© 2023 by the authors. Articles in this book are Open Access and distributed under the Creative Commons Attribution (CC BY) license. The book as a whole is distributed by MDPI under the terms and conditions of the Creative Commons Attribution-NonCommercial-NoDerivs (CC BY-NC-ND) license.

Contents

About the Editors ix

Preface xi

Jingpeng Li, Yun Lu and Huiqing Wang
Eco Polymeric Materials and Natural Polymer
Reprinted from: *Polymers* **2023**, *15*, 4021, doi:10.3390/polym15194021 1

Bingchao Duan, Minghui Yang, Quanchao Chao, Lan Wang, Lingli Zhang and Mengxing Gou et al.
Preparation and Properties of Egg White Dual Cross-Linked Hydrogel with Potential Application for Bone Tissue Engineering
Reprinted from: *Polymers* **2022**, *14*, 5116, doi:10.3390/polym14235116 11

Zhenbing Sun, Zhengjie Tang, Xiaoping Li, Xiaobao Li, Jeffrey J. Morrell and Johnny Beaugrand et al.
The Improved Properties of Carboxymethyl Bacterial Cellulose Films with Thickening and Plasticizing
Reprinted from: *Polymers* **2022**, *14*, 3286, doi:10.3390/polym14163286 24

Jihang Hu, Zongying Fu, Xiaoqing Wang and Yubo Chai
Manufacturing and Characterization of Modified Wood with In Situ Polymerization and Cross-Linking of Water-Soluble Monomers on Wood Cell Walls
Reprinted from: *Polymers* **2022**, *14*, 3299, doi:10.3390/polym14163299 38

Xuelian Li, Weizhong Zhang, Jingpeng Li, Xiaoyan Li, Neng Li and Zhenhua Zhang et al.
Optically Transparent Bamboo: Preparation, Properties, and Applications
Reprinted from: *Polymers* **2022**, *14*, 3234, doi:10.3390/polym14163234 51

Aiyue Huang, Qin Su, Yurong Zong, Xiaohan Chen and Huanrong Liu
Study on Different Shear Performance of Moso Bamboo in Four Test Methods
Reprinted from: *Polymers* **2022**, *14*, 2649, doi:10.3390/polym14132649 69

Huiling Yu, Chengsheng Gui, Yaohui Ji, Xiaoyan Li, Fei Rao and Weiwei Huan et al.
Changes in Chemical and Thermal Properties of Bamboo after Delignification Treatment
Reprinted from: *Polymers* **2022**, *14*, 2573, doi:10.3390/polym14132573 78

Shaoli Wang, Shengju Song, Xuping Yang, Zhengqi Xiong, Chaoxing Luo and Yongxiu Xia et al.
Effect of Preparation Conditions on Application Properties of Environment Friendly Polymer Soil Consolidation Agent
Reprinted from: *Polymers* **2022**, *14*, 2122, doi:10.3390/polym14102122 89

Yaohui Ji, Wencheng Lei, Yuxiang Huang, Jiangyuan Wu and Wenji Yu
Influence of Resin Content and Density on Water Resistance of Bamboo Scrimber Composite from a Bonding Interface Structure Perspective
Reprinted from: *Polymers* **2022**, *14*, 1856, doi:10.3390/polym14091856 103

Wenxuan Wang, Feihan Yu, Zhichen Ba, Hongbo Qian, Shuai Zhao and Jie Liu et al.
In-Depth Sulfhydryl-Modified Cellulose Fibers for Efficient and Rapid Adsorption of Cr(VI)
Reprinted from: *Polymers* **2022**, *14*, 1482, doi:10.3390/polym14071482 119

Tong Tang, Benhua Fei, Wei Song, Na Su and Fengbo Sun
Tung Oil Thermal Treatment Improves the Visual Effects of Moso Bamboo Materials
Reprinted from: *Polymers* **2022**, *14*, 1250, doi:10.3390/polym14061250 129

Hongyi Wu, Nitong Bu, Jie Chen, Yuanyuan Chen, Runzhi Sun and Chunhua Wu et al.
Construction of Konjac Glucomannan/Oxidized Hyaluronic Acid Hydrogels for Controlled Drug Release
Reprinted from: *Polymers* **2022**, *14*, 927, doi:10.3390/polym14050927 142

Zongying Fu, Hui Wang, Jingpeng Li and Yun Lu
Determination of Moisture Content and Shrinkage Strain during Wood Water Loss with Electrochemical Method
Reprinted from: *Polymers* **2022**, *14*, 778, doi:10.3390/polym14040778 155

Bo Liu, Lina Tang, Qian Chen, Liming Zhu, Xianwu Zou and Botao Li et al.
Lignin Distribution on Cell Wall Micro-Morphological Regions of Fibre in Developmental *Phyllostachys pubescens* Culms
Reprinted from: *Polymers* **2022**, *14*, 312, doi:10.3390/polym14020312 165

Zixuan Yu, Xiaofeng Zhang, Rong Zhang, Yan Yu and Fengbo Sun
Improving the Mould and Blue-Stain-Resistance of Bamboo through Acidic Hydrolysis
Reprinted from: *Polymers* **2022**, *14*, 244, doi:10.3390/polym14020244 177

Shuai Zhang, Yu Wan, Weijie Yuan, Yaoxiang Zhang, Ziyuan Zhou and Min Zhang et al.
Preparation of PVA–CS/SA–Ca^{2+} Hydrogel with Core–Shell Structure
Reprinted from: *Polymers* **2022**, *14*, 212, doi:10.3390/polym14010212 189

Kexia Jin, Zhe Ling, Zhi Jin, Jianfeng Ma, Shumin Yang and Xinge Liu et al.
Local Variations in Carbohydrates and Matrix Lignin in Mechanically Graded Bamboo Culms
Reprinted from: *Polymers* **2021**, *14*, 143, doi:10.3390/polym14010143 202

Edson Antonio dos Santos Filho, Carlos Bruno Barreto Luna, Danilo Diniz Siqueira, Eduardo da Silva Barbosa Ferreira and Edcleide Maria Araújo
Tailoring Poly(lactic acid) (PLA) Properties: Effect of the Impact Modifiers EE-g-GMA and POE-g-GMA
Reprinted from: *Polymers* **2021**, *14*, 136, doi:10.3390/polym14010136 213

Khurram Shahzad, Mohammad Rehan, Muhammad Imtiaz Rashid, Nadeem Ali, Ahmed Saleh Summan and Iqbal Muhammad Ibrahim Ismail
Sustainability Evaluation of Polyhydroxyalkanoate Production from Slaughterhouse Residues Utilising Emergy Accounting
Reprinted from: *Polymers* **2021**, *14*, 118, doi:10.3390/polym14010118 228

Sixuan Wei, Rujie Peng, Shilong Bian, Wei Han, Biao Xiao and Xianghong Peng
Facile and Scalable Synthesis and Self-Assembly of Chitosan Tartaric Sodium
Reprinted from: *Polymers* **2021**, *14*, 69, doi:10.3390/polym14010069 249

Yao Yao, Zhenbing Sun, Xiaobao Li, Zhengjie Tang, Xiaoping Li and Jeffrey J. Morrell et al.
Effects of Raw Material Source on the Properties of CMC Composite Films
Reprinted from: *Polymers* **2021**, *14*, 32, doi:10.3390/polym14010032 259

Qi Ye, Yingchun Gong, Haiqing Ren, Cheng Guan, Guofang Wu and Xu Chen
Analysis and Calculation of Stability Coefficients of Cross-Laminated Timber Axial Compression Member
Reprinted from: *Polymers* **2021**, *13*, 4267, doi:10.3390/polym13234267 274

Ziling Shen, Zhi Ye, Kailin Li and Chusheng Qi
Effects of Coupling Agent and Thermoplastic on the Interfacial Bond Strength and the Mechanical Properties of Oriented Wood Strand–Thermoplastic Composites
Reprinted from: *Polymers* **2021**, *13*, 4260, doi:10.3390/polym13234260 288

Xueyu Wang, Yong Zhong, Xiangya Luo and Haiqing Ren
Compressive Failure Mechanism of Structural Bamboo Scrimber
Reprinted from: *Polymers* **2021**, *13*, 4223, doi:10.3390/polym13234223 299

Phattarawadee Nun-Anan, Sunisa Suchat, Narissara Mahathaninwong, Narong Chueangchayaphan, Seppo Karrila and Suphatchakorn Limhengha
Study of *Aquilaria crassna* Wood as an Antifungal Additive to Improve the Properties of Natural Rubber as Air-Dried Sheets
Reprinted from: *Polymers* **2021**, *13*, 4178, doi:10.3390/polym13234178 311

Luthfi Hakim, Ragil Widyorini, Widyanto Dwi Nugroho and Tibertius Agus Prayitno
Performance of Citric Acid-Bonded Oriented Board from Modified Fibrovascular Bundle of Salacca (*Salacca zalacca* (Gaertn.) Voss) Frond
Reprinted from: *Polymers* **2021**, *13*, 4090, doi:10.3390/polym13234090 328

Chusheng Qi, Jinyue Wang and Vikram Yadama
Heat Transfer Modeling of Oriented Sorghum Fibers Reinforced High-Density Polyethylene Film Composites during Hot-Pressing
Reprinted from: *Polymers* **2021**, *13*, 3631, doi:10.3390/polym13213631 344

Lulu Liang, Yu Zheng, Yitian Wu, Jin Yang, Jiajie Wang and Yingjie Tao et al.
Surfactant-Induced Reconfiguration of Urea-Formaldehyde Resins Enables Improved Surface Properties and Gluability of Bamboo
Reprinted from: *Polymers* **2021**, *13*, 3542, doi:10.3390/polym13203542 358

Na Su, Changhua Fang, Hui Zhou, Tong Tang, Shuqin Zhang and Xiaohuan Wang et al.
Effect of Rosin Modification on the Visual Characteristics of Round Bamboo Culm
Reprinted from: *Polymers* **2021**, *13*, 3500, doi:10.3390/polym13203500 373

Rui Peng, Jingjing Zhang, Chungui Du, Qi Li, Ailian Hu and Chunlin Liu et al.
Investigation of the Release Mechanism and Mould Resistance of Citral-Loaded Bamboo Strips
Reprinted from: *Polymers* **2021**, *13*, 3314, doi:10.3390/polym13193314 387

Minzhen Bao, Neng Li, Yongjie Bao, Jingpeng Li, Hao Zhong and Yuhe Chen et al.
Outdoor Wood Mats-Based Engineering Composite: Influence of Process Parameters on Decay Resistance against Wood-Degrading Fungi *Trametes versicolor* and *Gloeophyllum trabeum*
Reprinted from: *Polymers* **2021**, *13*, 3173, doi:10.3390/polym13183173 402

About the Editors

Jingpeng Li

Jingpeng Li achieved his Ph.D. degree at the Chinese Academy of Forestry. He is currently an Associate Professor at the Laboratory of Efficient Utilization of Bamboo and Wood, China National Bamboo Research Center. His current work focuses on the functionalization of natural polymers for advanced applications. He is also interested in the use of natural materials in the manufacturing of bioplastics. He has published above 50 SCI-indexed journal articles and has gained an h-index of 21, according to the citation report from Google Scholar.

Yun Lu

Yun Lu, a distinguished recipient of the National Natural Science Foundation's Outstanding Youth Project, currently holds a post as a professor at the Research Institute of Wood Industry at the Chinese Academy of Forestry. She earned her Ph.D. in 2014 from Northeast Forestry University. Her research interests prominently encompass the fields of wood supramolecular science, wood cell wall engineering and bionic intelligent technology. She has published more than 70 articles in SCI journals such as ACS Energy Letters, Applied Catalysis B-Environmental. As the primary inventor, she has secured 12 national invention patents and earned 8 research grants. Professional affiliations include her membership in the Chinese Society of Forestry and the Chinese Papermaking Society. She also serves as a young editorial board member for the Journal of Forestry Engineering, Chinese Journal of Wood Science and Technology, Journal of Central South University of Forestry & Technology, the editorial board member of Scientific Reports, and the guest editor of Polymers and Journal of Renewable Materials. Her accolades include receiving the first prize for the Liang Xi Science and Technology Award in 2021, the Mao Yisheng Science and Technology Award for Wood Science in 2021, and the Youth Science and Technology Award of the National Forestry and Grassland Administration in 2022.

Huiqing Wang

Huiqing Wang graduated from the Department of Materials Science and Engineering, Beijing Institute of Technology, China, in 2013. Her postdoc was conducted at Wuhan University in the laboratory directed by Prof. Lina Zhang. She is currently an Associate Professor of Polymer Science in the School of Chemistry and Chemical Engineering at Hefei University of Technology, China. Her research group focuses on advanced functional cellulose/chitin materials, including new methods and novel structures of cellulose/chitin nanomaterials; the construction of fluorescent cellulose aerogels; high-strength antibacterial membranes; antibioadhesive coatings; porous carriers for drugs or enzymes; cellulose-induced chirality promotion; and nanocellulose for cell binding.

Preface

With increasing concern regarding the undesirable environmental and socioeconomic consequences of petrochemicals and limited fossil resources, biomass, bio-based polymers, and other renewable natural resources have increasingly become alternatives for the production of functional materials. Natural biomass, such as wood, bamboo, rattan, cellulose, bacterial cellulose, lignin, hemicellulose, chitin, alginate, silk, fibroin, starch, protein, collagen, gelatin, natural rubber, and their modified derivatives/composites, has been widely consumed for the preparation of bioplastics/biorubber in the form of film/member/hydrogel/foam/aerogels/fibers for various applications. Biobased synthetic polymers such as polyester, PLA, PHA, PBAT, PC, PBS, polyurethane, and so on can be derived from a variety of molecular biomasses such as straw glucose, plant oils, fatty acids, furan, terpenes, rosin acids, and amino acids. The use of such environmentally friendly or "green" polymer materials can avoid dependence on petroleum resources and reduce carbon emissions. Additionally, green solvent/process/technology for polymers and polymers for capturing pollution also contribute to the aim of global green and low-carbon transformation.

The presented reprint contains 30 high-quality original research and review papers by 202 authors from various research centers, including China, the USA, Australia, France, Brazil, Saudi Arabia, Thailand, and Indonesia. These papers were published in a Special Issue, "Eco Polymeric Materials and Natural Polymers", of the journal *Polymers*. These papers provide examples of the most recent developments in eco-polymeric materials and natural polymers. The Guest Editors would like to thank all authors who contributed to this Special Issue. The Guest Editors would also like to thank Special Issue Editor Jenny Hu for her overall professional attitude and kind assistance with the publications.

Jingpeng Li, Yun Lu, and Huiqing Wang
Editors

Editorial
Eco Polymeric Materials and Natural Polymer

Jingpeng Li [1,*], Yun Lu [2,*] and Huiqing Wang [3]

1. Key Laboratory of High Efficient Processing of Bamboo of Zhejiang Province, China National Bamboo Research Center, Hangzhou 310012, China
2. Research Institute of Wood Industry, Chinese Academy of Forestry, Beijing 100091, China
3. Department of Polymer Science and Engineering, School of Chemical Engineering, Hefei University of Technology, Hefei 230009, China; huiqing.wang@hfut.edu.cn
* Correspondence: lijp@caf.ac.cn (J.L.); y.lu@caf.ac.cn (Y.L.)

Citation: Li, J.; Lu, Y.; Wang, H. Eco Polymeric Materials and Natural Polymer. *Polymers* 2023, *15*, 4021. https://doi.org/10.3390/polym15194021

Received: 28 September 2023
Revised: 29 September 2023
Accepted: 30 September 2023
Published: 8 October 2023

Copyright: © 2023 by the authors. Licensee MDPI, Basel, Switzerland. This article is an open access article distributed under the terms and conditions of the Creative Commons Attribution (CC BY) license (https://creativecommons.org/licenses/by/4.0/).

With the increasing concern regarding the undesirable environmental and socioeconomic consequences of petrochemicals and limited fossil resources, biomass, bio-based polymers, and other renewable natural resources have increasingly become alternatives for the production of functional materials [1–3]. Natural biomasses, such as wood, bamboo, rattan, cellulose, bacterial cellulose, lignin, hemicellulose, chitin, alginate, silk, fibroin, starch, protein, collagen, gelatin, natural rubber, and their modified derivatives/composites, have been widely consumed for the preparation of bioplastics/biorubber in the form of films/members/hydrogels/foams/aerogels/fibers for various applications [4–6]. Biobased synthetic polymers such as polyester, poly(lactic acid) (PLA), polyhydroxyalkanoate (PHA), poly (butylene adipate-co-terephthalate) (PBAT), polycarbonate (PC), poly(butylene succinate) (PBS), polyurethane, and so on can be derived from a variety of molecular biomasses such as straw glucose, plant oils, fatty acids, furan, terpenes, rosin acids, and amino acids [7]. The use of such environmentally friendly or "green" polymer materials can avoid the dependence on petroleum resources and reduce carbon emissions [8,9]. Additionally, the green solvents, processes and technologies for polymers and the use of polymers for capturing pollution also contribute to the aim of global green and low-carbon transformation.

Nevertheless, eco polymeric materials and natural polymers face new challenges and problems every day. This Special Issue brings together different research works and reviews and attempts to cover the majority of the recent advances and applications of eco polymeric materials and natural polymers in the last few years.

This Special Issue gathers scientific works from research groups examining various eco polymeric materials, indicating advances in structural features, functions, and applications. The total number of manuscripts (30) published in this Special Issue indicates the importance of eco polymeric materials and natural polymers and the fact that many research groups and relevant members of the scientific community are thoroughly interested in the advancement of eco-friendly polymers and their advanced applications.

In their paper, Duan et al. (contribution 1) prepared an egg white dual cross-linked hydrogel through the induction of sodium hydroxide and the secondary cross-linking of protein chains by calcium ions. Characteristics of the dual cross-linked hydrogel were remarkably affected by the concentrations of calcium ions. The incorporation of calcium ions could benefit the thermal stability, swelling rate and texture of the hydrogels, while also reducing their swelling capacity. Calcium ions could impact the secondary structure of polypeptide chains and interact with protein chains, leading to more compact microstructure formation of the hydrogels. The results suggested that the egg white dual cross-linked hydrogels exhibited biocompatibility and cell-surface adhesion in vitro, indicating the potential for biomedical application.

In the study by Wu et al. (contribution 2), a stable composite hydrogel was prepared by incorporating konjac glucomannan (KGM) with oxidized hyaluronic acid (OHA), after which alkali processing and thermal treatment were conducted. The obtained hydrogel was pale yellow, smooth in surface, and had a favorable swelling capacity, which met the essential requirements for ideal drug-delivery applications. The OHA played an effective role in adjusting the swelling ratio and increasing the biodegradation rate. Furthermore, both the encapsulation efficiency of epigallocatechin gallate (EGCG) and the release properties of the hydrogels were significantly raised with the presence of OHA. The overall results suggest that the KGM/OHA hydrogel, loaded with EGCG, exhibited potential applications in controlled release.

In another study, Zhang et al. (contribution 3) prepared poly(vinyl alcohol)–chitosan/sodium alginate–Ca^{2+} (PVA–CS/SA–Ca^{2+}) core–shell hydrogels with a bilayer space by cross-linking PVA and CS to form a core structure and chelating SA and Ca^{2+} to form a shell structure to achieve multiple substance loading and multifunctional expression. The SA concentration and SA/Ca^{2+} cross-linking time show a positive correlation with the thickness of the shell structure; the PVA/CS mass ratio affects the structural characteristics of the core structure; and a higher CS content indicates the more obvious three-dimensional network structure of the hydrogel. Their optimal experimental conditions for the swelling degree of the core–shell hydrogel included an SA concentration of 5%; an SA/Ca^{2+} cross-linking time of 90 min; a PVA/CS mass ratio of 1:0.7; and a maximum swelling degree of 50 g/g.

Another piece of research, conducted by Wei et al. (contribution 4), presented a facile and scalable method to produce a mass of chitosan tartaric ester via solvent-evaporation causing crystallization, in which tartaric acid was used as the crystallization and the crosslinking agent. In their article, chitosan tartaric sodium was prepared via hydrolysis with NaOH aqueous solution. As a result, the acquired nanostructured chitosan tartaric sodium, which is dispersed in an aqueous solution 20–50 nm in length and 10–15 nm in width, shows both the features of carboxyl and amino functional groups. Moreover, morphology regulation of the chitosan tartaric sodium nanostructures can be easily achieved by adjusting the solvent evaporation temperature. This work proves that this is a simple route to prepare chitosan-based nanostructure patterns.

A study by Filho et al. (contribution 5) found that the properties of PLA can be tailored by adding small concentrations of ethylene elastomeric grafted with glycidyl methacrylate (EE-g-GMA) and poly(ethylene-octene) grafted with glycidyl methacrylate (POE-g-GMA), generating promising eco-friendly materials. The blends PLA/EE-g-GMA and PLA/POE-g-GMA showed better impact properties and thermal stability compared to pure PLA. The increase in crystallinity contributed to maintaining the thermomechanical strength, Shore D hardness, and shifting the thermal stability of the PLA/EE-g-GMA and PLA/POE-g-GMA blends to a higher temperature. The obtained results suggested a good interaction between PLA and the EE-g-GMA and POE-g-GMA systems, due to the glycidyl methacrylate functional group. In light of this, new environmentally friendly and semi-biodegradable materials can be manufactured for application in the packaging industry.

In their paper, Sun et al. (contribution 6) prepared a carboxymethyl bacterial cellulose-based composite film with good thermal stability and mechanical properties. For the composite films with the addition of 1.5% carboxymethyl bacterial cellulose (% v/v), 1% sodium alginate, and 0.4% glycerin, the tensile strength was 38.13 MPa, the elongation at break was 13.4%, the kinematic viscosity of the film solution was 257.3 mm^2/s, the opacity was 4.76 A/mm, the water vapor permeability was 11.85%, and the pyrolysis residue was 45%. Regression analysis of the data on mechanical properties yielded a significant correlation between thickeners and plasticizers regarding the tensile strength and elongation at break of the composite films.

In order to improve the survival rate of transplanted seedlings and improve the efficiency of seedling transplantation, Wang et al. (contribution 7) developed an environmentally friendly polymer konjac glucomannan (KGM)/chitosan (CA)/poly(vinyl alcohol) (PVA) ternary blend soil consolidation agent to consolidate the soil ball at the root of transplanted seedlings. They found that the film-forming performance of the adhesive was better when the KGM content was 4.5%, the CA content was in the range of 2–3%, the PVA content was in the range of 3–4%, and the preparation temperature was higher than 50 °C.

The polymer soil consolidation agent prepared under this condition has good application prospects in seedling transplanting.

As one of the hazardous heavy metal ion pollutants, Cr(VI) has attracted much attention in the sewage treatment research field due to its broad distribution range and serious toxicity [10]. Wang et al. (contribution 8) developed a simple and effective strategy for preparing cellulose fibers with a stable 3D network structure using dissolution, regeneration, wet spinning, and freeze drying. Based on the rich pore structure of cellulose fibers, thioglycolic acid was used to deeply sulfhydryl-modify them to obtain sulfhydryl-modified cellulose fibers for efficient and rapid adsorption of Cr(VI). The maximum adsorption capacity of sulfhydryl-modified cellulose fibers to Cr(VI) can reach 120.60 mg g^{-1}, the adsorption equilibrium can be achieved within 300 s, and its adsorption rate can reach 0.319 mg g^{-1} s^{-1}. The in-depth sulfhydryl-modified cellulose fibers are also available for other heavy metal ions. The low cost and environmentally friendly properties of the as-synthesized material demonstrate its potential for practical usage for the treatment of heavy metal ion pollution in wastewater.

The paper by Yao et al. (contribution 9) investigated the effects of five plant sources on the resulting properties of sodium carboxymethyl cellulose (CMC) and CMC/sodium alginate/glycerol composite films. The degree of substitution and resulting tensile strength tended to be 20% lower in leaf-derived CMC compared to those prepared from wood or bamboo. Microstructures of bamboo cellulose, bamboo CMC powder, and bamboo leaf CMC composites' films all differed from pine-derived material, but plant sources had no noticeable effect on the X-ray diffraction characteristics, Fourier transform infrared spectroscopy spectra, or pyrolysis properties of the CMC or composite films. The results highlighted the potential for using plant sources as a tool for varying CMC properties for specific applications.

In one study, Hakim et al. (contribution 10) examined the performance of citric acid-bonded orientation boards from a modified fibrovascular bundle salacca frond under NaOH + Na_2SO_3 treatment and the bonding mechanism between the modified fibrovascular bundle frond and citric acid. Their results found that the combination of 1% NaOH + 0.2% Na_2SO_3 treatment for 30 and 60 min immersion is successful in reducing the water absorption and thickness swelling of the orientation board. The findings of this study indicated that there is a reaction between the hydroxyl group in the modified fibrovascular bundle and the carboxyl group in citric acid.

A one-dimensional heat transfer model of natural fiber-reinforced thermoplastic composites during hot pressing was established by Qi et al. (contribution 11). The novelty of this study is that the apparent heat capacity of thermoplastics was first simulated and then coupled with the heat transfer model to simulate the temperature distribution of natural fiber-reinforced thermoplastics composites during hot pressing. Both the experimental and simulated data suggested that a higher temperature and/or a longer duration during the hot-pressing process should be used to fabricate oriented sorghum fiber-reinforced high-density polyethylene film composites as the high-density polyethylene content increases.

Another interesting paper by Shahzad et al. (contribution 12) evaluated the environmental burden of polyhydroxyalkanoate production from slaughtering residues by utilizing the Emergy Accounting methodology. The emergy intensity for polyhydroxyalkanoate production (seJ/g) shows a minor improvement ranging from 1.5% to 2% by changing only the electricity provision resources. This impact reaches up to 17% when electricity and heat provision resources are replaced with biomass resources. Similarly, the emergy intensity for polyhydroxyalkanoate production using electricity EU27 mix, coal, hydropower, wind power, and biomass is about 5% to 7% lower than the emergy intensity of polyethylene high density. In comparison, its value is up to 21% lower for electricity and heat provision from biomass.

Wood is a typical natural polymeric material. Wood drying is an essential step in wood processing, and it is also the most energy- and time-demanding step. Fu et al. (contribution 13) presented an electrochemical method to determine wood moisture content and shrinkage strain during drying. As the moisture content changed from 42% to 12%, the resistance increased from 1.0×10^7 Ω to 1.2×10^8 Ω. Both the shrinkage strain and resistance change rate increased with the decrease in wood moisture content, especially for the moisture content range of 23% to 8%, where the shrinkage strain and resistance change rate increased by 4% and 30%, respectively. This demonstrated the feasibility of the electrochemical method for measuring wood moisture content and shrinkage strain.

Wood modification can improve the dimensional stability, strength, and other properties of wood, and it has been extensively used. Hu et al. (contribution 14) improved the dimensional stability of wood via the in situ polymerization of water-soluble monomers in water. 2-Hydroxyethyl methacrylate and glyoxal were injected into the wood cell walls and activated cross-linking reactions to form interpenetrating polymer network structures. The polymer network blocked the partial pores and reduced wood hydroxyl, which simultaneously and significantly increased the wood's transverse connections, dimensions, and stability. This work advances fast-growing-wood modification by introducing a novel research strategy.

In another study, poplar veneer–thermoplastic composites and oriented strand–thermoplastic composites were fabricated by Shen et al. (contribution 15) using hot pressing. Their result found that the use of both KH550 and MDI as coupling agents improved the interfacial bond strength between wood and thermoplastics under dry conditions. The use of MDI resulted in a much greater increase in the interfacial bond strength than KH550 under both dry and wet conditions, while KH550 had a negative effect under wet conditions. The better interfacial bond strength between wood and thermoplastics provided oriented strand–thermoplastic composites with better mechanical properties and dimensional stability. The obtained results will guide the industry to produce high-performance wood–plastic composites using hot pressing for general applications.

During outdoor use, wood composites are susceptible to destruction by rot fungi. Bao et al. (contribution 16) investigated the effects of resin content and density on the resistance of outdoor wood mat-based engineering composites to fungal decay through fungal decay tests for a period of 12 weeks. The highest antifungal effects against *T. versicolor* (12.34% mass loss) and *G. trabeum* (19.43% mass loss) were observed at a density of 1.15 g/m^3 and a resin content of 13%. As a result of the chemical composition and microstructure measurements, the resistance of the outdoor wood mat-based engineering composite against *T. versicolor* and *G. trabeum* fungi was improved remarkably by increasing the density and resin content. The results of this study will provide a technical basis to improve the decay resistance of wood mat-based engineering composite in outdoor environments.

In the study by Ye et al. (contribution 17), the stability coefficient calculation theories in different national standards were analyzed and then the stability bearing capacity of cross-laminated timber elements with four slenderness ratios was investigated. Their results show that the average deviation between the fitting curve and calculated results of European and American standard was 5.43% and 3.73%, respectively, and the average deviation between the fitting curve and the actual test results was 8.15%. The stability coefficients calculation formulae could be used to reliably predict the stability coefficients of cross-laminated timber specimens with different slenderness ratios.

Another interesting paper by Nun-Anan et al. (contribution 18) investigated the effects of Aquilaria crassna wood (ACW) on the antifungal, physical and mechanical properties of natural rubber as air-dried sheets (ADS) and ADS filled with ACW. They found that the ACW-filled ADS had an increased Mooney viscosity, initial plasticity, and high thermo-oxidation plasticity (i.e., high plasticity retention index PRI). Additionally, superior green strength was observed for the ACW-filled ADS over the ADS without an additive because of chemical interactions between lignin and proteins in the natural rubber molecules eliciting greater gel formation. A significant inhibition of fungal growth on the natural rubber products during storage over a long period (5 months) was observed for ACW-filled ADS. The results suggested that these filled intermediate natural rubber products provide added value through an environmentally friendly approach, which is attractive to consumers.

Bamboo is a natural fiber-reinforced composite with excellent performance which is, to a certain extent, an alternative to the shortage of wood resources. In one study, the distribution of lignin components and lignin content in bamboo micro-morphological regions was measured by Liu et al. (contribution 19) at a semi-quantitative level according to age and radial location by means of visible-light microspectrophotometry coupled with the Wiesner and Maule reaction. They found that lignification develops with aging. Guaiacyl lignin units and syringyl lignin units were found in the cell wall of the fiber, parenchyma, and vessel. Differences in lignin content among different ages, different radial locations, and different micro-morphological regions of the cell wall were observed in this paper. It is considered that lignin plays an important role in cell-wall formation and the cell wall's mechanical properties. Lignin is related to the physical and mechanical properties of bamboo. Therefore, this study of the distribution of and change in lignin in bamboo development is conducive to the mastery and prediction of various properties in the bamboo development, and has guiding significance for bamboo and lignin industrial utilization.

In another study, combined microscopic techniques were used by Jin et al. (contribution 20) to non-destructively investigate the compositional heterogeneity and variation in cell wall mechanics in moso bamboo. Along the radius of bamboo culms, the concentration of xylan within the fiber sheath increased, while that of cellulose and lignin decreased gradually. At the cellular level, although the consecutive broad layer of fiber revealed a relatively uniform cellulose orientation and concentration, the outer broad layer with a higher lignification level has a higher elastic modulus (19.59–20.31 GPa) than that of the inner broad layer close to the lumen area (17.07–19.99 GPa). Comparatively, the cell corner displayed the highest lignification level, while its hardness and modulus were lower than that of the fiber broad layer, indicating that the cellulose skeleton is the prerequisite of cell-wall mechanics. The obtained cytological information is helpful to understand the origin of the anisotropic mechanical properties of bamboo.

Huang et al. (contribution 21) investigated the different shear performances of bamboo using four test methods: the tensile-shear, step-shear, cross-shear, and short-beam-shear methods. They indicated that the shear strength was significantly different in the four test methods and was highest in the step-shear-test method, but lowest in the tensile-shear-test method. The compound mode of compression and shear for the axial resulted in the maximum shear strength in the step-shear test, while the interface shear caused the tensile-shear strength to be the lowest. However, the shear changed the original fracture behavior of the tension, bending, and compression. Additionally, the axial-shear-test method caused typical interface-shear failure in the tensile-shear test and the overall tearing of fiber bundles in the step-shear test, while the parenchyma-cells collapsed in the cross-shear test. However, the short-beam-shear shearing characteristics resembled bending with the fiber bundle being pulled out. The findings of this study will inform the good use and manufacturing process of bamboo culm.

A study by Tang et al. (contribution 22) investigated the effects of tung oil thermal treatment on bamboo color at different temperatures and durations of time. The obtained results showed that the lightness (L^*) of bamboo decreased as the tung oil temperature or duration of time increased. The red–green coordinates (a^*) and color saturation (C^*) of bamboo were gradually increased as the tung oil temperature rose from 23 °C to 160 °C, while the a^* and C^* were gradually decreased when the temperature continued to rise from 160 °C to 200 °C. Eye movement data showed that the popularity of bamboo furniture was significantly improved at 23–100 °C and slightly improved at 160–180 °C with tung oil treatment. The findings of this study suggested that tung oil thermal treatment played a positive role in improving the visual effects and additional value of bamboo.

In another study, natural resin rosin was used by Su et al. (contribution 23) to treat round bamboo culm using the impregnation method. The obtained results showed that proper heating of the modified system was conducive to the formation of a continuous rosin film, which increased the gloss value. Heating decreased the brightness of the bamboo culm and changed the color from the green and yellow tones to red and blue. However, the heating temperature should not exceed 60 °C. They also used eye tracking technology to evaluate the users' preference for the visual characteristics of the bamboo culm surface. The findings of this study indicated that natural rosin resin could effectively improve the visual characteristics of bamboo culm, and different visual effects on bamboo surfaces were obtained in different temperature ranges.

Bamboo is easily attacked by fungus, resulting in a shorter service life and higher loss in storage and transportation [11]. The mold resistance of bamboo strips treated with low-molecular-weight organic acids and inorganic acid was first tested by Yu et al. (contribution 24), and then effect of citric acid with different concentrations was studied. Bamboo treated with acetic acid, propionic acid, oxalic acid, citric acid, and hydrochloric acid in a low concentration could improve their fungus growth rating from 4 in control samples to 2 or 3. Citric acid is effective in preventing mildew, and the mold resistance increased with the increased concentration of citric acid, and the fungus growth rating could reach 1 when the citric acid concentration was greater than 8%, while treating bamboo with citric acid in the concentration of 10% could control the infected area in the range of 10–17%. The improved mold and blue-stain resistance of treated bamboo could be attributed to the reduced nutrients in bamboo due to the hydrolysis of starch grains in parenchyma cells and the dissolution of soluble sugar.

In another paper, Peng et al. (contribution 25) synthesized the sustained-release system loading citral by using PNIPAm nanohydrogel as a carrier and analyzed its drug-release kinetics and mechanism. Their experimental results revealed that the release kinetics equation of the system conformed to the first order; the higher the external temperature, the better the match was. In the release process, PNIPAm demonstrated a good protection and sustained-release effect on citral. The laboratory mold control experiment results revealed that under the optimal conditions of release and impregnation time, the control efficiency of the bamboo treatment with pressure impregnation against the common bamboo molds, such as *P. citrinum*, *T. viride*, *A. niger*, and mixed mold reached 100% after 28 days, and the original colour of the bamboo was maintained during the mold control process.

Two manuscripts focusing on bamboo scrimber composites are also included in this Special Issue. In the first study, Ji et al. (contribution 26) prepared the bamboo scrimber composites using moso bamboo and phenol-formaldehyde resin, and the changes in the macroscopic and microscopic bonding interfaces before and after 28 h water-resistance tests were observed and analyzed. They showed that the water resistance of the bamboo scrimber composite increased with increasing resin content, with higher thickness swelling rates observed at higher densities. Obvious cracks were found at the macroscopic interface after 28 h tests, with higher resin contents leading to fewer and smaller cracks. With increasing density, the longitudinal fissures due to the defibering process decreased, having an effect on the width swelling rates. They suggested that the macroscopic and microscopic bonding interface structures of the bamboo scrimber composite are closely related to their water resistance.

The second study, by Wang et al. (contribution 27), investigated the influence of grain direction on the compression properties and failure mechanism of bamboo scrimber. They showed that the compressive load–displacement curves of bamboo scrimber in the longitudinal, tangential and radial directions contained elastic, yield and failure stages. The compressive strength and elastic modulus of the bamboo scrimber in the longitudinal direction were greater than those in the radial and tangential directions, and there were no significant differences between the radial and tangential specimens. The main failure mode of bamboo scrimber under longitudinal and radial compression was shear failure, and the main failure mode under tangential compression was interlayer separation failure. This study can provide benefits for the rational design and safe application of bamboo scrimber in practical engineering.

Moreover, Liang et al. (contribution 28) developed a facile strategy using the surfactant-induced reconfiguration of urea–formaldehyde (UF) resins to enhance the interface with bamboo and significantly improve its gluability. Through the coupling of a variety of surfactants, the viscosity and surface tension of the UF resins were properly regulated. The resultant surfactant-reconfigured UF resin showed much improved wettability and spreading performance to the surface of both green bamboo and yellow bamboo. Moreover, their reconfigured UF resin can reduce the amount of glue spread applied to bond the laminated commercial bamboo veneer products to 60 g m^{-2}, while the products prepared using the initial UF resin are unable to meet the requirements of the test standard, suggesting that this facile method is an effective way to decrease the application of petroleum-based resins and production costs.

In addition, in another study, bamboo delignification is a common method for studying its functional value-added applications. In Yu et al. (contribution 29)'s study, bamboo samples were delignified by means of treatment with sodium chlorite. They demonstrated that the lignin peak decreased or disappeared, and some hemicellulose peaks decreased, indicating that sodium chlorite treatment effectively removed lignin and partly decomposed hemicellulose, although cellulose was less affected. They suggested that delignified bamboo develops loose surfaces, increased pores, and noticeable fibers, indicating that alkali-treated bamboo has promising application potential due to its novel and specific functionalities.

Finally, in their review article, Li et al. (contribution 30) summarized the methods for preparing transparent bamboo, including delignification and resin impregnation. Potential applications of transparent bamboo are discussed using various functionalizations achieved through doping nanomaterials or modified resins to realize advanced energy-efficient building materials, decorative elements, and optoelectronic devices. Finally, challenges associated with the preparation, performance improvement, and production scaling of transparent bamboo are summarized, suggesting opportunities for the future development of this novel, bio-based, and advanced material.

This Special Issue has brought together experts that have studied and explored various aspects of eco polymeric materials and natural polymers. We would like to thank all researchers who have contributed to the production of this Special Issue of *Polymers*. In addition, I would like to express my gratitude to the Editorial Team who helped prepare the "Eco Polymeric Materials and Natural Polymer" Special Issue.

List of Contributions

1. Duan, B.; Yang, M.; Chao, Q.; Wang, L.; Zhang, L.; Gou, M.; Li, Y.; Liu, C.; Lu, K., Preparation and Properties of Egg White Dual Cross-Linked Hydrogel with Potential Application for Bone Tissue Engineering. *Polymers* 2022, 14(23), 5116; https://doi.org/10.3390/polym14235116.
2. Wu, H.; Bu, N.; Chen, J.; Chen, Y.; Sun, R.; Wu, C.; Pang, J., Construction of Konjac Glucomannan/Oxidized Hyaluronic Acid Hydrogels for Controlled Drug Release. *Polymers* 2022, 14(5), 927; https://doi.org/10.3390/polym14050927.
3. Zhang, S.; Wan, Y.; Yuan, W.; Zhang, Y.; Zhou, Z.; Zhang, M.; Wang, L.; Wang, R., Preparation of PVA–CS/SA–Ca^{2+} Hydrogel with Core–Shell Structure. *Polymers* 2022, 14(1), 212; https://doi.org/10.3390/polym14010212.
4. Wei, S.; Peng, R.; Bian, S.; Han, W.; Xiao, B.; Peng, X., Facile and Scalable Synthesis and Self-Assembly of Chitosan Tartaric Sodium. *Polymers* 2022, 14(1), 69; https://doi.org/10.3390/polym14010069.
5. dos Santos Filho, E. A.; Luna, C. B. B.; Siqueira, D. D.; Barbosa Ferreira, E. d. S.; Araujo, E. M., Tailoring Poly(lactic acid) (PLA) Properties: Effect of the Impact Modifiers EE-g-GMA and POE-g-GMA. *Polymers* 2022, 14(1), 136; https://doi.org/10.3390/polym14010136.
6. Sun, Z.; Tang, Z.; Li, X.; Li, X.; Morrell, J. J.; Beaugrand, J.; Yao, Y.; Zheng, Q., The Improved Properties of Carboxymethyl Bacterial Cellulose Films with Thickening and Plasticizing. *Polymers* 2022, 14(16), 3286; https://doi.org/10.3390/polym14163286.
7. Wang, S.; Song, S.; Yang, X.; Xiong, Z.; Luo, C.; Xia, Y.; Wei, D.; Wang, S.; Liu, L.; Wang, H.; Sun, L.; Du, L.; Li, S., Effect of Preparation Conditions on Application Properties

8. Wang, W.; Yu, F.; Ba, Z.; Qian, H.; Zhao, S.; Liu, J.; Jiang, W.; Li, J.; Liang, D., In-Depth Sulfhydryl-Modified Cellulose Fibers for Efficient and Rapid Adsorption of Cr(VI). *Polymers* 2022, 14(7), 1482; https://doi.org/10.3390/polym14071482.
9. Yao, Y.; Sun, Z.; Li, X.; Tang, Z.; Li, X.; Morrell, J. J.; Liu, Y.; Li, C.; Luo, Z. J. P., Effects of Raw Material Source on the Properties of CMC Composite Films. *Polymers* 2022, 14(1), 32; https://doi.org/10.3390/polym14010032.
10. Hakim, L.; Widyorini, R.; Nugroho, W. D.; Prayitno, T. A., Performance of Citric Acid-Bonded Oriented Board from Modified Fibrovascular Bundle of Salacca (*Salacca zalacca* (Gaertn.) Voss) Frond. *Polymers* 2021, 13(23), 4090; https://doi.org/10.3390/polym13234090.
11. Qi, C.; Wang, J.; Yadama, V., Heat Transfer Modeling of Oriented Sorghum Fibers Reinforced High-Density Polyethylene Film Composites during Hot-Pressing. *Polymers* 2021, 13(21), 3631; https://doi.org/10.3390/polym13213631.
12. Shahzad, K.; Rehan, M.; Rashid, M. I.; Ali, N.; Summan, A. S.; Ismail, I. M. I., Sustainability Evaluation of Polyhydroxyalkanoate Production from Slaughterhouse Residues Utilising Emergy Accounting. *Polymers* 2022, 14(1), 118; https://doi.org/10.3390/polym14010118.
13. Fu, Z.; Wang, H.; Li, J.; Lu, Y., Determination of Moisture Content and Shrinkage Strain during Wood Water Loss with Electrochemical Method. *Polymers* 2022, 14(4), 778; https://doi.org/10.3390/polym14040778.
14. Hu, J.; Fu, Z.; Wang, X.; Chai, Y., Manufacturing and Characterization of Modified Wood with In Situ Polymerization and Cross-Linking of Water-Soluble Monomers on Wood Cell Walls. *Polymers* 2022, 14(16), 3299; https://doi.org/10.3390/polym14163299.
15. Shen, Z.; Ye, Z.; Li, K.; Qi, C., Effects of Coupling Agent and Thermoplastic on the Interfacial Bond Strength and the Mechanical Properties of Oriented Wood Strand–Thermoplastic Composites. *Polymers* 2021, 13(23), 4260; https://doi.org/10.3390/polym13234260.
16. Bao, M.; Li, N.; Bao, Y.; Li, J.; Zhong, H.; Chen, Y.; Yu, Y., Outdoor Wood Mats-Based Engineering Composite: Influence of Process Parameters on Decay Resistance against Wood-Degrading Fungi *Trametes versicolor* and *Gloeophyllum trabeum*. *Polymers* 2021, 13(18), 3173; https://doi.org/10.3390/polym13183173.
17. Ye, Q.; Gong, Y.; Ren, H.; Guan, C.; Wu, G.; Chen, X., Analysis and Calculation of Stability Coefficients of Cross-Laminated Timber Axial Compression Member. *Polymers* 2021, 13(23), 4267; https://doi.org/10.3390/polym13234267.
18. Nun-Anan, P.; Suchat, S.; Mahathaninwong, N.; Chueangchayaphan, N.; Karrila, S.; Limhengha, S., Study of *Aquilaria crassna* Wood as an Antifungal Additive to Improve the Properties of Natural Rubber as Air-Dried Sheets. *Polymers* 2021, 13(23), 4178; https://doi.org/10.3390/polym13234178.
19. Liu, B.; Tang, L.; Chen, Q.; Zhu, L.; Zou, X.; Li, B.; Zhou, Q.; Fu, Y.; Lu, Y., Lignin Distribution on Cell Wall Micro-Morphological Regions of Fibre in Developmental *Phyllostachys pubescens* Culms. *Polymers* 2022, 14(2), 312; https://doi.org/10.3390/polym14020312.
20. Jin, K.; Ling, Z.; Jin, Z.; Ma, J.; Yang, S.; Liu, X.; Jiang, Z., Local Variations in Carbohydrates and Matrix Lignin in Mechanically Graded Bamboo Culms. *Polymers* 2022, 14(1), 143; https://doi.org/10.3390/polym14010143.
21. Huang, A.; Su, Q.; Zong, Y.; Chen, X.; Liu, H., Study on Different Shear Performance of Moso Bamboo in Four Test Methods. *Polymers* 2022, 14(13), 2649; https://doi.org/10.3390/polym14132649.
22. Tang, T.; Fei, B.; Song, W.; Su, N.; Sun, F., Tung Oil Thermal Treatment Improves the Visual Effects of Moso Bamboo Materials. *Polymers* 2022, 14(6), 1250; https://doi.org/10.3390/polym14061250.

23. Su, N.; Fang, C.; Zhou, H.; Tang, T.; Zhang, S.; Wang, X.; Fei, B., Effect of Rosin Modification on the Visual Characteristics of Round Bamboo Culm. *Polymers* 2021, 13(20), 3500; https://doi.org/10.3390/polym13203500.
24. Yu, Z.; Zhang, X.; Zhang, R.; Yu, Y.; Sun, F., Improving the Mould and Blue-Stain-Resistance of Bamboo through Acidic Hydrolysis. *Polymers* 2022, 14(2), 244; https://doi.org/10.3390/polym14020244.
25. Peng, R.; Zhang, J.; Du, C.; Li, Q.; Hu, A.; Liu, C.; Chen, S.; Shan, Y.; Yin, W., Investigation of the Release Mechanism and Mould Resistance of Citral-Loaded Bamboo Strips. *Polymers* 2021, 13(19), 3314; https://doi.org/10.3390/polym13193314.
26. Ji, Y.; Lei, W.; Huang, Y.; Wu, J.; Yu, W., Influence of Resin Content and Density on Water Resistance of Bamboo Scrimber Composite from a Bonding Interface Structure Perspective. *Polymers* 2022, 14(9), 1856; https://doi.org/10.3390/polym14091856.
27. Wang, X. Y.; Zhong, Y.; Luo, X. Y.; Ren, H. Q., Compressive Failure Mechanism of Structural Bamboo Scrimber. *Polymers* 2021, 13(23), 4223; https://doi.org/10.3390/polym13234223.
28. Liang, L.; Zheng, Y.; Wu, Y.; Yang, J.; Wang, J.; Tao, Y.; Li, L.; Ma, C.; Pang, Y.; Chen, H.; Yu, H.; Shen, Z., Surfactant-Induced Reconfiguration of Urea-Formaldehyde Resins Enables Improved Surface Properties and Gluability of Bamboo. *Polymers* 2021, 13(20), 3542; https://doi.org/10.3390/polym13203542.
29. Yu, H.; Gui, C.; Ji, Y.; Li, X.; Rao, F.; Huan, W.; Li, L., Changes in Chemical and Thermal Properties of Bamboo after Delignification Treatment. *Polymers* 2022, 14(13), 2573; https://doi.org/10.3390/polym14132573.
30. Li, X. L.; Zhang, W. Z.; Li, J. P.; Li, X. Y.; Li, N.; Zhang, Z. H.; Zhang, D. P.; Rao, F.; Chen, Y. H., Optically Transparent Bamboo: Preparation, Properties, and Applications. *Polymers* 2022, 14(16), 3234; https://doi.org/10.3390/polym14163234.

Author Contributions: Conceptualization, J.L., Y.L. and H.W.; Writing and editing, J.L., Y.L. and H.W. All authors have read and agreed to the published version of the manuscript.

Funding: This work was supported by the National Natural Science Foundation of China (grant nos. 32122058 and 32101604), the Fundamental Research Funds for the Central Non-Profit Research Institution of CAF (grant no. CAFYBB2022QA006 and CAFYBB2022XD004-02), and the Project of Forestry Science and Technology of the Zhejiang Province (grant no. 2021SY13).

Conflicts of Interest: The authors declare no conflict of interest.

References

1. Cywar, R.M.; Rorrer, N.A.; Hoyt, C.B.; Beckham, G.T.; Chen, E.Y.X. Bio-based polymers with performance-advantaged properties. *Nat. Rev. Mater.* **2022**, *7*, 83–103.
2. Garrison, T.F.; Murawski, A.; Quirino, R.L. Bio-Based Polymers with Potential for Biodegradability. *Polymers* **2016**, *8*, 262. [CrossRef] [PubMed]
3. Li, J.; Ma, R.; Yao, S.; Qin, D.; Lu, Y.; Chen, Y.; Jiang, Z.; Yang, D. Ultrafine Pd Nanoparticles Encapsulated in Mesoporous TiO_2 Region Selectively Confined in Bamboo Microchannels: An Ultrastable Continuous-Flow Catalytic Hydrogenation Microreactor. *Small Struct.* **2023**, 2300137. [CrossRef]
4. Brodin, M.; Vallejos, M.; Opedal, M.T.; Area, M.C.; Chinga-Carrasco, G. Lignocellulosics as sustainable resources for production of bioplastics–A review. *J. Clean. Prod.* **2017**, *162*, 646–664.
5. Nandakumar, A.; Chuah, J.-A.; Sudesh, K. Bioplastics: A boon or bane? *Renew. Sustain. Energy Rev.* **2021**, *147*, 111237.
6. Sun, S.; Zhang, J.; Zhang, L.; Liu, S.; Wang, R.; Kang, H.; Wang, Z. Research Progress in Biorubber. *Polym. Bull.* **2013**, *04*, 42–50.
7. Phadke, G.; Rawtani, D. Bioplastics as polymeric building blocks: Paving the way for greener and cleaner environment. *Eur. Polym. J.* **2023**, *199*, 112453.
8. Muelhaupt, R. Green Polymer Chemistry and Bio-based Plastics: Dreams and Reality. *Macromol. Chem. Phys.* **2013**, *214*, 159–174. [CrossRef]
9. Zhao, H.; Wang, J.; Meng, Y.; Li, Z.; Fei, B.; Das, M.; Jiang, Z. Bamboo and rattan: Nature-based solutions for sustainable development. *Innovation* **2022**, *3*, 100337. [CrossRef]

10. Mallik, A.K.; Moktadir, M.A.; Rahman, M.A.; Shahruzzaman, M.; Rahman, M.M. Progress in surface-modified silicas for Cr(VI) adsorption: A review. *J. Hazard. Mater.* **2022**, *423*, 127041.
11. Yan, J.; Niu, Y.; Wu, C.; Shi, Z.; Zhao, P.; Naik, N.; Mai, X.; Yuan, B. Antifungal effect of seven essential oils on bamboo. *Adv. Compos. Hybrid Mater.* **2021**, *4*, 552–561. [CrossRef]

Disclaimer/Publisher's Note: The statements, opinions and data contained in all publications are solely those of the individual author(s) and contributor(s) and not of MDPI and/or the editor(s). MDPI and/or the editor(s) disclaim responsibility for any injury to people or property resulting from any ideas, methods, instructions or products referred to in the content.

Article

Preparation and Properties of Egg White Dual Cross-Linked Hydrogel with Potential Application for Bone Tissue Engineering

Bingchao Duan *, Minghui Yang, Quanchao Chao, Lan Wang, Lingli Zhang, Mengxing Gou, Yuling Li, Congjun Liu and Kui Lu *

School of Chemical Engineering and Food Science, Zhengzhou University of Technology, Zhengzhou 450044, China
* Correspondence: duanbc100@163.com (B.D.); lukui126@126.com (K.L.)

Abstract: In this study, an egg white dual cross-linked hydrogel was developed based on the principle that the external stimulus can denature proteins and cause them to aggregate, forming hydrogel. The sodium hydroxide was used to induce gelation of the egg white protein, subsequently introducing calcium ions to cross-link with protein chains, thereby producing a dual cross-linked hydrogel. The characteristics of the dual cross-linked hydrogels—including the secondary structure, stability, microstructure, swelling performance, texture properties, and biosafety—were investigated to determine the effects of calcium ion on the egg white hydrogel (EWG) and evaluate the potential application in the field of tissue engineering. Results showed that calcium ions could change the β-sheet content of the protein in EWG after soaking it in different concentrations of $CaCl_2$ solution, leading to changes in the hydrogen bonds and the secondary structure of polypeptide chains. It was confirmed that calcium ions promoted the secondary cross-linking of the protein chain, which facilitated polypeptide folding and aggregation, resulting in enhanced stability of the egg white dual cross-linked hydrogel. Furthermore, the swelling capacity of the EWG decreased with increasing concentration of calcium ions, and the texture properties including hardness, cohesiveness and springiness of the hydrogels were improved. In addition, the calcium cross-linked EWG hydrogels exhibited biocompatibility and cell-surface adhesion in vitro. Hence, this work develops a versatile strategy to fabricate dual cross-linked protein hydrogel with biosafety and cell-surface adhesion, and both the strategy and calcium-egg white cross-linked hydrogels have potential for use in bone tissue engineering.

Keywords: egg white hydrogel; dual cross-linking; metal ions; secondary structure; biocompatibility

Citation: Duan, B.; Yang, M.; Chao, Q.; Wang, L.; Zhang, L.; Gou, M.; Li, Y.; Liu, C.; Lu, K. Preparation and Properties of Egg White Dual Cross-Linked Hydrogel with Potential Application for Bone Tissue Engineering. *Polymers* **2022**, *14*, 5116. https://doi.org/10.3390/polym14235116

Academic Editors: Jingpeng Li, Yun Lu and Huiqing Wang

Received: 26 September 2022
Accepted: 19 November 2022
Published: 24 November 2022

Publisher's Note: MDPI stays neutral with regard to jurisdictional claims in published maps and institutional affiliations.

Copyright: © 2022 by the authors. Licensee MDPI, Basel, Switzerland. This article is an open access article distributed under the terms and conditions of the Creative Commons Attribution (CC BY) license (https://creativecommons.org/licenses/by/4.0/).

1. Introduction

Natural macromolecules are a kind of polymer, existing in plants and animals, including the human body. They include polysaccharides, peptides, polynucleotides, and polyesters [1]. Natural polymers are an abundant resource for applications in the food and medical industries, due to that they have many advantages such as good biocompatibility, biodegradability, non-toxicity, and sustainability [2,3]. Hydrogel is a kind of three-dimensional network material. Hydrogels prepared from natural polymers should inherit the advantages of natural macromolecules, such as biocompatibility and biodegradability [4]. Thus, they are attracting attention particularly in biotechnology for uses such as drug delivery [5], biological sensing [6], wound dressing [7], and desalination [8,9].

Protein is a kind of natural polymer, abundant in nature, and with great prospects in polymer research [10,11]. In the field of polymer material science, the development of protein hydrogels is an important direction of current research [12–14]. Protein hydrogels retain a three-dimensional network structure similar to the extracellular matrix of animal tissue, featuring high water content. Hydrogels based on the proteins collagen [15], fibrin [16], elastin [17], and silk [18] have exhibited superior characteristics as compared to synthetic polymer-based hydrogels, such as biocompatibility, biodegradability, and low immunogenicity. Thus, these natural hydrogels are being extensively used for tissue engineering of bone and cartilage and in other biomedical application [19–22].

Egg white protein has a variety of functional characteristics, such as gelation, water holding capacity, foaming and emulsification. It is important in food manufacturing because it can improve the functionality, texture and flavor of food products [23]. Under heating, freezing, high pressure, extreme acid and alkali, ions, enzyme and other treatments, egg white protein is able to coagulate and form hydrogel with a three-dimensional network structure, which can strongly affect the structures, senses and flavors of egg white protein products [24,25]. The formation of protein hydrogels can be attributed to the formation of hydrogen bonds, disulfide bonds and electrostatic interaction, which induce the aggregation of protein molecules [26]. Proteins undergo denaturation in response to certain external stimuli (e.g., physical factors such as heating, ultraviolet light and pressure or pH, metal ions), and denaturation is one of the important mechanisms in the formation of protein hydrogels. The gelation of egg white protein involves multiple processes, including denaturation, aggregation and formation of gel network, and the gelling properties mainly depend on the medium conditions such as pH, ionic strength and salt type [27]. The interaction and the molecular conformation of protein chains will change during protein hydrogel formation. Studies have revealed that metal ions are able to impact the intermolecular and intramolecular interactions of protein chains and the conformation of protein molecules, further affecting the characteristics of protein gels [28,29].

In recent years, hydrogel materials have been extensively used in bone tissue engineering [30,31]; however, the application of protein hydrogels in bone tissue restoration and bone scaffold are severely limited due to their poor mechanical property [32]. Currently, tough and strong hydrogels are achieved by building dual network structures, compositing inorganic nanoparticles and introducing conductive materials and fibrous networks, improving the mechanical properties and the bone tissue repair ability of hydrogels [32]. Traditional hydrogels are fabricated through a single cross-linked mode, resulting in the lack of the energy dissipation pathways. Meanwhile, the dual crosslinking can improve the intermolecular interaction and cross-linking density in the hydrogel, providing an effective way to dissipate energy and increase the mechanical strength [33]. In this study, we prepared egg white hydrogel (EWG) according to the principle that proteins could be denatured by strong alkali, subsequently introducing calcium ions to induce aggregation and cross-linking with protein chains, thereby producing a dual cross-linked hydrogel (Figure 1a). Moreover, the secondary structure, stability, microstructure, swelling performance, texture and biocompatibility of the obtained hydrogels were investigated to research the effect of calcium ion on EWG.

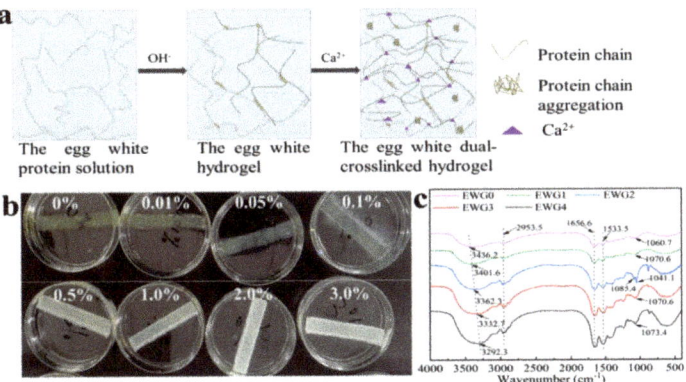

Figure 1. (a) Schematic of the EWG hydrogel preparation process; (b) photos of the hydrogel immersed in different concentrations of CaCl$_2$ solution. The number represents the concentration of CaCl$_2$ solution; (c) FT−IR spectra of the hydrogels.

2. Materials and Methods

2.1. Materials

Sodium hydroxide (NaOH) and calcium chloride (CaCl$_2$) were provided by Aladdin (Shanghai, China). Fresh eggs were purchased from a supermarket. The egg white was carefully separated from the yolk and then kept in a sealed chamber at 4 °C until further use. The Dulbecco's Modified Eagle's Medium (DMEM; glutamine, high glucose), fetal bovine serum (FBS), 3-[4,5-dimethylthiazol-2-yl]-2,5-diphenyl tetrazolium bromide (MTT) and the cell staining agent (Calcein AM and PI) were obtained from Sangon Biotech (Shanghai) Co., Ltd., Shanghai, China. All other chemicals were of analytical grade and used without further purification.

2.2. Preparation of the Egg White Hydrogel

NaOH solution (25 mg/mL) was added dropwise into the egg white (3:7 volume ratio) at room temperature with gentle stirring, then the solution was transferred to a mold (60 mm × 15 mm × 40 mm) and kept stationary 4 °C to release any air bubble present in the solution. After gelling, the egg white hydrogel was obtained and washed thoroughly with pure water to remove the residual sodium hydroxide.

CaCl$_2$ was dissolved in deionized water to obtain the sample solution at different concentrations. The egg white hydrogel samples (the samples were prepared in the size of 15 mm × 15 mm × 50 mm with a knife) were immersed in CaCl$_2$ solution (0.1%, 0.5%, 1%, 2%, 3% w/v), respectively, for 2 h at room temperature. After soaking, the hydrogel samples were taken and washed with pure water to remove the calcium chloride on the surface. The hydrogel samples soaked by pure water and CaCl$_2$ solution by 0.1%, 0.5%, 1% and 2% (w/v) were coded as EWG0, EWG1, EWG2, EWG3 and EWG4, respectively.

2.3. Characterization

The wet hydrogels were frozen in liquid nitrogen and snapped immediately, then freeze-dried using a vacuum freezing dryer (LEG-10C, Sihuan Furui Technology Development Co., Ltd., Hong Kong, China). The freeze-drying conditions were as follows: vacuum: 1Pa; cryo-temperature: −70 °C; material temperature: −50 °C; duration of drying: 24 h. The fracture sections of the freeze-dried samples were sputtered with gold for scanning electron microscopy (SEM, Zeiss, SIGMA, Roedermark, Germany) analysis.

The hydrogel samples were cut into particle-like size after being freeze-dried and vacuum-dried for 2 h at 60 °C. Then samples were transfer to a mortar and grind to a fine powder for measurements. The structural changes of the hydrogels were characterized by Fourier transform infrared spectroscopy (FT−IR, Spectrum3, Perkin Elmer, Waltham,

MA, USA), X-ray Diffraction Analysis (XRD, Regaku ultima IV, Japan). The data of amide I band (1700–1600 cm^{-1}) of FT-IR spectrum were analyzed by PeakFit software.

2.4. Swelling Tests

The thermal stability of the hydrogels was investigated by thermogravimetric analysis (TG, Discovery TGA 550, New Castle, DE, USA). The analysis was performed from room temperature to 800 °C with a heating rate of 15 °C/min in air atmosphere.

The swelling ratios of the hydrogels in pure water were tested through the gravimetric method. The freeze-dried hydrogels were immersed into pure water at 37 °C for modelling the body temperatures. Then the water in the surface of the hydrogels was gently wiped, and the weight of the samples was registered at predefined period. The swelling ratio (SR) was calculated as SR = (Ws − Wd)/Wd, where the Ws and Wd are the weight of the swollen and dried hydrogel, respectively. Similarly, the porosity ((Ws − Wd)/Ws) of different hydrogels after achieving swelling equilibrium was determined by using the identical measurement method of the mass swelling ratio [34].

2.5. Texture Tests

Texture analysis was performed using a texture analyzer (TA, BROOKFIELD CT3, USA) at room temperature. The hydrogel samples were prepared in the size of 10 mm × 10 mm × 15 mm with a knife, and the cylindrical probe TA4/1000 cylinder was selected for measurement. The clipped samples were compressed twice at 1 mm/s to 50% of their original height. The results were calculated with Texture Expert version 1.22 (Stable Micro Systems, Surrey, UK). All of these steps were performed six times.

2.6. Cell Experiments

HEK293 cells (human embryonic kidney-293 cells) were obtained from the China Center for Typical Culture Collection and cultured at 37 °C in a 5% CO_2 incubator. The culture medium containing 89.20% DMEM with 98 μg/mL penicillin/Streptomycin, 9.80% fetal bovine serum was changed every 2 days.

The cytotoxicity of the hydrogels was determined via an MTT assay. The hydrogel samples (100 mg) were sterilized under a UV lamp for 24 h, and then placed in DMEM containing 9.80% fetal bovine serum for 24 h at 37 °C to prepare the hydrogel extract solution. Thereafter, the resulting extracts were filtered using a 0.22-mm syringe for MTT tests. HEK293 cells (5×10^3 cells/well) were incubated in 96-wells plates for 12 h, and the extract solution of different hydrogels was added for 24 h incubation. Then the medium was aspirated and 10 μL of MTT (5 mg/mL) was added to each well for another 4-h incubation. After the culture medium was removed, 200 μL of DMSO was added in each well. After gently shaking for few minutes, the plates were carried out to measure the absorbance at 492 nm using a microplate reader (SparkTM 10M, Tecan). The cell viability was calculated based on the equation: Cell viability (%) = (the absorbance of sample-treated group/the absorbance of PBS-treated group) × 100%.

The live/dead cell staining assays were performed to evaluate the cell adhesion on the hydrogel surfaces and the biocompatibility. The hydrogel EWG4 sample with 10 mm diameter and a thickness of about 0.1 cm was sterilized under a UV lamp for 24 h, and then transferred to the bottom of 24-well plastic culture plates. HEK293 cells were seed to 24-well plastic culture plates on the sample (5×10^4 cells/well) for 48 h incubation. Then, the samples were washed by physiological saline and stained by the cell staining agent (Calcein AM and PI) for 30 min at 37 °C. Thereafter, the samples were washed again for three times with physiological saline and then observed using a fluorescence microscope (Olympus BX51, Olympus Corporation, Tokyo, Japan).

2.7. Statistical Analysis

All measurements were repeated at least three times independently and expressed as mean ± standard deviation (SD). All values reported in this study are expressed as mean

± standard deviation, and $p < 0.05$ (*), $p < 0.01$ (**) and $p < 0.001$ (***) signify significant and extremely significant differences, respectively.

3. Results and Discussion

3.1. Preparation and Morphology Analysis of the EWG Hydrogel

There are a variety of proteins in fresh egg white solution. In the liquid state, these proteins are stabilized via electrostatic interaction, hydrogen bonds and thiol ester bonds between protein chains. However, when heated or exposed to high pressure, alkali or acid, these chains are induced to unfold and rearrange into various forms. In this work, the egg white hydrogel was prepared by alkali induction and it was transparent, light yellow, smooth and ductile. Then, the egg white hydrogel was soaked in calcium chloride solution with different concentrations in order to induce secondary cross-linking between the calcium ions and the egg white hydrogel through the interaction of calcium ions with the particular amino acids on the polypeptide chain. As shown in Figure 1b, the hydrogel samples remained solid without breaking or decomposition after soaking in calcium chloride solution for 2 h. With increasing concentration, more calcium ions gradually infiltrated into the hydrogel, leading to deeper color and decreased transparency of the hydrogels.

3.2. FT–IR Analysis

The structural characteristics and conformational changes of the hydrogels were evaluated by Fourier transform infrared spectroscopy (FT–IR). As shown in Figure 1c, the absorption peaks at ~3350 and 2953 cm^{-1} were assigned to the stretching vibrations of N–H and O–H and of intermolecular hydrogen bonds, respectively [35]. With the increase of calcium concentration, the absorption band at 3436.2 cm^{-1} gradually shifted to 3292.3 cm^{-1} and the intensity of peaks at 2953 cm^{-1} increased significantly, suggesting that the introduction of calcium ions affected the hydrogen bonds between the amino and hydroxyl groups in the polypeptide chain and the protein molecular chain. The typical absorption peaks at 1656 and 1533 cm^{-1} were related to the amide I (C=O stretching) and amide II (N–H bending) modes of the protein chain structure, respectively [36]. The intensity changes of the peaks at 1533.5 cm^{-1} indicated that the amino and hydrogen bond in the protein molecular were impacted by calcium ions. Previous studies [35] have demonstrated that the amide I band (1700–1600 cm^{-1}) is the most characteristic spectral region related to the secondary structure of proteins and polypeptides, and the secondary structures and conformational changes could be investigated by quantitative analysis of the amide I band. Therefore, the absorption peak of this region was used to quantitatively calculate the specific proportion of each secondary structure (α-helices 1650–1658 cm^{-1}, β-sheets 1640–1610 cm^{-1}, β-turns 1700–1660 cm^{-1}, random coils 1650–1640 cm^{-1}) [35,37]. The results showed (Table 1 and Figure S1) that the β-sheets content of the hydrogel increased significantly as calcium ion concentration increased up to 1%, and then decreased at higher calcium concentrations. It had been indicated that the structure of β-sheets is prone to protein aggregation and particularly important for the hydrogel formation and stability [37]. The interaction of calcium ions and protein molecules could facilitate the polypeptides to fold the β-sheets; however, the cross-linking between calcium ion and protein chains along with the increasing of calcium ion concentration was dominated and disturbed gradually the hydrogen bonding in β-sheets structures of the protein, leading to the reduce of the β-sheets content in the hydrogels soaking with high concentrations of CaCl$_2$ solution. Meanwhile, the original conformation of EWG0 was changed after introducing calcium ion, resulting in the decrease of β-turns content and the increase of random coil structure. In addition, the changes of the band at 1030–1090 cm^{-1} assigned to C–O stretching vibration indicated that the structure of the polypeptide chain was changed by the calcium ions. Together, the results suggested that calcium ions were able to interact with the polypeptides and change secondary structure of the protein in the hydrogel.

Table 1. Effect of CaCl$_2$ addition on the secondary structures of egg white protein.

Concentrations of CaCl$_2$ Solution	Hydrogel Samples	Relative Content (%)			
		β-Sheets	Random Coil	α-Helices	β-Turns
0%	EWG0	23.56	14.93	14.56	46.96
0.1%	EWG1	34.54	12.35	11.41	41.71
0.5%	EWG2	33.21	13.55	13.34	39.91
1%	EWG3	36.38	14.54	15.11	33.98
2%	EWG4	21.98	31.12	15.81	31.08

3.3. XRD Analysis and Thermal Stability

To further study the influence of calcium ions on hydrogel structure, the hydrogels were characterized by XRD, TG and DTG. The XRD spectra of the EWG0, EWG1, EWG2, EWG3 and EWG4 are shown in Figure 2a. The hydrogels exhibited distinct peaks at 2θ = 20°, which was indicated to the β-sheets secondary structure of the egg white protein. The XRD diffraction intensity of the EWG1 (5885) and EWG2 (5713) was enhanced in comparison with that of EWG0 (5233), and then the intensity decreased along with the increase in the concentration of calcium ions (EWG3:4378; EWG4:3096), confirming that the β-sheets structure in the crystalline region of the protein was disturbed by the interaction of protein with calcium ions [38,39], which was consistent with the FT−IR results.

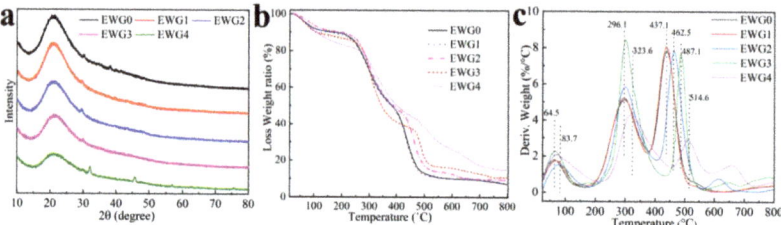

Figure 2. XRD patterns (a), TG curves (b) and DTG patterns (c) of hydrogels.

TG and DTG analyses were applied to investigate the thermal properties of the hydrogels. The TG curves (Figure 2b) revealed that the weight loss of EWG4 (17.3%) was higher than that of EWG0 (9.8%), and the temperature corresponding to the endothermic peak on the DTG curve of the hydrogels cross-linked by calcium ions (EWG1, EWG2, EWG3 and EWG4) (Figure 2c) was increased as compared with EWG0, indicating the water retention capacity and thermal stability of the hydrogel was enhanced after interacting with calcium ion. The second stage of weight loss was mainly due to the breaking of unstable non-covalent bonds of the protein chains and the covalent bonds of the small molecules in the protein backbone. As more calcium ions were added, the hydrogel exhibited an increased degradation temperature and decreased weight loss rate. Meanwhile, the weight loss decreased from 50.767% (EWG0) to 43.926% (EWG4), which was indicative of enhanced thermal stability. The explanation for this pattern is that the interaction of calcium ions with protein promoted secondary cross-linking of the protein chain and increased hydrogel cross-linking density, which could effectively inhibit heat conduction, thereby hindering thermal degradation of protein skeleton, and enhancing the thermal stability of the hydrogel [40,41]. In summary, the results indicated that the dual cross-linked structure involving the calcium ion formed a heat-stable system, which translated into improved stability of the hydrogel.

3.4. Microscopic Examination

SEM was used to study the microstructure changes of the hydrogels. As shown in Figure 3, EWG0 exhibited a loose and homogeneous three-dimensional structure with lots of pores. It had been reported that protein molecule chains were able to unfold and rearrange themselves to form an ordered three-dimensional protein network under strong alkaline condition [42]. After soaking in calcium chloride, the hydrogels exhibited inhomogeneous and rough microstructure with the disordered porous structure. When the calcium ion concentration reached 1.0% and 2.0%, the porous structure of EWG3 and EWG4 was significantly reduced, accompanied by the appearance of rough, flat edges and coarse fibers (Figure S2). The results suggested that the interaction with calcium cations induced the secondary crosslinking of the protein chains and improved the degree of crosslinking, leading to the formation of more compact microstructure of the hydrogel [43,44].

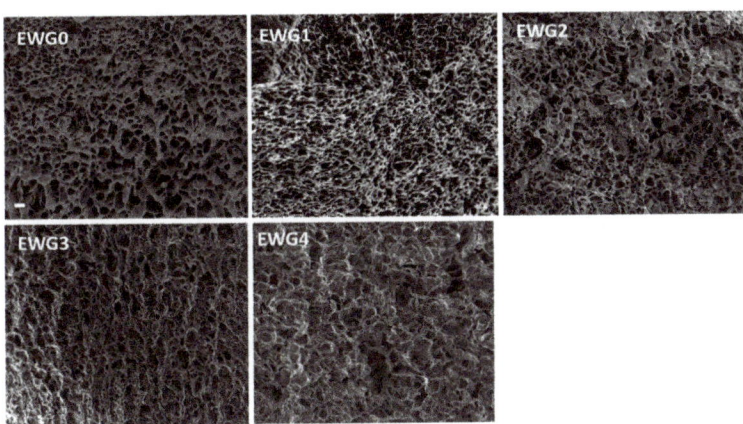

Figure 3. SEM images of the hydrogels. The scale bar is 10 μm.

3.5. Effect of Calcium Ions on Hydrogel Swelling Performance

The swelling performance of the hydrogels was strongly associated with hydrophilic groups and the pore network structure of the hydrogel, both of which are key to absorbing water. The swelling behavior of the hydrogel was tested in ultra-pure water. The results (Figure 4) show that the swelling rate increased rapidly with the extension of time and reached the swelling equilibrium after about 4 h. This pattern is common for hydrogels. Compared with the EWG0 group, EWG1, the hydrogel soaked in a low concentration of calcium ions (0.1%) showed a similar swelling rate, whereas EWG2, EWG3 and EWG4 showed gradual decreases with increasing concentration of calcium ions. The formation of the EWG0 hydrogel depended on the physical cross-linking with only weak binding between protein chains, giving water molecules easy access to the hydrogel interior and resulting in high expansibility. Moreover, it was found that the EWG0 hydrogel was gradually degraded and ruptured with extended soaking time, which was because the water molecules destroyed the three-dimensional structure with weak intermolecular forces [45]. It is well known that the degree of cross-linking directly affects the water absorption of a hydrogel [35]. When the egg white hydrogel was soaked in calcium chloride solution, calcium ions interacted with the amino acids of the protein chain to prompt the cross-linking and aggregation of the egg white protein and induce the formation of a tight three-dimensional network, resulting in significant decreases in water absorption and storage capacity of the hydrogels. Moreover, the cross-linked networks with tight structure could resist the destructive infiltration of moisture. The porosities of EWG0, EWG1, EWG2, EWG3 and EWG4 were 97.2 ± 2.34, 96.9 ± 3.16, 93.7 ± 4.24, 91.4 ± 4.94 and 82.4 ± 3.41, respectively (Figure S3). The porosities of the hydrogels decrease with increasing concentration of calcium ions, suggesting that the interaction between calcium ions and protein facilitate the cross-linking and aggregation of the protein chains, leading to the decrease of hydrogel porosity. It was found that the equilibrium swelling ratio of EWG0 was seven times higher than that of EWG4, indicating that the high concentration of calcium ions could effectively reduce the swelling capacity of the original egg white hydrogel. Therefore, it has great potential for preparing the sensitive double-layer hydrogel actuators using the EWG0 and EWG4 as the humidity responder and humidity inert layer, respectively (Figure S4) [46].

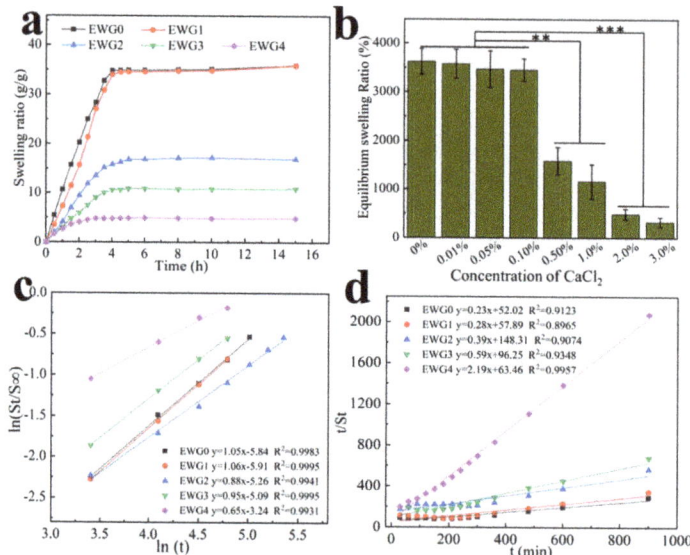

Figure 4. (a) Swelling kinetics of the hydrogels in distilled water at 37 °C; (b) equilibrium swelling ratio of the hydrogels in distilled water as a function of CaCl$_2$ concentration. ** $p < 0.01$, *** $p < 0.001$; (c) plots of $\ln(S_t/S_\infty)$ versus $\ln t$; and (d) t/S_t versus t for the hydrogels.

Swelling kinetic of the hydrogels is evaluated by Schott's second-order diffusion kinetic model and Fickian diffusional kinetic model [47,48]. The swelling data achieved from the first 60% of the fractional water uptake are fitted with the following equation to determine water diffusion mechanism of hydrogel samples: $\ln(S_t/S_\infty) = \ln k + n\ln t$, where S_t and S_∞ are the water uptake at time t and the equilibrium water uptake. The k parameter is a constant of the solvent-polymer system; the n parameter specifies the diffusion mechanism of water molecules. $n < 0.5$ indicates Fickian diffusion, $0.5 < n < 1$ indicates non-Fickian diffusion and $n = 1$ indicates that the diffusion mechanism is case-II. Figure 4c shows the plots of $\ln(S_t/S_\infty)$ versus $\ln t$, the slopes and intercepts of the plotted lines could be used to calculate n and k. The values of n for EW0 and EWG1 are close to 1, indicating that the water diffusion mechanism in EWG0 and EWG1 is case-II (relaxation-controlled) transport. The values of n for EWG2, EWG3 and EWG4 are greater than 0.5, implying the water diffusion mechanisms are non-Fickian diffusion type. The water diffusion mechanism changes of the hydrogels are caused by the secondary cross-linking of the protein chain and the increased crosslinking density of hydrogels, which limit the protein chains relaxation and hinder the diffusion of water [49]. The Schott's second-order diffusion kinetic model is used to get further information about the swelling rate: $t/S_t = A + Bt$, where $A = \frac{1}{K_s S_\infty^2}$ is the initial swelling rate of the hydrogel and K_s is the swelling rate constant, $B = 1/S_\infty$ is the converse of the equilibrium swelling. The plots of t/S_t versus t are plotted for the hydrogel samples (Figure 4d). The theoretical swelling equilibrium (shown in Table 2) of EWG0, EWG1, EWG2, EWG3 and EWG4 hydrogels are close to their corresponding experimental values. The swelling rate constants (K_s) of EWG3 and EWG4 are higher than that of EWG0, suggesting that the hydrogels with high crosslinking density possess the faster swelling rate and reach the swelling equilibrium in a shorter time, which is consistent with the experimental results.

Table 2. Second-order kinetic parameters for hydrogels.

Sample	EWG0	EWG1	EWG2	EWG3	EWG4
A	52.02	57.89	148.32	96.25	63.46
B	0.23	0.28	0.39	0.59	2.19
$K_s \times 10^{-3}$ (min^{-1})	1.08	1.39	1.09	3.68	75.83
S_∞ (%)	3969.3	3501.9	1675.5	1756.4	492.5
ESR (%)	5463.1	4786.1	2657.8	2219.7	518.1

ESR: experimental swelling equilibrium value.

3.6. Effect of Calcium Ions on Hydrogel Texture

Soaking the egg white hydrogel in different concentrations of calcium ions can change their microstructure, and, thereby, their properties. The textural properties of hardness, cohesiveness and springiness of the hydrogels were tested using a texture analyzer. As shown in Figure 5, the hardness of the hydrogels remained basically unchanged when the concentration of ions was less than 0.5% and when it was significantly enhanced as calcium ion concentration increased. Hardness is related to the structural strength of a hydrogel [50,51], and the overall structure of the protein is changed by the cross-linking of calcium ions with particular amino acids of adjacent peptides, resulting in the enhancement of the hydrogel hardness. Cohesiveness and springiness of the hydrogel were also affected by the addition of calcium. Previous studies [52,53] have shown that the conformation of proteins and polymerized protein chains are affected by divalent metal ions, leading to changes of the texture properties of protein hydrogels. In addition, cohesiveness and elasticity are influenced by the microstructure of the hydrogel. The SEM results showed that the introduction of calcium ions promoted a smaller three-dimensional pore structure and more compact microstructure of the hydrogel, which resulted in the enhancement of the cohesiveness and elasticity of the hydrogels. It could be found that the trends toward less hardness, cohesiveness and springiness of the hydrogels at low calcium concentration that might be because the calcium ions consumed hydroxyl ions, such that the three-dimensional network structure of the original hydrogel could not be maintained. As calcium concentration increased, the equilibrium between calcium and hydroxide ions was reached, and the interactions of the redundant calcium with the particular amino acids of the peptide chains gradually dominated and impacted the texture properties of the hydrogels.

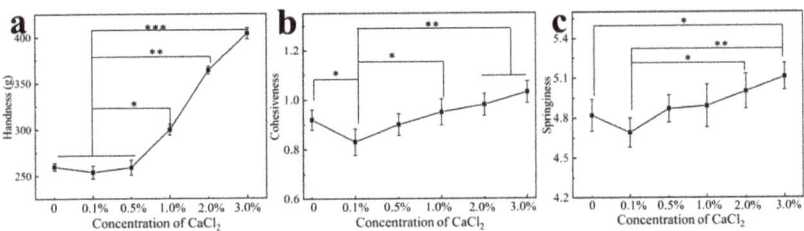

Figure 5. (a) Changes in hardness (a), cohesiveness (b) and springiness (c) of the hydrogels as a function of CaCl$_2$ concentration. * $p < 0.05$, ** $p < 0.01$, *** $p < 0.001$.

3.7. Cytocompatibility

To investigate the potential of the hydrogels in biomedical applications (e.g., wound healing and bone tissue repair requiring calcium ions [54]), the hydrogels were co-cultured with HEK-293 cells for assessing the cytocompatibility and cell adhesion on the hydrogel. The MTT results (Figure 6a) showed that EWG0 and EWG1 showed cell survival rates similar to the control. With the increase of calcium chloride concentration, the cell survival rates of EWG2, EWG3 and EWG4 groups decreased slightly but remained above 80%, indicating that all hydrogels possessed cytocompatibility. Compared to the egg-white-

/eggshell-based biomimetic hybrid hydrogels, the cells treated with EWG1 hydrogel for 24 h presented the similar proliferation rate [31], while the EWG2, EWG3 and EWG4 exhibited the lower proliferation rate, indicating that the high concentration calcium ions might be not advantageous to the cell proliferation. In addition, the live/dead cell staining assays (i.e., live cells stained fluorescent green, dead cells stained fluorescent red) were performed to study the cell adhesion and viability on the hydrogel surfaces. The results appear in Figure 6b. HEK-293 cells was able to survival normally and adhere to the EWG4 surface, demonstrating that the cross-linked hydrogels prepared with the highest calcium concentration were non-toxic, cytocompatible and adaptive for cell survival. In conclusion, calcium ion secondary cross-linked egg white gel showed excellent biocompatibility and biosafety. It should have great value for potential applications in the biomedical fields, particularly bone tissue engineering.

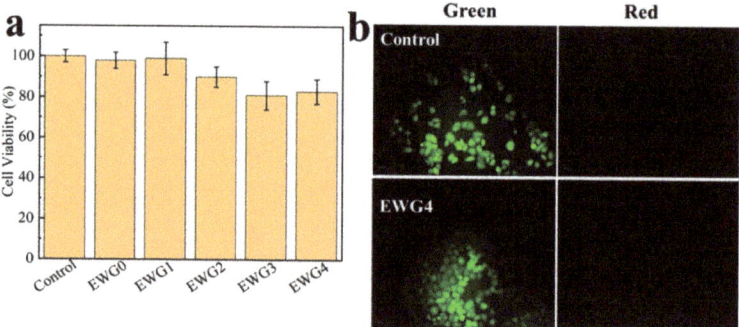

Figure 6. (**a**) Cell viability of HEK-293 cells on EWG0, EWG1, EWG2, EWG3 and EWG4 after 24 h culturing. (**b**) Live/dead staining florescent photographs of HEK-293 cells loaded with EWG4 for 48 h.

4. Conclusions

In summary, an egg white dual cross-linked hydrogel was prepared through the induction of sodium hydroxide and the secondary cross-linking of protein chains by calcium ions. Characteristics of the dual cross-linked hydrogel were remarkably affected by the concentrations of calcium ions. The incorporation of calcium ions could benefit thermal stability, swelling rate and texture of the hydrogels, while also reducing swelling capacity. Calcium ions could impact the secondary structure of polypeptide chains and interact with protein chains, leading to more compact microstructure formation of the hydrogels. Remarkably, the egg white dual cross-linked hydrogels exhibited biocompatibility and cell-surface adhesion in vitro, indicating the potential for biomedical application.

Supplementary Materials: The following supporting information can be downloaded at: https://www.mdpi.com/article/10.3390/polym14235116/s1, Figure S1. Normalized FT–IR spectra of hydrogels. Each peak represents a different secondary structure. (α-helices 1650–1658 cm^{-1}, β-sheets 1640–1610 cm^{-1}, β-turns 1700–1660 cm^{-1}, random coils 1650–1640 cm^{-1}). Figure S2. SEM images of the hydrogels. The scale bar is 5 μm. Figure S3. The porosities of the hydrogels. Figure S4. The photo of self-bending double layer hydrogel. The upper layer (white) is the hydrogel that soaked in calcium chloride, the lower layer (pale yellow) is the hydrogel that soaked without calcium chloride. The double layer hydrogel exhibited a smaller curvature that could be duo to the inapposite gel thickness.

Author Contributions: Formal analysis, investigation, B.D., L.W., L.Z. and M.G.; writing—original draft preparation, B.D., M.Y. and Q.C.; writing—review and editing, Y.L., C.L. and K.L.; resources, B.D. and K.L. All authors have read and agreed to the published version of the manuscript.

Funding: This work was supported by the Basic Research and Applied Basic Research Project of Zhengzhou Science and Technology Bureau (Grant No. zkz202111), the High-level Talent Fund start-up Project of Zhengzhou University of Technology (No. 22078), the National Natural Science

Foundation of China (No. 21572046), the Science and Technology Breakthrough Plan of Henan Province (Grant No. 212102310857), the Key Projects of Henan Provincial High School (Grant No. 21B150020, 21A550013, 22B150021), the College Students Innovation and Entrepreneurship Project of Henan Province (No. S202111068004), the College Students Innovation and Entrepreneurship Project of Zhengzhou University of Technology (Preparation and properties of Ion-induced egg white gel).

Data Availability Statement: The data presented in this study are available on request from the corresponding author.

Conflicts of Interest: The authors declare no conflict of interest.

References

1. Banu, J.R.; Kavitha, S.; Kannah, R.Y.; Devi, T.P.; Gunasekaran, M.; Kim, S.H.; Kumar, G. A review on biopolymer production via lignin valorization. *Bioresour. Technol.* **2019**, *290*, 121790. [CrossRef] [PubMed]
2. Liu, J.; Xie, L.; Wang, Z.; Mao, S.; Gong, Y.; Wang, Y. Biomass-derived ordered mesoporous carbon nano-ellipsoid encapsulated metal nanoparticles inside: Ideal nanoreactors for shape-selective catalysis. *Chem. Commun.* **2019**, *56*, 229–232. [CrossRef] [PubMed]
3. Gong, Y.; Li, D.; Luo, C.; Fu, Q.; Pan, C. Highly porous graphitic biomass carbon as advanced electrodematerials for supercapacitors. *Green Chem.* **2017**, *19*, 4132–4140. [CrossRef]
4. Karoyo, A.H.; Wilson, L.D. A review on the design and hydration properties of natural polymer-based hydrogels. *Materials* **2021**, *14*, 1095. [CrossRef]
5. Sun, Z.; Song, C.; Wang, C.; Hu, Y.; Wu, J. Hydrogel-based controlled drug delivery for cancer treatment: A review. *Mol. Pharm.* **2019**, *17*, 373–391. [CrossRef]
6. Tavakoli, J.; Tang, Y. Hydrogel based sensors for biomedical applications: An updated review. *Polymers* **2017**, *9*, 364. [CrossRef]
7. Liang, Y.; He, J.; Guo, B. Functional hydrogels as wound dressing to enhance wound healing. *ACS Nano* **2021**, *15*, 12687–12722. [CrossRef]
8. Wang, Z.; Wu, X.; Dong, J. Porifera-inspired cost-effective and scalable "porous hydrogel sponge" for durable and highly efficient solar-driven desalination. *Chem. Eng. J.* **2022**, *427*, 130905. [CrossRef]
9. Palanivelu, S.D.; Armir, N.A.Z.; Zulkifli, A.; Hair, A.H.A.; Salleh, K.M.; Lindsey, K.; Che-Othman, M.H.; Zakaria, S. Hydrogel Application in Urban Farming: Potentials and Limitations—A Review. *Polymers* **2022**, *14*, 2590. [CrossRef]
10. Mu, X.; Yuen, J.S.K., Jr.; Choi, J.; Zhang, Y.; Cebe, P.; Jiang, X.; Zhang, Y.S.; Kaplan, D.L. Conformation-driven strategy for resilient and functional protein materials. *Proc. Natl. Acad. Sci. USA* **2022**, *119*, e2115523119. [CrossRef]
11. Shen, Y.; Levin, A.; Kamada, A.; Toprakcioglu, Z.; Rodriguez-Garcia, M.; Xu, Y.; Tuomas, P.J.K. From protein building blocks to functional materials. *ACS Nano* **2021**, *15*, 5819–5837. [CrossRef] [PubMed]
12. Zustiak, S.P.; Wei, Y.; Leach, J.B. Protein–hydrogel interactions in tissue engineering: Mechanisms and applications. *Tissue Eng. Part B-Rev.* **2013**, *19*, 160–171. [CrossRef] [PubMed]
13. Davari, N.; Bakhtiary, N.; Khajehmohammadi, M.; Sarkari, S.; Tolabi, H.; Ghorbani, F.; Ghalandari, B. Protein-Based Hydrogels: Promising Materials for Tissue Engineering. *Polymers* **2022**, *14*, 986. [CrossRef] [PubMed]
14. Tang, Y.; Zhang, X.; Li, X.; Ma, C.; Chu, X.; Wang, L.; Xu, W. A review on recent advances of Protein-Polymer hydrogels. *Eur. Polym. J.* **2022**, *162*, 110881. [CrossRef]
15. Sarrigiannidis, S.O.; Rey, J.M.; Dobre, O.; Gonzalez-Garcia, C.; Dalby, M.J.; Salmeron-Sanchez, M. A tough act to follow: Collagen hydrogel modifications to improve mechanical and growth factor loading capabilities. *Mater. Today Bio.* **2021**, *10*, 100098. [CrossRef]
16. Nelson, D.W.; Gilbert, R.J. Extracellular Matrix-Mimetic Hydrogels for Treating Neural Tissue Injury: A Focus on Fibrin, Hyaluronic Acid, and Elastin-Like Polypeptide Hydrogels. *Adv. Healthc. Mater.* **2021**, *10*, 2101329. [CrossRef]
17. Mizuguchi, Y.; Mashimo, Y.; Mie, M.; Kobatake, E. Temperature-responsive multifunctional protein hydrogels with elastin-like polypeptides for 3-D angiogenesis. *Biomacromolecules* **2020**, *21*, 1126–1135. [CrossRef]
18. Kapoor, S.; Kundu, S.C. Silk protein-based hydrogels: Promising advanced materials for biomedical applications. *Acta Biomater.* **2016**, *31*, 17–32. [CrossRef]
19. Carvalho, M.S.; Cabral JM, S.; da Silva, C.L.; Vashishth, D. Bone matrix non-collagenous proteins in tissue engineering: Creating new bone by mimicking the extracellular matrix. *Polymers* **2021**, *13*, 1095. [CrossRef]
20. Yue, S.; He, H.; Li, B.; Hou, T. Hydrogel as a biomaterial for bone tissue engineering: A review. *Nanomaterials* **2020**, *10*, 1511. [CrossRef]
21. Pham, H.M.; Zhang, Y.; Munguia-Lopez, J.G.; Tran, S.D. Egg White Alginate as a Novel Scaffold Biomaterial for 3D Salivary Cell Culturing. *Biomimetics* **2021**, *7*, 5. [CrossRef]
22. Zhang, Y.; Pham, H.M.; Munguia-Lopez, J.G.; Kinsella, J.M.; Tran, S.D. The optimization of a novel hydrogel—Egg white-alginate for 2.5 D tissue engineering of salivary spheroid-like structure. *Molecules* **2020**, *25*, 5751. [CrossRef] [PubMed]
23. Luo, Y.; Li, M.; Zhu, K.X.; Guo, X.N.; Peng, W.; Zhou, H.M. Heat-induced interaction between egg white protein and wheat gluten. *Food Chem.* **2016**, *197*, 699–708. [CrossRef]

24. He, W.; Xiao, N.; Zhao, Y.; Yao, Y.; Xu, M.; Du, H.; Wu, N.; Tu, Y. Effect of polysaccharides on the functional properties of egg white protein: A review. *J. Food Sci.* **2021**, *86*, 656–666. [CrossRef] [PubMed]
25. Babaei, J.; Khodaiyan, F.; Mohammadian, M. Effects of enriching with gellan gum on the structural, functional, and degradation properties of egg white heat-induced hydrogels. *Int. J. Biol. Macromol.* **2019**, *128*, 94–100. [CrossRef]
26. Uman, S.; Dhand, A.; Burdick, J.A. Recent advances in shear-thinning and self-healing hydrogels for biomedical applications. *J. Appl. Polym. Sci.* **2020**, *137*, 48668. [CrossRef]
27. Lv, X.; Huang, X.; Ma, B.; Chen, Y.; Batool, Z.; Fu, X.; Jin, Y. Modification methods and applications of egg protein gel properties: A review. *Compr. Rev. Food Sci. Food Saf.* **2022**, *21*, 2233–2252. [CrossRef] [PubMed]
28. Stamboroski, S.; Boateng, K.; Lierath, J.; Kowalik, T.; Thiel, K.; Koppen, S.; Noeske, P.L.M.; Bruggemann, D. Influence of divalent metal ions on the precipitation of the plasma protein fibrinogen. *Biomacromolecules* **2021**, *22*, 4642–4658. [CrossRef]
29. Deng, C.; Shao, Y.; Xu, M.; Yao, Y.; Wu, N.; Hu, H.; Zhao, Y.; Tu, Y. Effects of metal ions on the physico-chemical, microstructural and digestion characteristics of alkali-induced egg white gel. *Food Hydrocoll.* **2020**, *107*, 105956. [CrossRef]
30. Wei, Z.Z.; Dong, X.; Zhang, Y.Q. A mechanically robust egg white hydrogel scaffold with excellent biocompatibility by three-step green processing. *Sci. China Technol. Sci.* **2022**, *65*, 1599–1612. [CrossRef]
31. Huang, K.; Hou, J.; Gu, Z.; Wu, J. Egg-white-/eggshell-based biomimetic hybrid hydrogels for bone regeneration. *ACS Biomater. Sci. Eng.* **2019**, *5*, 5384–5391. [CrossRef] [PubMed]
32. Zhao, H.; Liu, M.; Zhang, Y.; Yin, J.; Pei, R. Nanocomposite hydrogels for tissue engineering applications. *Nanoscale* **2020**, *12*, 14976–14995. [CrossRef] [PubMed]
33. Yin, S.; Zhang, W.; Zhang, Z.; Jiang, X. Recent advances in scaffold design and material for vascularized tissue-engineered bone regeneration. *Adv. Healthc. Mater.* **2019**, *8*, 1801433. [CrossRef] [PubMed]
34. Liu, S.; Wang, Y.N.; Ma, B.; Shao, J.; Liu, H.; Ge, S. Gingipain-responsive thermosensitive hydrogel loaded with SDF-1 facilitates in situ periodontal tissue regeneration. *ACS Appl. Mater. Inter.* **2021**, *13*, 36880–36893. [CrossRef] [PubMed]
35. Chang, C.; Duan, B.; Cai, J.; Zhang, L. Superabsorbent hydrogels based on cellulose for smart swelling and controllable delivery. *Eur. Polym. J.* **2010**, *46*, 92–100. [CrossRef]
36. Deng, X.; Attalla, R.; Sadowski, L.P.; Chen, M.; Majcher, M.J.; Urosev, I.; Yin, D.C.; Selvaganapathy, P.R.; Filipe, C.D.M.; Hoare, T. Autonomously Self-Adhesive Hydrogels as Building Blocks for Additive Manufacturing. *Biomacromolecules* **2018**, *19*, 62–70. [CrossRef]
37. Gao, X.; Yao, Y.; Wu, N.; Xu, M.; Zhao, Y.; Tu, Y. The sol-gel-sol transformation behavior of egg white proteins induced by alkali. *Int. J. Biol. Macromol.* **2020**, *155*, 588–597. [CrossRef]
38. Zhang, F.; Lu, Q.; Ming, J.; Dou, H.; Zuo, B.; Qin, M.; Li, F.; Kaplan, D.L.; Zhang, X. Silk dissolution and regeneration at the nanofibril scale. *J. Mater. Chem. B* **2014**, *2*, 3879–3885. [CrossRef]
39. Bai, Y.; Liu, X.; Shi, S.Q.; Li, J. A tough and mildew-proof soybean-based adhesive inspired by mussel and algae. *Polymers* **2020**, *12*, 756. [CrossRef]
40. Wang, Y.R.; Zhang, B.; Fan, J.L.; Yang, Q.; Chen, H.Q. Effects of sodium tripolyphosphate modification on the structural, functional, and rheological properties of rice glutelin. *Food Chem.* **2019**, *281*, 18–27. [CrossRef]
41. Li, K.; Jin, S.; Wei, Y.; Li, X.; Li, J.; Shi, S.Q.; Li, J. Bioinspired hyperbranched protein adhesive based on boronic acid-functionalized cellulose nanofibril and water-soluble polyester. *Compos. Part B-Eng.* **2021**, *219*, 108943. [CrossRef]
42. Farjami, T.; Babaei, J.; Nau, F.; Dupont, D.; Madadlou, A. Effects of thermal, non-thermal and emulsification processes on the gastrointestinal digestibility of egg white proteins. *Trends Food Sci. Technol.* **2021**, *107*, 45–56. [CrossRef]
43. Van den Berg, L.; Rosenberg, Y.; Van Boekel, M.A.J.S.; Rosenberg, M.; Velde, F. Microstructural features of composite whey protein/polysaccharide gels characterized at different length scales. *Food Hydrocoll.* **2009**, *23*, 1288–1298. [CrossRef]
44. Sheng, L.; Liu, Q.; Dong, W.; Cai, Z. Effect of high intensity ultrasound assisted glycosylation on the gel properties of ovalbumin: Texture, rheology, water state and microstructure. *Food Chem.* **2022**, *372*, 131215. [CrossRef]
45. Guo, L.; Niu, X.; Chen, X.; Lu, F.; Gao, J.; Chang, Q. 3D direct writing egg white hydrogel promotes diabetic chronic wound healing via self-relied bioactive property. *Biomaterials* **2022**, *282*, 121406. [CrossRef]
46. Chang, Q.; Darabi, M.A.; Liu, Y.; He, Y.; Zhong, W.; Mequanin, K.; Li, B.; Lu, F.; Xing, M.M.Q. Hydrogels from natural egg white with extraordinary stretchability, direct-writing 3D printability and self-healing for fabrication of electronic sensors and actuators. *J. Mater. Chem. A* **2019**, *7*, 24626–24640. [CrossRef]
47. Ganji, F.; Vasheghani, F.S.; Vasheghani, F.E. Theoretical description of hydrogel swelling: A review. *Iran. Polym. J.* **2010**, *19*, 375–398.
48. Olad, A.; Doustdar, F.; Gharekhani, H. Starch-based semi-IPN hydrogel nanocomposite integrated with clinoptilolite: Preparation and swelling kinetic study. *Carbohyd. Polym.* **2018**, *200*, 516–528. [CrossRef]
49. Zhang, B.; Xu, G.; Huang, Y. Prepration and characterization of silk fibroin-polyurethane composite hydrogels. *Acta Polym. Sin.* **2012**, *12*, 965–971. [CrossRef]
50. Lau, M.H.; Tang, J.; Paulson, A.T. Texture profile and turbidity of gellan/gelatin mixed gels. *Food Res. Int.* **2000**, *33*, 665–671. [CrossRef]
51. Chandra, M.V.; Shamasundar, B.A. Texture profile analysis and functional properties of gelatin from the skin of three species of fresh water fish. *Int. J. Food Prop.* **2015**, *18*, 572–584. [CrossRef]

52. Chakraborty, A.; Basak, S. Interaction with Al and Zn induces structure formation and aggregation in natively unfolded caseins. *J. Photochem. Photobiol. B* **2008**, *93*, 36–43. [CrossRef] [PubMed]
53. Shao, Y.; Zhao, Y.; Xu, M.; Chen, Z.; Wang, S.; Tu, Y. Effects of copper ions on the charteristics of egg white gel induced by strong alkali. *Poultry Sci.* **2017**, *96*, 4116–4123. [CrossRef] [PubMed]
54. Xue, X.; Hu, Y.; Deng, Y.; Su, J. Recent advances in design of functional biocompatible hydrogels for bone tissue engineering. *Adv. Funct. Mater.* **2021**, *31*, 2009432. [CrossRef]

Article

The Improved Properties of Carboxymethyl Bacterial Cellulose Films with Thickening and Plasticizing

Zhenbing Sun [1,†], Zhengjie Tang [1,†], Xiaoping Li [1,2,†], Xiaobao Li [1], Jeffrey J. Morrell [3,*], Johnny Beaugrand [4,*], Yao Yao [1] and Qingzhuang Zheng [1]

1. Yunnan Key Laboratory of Wood Adhesives and Glue Products, Southwest Forestry University, Kunming 650224, China
2. International Joint Research Center for Biomass Materials, Southwest Forestry University, Kunming 650224, China
3. National Centre for Timber Durability and Design Life, University of the Sunshine Coast, Brisbane, QLD 4102, Australia
4. Biopolymères Interactions Assemblages (BIA), INRA, Rue de la Géraudière, F-44316 Nantes, France
* Correspondence: jmorrell@usc.edu.au (J.J.M.); johnny.beaugrand@inrae.fr (J.B.)
† These authors contributed equally to this work.

Citation: Sun, Z.; Tang, Z.; Li, X.; Li, X.; Morrell, J.J.; Beaugrand, J.; Yao, Y.; Zheng, Q. The Improved Properties of Carboxymethyl Bacterial Cellulose Films with Thickening and Plasticizing. *Polymers* **2022**, *14*, 3286. https://doi.org/10.3390/polym14163286

Academic Editors: Jingpeng Li, Yun Lu and Huiqing Wang

Received: 21 July 2022
Accepted: 9 August 2022
Published: 12 August 2022

Publisher's Note: MDPI stays neutral with regard to jurisdictional claims in published maps and institutional affiliations.

Copyright: © 2022 by the authors. Licensee MDPI, Basel, Switzerland. This article is an open access article distributed under the terms and conditions of the Creative Commons Attribution (CC BY) license (https://creativecommons.org/licenses/by/4.0/).

Abstract: This study aims to improve the thermal stability and mechanical properties of carboxymethyl bacterial cellulose (CMBC) composite films. Experiments were conducted by preparing bacterial cellulose (BC) into CMBC, then parametrically mixing sodium alginate/starch/xanthan gum/gelatin and glycerin/sorbitol/PEG 400/PEG 6000 with CMBC to form the film. Scanning electron microscopy, X-ray diffractometry, infrared spectroscopy, mechanical tests, and thermogravimetric analysis showed that the composite films had better mechanical properties and thermal stability with the addition of 1.5% CMBC (% v/v), 1% sodium alginate, and 0.4% glycerin. Tensile strength was 38.13 MPa, the elongation at break was 13.4%, the kinematic viscosity of the film solution was 257.3 mm^2/s, the opacity was 4.76 A/mm, the water vapor permeability was 11.85%, and the pyrolysis residue was 45%. The potential causes for the differences in the performance of the composite films were discussed and compared, leading to the conclusion that CMBC/Sodium alginate (SA)/glycerin (GL) had the best thermal stability and mechanical properties.

Keywords: glycerin; sodium alginate; tensile strength; elongation at break

1. Introduction

Bacterial cellulose (BC) is a porous reticulated nanoscale biopolymer synthesized by microbial fermentation [1]. It is a straight-chain molecule consisting of β-D-glucose bound by β-1,4-glycosidic bonds and is also known as β-1,4-glucan [2]. It was first discovered by the British scientist A.J. Brown in 1886 and was identified by physical and chemical analysis as a hydrolyzed cellulose-like substance, with glucose as the main component of the hydrolysate [3]. Bacterial cellulose consists of unique filamentous fibers with a diameter of 0.01–0.10 μm, two to three orders of magnitude smaller than vegetable cellulose. Each filamentous fiber consists of several microfibers in a mesh structure [4]. In addition, bacterial cellulose has high crystallinity, good elasticity and mechanical properties [5], biodegradability, low cost and toxicity [6], and high biocompatibility [7], making it useful as a medical packaging material.

Carboxymethyl cellulose (CMC) is an industrially essential biopolymer resulting from the partial substitution of the 2, 3, and 6 hydroxyl groups on the glucose unit of cellulose by the carboxymethyl group [8,9]. To reduce the use of plastics, researchers have studied a wide range of materials such as CMC, gelatin, carrageenan, starch films, and other biosourced polymers. However, all pure polymer films have some drawbacks, such as the brittleness of gelatin [10,11], the poor thermal stability of carrageenan films [12,13], and the poor water solubility and weak mechanical properties of CMC and starch films [14–17]. The tensile strength of pure CMC films is 6–9 MPa [18–21], which cannot meet the requirements for most commercial applications. Modification of CMC-based composite films has been explored by adding different materials to produce composite films with improved mechanical properties [22], thermal stability [23], and degradability, potentially making these materials useful for food preservation [24].

Many studies have been carried out on CMC-based composite films. Rungsiri et al. prepared CMC from durian peel and combined it with rice starch to produce composite films [25]. The performance of these composite films varied between starch and CMC blends, as the amounts of straight-chain and branched-chain polymers in the starch affected functional properties and interactions with other materials [26]. The composite film was tested on tomatoes for its ability to inhibit spoilage and quality loss. Nazmi et al. showed that the addition of gelatine to CMC reduced the flexibility of the film but improved the tensile strength, puncture resistance, and thermal stability [27]. In addition, different gelatin sources had different effects on the mechanical properties of the films. In summary, adding starch, gelatine, and sodium alginate to most CMC films improves their mechanical properties and thermal stability [28,29]. However, BC is also a type of cellulose. There are few reports on the preparation of CMBC and its compounding with macromolecular polymers to produce films with improved mechanical properties and thermal stability.

2. Materials and Methods

2.1. BC Preparation

BC made from *Taonella mepensis* was supplied by Beinacruz Biotechnology Co., Ltd. (Suzhou, Jiangsu Province, China). A batch of 54 experimental bottles (cylindrical containers with a diameter of 6cm and a height of 8 cm) was cultivated by autoclaving, inoculating with the bacterium, incubating at 120 °C for 30 min, and finally incubating for 7 days at 37 °C in a medium containing: 20.0 g glucose (Tianjin Fuyu Fine Chemical Co., Ltd., Tianjin, China), 5.0 g yeast paste (Beijing Aoboxing Bio-tech Co., Ltd., Beijing, China), 1.0 g K_2HPO_4 (Shanghai Aladdin Biochemical Technology Co., Ltd., Shanghai, China), 15.0 g $MgSO_4$ (Tianjin Windship Chemical Reagent Technology Co., Ltd., Tianjin, China), 5 mL anhydrous ethanol (Tianjin Fuyu Fine Chemical Co., Ltd., Tianjin, China), and 1.0 L distilled water at pH 4.5. The wet BC (Figure 1A) was taken out and soaked in distilled water, and the distilled water was changed every hour for 12 consecutive times. The wet BC was cut into 3 cm-diameter pieces that were treated with 1% sodium hydroxide solution (Tianjin Fuyu Fine Chemical Co., Ltd., Tianjin, China) at 80 °C for 1 h. The treated BC was soaked in an excess of distilled water, which was changed every 2 h until the pH was 7. The BC gel was drained and vacuum-dried for 72 h at −80 °C before being crushed into a powder using a high-speed pulverizer (Yongkang Pu'ou Hardware Products Co., Ltd., Jinhua, Zhejiang Province, China). The ground material was passed through a 60-mesh sieve, and prepared for use.

Figure 1. *Cont.*

Figure 1. (**A**) Examples of BC (Contains water), (**B**) Sodium alginate, starch, xanthan gum, and gelatin, (**C**) Composite film solution, (**D**) CMBC/SA/GL composite film.

2.2. Carboxymethyl Cellulose BC (CMBC) Preparation

8 g BC, 160 mL of 95% ethanol (Tianjin Fuyu Fine Chemical Co., Ltd., Tianjin, China), and 40 mL of 30% NaOH (Tianjin Fuyu Fine Chemical Co., Ltd., Tianjin, China) solution were mixed and stirred for 60 min at 30 °C. Then 10 g of sodium chloroacetate (Shanghai Macklin Biochemical Co., Ltd., Shanghai, China) were added and the temperature was increased to 65 °C and stirred for 3 h. Glacial acetic acid (90%) (Tianjin Hengxing Chemical Preparation Co., Ltd., Tianjin, China) solution was added to reduce the pH of the mixture and then the samples were washed with ethanol (Tianjin Fuyu Fine Chemical Co., Ltd., Tianjin, China) until the pH was 7. The neutralized samples were oven-dried at 65 °C and stored for later use.

2.3. Preparation of CMBC Composite films

1.5 g of CMBC was placed in a heated magnetic mixer (Shanghai Lichenbang Instrument Technology Co., Ltd, DF-101Z, Shanghai, China) for 30 min at 45 °C at a stirring rate of 900–1000 r/min while different proportions of sodium alginate (Beijing Coolaber Technology Co., Ltd., Beijing, China) (Figure 1B), starch (SR) (Chengdu Jinshan Chemical Reagent Co., Ltd., Chengdu, Sichuan Province, China), xanthan gum (XG) (Shanghai Yuanye Biotechnology Co., Ltd., Shanghai, China), gelatin (GEL) (Tianjin Windship Chemical Reagent Technology Co., Ltd., Tianjin, China) were added to produce a final concentration of 1.5 percent solution (0.2%, 0.4%, 0.7%, 1%, 1.3%, all the above proportions are based on the mass of the solvent, the same below) along with different levels of glycerin (Shanghai Yuanye Biotechnology Co., Ltd., Shanghai, China), sorbitol (Tianjin Fuyu Fine Chemical Co., Ltd., Tianjin, China) (SO), PEG 400 (Beijing Coolaber Technology Co., Ltd., Beijing, China), and PEG 6000 (Beijing Coolaber Technology Co., Ltd., Beijing, China) (0.2%, 0.4%, 0.6%). The mixtures were stirred to dissolve the components and then the film solution (Figure 1C) was placed in an ultrasonic apparatus (Kunshan ultrasonic Intrasonic Co., Ltd., KF-101Z, Kunming, China) operated at 50 HZ for 12 min to remove air bubbles. The film was cast on a PTFE mold (Yangzhong Fuda Insulation Electric Co., Ltd., Yangzhong, Jiangsu Province, China), dried in a blast box (Shanghai Yiheng Scientific Instruments Co., Ltd., Shanghai, China) at 30 °C for 48 h, and then removed (Figure 1D).

2.4. Properties of the Composite Films

Tensile strength (MPa) and elongation at break (%) were measured on ten 0.089- to 0.098-mm by 150-mm-long dog-bone samples of each material on a Universal Testing Machine according to procedures described in GB/T 1040.1-20 06 (Plastics Determination of tensile properties). A load was applied to failure at a rate of 1 mm/min.

The opacity of the CMBC composite films was tested by cutting 10- by 40-mm-long samples and placing them on the inner surface on one side of a cuvette and then measuring absorbance at 600 nm on a 752# ultraviolet spectrophotometer (XP-Spectrum Company, Shanghai, China). Five tests were performed for each material [30].

The viscosity of the composite film solution was measured using a Nicolay rotational rheometer meter (MARS60, Thermo Fisher Scientific, Waltham, MA, USA).

Water vapor permeability examined the ability of water vapor to pass through the composite film in order to determine the suitability of each film for maintaining internal moisture.

The water vapor transmission coefficient of the specimen was calculated according to Equation (1) [31].

$$P = \frac{\Delta m \times d}{A \times t \times \Delta P} \quad (1)$$

where P is the water vapor transmission coefficient of the sample in grams of centimeters per square centimeter per second Pascal [g. cm/(cm². s. Pa)],

Δm is the amount of change in the mass of the sample in grams (g) during the period t,

A is the sample area through the water vapor in square meters (m^2),

t is the difference in time between two intervals after the mass change has stabilized in hours (h),

d is the thickness of the specimen in centimeters (cm), and

ΔP is the difference in water vapor pressure between the two sides of the specimen in Pascal (Pa).

2.5. Material Characterization

Microstructure: The composite films were placed on an aluminum grid and examined by field emission scanning electron microscopy on a Nova Nano SEM450 microscope (FEI, Hillsboro, OR, USA). At least five fields were examined for each material.

Fourier Transform Infrared Spectroscopy (FTIR): The composite films were analyzed on a Nicolet i50 FTIR Analyzer (Thermo Scientific, Waltham, MA, USA). The samples were subjected to 64 scans, and the resulting spectra were baseline-corrected and then analyzed for differences in spectra for different raw materials.

X-ray diffraction (XRD): Bacterial cellulose and carboxymethyl bacterial cellulose were examined by X-ray diffractometry on a Rigaku Ultima IV X-ray diffractometer (Rigaku Corp, Tokyo, Japan) (XRD, Ulti,) using a scanning angle from 10° to 40°, a step size of 0.026° (accelerating current = 30 mA and voltage = 40 kV), and Cu-Kα radiation of λ = 0.154 nm.

Thermogravimetric (TG) analysis: Approximately 5.0 to 6.0 mg of the Carboxymethyl BC composite films were ground to pass an 80-mesh to 120-mesh screen and placed into sample holders for analysis on a TGA 92 thermo gravimetric analyzer (KEP Technologies EMEA, Caluire, France). N_2 was used as the shielding gas and Al_2O_3 as the reference compound. The temperature was increased from room temperature (approx. 20–23 °C) to 600 °C at a rate of 20 °C/min to produce thermogravimetric curves.

3. Results and Discussion

3.1. TS and EB of the Composite Films

Tensile strength (TS) and elongation at break (EB) are the basic indicators for evaluating the film properties. Tables 1–4 show that the additions of glycerin, sorbitol, PEG 400, and PEG 6000 were positively correlated with the elongation at break of the films, and this result is consistent with the results of previous studies [32]. It is worth mentioning that the addition of glycerin had the most significant effect on the elongation at break of the composite films. In addition, the tensile strength of the composite films first increased and then decreased with the addition of sodium alginate, starch, xanthan gum, or gelatin.

Table 1. The effect of GI addition on tensile strength (TS) and elongation at break (EB) of the composite films.

GI Addition	GI (0.2%)		GI (0.4%)		GI (0.6%)	
properties	TS (MPa)	EB (%)	TS (MPa)	EB (%)	TS (MPa)	EB (%)
SA (0.2%)	12.2 (2.28)	29.7 (3.05)	10.6 (2.02)	48.2 (4.09)	6.70 (0.42)	52.5 (4.17)
SA (0.4%)	13.7 (1.19)	25.3 (2.04)	13.6 (2.01)	42.1 (2.89)	13.3 (1.19)	51.4 (3.34)
SA (0.7%)	19.2 (1.23)	23.9 (2.36)	14.9 (2.77)	37.6 (2.49)	12.1 (1.89)	45.6 (3.31)
SA (1.0%)	38.1 (13.4)	25.3 (2.04)	15.1 (1.18)	35.2 (3.35)	15.6 (1.82)	44.3 (3.57)
SA (1.3%)	21.1 (2.05)	10.3 (0.78)	24.6 (2.71)	31.5 (2.55)	15.2 (1.27)	35.4 (2.60)
SR (0.2%)	18.7 (1.71)	33.5 (4.15)	20.8 (3.48)	55.7 (1.68)	12.5 (1.83)	65.7 (9.07)
SR (0.4%)	24.9 (2.13)	25.0 (2.13)	30.1 (5.36)	53.6 (3.57)	20.9 (2.69)	65.0 (3.18)
SR (0.7%)	21.1 (3.01)	16.8 (1.22)	21.4 (3.13)	43.5 (2.24)	10.4 (1.82)	52.6 (2.92)
SR (1.0%)	18.5 (3.13)	10.6 (2.26)	21.2 (2.83)	39.9 (2.56)	10.5 (1.61)	50.0 (1.91)
SR (1.3%)	14.9 (1.26)	7.03 (1.03)	23.8 (1.90)	20.8 (2.81)	8.90 (0.72)	15.8 (1.59)
XG (0.2%)	20.6 (2.67)	6.98 (0.84)	29.7 (2.45)	29.1 (2.74)	10.8 (0.84)	32.0 (2.45)
XG (0.4%)	22.7 (3.65)	7.54 (0.76)	30.3 (5.66)	28.6 (2.38)	12.4 (2.18)	30.8 (2.76)
XG (0.7%)	25.2 (2.79)	8.35 (1.53)	26.6 (2.66)	25.4 (2.49)	14.2 (2.30)	29.3 (2.89)
XG (1.0%)	26.2 (2.93)	9.19 (1.57)	29.4 (4.39)	14.7 (1.43)	13.3 (2.24)	29.9 (2.81)
XG (1.3%)	29.1 (3.93)	6.79 (1.01)	30.1 (3.25)	13.0 (2.64)	16.1 (2.99)	21.5 (3.69)
GEL (0.2%)	15.3 (3.15)	23.9 (4.91)	10.8 (0.89)	25.9 (1.65)	11.9 (1.23)	27.0 (2.78)
GEL (0.4%)	30.6 (0.24)	46.2 (5.08)	13.2 (2.28)	49.8 (4.94)	10.7 (1.48)	57.8 (6.03)
GEL (0.7%)	15.6 (1.30)	18.7 (1.09)	19.7 (1.89)	19.0 (2.34)	21.3 (3.26)	35.7 (4.67)
GEL (1.0%)	7.01 (2.05)	10.8 (1.84)	15.1 (1.18)	32.2 (2.65)	15.8 (1.22)	23.6 (2.44)
GEL (1.3%)	10.7 (2.60)	14.6 (2.85)	27.0 (3.85)	38.4 (0.42)	14.1 (3.69)	13.6 (0.42)

Numbers in parentheses are standard deviations.

Table 2. The effect of SO addition on the tensile strength (TS) and elongation at break (EB) of the composite films.

SO Addition	SO (0.2%)		SO (0.4%)		SO (0.6%)	
properties	TS (MPa)	EB (%)	TS (MPa)	EB (%)	TS (MPa)	EB (%)
SA (0.2%)	24.6 (3.97)	24.4 (0.75)	24.0 (4.88)	30.5 (2.51)	22.1 (3.77)	45.8 (0.36)
SA (0.4%)	25.6 (4.65)	15.9 (1.90)	18.8 (1.94)	29.7 (2.11)	17.5 (0.27)	32.4 (2.93)
SA (0.7%)	26.2 (0.03)	13.7 (0.37)	24.1 (3.13)	20.5 (1.94)	16.9 (1.26)	29.5 (1.57)
SA (1.0%)	26.9 (2.67)	7.93 (0.89)	26.0 (3.71)	10.7 (1.06)	13.3 (0.99)	23.5 (1.12)
SA (1.3%)	26.4 (3.74)	6.09 (1.57)	16.6 (1.72)	4.24 (0.69)	14.0 (0.83)	16.9 (1.01)
SR (0.2%)	20.8 (1.71)	9.80 (1.58)	24.5 (2.89)	27.6 (2.65)	18.7 (1.85)	32.1 (5.68)
SR (0.4%)	26.6 (4.01)	7.56 (1.48)	27.0 (2.91)	21.1 (2.35)	26.7 (3.47)	26.8 (4.74)
SR (0.7%)	25.9 (4.27)	5.98 (0.34)	26.2 (2.30)	16.2 (1.65)	26.1 (1.93)	25.1 (3.90)
SR (1.0%)	21.2 (2.94)	4.35 (0.84)	24.5 (2.73)	9.01 (2.25)	20.8 (2.10)	13.9 (1.87)
SR (1.3%)	26.6 (3.00)	4.21 (1.67)	26.6 (2.19)	4.29 (0.60)	19.1 (2.14)	6.67 (0.84)
XG (0.2%)	18.6 (3.56)	7.09 (0.95)	20.8 (1.32)	8.08 (1.08)	16.9 (1.63)	8.78 (0.96)
XG (0.4%)	19.9 (0.92)	3.02 (0.54)	25.3 (2.05)	7.01 (1.38)	20.5 (2.08)	7.32 (0.95)
XG (0.7%)	16.5 (3.67)	1.53 (0.34)	20.6 (2.08)	3.00 (0.56)	19.4 (1.44)	4.88 (1.20)
XG (1.0%)	12.3 (2.30)	1.51 (0.29)	21.6 (0.55)	2.80 (0.13)	14.7 (1.24)	3.15 (0.91)
XG (1.3%)	12.8 (1.36)	1.16 (0.10)	19.6 (2.27)	1.83 (0.43)	25.1 (2.31)	2.29 (0.53)
GEL (0.2%)	14.1 (1.50)	13.8 (1.85)	16.0 (0.70)	18.1 (2.73)	7.55 (1.34)	25.1 (3.42)
GEL (0.4%)	9.11 (1.49)	6.45 (1.42)	9.67 (0.78)	8.13 (1.82)	10.7 (2.38)	20.5 (2.72)
GEL (0.7%)	10.5 (1.43)	5.87 (0.45)	18.9 (1.89)	8.67 (1.53)	11.9 (2.87)	9.10 (1.86)
GEL (1.0%)	8.10 (2.45)	5.56 (0.89)	25.8 (1.87)	6.77 (0.07)	16.1 (1.12)	3.62 (0.05)
GEL (1.3%)	7.42 (1.73)	2.75 (0.74)	29.1 (3.29)	4.37 (0.38)	23.6 (2.56)	5.29 (0.62)

Numbers in parentheses are standard deviations.

Table 3. The effect of PEG400 addition on the tensile strength (TS) and elongation at break (EB) of the composite films.

PEG 400 Addition	PEG 400 (0.2%)		PEG 400 (0.4%)		PEG 400 (0.6%)	
properties	TS (MPa)	EB (%)	TS (MPa)	EB (%)	TS (MPa)	EB (%)
SA (0.2%)	23.2 (0.99)	6.66 (1.31)	18.2 (3.79)	8.39 (1.66)	14.8 (3.32)	14.7 (3.38)
SA (0.4%)	16.4 (4.12)	3.67 (0.32)	18.0 (3.25)	5.14 (0.51)	11.2 (1.41)	10.7 (5.09)
SA (0.7%)	24.1 (1.30)	6.54 (0.95)	21.4 (3.45)	1.43 (0.42)	19.5 (1.53)	9.23 (1.28)
SA (1.0%)	28.0 (1.68)	3.09 (0.66)	26.5 (4.19)	5.42 (0.66)	17.9 (0.90)	6.22 (1.21)
SA (1.3%)	26.9 (1.18)	1.40 (0.58)	26.1 (1.67)	2.13 (0.34)	20.4 (3.77)	3.77 (0.39)
SR (0.2%)	17.9 (3.34)	13.4 (2.66)	18.4 (1.23)	10.0 (0.60)	15.1 (1.65)	3.94 (0.96)
SR (0.4%)	19.6 (0.66)	8.70 (0.46)	19.9 (0.93)	7.30 (1.05)	18.0 (2.05)	6.64 (0.54)
SR (0.7%)	17.6 (0.48)	2.50 (0.21)	17.0 (0.63)	2.50 (0.21)	17.1 (0.64)	2.33 (0.07)
SR (1.0%)	19.3 (0.65)	2.19 (0.70)	15.8 (1.21)	1.57 (0.23)	14.3 (2.89)	1.35 (0.29)
SR (1.3%)	20.1 (2.45)	1.06 (0.53)	18.04 (2.01)	1.50 (0.34)	19.3 (1.56)	1.21 (0.72)
XG (0.2%)	10.9 (0.53)	2.92 (0.20)	13.9 (0.54)	2.48 (1.08)	9.64 (1.31)	2.98 (0.11)
XG (0.4%)	8.04 (0.64)	1.15 (0.06)	17.8 (0.14)	1.39 (0.23)	16.6 (1.34)	1.50 (0.19)
XG (0.7%)	9.33 (0.81)	0.60 (0.04)	16.7 (1.05)	0.75 (0.06)	14.7 (0.19)	1.09 (0.31)
XG (1.0%)	9.52 (0.15)	0.51 (0.07)	10.7 (1.87)	0.54 (0.02)	8.76 (0.89)	0.98 (0.06)
XG (1.3%)	9.75 (0.75)	0.73 (0.21)	10.4 (1.14)	0.72 (0.54)	12.6 (1.32)	0.74 (0.24)
GEL (0.2%)	—	—	—	—	—	—
GEL (0.4%)	—	—	—	—	—	—
GEL (0.7%)	—	—	—	—	—	—
GEL (1.0%)	—	—	—	—	—	—
GEL (1.3%)	—	—	—	—	—	—

Numbers in parentheses are standard deviations. —: Indicates that some samples in this group did not form a composite film.

Table 4. The effect of PEG6000 addition on the tensile strength (TS) and elongation at break (EB) of the composite films.

PEG 6000 Addition	PEG 6000 (0.2%)		PEG 6000 (0.4%)		PEG 6000 (0.6%)	
properties	TS (MPa)	EB (%)	TS (MPa)	EB (%)	TS (MPa)	EB (%)
SA (0.2%)	13.6 (0.68)	9.74 (0.65)	31.8 (2.13)	3.59 (0.29)	23.9 (0.63)	1.77 (0.17)
SA (0.4%)	14.7 (0.83)	4.77 (0.36)	16.7 (1.56)	2.14 (0.54)	15.9 (2.08)	2.07 (0.20)
SA (0.7%)	23.4 (4.17)	2.53 (0.61)	16.5 (0.95)	2.34 (0.02)	18.4 (0.89)	2.33 (0.25)
SA (1.0%)	25.5 (1.10)	3.20 (0.30)	17.8 (0.93)	2.05 (0.05)	13.2 (0.41)	1.14 (0.19)
SA (1.3%)	20.5 (0.29)	2.07 (0.18)	18.8 (1.32)	2.04 (0.01)	10.9 (0.68)	0.75 (0.16)
SR (0.2%)	21.9 (1.67)	2.22 (0.25)	28.8 (1.17)	3.46 (0.38)	29.0 (1.93)	4.99 (0.69)
SR (0.4%)	25.6 (1.14)	2.11 (0.50)	30.1 (1.20)	2.97 (0.27)	21.4 (0.76)	3.36 (0.76)
SR (0.7%)	28.1 (2.05)	2.50 (0.37)	28.9 (0.61)	3.02 (0.55)	17.3 (1.90)	3.06 (0.38)
SR (1.0%)	26.2 (1.41)	1.38 (0.15)	22.1 (0.88)	2.04 (0.30)	10.6 (1.43)	2.11 (0.20)
SR (1.3%)	25.2 (0.51)	1.41 (0.20)	21.4 (2.10)	1.04 (0.34)	20.1 (2.34)	1.52 (0.23)

Numbers in parentheses are standard deviations.

Table 5 shows the p values and correlation coefficients of the binary regression equation with thickener and plasticizer as independent variables and tensile strength and elongation at break as dependent variables. Thickener and plasticizer were positively correlated with elongation at break of the composite film. The correlation coefficient was above 0.86, and the p value was >0.05. In addition, except for CMBC/GEL/GL (Figure 2D), CMBC/SA/PEG 6000 (Figure 2L), and CMBC/SR/PEG 6000 (Figure 2M), the addition of thickeners and plasticizers was positively correlated with tensile strength of the films, with a correlation coefficient above 0.72 and p value > 0.05.

Table 5. p value and Correlation coefficients for the effects of thickeners and plasticizers on tensile strength (TS) and elongation at break (EB) of composite films.

Composite Film	TS (MPa)		EB (%)	
	p Value	Correlation Coefficient	p Value	Correlation Coefficient
CMBC/SA/Gl	<0.0001	0.94	<0.0001	0.99
CMBC/SR/Gl	0.0019	0.95	<0.0001	0.99
CMBC/XG/Gl	0.0002	0.97	0.0001	0.98
CMBC/GEL/Gl	>0.05	—	0	0
CMBC/SA/SO	0.0008	0.91	<0.0001	0.99
CMBC/SR/SO	0.0164	0.90	<0.0001	0.98
CMBC/XG/SO	<0.0001	0.99	0.0002	0.98
CMBC/GEL/SO	0.0004	0.97	<0.0001	0.99
CMBC/SA/PEG400	<0.0001	0.93	0.0022	0.95
CMBC/SR/PEG400	0.0251	0.72	0.0003	0.97
CMBC/XG/PEG400	0.0114	0.91	<0.0001	0.99
CMBC//GEL/PEG400	—	—	—	—
CMBC/SA/PEG6000	>0.05	—	0.0057	0.86
CMBC/SR/PEG6000	>0.05	—	<0.0001	0.96

—: Indicates that when p value > 0.05, there is no correlation coefficient.

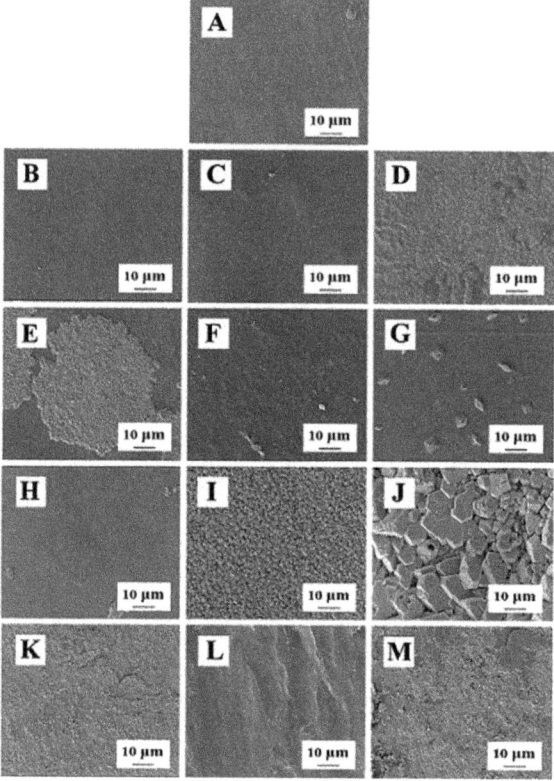

Figure 2. The microstructure of the composite films (**A**): CMBC/SA/Gl; (**B**): CMBC/SR/Gl; (**C**): CMBC/XG/Gl; (**D**): CMBC/GEL/Gl; (**E**): CMBC/SA/SO; (**F**): CMBC/SR/SO; (**G**): CMBC/XG/SO; (**H**): CMBC/GEL/SO; (**I**): CMBC/SA/PEG 400; (**J**): CMBC/SR/PEG 400; (**K**): CMBC/XG/PEG 400; (**L**): CMBC/SA/PEG 6000; (**M**): CMBC/SR/PEG 6000).

Table 6 shows that among the various composite films, the CMBC/SA/GL composite film had the best mechanical properties, with a tensile strength of 38.13 MPa and an elongation at break of 13.40%, at which time sodium alginate (1%) and propanetriol (0.2%) were added. The tensile strength of conventional plastics ranges from 8 MPa to 20 MPa [33]. In comparison, the tensile strength of the composite film was 38.13 MPa, indicating that the composite film had better mechanical properties than conventional plastics.

Table 6. The physical properties of CMBC composite films with difference additives.

Composite Film	Amount of Plasticizer Added (%)	Amount of Thickener Added (%)	The Kinematic Viscosity (mm^2/s)	Opacity (A/mm)	Water Vapor Permeability g·cm/(cm^2·s·Pa)
CMBC/SA/Gl	1	0.2	257.3 (3.34)	4.76 (1.11)	0.12 (0.02)
CMBC/SR/Gl	0.4	0.4	52.9 (5.38)	5.00 (0.38)	0.07 (0.01)
CMBC/XG/Gl	0.4	0.4	1018.1 (19.4)	5.93 (0.57)	0.11 (0.03)
CMBC/GEL/Gl	0.4	0.2	39.1 (3.82)	5.21 (0.31)	0.08 (0.02)
CMBC/SA/SO	1	0.2	821.0 (8.53)	5.76 (0.45)	0.10 (0.07)
CMBC/SR/SO	0.4	0.4	43.9 (2.32)	11.1 (0.53)	0.06 (0.01)
CMBC/XG/SO	0.4	0.4	1382.0 (21.70)	6.79 (0.14)	0.06 (0.03)
CMBC/GEL/SO	0.3	0.4	54.0 (4.03)	5.37 (0.43)	0.04 (0.01)
CMBC/SA/PEG400	1	0.2	471.1 (2.95)	4.70 (0.08)	0.07 (0.02)
CMBC/SR/PEG400	1.3	0.2	34.6 (5.34)	9.64 (1.03)	0.1 (0.04)
CMBC/XG/PEG400	0.4	0.4	1238.8 (10.60)	10.2 (1.92)	0.08 (0.03)
CMBC//GEL/PEG400	—	—	—	—	—
CMBC/SA/PEG6000	1	0.2	579.9 (4.31)	5.53 (0.32)	0.11 (0.03)
CMBC/SR/PEG6000	0.4	0.4	27.9 (3.33)	10.1 (0.94)	0.07 (0.02)

The numbers in parentheses are standard deviations. —: Indicates that some samples in this group did not form a composite film.

Table 6 also shows the kinematic viscosity, opacity, and water vapor permeability for each group of best-performing samples. Thickeners had a significant effect on the kinematic viscosity of the film solution for four different thickeners in the film solution in the order of xanthan gum, sodium alginate, starch, and gelatin composite film. The kinematic viscosity of the CMBC/XG/SO composite film was the largest at 1381.95 mm^2/s, indicating that the solution had poor advective properties. The film did not easily form a smooth surface when poured into the mold and appeared inhomogeneous, typical of non-Newtonian fluids. The kinematic viscosity of the CMBC/SR/PEG 6000 composite film was the smallest, only 27.87 mm^2/s.

The opacity of each group of composite films is shown in Table 6. The opacity of the laminated film ranged from 4.76 to 11.1 A/mm. Compared with previous studies [34], the opacity of the composite films was higher in all groups because the molecular arrangement of the CMBC was more ordered than that of the CMC. This blocked the passage of visible light [35], making the composite films more crystalline and opaque [36]. In addition, the opacity of the composite films with glycerin as the plasticizer was lower, probably because the glycerin produced composite films with a lower molecular structure that was initially arranged in an orderly manner, thus making it easier for light to pass through.

Crystallinity, molecular size, and other properties of macromolecular polymers all affect the water vapor permeability of the films [37]. The water vapor permeability of the composite films was between 0.04% and 0.12%, lower compared to previous studies [38].

These differences may be due to the more ordered molecular arrangement of CMBC compared to CMC.

3.2. Micro-Structure of the Composite Films

Figure 2 shows the composite films prepared with glycerin as the plasticizer, except for the CMBC/GEL/GL composite film, which had many fluid-like attachments on the surface (Figure 2D). The other three composite films had compact surfaces without cracks and bumps (Figure 2A–C), indicating that the components in the composite films were well-integrated and had good interaction forces. In addition, the composite films prepared with sorbitol as the plasticizer formed many round pie-shaped (Figure 2E), crumbled (Figure 2F), granular (Figure 2G), and powder-like substances (Figure 2F,H) on the surface, indicating that some thickeners or CMBC were not incorporated into the composite film system. The composite films prepared with PEG400 and PEG6000 as plasticizers formed a large number of spherical particles (Figure 2I), clay-like materials (Figure 2J), sand-like particles (Figure 2K,M), and ridge-like protrusions (Figure 2L) on the surface, indicating that these composite films had a loose structure and that the components were not sufficiently integrated.

3.3. FTIR Spectrum of Composite Films

The infrared spectra of the composite films are shown in Figure 3. The (O-H) stretching vibration peak of the CMBC composite membrane appears at 3280 cm^{-1}, while the (O-H) stretching vibration peak of the carboxymethyl cellulose appears at 3420 cm^{-1} [39], indicating the formation of extensive hydrogen bonds and strong interactions between the additives and the CMC [40]. As can be seen in Figure 3, the addition of sodium alginate/starch/xanthan gum/gelatin and glycerin/sorbitol/PEG 400/PEG 6000 did not change the chemical structure of the CMBC, as can be observed at 1600 cm^{-1} for carboxylate asymmetric stretching and 1420 cm^{-1} and 1310 cm^{-1} for carboxylate asymmetric stretching [41].

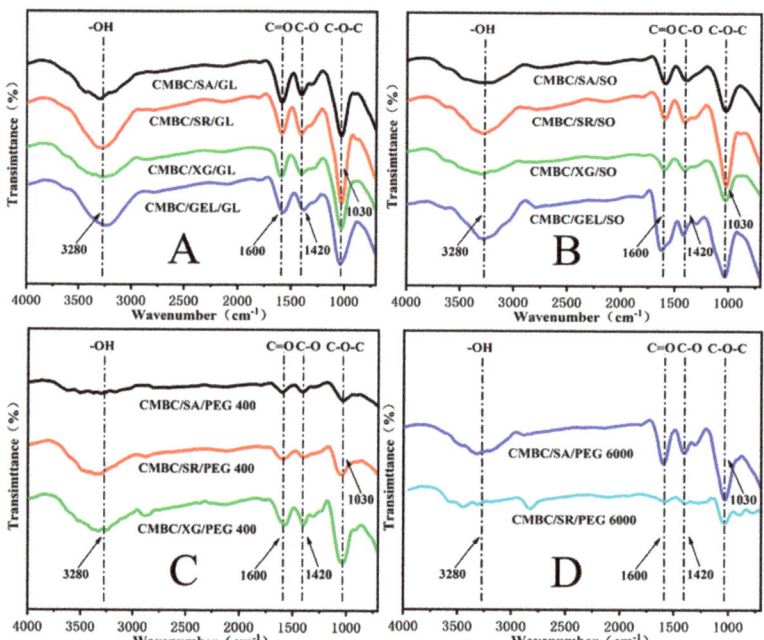

Figure 3. FTIR images of the composite films (**A**): CMBC/SA/G, CMBC/SR/Gl, CMBC/XG/Gl & CMBC/GEL/Gl; (**B**): CMBC/SA/SO, CMBC/SR/SO, CMBC/XG/SO & CMBC/GEL/SO; (**C**): CMBC/SA/PEG 400, CMBC/SR/PEG 400 & CMBC/XG/PEG 400; (**D**): CMBC/SA/PEG 6000 & CMBC/SR/PEG 6000).

3.4. XRD-Pattern Characteristics of the Composite Films

Figure 4 shows the XRD pattern of the composite film. The composite film with glycerol and PEG 6000 as plasticizers showed a broad diffraction peak at diffraction angles of 19.8° to 22.1°, indicating that the components of the composite film were well-bound and strongly interacted with each other (Figure 4A,D; Figure 4B,C). CMBC/CG/PEG 400 showed a broad diffraction peak at 11.3° and 20.6°, respectively [42], both indicating that sodium alginate and xanthan gum were not fully integrated into the composite film system.

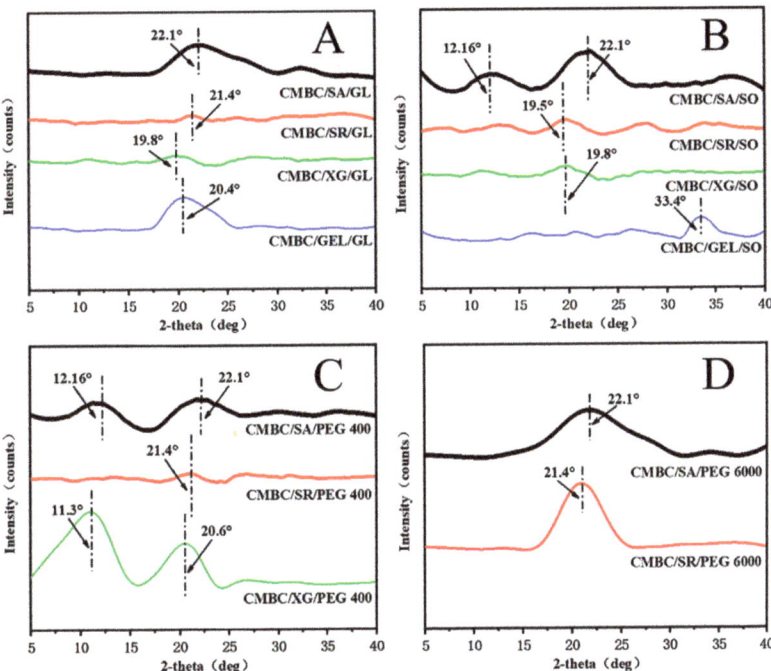

Figure 4. XRD patterns of the composite films (**A**): CMBC/SA/G, CMBC/SR/Gl, CMBC/XG/Gl & CMBC/GEL/Gl; (**B**): CMBC/SA/SO, CMBC/SR/SO, CMBC/XG/SO & CMBC/GEL/SO; (**C**): CMBC/SA/PEG 400, CMBC/SR/PEG 400 & CMBC/XG/PEG 400; (**D**): CMBC/SA/PEG 6000 & CMBC/SR/PEG 6000).

3.5. The Pyrolysis Characteristics of the Composite Films

Figure 5A–F illustrates the thermal performance of the films. The CMBC/GEL/GL, CMBC/SA/PEG 6000, and CMBC/SR/PEG 6000 composite films all showed two other pyrolysis peaks in addition to a pyrolysis peak of bound water near 100 °C [43], indicating that these three composite films had a nonuniform internal structure leading to poor thermal stability. However, the initiation temperature of the CMBC/SR/GL composite film was significantly lower than the temperatures of the other composite films. The main reason for this is that some of the starch was not dissolved during the preparation of the composite films and collected on the film surface. The TDG-TG diagram shows that the pyrolysis peak of the composite film was between 250 °C and 311 °C. In addition, from the DTG-TG curves in Figure 5 and related calculations, the following main parameters commonly used to reflect the pyrolysis characteristics can be obtained:

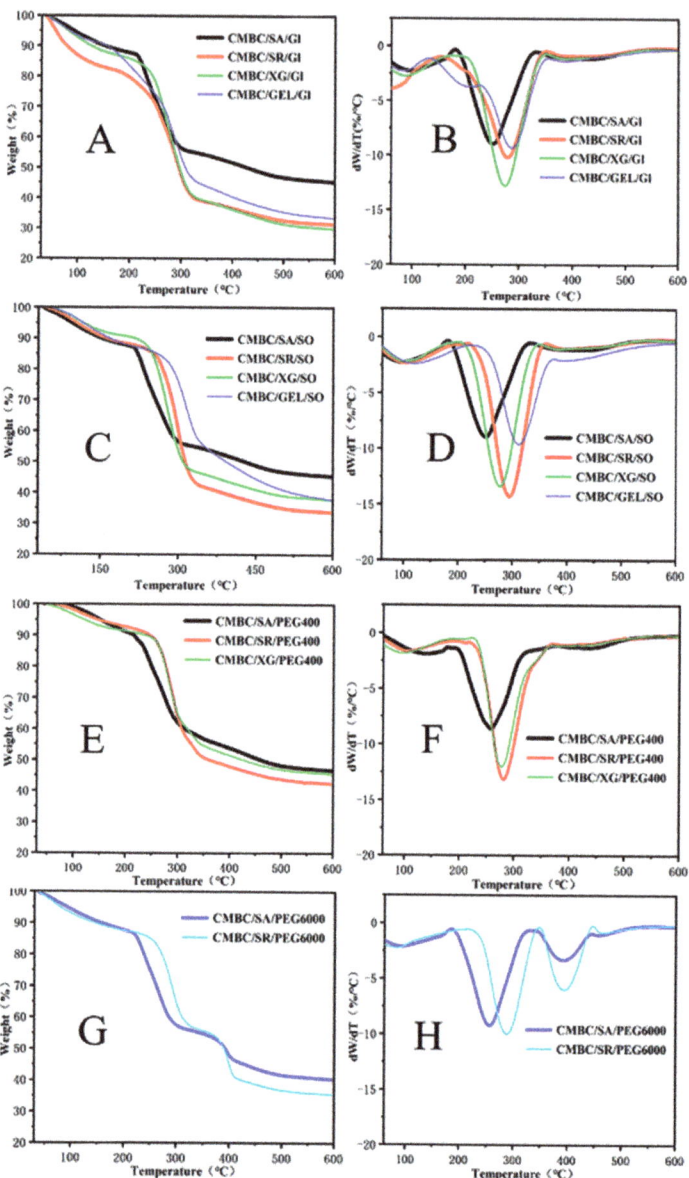

Figure 5. Pyrolysis curves of the composite films ((**A,C,E,G**) for TG; (**B,D,F,H**) for DTG).

the initial temperature of pyrolysis Ts;
the maximum rate of pyrolysis weight loss $(d_w/d_t)_{max}$;
the peak temperature Tmax corresponding to $(d_w/d_t)_{max}$;
the average rate of pyrolysis weight loss $(d_w/d_t)_{mean}$;
the pyrolysis maximum weight loss rate V_∞;
the temperature interval $\Delta T1/2$ corresponding to $(d_w/d_t)/(d_w/d_t)_{max} = 1/2$.

The data show that the initial temperature Ts of the pyrolysis of the composite films ranged from 213 °C to 277 °C (Table 6). The maximum weight loss rate V_∞ of the pyrolysis ranged from 53% to 72%, while the weight loss of the CMBC/SA/PEG400, CMBC/XG/PEG 400, and CMBC/SA/GL composite films was the smallest, at 53%, 54%, and 55%, respectively, indicating that these films had better thermal stability. The pyrolysis temperatures of conventional plastics ranged between 130 °C–145 °C [44]. In contrast, the pyrolysis temperatures of the composite films were 250 °C–300 °C, indicating that composite films have better thermal stability than traditional plastics. The above pyrolysis-related parameters can be combined to develop a comprehensive index D to characterize the degree of difficulty of the pyrolysis of composite films.

$$D = \frac{(d_w/d_t)_{max}(d_w/d_t)_{mean}V_\infty}{T_S T_{max} \Delta T_{1/2}} \quad (2)$$

Equation (2) shows that the lower the T_S (corresponding to a more significant value of D), the easier it is to pyrolyze, $(d_w/d_t)_{max}$ and $(d_w/d_t)_{mean}$, and the more intense the pyrolysis (corresponding to a more considerable D value). The lower values also indicate that the more significant the V_∞, the more pyrolysis (corresponding to a more considerable D value); the smaller the T_{max} and $\Delta T_{1/2}$, the earlier and more concentrated the pyrolysis peaks appear and the more favorable the pyrolysis and gasification. The D values calculated from the above equation were also included in Table 7, which showed that CMBC/SA/PEG 400 and CMBC/SA/GL had the smallest D values of 1.65×10^{-6} and 1.92×10^{-6}, indicating that these two composite films had the best thermal stability.

Table 7. Pyrolytic properties of the composite films.

Composite Film	$(d_w/d_t)_{max}$ (%/°C)	$(d_w/d_t)_{mean}$ (%/°C)	V_∞ (%)	T_S (°C)	T_{max} (°C)	$\Delta T_{1/2}$ (°C)	D
CMBC/SA/GL	−8.956	−1.59	55	213	250.42	76	1.92×10^{-6}
CMBC/SR/GL	−10.1714	−1.91	69	220	278.3	95.6	2.80×10^{-6}
CMBC/XG/GL	−12.7741	−1.99	70	232	273.9	85.4	3.96×10^{-6}
CMBC/SA/SO	−10.126	−1.83	61	225	264.0	69.8	2.68×10^{-6}
CMBC/SR/SO	−14.2546	−1.93	66	265	294.4	70.7	4.53×10^{-6}
CMBC/XG/SO	−13.3623	−1.81	72	246	276.2	73.1	4.58×10^{-6}
CMBC/GEL/SO	−9.5676	−1.88	62	277	310.4	73.6	2.59×10^{-6}
CMBC/SA/PEG400	−8.597	−1.54	53	219	256.3	90.1	1.65×10^{-6}
CMBC/SR/PEG 400	−13.1144	−1.65	58	251	280.1	74.2	2.28×10^{-6}
CMBC/XG/PEG400	−11.981	−1.58	54	252	277.0	73.2	2.09×10^{-6}

4. Conclusions

A carboxymethyl BC–based composite film was successfully prepared with good thermal stability and mechanical properties. The effects of the addition of sodium alginate/starch/xanthan gum/gelatin and propanetriol/sorbitol/PEG 400/PEG 6000 on the mechanical strength and thermal stability of the carboxymethyl BC–based composite film were investigated. The prepared composite films had suitable mechanical strength and thermal stability when 1.0% sodium alginate and 0.2% propanetriol were added. Regression analysis of the data on mechanical properties yielded a significant correlation between thickeners and plasticizers on the tensile strength and elongation at the break of the composite films (Table 5). The effects of the simultaneous addition of various thickeners on the mechanical properties and thermal stability of composite films will be further studied.

Author Contributions: Conceptualization, X.L. (Xiaoping Li) and J.J.M.; methodology, X.L. (Xiaoping Li) and J.B.; validation, Z.S., Z.T., X.L. (Xiaobao Li), Y.Y. and Q.Z.; formal analysis, X.L. (Xiaoping Li) and J.J.M.; investigation, Z.S., Z.T., X.L. (Xiaobao Li), Y.Y. and Q.Z.; resources, X.L. (Xiaoping Li) and Z.S.; data curation, Z.T.; writing—original draft preparation, Z.S.; writing—review and editing, J.J.M., X.L. (Xiaoping Li) and J.B.; supervision, X.L. (Xiaoping Li); project administration, X.L. (Xiaoping Li) and J.J.M.; funding acquisition, X.L. (Xiaoping Li) and Z.T. All authors have read and agreed to the published version of the manuscript.

Funding: This study was supported by the National Nature Science Foundation (31870551), the Top Young Talents in Yunnan Province (YNWR-QNBJ-2018-120), and the 111 Project (D21027).

Institutional Review Board Statement: Not applicable.

Informed Consent Statement: Not applicable.

Data Availability Statement: The data presented in this study are available from the listed authors.

Conflicts of Interest: The authors declare no conflict of interest.

References

1. Choi, S.; Rao, K.; Zo, S. BC and Its applications. *Polymers* **2020**, *14*, 1080. [CrossRef] [PubMed]
2. Ross, P.; Mayer, R.; Benziman, M. Cellulose biosynthesis and function in bacteria. *Microbiol. Rev.* **1991**, *55*, 35–58. [CrossRef] [PubMed]
3. Jia, S.; Ou, H.; Ma, X. Preliminary studies on the structure and properties of BC. *J. Cell. Sci. Technol.* **2002**, *3*, 25–30.
4. Iguchi, M.; Yamanaka, S.; Budhiono, A. BC—A masterpiece of nature's arts. *J. Mater. Sci.* **2000**, *35*, 261–270. [CrossRef]
5. Abdelhamid, H.; Mathew, A. Cellulose-based nanomaterials Advance Biomedicine: A Review. *Int. J. Mol. Sci.* **2022**, *23*, 5405. [CrossRef]
6. Tanpichai, S.; Boonmahitthisud, A.; Soykeabkaew, N. Review of the recent developments in all-cellulose nanocomposites: Properties and applications. *Carbohydr. Polym.* **2022**, *286*, 119192. [CrossRef]
7. Abdelhamid, H.; Mathew, A. Cellulose-based materials for water remediation. *Front. Chem. Eng.* **2021**, *74*, 790314. [CrossRef]
8. Hari, P.R.; Chandy, T.; Sharma, C.P. Chitosan/calcium alginate microcapsules for Intestinal delivery of nitrofurantoin. *J. Microencapsul.* **1996**, *13*, 319–329. [CrossRef]
9. Yuk, S.H.; Cho, S.H.; Lee, H.B. Electric current sensitive drug delivery systems using sodium alginate/polyacrylic acid composites. *Pharm. Res.* **1992**, *9*, 955–957. [CrossRef]
10. Kumar, R.; Ghoshal, G.; Goyal, M. Synthesis and functional properties of gelatin/CA-starch composite film: Excellent food packaging material. *J. Food Sci. Technol.* **2019**, *56*, 1954–1965. [CrossRef]
11. Pan, L.; Li, P.; Tao, Y. Preparation and properties of microcrystalline cellulose/fish gelatin composite film. *Materials* **2020**, *13*, 4370. [CrossRef] [PubMed]
12. Meng, F.; Zhou, Y.; Liu, J. Thermal decomposition behaviors and kinetics of carrageenan-poly vinyl alcohol bio-composite film. *Carbohydr. Polym.* **2018**, *201*, 96–104. [CrossRef] [PubMed]
13. Roy, S.; Rhim, J. Carrageenan/agar-based functional film integrated with zinc sulfide nanoparticles and Pickering emulsion of tea tree essential oil for active packaging applications. *Int. J. Biol. Macromol.* **2021**, *193*, 2038–2046. [CrossRef]
14. Lin, B.; Li, C.; Chen, F. Continuous blown film preparation of high starch content composite films with high ultraviolet aging resistance and excellent mechanical properties. *Polymers* **2021**, *13*, 3813. [CrossRef]
15. Liu, P.; Gao, W.; Zhang, X. Effects of ultrasonication on the properties of maize starch/stearic acid/ sodium carboxymethyl cellulose composite film. *Ultrason. Sonochem.* **2021**, *72*, 105447. [CrossRef] [PubMed]
16. Lin, D.; Zheng, Y.; Wang, X. Study on physicochemical properties, antioxidant and antimicrobial activity of okara soluble dietary fiber/sodium carboxymethyl cellulose/thyme essential oil active edible composite films incorporated with pectin. *Int. J. Biol. Macromol.* **2020**, *165*, 1241–1249. [CrossRef]
17. Roy, S.; Rhim, J. Carboxymethyl cellulose-based antioxidant and antimicrobial active packaging film incorporated with curcumin and zinc oxide. *Int. J. Biol. Macromol.* **2020**, *148*, 666–676. [CrossRef]
18. Hao, X.; Clifford, A.; Poon, R. Carboxymethyl cellulose and composite films prepared by electrophoretic deposition and liquid-liquid particle extraction. *Colloid Polym. Sci.* **2018**, *296*, 927–934.
19. Wang, K.; Du, L.; Zhang, C. Preparation of chitosan/curdlan/carboxymethyl cellulose blended film and its characterization. *J. Food Sci. Technol.* **2019**, *56*, 5396–5404. [CrossRef]
20. Xu, Y.; Li, Q.; Man, L. Bamboo-derived carboxymethyl cellulose for liquid film as renewable and biodegradable agriculture mulching. *Int. J. Biol. Macromol.* **2021**, *192*, 611–617. [CrossRef]

21. Jannatyha, N.; Shojaee-Aliabadi, S.; Moslehishad, M. Comparing mechanical, barrier and antimicrobial properties of nanocellulose/CMC and nanochitosan/CMC composite films. *Int. J. Biol. Macromol.* **2020**, *164*, 2323–2328. [CrossRef] [PubMed]
22. Liu, J.; Chen, P.; Qin, D. Nanocomposites membranes from cellulose nanofibers, SiO$_2$ and carboxymethyl cellulose with improved properties. *Carbohydr. Polym.* **2020**, *233*, 115818. [CrossRef] [PubMed]
23. Mandal, A.; Chakrabarty, D. Studies on mechanical, thermal, and barrier properties of carboxymethyl cellulose film highly filled with nanocellulose. *J. Thermoplast. Compos. Mater.* **2019**, *32*, 995–1014. [CrossRef]
24. Volpe, M.G.; Siano, F.; Paolucci, M.; Sacco, A.; Sorrentino, A.; Malinconico, M.; Varricchio, E. Active edible coating effectiveness in shelf-life enhancement of trout (*Oncorhynchusmykiss*) fillets. *LWT—Food Sci. Technol.* **2015**, *60*, 615–622. [CrossRef]
25. Suriyatem, R.; Auras, R.A.; Rachtanapun, P. Utilization of carboxymethyl cellulose from durian rind agricultural waste to improve physical properties and stability of Rice Starch-Based Film. *J. Polym. Environ.* **2019**, *27*, 286–298. [CrossRef]
26. Suriyatem, R.; Auras, R.; Rachtanapun, P. Biodegradable rice starch/carboxymethyl chitosan films with added propolis extract for potential use as active food packaging. *Polymers* **2018**, *10*, 954. [CrossRef]
27. Nazmi, N.; Isa, M.; Sarbon, N. Preparation and characterization of chicken skin gelatin/CMC composite film as compared to bovine gelatin film. *Food Biosci.* **2017**, *19*, 149–155. [CrossRef]
28. Li, H.; Huneault, M. Comparison of sorbitol and glycerin as plasticizers for thermoplastic starch in TPS/PLA blends. *J. Appl. Polym. Sci.* **2011**, *119*, 2439–2448. [CrossRef]
29. Azirah, M.; Isa, M.; Sarbon, N. Effect of xanthan gum on the physical and mechanical properties of gelatin-carboxymethyl cellulose film blends. *Food Packag. Shelf Life* **2016**, *9*, 55–63.
30. Han, J.; Floros, J. Casting antimicrobial packaging films and measuring their physical properties and antimicrobial activity. *J. Plast. Film Sheet.* **1997**, *13*, 287–298. [CrossRef]
31. Talja, R.; Helén, H.; Roos, Y.; Jouppila, K. Effect of type and content of binary polyol mixtures on physical and mechanical properties of starch-based edible films. *Carbohydr. Polym.* **2008**, *71*, 269–276. [CrossRef]
32. Lan, W.; Zhang, R.; Wang, Y. Preparation and characterization of carboxymethyl cellulose/sodium alginate/chitosan composite membrane. *China Plast. Ind.* **2017**, *45*, 144–149.
33. Lai, Y.; Xu, C.; Chen, J. Study on properfies of starch-filled polyethylene. *Plast. Sci. Technol.* **1996**, *6*, 7–10.
34. Hosseini, S.; Javidi, Z.; Rezaei, M. Efficient gas barrier properties of multi-layer films based on poly(lactic acid) and fish gelatin. *Int. J. Biol. Macromol.* **2016**, *92*, 1205–1214. [CrossRef] [PubMed]
35. Kanmani, P.; Rhim, J. Physical, mechanical and antimicrobial properties of gelatin based active nanocomposite films containing AgNPs and nanoclay. *Food Hydrocoll.* **2014**, *35*, 644–652. [CrossRef]
36. Zhao, M.; Sun, Z.; Hsu, C. Zinc oxide films with high transparency and crystallinity prepared by a low temperature spatial atomic layer deposition process. *Nanomaterials* **2020**, *10*, 459. [CrossRef]
37. Wang, B.; Jia, D.; Wang, K. Effects of NaOH treatment on structure and properties of blend films made by collagen, konjac, glucomanan and chitosan. *J. Funct. Mater.* **2005**, *36*, 1107–1115.
38. Oun, A.; Rhim, J. Preparation and characterization of sodium carboxymethyl cellulose/cotton linter cellulose nanofibril composite films. *Carbohydr. Polym.* **2015**, *127*, 101–109. [CrossRef]
39. Yao, Y.; Sun, Z.; Li, X. Effects of raw material source on the properties of CMC composite films. *Polymers* **2021**, *14*, 32. [CrossRef]
40. Liu, D.; Bian, Q.; Li, Y. Effect of oxidation degrees of graphene oxide on the structure and properties of poly (vinyl alcohol) composite films. *Compos. Sci. Technol.* **2016**, *129*, 146–152. [CrossRef]
41. Mansur, A.A.; de Carvalho, F.G.; Mansur, R.L.; Carvalho, S.M.; de Oliveira, L.C.; Mansur, H.S. Carboxymethylcellulose/ZnCdS fluorescent quantum dot nanoconjugates for cancer cell bioimaging. *Int. J. Biol. Macromol.* **2017**, *96*, 675–686. [CrossRef] [PubMed]
42. Hu, X.; Wang, K.; Yu, M. Characterization and antioxidant activity of a low-molecular-weight xanthan gum. *Biomolecules* **2019**, *9*, 730. [CrossRef] [PubMed]
43. Kapusniak, J.; Siemion, P. Thermal reactions of starch with long-chain unsaturated fatty acids. Part 2. Linoleic acid. *J. Food Eng.* **2007**, *78*, 323–332. [CrossRef]
44. Pöllänen, M.; Suvanto, M.; Pakkanen, T. Cellulose reinforced high density polyethylene composites—Morphology, mechanical and thermal expansion properties. *Compos. Sci. Technol.* **2013**, *76*, 21–28. [CrossRef]

Article

Manufacturing and Characterization of Modified Wood with In Situ Polymerization and Cross-Linking of Water-Soluble Monomers on Wood Cell Walls

Jihang Hu *, Zongying Fu, Xiaoqing Wang and Yubo Chai *

Research Institute of Wood Industry, Chinese Academy of Forestry, Beijing 100091, China
* Correspondence: hujihang88@caf.ac.cn (J.H.); chaiyubo@caf.ac.cn (Y.C.)

Abstract: Fast-growing plantation wood has poor dimensional stability and easily cracks, which limits its application. As wood modification can improve the dimensional stability, strength, and other properties of wood, it has been extensively used. In this study, 2-Hydroxyethyl methacrylate (HEMA) and glyoxal were applied to treat poplar wood (*Populus euramevicana cv.I-214*) by using vacuum pressure impregnation to improve its dimensional stability. The weight percentage gain (WPG), anti-swelling efficiency (ASE), water absorption rate (WAR), leachability (L), and other properties of modified wood were examined. Results showed that the modifier was diffused into the cell walls and intercellular space and reacted with the wood cell wall after heating to form a stable reticular structure polymer which effectively decreased the hydroxyl content in the wood and blocked the water movement channel; thus, further improving the physical performance of wood. These results were confirmed by scanning electron microscopy (SEM), energy-dispersive X-ray spectroscopy (EDX), X-ray diffraction (XRD), Fourier-transform infrared spectroscopy (FTIR), and nuclear magnetic resonance (NMR). When the ratio of the modifier was 80:20, the concentration of the modifier was 40%, and the curing temperature was 120 °C, the modified poplar had the best performance, which showed a low WAR (at its lowest 58.39%), a low L (at its lowest 10.44%), and a high ASE (of up to 77.94%).

Keywords: poplar; cell walls; cross-linking modification; wood properties

Citation: Hu, J.; Fu, Z.; Wang, X.; Chai, Y. Manufacturing and Characterization of Modified Wood with In Situ Polymerization and Cross-Linking of Water-Soluble Monomers on Wood Cell Walls. *Polymers* **2022**, *14*, 3299. https://doi.org/10.3390/polym14163299

Academic Editor: Antonios N. Papadopoulos

Received: 27 July 2022
Accepted: 9 August 2022
Published: 12 August 2022

Publisher's Note: MDPI stays neutral with regard to jurisdictional claims in published maps and institutional affiliations.

Copyright: © 2022 by the authors. Licensee MDPI, Basel, Switzerland. This article is an open access article distributed under the terms and conditions of the Creative Commons Attribution (CC BY) license (https://creativecommons.org/licenses/by/4.0/).

1. Introduction

Wood, as a significant natural and renewable material, has been extensively used in furniture, decoration, construction, and other fields because of its outstanding high strength-to-weight ratio, easy processing, beautiful appearance, and affordability [1,2]. Owing to the capability of carbon dioxide storage, wood and wood utilization are more friendly to the environment [3]. Fast-growing wood is considered an alternative source of natural wood due to its short planting time and is an ideal material for a sustainable society [4]. Nevertheless, the application of fast-growing wood is restricted by its poor dimensional stability, poor physical and mechanical properties, and easy cracking. However, wood modification technology could significantly improve the performance and quality of fast-growing wood, which is of great significance in broadening its application range and extending its service life [5]. Therefore, using modified fast-growing wood can relieve the shortage of wood supply and protect the over-exploitation of natural resources as an alternative means. Among numerous wood modification methods, chemical cross-linking modification has received more attention, with the advantages of a simple process, significant modification effect, and excellent stability [6,7]. It has been found that the strength and density of wood increase if the wood modifier is able to penetrate and enlarge the wood cell wall. If the modifier can react with the cell wall constituents, the modification impact is substantially improved, and the stability is extended [7–9]. The dimensional stability of wood can be enhanced by reducing the hydrophilic -OH group of cell wall components or bulking the cell wall with a modifier. Thus, it is a significant issue for the wood industry to ensure that the wood is non-toxic and environmentally sustainable during the treatment and use processes, and that it will not cause damage to the environment and the human body.

Hydroxyethyl methacrylate (HEMA) includes numerous active hydroxyl groups, which contribute to its biocompatibility. Additionally, since it is simple for the cross-linking agent to self-polymerize or create a network structure, it has excellent mechanical qualities. Due to the great strength and biocompatibility of Poly (2-hydroxyethyl methacrylate) (pHEMA) hydrogels, they have been used in a variety of biomedical applications [10–12]. Additionally, there have been several uses for wood modification in recent years [13]. Due to its high concentration of active hydroxyl groups, HEMA was selected as the main modification to increase the performance of wood cell walls [14]. It is expected that the hydroxyl groups will enhance the polymers' hydrogen bonding capabilities with wood components [15,16]. Glyoxal (GA) is a widely used cross-linking agent for the preparation of high-molecular-weight gel materials. It has excellent graft performance and biocompatibility. Due to its excellent cross-linking performance and low toxicity and volatility, it is widely used in the textile, paper, and chemical industries, etc. [17,18]. Modifying wood to enhance the dimensional stability and water resistance of wood using a cross-linking agent is an efficient modification strategy. As a result, treating wood with glyoxal may significantly enhance its physical and mechanical qualities [19]. Glyoxal has two interconnected aldehyde groups in its molecular structure and has high chemical activity and all the chemical properties of aldehydes. It is widely used in the wood industry to replace formaldehyde in the synthesis of resins [20–22]. Glyoxal is a well-studied bifunctional aldehyde which has the competent ability to react with hydroxyl groups of wood cell wall polymers and enhance dimensional stability, water repellence, and durability of wood [19]. However, these improvements are often accompanied by brittleness and strength losses in the wood [23].

The impregnation of wood cell walls with hydroxyethyl methacrylate and glyoxal resulted in the modification of the cell walls, and thermally induced in situ polymerization resulted in the formation of modified materials with outstanding characteristics. The purpose of this work is to enhance the quality of wood (dimensional stability and water absorption) by grafting polymers onto the cell walls of wood, expanding the cell walls, and converting hydrophilic hydroxyl groups into bigger hydrophobic groups. This work advances fast-growing-wood modification by introducing a novel research strategy. Modified wood is projected to be employed in furniture manufacture and interior decorating, broadening the spectrum of fast-growing wood's applications and enabling the replacement of natural forest wood with fast-growing wood.

2. Materials and Methods

2.1. Materials

The poplar wood (*Populus euramericana cv.I-214*) was collected from the Sun jia zhuang forest farm in Yi County, Hebei province, China. The average tree age and diameter at breast height are about 20 years and 25~33 cm, respectively. Additionally, the air-dry density is about 0.35 g/cm^3. Mature sapwood was randomly picked and sawed into 20 mm × 20 mm × 20 mm pieces, which needed to be completely dry before the impregnation experiment. All chemical reagents were analytical grade. HEMA (Kangboshunda Chemical Products Co., Ltd., Beijing, China), glyoxal of chemical grade (40%), analytical-grade calcium chloride ($CaCl_2$), hydrogen peroxide (30% aqueous solution, H_2O_2), and ammonium persulfate (($NH_4)_2S_2O_8$) (Aladdin Co., Ltd., Shanghai, China; Beijing, China).

2.2. Wood Modification Methods

Modifiers included HEMA and glyoxal, and the compound initiator consisted of $CaCl_2$, H_2O_2, and $(NH_4)_2S_2O_8$. The modifiers were prepared with varied HEMA to glyoxal ratios (HEMA:GA = 100:0, 90:10, 80:20, 70:30, 60:40) and different concentrations (10%, 20%,

30%, 40%). The amount of compound initiator was 3% of the mass of the modifier, and the mass ratio of $CaCl_2$, H_2O_2, and $(NH_4)_2S_2O_8$ is 1:1:1. $CaCl_2$, H_2O_2, and $(NH_4)_2S_2O_8$. One parameter was taken as the variable and the others were left unchanged. Specific parameters have been shown in Table 1.

Table 1. Parameters of treatments for each group of wood samples.

The Ratio of HEMA and GA	$CaCl_2/H_2O_2/(NH_4)_2S_2O_2$ (%)	The Content of HEMA/GA (%)	Curing Temperature (°C)
100:0	3	10	80
90:10	3	20	100
80:20	3	30	120
70:30	3	40	140
60:40	3	-	-

Wood samples were impregnated respectively with the above prepared solution, through the vacuum-pressure method as vacuuming (0.1 MPa, 1 h) → liquid injection → pressurization (1.0 MPa, 8 h) → pressure relief. After releasing the pressure, the wood sample was air-dried for 5 days and then heated to 40 °C for 24 h to initiate the reaction between the modifier and the wood cell wall. They were then heated for 6 h at 80 °C, 100 °C, 120 °C, and 140 °C, respectively, to complete the polymerization process (Figure 1). Finally, all the samples were dried at 103 ± 2 °C to achieve the same quality, except for the samples in 120 °C and 140 °C conditions. Each group had 10 duplicate samples, and had its statistical average value taken.

Figure 1. Schematic diagram of the wood modification process.

2.3. Characterization

Scanning electron microscope (SEM) and Energy-dispersive X-ray spectroscopy (EDX). SEM (Hitachi SU8020, Toyko, Japan) was used to investigate the distribution of modifiers in wood cell walls at an accelerating voltage of 10 kV. Samples of tangential modified and unmodified wood (5 mm × 5 mm × 5 mm) were coated with a platinum layer on an aluminum SEM stub. Additionally, EDX coupled with SEM was utilized to detect the concentration of the characteristic elements of wood samples.

X-ray diffraction (XRD) analysis was performed by a Bruker D8 Advance diffractogram from Germany. Cu-Kα radiation with graphite monochromator, 40 kV, 40 mA, and 2θ scan range of 5–45° (2θ) with a scanning speed of 2°/min were preset parameters. On the scanning curve, there was a maximum peak of (002) diffraction near 2θ = 22° and a minimum near 2θ = 18°. The empirical formula proposed by Segal et al. [24] was used to calculate the relative crystallinity. The crystallinity index (CrI) was calculated as follows:

$$\text{CrI } (\%) = \frac{100(I_{002} - I_{am})}{I_{002}} \qquad (1)$$

where I_{002} is the intensity of the diffraction angle of (002) crystal plane (2θ = 22°), and I_{am} is the scattering intensity of the amorphous background diffraction degrees (2θ = around 18°). The microfilament angle of the sample was calculated according to the Cave 0.6T method [24].

Fourier-Transform Infrared Spectroscopy (FTIR) was used to characterize the chemical composition of the modified wood samples. All samples (three samples each for modified and unmodified wood) were crushed to 100 mesh and ground using KBr pellets. A FTIR analysis was performed using a FTIR Spectrometer (Nicolet Nexus 6700, Madison, WI, USA) in the range of 400 to 4000 cm^{-1} in diffuse reflection mode with a 4 cm^{-1} resolution and 64 scans.

Nuclear magnetic resonance spectroscopy (NMR). ^{13}C NMR spectra of the modified wood samples were recorded on a Bruker Avance AVIII 400 MHz NMR spectrometer. The modified and unmodified poplar wood flour with 160 mesh was prepared for an NMR. The solid-state ^{13}C magic angle spinning (MAS) NMR spectra of the modified wood samples were examined and compared with unmodified wood samples. The MAS speed was set at 5000 Hz and scanned 8318 times at room temperature.

2.4. Wood Performance Evaluation

The samples with a size of 20 mm × 20 mm × 20 mm (longitudinal × radial × tangential) were prepared for weight percentage gain (WPG), bulking effect (BE), anti-swelling efficiency (ASE), water absorption (WAR), and leachability (L) tests. Ten replicates were conducted for each group, and the statistical average value was taken.

The weight percentage gain (WPG) was calculated according to the absolute dry mass before and after modification, as seen below (Formula (2)).

$$\text{WPG (\%)} = \frac{100(W_1 - W_0)}{W_0} \quad (2)$$

where W_0 and W_1 are the weight of oven-dried wood samples before and after treatment, respectively.

The bulking effect (BE) test is used to characterize the size changes of wood before and after modification. The BE was calculated (see Formula (3)).

$$\text{BE (\%)} = \frac{100(V_1 - V_0)}{V_0} \quad (3)$$

where V_0 and V_1 are the volume of oven-dried wood samples before and after treatment, respectively.

By completely immersing the modified and unmodified samples for 120 h, the volumes of the modified and unmodified samples before and after immersion were measured. From these volumes, ASE can be calculated as follows:

$$VSE\ (\%) = \frac{100(V_a - V_b)}{V_b} \quad (4)$$

where V_a and V_b are the volumes of wood samples after and before immersion, respectively.

$$\text{ASE (\%)} = \frac{100(VSE_u - VSE_t)}{VSE_u} \quad (5)$$

where VSE_u and VSE_t are the volumetric swelling efficiencies of untreated and treated wood samples, respectively.

The water absorption rate was calculated according to the "Determination method of water absorption of wood" (GB/T 1934.1-2009). First, the modified samples were numbered and weighed. The specimens were placed at least 50 mm below the surface of the water and soaked for 10 days, and then the samples were removed from the water, and the surface was wiped with a tissue and weighed again. The WAR was calculated as:

$$\text{WAR (\%)} = \frac{100(W_2 - W_1)}{W_1} \tag{6}$$

where W_1 and W_2 are the absolute dry mass after modified and the mass after soaking of the modified wood specimens, respectively.

After calculating the WAR, all samples were oven-dried at 103 ± 2 °C to a constant weight. The leaching rate due to water immersion was calculated as follows:

$$\text{L (\%)} = \frac{100(W_1 - W_3)}{(W_1 - W_0)} \tag{7}$$

where W_3 is the absolute dry weight of a modified wood sample after a 10 d water immersion.

3. Results and Discussion

3.1. Distribution of Modifiers

Cross-sectional microphotographs of both treated and untreated wood were obtained by SEM and EDX. As illustrated in Figure 2a,b, the cell walls in modified wood were thicker than that of unmodified wood, and unmodified wood had distinct gaps between the cell walls that were not present in modified wood. The untreated cell walls had a mean thickness of 5.04 ± 0.96 μm, and the treated cell walls had a mean thickness of 7.74 ± 0.57 μm (Table 2). The resulting polymer from the copolymerization of HEMA and glyoxal was grafted onto the wood cell wall, thereby improving the interfacial compatibility between the polymer and the wood cell wall without obvious gaps. According to the energy spectrum analysis (Figure 2e,f), the carbon, oxygen, and nitrogen contents of the modified wood cell walls were greater than those of the unmodified wood. This result indicates that the modifier was predominantly distributed in the cell walls and, to a lesser extent, in the cell lumen.

Table 2. The data of untreated and treated wood cell wall thickness.

Labels	Total	Max./μm	Min./μm	Mean./μm
Untreated wood	120	5.34	4.38	5.04
Treated wood	120	8.40	7.20	7.74

3.2. XRD Analysis

As seen in Table 3 and Figure 3b, the peak width of the X-ray diffraction pattern of poplar increased from 1.74 to 2.49, and the microfibril angle increased from 21.86° to 29.70°. In this hypothesis, it is believed that when the modifier causes the cell wall to swell, some of the amorphous region components are dissolved, resulting in the original arrangement of the microfilaments changing.

The XRD spectrum (Figure 3a) showed that the maximum diffraction values of the crystalline region of the modified wood (002) were around $2\theta = 22°$, which implies that the treatment had no effect on the crystalline region, and thus there was no change in the distance between the crystal layers. In terms of diffraction intensity, the treated wood increased a little, indicating a slight effect on the amorphous region of the wood microfibrils. Troughs appeared near $2\theta = 18°$, and the difference from untreated wood was less than 9%. Therefore, from the perspective of a crystal structure, there is no difference in the crystal structure of the cellulose of treated wood, and it is still cellulose I-type structure [25].

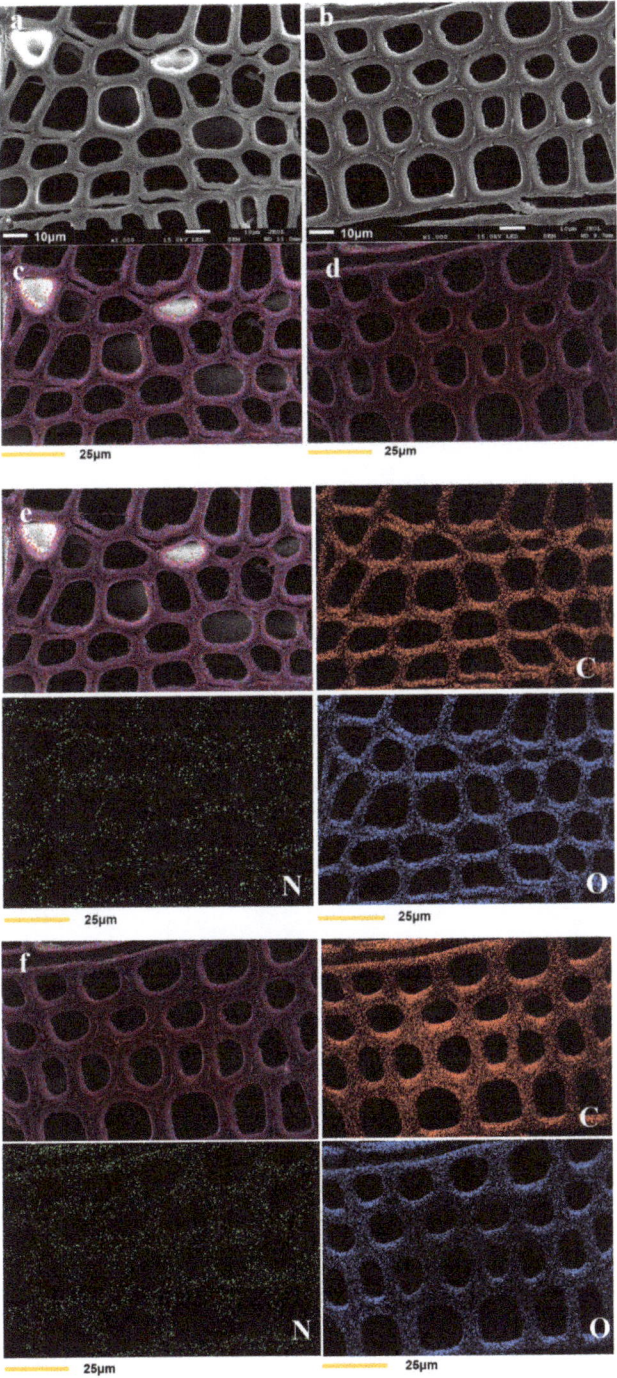

Figure 2. SEM morphologies of unmodified (**a**) and modified (**b**) wood samples, (**c**,**d**) are energy spectrum figures of untreated and treated wood, respectively. Samples (**e**,**f**) show the distribution of C, N, and O elements in the cell walls of unmodified and modified wood.

Table 3. Parameters of treatments for each wood sample.

	Crystallinity	Microfibril Angle	Crystal Breadth
Untreated wood	43.15%	21.86°	1.74
Treated wood	54.77%	29.70°	2.49

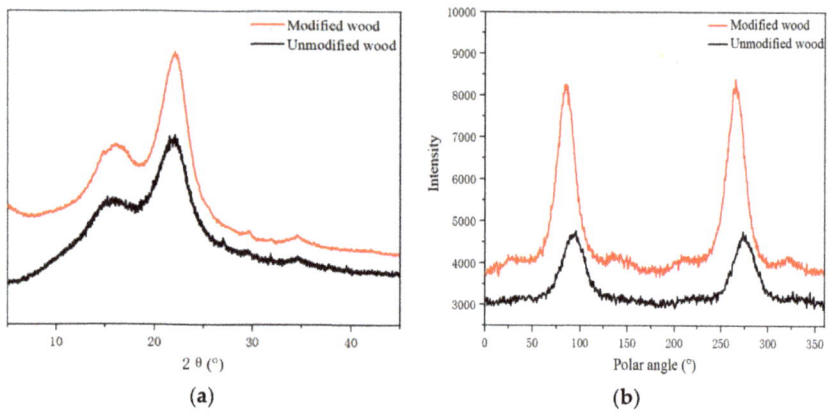

Figure 3. XRD patterns (a) and MFA (b) of modified and unmodified wood.

The relative crystallinity of wood is a critical parameter for describing the supramolecular structure of cellulose, which is intimately related to its physical and chemical properties. Additionally, relative crystallinity is indispensable for comprehending wood alteration, processing, and utilization [26]. The relative crystallinity of modified wood increased from 43.15% to 54.77% (Table 3) because the modifiers entered the cell wall of the wood and affected the crystalline structure of the cellulose molecular chain. When the amorphous fiber recombination occurred after the temperature curing, a stronger bond was established. As the relative crystallinity improved, the dimensional stability of the wood also improved, leading to further improvements in its physical properties. With the increase in wood crystallinity, the dimensional stability and nuclear density of wood also increased, which is consistent with the results of the physical properties evaluation of modified wood.

Despite the slight change, high-temperature curing alters the width of the crystallization region in wood. The half-peak width of the diffraction peak of the (002) crystal plane was used to calculate the width of the crystalline region of the wood [27]. When the reaction temperature reached a certain level, the crystalline zone and amorphous zone of the fibers changed, and the width of the crystalline zone of the modified materials increased. This was consistent with the findings of the microfibril angle and relative crystallinity investigation.

3.3. FTIR Analysis

The FTIR spectra in Figure 4 clearly show that the absorption peaks near 3373 cm^{-1} became weaker after grafting with HEMA and glyoxal, and the broad bands at 3200–3600 cm^{-1} were attributed to primary amine O-H bands, which indicated that the -OH on the wood was involved in the grafting reaction [28,29]. The strong absorption peaks of C=C and C=O appeared at 1666 cm^{-1}, and 1627 cm^{-1}, mainly caused by the stretching vibration of the carbonyl and aldehyde groups in HEMA and glyoxal. As shown in Figure 4, the peak at 1594 cm^{-1} corresponds to $-$C=O stretching and disappears from the spectrum of the modified material. This is due to the oxidation reaction between the modifier and the wood. In addition, the 898 cm^{-1} absorption peak in the modified wood was weakened, indicating that the modifier had a certain influence on the cellulose structure of the wood [30,31]. FTIR spectroscopy supported the XRD results.

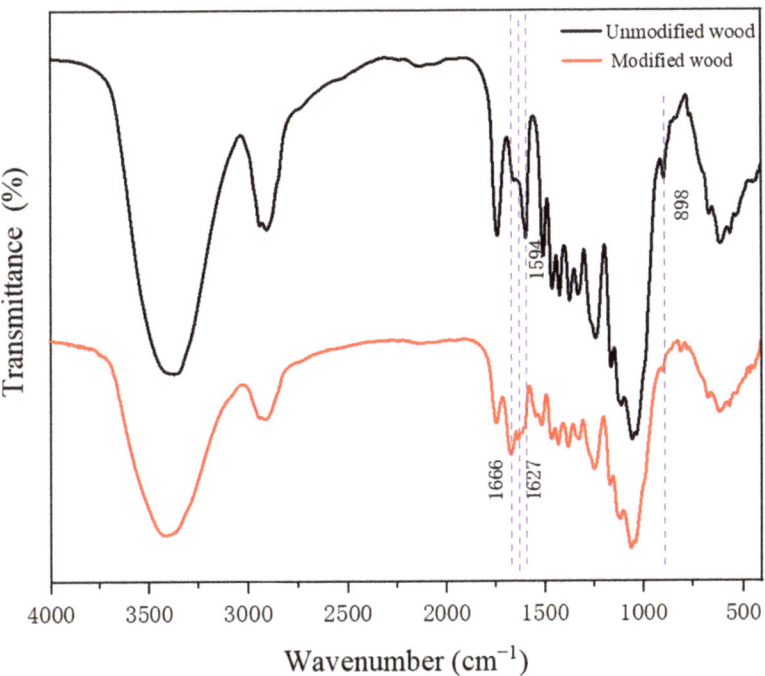

Figure 4. The FTIR spectra of the treated wood and untreated wood.

Both HEMA and glyoxal are highly reactive, and this functional group can react with hydroxyl groups in wood. Under high temperature and catalyst action, the hydroxyl groups between the wood modifiers can be cross-linked to form a three-dimensional network polymer under heating conditions. It is because of the chemical cross-linking reaction of the reactive modifiers in the wood and the uniform distribution in the wood that the physical and chemical properties of wood were comprehensively improved.

3.4. ^{13}C NMR Analysis

In order to further study whether wood reacts with HEMA/GA, and confirm the conclusion of FTIR, NMR (Figure 5) was used to study whether there is a new chemical shift in the modified wood and whether there is a chemical reaction in the modified wood.

The absorption peaks at 88 and 83 ppm were enhanced, indicating that the chemical cross-linking reaction occurred between the modifier and C4 in crystalline and non-crystalline areas, which rearranged the molecules in the non-crystalline area and formed a new crystalline area, thus increasing the crystallinity of the wood. This result is the same as the XRD result. The characteristic peaks at 55 ppm and 120–170 ppm have noticeable chemical shifts compared with the unmodified materials, and here represent the methoxy carbon and aromatic carbon components of lignin, indicating that a cross-linking reaction between the modifier and the carbon in lignin has occurred [32–34].

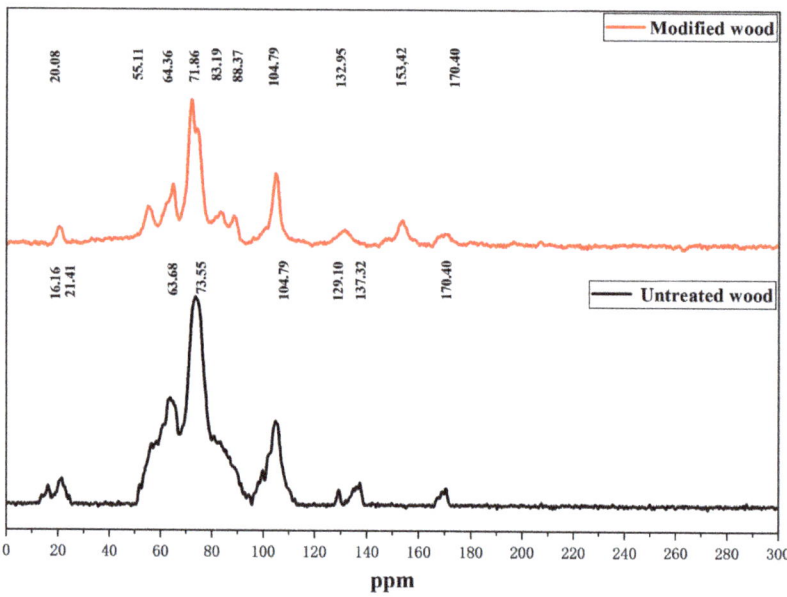

Figure 5. NMR spectra of untreated and treated wood.

3.5. Effect of the Ratio of HEMA and Glyoxal on Properties of Modified Wood

Due to the porous nature of wood and the higher concentration of free hydroxyl groups, it is excellent at storing water molecules and has high hygroscopicity. When wood absorbs water, water molecules combine with free hydroxyl groups, causing hemicellulose and cellulose to swell, their intermolecular hydrogen bonds to break, and more hydrophilic active sites to be exposed, thereby increasing water absorption and swelling of wood and resulting in deformation [35]. The main objectives of wood modification in this study are to reduce the number of hydroxyl groups and channels of water movement.

Figure 6 reflects that the ratio of HEMA and glyoxal has an influence on wood bulking effects BE, ASE, WPG, WAR, and L, while other factors remain unchanged (the modifier concentration was 20%; the curing temperature was 120 °C).

WPG and BE increased with the addition of glyoxal, demonstrating that glyoxal promotes the modifier into the wood cell wall. The BE reached its maximum when the HEMA to glyoxal ratio was 80:20, which indicates that the addition of glyoxal induces more modifiers into the wood cell wall and that the swelling of the cell wall by the modifier has reached its limit. Although WPG increased with increasing glyoxal (HEMA:GA = 70:30), BE decreased slightly, indicating that the modifier entered the cell cavity. When the ratio was 60:40, WPG decreased slightly, which could be attributed to an overabundance of glyoxal, which caused self-polymerization of glyoxal, thus preventing the modifier from entering the wood. When the HEMA to glyoxal ratio was 70:30, the variation trend was consistent with that of WPG, ASE, and WA, where the best values were 31.88%, 74.18%, and 107.44%, respectively. This is because modifiers can penetrate wood cell walls and react with the cell wall components of the wood to improve cell wall performance. Glyoxal is a reactive cross-linking agent for bifunctional groups, while HEMA is a cross-linking agent for bifunctional groups. By formulating polymers with the two ingredients, HEMA and glyoxal, and using hydroxyl cross-linking reactions, vinyl and hydroxyl groups can polymerize to form a three-dimensional structure that helps improve the cohesion strength of polymers, thus increasing the physical properties of wood [36]. Another approach is for the modifier to polymerize and block the transport channels of water in the wood; therefore, to improve the dimensional stability of the wood, these pores include vessels, pits, etc. Because of this, the water absorption rate of wood was reduced, which ultimately benefited the product's overall shelf life.

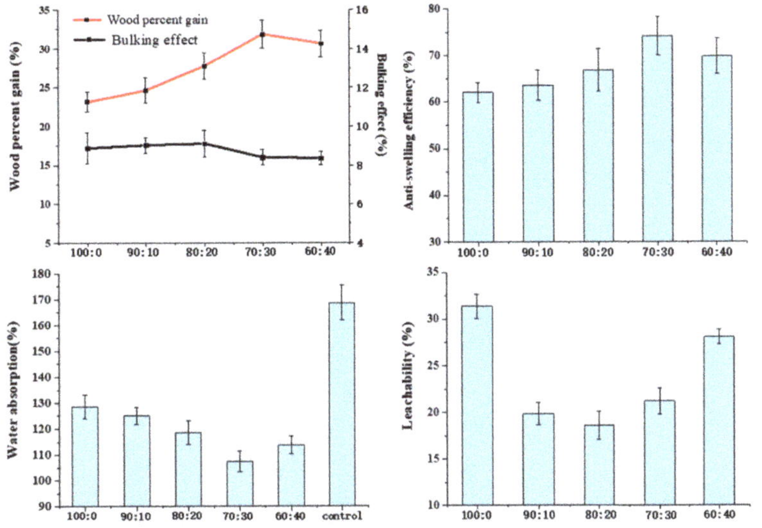

Figure 6. Effects of the ratio of HEMA and glyoxal on properties of wood.

Furthermore, both HEMA and glyoxal can react with hydrophilic hydroxyl groups on wood, thus reducing the ability of wood to absorb water. ASE increased to 74.18% when the ratio was 70:30. Compared to untreated wood (168.60%), the water absorption of WA (107.44%) decreased by 36.28%. Basically, the changing trend in leachability was the same as BE. The leachability was best when the ratio of 80 to 20 showed the modifier had the best retention ratio on the cell wall. A comprehensive evaluation of the physical properties of ASE, WA, and L requires an optimal modifier ratio of 70:30.

3.6. Effect of Modifier Concentration on Properties of Modified Wood

Figure 7 reflects that the modifier concentration influences the bulking effect, dimensional stability, weight gain, water absorption rates, and leachability of wood, while other factors remained unchanged (the ratio was 80 to 20; the curing temperature was 120 °C).

As the modifier concentration increased, WPG and BE both exhibited a positive correlation trend. The results indicated that the higher the concentration, the more modifier was infused into the wood and the greater the swelling of the cell wall. In Figure 7, an increase in modifier concentration results in a higher WPG. When the modifier concentration was 10%, the WPG was 14.22%, while ASE was 56.59%, indicating the high efficiency of HEMA/GA in improving the dimensional stability of poplar wood. When the concentration reached 40%, ASE and WPG increased to 77.94% and 60.19%. In addition, the cross-linking degree increased with increasing modifier concentration, and the oligomer decreased, reducing the modifier loss rate from 33.10% (the modifier concentration was 10%) to 10.38% (the modifier concentration was 40%). The initiator successfully grafted the modifier onto the hydroxyl groups in wood, thereby reducing the number of hydroxyl groups and the hygroscopic rate of the wood. Obviously, when the concentration was changed, the water absorption of wood decreased by 65.4% compared to untreated wood. The modifier was fixed into the wood, blocking water flow, causing a decrease in the water absorption of the wood. After a comprehensive inspection of ASE, WA, and L, it was determined that the best performance was obtained at a modifier concentration of 40%. The ASE obtained in this work was much higher than in other studies [7,17,37,38].

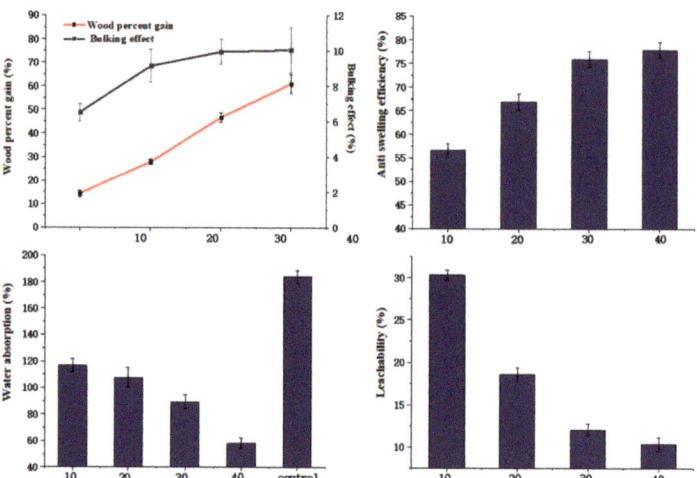

Figure 7. Effects of modifier concentrations on properties of wood.

3.7. Effects of Curing Temperature on Wood Properties

The effects of different curing temperatures on the properties of modified poplar wood were investigated. Other technological parameters, including HEMA to glyoxal ratio (80:20) and modifier concentration (20%), remained unchanged. The modification results of wood are impacted by temperature, which is directly related to the activity of the catalyst and the degree of cross-linking of the modifier. As can be seen from Figure 8, the curing temperature has no significant effect on BE. With a temperature exceeding 120 °C, WPG declined slightly due to the rapid volatilization of glyoxal and the degradation of wood cell wall components. The reactivity of HEMA and glyoxal, as well as the cross-linking with the wood composition, increased as the temperature increased. To ensure that the synthetic property of wood was the best, the curing temperature was set at 120 °C. At a curing temperature of 120 °C, the ASE of wood can reach 68.50%, with a leachability of 19.33%.

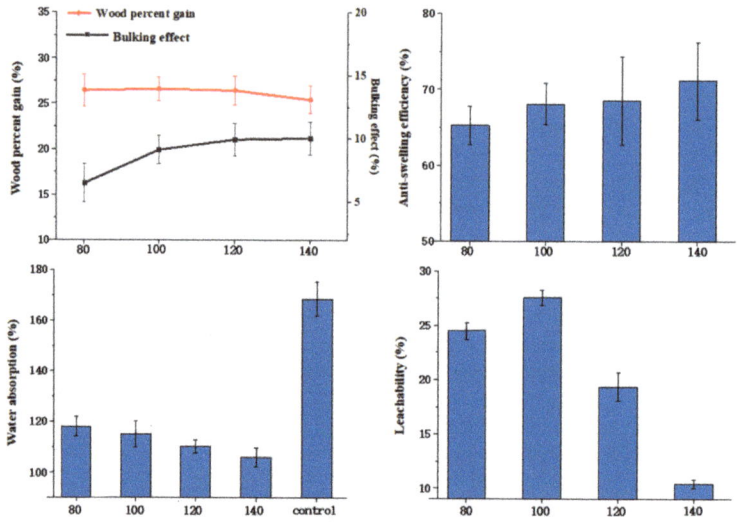

Figure 8. Effects of curing temperature on properties of wood.

4. Conclusions

In this study, the dimensional stability of wood was improved by the in situ polymerization of water-soluble monomers in water. A water-soluble monomer and cross-linker were injected into the wood cell walls and activated cross-linking reactions to form interpenetrating polymer network structures. The polymer network blocked the partial pore and reduced wood hydroxyl, which simultaneously and significantly increased the wood's transverse connection, dimension, and stability. The results showed that the best performance of ASE, WAR, and L was achieved at 77.94%, 58.39%, and 10.44%, with the experimental parameters of a 70:30 modifier ratio, 40% modifier concentration, and 120 °C curing temperature, respectively. Compared to unmodified wood, modified wood has thicker cell walls and smaller cell gaps. In addition, crystallinity, crystalline zone width, and the microfiber angle were all greater in modified wood, indicating that the modifiers are efficiently dispersed into the cell walls. The FTIR analysis reveals that the -OH of the modified wood, which is involved in the reaction of HEMA and glyoxal, is significantly reduced. Furthermore, the characteristic chemical bonds C=C and C=O in the modifiers were significantly increased, indicating that the modifiers strongly bonded to the wood. The results from FTIR and NMR analyses indicate that the modifiers existed in the wood cell wall and reacted with wood components. The enhanced absorption peaks at 88 and 83 ppm indicate the formation of a new crystallization zone, and consequently, the crystallinity of the modified material increased. Compared with the unmodified wood, the characteristic peaks of 55 ppm and 120–170 ppm showed significant chemical shifts, indicating a cross-linking reaction between the modifier and the carbon in the lignin. The results of the NMR analysis verified the results of FTIR and XRD.

Author Contributions: Data curation—writing and editing: J.H.; Supervision: Z.F. and X.W.; Project administration: Y.C. All authors have read and agreed to the published version of the manuscript.

Funding: This research was funded by the Fundamental Research Funds for the Central Non-profit Research Institution of CAF (CAFYBB2020ZC003).

Conflicts of Interest: The authors declare no conflict of interest.

References

1. Wu, Y. Newly advances in wood science and technology. *J. Cent. South Univ. For. Technol.* **2021**, *41*, 1–28.
2. Dong, Y.; Altgen, M.; Mäkelä, M.; Rautkari, L.; Hughes, M.; Li, J.; Zhang, S. Improvement of interfacial interaction in impregnated wood via grafting methyl methacrylate onto wood cell walls. *Holzforschung* **2020**, *74*, 967–977. [CrossRef]
3. Dong, Y.; Zhang, W.; Hughes, M.; Wu, M.; Zhang, S.; Li, J. Various polymeric monomers derived from renewable rosin for the modification of fast-growing poplar wood. *Compos. Part B Eng.* **2019**, *174*, 106902. [CrossRef]
4. Bi, X.; Zhang, Y.; Li, P.; Wu, Y.; Yuan, G.; Zuo, Y. Research Progress of Poplar Impregnation Modification and Its Application in Furniture. *Mater. Rep.* **2022**, *21*, 1–25.
5. Thybring, E.; Fredriksson, M. Wood modification as a tool to understand moisture in wood. *Forests* **2021**, *12*, 372. [CrossRef]
6. Dong, Y.; Liu, X.; Liu, R.; Wang, W.; Li, J. Research progress and development trend of wood chemical cross-linking modification. *China For. Prod. Ind.* **2020**, *57*, 10–14.
7. Huang, Y.; Li, G.; Chu, F. Modification of wood cell wall with water-soluble vinyl monomer to improve dimensional stability and its mechanism. *Wood Sci. Technol.* **2019**, *53*, 1051–1060. [CrossRef]
8. Chai, Y.B.; Liu, J.L.; Sun, B.L.; Qin, T.F.; Chu, F.X. Processing and properties of acetylated poplar treated without catalysts. *China Wood Ind.* **2015**, *29*, 5–9.
9. Xu, W.; Mao, W.G.; Wu, Z.H.; Huang, Q.T.; Lin, S.X. Research of impregnation modification and high-heat treatment of fast-growing wood for furniture. *China For. Sci. Technol.* **2015**, *29*, 9–21.
10. Keplinger, T.; Cabane, E.; Chanana, M.; Hass, P.; Merk, V.; Gierlinger, N.; Burgert, I. A versatile strategy for grafting polymers to wood cell walls. *Acta Biomater.* **2015**, *11*, 256–263. [CrossRef]
11. Dong, Y.; Wang, K.; Yan, Y.; Zhang, S.; Li, J. Grafting polyethylene glycol dicrylate (PEGDA) to cell walls of poplar wood in two steps for improving dimensional stability and durability of the wood polymer composite. *Holzforschung* **2016**, *70*, 919–926. [CrossRef]
12. Giridhar, N.; Pandey, K.K.; Prasad, B.E.; Bisht, S.S.; Vagdevi, H.M. Dimensional stabilization of wood by chemical modification using isopropenyl acetate. *Maderas. Cienc. Y Tecnol.* **2017**, *19*, 15–20.
13. Kurkowiak, K.; Emmerich, L.; Militz, H. Wood chemical modification based on bio-based polycarboxylic acid and polyols—Status quo and future perspectives. *Wood Mater. Sci. Eng.* **2021**, 1–15. [CrossRef]

14. Guo, D.; Shen, X.; Yang, S.; Huang, Y.; Li, G.; Chu, F. The mechanism to improve the dimensional stability of wood modification of water-soluble vinyl monomers. *Scientia Silvae Sinacae.* **2021**, *57*, 158–165.
15. Huang, Y.; Li, G.; Chu, F. *In situ* polymerization of 2-hydroxyethyl methacrylate (HEMA) and 3-(methacryloxy)propyltrimethoxysilane (MAPTES) in poplar cell wall to enhance its dimensional stability. *Holzforschung* **2019**, *73*, 469–474. [CrossRef]
16. Lei, Z.; Gao, J.; Liu, X.; Liu, D.; Wang, Z. Poly(glycidyl methacrylate-co-2-hydroxyethyl methacrylate) brushes as peptide/protein microarray substrate for improving protein binding and functionality. *ACS Appl. Mater. Interfaces* **2016**, *8*, 10174–10182. [CrossRef]
17. Qiu, H.; Yang, S.; Han, Y.; Shen, X.; Fan, D.; Li, G.; Chu, F. Improvement of the Performance of Plantation Wood by Grafting Water-Soluble Vinyl Monomers onto Cell Walls. *ACS Sustain. Chem. Eng.* **2018**, *6*, 14450–14459. [CrossRef]
18. Macková, H.; Plichta, Z.; Hlídková, H.; Sedláček, O.; Konefal, R.; Sadakbayeva, Z.; Dušková-Smrčková, M.; Horák, D.; Kubinová, Š. Reductively degradable poly(2-hydroxyethyl methacrylate) hydrogels with oriented porosity for tissue engineering applications. *ACS Appl. Mater. Interfaces* **2017**, *9*, 10544–10553. [CrossRef]
19. Yan, Y.; Dong, Y.; Chen, H.; Zhang, S.; Li, J. Effect of Catalysts and Sodium Hydroxide on Glyoxal-treated Wood. *BioResources* **2014**, *9*, 4540–4551. [CrossRef]
20. Ping, L.; Chai, Y.; Zhang, F.; Sun, B.; Liu, J. In Polymerization of Environment Friendly Melamine-Urea-Glyoxal Resin in Rubber Wood for Improved Physical and Mechanical Properties. *Int. J. Polym. Sci.* **2021**, *2021*, 8510571. [CrossRef]
21. Cao, L.; Pizzi, A.; Zhang, Q.; Tian, H.; Lei, H.; Xi, X.; Du, G. Preparation and characterization of a novel environment-friendly urea-glyoxal resin of improved bonding performance. *Eur. Polym. J.* **2022**, *162*, 110915. [CrossRef]
22. Xi, X.; Liao, J.; Pizzi, A.; Gerardin, C.; Amirou, S.; Delmotte, L. 5-Hydroxymethyl furfural modified melamine glyoxal resin. *J. Adhes.* **2020**, *96*, 1167–1185. [CrossRef]
23. Sun, W.; Shen, H.; Cao, J. Modification of wood by glutaraldehyde and poly (vinyl alcohol). *Mater. Des.* **2016**, *96*, 392–400. [CrossRef]
24. Yan, Y. *Study on Performance and Enhancement Mechanism of Glyoxal Compound Modified Fast-Growing Poplar Wood*; Beijing Forestry University: Beijing, China, 2017.
25. Segal, L.; Creely, J.J.; Martin, A.E., Jr.; Conrad, C.M. An Empirical Method for Estimating the Degree of Crystallinity of Native Cellulose Using the X-Ray Diffractometer. *Text. Res. J.* **1959**, *29*, 786–794. [CrossRef]
26. Cave, I.D. Theory of X-ray measurement of microfibril angle in wood. *Wood Sci. Technol.* **1997**, *31*, 225–234. [CrossRef]
27. Xiong, F.Q.; Zhou, L.; Liu, S.Q.; Qian, L.C.; Wang, Z.B.; Zhao, L. Effects of liquid nitrogen milling on particle structure of eucalyptus powder. *J. Nanjing For. Univ. Nat. Sci. Ed.* **2014**, *38*, 125–129.
28. Zhenhua, X.; Guangjie, Z. Influence of different treatments on wood crystal properties. *J. Northwest For. Univ.* **2007**, *22*, 169–171.
29. Li, Y. *Mechanism and Application of Compressive Deformation Fixation of Poplar by Glycerin Pretreatment*; Beijing Forestry Universiy: Beijing, China, 2010.
30. Shu, Z.; Liu, S.; Zhou, L.; Li, R.; Qian, L.; Wang, Y.; Wang, J.; Luo, S. Effects of impregnation temperature on the microfibril angle and crystal properties of modified poplar veneers based on [Bmim] Cl. *J. Northwest A F Univ.-Nat. Sci.* **2017**, *45*, 81–86.
31. Ding, Z.J. Research on Poplar Veneer Dyeing Processing and Lightfastness. Master's Thesis, Nanjing Forestry University, Nanjing, China, 2011.
32. Liu, Y.; Hu, J.H.; Gao, J.M.; Guo, H.W.; Chen, Y.; Cheng, Q.Z.; Via, B.K. Wood veneer dyeing enhancement by ultrasonic-assisted treatment. *BioResources* **2015**, *10*, 1198–1212. [CrossRef]
33. Jin, C.; Song, W.; Liu, T.; Xin, J.N.; William, C.H.; Zhang, J.W.; Liu, G.F.; Kong, Z.W. Temperature and pH responsive hydrogels using methacrylated lignosulfonate cross-linker: Synthesis, characterization, and properties. *ACS Sustain. Chem. Eng.* **2018**, *6*, 1763–1771. [CrossRef]
34. Wang, K.; Dong, Y.; Yan, Y.; Zhang, S.; Li, J. Improving dimensional stability and durability of wood polymer composites by grafting polystyrene onto wood cell walls. *Polym. Compos.* **2018**, *39*, 119–125. [CrossRef]
35. Yu, Y. Study on Quantitative Determination of Phenol Hydroxy Structure of Spruce Lignin by Nuclear Magnetic Resonance Technology. Master's Thesis, Guangxi University, Nanning, China, 2020.
36. Qin, Y.; Dong, Y.; Li, J. Effect of Modification with Melamine–Urea–Formaldehyde Resin on the Properties of Eucalyptus and Poplar. *J. Wood Chem. Technol.* **2019**, *39*, 360–371. [CrossRef]
37. Liu, Q.; Du, H.; Lyu, W. Physical and Mechanical Properties of Poplar Wood Modified by Glucose-Urea-Melamine Resin/Sodium Silicate Compound. *Forests* **2021**, *12*, 127. [CrossRef]
38. Wang, X.; Chen, X.; Xie, X.; Cai, S.; Yuan, Z.; Li, Y. Multi-Scale Evaluation of the Effect of Phenol Formaldehyde Resin Impregnation on the Dimensional Stability and Mechanical Properties of Pinus Massoniana Lamb. *Forests* **2019**, *10*, 646. [CrossRef]

Review

Optically Transparent Bamboo: Preparation, Properties, and Applications

Xuelian Li [1], Weizhong Zhang [1], Jingpeng Li [2], Xiaoyan Li [2], Neng Li [2], Zhenhua Zhang [1], Dapeng Zhang [1], Fei Rao [1,*] and Yuhe Chen [2]

[1] School of Art and Design, Zhejiang Sci-Tech University, Hangzhou 310018, China
[2] Key Laboratory of High Efficient Processing of Bamboo of Zhejiang Province, China National Bamboo Research Center, Hangzhou 310012, China
* Correspondence: raofei@zstu.edu.cn

Abstract: The enormous pressures of energy consumption and the severe pollution produced by non-renewable resources have prompted researchers to develop various environmentally friendly energy-saving materials. Transparent bamboo represents an emerging result of biomass material research that has been identified and studied for its many advantages, including light weight, excellent light transmittance, environmental sustainability, superior mechanical properties, and low thermal conductivity. The present review summarizes methods for preparing transparent bamboo, including delignification and resin impregnation. Next, transparent bamboo performance is quantified in terms of optical, mechanical, and thermal conductivity characteristics and compared with other conventional and emerging synthetic materials. Potential applications of transparent bamboo are then discussed using various functionalizations achieved through doping nanomaterials or modified resins to realize advanced energy-efficient building materials, decorative elements, and optoelectronic devices. Finally, challenges associated with the preparation, performance improvement, and production scaling of transparent bamboo are summarized, suggesting opportunities for the future development of this novel, bio-based, and advanced material.

Keywords: transparent bamboo; optical properties; mechanical properties; functionalization and application

1. Introduction

As environmental concerns and the desire to establish a sustainable civilization become more urgent, bamboo has been identified as a potential replacement for materials based on non-renewable resources. There are about 1500 species of bamboo and 36 million hectares of bamboo planting area widely distributed across America, Asia, and Africa [1]. Indeed, bamboo is an important forest resource, having a higher yield, more rapid growth rate, and better mechanical properties than wood, as well as a high aspect ratio and excellent biodegradability. In terms of growth rate, bamboo has a short growth cycle of 3–5 years, whereas wood has a growth cycle of 20–60 years. Furthermore, single bamboo fiber has average tensile strength and modulus of 1.6 GPa and 33 GPa, respectively [2], which is significantly higher than other known natural fibers, such as cotton, coir, henequen, and ramie [3]. Therefore, bamboo has been widely used to fabricate various structural composites, including bamboo scrimber composites [4,5], laminated bamboo lumber [6,7], and bamboo-fiber-reinforced epoxy composites [8].

Recently, interest has arisen in the use of transparent wood (TW) to reduce energy consumption in buildings by capitalizing on its transparency, high mechanical properties, and excellent thermal insulation performance [9,10]. Fabricated by extracting lignin and filling the remaining wood template with a resin, TW exhibits excellent optical transmittance across a wide range of wavelengths [11]. Typically, TW is formed by combining the skeleton structure of wood with transparent organic resins such as an epoxy polymer (EP) or polymethyl methacrylate (PMMA) [12,13]. Using this approach, TW inherits the excellent characteristics of native wood, including a high modulus, low density, high strength, and high toughness [14]; similarly, its transparency depends on the composition and structure of the specific wood employed [15].

As a natural composite material, wood consists of cellulose, hemicellulose, and lignin, which are arranged in a multi-layered and hierarchical structure. Wood is naturally opaque and always has a certain color owing to two aspects. On the one hand, the chromophoric groups in lignin and other substances have a strong ability to absorb visible light at wavelengths of 380–780 nm [16,17], accounting for 80–95% of the total light absorption of wood and constituting the fundamental reason for its color [18]. On the other hand, many pores in the wood are primarily filled with air and water. This mesoporous structure scatters a great deal of light in the visible wavelength range [19,20]. Moreover, the different components in the wood structure create interfaces between materials with different refractive indexes that induce light refraction, scattering, and reflection when light propagates across them [21]. Therefore, removing the colored substances in wood and filling the cavities with a resin providing a refractive index approaching cellulose can make wood transparent.

Bamboo and wood share many characteristics: they have a highly aligned hierarchy and contain similar primary components [22], including cellulose, hemicellulose, and lignin. Therefore, bamboo also has the potential to be prepared as a transparent material. As mentioned above, bamboo has many advantages compared with wood. However, bamboo's high density and low porosity present notable challenges in preparing transparent bamboo (TB) [23]. For example, the high density of bamboo reduces its permeability during treatment; the density of mature bamboo (~0.65 g/cm^3) is much higher than that of the low-density wood species such as balsa, basswood, and poplar (normally ~0.1–0.4 g/cm^3), which are typically used to produce TW [23].

Furthermore, the poor permeability of bamboo increases the time and quantity of chemicals required to remove the lignin. Poor permeability also affects the resin filling rate, potentially reducing TB's light transmittance and mechanical properties [24]. As a result, recent studies have shown that applying the TW preparation process directly to the preparation of TB results in light transmittance of less than 10% [25]. The preparation of transparent bamboo composites (TBCs), which comprise TB combined with other materials to improve performance, faces many of the same challenges.

Therefore, this review summarizes recent advances in the production of functional TB to identify potential directions for future development. First, several simple and efficient TB preparation strategies are introduced based on delignification/lignin modification and resin impregnation. The optical and mechanical properties of TB prepared using different methods are then compared, and finally, various potential applications of TB in energy-efficient buildings, decorative elements, and optoelectronic devices are discussed.

2. Basics of Light–Bamboo Interaction

As light travels through the air and interacts with solids, its propagation can continue in the forward direction when refracted and/or absorbed, or it can be reflected backward at the air–solid interface (Figure 1a). In order to effectively discuss the optical properties of transparent materials, it is necessary to define the corresponding terms [26]. The total optical transmittance of an object (often referred to simply as its transmittance) is the ratio of the transmitted light intensity (including the intensities of the directly transmitted light $I_{T,direct}$ and diffused transmitted light $I_{T,diffuse}$) to the incident light intensity I_{I0}; the total transmittance is, therefore $(I_{T,direct} + I_{T,diffuse})/I_{I0}$. The optical haze is the ratio of $I_{T,diffuse}$ to the total transmitted light and is therefore defined as $I_{T,diffuse}/(I_{T,direct} + I_{T,diffuse})$, as shown in Figure 1a. Owing to the difference between refractive indexes, when light passes through an object, it will be refracted at an angle obeying Snell's law, expressed as $n_1 \sin\theta_1 = n_2 \sin\theta_2$, where n is the refractive index of each material and θ is the incident light angle in that material. Multiple factors can influence an object's optical transmittance and haze, including its surface roughness, thickness, refractive index, pore size distribution, porosity, etc. [27–29]. For two-phase materials, such as microscale composites, the higher the refractive index (RI) ratio between the two media, the stronger the light scattering, which corresponds to a larger proportion of reflected light and thus a lower transmittance. The attenuation of light occurs when light is transformed into other types of energy, such as heat. The more solid–solid interfaces in a composite or the greater its thickness, the lighter attenuation that occurs, thus reducing transmittance. Thus, a transparent composite material can be realized by providing a low RI ratio between phases, low light attenuation, and lesser thickness, resulting in a higher transmittance.

Bamboo is opaque because of its optically heterogeneous nature, a result of its microscale porous structure, different chemical components with different RIs in the cell walls, and contents of strongly light-absorbing chemical entities. Figure 1b illustrates the microstructure of bamboo. When light interacts with bamboo, a combination of scattering/reflection, refraction, transmission, and absorption occurs. Light scattering takes place at all interfaces between the cell walls (which have an RI of ~1.56) and air (which have an RI of 1.0). Inside the cell walls, the mismatch between the RIs of the main chemical components of lignin (1.61), cellulose (1.53), and hemicellulose (1.53) leads to additional light scattering [30].

Furthermore, lignin exhibits particularly strong light absorption among these components, accounting for 80–95% of all light absorbed by bamboo [18]. The details of this interaction depend on the light's wavelength and the bamboo's properties, such as its density, chemical compositions, and fiber direction. Generally, to make bamboo transparent, light absorption by chemical entities and light scattering at the air/cell wall interfaces and inside the cell walls must be reduced or eliminated.

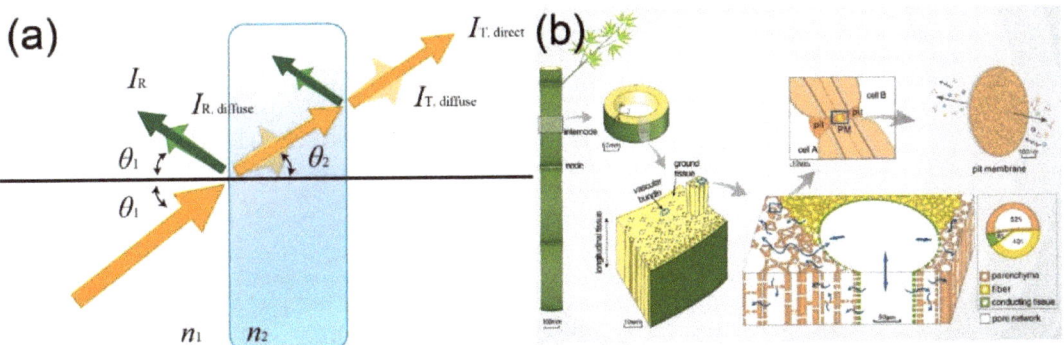

Figure 1. (a) Sketch illustrating light propagation through a solid object in a medium (i.e., air), where I_{I0} is the incident light intensity, $I_{T,direct}$ is the intensity of directly transmitted light, $I_{T,diffuse}$ is the intensity of diffusely transmitted light, I_R is the intensity of reflected light, $I_{R,diffuse}$ is the intensity of diffusely reflected light, θ_1 is the incident angle in the medium, θ_2 is the refracted angle in the solid, and n_1 and n_2, respectively, represent the refractive indices of the medium and solid. (b) Schematic of the hierarchical pore network in bamboo, where PM indicates the pit membrane [31]. ((a) Adapted from Li et al. [32] Copyright 2017 the Royal Society; (b) Reproduced with permission from Liu et al. [31]. Copyright 2021. Elsevier).

3. Preparation of TB

The preparation of TB can be divided into two steps. First, the color-producing compounds in bamboo are removed or modified (decolorization treatment); second, the bamboo is impregnated with a RI-matched resin. Removing or modifying the lignin and chromophoric groups of bamboo is a particularly crucial step. However, the delignified or lignin-modified bamboo is still opaque as the RIs of the substrate and air still do not match. Therefore, it is essential to impregnate the bamboo with a transparent resin providing matching RI to make the bamboo transparent.

3.1. Preparation of the Bamboo Template

The first key step in the preparation of TB is to remove or chemically modify the chromogenic substances in the bamboo (primarily lignin) to achieve a decolorized template. Decolorization is commonly achieved in wood and bamboo using acid delignification, alkali delignification, lignin modification, or enzyme delignification methods.

3.1.1. Acid Delignification Method

The preparation of delignified bamboo templates using the relatively simple acid delignification method was demonstrated by Wu et al. [25], who employed a certain concentration of sodium chlorite ($NaClO_2$) mixed with water and acetic acid (CH_3COOH) to treat bamboo at a pH value of 4.6 and a temperature of 80–90 °C until it turned white. The time required for delignification differed according to the size of the bamboo sample. Under this method, the $NaClO_2$ forms unstable chlorous acid in an acidic environment; this chlorous acid decomposes into Cl_2, ClO_2, and H_2O, which interact with the benzene ring structures in the conifer aldehyde and aromatic ketone in the lignin through an oxidative ring-opening reaction to form an acidic group, causing the material to degrade and dissolve in water. Furthermore, the hypochlorous acid produced by the reaction of Cl_2 with water is also a strong oxidant that reacts with lignin to finally produce o-quinone, small molecules (e.g., carboxylic acid) and corresponding alcohols. Lignin macromolecules are broken and dissolved through these reactions, thereby achieving the purpose of decolorization. When sodium hypochlorite (NaClO) is used to remove lignin instead, the main reactions include chlorination and oxidation, and the main reaction objects are the benzoquinone structure of the lignin benzene ring and the conjugated double bond of the side chain. These reactions are able to generate small molecules (such as CO_2 and carboxylic acid) and thereby remove lignin and other coloring substances from the sample [33,34].

In addition, the use of 1% sodium hydroxide (NaOH) to pretreat natural bamboo prior to acid delignification has been observed to help prepare TB with high optical transmittance [23]. Figure 2a summarizes the associated bamboo template preparation stages. The transmittance of TB prepared using NaOH and acid delignification was found to be ~8% lower than that of TW, but its cellulose volume fraction was ~400% higher, resulting in a tensile strength of 92 MPa, roughly twice that of TW. Furthermore, the developed TB exhibited a low heat conductivity of 0.203 $Wm^{-1} K^{-1}$, indicating a reduction of ~10% and ~80% compared to TW and traditional glass, respectively. Indeed, the pretreatment of bamboo with NaOH results in a more microporous cell wall structure, primarily owing to the partial removal of extract, hemicellulose, and lignin from the cell walls to form small cavities. These cavities' existence increases the cell wall's contraction and thereby decreases its thickness, and realizing a porous structure more conducive to the infiltration of lignin-removal chemicals.

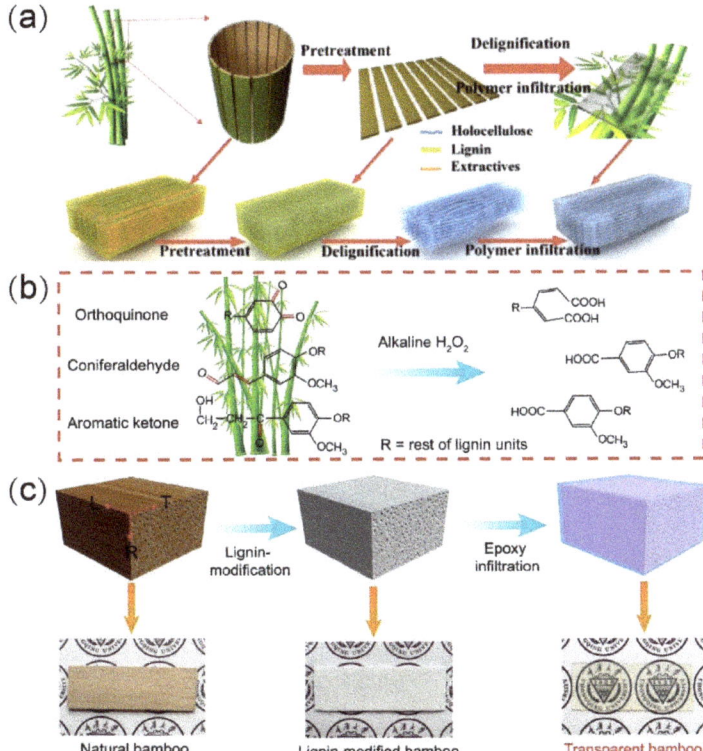

Figure 2. Preparation of transparent bamboo (TB) using the (**a**) acid delignification (including pretreatment) [23], (**b**,**c**) lignin modification [35] methods. ((**a**) Reproduced with permission from Wang et al. [23]. Copyright 2021 ACS Publications; (**b**,**c**) Reproduced with permission from Wang et al. [35]. Copyright 2022 Elsevier).

3.1.2. Alkali Delignification Method

Researchers have also prepared TW templates by treating wood samples with a mixed solution of NaOH and sodium sulfite (Na_2SO_3) at 100 °C for 12 h, as SO_3^{2-} can sulfonate lignin in alkaline conditions [36,37]. The purpose of this lignosulfonalization is to introduce sulfonic acid groups into the side chain of the benzene rings in lignin. The reaction products are not only soluble in water but can also break the bonds of various ethers. Considering that the quinone structure formed during alkaline sulfite treatment will darken the color of the sample, further bleaching with hydrogen peroxide (H_2O_2) is required [15]. Notably, H_2O_2 bleaching is also a delignification process. First, H_2O_2 is dissociated in an alkaline solution to form hydrogen peroxide anion (HOO^-), which can decompose the quinone structure and even degrade it into small molecule esters. It should be noted that the bleaching effect of HOO^- is closely related to the pH of the solution. When the pH increases, the concentration of HOO^- increases, leading to an improved bleaching effect. However, when the pH exceeds 10.5, HOO^- is easily decomposed into O_2, and the bleaching effect will deteriorate [15]. Critically, the alkali delignification method is complex and can readily cause sample deformation [38].

3.1.3. Lignin Modification (Lignin Retaining) Method

The lignin modification method has been used to prepare bamboo templates for TB by retaining the lignin while modifying the chromophoric groups within, as shown in Figure 2b. Because the lignin composition is retained, the mechanical strength of the resulting TB is typically higher than that of TB prepared by removing the lignin [35]. Moreover, the lignin modification method also avoids wasting the lignin components. For example, Wang et al. [35] removed the light-absorbing chromospheres (aromatic ketone, coniferaldehyde, and orthoquinone) of lignin using an alkali H_2O_2 treatment while retaining the aromatic skeleton lignin structure (Figure 2b). To do so, sodium silicate (3.0 wt%), sodium hydroxide (3.0 wt%), diethylenetriamine pentaacetate (0.1 wt%), magnesium sulfate (0.1 wt%), and H_2O_2 (4.0 wt%) were dissolved in deionized water to prepare the lignin modification solution in which the bamboo samples were immersed at 70 °C until they turned completely white. The lignin-modified bamboo was then thoroughly rinsed three times with boiled deionized water to remove any residual chemicals. Finally, the resulting bamboo templates were solvent exchanged using acetone and ethanol. This lignin modification method was confirmed to selectively remove or react the chromogenic groups in the bamboo while retaining most of the lignin. The resulting high-lignin-content TB had a transmittance of 87% and haze of 90%.

3.1.4. Biological Enzyme Delignification

Delignification using the biological enzyme method represents an environmentally friendly process, as the need for harmful chemicals is minimized. Jichun and Yan [39] used biological enzymes to degrade lignin and thereby achieve the decolorization of wood by employing the following procedure. The dried wood samples, pure water, biological enzyme (synthetic laccase/xylanase system at a dosage of 10 IU/g), and glacial acetic acid were combined in a 1:30–40 mass ratio of sample to water, and the pH was adjusted to 3–5 by adding trace quantities of hydrogen peroxide (up to 4% of the sample mass). A treatment temperature of 35–50 °C was then applied for 1–2 h, after which the samples were washed with deionized water. Next, the samples were extracted using 30 wt% dioxygen water and 25 wt% ammonia water at a volume ratio of 10:1. The extracted samples were then washed with deionized water and dehydrated by ultrasonic extraction to obtain the TW templates.

4. Properties of TB

4.1. Optical Properties

Ongoing research has produced TB with excellent optical transmittance and high haze (Table 1). There are three main interactions with light to consider when evaluating the optical properties of TB: (1) reflection at the outer gas/TB interface, (2) scattering in the form of reflection and refraction, and (3) absorption inside the TB. In TB, light scattering mainly occurs at the interface between the bamboo tissue and the polymer. The lower the difference between the RIs of the bamboo template and polymer, the less scattering will occur at their interface. High haze is primarily a result of collective scattering inside the composite material. Finally, light absorption is primarily caused by the presence of lignin in the bamboo. Delignification and lignin modification have accordingly been applied, as discussed in Section 3, to remove the components containing chromophores and thereby minimize light absorption.

As shown in Table 1, the optical properties of TB prepared using the same resin differ substantially according to the preparation method. As discussed in Section 1, the light transmittance obtained by applying the TW preparation process directly to the preparation of TB was not ideal [25]. However, the light transmittance of TB was effectively increased to 80% by pretreating with 1% NaOH before acid-chlorite delignification, as this increased the contraction of the cell wall and decreased its thickness through the partial removal of extract, lignin, and hemicellulose [23]. In addition, the lignin modification method produced TB with excellent light transmittance as high as 87% by removing the light-absorbing chromospheres while preserving the aromatic skeletal lignin structure [35].

The most frequently used resin for the preparation of TB is shown in Table 1 to be EP, primarily owing to the similarity of its RI to that of the bamboo template. This reduces the degree of scattering to obtain better optical performance. Factors such as the fiber volume fraction and thickness of the bamboo template will also affect the optical properties of TB. The volume fraction of bamboo fibers naturally increases from the inner side to the outer side of the bamboo culm; TB with the desired fiber volume fraction can be obtained for a given thickness by compressing the bamboo template prior to resin impregnation.

As shown in Figure 3a, the transmittance of TB generally decreases with increasing fiber volume fraction, primarily because a higher fiber volume fraction hinders the infiltration of the lignin-modification chemicals into the bamboo, thereby leaving more chromospheres in the lignin [35]. Figure 3a also shows the transmittance of the inner layer with a low fiber volume fraction is higher than that of the outer layer with high fiber volume fraction. Furthermore, as the bamboo thickness increases, the transmittance decreases and haze increases, primarily owing to the longer light pathway, which corresponds to an increasing degree of light attenuation and an increasing number of polymer/bamboo interfaces, leading to more light scattering. Indeed, as reported in Figure 3b, as the TB thickness increased from 0.3 to 2.9 mm, the transmittance decreased from 92.4% to 23.7%, and the haze increased from 43.5% to 82.95% [24]. Figure 3b also shows that the transmittance of multi-layer TB is far higher than that of single-layer TB with the same thickness because multi-layer TB enables more uniform epoxy resin impregnation through the laminations. For example, the transmittance of 1.2 mm thick multi-layer TB was found to be 78.6, whereas that of 1.2 mm thick single-layer TB was 10.4 [24].

In addition, TB also demonstrates an anisotropic optical property attributed to cellulose orientation-induced birefringence ($\triangle n \approx 0.074$–$0.08$) and the anisotropic structure of natural bamboo [40,41]. Indeed, the haze of TBC made from TB is higher in the radial direction than in the longitudinal direction (Figure 3c) because TBC has a lower density of polymer/cellulose interfaces in the longitudinal direction owing to the hollow cylindrical shape of bamboo cells [35]. On the one hand, the light scattering is highly anisotropic when the light is perpendicular to the longitudinal direction of the TBC; on the other hand, the light scattering is almost isotropic when the light is oriented along the longitudinal direction of the TBC (Figure 3d) [35]. Thus, different relationships between the orientation of the bamboo fibers and the direction of light (perpendicular and parallel) will result in different scattering patterns.

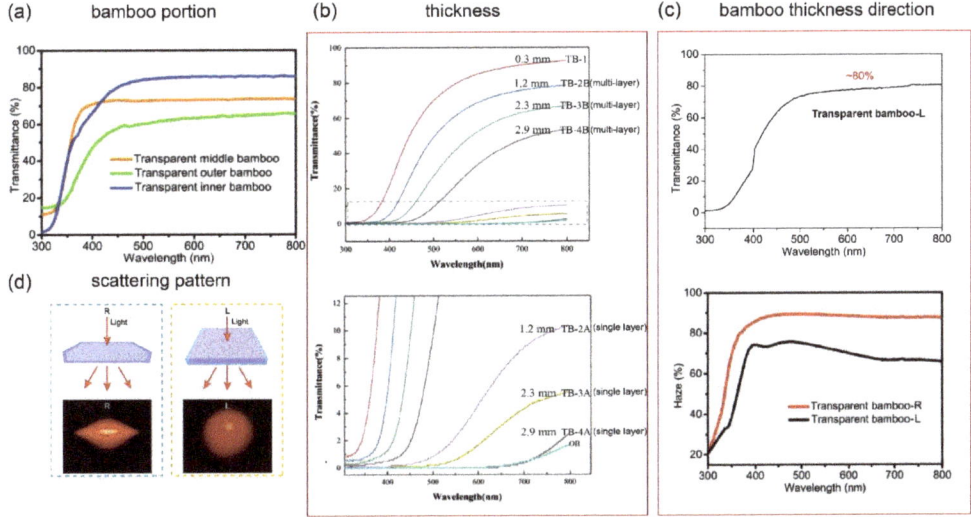

Figure 3. Transmittance and haze data for TB according to (**a**) bamboo fiber location [35], (**b**) bamboo thickness [24], and (**c**) bamboo thickness direction [35], as well as the (**d**) scattering pattern of TB according to bamboo thickness direction (anisotropy effect) [35]. ((**a**,**c**,**d**) Adapted from Wang et al. [35]; (**b**) Adapted from Wu et al. [24]).

Table 1. Summary of TB preparations and properties.

Ref.	Template Preparation Method; Temperature; Time	Polymer	Variable	t (mm)	Optical Property Tr (%)	haze (%)	Mechanical Property l × w × t (mm³)	TS (MPa)
Wu et al. [25]	Delignification: NaClO₂; 80–90 °C; 2–4 h	Epoxy (E51) (RI = approximately 1.5)	inner	1.1	approximately 12	—	3 × 4.4 × 1.1	35.31
			outer	1.8	approximately 2	—	3 × 7.8 × 1.8	82.18
Wang et al. [23]	Delignification: preconditioned in NaOH + 10 h; NaClO₂; 85 °C; 3 h	Two-part epoxy resin (Clearcast 7000) (RI = approximately 1.5)	—	1	approximately 80	80	165 × 13 × 1	92
			—	1.5	approximately 75	80	165 × 13 × 1.5	—
Wu et al. [24]	Delignification: NaClO₂; 80–90 °C; 2–3 h	Epoxy (E51) (RI = 1.52)	—	0.3	92.4	43.5	40 × 20 × 0.3	47.1
			multi-layer (3 layers)	1.2	78.6	70	40 × 20 × 1.2	61.89
			multi-layer (5 layers)	2.3	67.1	70.55	40 × 20 × 2.3	approximately 60
			multi-layer (7 layers)	2.9	23.7	82.95	40 × 20 × 2.9	approximately 60
			single layer	1.2	10.4	97.02	40 × 20 × 1.2	61.15
			single layer	2.3	5.5	~100	40 × 20 × 2.3	approximately 30
			single layer	2.9	1.7	~100	40 × 20 × 2.9	approximately 10
Wang et al. [35]	Lignin modification: lignin-modification solution; 70 °C; until become white	Epoxy (E51) (RI = approximately 1.5)	inner	1.5	87	—	—	—
			middle	1.5	74	—	—	—
			outer	1.5	66	—	—	—
			radial	1.5	—	90	20 × 8 × 1.5	30
			longitudinal	1.5	80	70	50 × 10 × 1.5	118

4.2. Mechanical Properties

The mechanical properties of TB are affected by the preparation method, bamboo properties such as cellulose content, cell structure morphology, and density; bamboo structural anisotropy; and the interfacial bond between the bamboo template and the impregnating resin. In addition, a summary including mechanical and physical properties of natural fibres is shown in Table 2.

Table 2. Summary of mechanical and physical properties of natural plant fibres.

Ref.	Type of Fibre	Density (g/cm³)	Tensile Strength (MPa)	Young's Modulus (MPa)
Abdul Khalil et al. [42]	Moso bamboo (*Phyllostachys pubescens*)	1.2–1.5	500–575	27–40
Liu et al. [3]	Oil palm	0.7–1.6	248	3.2
Liu et al. [3]	Pineapple	0.8–1.6	1.44	34.5–82.5
Cai et al. [43]	Abaca	1.5	717	18.6
Vijaya Ramnath et al. [44]	Jute	1.3–1.49	393–800	13–26.5
Ramesh et al. [45]	Sisal	1.41	350–370	12.8
Mohamad et al. [46]	Kenaf	1.2	282.6	7.13
Asim et al. [47]	Coconut	1.2	140–225	3–5

As shown in Table 1, the mechanical properties of TB produced using the same resin exhibit substantial differences based on the applied preparation method. Notably, as the lignin modification method removes the light-absorbing chromophore groups without entirely destroying the aromatic structure of lignin, the mechanical properties of the resulting TB are much better. As a result, the tensile strength of TB obtained by lignin modification can be as high as 118 MPa [35].

The mechanical properties of TB are strongly dependent on the mechanical properties of the bamboo template. Density, which mainly depends on the fiber diameter, fiber content, and cell wall thickness, exerts a considerable influence on the mechanical properties of bamboo. The bamboo fiber density increases with increasing fiber content. As mentioned in Section 4.1, the bamboo fiber volume fraction increases from the inner side to the outer side of the bamboo culm, resulting in different bamboo layer densities according to their original locations within the stalk. Thus, TB made from an outer bamboo layer is often stronger than TB made from an inner layer, as shown in Table 1. Note that Table 1 also indicates that the tensile strength of multi-layer TB is higher than that of single-layer TB with the same thickness. In the beginning, the tensile strength of 1.2 mm thick multi-layer TB was found to be 61.89 MPa, whereas that of 1.2 mm thick single-layer TB was found to be 61.15 MPa [24]. As the TB thickness increases, the difference in tensile strength grows larger. This result is related to interfacial compatibility. As mentioned in Section 4.1, multi-layer TB enables more uniform resin impregnation through the laminations, whereas it is difficult for the resin to infiltrate into the bamboo cells of single-layer TB. Therefore, the tensile strength of multi-layer TB is greater, making it more suitable as a structural material in electronics, household, and construction applications. Furthermore, the tensile strength of TB first increases and then decreases with increasing bamboo template thickness. A thicker bamboo template makes it difficult to completely remove lignin but and hinders the ability of the resin to infiltrate into the template. In contrast, a relatively thinner bamboo template exhibits more extensive lignin removal and can easily be impregnated with resin. The properties of TB can thus be tailored by compression of the bamboo template to provide different densities and thicknesses. Furthermore, owing to the anisotropic structure of bamboo, which has no radial cell elements, the tensile strength of TB is much higher in the longitudinal direction (about 118 MPa) than in the radial direction (about 30 MPa) [35].

4.3. Thermal Conductivity

Excellent thermal insulation performance with low heat conductivity is critical to realizing energy-efficient building materials [48,49]. The thermal conductivity of TB (0.203 W m^{-1} K^{-1}) is accordingly compared with that for common glass (0.974 W m^{-1} K^{-1}), TW (0.225 W m^{-1} K^{-1}) [23], and other materials in Figure 4a. Notably, the figure indicates that TB has a lower thermal conductivity than TW. The low thermal conductivities of both TB and TW can be attributed to the interfacial heat resistance and the phonon dispersion between the air and cell walls [50]. Thermal conduction in TB is also anisotropic, like its optical and mechanical properties. Indeed, the thermal conductivities of TB in the longitudinal and radial directions were found to be about 0.44 and 0.33 W m^{-1} K^{-1} (Figure 4b), respectively [35]. This indicates that the thermal energy tends to spread more in the direction parallel to the direction of bamboo growth owing to the orientation of the cellulose nanofibers [51]. Because of its low thermal conductivity yet relatively high light transmittance, TB represents a promising window material that can prevent heat dissipation and thus reduce the energy consumption of buildings.

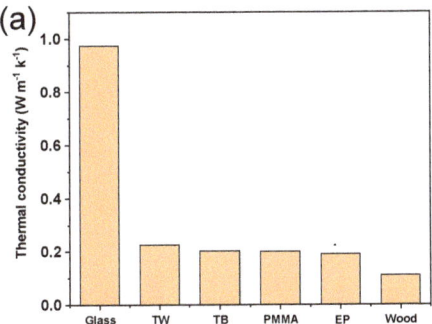
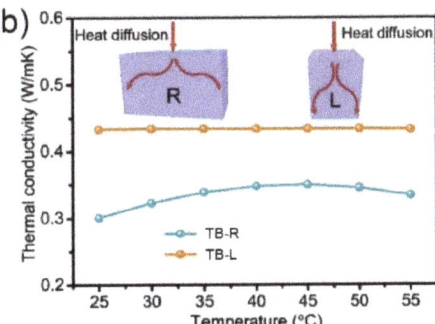

Figure 4. Thermal conductivity of TB: (**a**) comparison of the thermal conductivities of common glass, transparent wood (TW), TB, polymethyl methacrylate (PMMA), epoxy polymer (EP), and wood [23,52,53]; (**b**) thermal conductivities of TB in the longitudinal (L) and radial (R) directions [35]. ((**b**) adapted from Wang et al. [35]).

5. Functionalized TB and its Potential Applications

Bamboo has considerable multifunctional potential owing to its multi-scale pores and composition of lignin, hemicellulose, and cellulose. Typical modification strategies to improve the functionality of TB include cell wall modification, cell wall/lumen interface modification, and lumen filling, as shown in Figure 5. Notably, the bamboo template used to produce TB has been shown to demonstrate higher porosity than the original bamboo, potentially facilitating additional modification [23]. One method for TB modification is to fill the lumen space with polymer liquids decorated with functional nanoparticles, followed by polymerization and curing. Another TB modification method is to first modify the bamboo template, then apply monomer impregnation and polymerization. Taking TW as an example, successful modifications have realized ultraviolet (UV)-stabilized wood [54,55], magnetic wood [56,57], conductive wood [58–60], and stimuli-responsive wood [61]. Bamboo also has the potential to be prepared with the same functionalities owing to its similar structure and composition. Such functionalizations can enable the wide application of TB in fields such as energy-saving windows, decorative elements, and optoelectronic devices.

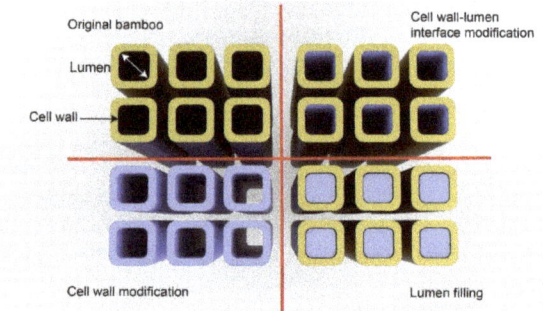

Figure 5. A selection of structural modification strategies to improve TB functionality (inspired by Burgert et al. [62]).

5.1. Energy-Saving Windows

Reducing building energy consumption has been the focus of much research. Windows are essential building components considering the requirements of human visual comfort and health; however, they are also the least energy-efficient elements of a building since energy is always transferred through a window in the direction opposite to that desired. In summer, when the outdoor temperature is higher than the indoor temperature, heat is transferred to the indoor environment through the windows; in winter, when the outdoor temperature is lower than the indoor temperature, the heat inside is released into the outdoor environment. Both phenomena increase the energy consumption required to maintain a comfortable indoor environment. Therefore, smart windows and low emissivity (low-E) glass have been designed to alleviate these problems [63–66]. However, there remain several issues with conventional low-E glass. Most importantly, low-E coatings are produced using either physical vapor deposition (PVD) or chemical vapor deposition (CVD), which require expensive, high-vacuum setups [67–69].

To address these problems, two kinds of TWCs (transparent wood composites) with anisotropic structures were developed to serve as energy-saving windows. These TWCs employed the infiltration of an EP resin dispersion containing 0.04 wt% tungsten-doped vanadium dioxide nanoparticles (W/VO$_2$ NPs) into a delignified wood template in the longitudinal or radial directions (W/VO$_2$-TPW-R or W/VO$_2$-TPW-L, respectively), followed by subsequent polymerization, as shown in Figure 6a [70]. Both TWCs exhibited excellent temperature regulation when used as windows: after 5 min of solar radiation, the temperature of a simulated room with a glass window rose sharply from 21.0 to 60.7 °C, whereas that of a room with an EP-resin impregnated TW window increased from 21.0 to 47.2 °C, that of a room with the W/VO$_2$-TPW-R window only increased from 21.0 to 40.6 °C, and that of a room with the W/VO$_2$-TPW-L window only increased from 21.0 to 39.1 °C. Thus, W/VO$_2$-TPW-R/L provided the most effective thermal insulation among the investigated window materials. The superior effectiveness of this material can be attributed to the large quantity of heat reflected by the VO$_2$ NPs. As a result, the indoor temperature increased at a significantly slower rate than when using ordinary glass panels. This novel TWC combining a low thermal conductivity TW template with thermochromic VO$_2$ NPs provided a potential solution for replacing expensive, heavy, high thermal conductivity and infrared transparent glass.

Similarly, energy-saving windows made of TB-based materials combining a low-E coating with radiative cooling (RC) have been reported. A material capable of RC spontaneously cools its surface by reflecting sunlight and radiating long wavelength infrared (LWIR) to the cold outer space. This technique can deliver effective cooling power through highly emissive (ε_{LWIR}) materials, though applications of related technology to windows have not been extensively explored [71]. To achieve the maximum cooling effect, the side of the window facing the outside must have a high ε_{LWIR} to promote RC, while the other side must have a low ε_{LWIR} to prevent heat exchange [72–74]. However, though the RC technique can reduce the energy consumption required for cooling in summer, it also increases the heating burden in winter owing to its heat-releasing characteristic. Therefore, manufacturing a TBC with enhanced RC for application in energy-saving windows could reduce the energy consumption when cooling is required.

Cellulose-based glass panes with high optical transparency were accordingly obtained by delignification of bamboo slices followed by epoxy infiltration, then depositing silver nanowire on one face to realize RC cellulose glass [70]. A low indoors-facing ε_{LWIR} of 0.3 was thereby achieved, together with a large ε_{LWIR} difference of 0.65 between the two faces. A schematic demonstration of the working mechanism of this RC cellulose glass is shown in Figure 6b. The RC performance was considerably enhanced by the suitable difference between the two faces ε_{LWIR}. Critically, compared to conventional low-E glass, RC cellulose can be produced using a solution-based process rather than complex PVD or CVD processes. Thus, low-cost and eco-friendly, energy-saving windows derived from bamboo demonstrate the considerable potential of TB functionalization in architecture.

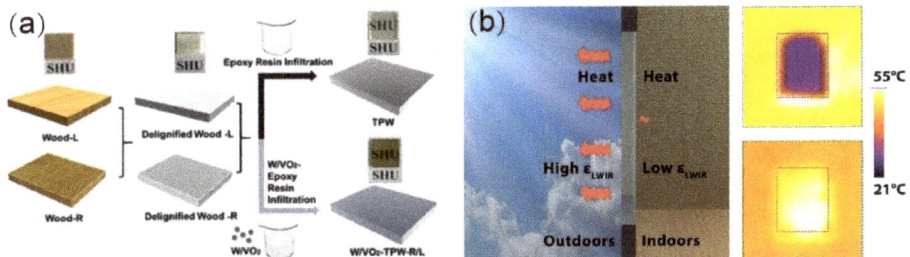

Figure 6. Functional transparent composite materials for energy-saving windows. (**a**) Schematic description of the synthesis route of tungsten-doped vanadium dioxide nanoparticles impregnated into a delignified wood template in the longitudinal or radial directions (W/VO$_2$-TPW-R/L) [70]. (**b**) Schematic demonstration of the working mechanism of radiative cooling (RC) cellulose glass [75]. ((**a**) Reproduced with permission from Zhang et al. [70]. Copyright 2020 ACS Publications; (**b**) Reproduced with permission from Zhou et al. [75]. Copyright 2021 ACS Publications).

5.2. Decorative Applications

The most obvious advantage of TB is that it conforms to the concept of sustainable and environmentally friendly development. Compared with glass, plastic, and other transparent household materials, TB is derived from renewable biomass materials and is produced using a low-cost, non-toxic process. Furthermore, TB has high light transmittance (over 80%) and haze (over 80%) as well as excellent mechanical properties with a higher tensile strength than wood. Though the tensile strength of bamboo is weakened when its internal fibers are destroyed during the delignification process, the tensile strength of the TB is increased after resin impregnation. Therefore, TB is not only useful for household decoration but also in the load-bearing components of household products.

Indeed, TB can serve as a replacement for transparent materials such as glass or plastic in many decorative applications. Furthermore, the application of TB in household products can increase the sense of transparency and mystery [76]. It has been found that overlapping certain thicknesses of TB can produce different effects of transmittance and haze, providing the possibility of forming different visual effects through weaving [77]. For example, a screen woven from TB is shown in Figure 7a. This screen design not only combines TB and wood but also highlights the transparency and haze of TB and reflects aspects of traditional Chinese culture using new materials. Lamp design represents another crucial part of the home decoration industry and should consider the application of new materials and technologies. Indeed, TB can address many deficiencies in the field of lamp design in China, including a serious homogeneity among products, lack of innovation, and design obsolescence. For example, the upper and lower lampshades shown in Figure 7b are made of TB, softening the light. The advantages of TB can thus be fully realized in the design of household decorative products while collocating it in mixed applications with other materials to achieve multiple styles. It is expected that designers will use the patterns present in TB or dope TB with multifarious guest materials to develop next-generation decorative materials and even furniture that possess high transparency, haze, and fascinating patterns to serve various interesting functions.

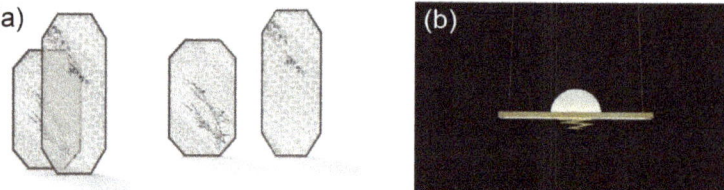

Figure 7. Decorative applications of TB: (**a**) woven composite screens; (**b**) sunset chandelier [76]. ((**a**,**b**) Reproduced with permission from ref. [76]. Copyright 2021 Editorial Department of *Journal of furniture*.).

5.3. Optoelectronic Devices

Transparent biomass materials offer a promising base for optoelectronic devices being developed for a wide range of applications such as mobile phone cameras, new energy conversion systems, and medical testing. Biomass-based transparent materials also have potential as structural materials for photovoltaic equipment such as solar cells [78–80]. Indeed, solar cell efficiency has been improved by 18.02% by placing TW on the top surface (Figure 8a) [78], as its high haze leads to large scattering angles that increase the light path length within the solar cell below. Magnetic TW has been obtained by adding magnetic nanoparticles during preparation [81,82]. Furthermore, TW chips treated with luminescent γ-Fe2O3@YVO4:Eu3+ nanoparticles have been shown to emit light under UV excitation (Figure 8b). Such TW functionality has a wide range of potential applications in luminous magnetic switches, lighting equipment, and anti-counterfeiting equipment.

In addition, a novel, stretchable, transparent, and electroconductive wood material has been fabricated by filling the channels and pores of delignified wood with poly (PDES (polymerizable deep eutectic solvents)) via in situ photopolymerization [83]. The combination of poly (PDES) with the wood matrix imbued the material with excellent transmittance (about 90%), stretchability (up to 80% strain), and superior electrical conductivity (up to 0.16 S m^{-1}).

A transparent conductive wood (TCW) slice can clearly show the parts beneath or make an embedded LED glow (Figure 8c), demonstrating its excellent transparent and conductive characteristics. Furthermore, its unique and outstanding performance characteristics mean that TCW exhibits excellent functional abilities to sense external stimuli, especially in terms of strain and touch. This allows TCW-based sensors to detect subtle human bending release activities and other weak pressures (Figure 8d).

Moreover, a completely wood-based flexible electronic circuit has been reported in which the substrate of the circuit—a strong, flexible, and transparent wood film (TWF)—was printed with a sustainable, bio-based lignin-derived carbon nanofiber conductive ink (Figure 8e) [84]. Notably, the TWF fabrication process maintained the original alignment of the cellulose nanofibers and promoted their combination in the process of removing lignin and hemicellulose. The Young's modulus and tensile strength of this TWF were determined to be 49.9 GPa and 469.9 MPa, respectively, which are greater than those of most natural fibers, plastics, polymers, and metal or engineered alloys, as shown in Figure 8f [84]. The combination of TWF and bio-based conductive ink can therefore produce environmentally friendly and sustainable wood-based electronic products for potential applications such as flexible circuits and sensors. Indeed, the development of flexible and/or transparent electrical devices fabricated from natural wood using environmentally friendly design methods is expected to open up new possibilities in biomedical devices, smart electronics, and other fields. Considering the similarity between the structure and composition of bamboo and wood, the use of bamboo in such applications is not only feasible but could also improve sustainability.

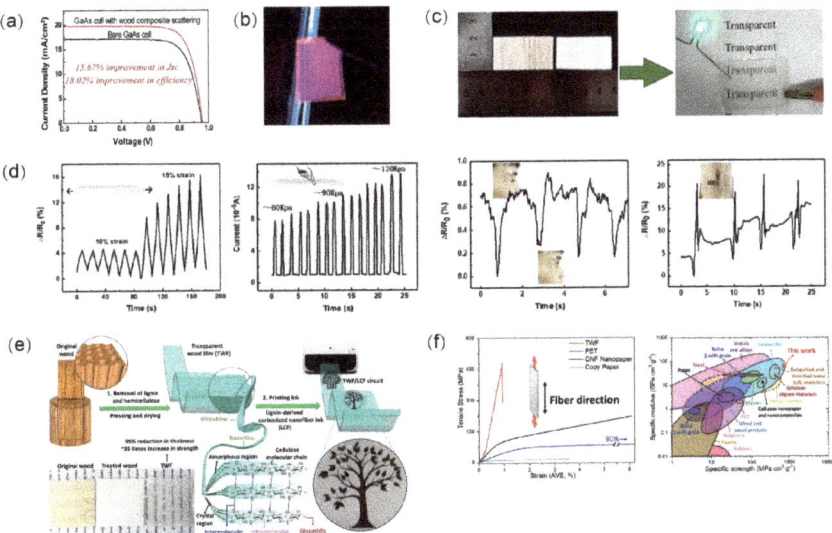

Figure 8. Application of TWC materials in optoelectronics: (**a**) current density versus voltage characteristics for a bare GaAs cell (black) and a GaAs cell with light management wood coating (red) [78]; (**b**) TW glowing under ultraviolet (UV) light [81]; (**c**) digital photograph of transparent conductive wood (TCW) with LED lighting [83]; (**d**) TCW used a strain/touch sensor to monitor human activities [83]; (**e**) processing of transparent wood film (TWF) for application in flexible electronics [84]; (**f**) mechanical properties of TWF compared with various other materials [84]. ((**a**) Adapted from Zhu et al. [78]; (**b**) Adapted from Gan et al. [81]; (**c**,**f**) Adapted from Fu et al. [84]; (**d**) Adapted from Wang et al. [83]).

6. Conclusions and Outlook

This review summarized the recent progress in TB, including different methods for preparing TB with improved optical, mechanical, and thermal conductivity properties; the realization of TB functionalization in energy-efficient building components; and discussions of potential TB use in decorative applications and optoelectronic devices.

1. The delignification process used to prepare TB will affect its optical and mechanical properties. Previous experimental results indicate that TB prepared by lignin modification has better properties than the TB prepared using other methods. However, lignin modification requires more chemical solvents than most other methods. Thus, several sustainability issues related to delignification technology optimization should be addressed in future research. For example, environmentally friendly "green" chemistry approaches are necessary to minimize the use of solvents, reduce reaction time and waste streams in the preparation of TB, and develop TB-specific functionalization methodologies. Proper resin selection is also essential to realizing the desired TB properties. However, as there have been few attempts to prepare TB using different resin types, different resins and resin modifications should be studied to increase the functionality of TB.
2. Scaling up the production of TB is expected to be challenging, although some strategies targeting TB have been explored with promising results. Furthermore, the results reported to date have focused on small and thin TB samples. As an increase in thickness causes the light to travel longer and decay inside the TB, increasing thickness while maintaining sufficient transparency represents a significant challenge to the successful implementation of this technology at the industrial scale. Various technical optimization approaches are therefore worthy of investigation, such as the effect of parallel lamination and cross lamination on TB performance. The potential functionalizations of TB include its development as a: (i) thermal insulation material, (ii) decorative material, (iii) conductive material, and (iv) or magnetic material. To date, TB has typically been singly functionalized for a specific scenario. The versatility of TB should therefore be further evaluated, such as the simultaneous combination of thermal insulation and magnetic functions to address a variety of situations using the same material. The functional utilization of bamboo is far less than that of wood at present, so there remains a wide range of strategies to be explored to develop novel functionalized bamboo materials.
3. The functionalization of TB enables its application to diverse fields such as: (i) energy-saving windows, (ii) decorative applications, and (iii) optoelectronic devices. Several applications targeting TB have been explored with promising results. As the research community pays increasing attention to sustainable development, it is likely that functional TB technology will continue to develop over the next several years.

It is a pleasant process to imagine the future integrated application of TB (Figure 9). TB with good optical and mechanical properties can find potential applications as windows and ceilings in a house or museum, which can replace the glass. The TB-based energy-saving window can improve energy efficiency due to its great thermal insulation properties. Simultaneously, this kind of window can scatter light, which makes the light distribution more uniform compared with normal glass. Similarly, the furniture and household, such as lampshades, desktops, and byobu made of TB, can soften the light and add a calm and tranquil scenery to the room. Different transparency and haze produced after the superposition of TB can form different visual effects. The bamboo-based furniture and household design reflect elements of Chinese traditional culture. In the design of optoelectronic devices, such as wearable sensors, while giving full play to the advantages of TB, more attention should be paid to combine theoretical calculation with theoretical simulation to realize the rapid development of modern bamboo science.

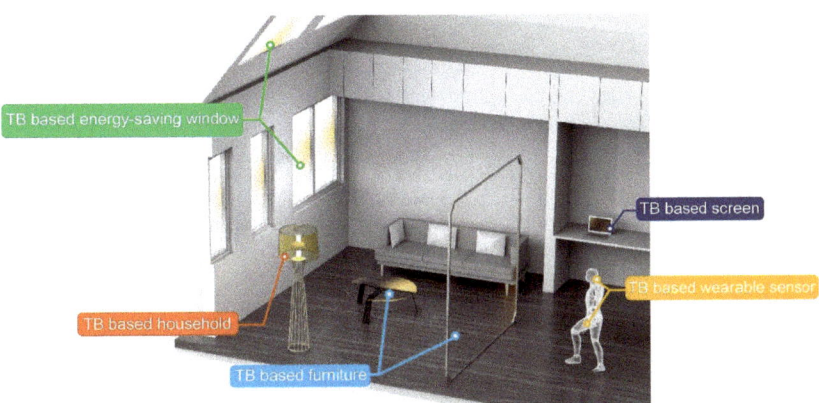

Figure 9. Design of a future integrated application of TB.

Author Contributions: Conceptualization, writing—original draft preparation, X.L. (Xuelian Li); formal analysis, investigation, writing—original draft preparation, W.Z.; investigation, writing—review and editing, J.L., X.L. (Xiaoyan Li), N.L., Z.Z., D.Z. and F.R.; project administration, F.R and Y.C. All authors have read and agreed to the published version of the manuscript.

Funding: This work was supported by the Key Research & Development Program of Zhejiang Province (2021C02012); the Science Foundation of Zhejiang Provincial Department of Education (113429A4F21070); and the Science Foundation of Zhejiang Sci-Tech University (11340031282014 & 11343132612052).

Institutional Review Board Statement: Not applicable.

Informed Consent Statement: Not applicable.

Data Availability Statement: Not applicable.

Conflicts of Interest: The authors declare no conflict of interest.

References

1. Scurlock, J.M.O.; Dayton, D.C.; Hames, B. Bamboo: An Overlooked Biomass Resource? *Biomass Bioenergy* **2000**, *19*, 229–244. [CrossRef]
2. Yu, Y.; Jiang, Z.; Fei, B.; Wang, G.; Wang, H. An Improved Microtensile Technique for Mechanical Characterization of Short Plant Fibers: A Case Study on Bamboo Fibers. *J. Mater. Sci.* **2011**, *46*, 739–746. [CrossRef]
3. Liu, D.; Song, J.; Anderson, D.P.; Chang, P.R.; Hua, Y. Bamboo Fiber and Its Reinforced Composites: Structure and Properties. *Cellulose* **2012**, *19*, 1449–1480. [CrossRef]
4. Yu, Y.; Liu, R.; Huang, Y.; Meng, F.; Yu, W. Preparation, Physical, Mechanical, and Interfacial Morphological Properties of Engineered Bamboo Scrimber. *Constr. Build. Mater.* **2017**, *157*, 1032–1039. [CrossRef]
5. Chung, M.J.; Wang, S.Y. Mechanical Properties of Oriented Bamboo Scrimber Boards Made of Phyllostachys Pubescens (Moso Bamboo) from Taiwan and China as a Function of Density. *Holzforschung* **2018**, *72*, 151–158. [CrossRef]
6. Lin, Q.; Huang, Y.; Li, X.; Yu, W. Effects of Shape, Location and Quantity of the Joint on Bending Properties of Laminated Bamboo Lumber. *Constr. Build. Mater.* **2020**, *230*, 117023. [CrossRef]
7. Chen, G.; Yu, Y.; Li, X.; He, B. Mechanical Behavior of Laminated Bamboo Lumber for Structural Application: An Experimental Investigation. *Eur. J. Wood Wood Prod.* **2020**, *78*, 53–63. [CrossRef]
8. Huang, J.K.; Young, W.B. The Mechanical, Hygral, and Interfacial Strength of Continuous Bamboo Fiber Reinforced Epoxy Composites. *Compos. Part. B Eng.* **2019**, *166*, 272–283. [CrossRef]
9. Wu, J.; Wu, Y.; Yang, F.; Tang, C.; Huang, Q.; Zhang, J. Impact of Delignification on Morphological, Optical and Mechanical Properties of Transparent Wood. *Compos. Part. A Appl. Sci. Manuf.* **2019**, *117*, 324–331. [CrossRef]
10. Li, Y.; Vasileva, E.; Sychugov, I.; Popov, S.; Berglund, L. Optically Transparent Wood: Recent Progress, Opportunities, and Challenges. *Adv. Opt. Mater.* **2018**, *6*, 1–14. [CrossRef]
11. Li, Y.; Fu, Q.; Yu, S.; Yan, M.; Berglund, L. Optically Transparent Wood from a Nanoporous Cellulosic Template: Combining Functional and Structural Performance. *Biomacromolecules* **2016**, *17*, 1358–1364. [CrossRef] [PubMed]
12. Subba Rao, A.N.; Nagarajappa, G.B.; Nair, S.; Chathoth, A.M.; Pandey, K.K. Flexible Transparent Wood Prepared from Poplar Veneer and Polyvinyl Alcohol. *Compos. Sci. Technol.* **2019**, *182*, 107719. [CrossRef]

13. Wu, Y.; Zhou, J.; Huang, Q.; Yang, F.; Wang, Y.; Liang, X.; Li, J. Study on the Colorimetry Properties of Transparent Wood Prepared from Six Wood Species. *ACS Omega* **2020**, *5*, 1782–1788. [CrossRef]
14. Chen, L.; Xu, Z.; Wang, F.; Duan, G.; Xu, W.; Zhang, G.; Yang, H.; Liu, J.; Jiang, S. A Flame-Retardant and Transparent Wood/Polyimide Composite with Excellent Mechanical Strength. *Compos. Commun.* **2020**, *20*, 100355. [CrossRef]
15. Wan, C.; Liu, X.; Huang, Q.; Cheng, W.; Su, J.; Wu, Y. A Brief Review of Transparent Wood: Synthetic Strategy, Functionalization and Applications. *Curr. Org. Synth.* **2021**, *18*, 615–623. [CrossRef]
16. Chen, D.; Gao, A.; Cen, K.; Zhang, J.; Cao, X.; Ma, Z. Investigation of Biomass Torrefaction Based on Three Major Components: Hemicellulose, Cellulose, and Lignin. *Energy Convers. Manag.* **2018**, *169*, 228–237. [CrossRef]
17. Cogulet, A.; Blanchet, P.; Landry, V. Wood Degradation under UV Irradiation: A Lignin Characterization. *J. Photochem. Photobiol. B Biol.* **2016**, *158*, 184–191. [CrossRef] [PubMed]
18. Schwanninger, M.; Steiner, M.; Zobl, H. Yellowing and IR-Changes of Spruce Wood as Result of UV-Irradiation. *J. Photochem. Photobiol. B Biol.* **2003**, *69*, 97–105.
19. Emaminasab, M.; Tarmian, A.; Oladi, R.; Pourtahmasi, K.; Avramidis, S. Fluid Permeability in Poplar Tension and Normal Wood in Relation to Ray and Vessel Properties. *Wood Sci. Technol.* **2016**. [CrossRef]
20. Zauer, M.; Hempel, S.; Pfriem, A. Investigations of the Pore-Size Distribution of Wood in the Dry and Wet State by Means of Mercury Intrusion Porosimetry. *Wood Sci Technol* **2014**, 1229–1240. [CrossRef]
21. Wang, L.; Liu, Y.; Zhan, X.; Luo, D.; Sun, X. Correction: Photochromic Transparent Wood for Photo-Switchable Smart Window Applications. *J. Mater. Chem. C* **2019**, 518055. [CrossRef]
22. Sun, Y.; Cheng, J. Hydrolysis of Lignocellulosic Materials for Ethanol Production: A Review. *Bioresour. Technol.* **2002**, *83*, 1–11. [CrossRef]
23. Wang, X.; Shan, S.; Shi, S.Q.; Zhang, Y.; Cai, L.; Smith, L.M. Optically Transparent Bamboo with High Strength and Low Thermal Conductivity. *ACS Appl. Mater. Interfaces* **2021**, *13*, 1662–1669. [CrossRef] [PubMed]
24. Wu, Y.; Wang, J.; Wang, Y.; Zhou, J. Properties of Multilayer Transparent Bamboo Materials. *ACS Omega* **2021**, *6*, 33747–33756. [CrossRef] [PubMed]
25. Wu, Y.; Wang, Y.; Yang, F.; Wang, J.; Wang, X. Study on the Properties of Transparent Bamboo Prepared by Epoxy Resin Impregnation. *Polymers* **2020**, *12*, 863. [CrossRef]
26. Zhu, H.; Parvinian, S.; Preston, C.; Vaaland, O.; Ruan, Z.; Hu, L. Transparent Nanopaper with Tailored Optical Properties. *Nanoscale* **2013**, *5*, 3787–3792. [CrossRef]
27. Hsu, P.C.; Song, A.Y.; Catrysse, P.B.; Liu, C.; Peng, Y.; Xie, J.; Fan, S.; Cui, Y. Radiative Human Body Cooling by Nanoporous Polyethylene Textile. *Science.* **2016**, *353*, 1019–1023. [CrossRef]
28. Larena, A.; Millán, F.; Pérez, G.; Pinto, G. Effect of Surface Roughness on the Optical Properties of Multilayer Polymer Films. *Appl. Surf. Sci.* **2002**, *187*, 339–346. [CrossRef]
29. Nogi, M.; Yano, H. Optically Transparent Nanofiber Sheets by Deposition of Transparent Materials: A Concept for a Roll-to-Roll Processing. *Appl. Phys. Lett.* **2009**, *94*. [CrossRef]
30. Fink, S. Transparent Wood—A New Approach in the Functional Study of Wood Structure. *Holzforschung* **1992**, *46*, 403–408. [CrossRef]
31. Liu, R.; Zhang, S.; Semple, K.; Lian, C.; Chen, M.; Luo, J.; Yang, F.; Dai, C.; Fei, B. Precise Microcasting Revealing the Connectivity of Bamboo Pore Network. *Ind. Crops Prod.* **2021**, *170*, 113787. [CrossRef]
32. Li, Y.; Fu, Q.; Yang, X.; Berglund, L. Transparent Wood for Functional and Structural Applications. *Philos. Trans. R. Soc. A Math. Phys. Eng. Sci.* **2018**, *376*. [CrossRef] [PubMed]
33. Zou, W.; Sun, D.; Wang, Z.; Li, R.; Yu, W.; Zhang, P. Eco-Friendly Transparent Poplar-Based Composites That Are Stable and Flexible at High Temperature. *RSC Adv.* **2019**, *9*, 21566–21571. [CrossRef] [PubMed]
34. Li, Y.; Yang, X.; Fu, Q.; Rojas, R.; Yan, M.; Berglund, L. Towards Centimeter Thick Transparent Wood through Interface Manipulation. *J. Mater. Chem. A* **2018**, *6*, 1094–1101. [CrossRef]
35. Wang, Y.; Guo, F.; Li, Y.; Zhu, W.; Huang, P.; Hu, N.; Fu, S. High Overall Performance Transparent Bamboo Composite via a Lignin-Modification Strategy. *Compos. Part. B Eng.* **2022**, *235*, 109798. [CrossRef]
36. Wu, Y.; Wu, J.; Yang, F.; Tang, C.; Huang, Q. Effect of H_2O_2 Bleaching Treatment on the Properties of Finished Transparent Wood. *Polymers* **2019**, *11*, 776. [CrossRef]
37. Shi, J.; Peng, J.; Huang, Q.; Cai, L.; Shi, S.Q. Fabrication of Densified Wood via Synergy of Chemical Pretreatment, Hot-Pressing and Post Mechanical Fixation. *J. Wood Sci.* **2020**, *66*. [CrossRef]
38. Yang, R.; Cao, Q.; Liang, Y.; Hong, S.; Xia, C.; Wu, Y.; Li, J.; Cai, L.; Sonne, C.; Van Le, Q.; et al. High Capacity Oil Absorbent Wood Prepared through Eco-Friendly Deep Eutectic Solvent Delignification. *Chem. Eng. J.* **2020**, *401*, 126150. [CrossRef]
39. Zhou, J.; Wu, Y. Research Progress on Preparation and Application of Transparent Wood. *China For. Prod. Ind.* **2020**, *57*, 17–22. [CrossRef]
40. Lasseuguette, E.; Roux, D.; Nishiyama, Y. Rheological Properties of Microfibrillar Suspension of TEMPO-Oxidized Pulp. *Cellulose* **2008**, *15*, 425–433. [CrossRef]
41. Jacucci, G.; Schertel, L.; Zhang, Y.; Yang, H.; Vignolini, S. Light Management with Natural Materials: From Whiteness to Transparency. *Adv. Mater.* **2021**, *33*. [CrossRef] [PubMed]

42. Abdul Khalil, H.P.S.; Bhat, I.U.H.; Jawaid, M.; Zaidon, A.; Hermawan, D.; Hadi, Y.S. Bamboo Fibre Reinforced Biocomposites: A Review. *Mater. Des.* **2012**, *42*, 353–368. [CrossRef]
43. Cai, M.; Takagi, H.; Nakagaito, A.N.; Katoh, M.; Ueki, T.; Waterhouse, G.I.N.; Li, Y. Influence of Alkali Treatment on Internal Microstructure and Tensile Properties of Abaca Fibers. *Ind. Crops Prod.* **2015**, *65*, 27–35. [CrossRef]
44. Vijaya Ramnath, B.; Manickavasagam, V.M.; Elanchezhian, C.; Vinodh Krishna, C.; Karthik, S.; Saravanan, K. Determination of Mechanical Properties of Intra-Layer Abaca-Jute-Glass Fiber Reinforced Composite. *Mater. Des.* **2014**, *60*, 643–652. [CrossRef]
45. Ramesh, M.; Palanikumar, K.; Reddy, K.H. Mechanical Property Evaluation of Sisal-Jute-Glass Fiber Reinforced Polyester Composites. *Compos. Part. B Eng.* **2013**, *48*, 1–9. [CrossRef]
46. Mohamad, M.; Ting, S.W.; Bakar, M.B.A.; Osman, W.H.W. The Effect of Alkaline Treatment on Mechanical Properties of Polylactic Acid Reinforced with Kenaf Fiber Mat Biocomposite. *IOP Conf. Ser. Earth Environ. Sci.* **2020**, *596*. [CrossRef]
47. Asim, M.; Jawaid, M.; Abdan, K.; Ishak, M.R. Effect of Alkali and Silane Treatments on Mechanical and Fibre-Matrix Bond Strength of Kenaf and Pineapple Leaf Fibres. *J. Bionic Eng.* **2016**, *13*, 426–435. [CrossRef]
48. Li, T.; Song, J.; Zhao, X.; Yang, Z.; Pastel, G.; Xu, S.; Jia, C.; Dai, J.; Dai, C.; Gong, A.; et al. Anisotropic, Lightweight, Strong, and Super Thermally Insulating Nanowood with Naturally Aligned Nanocellulose. *Sci. Adv.* **2018**, *4*, 1–10. [CrossRef]
49. Simona, P.L.; Spiru, P.; Ion, I.V. Increasing the Energy Efficiency of Buildings by Thermal Insulation. *Energy Procedia* **2017**, *128*, 393–399. [CrossRef]
50. Li, T.; Zhu, M.; Yang, Z.; Song, J.; Dai, J.; Yao, Y.; Luo, W.; Pastel, G.; Yang, B.; Hu, L. Wood Composite as an Energy Efficient Building Material: Guided Sunlight Transmittance and Effective Thermal Insulation. *Adv. Energy Mater.* **2016**, *6*. [CrossRef]
51. Zhao, X.; Huang, C.; Liu, Q.; Smalyukh, I.I.; Yang, R. Thermal Conductivity Model for Nanofiber Networks. *J. Appl. Phys.* **2018**, *123*. [CrossRef]
52. Qiu, Z.; Xiao, Z.; Gao, L.; Li, J.; Wang, H.; Wang, Y.; Xie, Y. Transparent Wood Bearing a Shielding Effect to Infrared Heat and Ultraviolet via Incorporation of Modified Antimony-Doped Tin Oxide Nanoparticles. *Compos. Sci. Technol.* **2019**, *172*, 43–48. [CrossRef]
53. Huang, X.; Jiang, P.; Tanaka, T. A Review of Dielectric Polymer Composites with High Thermal Conductivity. *IEEE Electr. Insul. Mag.* **2011**, *27*, 8–16. [CrossRef]
54. Gan, W.; Gao, L.; Sun, Q.; Jin, C.; Lu, Y.; Li, J. Multifunctional Wood Materials with Magnetic, Superhydrophobic and Anti-Ultraviolet Properties. *Appl. Surf. Sci.* **2015**, *332*, 565–572. [CrossRef]
55. Guo, H.; Fuchs, P.; Cabane, E.; Michen, B.; Hagendorfer, H.; Romanyuk, Y.E.; Burgert, I. UV-Protection of Wood Surfaces by Controlled Morphology Fine-Tuning of ZnO Nanostructures. *Holzforschung* **2016**, *70*, 699–708. [CrossRef]
56. Trey, S.; Olsson, R.T.; Ström, V.; Berglund, L.; Johansson, M. Controlled Deposition of Magnetic Particles within the 3-D Template of Wood: Making Use of the Natural Hierarchical Structure of Wood. *RSC Adv.* **2014**, *4*, 35678–35685. [CrossRef]
57. Merk, V.; Chanana, M.; Keplinger, T.; Gaan, S.; Burgert, I. Hybrid Wood Materials with Improved Fire Retardance by Bio-Inspired Mineralisation on the Nano- and Submicron Level. *Green Chem.* **2015**, *17*, 1423–1428. [CrossRef]
58. Zhang, Y.; Luo, W.; Wang, C.; Li, Y.; Chen, C.; Song, J.; Dai, J.; Hitz, E.M.; Xu, S.; Yang, C.; et al. High-Capacity, Low-Tortuosity, and Channel-Guided Lithium Metal Anode. *Proc. Natl. Acad. Sci. USA* **2017**, *114*, 3584–3589. [CrossRef]
59. Chen, C.; Zhang, Y.; Li, Y.; Dai, J.; Song, J.; Yao, Y.; Gong, Y.; Kierzewski, I.; Xie, J.; Hu, L. All-Wood, Low Tortuosity, Aqueous, Biodegradable Supercapacitors with Ultra-High Capacitance. *Energy Environ. Sci.* **2017**, *10*, 538–545. [CrossRef]
60. Hassel, B.I.; Trey, S.; Leijonmarck, S.; Johansson, M. A Study on the Morphology, Mechanical, and Electrical Performance of Polyaniline-Modified Wood—A Semiconducting Composite Material. *BioResources* **2014**, *9*, 5007–5023. [CrossRef]
61. Li, Y.; Fu, Q.; Rojas, R.; Yan, M.; Lawoko, M.; Berglund, L. Lignin-Retaining Transparent Wood. *ChemSusChem* **2017**, *10*, 3445–3451. [CrossRef] [PubMed]
62. Burgert, I.; Cabane, E.; Zollfrank, C.; Berglund, L. Bio-Inspired Functionalwood-Basedmaterials-Hybrids and Replicates. *Int. Mater. Rev.* **2015**, *60*, 431–450. [CrossRef]
63. Torres-Pierna, H.; Ruiz-Molina, D.; Roscini, C. Highly Transparent Photochromic Films with Tunable and Fast Solution-like Response. *Mater. Horizons* **2020**, 1–11. [CrossRef]
64. Zhang, S.; Cao, S.; Zhang, T.; Lee, J.Y. Plasmonic Oxygen-Deficient TiO_{2-x} Nanocrystals for Dual-Band Electrochromic Smart Windows with Efficient Energy Recycling. *Adv. Mater.* **2020**, *2004686*, 2–9. [CrossRef] [PubMed]
65. Zhou, Y.; Wang, S.; Peng, J.; Tan, Y.; Li, C.; Yin, F.; Zhou, Y.; Wang, S.; Peng, J.; Tan, Y.; et al. Article Liquid Thermo-Responsive Smart Window Derived from Hydrogel Liquid Thermo-Responsive Smart Window Derived from Hydrogel. *Joule* **2020**, *4*, 1–17. [CrossRef]
66. Ke, Y.; Zhang, Q.; Wang, T.; Wang, S.; Li, N.; Lin, G. Cephalopod-Inspired Versatile Design Based on Plasmonic VO_2 Nanoparticle for Energy-Efficient Mechano-Thermochromic Windows. *Nano Energy* **2020**, *73*. [CrossRef]
67. Chen, R.; Das, S.R.; Jeong, C.; Khan, M.R.; Janes, D.B.; Alam, M.A. Co-Percolating Graphene-Wrapped Silver Nanowire Network for High Performance, Highly Stable, Transparent Conducting Electrodes. *Adv. Funct. Mater.* **2013**, 1–9. [CrossRef]
68. Lee, D.; Lee, H.; Ahn, Y.; Jeong, Y.; Lee, D.-Y.; Lee, Y. Highly Stable and Flexible Silver Nanowire—Graphene Hybrid Transparent Conducting Electrodes for Emerging Optoelectronic Devices. *Nanoscale* **2013**, *5*, 7750. [CrossRef]
69. Yuan, Z.; Dryden, N.H.; Vittal, J.J.; Puddephatt, R.J. Chemical Vapor Deposition of Silver. *Chem. Mater.* **1995**, *7*, 1696–1702. [CrossRef]

70. Zhang, L.; Wang, A.; Zhu, T.; Chen, Z.; Wu, Y.; Gao, Y. Transparent Wood Composites Fabricated by Impregnation of Epoxy Resin and W-Doped VO2 Nanoparticles for Application in Energy-Saving Windows. *ACS Appl. Mater. Interfaces* **2020**, *12*, 34777–34783. [CrossRef]
71. Ke, Y.; Chen, J.; Lin, G.; Wang, S.; Zhou, Y.; Yin, J.; Lee, P.S.; Long, Y. Smart Windows: Electro-, Thermo-, Mechano-, Photochromics, and Beyond. *Adv. Energy Mater.* **2019**, *1902066*, 1–38. [CrossRef]
72. Heidarinejad, M.; Dalgo, D.; Mi, R.; Zhao, X.; Song, J. A Radiative Cooling Structural Material. *Struct. Mater.* **2019**, *763*, 760–763.
73. Raman, A.P.; Anoma, M.A.; Zhu, L.; Rephaeli, E.; Fan, S. Passive Radiative Cooling below Ambient Air Temperature under Direct Sunlight. *Nature* **2014**. [CrossRef] [PubMed]
74. Zhao, B.; Hu, M.; Ao, X.; Chen, N.; Pei, G. Radiative Cooling: A Review of Fundamentals, Materials, Applications, and Prospects. *Appl. Energy.* **2019**, *236*, 489–513. [CrossRef]
75. Zhou, C.; Julianri, I.; Wang, S.; Chan, S.H.; Li, M.; Long, Y. Transparent Bamboo with High Radiative Cooling Targeting Energy Savings. *ACS Mater. Lett.* **2021**, *3*, 883–888. [CrossRef]
76. Wang, J.; Wu, Y.; Wang, Y.; Zheng, Y. Research Progress of Transparent Bamboo and Home Product Design. *Furniture* **2021**, *42*, 52–56. [CrossRef]
77. Zhao, B.; Yu, L.; Li, H.; Fei, B. The Research Progress and Application of Bamboo Packaging Materials. *CHINA For. Prod. Ind.* **2018**. [CrossRef]
78. Zhu, M.; Li, T.; Davis, C.S.; Yao, Y.; Dai, J.; Wang, Y.; AlQatari, F.; Gilman, J.W.; Hu, L. Transparent and Haze Wood Composites for Highly Efficient Broadband Light Management in Solar Cells. *Nano Energy.* **2016**, *26*, 332–339. [CrossRef]
79. Nogi, M.; Karakawa, M.; Komoda, N.; Yagyu, H.; Nge, T.T. Transparent Conductive Nanofiber Paper for Foldable Solar Cells. *Sci. Rep.* **2015**, *5*, 17254. [CrossRef]
80. Li, Y.; Cheng, M.; Jungstedt, E.; Xu, B.; Sun, L.; Berglund, L. Optically Transparent Wood Substrate for Perovskite Solar Cells. *ACS Sustain. Chem. Eng.* **2019**, *7*, 6061–6067. [CrossRef]
81. Gan, W.; Xiao, S.; Gao, L.; Gao, R.; Li, J.; Zhan, X. Luminescent and Transparent Wood Composites Fabricated by Poly(Methyl Methacrylate) and γ-Fe2O3@YVO4:Eu3+ Nanoparticle Impregnation. *ACS Sustain. Chem. Eng.* **2017**, *5*, 3855–3862. [CrossRef]
82. Gan, W.; Gao, L.; Xiao, S.; Zhang, W.; Zhan, X.; Li, J. Transparent Magnetic Wood Composites Based on Immobilizing Fe_3O_4 Nanoparticles into a Delignified Wood Template. *J. Mater. Sci.* **2017**, *52*, 3321–3329. [CrossRef]
83. Wang, M.; Li, R.; Chen, G.; Zhou, S.; Feng, X.; Chen, Y.; He, M.; Liu, D.; Song, T.; Qi, H. Highly Stretchable, Transparent, and Conductive Wood Fabricated by in Situ Photopolymerization with Polymerizable Deep Eutectic Solvents. *ACS Appl. Mater. Interfaces* **2019**, *11*, 14313–14321. [CrossRef] [PubMed]
84. Fu, Q.; Chen, Y.; Sorieul, M. Wood-Based Flexible Electronics. *ACS Appl. Mater. Interfaces* **2020**. [CrossRef]

Article

Study on Different Shear Performance of Moso Bamboo in Four Test Methods

Aiyue Huang [1,2,†], **Qin Su** [1,2,†], **Yurong Zong** [1,2,†], **Xiaohan Chen** [1,2] **and Huanrong Liu** [1,2,*]

1. Department of Biomaterials, International Center for Bamboo and Rattan, Beijing 100102, China; huangaiyue1124@163.com (A.H.); 15956609529@163.com (Q.S.); zyrong666@163.com (Y.Z.); 18664331398@163.com (X.C.)
2. Key Laboratory of National Forestry and Grassland Administration/Beijing for Bamboo & Rattan Science and Technology, Beijing 100102, China
* Correspondence: liuhuanrong@icbr.ac.cn
† Co-author. These authors contributed equally to this work.

Abstract: Bamboo is recognized as a potential and sustainable green material. The longitudinal-splitting and shear strengths of bamboo are weak but critical to its utilizations. To discuss the different shear performances of bamboo, the shear strength and behaviors of bamboo culm were investigated by four test methods: the tensile-shear, step-shear, cross-shear, and short-beam-shear methods. Then, the different shear performance and mechanisms were discussed. Results indicated that the shear strength was significantly different in the four test methods and was highest in the step-shear-test method but lowest in the tensile-shear-test method. Moreover, the typical load-displacement curves were different across the shear methods but were similar to the curves of the respective loading modes. The axially aligned fiber bundles played an important role in all the shear performances. In the tensile-shear method, specimens fractured at the interface of the bamboo-fiber bundles. However, compress-shear behaviors were a combination of compression and shear. Then, the cross-shear method, in compress-shear, was lower than that of the step-shear method because of oval-shaped bamboo culm sections of different thickness. In the short-beam shear method, the behaviors and shearing characteristics were like bending with the fiber bundle pulled out.

Keywords: shear performance; test methods; moso bamboo; fracture behavior; shearing characteristics

Citation: Huang, A.; Su, Q.; Zong, Y.; Chen, X.; Liu, H. Study on Different Shear Performance of Moso Bamboo in Four Test Methods. *Polymers* **2022**, *14*, 2649. https://doi.org/10.3390/polym14132649

Academic Editor: Jingpeng Li

Received: 2 June 2022
Accepted: 26 June 2022
Published: 29 June 2022

Publisher's Note: MDPI stays neutral with regard to jurisdictional claims in published maps and institutional affiliations.

Copyright: © 2022 by the authors. Licensee MDPI, Basel, Switzerland. This article is an open access article distributed under the terms and conditions of the Creative Commons Attribution (CC BY) license (https://creativecommons.org/licenses/by/4.0/).

1. Introduction

Bamboo is widely used for buildings and other structural applications all over the world, not only for its fast growth, low cost and environmental friendliness, but also for its outstanding mechanical and physical properties [1–4]. Moso bamboo in China and Guadua bamboo in Colombia, for example, are capable of being used for dwellings and wide-span bridges [5]. However, due to the longitudinal arrangement of the fibers, bamboo has excellent tensile, bending, and compression properties in axial performance, except for its splitting and shearing properties. Longitudinal splitting is the dominant mode of failure for most bamboo in structural applications [6]. Shear failure, which is instantaneous and catastrophic, occurs in brittle materials. Therefore, the resistance to longitudinal splitting (cracking) is of fundamental importance in bamboo constructions.

Shear strength is determined by the anatomical structure of bamboo [7]. Bamboo culm, which consists of about 50% parenchyma cells, 40% fibers, and 10% vessels, is typically a bio-composite material [8]. The stiff fiber bundles embedded into the soft parenchyma tissues are arranged exclusively in an axial manner, which strongly contribute to the axial performances, but not the shearing properties, of bamboo. The axial shear strength of the internode of moso bamboo is just 13.5 MPa, only 9% of the tensile strength. In addition, it has been demonstrated that the density affects the axial-shear properties of bamboo [9,10]. The shear strengths of four-year-old moso bamboo with densities of

0.763 g/cm³, 0.798 g/cm³, and 0.853 g/cm³ are 21.7 MPa, 23.2 MPa, and 24.2 MPa, respectively [11]. Moreover, shear strength depends on the age of bamboo [12,13]. Shear strength first increases and then decreases with age, which could be related to modifications in density or the anatomical structure [13]. However, aside from age, shear strength also varies among species. For instance, the shear strength of *Bambusa pervariabilis* was 18.7 MPa, but that of *Dendrocalamopisis vario-striata* was 12.9 MPa [14].

Besides, the test methods and loading speed significantly affect shear-strength findings [15]. The shear strength of moso bamboo at a loading speed of 0.2 mm/min is 21.5 MPa, but it increases by 31% when the loading speed is increased to 2 mm/min. More elastic deformation occurred in bamboo at high loading speed, which would absorb part of the energy of external force and make bamboo exhibit higher shear strength. At the same time, the shape of the bamboo specimen affects its mechanical strength. According to previous studies, the tensile-shear strength, step-shear strength, and cross-shear strength of moso bamboo are 2.5–7 MPa [16–18], 15–21 MPa [10,11], and 11–15 MPa [19], respectively. Therefore, the shear strength varies under different test methods. In addition, most studies have only reported the shear strength of bamboo based on one method. Moreover, the fracture behavior and shearing characteristics of bamboo have not been studied in depth.

Therefore, this study compared the shear performance and shear mechanisms of moso bamboo under four test methods. Shear strengths and behaviors were also analyzed. Moreover, the shear-fracture characteristics were observed using a field emission-scanning electron microscope (FESEM). The different shear performances and fracture mechanisms of bamboo, based on the anisotropic and two-phase composite structure of bamboo culm, were also discussed.

2. Materials and Methods

2.1. Materials

Ten four-year-old moso bamboo (*Phyllostachys edulis*) culms were obtained from Huoshan, Anhui, China. Their average diameter at breast height was 100–130 mm, whereas the culm-wall thickness was 10–12 mm. The culms were air-dried for one year before the experiments. Samples were prepared from the internode sections of the whole bamboo wall, at a height of 1.5–2.0 m.

To avoid the effects of specimen's variety on test results, the specimens for shearing performance comparisons in four test methods were prepared from one internode. The referenced standards, shapes, dimensions of specimens, test setups, and number of specimens in different test methods are listed in Table 1. Prior to tests, all samples were conditioned in a climate-controlled chamber at 20 °C under 65% relative humidity for three weeks. Additionally, except for the short-beam-shear method, where the load was applied in the T direction, the loading direction in the other three methods was in the L direction.

2.2. Methods

A tester (Instron series 5582, Norwood, MA, USA) equipped with a load cell with capacity of 10 kN was applied for tensile-shear and short-beam-shear tests, but a load cell with 100 kN capacity was used for the cross-shear test. Another tester (Shijin, series WDW-E100D, JiNan, China) equipped with a load-cell capacity of 10 kN was applied for the step-shear test. The test speed was controlled by displacement, in which the specimen failed in 60–90 s. After mechanical-shear tests, the fracture surfaces of each tested specimen were cut off. The fracture characteristics were investigated in detail using a field emission-scanning electron microscopy (XL30 FEG-SEM, FEI, Hillsboro, OR, USA) after the sputter coating of the specimens with gold in a vacuum chamber.

Table 1. The shear-strength-test methods used in the present research.

Test Methods	Reference Standards	Quantity	Specimen Shape and Dimensions (mm)	Test Setup
Tensile-shear	ASTM D906 (Standard Test Method for Strength Properties of Adhesives in Plywood Type Construction in Shear by Tension Loading)	40		
Short-beam shear	ASTM D2344/2344M-2016 (Standard Test Method for Short-Beam Strength of Polymer Matrix Composite Materials and Their Laminates)	40		
Step-shear	GB/T 15780-1995 (Testing methods for physical and mechanical properties of bamboos)	60		
Cross-shear	ISO/TR 22157-2-2004 (Bamboo—Determination of physical and mechanical properties—Part 2: Labora-tory manual)	10		

3. Results and Discussion

3.1. Different Shear Strengths in Four Shear Test Methods

The shear strengths of the bamboo culm were considerably different in the four test methods. As shown in Figure 1, the shear strength under the step-shear method in compression mode was the highest. In contrast, the lowest shear strength under the tensile-shear method was only 8.65 MPa, which is 48% lower than the step-shear strength. In addition, the cross-shear strength and short-beam-shear strength were between them.

Besides the structure, the different bamboo strengths are mainly affected by the test methods and loading modes. According to ASTM D906, the tensile-shear test was performed on tension mode. Specimens were prepared with two notches. The shear surface was the L–R section in the axial direction at the end of both notches (Figure 2a). The shear-stress concentration at the pre-notched ends increased with the tensile loading, causing shear-plane failure. The tensile-shear strength was the lowest in this way.

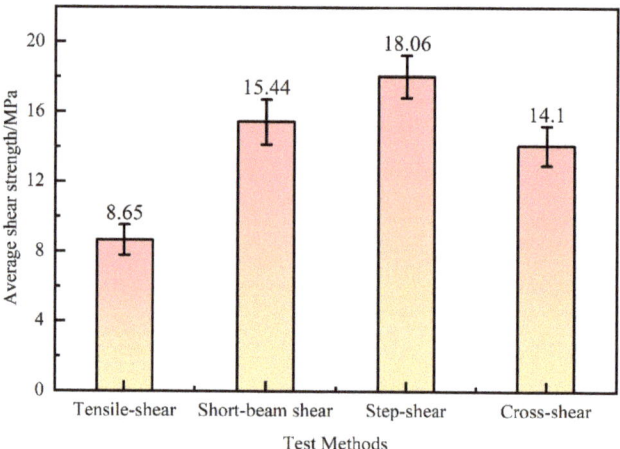

Figure 1. Shear strengths under different test methods.

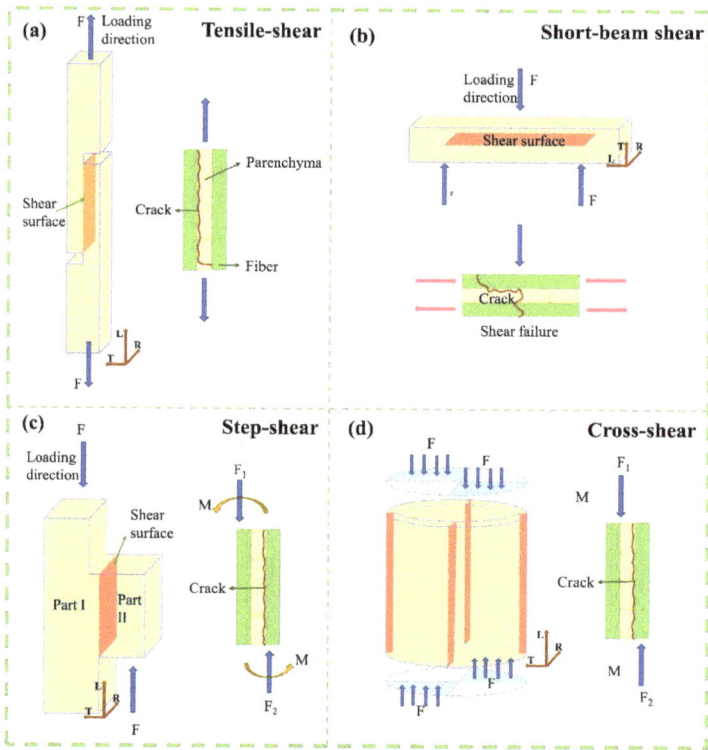

Figure 2. Loading modes of four test methods. (**a**) Tensile-shear test, (**b**) short-beam-shear test, (**c**) step-shear test, (**d**) cross-shear test.

However, based on the ISO/TR 22157-2:2004(E) and GB/T 15780-1995 test methods, cross-shear and step-shear tests under compression were applied on round bamboo culms and bamboo strips, respectively. The axially arranged fiber bundles were beneficial to the first compression stage and played an important role on the compression shear mode.

The circular bamboo culm was oval-shaped and with varied wall thickness. The shear failure occurred on the weakest side of the round bamboo culm (Figure 2d). Therefore, the cross-shear strength was only 14.1 MPa, 22% lower than the step-shear strength. The specimen dimensions of the step-shear method are shown in Table 1, where shear surface was in the gap between the top and bottom widths (Figure 2c), which caused tearing at the bottom of the specimen. This resulted in the highest shear strength (16.1 MPa). ASTM D 2344/2344M-2016 was referred to in the short-beam-shear test. Since three-point loading was applied to the specimen with a small span-thickness ratio, there was a compound-load mode of bending and shear (Figure 2b). Hence, the short-beam-shear strength was 15.4 MPa, only lower than the step-shear strength.

3.2. Fracture Behaviors of Bamboo in Four Test Methods

The typical load-displacement curves in the four shear methods are shown in Figure 3. The specimens showed different fracture behaviors under the test modes, which were dominantly related to the loading direction and mode. Bamboo is a typical gradient material, with fiber bundles and parenchyma cells that are axially arranged. Except for the short-beam-shear test, the loading direction of the other test modes were all parallel to the bamboo longitudinally. Bamboo lacks transverse-fiber bundles that could stop the longitudinal propagation of the crack in the tensile-shear, step-shear, and cross-shear tests, so it rapidly presented a load decrease once it reached the peak load. Additionally, although the load-displacement curves for the tensile-shear and compression-shear tests were similar for pure tension and compression. The shear changed the original fracture behavior, causing the curves to be different. In the tensile-shear test, the load-displacement curve continuously increased before failure but did not display increase and decrease patterns like for the pure-tension curves [20]. In the compression-shear, including step-shear, and cross-shear tests, the plastic-plateau stage was not notable, which was different from the pure-compression test [21]. The shear failure occurred as soon as the load peaked, followed by a sharp drop in the load. However, in the short-beam shear test, the loading direction was perpendicular to the fiber. The strong fiber prevented the propagation of the radial crack, which occurred in a stepwise manner. Thus, the fracture behavior of the short-beam shear test was like the pure-bending test [22].

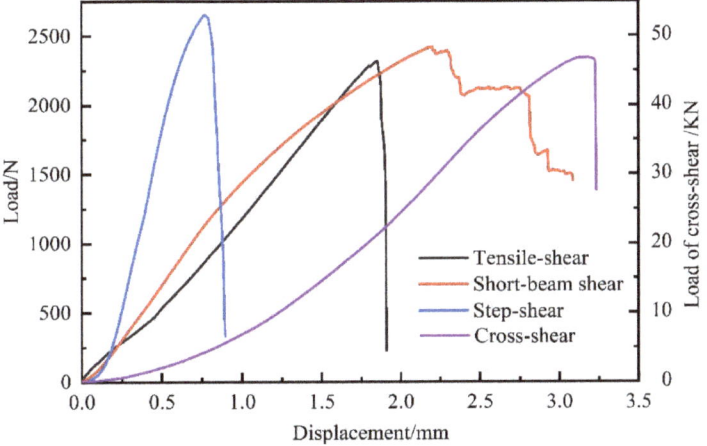

Figure 3. Typical load-displacement curves in four test methods.

3.3. Shearing Characteristics and Failure Mechanisms in Four Test Methods

The shearing characteristics and mechanisms of bamboo in the four test methods were investigated and analyzed from the macroscopic scale to the microscopic scale. The shearing in the four methods could be divided into axial and transverse. Axial shearing included the tensile-shear, step-shear, and cross-shear tests, whereas the transverse shear method was for the short-beam shear test. At the macroscopic scale, the shearing characteristics of the axial shear specimens were similar. As shown in Figures 4a, 6a and 7a, the crack paths at the outer surface in the axial direction were straight but serrated at the inner surface (Figures 4b, 6b and 7b). These were caused by the horizontally arranged cells on the inner surface and the longitudinally arranged cells on the outer surface [23]. Additionally, the density of fibers decreased gradually in the radial direction, from the outer to the inner culm wall, which caused the shear-failure surface to change from smooth to rough, as shown in Figures 4c, 6d and 7c. Furthermore, the main shearing characteristics were different under each test method. In the step-shear test, the tooth shape of the serrated crack at the inner surface was larger than that of the other two samples. Meanwhile, the bottom opening of the crack was significantly larger than that of the top. In the cross-shear test, the fracture always took place in one or two shear-failure surfaces (Figure 7a,c). Additionally, the crack deflected in the bamboo-culm wall (Figure 7a), accompanied by the buckled splitting of the bamboo-culm wall. From the top view, as in Figure 7c, twisting occurred at the failure position. As for the short-beam shear test, radial fracture occurred in the specimen in a manner similar to three-point bending. The crack was a zigzag at the bottom of the specimen (Figure 5b,e). Due to the gradient content of the fiber bundle, the crack was broader on the inner surface than on the outer surface (Figure 5e). In addition, there was no crack on the outer surface except the indention, which was different from the visible crack on the inner surface (Figure 5c,d).

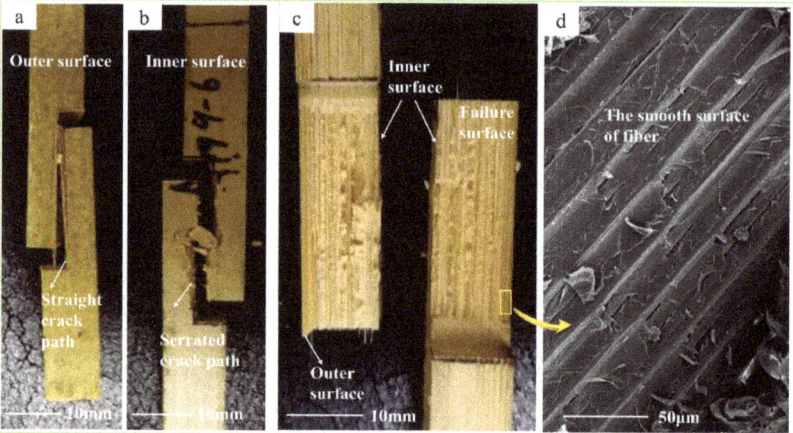

Figure 4. Shearing characteristics of tensile-shear method: (**a**) straight crack path at outer surface; (**b**) serrated crack path at inner surface; (**c**) shear-failure surface; (**d**) SEM image of smooth shear-failure surface.

At the microscopic scale, the shearing characteristics in the four test methods were also different. Except for the fiber pulled out in the short-beam-shear test (Figure 5f), the other test methods showed similar but different shearing characteristics. For the tensile-shear test, the fiber surface was smooth and only slightly peeled (Figure 4d). However, in the step-shear test, tearing of the fiber bundles and parenchyma cells at the bottom of the specimens occurred (Figure 6e,f). Moreover, the fibers and parenchymal cells collapsed in the cross-shear test (Figure 7d,e).

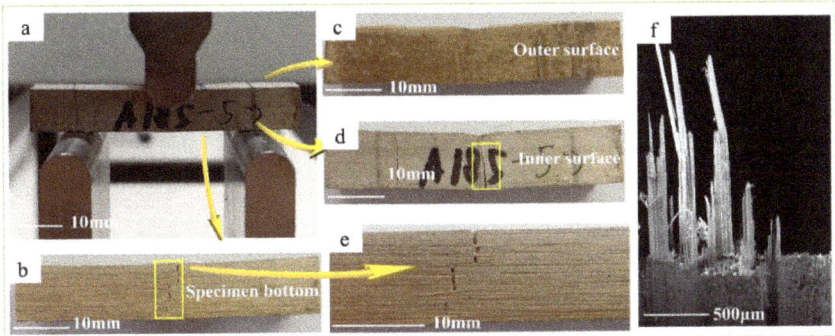

Figure 5. Shear-failure characteristics of short-beam-shear method: (**a**) the respective image of short-beam shear testing; (**b**) fracture in a zigzag manner at the bottom of specimen; (**c**) crack at outer surface; (**d**) straight crack path at inner surface; (**e**) a close-up look at zigzag crack path at specimen bottom; (**f**) SEM image of the pulled-out fiber bundle.

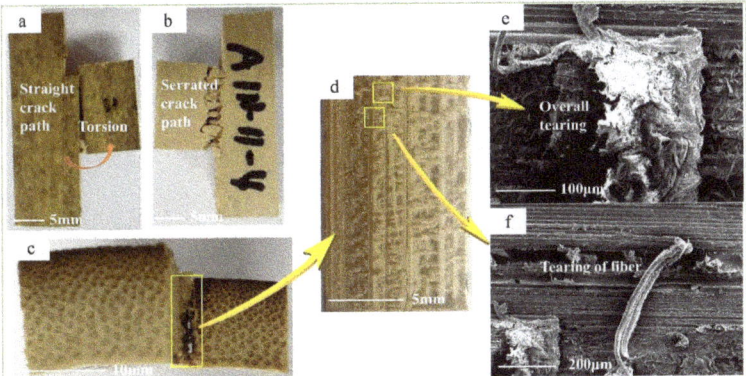

Figure 6. Shear-failure characteristics of step-shear method: (**a**) straight crack path at outer surface; (**b**) serrated crack path at inner surface; (**c**) Top view of failure specimen; (**d**) Shear failure surface; (**e**) SEM image of the overall tearing parenchyma cells; (**f**) SEM image of fiber tearing.

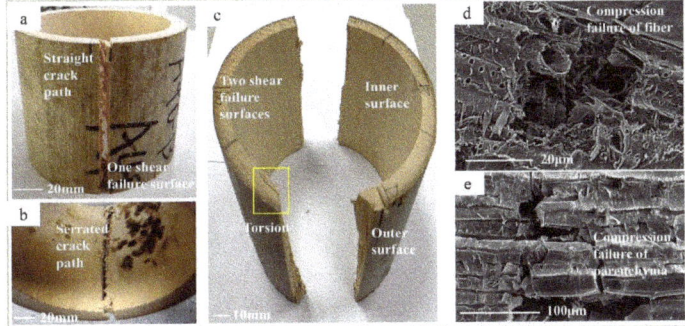

Figure 7. Shear-failure characteristics of cross-shear method: (**a**) failure specimen with one failure surface; (**b**) serrated crack path at inner surface; (**c**) failure specimen with two failure surfaces; (**d**) SEM image of the fiber fracture; (**e**) SEM image of the parenchyma-cell fracture.

4. Conclusions

This study focused on the different shear behaviors and fracture mechanisms of bamboo in four different shear-test methods. The findings of this study will inform the good use and manufacturing process of bamboo culm. Results suggested that the test methods have a considerable effect on the shear strength and behaviors. The compound mode of compression and shear for the axial resulted in the maximum shear strength in the step-shear test, while the interface shear caused the tensile-shear strength to be the lowest. The loading direction also affected the shear behavior under the four test methods. The typical load-distance curves of the ensile-shear, short-beam shear, and compression-shear tests were similar to the respective loading-modes' curves. However, the shear changed the original fracture behavior of the tension, bending, and compression. Additionally, the axial-shear-test methods caused typical interface-shear failure in the tensile-shear test and the overall tearing of fiber bundles in the step-shear test, while the parenchyma-cells collapsed in the cross-shear test. However, the short-beam-shear shearing characteristics were like bending with the fiber bundle pulled-out.

Author Contributions: Conceptualization, H.L.; methodology, Y.Z.; software, X.C.; data curation, A.H.; writing—original draft preparation, A.H. and Q.S.; writing—review and editing, A.H. and Q.S. All authors have read and agreed to the published version of the manuscript.

Funding: This work was financially supported by Fundamental Research Funds for the International Center for Bamboo and Rattan (Grant No. 1632021013).

Institutional Review Board Statement: Not applicable.

Informed Consent Statement: Not applicable.

Data Availability Statement: Not applicable.

Conflicts of Interest: The authors declare no conflict of interest.

References

1. Chen, M.L.; Ye, L.; Li, H.; Wang, G.; Chen, Q.; Fang, C.H.; Dai, C.P.; Fei, B.H. Flexural strength and ductility of moso bamboo. *Constr. Build. Mater.* **2020**, *246*, 118418. [CrossRef]
2. Wang, X.; Keplinger, T.; Gierlinger, N.; Burgert, I. Plant material features responsible for bamboo's excellent mechanical performance: A comparison of tensile properties of bamboo and spruce at the tissue, fiber and cell wall levels. *Ann. Bot.* **2014**, *114*, 1627–1635. [CrossRef] [PubMed]
3. Manandhar, R.; Kim, J.H.; Kim, J.T. Environmental, social and economic sustainability of bamboo and bamboo-based construction materials in buildings. *J. Asian Archit. Build. Eng.* **2019**, *18*, 49–59. [CrossRef]
4. Li, J.; Ma, R.; Lu, Y.; Wu, Z.; Liu, R.; Su, M.; Jin, X.; Zhang, R.; Bao, Y.; Chen, Y.; et al. Bamboo-inspired design of a stable and high-efficiency catalytic capillary microreactor for nitroaromatics reduction. *Appl. Catal. B Environ.* **2022**, *310*, 121297. [CrossRef]
5. Trujillo, D. Bamboo structures in Colombia. *Struct. Eng.* **2007**, *85*, 25–30.
6. Tang, T.K.H.; Welling, J.; Liese, W. Kiln drying for bamboo culm parts of the species Bambusa stenostachya, Dendrocalamus asper and Thyrsostachys siamensis. *J. Indian Acad. Wood Sci.* **2013**, *10*, 26–31. [CrossRef]
7. Habibi, M.K.; Lu, Y. Crack propagation in bamboo's hierarchical cellular structure. *Sci. Rep.* **2014**, *4*, 1–7. [CrossRef] [PubMed]
8. Liese, W.; Köhl, M. (Eds.) *Bamboo: The Plant and Its Uses*; Springer: Berlin/Heidelberg, Germany, 2015.
9. Arce-Villalobos, O.A. Fundamentals of the Design of Bamboo Structures. Ph.D. Thesis, Technische Universiteit Eindhoven, Eindhoven, The Netherlands, 1993.
10. Janssen, J.J. Designing and building with bamboo. *Neth. Int. Netw. Bamboo Ratt.* **2000**, *8*, 130–133.
11. Wang, Z.H. The Study on Variability in Bamboo Timber Properties and Relationship with Its Processing. Ph.D. Thesis, Chinese Academy of Forestry, Beijing, China, 2001.
12. Bautista, B.E.; Garciano, L.E.; Lopez, L.F. Comparative Analysis of Shear Strength Parallel to Fiber of Different Local Bamboo Species in the Philippines. *Sustainability* **2021**, *13*, 8164. [CrossRef]
13. Liese, W.; Weiner, G. Ageing of bamboo culms. A review. *Wood Sci. Technol.* **1996**, *30*, 77–89. [CrossRef]
14. Chen, G.J. The Interspecific Difference of Bamboo Mechanical Properties and Its Influencing Factors. Master's Thesis, Chinese Academy of Forestry, Beijing, China, 2019.
15. Gao, L.; Wang, Z.; Lin, T.; Cheng, H.T. Influence of test methods on moso bamboo shear strength parallel to grain. *China Wood Ind.* **2012**, *26*, 48–54.
16. Diao, Q.Q. Specific Modulus Grade of Dimension Bamboo Strips and Manufacture and Evaluation of Bamboo Lamination. Master's Thesis, Chinese Academy of Forestry, Beijing, China, 2018.

17. Song, G.N. Design, Manufacture and Evaluation of Graded Glued Bamboo Lamination for Marine Use. Master's Thesis, Chinese Academy of Forestry, Beijing, China, 2016.
18. Li, J.; Sun, Z.J. Effects of anisotropy and radial gradient variation on tensile shear strength of Moso bamboo. *J. Cent. South Univ. For. Technol.* **2013**, *33*, 120–123.
19. Zhang, W.F. Mechanical Properties and Brooming Processing of Bamboo-Culm. Master's Thesis, Chinese Academy of Forestry, Beijing, China, 2012.
20. Liu, H.R.; Jiang, Z.H.; Fei, B.H.; Hse, C.Y.; Sun, Z.J. Tensile behaviuor and fracture mechanism of moso bamboo (Phyllostachys pubescens). *Holzforschung* **2015**, *69*, 47–52. [CrossRef]
21. Zhang, X.X.; Li, J.H.; Yu, Z.X.; Yu, Y.; Wang, H.K. Compressive failure mechanism and buckling analysis of the graded hierarchical bamboo structure. *J. Mater. Sci.* **2017**, *52*, 6999–7007. [CrossRef]
22. Liu, H.R.; Wang, X.Q.; Zhang, X.B.; Sun, Z.J.; Jiang, Z.H. In situ detection of the fracture behaviour of moso bamboo (Phyllostachys pubescens) by scanning electron microscopy. *Holzforschung* **2016**, *70*, 1183–1190. [CrossRef]
23. Wang, X.; Zhang, S.; Chen, L.; Huang, B.; Fang, C.; Ma, X.; Liu, H.; Sun, F.; Fei, B. Effects of pith ring on the hygroscopicity and dimensional stability of bamboo. *Ind. Crops Prod.* **2022**, *184*, 115027. [CrossRef]

Article

Changes in Chemical and Thermal Properties of Bamboo after Delignification Treatment

Huiling Yu [1], Chengsheng Gui [2], Yaohui Ji [3], Xiaoyan Li [4], Fei Rao [5], Weiwei Huan [6] and Luming Li [6,*]

1. College of Engineering, Yantai Nanshan University, Yantai 265713, China; yhl053558@163.com
2. Zhejiang Shenghua Yunfeng New Material Co., Ltd., Huzhou 313200, China; gcs19882006@126.com
3. Research Institute of Wood Industry, Chinese Academy of Forestry, Beijing 100091, China; jiyaohui1994@gmail.com
4. China National Bamboo Research Center, Department of Efficient Utilization of Bamboo and Wood, Wenyi Road 310, Hangzhou 310012, China; xiaoyanli1995@sina.com
5. School of Art and Design, Zhejiang Sci-Tech University, Hangzhou 310018, China; raofei@zstu.edu.cn
6. College of Chemistry and Materials Engineering, Zhejiang A&F University, Hangzhou 311300, China; 20130874@zafu.edu.cn
* Correspondence: llm@zafu.edu.cn

Citation: Yu, H.; Gui, C.; Ji, Y.; Li, X.; Rao, F.; Huan, W.; Li, L. Changes in Chemical and Thermal Properties of Bamboo after Delignification Treatment. *Polymers* 2022, 14, 2573. https://doi.org/10.3390/polym14132573

Academic Editor: Sergiu Coseri

Received: 21 May 2022
Accepted: 21 June 2022
Published: 24 June 2022

Publisher's Note: MDPI stays neutral with regard to jurisdictional claims in published maps and institutional affiliations.

Copyright: © 2022 by the authors. Licensee MDPI, Basel, Switzerland. This article is an open access article distributed under the terms and conditions of the Creative Commons Attribution (CC BY) license (https://creativecommons.org/licenses/by/4.0/).

Abstract: Bamboo delignification is a common method for studying its functional value-added applications. In this study, bamboo samples were delignified by treatment with sodium chlorite. The effects of this treatment on the bamboo's microstructure, surface chemical composition, and pyrolysis behaviour were evaluated. Field-emission scanning electron microscopy (FE-SEM), Fourier-transform infrared (FTIR) spectroscopy, X-ray photoelectron spectroscopy (XPS), and X-ray diffraction (XRD) were conducted to evaluate these parameters. The FTIR results demonstrated that the lignin peak decreased or disappeared, and some hemicellulose peaks decreased, indicating that sodium chlorite treatment effectively removed lignin and partly decomposed hemicellulose, although cellulose was less affected. The XPS results showed that, after treatment, the oxygen-to-carbon atomic ratio of delignified bamboo increased from 0.34 to 0.45, indicating a lack of lignin. XRD revealed increased crystallinity in delignified bamboo. Further pyrolysis analysis of treated and untreated bamboo showed that, although the pyrolysis stage of the delignified bamboo did not change, the maximum thermal degradation rate (R_{max}) and its corresponding temperature (from 353.78 to 315.62 °C) decreased significantly, indicating that the pyrolysis intensity of the bamboo was weakened after delignification. Overall, this study showed that delignified bamboo develops loose surfaces, increased pores, and noticeable fibres, indicating that alkali-treated bamboo has promising application potential due to its novel and specific functionalities.

Keywords: chemical change; thermal property; bamboo; delignification; pyrolysis; bamboo microstructure

1. Introduction

Global ecological deterioration has shifted researchers' focus onto natural materials, and issues such as environmental friendliness and recyclability are becoming increasingly important in the development of new materials. Consequently, cellulose, which is the most abundant biopolymer on Earth with these properties, has been widely used as a source of raw materials. Natural fibres in wood and bamboo materials are abundant, biodegradable, and eco-friendly resources. Hence, they are considered high-quality alternatives [1]. Consequently, these materials have become research hotspots in recent years. Several functional studies have recently been conducted on various wood and bamboo fibre materials, such as transparent wood, aesthetic wood, flame retardant fibre materials, oil–water separation sponges, supercapacitor electrodes, and bioplastics [2–7]. Naturally occurring lignocellulosic materials, such as wood and bamboo, are porous materials. However, to further increase wood's porosity and susceptibility to functionalization, the most common strategy is delignification [8], which is the first and key step in studying the functionalization of lignocellulosic materials. As a result of delignification, nanopores in the cell wall structure become exposed. Additionally, delignification affects the chemical composition of wood, as it removes most of the lignin and part of the hemicellulose and facilitates the insertion of polymer and inorganic materials, thereby improving the wood's mechanical properties, hydrophobicity, magnetism, insulation, or transparent properties [9,10]. Previous studies obtained cellulose scaffolds using delignification treatment, which helped to expand the range of application of cellulose. Therefore, delignification treatment as a method of extracting cellulose fibres from wood and bamboo has recently garnered substantial interest. To date, several pre-treatment methods, including physico-chemical [11], chemical [12,13], and biological [14,15] methods, have been investigated to remove lignin and separate cellulose in wood and bamboo. The chemical method primarily involves acid or alkali solution impregnation treatment, whereas the physico-chemical method mainly uses acid or alkali solution steam treatment. Although physico-chemical methods improve the efficiency of delignification, they are associated with high energy consumption and cost. The biological method uses fungi for biodegradation, which is a green preparation method; however, the efficiency of the method is low. Therefore, the chemical method is a better choice with respect to cost and efficiency. Lignocellulosic scaffolds after delignification have been widely used in advanced functional materials, such as modified materials with excellent mechanical properties [16], low thermal conductivity [17,18], and thermal radiation cooling [19]. This suggests that delignification can expand the functional utilization of wood resources.

Bamboo is a lignocellulosic material composed of cellulose, hemicellulose, and lignin and is abundant, fast growing, sustainable, and renewable. Thus, bamboo fibres are considered potential raw resources for the fabrication of fibrous cellulose [20]. In bamboo composition, cellulose, hemicellulose, and lignin act as the skeleton, matrix, and encrusting materials, respectively [21]. Consequently, lignin removal causes lignocellulosic materials to become scaffolds that can be further decorated/grafted with different chemical functional groups, bringing novel and specific functionalities. Moreover, the mild conditions of delignification reactions (temperature; 80–120 °C and atmospheric pressure) do not impact the macrostructure substantially, preserving its original three-dimensional structure.

Although numerous studies have been conducted on the delignification of wood and bamboo and the functional transformation of treated materials, little information exists on changes in the material itself after delignification [22]. Alkali treatment is a common method for bamboo delignification. This study aimed to evaluate the microstructure, chemical change, and thermal degradation characteristics of bamboo delignified with sodium chlorite by comparing these characteristics before and after treatment. To the best of our knowledge, the microstructure, surface chemical composition, and pyrolysis behaviour of bamboo after sodium chlorite treatment have not been systematically studied yet. Hence, this study is expected to provide a reference basis for the functional value-added application of delignified bamboo.

2. Materials and Methods

2.1. Raw Materials

Five-year-old moso bamboo (*Phyllostachys pubescens* Mazel) was cut in Linan District, Zhejiang province. The samples were sawed 1.3 m from the base into 1 m long culms and divided into 20 mm wide and 0.5 mm thick strips. Sodium chlorite (2 wt.%) was obtained from Sinopharm Chemical Reagent Co., LTD (Huangpu, Shanghai, China), while acetic acid (AR grade) and ethanol were provided by Beijing Chemical Plant (DaXing, Beijing, China). Deionised water (DI) was made in the laboratory.

2.2. Delignification Treatment

First, 2000 mL of 2% sodium chlorite solution was prepared and the pH was adjusted to approximately 4.5 using acetic acid. Thereafter, the bamboo samples were completely soaked in sodium chlorite solution for 4 h at 50 °C. Subsequently, the samples were rinsed several times with DI, until the pH was neutral, to remove residual chemical substances. To better remove the residual reagent in the bamboo interior, the obtained samples were subjected to gradient dehydration using different ethanol concentrations (25, 50, 75, and 99.5 wt.%). The samples were immersed in each concentration of ethanol for 5 min in ascending order. Finally, all samples were washed with DI for 2 h to complete the bamboo delignification. All samples were stored in a room at 25 °C and 50% relative humidity for 15 days before the experimental study. The bamboo was categorised as: natural bamboo or NB (untreated bamboo) and delignified bamboo or DB (treated bamboo).

2.3. Characterization

The morphologies of NB and DB samples were measured using field-emission scanning electron microscopy (FE-SEM) (model SU8010, Hitachi, Japan). All samples were cut into tangential sections and coated with a gold layer.

For chemical analysis, Fourier transform infrared attenuated total internal reflectance (FTIR-ATR) spectra were obtained directly from the specimen surface using a Nicolet iS10 FTIR spectrometer (Thermo Scientific, Waltham, MA, USA) equipped with a diamond crystal ATR accessory (Smart iTX, Thermo Scientific, Waltham, MA, USA). For each measurement, 64 scans were conducted within 400–4000 cm^{-1} at 4 cm^{-1} resolution. Similarly, X-ray photoelectron spectroscopy (XPS) was performed at 150 W using a Thermo Scientific ESCALAB 250Xi spectrophotometer (Thermo Scientific, Waltham, MA, USA) equipped with a monochromatic Al Kα X-ray source (hv = 1486.6 eV), a 650 μm spot size, and pollutant carbon C1s = 284.8eV for charge correction. X-ray diffraction (XRD) data were obtained using a D8 ADVANCE X-ray instrument (Bruker, Karlsruhe, Germany) with Cu Kα radiation (wavelength, 1.5406 angstrom) at 40 kV, 40 mA, and a scan speed of 6°/min in the 2θ range of 5–50°. Pyrolysis characteristics were determined using a TG analyser (NETZSCH TG 209F1 Libra, Selb, Germany). A test temperature of 30–600 °C and linear heating rate of 10 °C/min^{-1} were employed under a N_2 flow of 10 mL/min. Each sample weighed approximately 10 mg.

3. Results and Discussion

3.1. Surface Microstructure Morphology

As microstructural changes can affect the functional application of delignified bamboo, SEM was used for observation and analysis in the present study. The surface microstructure morphology of bamboo before and after sodium chlorite treatment is shown in Figure 1.

The SEM images of NB reveal that the radial section is laevigatus; moreover, the smooth surfaces of the cell wall are clearly visible (Figure 1a). The cross section shows the orderly arrangement of pores (Figure 1c). A comparison between Figure 1a,c and Figure 1b,d shows that the compact cell walls in the radial section and cross section became loose after delignification treatment. Additionally, many micron-scale pores and cellulose nanofibres were observed on the DB cell walls (red lines in Figure 1d). The present study's findings confirm previous reports that delignification causes fibre cells to separate and jump out, resulting in a slight increase and decrease in the porosity and density of the treated bamboo, respectively [23]. Furthermore, the results confirm that alkali treatment causes a loss of bamboo matrix and the agglomeration of microfibres, which lead to size changes, surface roughness, cracking, and the loss of the mechanical strength of bamboo fibre [24].

3.2. Chemical Functional Groups

To further study which chemical composition changes in bamboo after sodium chlorite treatment led to microstructural changes, the results of the FTIR on DB and NB samples were compared and analysed. Figure 2 shows the FTIR spectra in the region ranging from 1800 to 800 cm^{-1}, which reflects the entire molecule's characteristics and is considered the fingerprint region of bamboo functional groups [22]. The FTIR spectra absorption peaks of the NB samples are defined in Table 1.

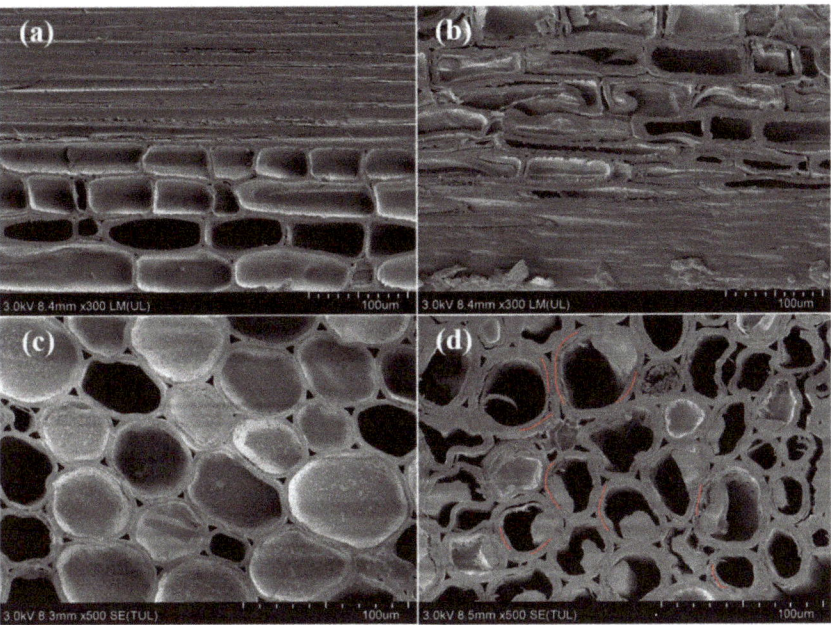

Figure 1. Scanning electron microscopy (SEM) images of (**a**,**c**) natural and (**b**,**d**) delignified bamboo. (**a**,**b**) Radial sections of natural bamboo (NB) and delignified bamboo (DB). (**c**,**d**) Cross sections of NB and DB.

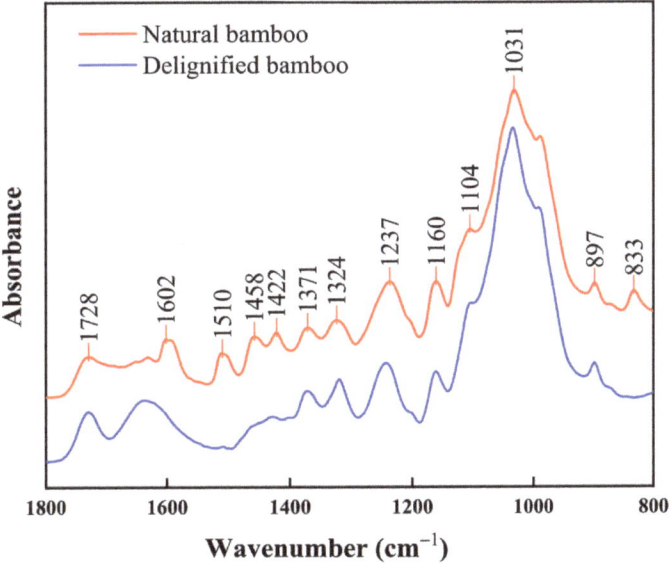

Figure 2. Fourier transform infrared (FTIR) spectra of NB and DB.

Table 1. Assignment of FTIR spectra absorption peaks of natural bamboo [22,25–27].

Wavenumber (cm^{-1})	Functional Group	Assignment
1728	C=O	Non-conjugated C=O in hemicellulose (xylans)
1602	C=C	C=C unsaturated linkages, aromatic skeletal vibration in lignin
1510	C=C	Aromatic skeletal vibration (C=C) in lignin
1458	C–H, O–H	Asymmetric bending in CH$_3$ (lignin)
1422	CH$_2$	Aromatic skeletal vibrations (lignin) and C–H deformation in plane (cellulose)
1371	C–H	C–H deformation in cellulose and hemicellulose
1324	O–H	phenol group (cellulose)
1237	C–O	Syringyl ring and C–O stretch in lignin and xylan
1160	C–O–C	C–O–C vibration in cellulose and hemicellulose
1104	C–H	Guaiacyl and syringyl (lignin)
1031	C–O, C–H	C–O stretch in cellulose and hemicelluloseC–H stretch in lignin
897	C–H	C–H deformation in cellulose
833	C–H	C–H vibration in guaiacyl derivatives

The regional range from 1800 to 800 cm^{-1} consisted of 13 absorption peaks (Figure 2). Seven peaks were attributed to the aromatic framework or main functional groups of lignin. The peaks at 1602, 1510, and 1422 cm^{-1} were attributed to the vibration contributions of C=C unsaturated linkages in the aromatic lignin skeleton, while the peak at 1458 cm^{-1} resulted from asymmetric bending in the CH$_3$ of lignin. The absorption peak at 1237 cm^{-1} was attributed to the syringyl rings and C–O stretch in lignin and xylan. Additionally, the 1104 cm^{-1} peak value was due to the structural contributions of guaiacyl and syringyl in lignin. The curve of the DB in Figure 2 revealed that after sodium chlorite treatment, the intensity of the absorption peaks at 1602, 1510, 1458, 1422, 1237, and 1104 cm^{-1} weakened to various degrees or even disappeared. This finding signified that the lignin was substantially decomposed after the alkali treatment. Additionally, the peak at 833 cm^{-1} was attributed to C–H vibration in guaiacyl derivatives in lignin. The peak disappeared completely after delignification treatment, which further demonstrates that the lignin was decomposed.

In addition to lignin decomposition, the peak strengths (1728, 1371, 1324, and 897 cm^{-1}) attributed to cellulose and hemicellulose showed only minor changes after the alkali treatment, in agreement with previous reports that chemical reactions between alkali and cellulose rarely occur [28]. However, the intensity of the peaks at 1160 and 1031 cm^{-1} slightly weakened, indicating that some hemicellulose may be degraded after treatment.

3.3. Chemical Composition

X-ray energy spectrum analysis (XPS) is a practical method for obtaining chemical and structural information on wood material surfaces [29]. Similar to wood, the chemical composition of bamboo comprises cellulose, hemicellulose, lignin, and small extract amounts, with carbon (C), hydrogen (H), and oxygen (O) as the main components. Therefore, the chemical properties of bamboo surfaces can be determined using XPS. In the present study, only the C and O elements in bamboo were assessed, as XPS cannot detect H elements. The main objects of XPS detection and analysis are 1s electrons in the inner shells of C and O atoms. Information on the chemical properties of the bamboo surfaces was obtained based on the C1s and O1s peak intensities and chemical shifts. The peaks of C_1 and O_1 were composed of components related to C and O functional groups in bamboo, respectively, and categorized as C_1 (C–C, C–H), C_2 (C–O), C_3 (O–C–O, C=O), and C_4 (O–C=O), and O_1 (O–C=O) and O_2 (C–O), according to the binding energy level. The C and O in different atomic binding states come from different sources [30,31]. The structural characteristics and chemical shifts of each are shown in Table 2.

High-resolution C1s and O1s XPS spectra of natural bamboo and delignified bamboo were processed and are presented in Figure 3. Studies have shown that the degradation of cellulosic materials can be detected through a change in the oxygen-to-carbon (O/C) atomic ratio [22]. Quantitative measurements of O/C atomic ratios were calculated from

Figure 3 using the total area of C and O peaks and their respective photoemission cross sections. The results are shown in Table 3.

Table 2. Classification of carbon (C) and oxygen (O) peak components of bamboo [32–35].

Element Component	Binding Energy (eV)	Binding Type	Main Resources
C_1	284.5	C–C, C–H	Lignin and extracts
C_2	285.5	C–O	Cellulose and hemicellulose
C_3	286.5	O–C–O, C=O	Cellulose
C_4	288.3	O–C=O	Hemicellulose and extracts
O_1	532.8	O–C=O	Lignin
O_2	534.1	C–O	Cellulose and hemicellulose

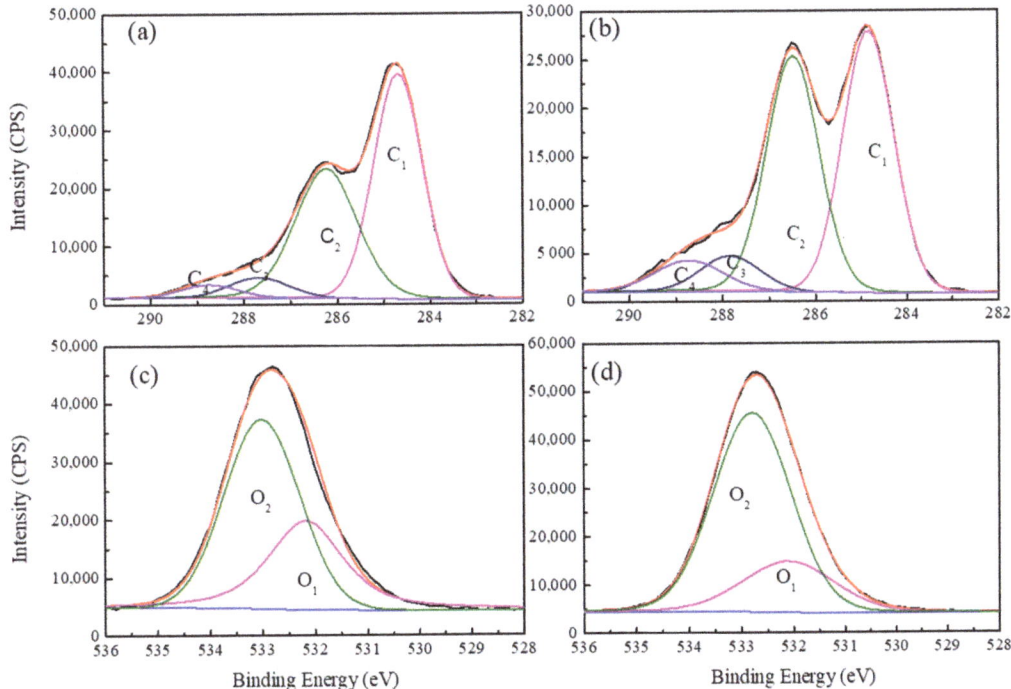

Figure 3. High-resolution X-ray photoelectron spectroscopy (XPS) spectra of (a,b) carbon and (c,d) oxygen peaks in (a,c) NB and (b,d) DB.

Table 3. Summary of XPS spectral parameters of NB and DB.

Samples	O/C Atomic Ratios	C (%)				O (%)	
		C_1	C_2	C_3	C_4	O_1	O_2
Natural bamboo	0.34	51.71	39.22	5.59	3.48	37.48	62.52
Delignified bamboo	0.45	44.88	41.65	6.73	6.73	23.96	76.04

In the C1s spectra, the peak components of C_1 were primarily from lignin and extracts (C–C, C–H), while those of C_2, C_3, and C_4 mainly originated from cellulose and hemicellulose (Table 2). After delignification treatment, the C_1 content decreased from 51.71% to 44.88%, which showed that the lignin was effectively decomposed by sodium chlorite (Table 3). Additionally, the O/C atomic ratios given in Table 3 for NB and DB were 0.34 and 0.45, respectively. It has been reported that the O/C atomic ratios of cellulose, hemicellulose, lignin, and extracts (mainly Lipophilic compounds) are about 0.83, 0.8, 0.33, and 0.1 [29,35,36], respectively. High O/C ratios reflect high carbohydrate content, while low ratios indicate the presence of more lignin and extracts on the bamboo surface [37]. Therefore, the increased O/C atomic ratio in this study after delignification treatment further confirmed the degradation effect of sodium chlorite on lignin.

In the O1s spectra, the chemical components representing the O_1 peak were lignin and extracts, and the O_2 peak was associated with hemicelluloses and cellulose [38,39]. An analysis of the delignification treatment of the O1s spectra of bamboo samples revealed that O_1 components decreased (from 37.48% to 23.96%) and O_2 components increased (from 62.52% to 76.04%) (Table 3), which is similar to reports in previous studies [40].

3.4. Crystalline Structure

XRD can be used to investigate the supramolecular structure of biomass materials [41]. For example, using XRD techniques to analyse the crystallinity and crystal width provides insights into the effects of different treatments on the cellulose, hemicellulose, and lignin content of bamboo [42,43]. In the present study, XRD testing of NB and DB samples was conducted using an X-ray generator, and the results are illustrated in Figure 4.

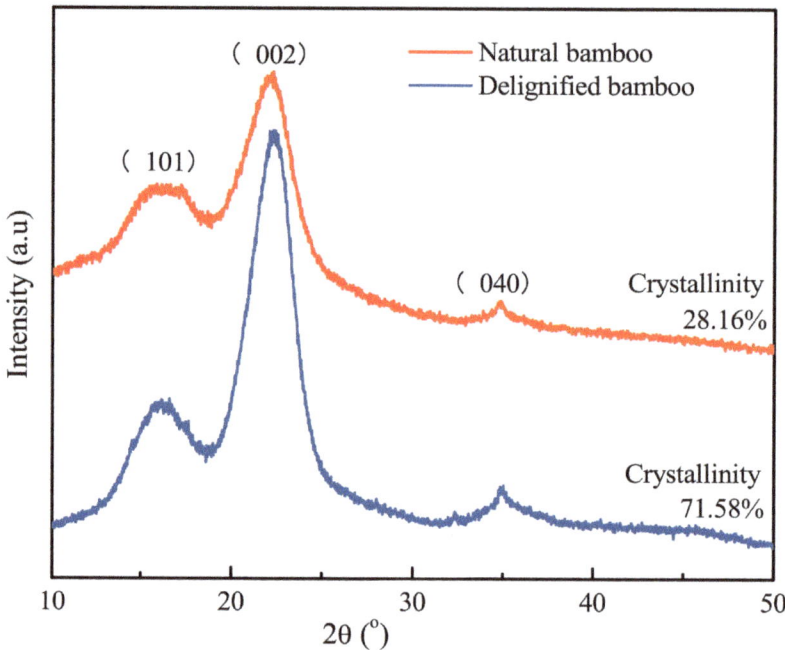

Figure 4. X-ray diffraction (XRD) patterns of NB and DB.

Three typical peaks corresponding to the (040), (002), and (101) lattice planes of cellulose I were 34.66°, 22.26°, and 16.12°, respectively [44]. The intensity of the (040) peak was low; hence, the two reflections (101) and (002) were used to reflect the crystalline structure. Among them, the peak reflecting the crystal zone width was (002), and that reflecting the zone length was (040). The characteristic peaks of bamboo (101) and (002) corresponded to those of delignified bamboo, and their positions were almost unchanged, indicating that sodium chlorite treatment has almost no effect on the crystalline region of cellulose. However, the peak intensities of (101) and (002) of delignified bamboo were stronger than those of natural bamboo. It was observed from the calculation results that the crystallinity indices, after sodium chlorite treatment, were much higher than those of the natural bamboo (from 28.16% to 71.58%). This is because amorphous hemicellulose and lignin are selectively removed by sodium chlorite in the delignification process [45,46]. Simultaneously, the hydroxyl group of the amorphous microfibrils is exposed, forming hydrogen bonds with microfibrils on the surface of the crystallization region, thus improving the crystallinity.

3.5. Pyrolysis Properties

The pyrolysis of bamboo can be regarded as the superposition of the pyrolysis process of three main components (cellulose, hemicellulose, and lignin) [47]. Investigating the pyrolysis characteristics of bamboo before and after delignification treatment will help to better design and prepare functional bamboo composites by thermochemical conversion methods, such as gasification and pyrolysis [42]. Thermogravimetric analysis (TGA) was performed to study the alkali treatment effect on the pyrolysis characteristics of bamboo. The TGA and derivative thermogravimetry (DTG) curves of the samples are shown in Figure 5. Table 4 lists the corresponding data.

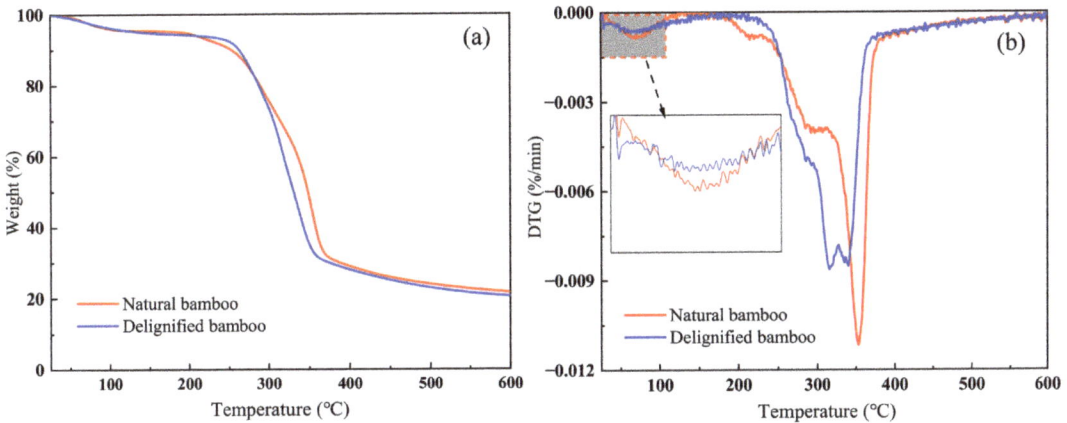

Figure 5. (a) Thermogravimetric analysis (TGA) and (b) derivative thermogravimetry (DTG) curves of NB and DB.

Table 4. TGA data for NB and DB.

Sample	T_{max} (°C)		R_{max} (%/(°C))		Residues (wt.%)
	Stage 1	Stage 2	Stage 1	Stage 2	
Natural bamboo	70.12	353.78	8.39×10^{-4}	111.71×10^{-4}	21.68
Delignified bamboo	69.11	315.62	6.02×10^{-4}	86.42×10^{-4}	20.41

T_{max}: temperature at the maximum weight-loss rate; R_{max}: maximum decomposition rate.

The difference between the NB and DB in the decomposition temperature was minor (Figure 5a). Notably, the pyrolysis process of sodium-chlorite-treated and -untreated bamboo was divided into two stages (Figure 5b). In the first stage, the temperature ranged from the initial temperature to approximately 140 °C. Small losses in sample mass were mainly caused by moisture loss, and the duration of when these losses occurred was identified as the water evaporation stage [48] (stage 1). A small weightlessness peak (fastest evaporation rate) appeared at 70.12 °C for the untreated bamboo and at 69.11 °C for the treated bamboo. The enlarged image in Figure 5b shows that the difference in the R_{max} values of these two temperature points is not large, and the slight difference can likely be attributed to the difference in initial conditions. In the natural state, the moisture content of different samples may be slightly different, which explains why the pyrolysis curves of the two samples in stage 1 are slightly different. In the second stage, which is the main reaction stage, the temperature ranged from the initial temperature to approximately 140 °C to 450 °C, and sample weightlessness was evident. In Table 4, we can see that the temperature of the treated bamboo samples was significantly lower than that of the untreated samples at the maximum thermal degradation rate (R_{max}), indicating that sodium chlorite effectively removes lignin in bamboo fibre so that the remaining cellulose and hemicellulose are easily decomposed by heat. Additionally, the R_{max} value of the treated samples was significantly lower than that of the untreated bamboo (Figure 5b and Table 4), indicating that under similar pyrolysis conditions, lignin presence intensifies bamboo pyrolysis. Studies have shown that it is difficult to decompose lignin, that its weight loss occurs over a wide temperature range (160–900 °C), and that the amount of generated solid residue is very high (approximately 40 wt.%) [49]. Consequently, at the end of the experiment (the temperature reached 600 °C), the residual weight of the untreated bamboo samples was slightly larger than that of the treated bamboo, which was mainly caused by the presence of lignin.

4. Conclusions

Alkali treatment exhibited clear effects on the chemical composition of bamboo. Compared with the microstructure of untreated bamboo, the structure of bamboo treated with sodium chlorite was notably looser and coarser, with increased pores, decreased density, and the presence of fibres. The FTIR analysis results revealed that after bamboo treatment, the lignin-related peaks at 1602, 1510, 1458, 1422, 1237, 1104, and 833 cm^{-1} weakened or disappeared, meaning that the lignin was substantially decomposed. Similarly, the intensity of peaks at 1160 and 1031 cm^{-1} weakened slightly, indicating that some hemicellulose may be degraded after treatment. These changes directly result in the loose appearance of the microstructure. Additionally, the XPS analysis results revealed that the proportion of C_1 and O_1 attributed to lignin decreased from 51.71% to 44.88% and 37.48% to 23.96%, respectively, and that the O/C atomic ratio increased from 0.34 to 0.45, indicating that the surfaces of treated bamboo contain less lignin and fewer extracts. The XRD results showed that although the sodium chlorate treatment had little effect on the crystallization zone of bamboo cellulose, the crystallinity of treated bamboo increased significantly (from 28.16% to 71.58% at the end of the treatment). This was attributed to the fact that amorphous hemicellulose and lignin were removed during the treatment. Furthermore, the hydroxyl group of the amorphous microfibre was exposed and formed hydrogen bonds with the microfibre on the surface of the crystallization region, leading to increased crystallinity. TG analysis showed that bamboo pyrolysis before and after treatment could be divided into two stages: water evaporation and main decomposition. In the main decomposition stage, the maximum pyrolytic rate of treated bamboo was 315.62 °C, which was significantly lower than that of untreated bamboo, and the maximum pyrolysis rate was lower than that of untreated bamboo, indicating that after delignification, the intensity of bamboo pyrolysis decreases.

Author Contributions: Methodology, software, investigation, data curation, and writing—original draft, H.Y.; supervision, resources, and writing—review and editing, L.L.; visualization and investigation, Y.J. and W.H.; investigation, C.G., X.L. and F.R. All authors have read and agreed to the published version of the manuscript.

Funding: This research was funded by the Talent Launching Project of Zhejiang A & F University, grant number 2021LFR037.

Institutional Review Board Statement: Not applicable.

Informed Consent Statement: Not applicable.

Data Availability Statement: The data presented in this study are available on request from the corresponding author.

Conflicts of Interest: The authors declare no conflict of interest.

References

1. Takatani, M.; Ito, H.; Ohsugi, S.; Kitayama, T.; Saegusa, M.; Kawai, S.; Okamoto, T. Effect of Lignocellulosic Materials on the Properties of Thermoplastic Polymer/Wood Composites. *Holzforschung* **2000**, *54*, 197–200. [CrossRef]
2. Li, Y.; Fu, Q.; Yu, S.; Yan, M.; Berglund, L. Optically Transparent Wood from a Nanoporous Cellulosic Template: Combining Functional and Structural Performance. *Biomacromolecules* **2016**, *17*, 1358–1364. [CrossRef] [PubMed]
3. Mi, R.; Chen, C.; Keplinger, T.; Pei, Y.; He, S.; Liu, D.; Li, J.; Dai, J.; Hitz, E.; Yang, B.; et al. Scalable Aesthetic Transparent Wood for Energy Efficient Buildings. *Nat. Commun.* **2020**, *11*, 3836. [CrossRef] [PubMed]
4. Ghanadpour, M.; Carosio, F.; Larsson, P.T.; Wågberg, L. Phosphorylated Cellulose Nanofibrils: A Renewable Nanomaterial for the Preparation of Intrinsically Flame-Retardant Materials. *Biomacromolecules* **2015**, *16*, 3399–3410. [CrossRef]
5. Guan, H.; Cheng, Z.; Wang, X. Highly Compressible Wood Sponges with a Spring-like Lamellar Structure as Effective and Reusable Oil Absorbents. *ACS Nano* **2018**, *12*, 10365–10373. [CrossRef]
6. Wang, F.; Liu, X.; Duan, G.; Yang, H.; Cheong, J.Y.; Lee, J.; Ahn, J.; Zhang, Q.; He, S.; Han, J.; et al. Wood-Derived, Conductivity and Hierarchical Pore Integrated Thick Electrode Enabling High Areal/Volumetric Energy Density for Hybrid Capacitors. *Small* **2021**, *17*, 2102532. [CrossRef]
7. Chen, G.; Wu, Z.; Shen, Z.; Li, H.Y.; Li, J.; Lü, B.; Song, G.; Gong, X.; Qin, M.; Yao, C.L.; et al. Scalable, Strong and Water-Stable Wood-Derived Bioplastic. *Chem. Eng. J.* **2022**, *439*, 135680. [CrossRef]
8. Beims, R.F.; Arredondo, R.; Sosa Carrero, D.J.; Yuan, Z.; Li, H.; Shui, H.; Zhang, Y.; Leitch, M.; Xu, C.C. Functionalized Wood as Bio-Based Advanced Materials: Properties, Applications, and Challenges. *Renew. Sustain. Energy Rev.* **2022**, *157*. [CrossRef]
9. Berglund, L.A.; Burgert, I. Bioinspired Wood Nanotechnology for Functional Materials. *Adv. Mater.* **2018**, *30*, 1704285. [CrossRef]
10. Chen, C.; Hu, L. Nanoscale ion regulation in wood-based structures and their device applications. *Adv. Mater.* **2020**, *33*, 2002890. [CrossRef]
11. Liang, Y.; Zheng, G.; Xia, C.; Zuo, S.; Ge, S.; Yang, R.; Ma, X.; Fei, B.; Li, J.; Cheng, C.K.; et al. Synthesis of ultra-high strength structured material from steam-modified delignification of wood. *J. Clean. Prod.* **2022**, *351*, 131531. [CrossRef]
12. Zhu, M.; Song, J.; Li, T.; Gong, A.; Wang, Y.; Dai, J.; Yao, Y.; Luo, W.; Henderson, D.; Hu, L. Henderson, and L-B. Hu: Highly anisotropic, highly transparent wood composites. *Adv. Mater.* **2016**, *28*, 5181. [CrossRef]
13. Frey, M.; Widner, D.; Segmehl, J.S.; Casdorff, K.; Keplinger, T.; Burgert, I. Delignified and densified cellulose bulk materials with excellent tensile properties for sustainable engineering. *ACS Appl. Mater. Interfaces* **2018**, *10*, 5030. [CrossRef]
14. Lee, J. Biological conversion of lignocellulosic biomass to ethanol. *J. Biotechnol.* **1997**, *56*, 1–24. [CrossRef]
15. Bak, J.S.; Ko, J.K.; Choi, I.G.; Park, Y.C.; Seo, J.H.; Kim, K.H. Fungal pretreatment of lignocellulose by Phanerochaete chrysosporium to produce ethanol from rice straw. *Biotechnol. Bioeng.* **2009**, *104*, 471–482. [CrossRef]
16. Chen, C.; Li, Z.; Mi, R.; Dai, J.; Xie, H.; Pei, Y.; Li, J.; Qiao, H.; Tang, H.; Yang, B.; et al. Rapid Processing of Whole Bamboo with Exposed, Aligned Nanofibrils toward a High-Performance Structural Material. *ACS Nano* **2020**, *14*, 5194–5202. [CrossRef]
17. Wang, Y.Y.; Wang, X.Q.; Li, Y.Q.; Huang, P.; Yang, B.; Hu, N.; Fu, S.Y. High-Performance Bamboo Steel Derived from Natural Bamboo. *ACS Appl. Mater. Interfaces* **2021**, *13*, 1431–1440. [CrossRef]
18. Wang, X.; Shan, S.; Shi, S.Q.; Zhang, Y.; Cai, L.; Smith, L.M. Optically Transparent Bamboo with High Strength and Low Thermal Conductivity. *ACS Appl. Mater. Interfaces* **2021**, *13*, 1662–1669. [CrossRef]
19. Zhou, C.; Julianri, I.; Wang, S.; Chan, S.H.; Li, M.; Long, Y. Transparent Bamboo with High Radiative Cooling Targeting Energy Savings. *ACS Appl. Mater. Interfaces* **2021**, *3*, 883–888. [CrossRef]
20. Lin, Q.; Huang, Y.; Yu, W. Effects of Extraction Methods on Morphology, Structure and Properties of Bamboo Cellulose. *Ind. Crops Prod.* **2021**, *169*. [CrossRef]
21. Marchessault, R. Wood Chemistry, Fundamentals and Applications. 2nd Edn., by Ero Sjöström, Academic Press, NY, 1993, 250 Pages of Text and 42 Pages of Bibliography and Index (ISBN 0-12-647481-8). *Carbohydr. Res.* **1994**, *252*, C1. [CrossRef]
22. Yu, H.; Zheng, H.; Zhan, M.; Zhang, W.; Wang, J.; Pan, X.; Zhuang, X. Wei Surface Characterization and Biodegradability of Sodium Hydroxide-Treated Moso Bamboo Substrates. *Eur. J. Wood Wood Prod.* **2021**, *79*, 443–451. [CrossRef]
23. Wu, M.B.; Hong, Y.M.; Liu, C.; Yang, J.; Wang, X.P.; Agarwal, S.; Greiner, A.; Xu, Z.K. Delignified Wood with Unprecedented Anti-Oil Properties for the Highly Efficient Separation of Crude Oil/Water Mixtures. *J. Mater. Chem. A* **2019**, *7*, 16735–16741. [CrossRef]
24. Chen, H.; Yu, Y.; Zhong, T.; Wu, Y.; Li, Y.; Wu, Z.; Fei, B. Effect of Alkali Treatment on Microstructure and Mechanical Properties of Individual Bamboo Fibres. *Cellulose* **2017**, *24*, 333–347. [CrossRef]
25. Wu, J.; Wu, Y.; Yang, F.; Tang, C.; Huang, Q.; Zhang, J. Impact of Delignification on Morphological, Optical and Mechanical Properties of Transparent Wood. *Compos. Part A Appl. Sci. Manuf.* **2019**, *117*, 324–331. [CrossRef]
26. Xu, G.; Wang, L.; Liu, J.; Wu, J. FTIR and XPS Analysis of the Changes in Bamboo Chemical Structure Decayed by White-Rot and Brown-Rot Fungi. *Appl. Surf. Sci.* **2013**, *280*, 799–805. [CrossRef]
27. Wang, X.; Ren, H. Comparative Study of the Photo-Discoloration of Moso Bamboo (Phyllostachys Pubescens Mazel) and Two Wood Species. *Appl. Surf. Sci.* **2008**, *254*, 7029–7034. [CrossRef]
28. Yuan, Z.; Kapu, N.S.; Beatson, R.; Chang, X.F.; Martinez, D.M. Effect of Alkaline Pre-Extraction of Hemicelluloses and Silica on Kraft Pulping of Bamboo (Neosinocalamus Affinis Keng). *Ind. Crops Prod.* **2016**, *91*, 66–75. [CrossRef]
29. Nguila Inari, G.; Petrissans, M.; Lambert, J.; Ehrhardt, J.J.; Gérardin, P. XPS Characterization of Wood Chemical Composition after Heat-Treatment. *Surf. Interface Anal.* **2006**, *38*, 1336–1342. [CrossRef]

30. Du Guanben, Application of X-ray Photoelectron Spectroscopy in Wood Science and Technology. *China Wood Ind.* **1999**, *13*, 17–20+29. [CrossRef]
31. Watling, K.M.; Parr, J.F.; Rintoul, L.; Brown, C.L.; Sullivan, L.A. Raman, infrared and XPS study of bamboo phytoliths after chemical digestion. *Spectrochim. Acta Part A Mol. Biomol. Spectrosc.* **2011**, *80*, 106–111. [CrossRef]
32. Wang, X.Q.; Ren, H.Q.; Zhao, R.J.; Cheng, Q.; Chen, Y.P. FTIR and XPS spectroscopic studies of photodegradation of Moso bamboo (Phyllostachys Pubescens Mazel). *Spectrosc. Spectr. Anal.* **2009**, *29*, 1864–1867. [CrossRef]
33. Gérardin, P.; Petrič, M.; Petrissans, M.; Lambert, J.; Ehrhrardt, J.J. Evolution of Wood Surface Free Energy after Heat Treatment. *Polym. Degrad. Stab.* **2007**, *92*, 653–657. [CrossRef]
34. Huang, X.; Kocaefe, D.; Kocaefe, Y.; Boluk, Y.; Pichette, A. Study of the Degradation Behavior of Heat-Treated Jack Pine (Pinus Banksiana) under Artificial Sunlight Irradiation. *Polym. Degrad. Stab.* **2012**, *97*, 1197–1214. [CrossRef]
35. Nguila Inari, G.; Pétrissans, M.; Dumarcay, S.; Lambert, J.; Ehrhardt, J.J.; Šernek, M.; Gérardin, P. Limitation of XPS for Analysis of Wood Species Containing High Amounts of Lipophilic Extractives. *Wood Sci. Technol.* **2011**, *45*, 369–382. [CrossRef]
36. Kocaefe, D.; Huang, X.; Kocaefe, Y.; Boluk, Y. Quantitative Characterization of Chemical Degradation of Heat-Treated Wood Surfaces during Artificial Weathering Using XPS. *Surf. Interface Anal.* **2013**, *45*, 639–649. [CrossRef]
37. Meng, F.D.; Yu, Y.L.; Zhang, Y.M.; Yu, W.J.; Gao, J.M. Surface Chemical Composition Analysis of Heat-Treated Bamboo. *Appl. Surf. Sci.* **2016**, *371*, 383–390. [CrossRef]
38. Hua, X.; Kaliaguine, S.; Kokta, B.; Adnot, A. Surface Analysis of Explosion Pulps by ESCA Part 1. Carbon (1s) Spectra and Oxygen-to-Carbon Ratios. *Wood Sci. Technol.* **1993**, *27*, 449–459. [CrossRef]
39. Kamdem, D.P.; Riedl, B.; Adnot, A.; Kaliaguine, S. ESCA Spectroscopy of Poly(Methyl Methacrylate) Grafted onto Wood Fibres. *J. Appl. Polym. Sci.* **1991**, *43*, 1901–1912. [CrossRef]
40. Koubaa, A.; Riedl, B.; Koran, Z. Surface analysis of press dried-CTMP paper samples by electron spectroscopy for chemical analysis. *J. Appl. Polym. Sci.* **1996**, *61*, 545–552. [CrossRef]
41. Liu, Y.; Hu, H. X-ray diffraction study of bamboo fibres treated with NaOH. *Fibres Polym.* **2009**, *9*, 735–739. [CrossRef]
42. Jiang, Z.; Liu, Z.; Fei, B.; Cai, Z.; Yu, Y.; Liu, X. The Pyrolysis Characteristics of Moso Bamboo. *J. Anal. Appl. Pyrolysis* **2012**, *94*, 48–52. [CrossRef]
43. Sun, R.H.; Li, X.J.; Liu, Y.; Hou, R.G.; Qiao, J.Z. Effects of high temperature heat treatment on FTIR and XRD characteristics of bamboo bundles. *J. Cent. South Univ. For. Technol.* **2013**, *33*, 97–100. [CrossRef]
44. Yang, L.; Wu, Y.; Yang, F.; Wang, W. Study on the Preparation Process and Performance of a Conductive, Flexible, and Transparent Wood. *J. Mater. Res. Technol.* **2021**, *15*, 5396–5404. [CrossRef]
45. Wang, Y.Y.; Li, Y.Q.; Xue, S.S.; Zhu, W.B.; Wang, X.Q.; Huang, P.; Fu, S.Y. Superstrong, Lightweight, and Exceptional Environmentally Stable SiO_2@GO/Bamboo Composites. *ACS Appl. Mater. Interfaces* **2022**, *14*, 7311–7320. [CrossRef]
46. Lin, J.; Yang, Z.; Hu, X.; Hong, G.; Zhang, S.; Song, W. The Effect of Alkali Treatment on Properties of Dopamine Modification of Bamboo Fibre/Polylactic Acid Composites. *Polymers* **2018**, *10*, 403. [CrossRef]
47. Raveendran, K. Pyrolysis Characteristics of Biomass and Biomass Components. *Fuel* **1996**, *75*, 987–998. [CrossRef]
48. Li, L.; Chen, Z.; Lu, J.; Wei, M.; Huang, Y.; Jiang, P. Combustion Behavior and Thermal Degradation Properties of Wood Impregnated with Intumescent Biomass Flame Retardants: Phytic Acid, Hydrolyzed Collagen, and Glycerol. *ACS Omega* **2021**, *6*, 3921–3930. [CrossRef]
49. Yang, H.; Yan, R.; Chen, H.; Lee, D.H.; Zheng, C. Characteristics of Hemicellulose, Cellulose and Lignin Pyrolysis. *Fuel* **2007**, *86*, 1781–1788. [CrossRef]

Article

Effect of Preparation Conditions on Application Properties of Environment Friendly Polymer Soil Consolidation Agent

Shaoli Wang [1], Shengju Song [2], Xuping Yang [3], Zhengqi Xiong [4], Chaoxing Luo [4], Yongxiu Xia [1], Donglu Wei [4], Shaobo Wang [5], Lili Liu [4], Hong Wang [4], Lifang Sun [1], Lichao Du [2] and Shaofeng Li [1,*]

1. State Key Laboratory of Tree Genetics and Breeding, Experimental Center of Forestry in North China, National Permanent Scientific Research Base for Warm Temperate Zone Forestry of Jiulong Mountain in Beijing, Chinese Academy of Forestry, Beijing 100091, China; wshaoli@iccas.ac.cn (S.W.); yxxia@caf.ac.cn (Y.X.); slf2014@163.com (L.S.)
2. R & D Department, China Academy of Launch Vehicle Technology, Beijing 100076, China; songshengju99@163.com (S.S.); duchao0620@163.com (L.D.)
3. Chinese Academy of Forestry, Beijing 100091, China; yxp@caf.ac.cn
4. College of Material and Chemical Engineering, Heilongjiang Institute of Technology, Harbin 150050, China; xiongzhengqi1020@163.com (Z.X.); luochaoxing525@163.com (C.L.); mit_ing@163.com (D.W.); liulili760802@126.com (L.L.); wanghongg357@126.com (H.W.)
5. Beijing Yangsheng New Material Technology Co., Ltd., Beijing 102299, China; wshaobo2009@163.com
* Correspondence: Lisf@caf.ac.cn

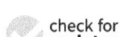

Citation: Wang, S.; Song, S.; Yang, X.; Xiong, Z.; Luo, C.; Xia, Y.; Wei, D.; Wang, S.; Liu, L.; Wang, H.; et al. Effect of Preparation Conditions on Application Properties of Environment Friendly Polymer Soil Consolidation Agent. *Polymers* 2022, 14, 2122. https://doi.org/10.3390/polym14102122

Academic Editors: Jesús-María García-Martínez, Jingpeng Li, Yun Lu and Huiqing Wang

Received: 14 April 2022
Accepted: 18 May 2022
Published: 23 May 2022

Publisher's Note: MDPI stays neutral with regard to jurisdictional claims in published maps and institutional affiliations.

Copyright: © 2022 by the authors. Licensee MDPI, Basel, Switzerland. This article is an open access article distributed under the terms and conditions of the Creative Commons Attribution (CC BY) license (https:// creativecommons.org/licenses/by/ 4.0/).

Abstract: In order to improve the survival rate of transplanted seedlings and improve the efficiency of seedling transplantation, we developed an environmental friendly polymer konjac glucomannan (KGM)/chitosan (CA)/polyvinyl alcohol (PVA) ternary blend soil consolidation agent to consolidate the soil ball at the root of transplanted seedlings. In the previous research, we found that although the prepared KGM/CA/PVA ternary blend soil consolidation agent can consolidate the soil ball at the root of the seedling, the medium solid content of the adhesive was high, which affects its spraying at the root of the seedling. At the same time, the preparation temperature of the KGM/CA/PVA ternary blend was also high. Therefore, to reduce the energy consumption and the cost of the KGM/CA/PVA ternary blend soil consolidation agent in the preparation process, this paper studied the influence of preparation conditions on the application performance of the environmental friendly polymer soil consolidation agent. We aimed to reduce the highest value CA content and preparation temperature of the KGM/CA/PVA ternary blend adhesive on the premise of ensuring the consolidation performance of the KGM/CA/PVA ternary blend adhesive on soil balls. It was prepared for the popularization and application of the environmental friendly polymer KGM/CA/PVA ternary blend soil consolidation agent in seedling transplanting. Through this study, it was found that the film-forming performance of the adhesive was better when the KGM content was 4.5%, the CA content was in the range of 2–3%, the PVA content was in the range of 3–4%, and the preparation temperature was higher than 50 °C. The polymer soil consolidation agent prepared under this condition has a good application prospect in seedling transplanting.

Keywords: polymer soil consolidation agent; preparation conditions; consolidated adhesive film; compressive strength; seedling transplanting

1. Introduction

Both barren mountain afforestation and urban greening involve seedling transplanting [1]. In order to improve the survival rate of transplanted seedlings, the most important thing is to ensure the integrity of the soil ball at the root of transplanted seedlings. The diameter of the soil ball at the root of the seedling is generally required to be 5–10 times the diameter at breast height (DBH) of the tree itself, and the height of the soil ball is generally about 2/3 of its diameter. After successful lifting from the nursery, the soil ball is very easy to break due to squeezing, bumping, and mutual impact during its transportation. To ensure the integrity of transplanted seedlings, the commonly used methods are wrapping, binding, or wooden box packaging [1,2]. There are two problems in the current method to ensure the integrity of the soil ball. First, some materials wrapped and bound, such as plastic rope, plastic cloth, and iron wire, do not degrade easily and can cause environmental pollution. Second, the technical requirements for soil ball wrapping, binding, and seedling raising are high, so it not only consumes a lot of manpower and material resources but also has low seedling raising efficiency. The key is that it is difficult to ensure the integrity of the soil ball in the end, especially in the process of loading, unloading, and transporting seedlings where the mother soil ball is often damaged.

In view of the demand for seedling transplanting and maintaining the integrity of the mother soil ball, our research team prepared KGM/CA binary blend soil consolidation agents and KGM/CA/PVA ternary blend soil consolidation agents with KGM, CA, and PVA as the main raw materials in the previous research work [3–5]. The soil consolidation agent was evenly sprayed on the surface of the soil ball at the root of the transplanted seedlings. After the consolidation agent was dry, it could form a hard film on the surface of the soil ball. In the previous research, it was found that the film formed on the surface of the soil ball by KGM/CA and KGM/CA/PVA blend soil consolidation agents can protect the soil ball in the process of transportation, which reduced wear and improved its resistance to transportation oscillation [3–5]. The application of the KGM/CA and KGM/CA/PVA blend consolidation agent also had no adverse effect on the growth of the seedlings.

KGM, the raw material for preparing the soil consolidation agent, is a rich natural macromolecular polysaccharide, which can be extracted from amorphophallus konjac plant tubers [6–10]. KGM is mainly composed of D-mannose and D-glucose units through β-1,4-linkages in a mole ratio of 1.6:1 [11–13]. Due to the presence of active primary hydroxyl ($-CH_2OH$) at the C(6) position of each sugar unit, KGM can participate in many chemical reactions, such as nitration, etherification, and graft polymerization [14]. The O-acetyl ($-OCH_3$) group located at the C(6) position of the sugar residues contributes to the solubility and swelling of KGM [15,16]. KGM is water-soluble and has high viscosity even at low concentrations. It can also form a gel network structure through extensive hydrogen bonding and entanglement [17]. Therefore, KGM is very suitable for preparing soil consolidation agents. However, the glue solution prepared by a single KGM can not meet the consolidation of soil. Considering the convenience and economy of preparing soil consolidation agents, it is wise to use physical blending modification technology to improve the consolidation performance of soil.

There are two directions in the physical blending modification of KGM. One is to blend with other gel polysaccharides, chitosan, starch, and other substances, so as to improve the viscosity or gel strength of products [18–21]. Second, KGM is blended with other synthetic polymers to give it new functions. In the process of blending, the addition of other polymer materials can greatly improve the hydrogen bonding in KGM molecules and form a new spatial network structure [22,23].

PVA contains a large number of hydroxyl groups in its molecular chain, which can be crosslinked to form a macromolecular network. Meanwhile, PVA is a water-soluble polymer material with good biocompatibility and film-forming properties [24,25]. After blending with PVA, the mechanical properties of the KGM film can be improved.

CA is the product of deacetylation of chitin macromolecule, and its structure is similar to cellulose [26,27]. Many hydroxyl and amino groups are distributed on the macromolecular chain of chitosan, which gives it good solubility and reactivity. CA also has good biocompatibility, adsorption, film-forming, permeability, moisture retention, and biodegradability properties [27,28]. Blending CA with KGM and PVA can improve the adhesion of the glue solution on the soil surface, film-forming, moisturizing, and air permeability of the glue film. However, because the price of CA is higher than that of KGM and PVA, it is necessary to explore the influence of the solid content of CA in the KGM/CA/PVA ternary blend on the application performance of the soil consolidation agent, which is of great significance for the market application of KGM/CA/PVA ternary blend soil consolidation agents in the future.

In the preparation process of the soil consolidation agent, it was found that the preparation temperature of the adhesive had an obvious influence on the viscosity and film-forming of the KGM/CA/PVA ternary blend adhesive, as well as the consolidation of soil and the anti-compression and anti-transport oscillation of the consolidated soil column. The higher the preparation temperature of the KGM/CA/PVA ternary blend adhesive, the higher the preparation cost and an increase in energy consumption, which is not conducive to reducing carbon emission. At the same time, the content of CA and PVA can not only affect the viscosity, fluidity, film-forming, and consolidation of the soil ball but also affect the preparation cost of the KGM/CA/PVA consolidation agent.

Therefore, in order to reduce the energy consumption and cost of the KGM/CA/PVA ternary blend soil consolidation agent in the preparation process, this paper studied the influence of its preparation conditions, such as preparation temperature, CA and PVA content, on the application performance of environmental friendly polymer KGM/CA/PVA soil consolidation agent. It can also reduce the content of CA in the adhesive and the preparation temperature of polymer soil consolidation agent on the premise of ensuring the consolidation performance of KGM/CA/PVA ternary blend soil consolidation agent on soil balls. It is prepared for the popularization and application of the environmental friendly polymer KGM/CA/PVA ternary blend soil consolidation agent in seedling transplanting.

2. Materials and Methods

2.1. Materials and Experimental Instruments

(1) Materials: The main raw material konjac flour (KGM, 200 g/bottle) was provided by Bozhou Baofeng bio-technology limited company. Chitosan (CA, chemical pure), polyvinyl alcohol (PVA, superior-grade pure), acetic acid (excellent-grade pure), sodium hydroxide (analytical purity), and other compounds were supplied by Sinopharm Chemical Reagent Co., Ltd. of China, Shanghai, China.

(2) Experimental instruments: Water bath heating pot; Mechanical agitator; Automatic film-coating machine; Stereomicroscope (Leica DFC425C); Mechanical testing machine INSTRON 5582; Simulated transportation vibration test bench hk-120 with a payload of 300 kg.

2.2. Preparation and Viscosity Test Method of Environmental Friendly Soil Consolidation Agent

(1) Preparation of polyvinyl alcohol (PVA) solution: Add 360 g of ultrapure water to the neck mouth flask, add 40 g of PVA (molecular weight: 1840 g/mol) to the four mouth flask, and then put the flask into a water bath, heat it to the temperature of 95 °C, and stir mechanically (rotating speed 150 r/min) for 2 h to obtain a PVA solution with a mass fraction of 10%.

(2) Preparation of chitosan solution: Weigh a certain amount of CA into a four neck flask, add a pre-configured acetic acid solution with a mass fraction of 20%, stir until it is evenly dissolved to obtain a dilute acid solution of CA.

(3) Preparation of KGM/CA/PVA ternary blending soil consolidation agent: Weigh a certain amount of KGM into a four neck flask, add an appropriate amount of CA dilute acid solution according to the preset proportion, and mechanically stir (400 r/min) for a certain time at the target temperature until KGM and CA in the system are completely dissolved. Then the NaOH was added to the system to adjust the pH value of the reaction solution to 4.2–4.5. Finally, KGM/CA/PVA ternary blend adhesive was obtained by adding 10% polyvinyl alcohol solution prepared in advance and stirring for 1 h. Then, according to the actual needs, tackifier and preservative can be added to KGM/CA/PVA ternary blend adhesive. Finally, transfer the KGM/CA/PVA ternary blend adhesive to a wide mouth bottle and seal it for standby after it is reduced to room temperature.

(4) Test method for viscosity of KGM/CA/PVA ternary blend adhesive: Using Brooke digital display DV3 viscometer, immerse the rotor into the ternary blend adhesive of different formulations at room temperature, select rotor No. 64, set the speed to 10 r/min to obtain the viscosity of the ternary blend adhesive.

2.3. Preparation and Test of KGM/CA/PVA Ternary Blend Film

(1) Preparation of KGM/CA/PVA ternary blend film: Weigh 5 g KGM/CA/PVA ternary blend adhesive, and then pour the adhesive directly onto the substrate in the middle of the automatic coating machine console. Select the appropriate scraper and the appropriate mode. The scraper will move back and forth on the track at a certain speed until the KGM/CA/PVA ternary blend adhesive is evenly coated on the substrate. Then, the substrate with the adhesive film is translated to a dry position and naturally dried to form a film.

(2) The SEM image of KGM/CA/PVA ternary blend film: After the prepared KGM/CA/PVA ternary blend film was dried in a dryer for 24 h, the film with uniform appearance was selected and cut into small strips, and the sample for observing the upper surface was prepared. In addition, the membrane at the same part were brittle broken after liquid nitrogen freezing, and the cross-section port was scanned. Place the pasted sample on the copper table, spray gold under 13.3 Pa vacuum for 20–30 s, and the thickness is about 690 nm. Under the condition of accelerating voltage of 20 kV, the surface and cross-section morphology of the film samples were observed by scanning electron microscope (SEM).

2.4. Preparation and Tests of Soil Column Samples

(1) Preparation of soil column samples: Two different types of consolidated soil columns were prepared from cinnamon soil (loam) with different formulations of KGM/CA/PVA ternary blend adhesive. The preparation method is similar to the original [3].

(2) Optical characterization of KGM/CA/PVA ternary blend adhesive film on the surface of consolidated soil column: Referring to the original characterization method [3,5], Leica DFC425C stereoscope was used to observe the surface morphology of the bonding film on the upper surface of the consolidated soil column. Obtain the best image of the consolidated adhesive film on the upper surface of the soil column sample by adjusting the "light source" and "focusing/zoom", and transmit the image to the picture window of the software for saving.

(3) Test method of compressive strength of consolidated soil column: Referring to the original test method [3,5], the compressive strength of consolidated soil column is tested by INSTRON 5582 universal testing machine. The test conditions are that the time interval is 0.5 s and the compression rate is 1 mm/min.

(4) Test method of anti-transport oscillation of consolidated soil column: Referring to the original test method [3], according to the American Transportation Association standard (ISTA) and American Society of materials standard (ASTM), the transportation oscillation resistance of consolidated soil column is tested by simulated transportation vibration test bench HK-120, as shown in Table 1.

Table 1. ISTA/ASTM Standard Test Method for Simulating Transportation Vibration.

Oscillation Test Sequence	Test Speed (r.min^{-1})	Corresponding Frequency (Hz)	Test Time t (min)
Test 1	180	3.0	79
Test 2	210	3.5	66
Test 3	240	4.0	60

2.5. Method of Transplanting Seedlings with KGM/CA/PVA Ternary Blend Adhesive in Seedling Transplanting

The application of ternary blend soil consolidation agent for seedling transplanting adopts a relatively simple method: Take loam as the main consolidation object and seedlings with DBH of 5–10 mm growing outdoors as the main transplanting object. Firstly, circle a circle with a diameter of about 10–15 cm with the transplanted seedling as the center at the root of the seedling, and then gradually clean the soil outside the circle from top to bottom until a conical soil ball is formed at the root of the transplanted seedling. There is about 1 cm contact surface between the bottom of the conical soil ball and the earth. Then, the soil consolidation agent is sprayed on the upper and side surfaces of the conical soil ball. After about 1 day of consolidation, the glue liquid on the surface of the soil ball will form a hard shell on the surface of the soil ball together with the soil on the surface of the soil ball. After the soil ball is lifted, it can be directly transported to a new planting site for transplanting. When transplanting seedlings in sandy soil with the soil consolidation agent, except that the method of preparing the soil ball is slightly different, the transplanting steps are similar to those in loam [3–5].

3. Results

3.1. Effect of Preparation Conditions on Viscosity of KGM/CA/PVA Ternary Blend Adhesive

In previous studies, it was found that the preparation temperature and solid content of the blend adhesive had a great influence on the viscosity and fluidity of the blend adhesive [3–5]. In order to reduce the content of CA in the ternary blend solution and ensure the viscosity of the solution to the consolidation of soil columns, the effect of CA content on the viscosity of the solution at different temperatures was explored.

It can be seen from Figure 1a,b that the viscosity of the KGM/CA/PVA blend adhesive increases gradually with the increase of CA content in the glue at different temperatures. It was found that the viscosity increased slowly when the CA content was less than 1.5% at the preparation temperature of 40 °C; when the CA content was higher than 1.5%, the viscosity of the adhesive increased rapidly, as shown in Figure 1a. When the preparation temperature was 50 °C, and the CA content of the KGM/CA/PVA blend adhesive was lower than 3%, the viscosity increased rapidly; when the CA content was higher than 3%, the increase of adhesive viscosity slowed down, as shown in Figure 1b. Therefore, we explored the effect of CA content (2–3%) in the glue solution on the viscosity of the blend adhesive system at different temperatures, as shown in Figure 1c. It was found that when the CA content was fixed, with the increase of the preparation temperature, the viscosity of the adhesive first increased and then decreased, and the viscosity was highest at 60 °C. When the CA content in the blend adhesive was 3%, the viscosity in the blend adhesive system was highest as a whole. When the preparation temperature of the glue was at 60 °C, the viscosity of the adhesive reached a maximum value. It can be seen that the preparation temperature and the CA content in the adhesive have a great influence on the viscosity of the KGM/CA/PVA blend adhesive. Excessive viscosity of the KGM/CA/PVA blend adhesive can affect its fluidity, so it is necessary to select the appropriate preparation conditions according to the consolidation demand of the soil.

3.2. Effect of Preparation Conditions on Properties of KGM/CA/PVA Ternary Blend Film

The preparation conditions of the KGM/CA/PVA ternary blend adhesive, such as preparation temperature, CA content, and PVA content, have a great impact not only on the viscosity but also on the film-forming property, strength, toughness, and internal structure of the film.

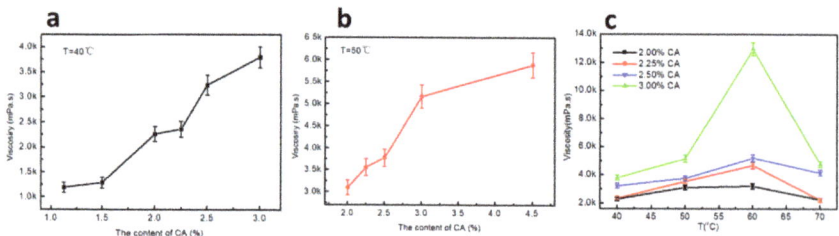

Figure 1. (**a**) Effect of CA content on viscosity of ternary blend adhesive at 40 °C; (**b**) Effect of CA content on viscosity of ternary blend adhesive at 50 °C; (**c**) Effect of CA content on viscosity of ternary blend adhesive at different temperatures.

3.2.1. Effect of Preparation Conditions of KGM/CA/PVA Ternary Blend Adhesive on Film-Forming Properties

The prepared ternary blend adhesive was prepared into a thin film by an automatic coating machine. When the KGM/CA/PVA ternary blend adhesive with different CA content was prepared into a glue film at 40 °C, it was found that when the CA content was 0%, the formed thin film was relatively brittle and it was difficult to form a whole film, as shown in Figure 2a. From Figure 2b,c, it was found that the brittleness of the film was gradually improved, and the formed film was gradually complete with the gradual increase of CA content. However, due to the low preparation temperature of the KGM/CA/PVA ternary blend adhesive, the quality of the formed adhesive film was relatively poor. With the increase in temperature, the viscosity of the KGM/CA/PVA ternary blend adhesive increases, and the toughness of the film increased to a certain extent. As shown in Figure 2d, the adhesive film formed by the adhesive prepared under the condition of 50 °C had a surface of well-distributed pore, which met the application requirements of the consolidated soil ball. At the same time, the toughness of the adhesive film was good, and the complete membrane can be cut evenly with a die. As the temperature continued to rise to 60 °C, the formed adhesive film was more flexible and better, as shown in Figure 2e. When the temperature reached to 70 °C, the formed adhesive film can be bent at multiple angles, as shown in Figure 2f.

3.2.2. Effect of Preparation Conditions of KGM/CA/PVA Ternary Blend Adhesive on the Structure of Film

The effect of temperature on the internal structure of the film can be seen from the surface and cross-section of the film. The adhesive film formed by the KGM/CA/PVA ternary blend adhesive was prepared at 40 °C. As shown in Figure 3a, it can be seen that the adhesive film was not densely arranged, was unevenly distributed, and had many cavities. As shown in Figure 3b, it can be seen that the links between the cured products on the surface of the adhesive film were weak, the pores formed were large, and there was no cross-linking between the molecular chains, so the adhesive film formed was brittle. From Figure 3c,d, it can be seen that when the temperature for preparing the ternary blend adhesive was increased to 70 °C, the surface of the adhesive film was densely arranged, and there were many short protrusions on the surface and connected into a network structure. At the same time, there were a small number of pores on the surface of the formed adhesive film, which can meet the application requirements of the transplanted seedlings consolidated soil ball. Through Figure 3e,f, it can be seen that KGM, CA, and PVA were cross-linked together, permeated, and interspersed with each other, which reflected the enhanced interaction between KGM, CA, and PVA. It can be seen from Figure 3f that an interpenetrating structure was formed between KGM, CA, and PVA, indicating that KGM, CA, and PVA have good compatibility. This morphology not only ensured the bonding strength but also improved the water resistance to a certain extent.

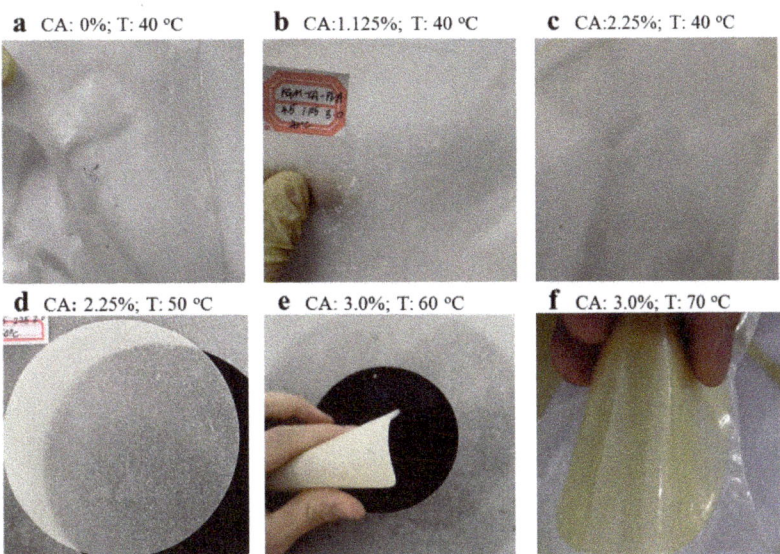

Figure 2. Film-forming properties of different KGM/CA/PVA ternary blend adhesive. Preparation conditions of KGM/CA/PVA ternary blend adhesive: the content of KGM and PVA were 4.5% and 3.0% respectively, the pH value is 5.0.

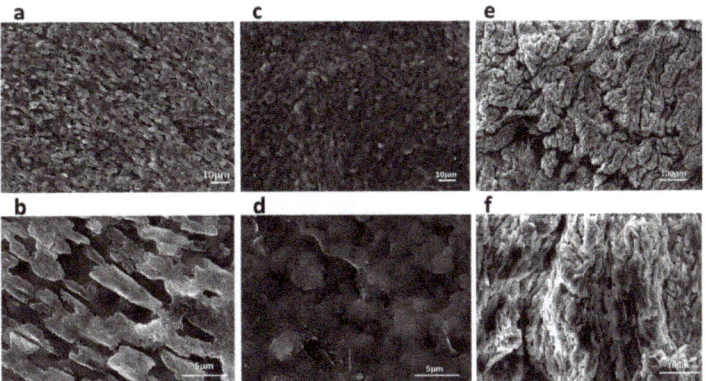

Figure 3. The SEM image of adhesive film. (**a–d**) Surface morphology of adhesive film; (**e,f**) Cross-section morphology of adhesive film. Preparation conditions of KGM/CA/PVA ternary blend adhesive: the content of KGM and PVA were 4.5% and 3.0% respectively, the pH value is 5.0. (**a,b**) the content of the CA was 2.5%, the preparation temperature was 40 °C; (**c–f**) the content of the CA was 3%, the preparation temperature was 70 °C.

3.3. Effect of Preparation Conditions on the Morphology of Consolidated Adhesive Film

The loam in cinnamon soil was prepared into the soil column by mold, and KGM/CA/PVA ternary blend adhesive was evenly sprayed on the surface of the soil column to form a uniform glue film. The soil column at this stage contains a certain amount of water. During the drying of the soil column, with the volatilization of water, the morphology of the polymer consolidated film covering the surface of the soil column can change, and produce bubbles similar to fish eye. The moisture permeability of the outer polymer film can be directly judged by observing the morphology of the solid film on the surface of the soil column through the body microscope. It was found that the preparation temperature and the content of CA in the KGM/CA/PVA ternary blend can affect the surface morphology of the consolidated film. When the preparation temperature was 50 °C, relatively large holes were easier to form on the surface of the consolidated adhesive film, as shown in Figure 4a–d. When the CA content was low, a reticular membrane with large pores was formed on the surface of the soil column, as shown in Figure 4a. As the CA content gradually increased from 1% to 3%, the holes on the surface of the consolidated adhesive film gradually shrank and formed many small bubbles, as shown in Figure 4b–d. When the preparation temperature of the glue solution rose to 60 °C and 70 °C, it was found that with the increase of CA content, the bubbles on the surface of the consolidated adhesive film were also gradually decreasing, as shown in Figure 4e–l. It can be seen that in the KGM/CA/PVA ternary blend adhesive, with the increase of CA content, the film-forming performance of the soil column consolidated adhesive film was better. Therefore, under the same humidity conditions, its bubbles gradually decreased.

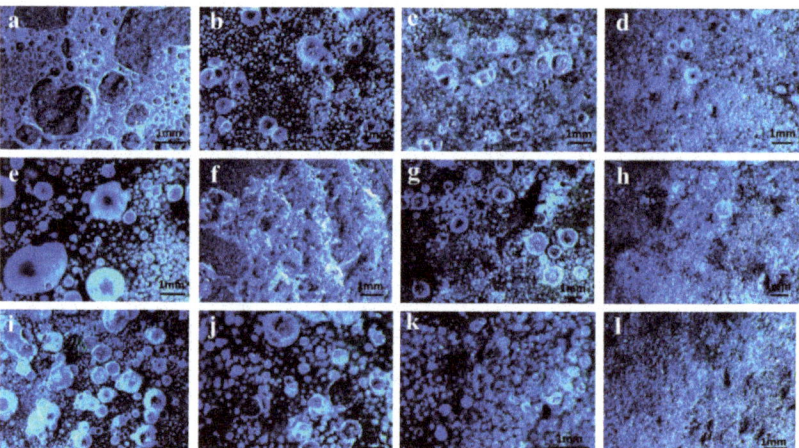

Figure 4. Surface morphology of consolidated adhesive film. (**a–d**) Preparation conditions of KGM/CA/PVA ternary blend adhesive: the preparation temperature was 50 °C; The contents of CA were 1%, 2%, 2.5%, 3%, respectively; (**e–h**) Preparation conditions of KGM/CA/PVA ternary blend adhesive: the preparation temperature was 60 °C; The contents of CA were 1%, 2%, 2.5%, 3%, respectively; (**i–l**) Preparation conditions of KGM/CA/PVA ternary blend adhesive: the preparation temperature was 70 °C; The contents of CA were 1%, 2%, 2.5%, 3%, respectively. Other preparation conditions of (**a–l**) KGM, PVA content were 4.5% and 3%, respectively; The pH value of the ternary blend was 5.0; pH of cinnamon soil was 8.5; Particle size of soil was 1 mm.

It was found that when the CA content in the KGM/CA/PVA ternary blend adhesive was the same, the bubbles on the surface of the consolidated film also showed a gradually decreasing trend with the preparation temperature of the solution rising from 50 °C to 70 °C, as shown in Figure 4a,e,i, and Figure 4b,f,j. This was because, with the increase in temperature, CA dissolved more thoroughly in the blend adhesive. CA formed a short cross-linking network structure after blending with KGM and PVA. The film-forming performance of the KGM/CA/PVA ternary blend was better. Therefore, under the same humidity conditions, the bubbles on the surface of the consolidated film were small.

3.4. Effect of Preparation Conditions on Compressive Properties of Consolidated Soil Columns

It was found that the preparation temperature and solid content of the KGM/CA/PVA ternary blend adhesive had great effects on the compressive properties of the consolidated soil columns. The film-forming performance of PVA is good, but when the concentration of PVA is large, the viscosity is too large, which is not conducive to the fluidity of glue solution. Therefore, it is necessary to study the influence of PVA content on the compression resistance of consolidated soil columns. It was found that the compressive strength of consolidated soil columns increased with the increase of PVA content from 1% to 5% at the same temperature, as shown in Figure 5a. At the same time, it was also found that when the PVA content was fixed, the compressive strength of the consolidated soil column increased with the increase of the preparation temperature from 40 °C to 70 °C, especially when the temperature was higher than 50 °C, the compressive strength of the consolidated soil column increased rapidly, and the maximum compressive strength reached 4.57 MPa. It can be seen that the preparation temperature and PVA content of the KGM/CA/PVA ternary blend adhesive can directly affect the compressive strength of the consolidated soil column.

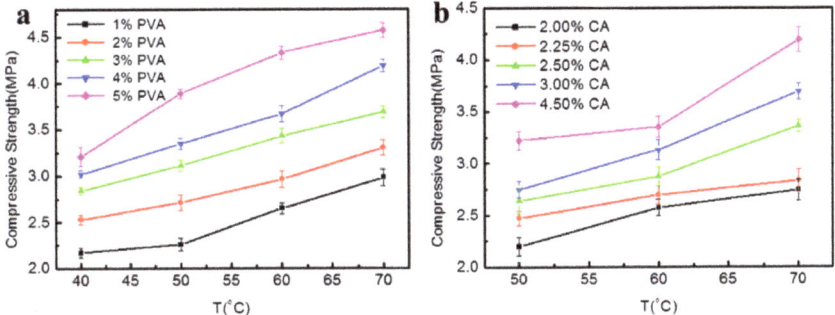

Figure 5. Effect of preparation conditions on compressive properties of consolidated soil columns Preparation conditions of KGM/CA/PVA ternary blend adhesive: the content of KGM was 4.5%, the pH value is 5.0, pH of cinnamon soil was 8.5; Particle size of soil was 1 mm. (**a**) The content of the CA was 4.5%; (**b**) The content of the PVA was 4.0%.

Due to the high price of CA, in order to reduce the preparation cost of KGM/CA/PVA ternary blend soil consolidation agent, it is necessary to study the effect of the reduction of CA content in the glue on the compressive properties of consolidated soil columns. It was found that at the same temperature, with the decrease of CA content, the compressive strength of the consolidated soil column also showed a decreasing trend, as shown in Figure 5b. However, at 60 °C, with the decrease of CA content, the compressive strength of the consolidated soil column decreased relatively little. At the same time, it was also found that the compressive strength of the consolidated soil column was higher with the increase in the preparation temperature of the glue solution. Therefore, it is necessary to select the appropriate preparation temperature and CA content according to the actual application requirements.

3.5. Effect of Preparation Conditions on Anti-Transport Oscillation of Consolidated Soil Column

During the transportation of transplanted seedlings, the consolidated soil ball needs to bear various oscillations during the transportation from the nursery to the planting point. According to the effect of the preparation conditions of the KGM/CA/PVA ternary blend adhesive on the viscosity of glue and the anti-compression performance of the consolidated soil column, we selected three better conditions for preparing KGM/CA/PVA adhesive, as shown in Table 2. In order to study the influence of different preparation conditions on the anti-transportation oscillation of the consolidated soil column, we used the simulated transportation vibration test-bed HK-120 to test the anti-transportation oscillation of the consolidated soil column. The oscillation conditions were shown in Table 1. As shown in Figure 6, in oscillation experiment 1 (180 rpm, 3.0 Hz, 79 min), it can be seen that samples a_2, b_2, c_2 and d_2 have almost no wear; In oscillation experiment 2 (210 rpm, 3.5 Hz, 66 min), compared with the control sample a_3, it can be found that the wear of sample b_3 was very small; The wear of samples c_3 and d_3 were larger than that of b_3 but smaller than that of sample a_3; In oscillation test 3 (240 rpm, 4.0 Hz, 60 min), with the continuous increase of amplitude, the surface of the four soil column samples showed obvious wear. However, it can be clearly seen that the wear degree of soil column samples b_4, c_4, and d_4 consolidated with glue was less than that of the control sample a_4. It can be seen that the KGM/CA/PVA ternary blend film formed on the surface of the consolidated soil column has a good protective effect on the soil column in the process of transportation oscillation.

Table 2. Details of Consolidated Soil Column Samples.

Sample No.	Preparation Conditions of Glue Solution
Sample a_1–a_4	Blank control
Sample b_1–b_4	50 °C, KGM, CA and PVA(4.5%, 4%, 4%), pH4.5
Sample c_1–c_4	60 °C, KGM, CA and PVA(4.5%, 4%, 3%), pH 4.5
Sample d_1–d_4	70 °C, KGM, CA and PVA(4.5%, 3%, 4%), pH4.5

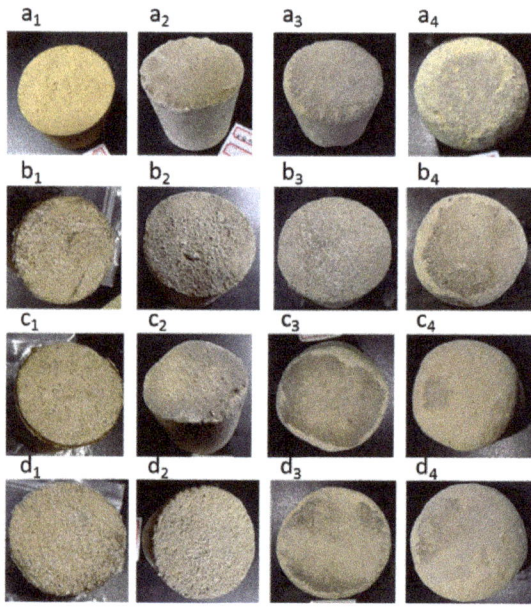

Figure 6. Wear condition of consolidated soil column samples. (**a1–a4**) was the blank control group; (**a1,b1,c1,d1**) formed the control group without any oscillation test; The preparation conditions of (**b1–b4**) spraying KGM/CA/PVA ternary blend adhesive were 50 °C, the contents of KGM, CA and PVA are 4.5%, 4%, and 4%, respectively, the pH value is 4.5; The preparation conditions of (**c1–c4**) spraying KGM/CA/PVA ternary blend adhesive were 60 °C, the contents of KGM, CA and PVA are 4.5%, 4%, and 3%, respectively, the pH value is 4.5; The preparation conditions of (**d1–d4**) spraying KGM/CA/PVA ternary blend adhesive were 70 °C, the contents of KGM, CA and PVA were 4.5%, 3%, and 4% respectively, the pH value was 4.5.

3.6. Preliminary Application of Polymer Soil Consolidation Agent in Seedling Transplanting

In order to verify that the KGM/CA/PVA ternary blend adhesive can indeed consolidate the loam soil ball at the root of seedlings, we conducted a preliminary application study on the prepared polymer soil consolidation agent in the actual transplanting process of seedlings. Firstly, during the transplanting process of sierra salvia and Euonymus japonicas, we sprayed the KGM/CA/PVA ternary blend adhesive on the soil ball at the root of sierra salvia or Euonymus japonicas. After the blend adhesive dried, a layer of consolidated film formed on the surface of the soil ball. At this time, the plant sierra salvia or Euonymus japonicas can be transplanted directly with the soil ball, as shown in Figure 7a,b. It was found that the blend adhesive can consolidate the soil ball with a diameter of about 20 cm. After observation, the consolidated glue had no adverse effect on the growth of sierra salvia and Euonymus japonicas. It can be concluded that the KGM/CA/PVA ternary blend adhesive prepared by us has a good consolidation effect on loam and brings convenience to the process of seedling transplantation.

Figure 7. Preliminary application of KGM/CA/PVA ternary blend in seedling transplanting. (**a**) The diameter at breast height (DBH) of sierra salvia was 5 mm; (**b**) The DBH of Euonymus japonicas was 12 mm; (**c**) The DBH of Juniperus Sabina was 5 mm; (**d**) The DBH of Euonymus japonicas was 8 mm.

In order to further verify the consolidation effect of the KGM/CA/PVA ternary blend adhesive on sandy soil, we sprayed the adhesive on the surface of the sandy soil ball during the transplanting of Juniperus sabina. After drying, a layer of consolidation glue film was formed on the surface of the sandy soil ball, as shown in Figure 7c. At this time, you can directly carry the Juniperus Sabina seedlings and soil balls to transplant Juniperus Sabina to the planting site. To further verify the consolidation effect of the glue on sandy soil, we sprayed the glue on the sandy soil ball with a diameter of about 15–30 cm at the root of Euonymus japonicus. After drying, the glue formed a relatively hard transparent polymer glue film on the surface of the sandy soil ball. Due to the protective effect of the adhesive film, Euonymus japonicus can be directly transported to the planting site with consolidated soil balls, as shown in Figure 7d. After transplanting Euonymus japonicus, it was found that it grew well. It can be seen that the KGM/CA/PVA ternary blend adhesive also has a good consolidation effect on sandy soil balls, and has no adverse effect on the growth of seedlings.

4. Discussion

Because the molecular structure of chitosan contains active groups such as amino, hydroxyl, acetylamino, and electronic pyran ring, it can undergo chemical reactions such as hydrolysis, acylation, carboxymethylation, condensation, and complexation under specific conditions, and can produce derivatives with various physical and chemical functions. Therefore, CA plays an important role in KGM/CA/PVA ternary blend adhesive. However, due to its high value, in order to improve the fluidity of the adhesive and reduce the preparation cost of the adhesive, it is necessary to study the effect of CA content on the consolidation properties of the KGM/CA/PVA ternary blend adhesive. During the study, it was found that the viscosity of the adhesive increased with the increase of CA content at different temperatures. Especially at 60 °C, the viscosity of the adhesive reached a maximum. Therefore, it is necessary to screen the appropriate CA content in combination with the consolidation performance of the consolidated adhesive film on the soil ball.

For the prepared film, it was found that at the same temperature, with the increase of CA content, the film-forming property and toughness of the KGM/CA/PVA ternary blend adhesive were significantly improved. When the temperature was higher than 50 °C, the bending performance of the prepared adhesive film was better as a whole. The SEM micrograph of the adhesive film showed that when the temperature was lower than 50 °C, the links between the cured products on the surface of the adhesive film were weak, the pores formed were large, and there was no cross-linking between the molecular chains, so the adhesive film was brittle. At 70 °C, the surface of the film was densely arranged, and there were many short protrusions on the surface connected to a network structure. It can be seen that the consolidation performance of the film was better when the preparation temperature of the adhesive was higher.

After the KGM/CA/PVA ternary blend adhesive was consolidated on the surface of the soil column, the consolidated film on the surface of the soil column was observed. It was proved again that the film-forming performance of the consolidated adhesive film of the soil column was better with the increase of CA content. Therefore, its bubbles can gradually decrease under the same humidity conditions. When the CA content of the KGM/CA/PVA ternary blend adhesive was the same, the bubbles on the surface of the consolidated film also showed a gradually decreasing trend with the increase of the preparation temperature of the solution. This was because, with the increase in temperature, CA dissolved more thoroughly in the KGM/CA/PVA ternary blend adhesive. CA formed a short cross-linking network structure after blending with KGM and PVA. The film-forming performance of the KGM/CA/PVA ternary blend adhesive was better.

In the test of the compressive strength of the consolidated soil column, it was found that the compressive strength of the consolidated soil column can decrease with the decrease of CA content at the same temperature. At the same time, it was also found that the compressive strength of the consolidated soil column was higher with the increase in adhesive preparation temperature. It was proved again that when the preparation temperature of the adhesive was high, CA was completely dissolved in the adhesive, and blended with KGM and PVA to form a temporary cross-linked network structure, which had a good protective effect on the consolidated soil column. In the study of anti-transport oscillation of consolidated soil column, it was found that compared with the soil column not consolidated with adhesive film, the surface wear of the consolidated soil column was less when the surface of the soil column was protected by adhesive film.

During the exploration of seedling transplanting application, it was found that the KGM/CA/PVA ternary blend adhesive had a good consolidation effect on loam and sandy soil. With the increase of soil ball diameter and the change of soil texture, the requirements for the properties of ternary blend adhesive were slightly different. Therefore, KGM/CA/PVA ternary blend adhesive with different consolidation properties can be prepared according to the actual needs of seedling transplantation.

5. Conclusions

This work was performed primarily to study the influence of the preparation conditions of KGM/CA/PVA ternary blend adhesive on the consolidation performance of soil consolidation agents. Its main purpose was to study the effects of the preparation temperature of the glue solution and the content of CA and PVA on the viscosity of the adhesive, the film-forming property of the ternary blend adhesive, the structure of the consolidated film, and its influence on the consolidation performance of the soil column. It was found that the viscosity of the KGM/CA/PVA ternary blend adhesive increased with the increase of temperature and CA content. However, the high viscosity of the adhesive can lead to the decline of the flow performance of the adhesive and then result in the decline of the film-forming performance of the KGM/CA/PVA ternary blend adhesive. Through this study, it was found that the film-forming performance of the adhesive was better when the calcium content was 2–3% and the PVA content was 3–4%. When the preparation temperature was higher than 50 °C, not only the film-forming performance of KGM/CA/PVA adhesive, the brittleness, the moisture, and air permeability of KGM/CA/PVA adhesive film were significantly improved, but also the protection of the adhesive film to the consolidated soil ball (soil column) was enhanced.

In the process of seedling transplanting, it was found that the KGM/CA/PVA ternary blend adhesive had a good consolidation effect on the loam and sandy soil at the root of transplanted seedlings. In the practical application of seedling transplanting, due to the need to consider the preparation cost of the KGM/CA/PVA ternary blend adhesive, while ensuring the soil consolidation effect of the KGM/CA/PVA consolidation agent, it was particularly necessary to consider reducing the content of CA and the preparation temperature of the ternary blend adhesive, so as to achieve the purpose of reducing the preparation energy consumption and cost. Therefore, in the process of practical application, it is necessary to select the polymer soil consolidation agent with the appropriate performance according to the size of the soil ball at the root of seedlings and the texture of the soil.

Author Contributions: Conceptualization, S.S., L.D. and L.L.; methodology, S.W. (Shaoli Wang) and Y.X.; software, S.W. (Shaoli Wang) and Z.X.; validation, Z.X., C.L., D.W. and H.W.; formal analysis, X.Y. and Y.X.; investigation, Z.X. and C.L.; resources, S.W. (Shaoli Wang), X.Y. and Y.X.; data curation, S.S., H.W. and L.S.; writing—original draft preparation, S.W. (Shaoli Wang) and S.W. (Shaobo Wang); writing—review and editing, S.S. and L.S.; visualization, S.W. (Shaoli Wang) and S.W. (Shaobo Wang); supervision, S.S. and S.L.; project administration, S.W. (Shaoli Wang); funding acquisition, S.L. All authors have read and agreed to the published version of the manuscript.

Funding: This research was funded by the Chinese Academy of Forestry-Special funds for basic scientific research service expenses of the central level public welfare research institutes [CAFYBB2020QD001].

Institutional Review Board Statement: Not applicable.

Informed Consent Statement: Not applicable.

Data Availability Statement: This will be made available upon request through the corresponding author.

Acknowledgments: Jiasong He encouraged and enlightened us with his profound knowledge and professional spirit so that the project was successfully completed. Therefore, we express our sincere gratitude. Furthermore, associate researcher Yong Zhong at the Research Institute of Wood Industry, Chinese Academy of Forestry is greatly acknowledged for his support in conducting the compression test of the consolidated soil columns.

Conflicts of Interest: The authors declare no conflict of interest.

References

1. Arnold, M.A. Challenges and benefits of transplanting large trees: An introduction to the workshop. *Hort Technol.* **2005**, *15*, 115–117. [CrossRef]
2. Sun, Y. Big Tree Transplanting Techniques in Urban Road Renovation: A Case Study of Big Tree Transplanting in Urban Road Renovation of Jinchang City, Gansu Province. *J. Landsc. Res.* **2015**, *7*, 1–2.
3. Wang, S.; Yang, X.; Luo, C.; Xiong, Z.; Wang, S.B.; Song, S.; Liu, L.; Li, S. Effects of Polymer Soil Consolidation Agent on Soil Consolidation Performance and Its Application in Seedling Transplantation. *Sci. Silvae Sin.* **2022**, *58*, 151–161.
4. Wang, S.; Wang, R.; Sun, L.; Wang, H.; Xin, X.; Wei, D.; Yang, X. A Degradable Soil Consolidation Agent and Its Preparation Method and Transplantation Method of Plant Seedlings. Chinese Patent CN 108774530 A, 9 November 2018.
5. Wang, S.; Wei, D.; Yang, X.; Song, S.; Sun, L.; Xin, X.; Zheng, G.; Wang, R.; Liu, L.; Sun, J.; et al. Study on the effects of polymer-type soil consolidation agent for seedling transplantation. *Sci. Rep.* **2021**, *11*, 5575. [CrossRef] [PubMed]
6. Chua, M.; Baldwin, T.C.; Hocking, T.J.; Chan, K. Traditional uses and potential health benefits of Amorphophallus konjac K. Koch ex N.E.Br. *J. Ethnopharmacol.* **2010**, *128*, 268–278. [CrossRef]

7. Katsuraya, K.; Okuyama, K.; Hatanaka, K.; Oshima, K.; Sato, T.; Matsuzaki, K. Constitution of konjac glucomannan: Chemical analysis and 13C NMR spec-troscopy. *Carbohydr. Polym.* **2003**, *53*, 183–189. [CrossRef]
8. Agoub, A.A.; Smith, A.M.; Giannouli, P.; Richardson, R.K.; Morris, E.R. "Melt-in-the-mouth" gels from mixtures of xanthan and konjac glucomannan under acidic conditions: A rheological and calorimetric study of the mechanism of synergistic gelation. *Carbohydr. Polym.* **2007**, *69*, 713–724. [CrossRef]
9. Fang, W.X.; Wu, P.W. Variations of Konjac glucomannan (KGM) from Amorphophallus konjac and its refined powder in China. *Food Hydrocoll.* **2004**, *18*, 167–170. [CrossRef]
10. Li, B.; Xie, B.J.; Kennedy, J.F. Studies on the molecular chain morphology of konjac glucomannan. *Carbohydr. Polym.* **2006**, *64*, 510–515. [CrossRef]
11. Cescutti, P.; Campa, C.; Delben, F.; Rizzo, R. Structure of the oligomers obtained by enzymatic hydrolysis of the glucomannan produced by the plant Amorphophallus konjac. *Carbohydr. Res.* **2002**, *337*, 2505–2511. [CrossRef]
12. Dave, V.; Sheth, M.; McCarthy, S.P.; Ratto, J.A.; Kaplan, D.L. Liquid crystalline, rheological and thermal properties of konjac glucomannan. *Polymer* **1998**, *39*, 1139–1148. [CrossRef]
13. Yeh, S.L.; Lin, M.S.; Chen, H.L. Partial hydrolysis enhances the inhibitory effects of konjac glucomannan from Amorphophallus konjac C. Koch on DNA damage induced by fecal water in Caco-2 cells. *Food Chem.* **2010**, *119*, 614–618. [CrossRef]
14. Xu, C.G.; Luo, X.G.; Lin, X.Y.; Zhuo, X.R.; Liang, L.L. Preparation and characterization of polylactide/thermoplastic konjac glucomannan blends. *Polymer* **2009**, *50*, 3698–3705. [CrossRef]
15. Chen, Z.G.; Zong, M.H.; Li, G.J. Lipase-catalyzed acylation of konjac glucomannan in organic media. *Process Biochem.* **2006**, *41*, 1514–1520. [CrossRef]
16. Alvarez-Manceñido, F.; Mariana, L.; Martínez-Pacheco, R. Konjac gluco-mannan/xanthan gum enzyme sensitive binary mixtures for colonic drug delivery. *Eur. J. Pharm. Biopharm.* **2008**, *69*, 573–581. [CrossRef]
17. Luo, X.; He, P.; Lin, X. The mechanism of sodium hydroxide solution promoting the gelation of Konjac glucomannan (KGM). *Food Hydrocoll.* **2013**, *30*, 92–99. [CrossRef]
18. Zhu, F. Modifications of konjac glucomannan for diverse applications. *Food Chem.* **2018**, *256*, 419–426. [CrossRef]
19. Chen, J.; Liu, C.; Chen, Y.; Chen, Y.; Chang, P.R. Structural characterization and properties of starch/konjac glucomannan blend films. *Carbohydr. Polym.* **2008**, *74*, 946–952. [CrossRef]
20. Cheng, L.H.; Abd Karim, A.; Norziah, M.H.; Seow, C.C. Modification of the microstructural and physical properties of konjac glucomannan-based films by alkali and sodium carboxymethylcellulose. *Food Res. Int.* **2002**, *35*, 829–836. [CrossRef]
21. Li, B.; Kennedy, J.F.; Peng, J.L.; Yie, X.; Xie, B.J. Preparation and performance evaluation of glucomannan–chitosan–nisin ternary antimicrobial blend film. *Carbohydr. Polym.* **2006**, *65*, 488–494. [CrossRef]
22. Qiao, D.; Shi, W.; Luo, M.; Jiang, F.; Zhang, B. Polyvinyl alcohol inclusion can optimize the sol-gel, mechanical and hydrophobic features of agar/konjac glucomannan system. *Carbohydr. Polym.* **2022**, *277*, 118879. [CrossRef] [PubMed]
23. Gu, R.; Mu, B.; Yang, Y. Bond performance and structural characterization of polysaccharide wood adhesive made from Konjac glucomannan/chitosan/polyvinyl alcohol. *BioResources* **2016**, *11*, 8166–8177. [CrossRef]
24. Ma, Q.; Du, L.; Yang, Y.; Wang, L. Rheology of film-forming solutions and physical properties of tara gum film reinforced with polyvinyl alcohol (PVA). *Food Hydrocoll.* **2017**, *63*, 677–684. [CrossRef]
25. Abral, H.; Atmajaya, A.; Mahardika, M.; Hafizulhaq, F.; Handayani, D.; Sapuan, S.M.; Ilyas, R.A. Effect of ultrasonication duration of polyvinyl alcohol (PVA) gel on characterizations of PVA film. *J. Mater. Res. Technol.* **2020**, *9*, 2477–2486. [CrossRef]
26. Wang, W.; Meng, Q.; Li, Q.; Liu, J.; Zhou, M.; Jin, Z.; Zhao, K. Chitosan derivatives and their application in biomedicine. *Int. J. Mol. Sci.* **2020**, *21*, 487. [CrossRef] [PubMed]
27. Mourya, V.K.; Inamdar, N.N. Chitosan-modifications and applications: Opportunities galore. *React. Funct. Polym.* **2008**, *68*, 1013–1051. [CrossRef]
28. Bakshi, P.S.; Selvakumar, D.; Kadirvelu, K.; Kumar, N.S. Chitosan as an environment friendly biomaterial—A review on recent modifications and applications. *Int. J. Biol. Macromol.* **2020**, *150*, 1072–1083. [CrossRef]

Article

Influence of Resin Content and Density on Water Resistance of Bamboo Scrimber Composite from a Bonding Interface Structure Perspective

Yaohui Ji, Wencheng Lei, Yuxiang Huang , Jiangyuan Wu and Wenji Yu *

Research Institute of Wood Industry, Chinese Academy of Forestry, Beijing 100091, China; jiyaohui1994@gmail.com (Y.J.); leiwenc@163.com (W.L.); yxhuang@caf.ac.cn (Y.H.); wjy125616214@163.com (J.W.)
* Correspondence: chinayuwj@126.com

Abstract: As a new type of green environmental protection material for outdoor use, the water resistance of bamboo scrimber composite (BSC) is crucial—the primary reason for a decrease in water resistance being bonding interface failure. From a bonding interface structure perspective, the influence mechanism of the resin content and density on the water resistance of BSCs remains unknown. Therefore, in this study, BSCs were prepared using Moso bamboo and phenol-formaldehyde resin, and the changes in the macroscopic and microscopic bonding interfaces before and after 28-h water-resistance tests were observed and analyzed. The results showed that the water resistance of the BSC increased with increasing resin content, with higher thickness swelling rates (TSRs) observed at higher densities. Obvious cracks were found at the macroscopic interface after 28-h tests, with higher resin contents leading to fewer and smaller cracks. With increasing density, the longitudinal fissures due to defibering process decreased, having an effect on width swelling rates (WSRs). Furthermore, porosity measurements revealed changes in the microscopic bonding interface; the difference in porosity before and after testing (D-value) showed the same trend as water resistance. Generally, we conclude that the macroscopic and microscopic bonding interface structures are closely related to BSC water resistance.

Keywords: water resistance; bamboo scrimber composite; bonding interface

Citation: Ji, Y.; Lei, W.; Huang, Y.; Wu, J.; Yu, W. Influence of Resin Content and Density on Water Resistance of Bamboo Scrimber Composite from a Bonding Interface Structure Perspective. *Polymers* **2022**, *14*, 1856. https://doi.org/10.3390/polym14091856

Academic Editor: Beom Soo Kim

Received: 11 April 2022
Accepted: 29 April 2022
Published: 30 April 2022

Publisher's Note: MDPI stays neutral with regard to jurisdictional claims in published maps and institutional affiliations.

Copyright: © 2022 by the authors. Licensee MDPI, Basel, Switzerland. This article is an open access article distributed under the terms and conditions of the Creative Commons Attribution (CC BY) license (https://creativecommons.org/licenses/by/4.0/).

1. Introduction

With the continued growth in green buildings and sustainable urban construction, it has become necessary to develop green, renewable biomaterials to replace conventional structural materials—which are usually high density or polluting materials—such as metals, ceramics, and polymer composites [1–4]. Moso bamboo (*Phyllostachys pubescens* Mazel) accounts for 67% of all bamboo forests in China, and is therefore used more extensively than other bamboo species [5]. The findings of previous studies have shown that Moso bamboo fibers offer superior mechanical properties compared to wood fibers [6]. To fully exploit high-strength bamboo fibers, a novel strategy to use a Moso bamboo scaffold as a matrix reinforced by phenol-formaldehyde (PF) resin has been developed. Bamboo scrimber composite (BSC)—a renewable, engineered material—consists of crushed bamboo bundles assembled along the grain direction after PF resin immersion and pressed into a dense block. With its multi-level and multi-scale structure, BSC has superb physicomechanical performance, high utilization ratios (over 90%), excellent carbon sequestration, and is widely used in various fields, including furniture, flooring, garden landscaping, construction, and wind turbine blades [7–10]. It also overcomes the defects of anisotropy and the inhomogeneous density of natural bamboo.

In recent years, most studies of BSCs have focused on their physical and mechanical properties. Kumar et al. investigated the effects of density on the mechanical and water absorption properties of BSCs prepared via cold-pressing and hot-curing. They found that the water absorption and all mechanical properties of BSCs—such as tensile strength, compression strength, shear strength, and flexural strength—increased with increasing density [11]. Sharma et al. compared the mechanical properties of two types of BSCs and laminated bamboo sheets [12]. Subsequently, Shangguan et al. investigated the mechanical properties and chemical composition of heat-treated BSCs [13]. Furthermore, Rao et al. found that, as the resin content increased, the water resistance of the BSCs increased, shear strength increased at first and then decreased, and compressive strength decreased [5]. Rao et al. also investigated the effects of different units and resin molecular weights on the mechanical performance and water resistance [14,15]. However, most reports have included only superficial discussions of the water resistance of BSCs, and few studies have been conducted on its influence mechanism.

Although qualified BSCs can be produced by most enterprises, some problems caused by poor water resistance still exist, such as cracking, fuzzing, and deformation under high-humidity conditions. Resin content and density are two vital factors that determine dimensional stability and water repellency [16]. As the resin content increases, the water resistance of composites increases gradually [17–19]. Rao et al. demonstrated the optimum resin content of BSC to be 20% [5]. Based on previous studies [17,20], density has a major influence on the water absorption properties of composites. For example, the mechanical properties increased with increasing density [21], whereas the water absorption was reduced [11,22]. Research on the water resistance of BSCs has focused primarily on performance but has not established the essential influences, especially on the bonding interface changes.

Improvement in water resistance can be achieved by ameliorating the compatibility of the bonding interface [23]. The fundamental reason for the decline in dimensional stability is the interface failure caused by uneven adhesive penetration and poor bonding [17,24–26]. Macroscopic interface failure refers to interlaminar cracking, which can be caused by weak gluing between two fluffed bamboo mats (FBMs). Microscopic interface failure refers to the springback and swelling deformation of the compressed cell cavity in a humid environment at a position with less effective bonding. Therefore, it is of vital importance to understand how cooperative coupling between the macroscopic and microscopic structures at the bonding interface affect the water resistance of BSCs. Meng et al. investigated the bonding interface structure via fluorescence microscopy, scanning electron microscopy coupled with energy-dispersive X-ray spectroscopy (SEM-EDX), and transmission electron microscopy (TEM). The results suggested that cured PF resin not only entered the cell lumina but also was deposited on the bamboo cell walls surface [23]. Yu et al. also confirmed that the cells in BSC were filled with PF resin through SEM and TEM observation [17]. However, few studies have been conducted on the response relationship between the bonding interface failure and water resistance of BSCs.

In addition, it is worth mentioning that researchers usually choose to increase the concentrations of PF resin (appropriately) to solve the problem that the FBMs cannot absorb adequate, predetermined adhesive volumes because of complete saturation during the impregnation process, which can result in experimental errors due to differences in the permeabilities of PF resins with inconsistent resin solid concentrations.

In this study, PF resin with the same solid concentration was used, with the procedure from dipping to drying being repeated several times until the stated dose had been reached. We examined the influence of the resin content and density on the water resistance of BSCs from an interface failure perspective. Furthermore, the macroscale and microscale bonding interface structures of untreated samples under different preparation conditions were observed using laser scanning microscopy (LSM) and SEM. The macroscopic interface failure morphology was observed after water-resistance tests, using ultra-depth electron microscopy (UDEM), the microscopic interface failure being reflected by changes in the porosity and pore-diameter distribution before and after the test.

2. Material and Methods

2.1. Materials

Four-year-old Moso bamboo (*Phyllostachys pubescens* Mazel) was obtained from Huzhou, Zhejiang Province, China. PF resin of 47.5 wt%, the viscosity of 40 cps, Mw of 919, and pH of 10.04 was supplied by the Beijing Dynea Chemical Industry Co., Ltd., Beijing, China. Safranine T (Macklin) and toluidine blue O (Macklin) were used as received.

2.2. Preparation of BSC

The method used to produce the BSC is shown in Figure 1. The bamboo was defibered using a crushing machine to obtain FBMs with a series of uniform linear cracks along the grain direction (Figure 1a). Bamboo is divided into three parts in the enlarged schematic diagram of a bamboo section: bamboo epidermis (BE), bamboo middle (BM), and bamboo pit ring (BPR), and some parts of the BS and BPR are removed during defibering. The side close to the BS in an FBM is referred to as "the outer surface", while that close to the BPR is "the inner surface". The FBMs were then dipped into the PF resin, which was diluted into 15 wt% solutions by adding distilled water (Figure 1b). After air-drying to 10% MC, the FBMs with target resin contents were assembled into the mold along the grain based on the desired density, as summarized in Table 1. After pressing in a hot-pressing machine (Carver 3925, CARVER Inc., Wabash, IL, USA) at 150 °C for 10 min and unloaded while cooling at 30–40 °C, BSCs of dimensions 300 × 120 × 10 mm (longitudinal × width × thickness) were obtained (Figure 1c,d). We screened the samples with density ranges less than ±0.02 g/cm^3 from the set value and considered them to have the same density gradient. Finally, all BSCs were conditioned in a chamber at 20 ± 2 °C and relative humidity (RH) of 65 ± 5% for 2 weeks before being tested.

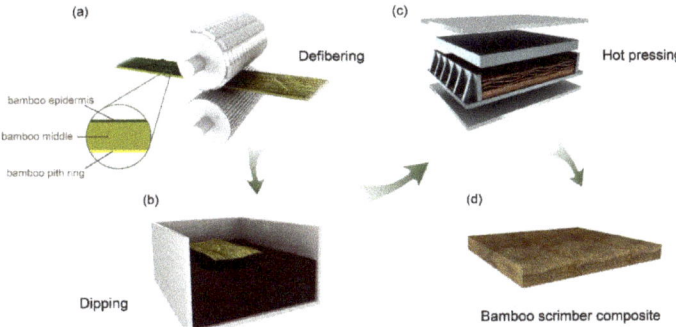

Figure 1. Schematic illustration of BSC preparation. Note: (**a**) Defibering; (**b**) Dipping into the PF resin; (**c**) Hot pressing; (**d**) BSC.

Table 1. BSC preparation conditions.

Sample	Preparation Conditions	
	Resin Content (%)	Density (g/cm^3)
BSC-10-1.0	10	1.00
BSC-10-1.15	10	1.15
BSC-10-1.3	10	1.30
BSC-15-1.0	15	1.00
BSC-15-1.15	15	1.15
BSC-15-1.3	15	1.30
BSC-20-1.0	20	1.00
BSC-20-1.15	20	1.15
BSC-20-1.3	20	1.30

2.3. Characterizations

2.3.1. Water Resistance

The water absorption rate (WAR), width swelling rate (WSR), and thickness swelling rate (TSR) of the BSC samples were determined based on the GB/T 40247-2021 standard. The 28-h cycle hydrothermal treatment included 4 h of immersion in boiling water at $100 \pm 2\,°C$, 20 h of oven-drying at $63 \pm 3\,°C$, and another 4 h of immersion in boiling water at $100 \pm 2\,°C$. The sample surface was placed parallel to the horizontal plane in a water bath kettle. Four samples of dimensions $50 \times 50 \times 10$ mm^3 were tested in each case. Then, the weight and size changes were measured after carefully removing any excess water from the specimen surfaces. The WAR, WSR, and TSR were calculated using Equations (1)–(3):

$$\text{WAR}(\%) = \frac{m_1 - m_0}{m_0} \times 100 \tag{1}$$

$$\text{WSR}(\%) = \frac{b_1 - b_0}{b_0} \times 100 \tag{2}$$

$$\text{TSR}(\%) = \frac{t_1 - t_0}{t_0} \times 100 \tag{3}$$

where m_0, b_0, and t_0 are the weight, width, and thickness of the pristine samples, respectively, and m_1, b_1, and t_1 are the weight, width, and thickness of the 28-h cycle-treated samples, respectively.

Based on the experimental data of the water-resistance tests, the optimum process conditions of BSCs—resin content: 20%, density: 1.15 g/cm^3 (vide infra)—were selected to determine the value of the other fixed variable when one of the variables changes. Therefore, BSC samples obtained under five conditions with different resin contents and densities, namely BSC-10-1.15, BSC-15-1.15, BSC-20-1.0, BSC-20-1.15, and BSC-20-1.3, where "BSC-x-y" refers to a resin content of x% and a density of y g/cm^3, were prepared for the following tests.

2.3.2. Morphological Characterizations

Cross-sections of BSC samples $10 \times 10 \times 10$ mm in size, prepared using the five above-mentioned process conditions, were cut with a slicer (RM2245, LEICA, Wetzlar, Germany). To analyze the distribution of the PF resin through BSCs of different resin contents, LSM (LSM 980 with Airyscan 2, Zeiss, Oberkochen, Germany) was performed. Specimens were stained with 0.5% toluidine blue O prior to LSM testing to suppress the auto-fluorescence of the bamboo lignin. The cell deformation of the BSCs at different densities was examined using SEM (SU8020, Hitachi, Ltd., Tokyo, Japan). The failure morphologies of the cross-sectional surfaces at the macroscopic bonding interfaces of the BSC samples were observed by UDEM (VHX-6000, Keyence, Osaka, Japan) in transmission mode following completion of the 28-h cycle treatment.

2.3.3. Mercury Intrusion Porosimetry (MIP)

MIP testing was performed using a Mercury Porosimeter (AutoPore V 9600, Micrometrics Inc., Norcross, GA, USA) in the 0.0007–420.5950 Mpa pressure range. Because mercury is a non-wetting fluid—which means it cannot penetrate a porous solid through capillary forces—we approximated bamboo as a bundle of small cylindrical capillaries to simplify the calculations [27]. The pore volume was obtained based on the quantity of mercury intruded; hence, the total internal volume could be calculated. In addition, based on the Washburn equation [28], the pore radius (r), porosity (φ), and aperture size distribution ($D(r)$) can be determined as follows:

$$r = -\frac{2\gamma \cdot \cos\theta}{P} \tag{4}$$

$$\varphi = \frac{V_T}{V_S} \tag{5}$$

$$D(r) = \frac{P}{r}\frac{dV}{dP} \tag{6}$$

where P is the measured pressure, γ is the surface tension of mercury (0.48 N/m), $\theta = 140°$ is the wetting angle of mercury, V_T is the total volume of mercury, and V_S is the total sample volume.

In addition, in order to compare the change in porosity before and after the water resistance test, the porosity difference is defined as D-value (D), which is calculated by the following equation:

$$D = \varphi_1 - \varphi_0 \tag{7}$$

where φ_0 is the porosity of untreated BSC and φ_0 is the porosity of BSC after 28-cycle treatment, respectively.

2.4. Statistical Analysis

To detect the difference in water resistance among the BSC samples prepared using different resin contents and densities, analysis of variance (ANOVA) was performed using SPSS (IBM SPSS software version 24, SPSS Inc., Chicago, IL, USA)—F being the outcome of the F test, which equals the ratio of the mean square between groups and within groups. Additionally, p can determine the significance of differences and represent probabilities under the corresponding F values, while defining 5% as the significance level ($p < 0.05$).

3. Results and Discussion

3.1. Dimensional Stability

Resin content and density are two main factors that affect BSC water resistance. Figure 2 shows the WSR, TSR, and WAR of the BSCs after 28-h cycle treatment. The WSR and WAR of BSC-20-1.3 were significantly lower than those of the other samples, which was the minimum value among the BSCs in this study. For BSC-20-1.0, the TSR was as low as 6.68%, which was the lowest value among the BSCs in this study. The high standard deviations of the WSR and WAR for BSC-20-1.0 were due to the high variability among the samples. At the set levels, it had the highest resin content (20%). Correspondingly, the percentage of bamboo in BSC-20-1.0 was the smallest with the same density, so it was more dispersed and had more voids in the lay-up process. The WSR and WAR of BSC-10-1.0 had high standard deviations, indicating that the lower resin content can also lead to higher variables due to uneven permeation of PF resin. Meanwhile, these findings are attributed to the fact that both BSC-10-1.0 and BSC-20-1.0 had the lowest hot-pressing pressure (smallest density) and the resin did not flow uniformly to all parts of the samples; thus, the performance gaps among BSC samples at 1.00 g/cm^3 density were large.

As the resin content of the BSC increased, the WSR, TSR, and WAR gradually decreased during the water-soaking test. This is attributable to the fact that water can penetrate bamboo parenchymal cells, vessels, or intercellular spaces, and be absorbed through hydrogen bonds due to the many free hydroxyl groups in the bamboo cell wall [8,29]. Moreover, the impregnation of the PF resin increased after the defibering process. PF resin—which can penetrate the cell lumens, swell the cell wall, and form a rigid cross-linked hydrophobic network during curing—helps to improve the dimensional stability of bamboo by blocking the impregnation path of water [7,30–32]. Therefore, the ability of the network system to effectively prevent the hydroxyl groups of bamboo from interacting with water molecules is strengthened with an increase in the resin content.

With a density of 1.00 g/cm^3, the WSR, TSR, and WAR of BSC-10-1.0 were 2.39, 3.29, and 3.15 times those of BSC-20-1.0, respectively. By contrast, when the density increased to 1.30 g/cm^3, the WSR, TSR, and WAR of BSC-10-1.3 were 1.67, 2.53, and 2.74 times those of BSC-20-1.3, respectively. This phenomenon showed that the higher resin content BSCs had higher water resistance—consequently, 20% was selected as the optimum resin content condition. Moreover, a higher density could decrease the effect of resin content on the water resistance of BSCs. Based on the ANOVA results shown in Table 1, the resin content significantly affected all values of water resistance ($p < 0.05$).

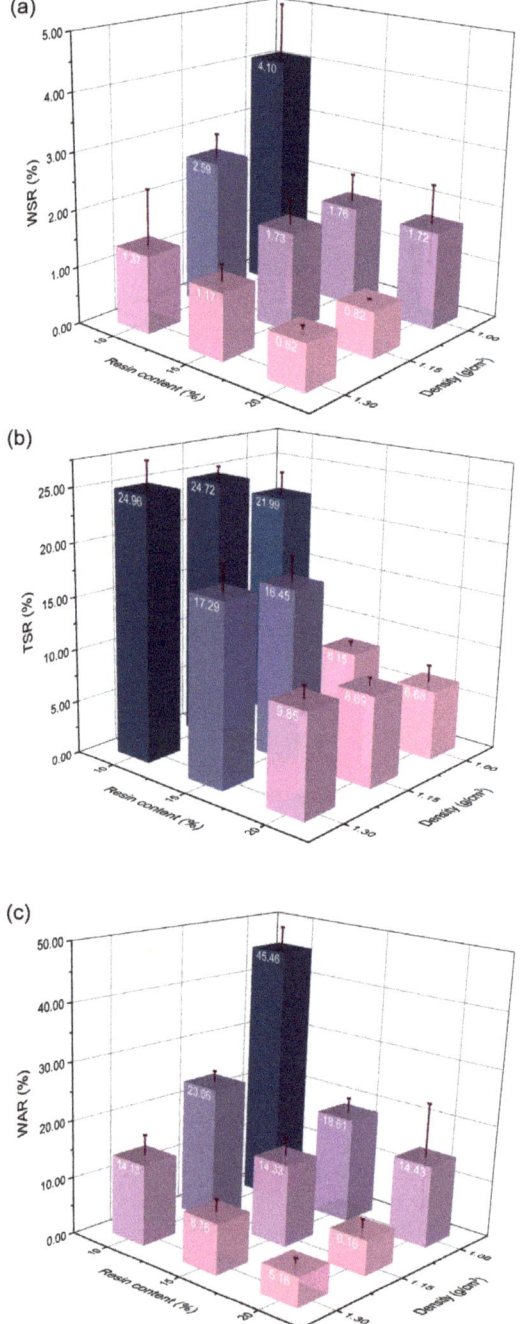

Figure 2. Water resistance of BSC samples including (**a**) WSR, (**b**) TSR, and (**c**) WAR.

The TSR—which represents the swelling of the bamboo cell wall plus the swelling caused by the compressive release—showed an upward trend with increasing BSC density after the water-soaking test (Figure 2b). The hydrothermal cycle treatment was more efficient or faster in inducing the TSR in higher-density BSCs. The TSR of BSC-20-1.3 was 47.46% higher than that of BSC-20-1.0. However, a high density had a positive effect on the WSR and WAR of the BSC. These findings could be ascribed to the closure of the water infiltration path due to the formation of interlocked structures between the dense cells and PF resin under high pressure [22]. Based on the ANOVA analysis (Table 1), the density showed clear effects on the WSR and WAR of BSCs ($p < 0.05$), except for the TSR ($p = 0.291 > 0.05$). At a resin content of 20%, the WSR and WAR of BSC-20-1.3 were 52.33% and 64.24% lower than BSC-20-1.0, respectively.

It should be noted that the water resistance of a material is good only when all three values (WSR, TSR, and WAR) are low. As mentioned above, 20% was chosen to be the optimum resin content. At 20% resin content, the WAR and WSR of BSC-20-1.3 were the lowest and the TSR of BSC-20-1.0 was the smallest. Considering all water resistance indicators, 1.15 g/cm^3 (the middle value) was confirmed to be the best density. Therefore, the best process was that with 20% resin content and 1.15 g/cm^3 density. Besides the samples under the optimum process (BSC-20-1.15), BSC-10-1.15, BSC-15-1.15, BSC-20-1.0, and BSC-20-1.3 were also selected for the subsequent characterization to study the effects of the resin content and density on the water resistance of BSCs. Moreover, the interaction between resin content and density had a significant influence on the WSR, TSR, and WAR ($p < 0.05$), which also revealed that they were not independent factors and had interactive effects on each other (Table 2).

Table 2. Water-resistance data analyzed by ANOVA.

Factor	WSR		TSR		WAR	
	F	p	F	p	F	p
resin content	7.057	0.004	55.425	0.000	8.770	0.001
density	4.881	0.017	1.301	0.291	6.260	0.006
interaction	3.293	0.034	3.912	0.019	9.395	0.000

3.2. Bonding Interface Morphology

Two FBMs, separately dyed with Safranine T and Toluidine Blue O, were hot-pressed to prepare BSCs under the optimum process conditions. As shown in Figure 3, uneven dyeing outlined the shape of the bamboo fiber bundles, where the cracks generated by the defibering process were compressed and cross-linked with the PF resin. In addition, the yellow curve—that is, the boundary between the two FBMs after hot pressing—shows a comb-meshing structure. Compared with other bamboo/wood composites, this bonding structure is unique and stable and provides BSCs with higher strength and superior water resistance.

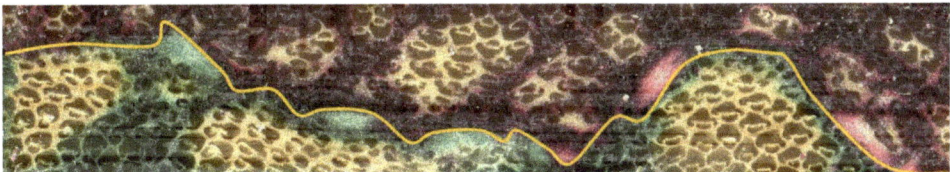

Figure 3. The comb-meshing structure of the macro bonding interface.

LSM can reveal the distribution of the PF resin—hence, BSC-10-1.15, BSC-15-1.15, and BSC-20-1.15 were selected for observation. The laser wavelengths were set to 405 nm (green channel) and 488 nm (red channel).

In Figure 4, green represents lignin, while orange and red represent the PF resins. It is worth mentioning that orange is formed by the superposition of the red and green parts in the dual-channel mode. The yellow arrows refer to the residual parts of the BPR—that is, stone cells. In contrast to other bamboo composites, the macroscopic bonding interface of the BSC was in the form of point-to-plane bonding rather than linear [33,34]. However, when the resin content was less than 15%, the bonding interfaces were spotted structures (Figure 4(a-2,b-2)), with planar structures appearing at resin contents of up to 20% (Figure 4(c-2)). When the resin content increased from 10% to 20%, the area of the red part increased, which meant that the PF resin was better distributed (Figure 4a–c). It was also determined that the penetration of the PF resin could be affected by the resin content.

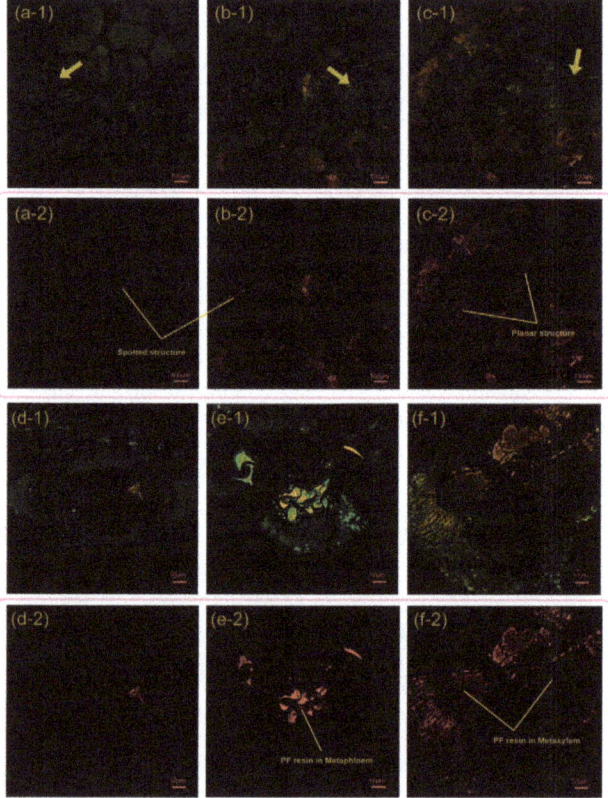

Figure 4. The LSM images of the macroscopic bonding interface and the cured PF resin trapped in the bamboo vascular tissue: (**a,d**) BSC-10-1.15, (**b,e**) BSC-15-1.15, (**c,f**) BSC-20-1.15. Note: 1 and 2 denote dual-channel superposition diagrams and single-channel diagrams, respectively.

Although a few fiber cells were not completely shielded by the toluidine blue O, a general trend could still be observed (Figure 4). From Figure 4d–f, the distribution of PF resin in the vessels reflected the growth of resin content. Cured PF resin filled the metaphloem and metaxylem at 20% resin content (Figure 4(f-2)).

The parenchyma tissues and vessels of the FBMs were densified during hot pressing. As the density increased, the degree of deformation increased, according to the SEM diagrams shown in Figure 5. It could be seen that, when the density was less than 1.15 g/cm^3, more PF resin remained in the lumens. At 1.30 g/cm^3, the cell cavity and vessels were almost closed, causing large internal stress, with less PF resin present—which meant that the binding force of the PF resin on cell rebound would be reduced if the samples were in a hydrothermal environment, resulting in a decline in diameter stability. This inference is consistent with the previous water-resistance test results. In addition, many nanopores with diameters ranging from tens to hundreds of nanometers could be observed in the cured PF resin (Figure 5g–i).

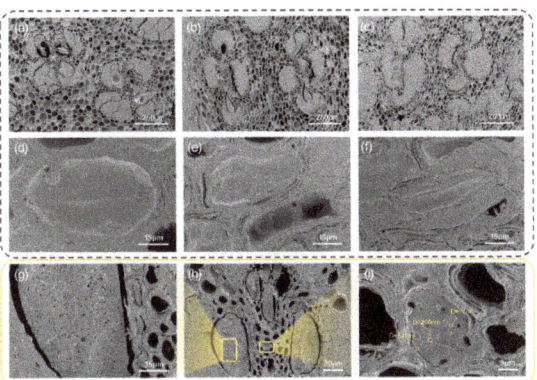

Figure 5. SEM micrographs of BSC-20-1.0 (**a**,**d**), BSC-20-1.15 (**b**,**e**), BSC-20-1.3 (**c**,**f**), and cured PF resin (**g**–**i**). Note: D denotes the diameter of the pore.

3.3. Macroscopic Interface Failure Morphology

The FBM has two sides, namely the outer and inner surfaces, and two BSC assembly patterns (outer-to-inner and inner-to-inner) were observed (Figure 6). Figure 7a,g show the original 2D morphologies of the bamboo fiber bundles in the FBMs. Bamboo has no radial transfer structure that prevents the water-soluble PF resin from entering internal cells [35]. Consequently, a defibering process was adopted to create dot- and linear-shaped fissures that disrupt few fibers—which is not only conducive to the penetration of the adhesive but also maintains the outstanding strength of the natural bamboo. The brittle parts of the BE and BPR fell off easily during the defibering process. In the 3D images, the red-shaded areas correspond to the residual BE waxy layer (Figure 7b) and stone cells in the BPR (Figure 7h) of the bamboo after mechanical fluffing. In addition, the maximum linear-shaped crack depths of the outer and inner surfaces were determined to be 3740.19 and 3916.20 µm, respectively, which were sufficient to open the internal impregnation channels.

Figure 6. Two BSC assembly patterns.

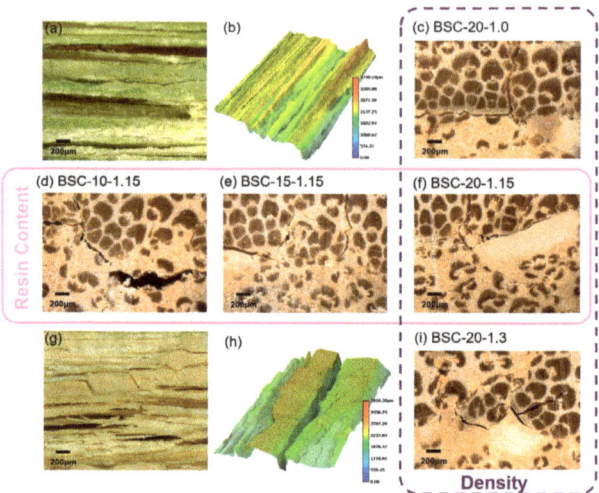

Figure 7. 2D (**a**,**g**) and 3D (**b**,**h**) images of the outer (**a**,**b**) and inner (**g**,**h**) surfaces of the FBM and failure morphologies at the outer-to-inner bonding interfaces of the BSCs (**c**–**f**,**i**) after the 28-h cycle treatment. Note: The pink frame represents BSCs at various resin contents; the purple frame represents BSCs at various densities.

Compared to specimens of different densities (Figure 8a), it can be seen from the photos that changes in the thickness of BSCs with different resin contents before and after the 28-h cycle treatment were more substantial (Figure 8f), indicating that the resin content had a greater impact on the TSR than the density. This was also proven by the previous ANOVA results. Furthermore, the BSC-20-1.3 sample had more cracks in its macroscopic appearance than BSC-20-1.0, after the 28-h cycle treatment (Figure 8a).

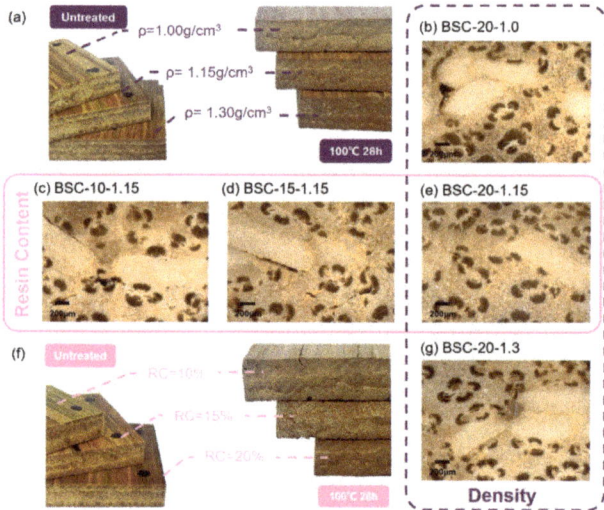

Figure 8. Images of BSC samples at various density (**a**) and resin content (**f**) conditions before and after the water-resistance test and failure morphologies at the inner-to-inner bonding interfaces of the BSCs (**b**–**e**,**g**) after the 28-h cycle treatment. Note: The pink frame represents BSCs at various resin contents; the purple frame represents BSCs at various densities.

To explore the effect of the resin content and density on the water resistance, the failure area of the macroscopic bonding interface of the BSCs was observed along a transverse section. By means of UDEM images, we established that the outer-to-inner bonding interfaces were always damaged after treatment when the density was raised. However, increasing the resin content contributed to clear improvements (Figure 7). Destruction of the macroscopic bonding interface mostly occurred where the BE and BPR remained. The width and number of transverse cracks decreased with increasing resin content (Figure 7d–f). The TSR showed the same trend with changing resin content (Figure 2b). In the case of BSC-20-1.3 (Figure 7i), new thin cracks in the transverse direction appeared after the water-resistance test as a result of the stress stored via high compression, which led to a larger springback [36], but generally speaking, no obvious transverse fracture trend was observed as the density was varied. However, the longitudinal cracks due to the defibering process decreased with an increase in density (Figure 7c,f,i) because the PF resin increased its fluidity under the action of high temperature and high pressure, making it easier to penetrate these large fissures. This also explains why WSR correlates negatively with density, as shown in Figure 2a. There were visible cracks between the inner and inner surfaces of BSC-10-1.15 and BSC-15-1.15, whereas there were no fractures in the others (Figure 8). In general, the thickness modification depends on the bonding strength of the outer-to-inner interface. Overall, at either bonding interface, there was an inverse correlation between the resin content and the number of cracks, indicating that the increase in PF resin adhesive improved the interfacial bonding strength.

3.4. Analysis of Porosity

It was difficult to directly observe the failure morphology of the microscopic bonding interface; therefore, the pore information was measured using MIP tests to indirectly reflect the interface. Prior to analyzing the pore size distribution, the porosity of the BSC samples under different preparation conditions should first be discussed. Figure 9 shows the relationship between the resin content/density and porosity of the BSCs before and after the 28-h cycle treatment. As observed from the data of untreated and treated BSCs, the increase in density led to a decrease in porosity owing to the compression and closure of partial lumens, which was confirmed by previous morphological observations. BSC D-values at various densities (Table 3) were calculated, which revealed that BSC-20-1.3 has the largest D-value, consistent with the previous conclusion about the correlation between density and TSR, which indicated that changes in the microscopic bonding interface caused by the hydrothermal treatment might be the reason for the change in the TSR. Furthermore, the porosity of the untreated samples exhibited a positive correlation with the resin content, whereas the porosity of the treated group was negatively correlated. The D-value of BSC-20-1.15 was considerably less than that of BSC-10-1.15; consequently, the parts of the pores that increased due to growth in resin content did not change much under hydrothermal action. Combined with the above morphological analysis, we speculate that the increased parts of pores are formed via the curing of the PF resin. The D-value of BSC-20-1.15 was the smallest of all samples, consistent with the best conditions selected using the water-resistance test mentioned above. This also confirmed that the failure of the microscopic bonding interface was closely related to water resistance.

Table 3. D-values of BSC samples.

Sample	D-Value/%
BSC-10-1.15	12.3275
BSC-15-1.15	4.9813
BSC-20-1.0	4.0425
BSC-20-1.15	2.6535
BSC-20-1.3	13.4705

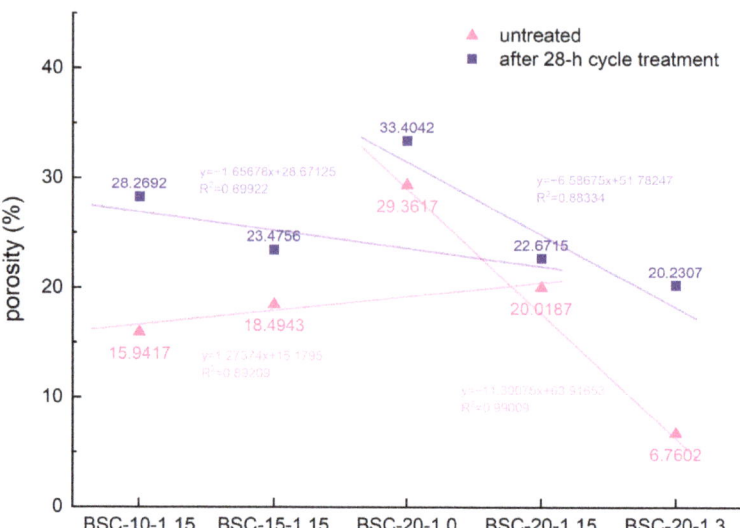

Figure 9. Porosity of the BSC samples before and after 28-h cycle treatment.

Typical MIP curves for the log differential intrusion versus pore size diameter for the BSCs are shown in Figure 10. Based on the peak positions, it can be concluded that the pore distribution in the BSCs could be divided into two parts—that is, pores of diameters of 10–100 nm and those >1000 nm. The bamboo structure consisted of parenchyma, fibers, and vessel cells. These component cells had lumens with diameters in the range of 21–40 μm [37,38] for parenchyma, 0.26–24.96 μm for fibers [39,40], and 40.17–259.91 μm for vessels [21].

Pores with diameters ranging from 10–100 nm could be pit-membrane voids or other small voids [41]. With an increase in the resin content, the number of pores (10–100 nm) in the untreated samples increased (Figure 10a,b,d). After the 28-h cycle treatment, there were fewer nanopores (10–100 nm) of BSC-15-1.15 and BSC-20-1.15 than those of the untreated group, which might have been due to the PF resin filling the void space of the pits. In addition, it also showed that some nanopores produced due to the increase in resin content had excellent hydrophobicity. This might be because of the pores formed by water evaporation in the PF resin during curing, further confirming earlier speculation. In contrast to BSC-20-1.0 and BSC-20-1.3, BSC-20-1.15 exhibited fewer pores (10–100 nm) before treatment than after treatment (Figure 10c–e), suggesting that the PF resin could fully penetrate and fill the nanopores under the proper stress only at a density of 1.15 g/cm^3.

From Figure 10, it can be seen that the 28-h cycle treatment led to an increase in the number of pores of diameters greater than 1000 nm. Micron pores (diameters greater than 1000 nm) contain cavities of various cells, cell gaps, cracks generated by defibering, gaps between FBMs, etc. The micron pores (>1000 nm) of BSC-20-1.15 were smaller than those of BSC-10-1.15. Compared with the pore size distribution before and after the 28-h cycle treatment, the distribution of micron pores (>1000 nm) in the samples after the 28-h cycle treatment decreased as the resin content increased (Figure 10a,b,d)—that is, as the resin content increased, the permeability of the PF resin increased, making it easier for it to enter the larger pores. Meanwhile, with the increase in density, the micron pore distribution of untreated BSC decreased considerably, caused by the higher compaction ratios (Figure 10c–e). Based on the untreated group, the changes in the micron pores in the samples after 28-h of cycle treatment increased as the density increased, revealing that excessive pressure weakened the binding force of the PF on micron pores, which rebounded more during the water-resistance test.

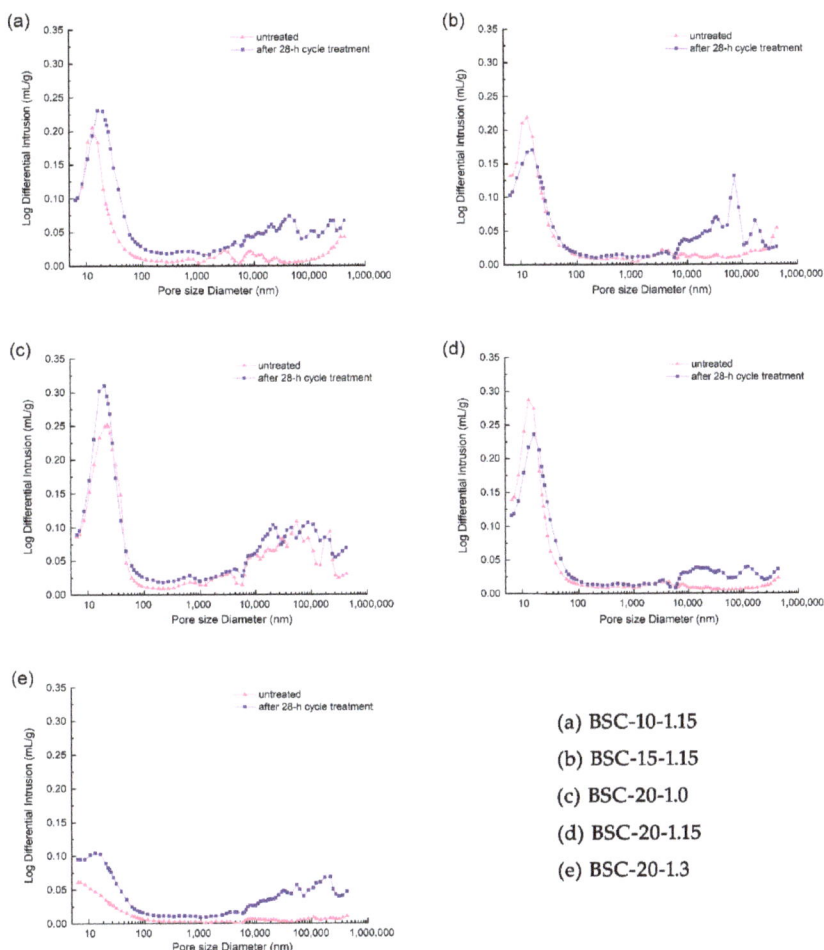

Figure 10. Pore size distribution of the BSC samples before and after 28-h cycle treatment. Note: (**a**) BSC-10-1.15; (**b**) BSC-15-1.15; (**c**) BSC-20-1.0; (**d**) BSC-20-1.15; (**e**) BSC-20-1.3.

4. Conclusions

In this study, the water resistance and macroscopic and microscopic bonding interface characterization of BSCs with different resin contents (10%, 15%, and 20%) and densities (1.00 g/cm^3, 1.15 g/cm^3, and 1.30 g/cm^3) were investigated. With an increase in density, the TSR increased, while the WSR and WAR decreased. At the same time, the resin content was inversely proportional to the three water-resistance indices of the BSCs. The BSC water resistance at a resin content of 20% and density of 1.15 g/cm^3 was the best.

It could be concluded that there were nonlinear and point-to-plane structures in the macroscopic bonding interface, and that the PF resin was more evenly and widely distributed with an increase in resin content. Moreover, as the density increased, the compression of the cell cavity and vessels increased.

To analyze the influence mechanism of the resin content and density on the water resistance of the BSCs from a bonding interface perspective, the failure morphology of the macroscopic bonding interface after the water-resistance test was observed, which showed that the failure of the outer-to-inner interface had a more obvious effect on water resistance than that of the inner-to-inner interface. The transverse and longitudinal fissures at the macroscopic bonding interfaces on the cross-sections of the BSCs decreased considerably with increasing resin content, which was consistent with the trends of WSR and TSR. With increasing density, only longitudinal fissures due to defibering decreased, which affected the WSR. In conclusion, the influence of the resin content on all water-resistance indices of the BSCs could be reflected by the macroscopic interface failure, which could only explain the dimensional changes in the width direction concerning the density factor.

The microscopic bonding interface failure was measured using a MIP test, and the changes in porosity and pore distribution before and after 28-h cycle treatment were compared. As the resin content increased, the PF resin filled more nano- and micron-sized pores. The D-value was the smallest when the resin content was up to 20%. When the density increased, the porosities of the untreated and treated samples decreased. The D-value at a density of 1.30 g/cm^3 was the largest. The microscopic bonding interface structure affected all water resistance indicators, with the TSR being more influenced than the other indicators. Therefore, the failure of the microscopic bonding interface could suggest the influence mechanism of the two factors on the BSC water resistance.

Water resistance plays an important role in BSCs and depends on the lifespan of the related products. Research on the water repellency and dimensional stability of BSCs from a bonding interface structure perspective is innovative and has theoretical significance for guiding performance improvements. Further research should be conducted to investigate novel methods that could be used to determine the influence mechanism of water resistance from a bonding interface perspective at other scales, such as ultramicroscopic and molecular scales.

Author Contributions: Data curation, investigation, and writing—original draft, Y.J.; Investigation, W.L. and J.W.; validation, Y.H.; writing—review and editing, W.Y. All authors have read and agreed to the published version of the manuscript.

Funding: This work was supported by the National Natural Science Foundation of China (No. 31,971,738).

Data Availability Statement: The data presented in this study are available on request from the corresponding author.

Acknowledgments: We wish to thank Beijing Zhongkebaice Technology Service Co., Ltd. for the equipment support.

Conflicts of Interest: The authors declare no competing interests.

References

1. Wang, Y.Y.; Wang, X.Q.; Li, Y.Q.; Huang, P.; Yang, B.; Hu, N.; Fu, S.Y. High-Performance Bamboo Steel Derived from Natural Bamboo. *ACS Appl. Mater. Interfaces* **2021**, *13*, 1431–1440. [CrossRef] [PubMed]
2. Cheng, Z.; Zhou, H.; Lu, Q.; Gao, H.; Lu, L. Extra Strengthening and Work Hardening in Gradient Nanotwinned Metals. *Science* **2018**, *362*, eaau1925. [CrossRef] [PubMed]
3. Ramage, M.H.; Burridge, H.; Busse-Wicher, M.; Fereday, G.; Reynolds, T.; Shah, D.U.; Wu, G.; Yu, L.; Fleming, P.; Densley-Tingley, D.; et al. The Wood from the Trees: The Use of Timber in Construction. *Renew. Sustain. Energy Rev.* **2017**, *68*, 333–359. [CrossRef]
4. Orsini, F.; Marrone, P. Approaches for a Low-Carbon Production of Building Materials: A Review. *J. Clean. Prod.* **2019**, *241*, 118380. [CrossRef]
5. Rao, F.; Ji, Y.; Li, N.; Zhang, Y.; Chen, Y.; Yu, W. Outdoor Bamboo-Fiber-Reinforced Composite: Influence of Resin Content on Water Resistance and Mechanical Properties. *Constr. Build. Mater.* **2020**, *261*, 120022. [CrossRef]
6. Guo, W.; Kalali, E.N.; Wang, X.; Xing, W.; Zhang, P.; Song, L.; Hu, Y. Processing Bulk Natural Bamboo into a Strong and Flame-Retardant Composite Material. *Ind. Crops Prod.* **2019**, *138*, 111478. [CrossRef]
7. Anwar, U.M.K.; Paridah, M.T.; Hamdan, H.; Sapuan, S.M.; Bakar, E.S. Effect of Curing Time on Physical and Mechanical Properties of Phenolic-Treated Bamboo Strips. *Ind. Crops Prod.* **2009**, *29*, 214–219. [CrossRef]
8. Abdul Khalil, H.P.S.; Bhat, I.U.H.; Jawaid, M.; Zaidon, A.; Hermawan, D.; Hadi, Y.S. Bamboo Fibre Reinforced Biocomposites: A Review. *Mater. Des.* **2012**, *42*, 353–368. [CrossRef]
9. Yu, W.; Yu, Y. Development and Prospect of Wood and Bamboo Scrimber Industry in China. *China Wood Ind.* **2013**, *27*, 5–8.

10. Huang, Y.; Ji, Y.; Yu, W. Development of Bamboo Scrimber: A Literature Review. *J. Wood Sci.* **2019**, *65*, 25. [CrossRef]
11. Kumar, A.; Vlach, T.; Laiblova, L.; Hrouda, M.; Kasal, B.; Tywoniak, J.; Hajek, P. Engineered Bamboo Scrimber: Influence of Density on the Mechanical and Water Absorption Properties. *Constr. Build. Mater.* **2016**, *127*, 815–827. [CrossRef]
12. Sharma, B.; Gatóo, A.; Bock, M.; Ramage, M. Engineered Bamboo for Structural Applications. *Constr. Build. Mater.* **2015**, *81*, 66–73. [CrossRef]
13. Shangguan, W.; Gong, Y.; Zhao, R.; Ren, H. Effects of Heat Treatment on the Properties of Bamboo Scrimber. *J. Wood Sci.* **2016**, *62*, 383–391. [CrossRef]
14. Rao, F.; Rao, F.; Rao, F.; Zhu, X.; Zhang, Y.; Ji, Y.; Lei, W.; Li, N.; Zhang, Z.; Chen, Y.; et al. Construction and Building Materials Water Resistance and Mechanical Properties of Bamboo Scrimber Composite Made from Different Units of Bambusa Chungii as a Function of Resin Content. *Constr. Build. Mater.* **2022**, *335*, 127250. [CrossRef]
15. Rao, F.; Ji, Y.; Huang, Y.; Li, N.; Zhang, Y.; Chen, Y.; Yu, W. Influence of Resin Molecular Weight on Bonding Interface, Water Resistance, and Mechanical Properties of Bamboo Scrimber Composite. *Constr. Build. Mater.* **2021**, *292*, 123458. [CrossRef]
16. Shams, M.I.; Yano, H.; Endou, K. Compressive Deformation of Wood Impregnated with Low Molecular Weight Phenol Formaldehyde (PF) Resin I: Effects of Pressing Pressure and Pressure Holding. *J. Wood Sci.* **2004**, *50*, 337–342. [CrossRef]
17. Yu, Y.; Liu, R.; Huang, Y.; Meng, F.; Yu, W. Preparation, Physical, Mechanical, and Interfacial Morphological Properties of Engineered Bamboo Scrimber. *Constr. Build. Mater.* **2017**, *157*, 1032–1039. [CrossRef]
18. Du, C.G.; Li, R. Study on Bamboo Particleboard Bambooceramics—PF Resin Content Influence on Properties of Bambooceramics. *Adv. Mater. Res.* **2014**, *1035*, 41–44. [CrossRef]
19. Grinins, J.; Biziks, V.; Irbe, I.; Rizikovs, J. Water Related Properties of Birch Wood Modified with Phenol-Formaldehyde (PF) Resins. *Key Eng. Mater.* **2019**, *800*, 246–250. [CrossRef]
20. Wei, J.; Rao, F.; Huang, Y.; Zhang, Y.; Qi, Y.; Yu, W.; Hse, C.Y. Structure, Mechanical Performance, and Dimensional Stability of Radiata Pine (Pinus Radiata D. Don) Scrimbers. *Adv. Polym. Technol.* **2019**, *2019*, 5209624. [CrossRef]
21. Sun, Y.; Zhang, Y.; Huang, Y.; Wei, X.; Yu, W. Influence of Board Density on the Physical and Mechanical Properties of Bamboo Oriented Strand Lumber. *Forests* **2020**, *11*, 567. [CrossRef]
22. Bao, M.; Huang, X.; Zhang, Y.; Yu, W.; Yu, Y. Effect of Density on the Hygroscopicity and Surface Characteristics of Hybrid Poplar Compreg. *J. Wood Sci.* **2016**, *62*, 441–451. [CrossRef]
23. Meng, F.; Liu, R.; Zhang, Y.; Huang, Y.; Yu, Y.; Yu, W. Improvement of the Water Repellency, Dimensional Stability, and Biological Resistance of Bamboo-Based Fiber Reinforced Composites. *Polym. Compos.* **2019**, *40*, 506–513. [CrossRef]
24. Yu, Y.; Huang, Y.; Zhang, Y.; Liu, R.; Meng, F.; Yu, W. The Reinforcing Mechanism of Mechanical Properties of Bamboo Fiber Bundle-Reinforced Composites. *Polym. Compos.* **2019**, *40*, 1463–1472. [CrossRef]
25. Zhang, Y.H.; Zhu, R.X.; Wen-Ji, Y.U.; Ren, D.H. Performance of Exterior Crushed Bamboo-Mat Composite after Accelerated Aging Test. *China Wood Ind.* **2012**, *26*, 6–8.
26. Doan, T.T.L.; Gao, S.L.; Mäder, E. Jute/Polypropylene Composites I. Effect of Matrix Modification. *Compos. Sci. Technol.* **2006**, *66*, 952–963. [CrossRef]
27. Vitas, S.; Segmehl, J.S.; Burgert, I.; Cabane, E. Porosity and Pore Size Distribution of Native and Delignified Beech Wood Determined by Mercury Intrusion Porosimetry. *Materials* **2019**, *12*, 416. [CrossRef]
28. Gardner, W. Note on the Dynamics of Capillary Flow. *Phys. Rev.* **1921**, *18*, 206–209. [CrossRef]
29. Abdullah, C.K.; Jawaid, M.; Abdul Khalil, H.P.S.; Zaidon, A.; Hadiyane, A. Oil Palm Trunk Polymer Composite: Morphology, Water Absorption, and Thickness Swelling Behaviours. *BioResources* **2012**, *7*, 2948–2959. [CrossRef]
30. Gabrielli, C.P.; Kamke, F.A. Phenol-Formaldehyde Impregnation of Densified Wood for Improved Dimensional Stability. *Wood Sci. Technol.* **2010**, *44*, 95–104. [CrossRef]
31. Abdul Khalil, H.P.S.; Bhat, A.H.; Jawaid, M.; Amouzgar, P.; Ridzuan, R.; Said, M.R. Agro-Wastes: Mechanical and Physical Properties of Resin Impregnated Oil Palm Trunk Core Lumber. *Polym. Compos.* **2010**, *31*, 638–644. [CrossRef]
32. Deka, M.; Saikia, C.N.; Baruah, K.K. Treatment of Wood with Thermosetting Resins: Effect on Dimensional Stability, Strength and Termite Resistance. *Indian J. Chem. Technol.* **2000**, *7*, 312–317.
33. Guan, M.; Huang, Z.; Zeng, D. Shear Strength and Microscopic Characterization of a Bamboo Bonding Interface with Phenol Formaldehyde Resins Modified with Larch Thanaka and Urea. *BioResources* **2016**, *11*, 492–502. [CrossRef]
34. Huang, Y.; Lin, Q.; Yang, C.; Bian, G.; Zhang, Y.; Yu, W. Multi-Scale Characterization of Bamboo Bonding Interfaces with Phenol-Formaldehyde Resin of Different Molecular Weight to Study the Bonding Mechanism. *J. R. Soc. Interface* **2020**, *17*, 20190755. [CrossRef] [PubMed]
35. Yu, Y.; Huang, X.; Yu, W. A Novel Process to Improve Yield and Mechanical Performance of Bamboo Fiber Reinforced Composite via Mechanical Treatments. *Compos. Part B Eng.* **2014**, *56*, 48–53. [CrossRef]
36. Yu, Y.; Zhu, R.; Wu, B.; Hu, Y.; Yu, W. Fabrication, Material Properties, and Application of Bamboo Scrimber. *Wood Sci. Technol.* **2015**, *49*, 83–98. [CrossRef]
37. Yu, Y.; Yu, W. Manufacturing Technology of Bamboo-Based Fiber Composites with High-Performance. *World Bamboo Ratt.* **2013**, *11*, 6–10.

38. Lian, C.P.; Liu, R.; Zhang, S.Q.; Luo, J.J.; Fei, B.H. Research Progress on Anatomical Structure of Bamboo Vascular Bundles. *China For. Prod. Ind.* **2018**, *45*, 8–12.
39. Liu, X.E.; Liu, Z.; Wang, Y.H.; Fei, B.H.; Zhou, X.; Zhang, L.F.; Gao, L.Y.; Jin, L.U. Investigation on Main Anatomical Characteristics of Dendrocalamus Giganteus. *J. Anhui Agric. Univ.* **2012**, *39*, 890–893.
40. Wang, Y.; Zhan, H.; Ding, Y.; Wang, S.; Lin, S. Variability of Anatomical and Chemical Properties with Age and Height in Dendrocalamus Brandisii. *BioResources* **2016**, *11*, 1202–1213. [CrossRef]
41. Lian, C.; Liu, R.; Xiufang, C.; Zhang, S.; Luo, J.; Yang, S.; Liu, X.; Fei, B. Characterization of the Pits in Parenchyma Cells of the Moso Bamboo [Phyllostachys Edulis (Carr.) J. Houz.] Culm. *Holzforschung* **2019**, *73*, 629–636. [CrossRef]

Article

In-Depth Sulfhydryl-Modified Cellulose Fibers for Efficient and Rapid Adsorption of Cr(VI)

Wenxuan Wang [1], Feihan Yu [1], Zhichen Ba [1], Hongbo Qian [1], Shuai Zhao [1], Jie Liu [2], Wei Jiang [3], Jian Li [1] and Daxin Liang [1,*]

[1] Key Laboratory of Bio-Based Material Science and Technology (Ministry of Education), Northeast Forestry University, Harbin 150040, China; wxwang@nefu.edu.cn (W.W.); yufeihan@nefu.edu.cn (F.Y.); bazc_@nefu.edu.cn (Z.B.); 1961026882@nefu.edu.cn (H.Q.); zs861944571@163.com (S.Z.); lijiangroup@163.com (J.L.)
[2] Department of Military Facilities, Army Logistics Academy, Chongqing 401331, China; liujiely@homail.com
[3] State Key Laboratory of Bio-Fibers and Eco-Textiles, Qingdao University, Qingdao 266071, China; wejiangqd@qdu.edu.cn
* Correspondence: daxin.liang@nefu.edu.cn; Tel.: +86-18182806182

Citation: Wang, W.; Yu, F.; Ba, Z.; Qian, H.; Zhao, S.; Liu, J.; Jiang, W.; Li, J.; Liang, D. In-Depth Sulfhydryl-Modified Cellulose Fibers for Efficient and Rapid Adsorption of Cr(VI). *Polymers* **2022**, *14*, 1482. https://doi.org/10.3390/polym14071482

Academic Editor: Dimitrios Bikiaris

Received: 28 February 2022
Accepted: 28 March 2022
Published: 6 April 2022

Publisher's Note: MDPI stays neutral with regard to jurisdictional claims in published maps and institutional affiliations.

Copyright: © 2022 by the authors. Licensee MDPI, Basel, Switzerland. This article is an open access article distributed under the terms and conditions of the Creative Commons Attribution (CC BY) license (https://creativecommons.org/licenses/by/4.0/).

Abstract: As one of the hazardous heavy metal ion pollutants, Cr(VI) has attracted much attention in the sewage treatment research field due to its wide distribution range and serious toxicity. In this paper, cellulose fibers were prepared by wet spinning and followed by freeze drying, resulting in large porosity. Subsequently, in-depth sulfhydryl modification was applied with cellulose fibers for efficient and rapid adsorption of Cr(VI). The maximum adsorption capacity of sulfhydryl-modified cellulose fibers to Cr(VI) can reach 120.60 mg g^{-1}, the adsorption equilibrium can be achieved within 300 s, and its adsorption rate can reach 0.319 mg g^{-1} s^{-1}. The results show that the in-depth sulfhydryl-modified cellulose fibers perform excellent adsorption capacity for chromium, and are also available for other heavy metal ions. At the same time, the low cost and environmentally friendly property of the as-synthesized material also demonstrate its potential for practical usage for the treatment of heavy metal ion pollution in waste water.

Keywords: cellulose fibers; adsorption; Cr(VI); sulfhydryl-modified

1. Introduction

Chromium (Cr) is commonly used in tanning, metalworking, mining, and electroplating industries, causing serious pollution of water, soil, and plant resources worldwide [1,2]. Cr in industrial wastewater is mainly hexavalent compounds, such as chromate ions; Cr(VI) toxicity is approximately 100 times higher than Cr(III), and it can cause lung cancer and nasopharyngeal cancer [3]. The World Health Organization (WHO) stipulates that the maximum allowable value of Cr(VI) in drinking water is 50 µg/L [4]. Traditional treatment methods, such as membrane separation [5], chemical reduction [6], adsorption [7,8], as well as plasma method [9], have been used to address Cr(VI) in wastewater. Currently, many materials (such as graphene, MXene, microalgal, etc.) are used in the field of adsorption, which is a method with the advantages of reusability, high selectivity, and simple application [10–12]. Unfortunately, several problems are associated with these traditional adsorbents, such as low removal efficiency, weak selectivity, and high cost. Therefore, fabricating an excellent adsorbent to remove Cr(VI) in wastewater has become inevitable.

Cellulose is the most widespread biomass material in nature, which is nontoxic, nonpolluting, and easily degraded, making it an environmentally friendly material [13,14]. The dissolution and regeneration of cellulose are necessary for the wide application of cellulose, and different forms of cellulose materials (cellulose microspheres, cellulose fibers, cellulose films, cellulose aerogels, foams, etc.) can be prepared through different solvent systems [15]. The ionic liquid is an excellent cellulose dissolving system, which makes the reshaping of cellulose easy to operate. However, the ionic liquid is expensive and not suitable for large-scale applications. Fortunately, ionic liquids can be recycled and then cellulose dissolved, which greatly reduces the cost [16]. Cellulose fiber (CF) is a material with the advantages of large specific surface area, low density, high mechanical strength, reticular porosity, good hydrophilicity, etc., making it a new type of functional material that is easy to be modified [17]. CFs can be prepared by wet spinning, and pure cellulose fibers also have a certain adsorption effect [18]. Many studies have been carried out on the modification of cellulose fibers to further improve their adsorption efficiency. Song et al. [19] used polyethyleneimine-cellulose fibers for the fast recovery of Au(I) from alkaline e-waste leachate, and Ali et al. [20] used cellulose fiber yarns with high wet strength and their affinity for chitosan combined with the immobilization of tempo-oxidized cyclodextrins to capture 17α-vinyl estradiol from aqueous solutions. These modification methods can improve the adsorption performance of cellulose fibers; unfortunately, the methods are too complex to be implemented and too costly for large-scale applications. Sulfhydryl (–SH) is a common adsorption group in pollutant treatment [21]. Cellulose contains many hydroxyls, and thioglycolic acid is used to esterify it to achieve sulfhydrylation [22]. The adsorption performance of sulfhydryl-modified cellulose is greatly improved, and it can adsorb inorganic/organic pollutants.

Here, we proposed a simple and effective strategy for preparing CFs with a stable 3D network structure using dissolution, regeneration, wet spinning, and freeze drying. Based on the rich pore structure of CFs, thioglycolic acid was used to deeply sulfhydryl-modified them to obtain sulfhydryl-modified cellulose fibers (CFs–SH). The adsorption of Cr(VI) on the CFs–SH has the characteristics of high adsorption capacity and high adsorption efficiency because of being deeply sulfhydryl-modified; simultaneously, the CFs–SH can adsorb more kinds of metal ions, making CFs–SH have a wide application space.

2. Materials and Methods

2.1. Materials

All chemicals were used without further purification. Potassium dichromate ($K_2Cr_2O_7$, 99%), sodium hydroxide (NaOH, 99%), cobalt nitrate ($Co(NO_3)_2 \bullet 6H_2O$, 99%), cadmium nitrate ($Cd(NO_3)_2 \bullet 4H_2O$, 99%), thioglycolic acid ($C_2H_4O_2S$, 99.5%), diphenylcarbazide ($C_{13}H_{14}N_4O$, 99.7%) and 1-Butyl-3-methyl imidazole chloride ([Bmim]Cl 99%) were obtained from Sigma-Aldrich Co. Ltd. (Shanghai, China). Cupric nitrate ($Cu(NO_3)_2 \bullet 3H_2O$, 99%), nickel nitrate ($Ni(NO_3)_2 \bullet 6H_2O$, 99%), zinc nitrate ($Zn(NO_3)_2 \bullet 6H_2O$, 99%) and lead nitrate ($Pb(NO_3)_2$, 99%) were purchased from Tianjin Kemiou Chemical Reagent Co., Ltd. (Tianjin, China). Hydrochloric acid (HCl, 37%), nitric acid (HNO_3, 37%), and ethanol (C_2H_6O, 99%) were sourced from a local supplier.

2.2. Characterization

The morphology and structural characteristics of samples were observed using a scanning electron microscope equipped with an energy-dispersive X-ray spectrometer (SEM-EDS; TM3030, Tokyo Prefecture, Japan). Thermal stability was evaluated using a simultaneous thermal analyzer (Netzsch STA 449F3, Frankfurt, Main, Germany) based on the analysis of thermal gravimetric analysis (TG) and differential scanning calorimetry (DSC), at a temperature range of 50–600 °C with a heating rate of 10 °C min^{-1} under N_2 atmosphere. The crystal structures of pure CFs and CFs–SH were analyzed using an X-ray diffractometer (XRD; D/max-2200VPC, Tokyo Prefecture, Japan). Fourier transform infrared spectroscopy (FTIR; Frontier, Perkin Elmer, Waltham, MA, USA) was used to characterize the abundant functional group in the pure CFs and CFs–SH. The Cr(VI) concentration was monitored using a UV–vis spectrophotometer (UV-1800, Shimadzu, Kyoto Prefecture, Japan) at 540 nm. The adsorption capacity of CFs-SH toward the adsorption of other heavy metal ions by inductively coupled plasma mass spectrometry (ICP-MS; NEXILN350D, PerkinElmer, Waltham, MA, USA).

2.3. Preparation of Pure CFs and CFs–SH

The α-Cellulose (5 wt%) was initially dissolved in [Bmim]Cl at 90 °C for 1 h, the cellulose solution was vacuumed at 80 °C to remove air bubbles, and the solution was injected into ethanol using a 10 mL syringe (needle diameter 800 μm) under the action of a syringe pump. The obtained fibers were washed with water and then prefrozen at −12 °C for 12 h and subsequently freeze-dried for 48 h under high-vacuum conditions (0.010 mbar) at −56 °C to obtain pure CFs. Pure CFs (5 g) were further immersed in thioglycolic acid solution (200 mL, the volume ratio of H_2SO_4 and thioglycolic acid was 1:100) for 4 h, and washed 3–5 times in distilled water until the pH of solution was neutral, then finally, dried at 45 °C for 24 h.

2.4. Adsorption of Cr(VI)

The dried pure CFs and CFs–SH (50 mg, the optimal adsorbent quality selection was shown in Figure S1) were immersed in 20 mL of $K_2Cr_2O_7$ aqueous solution (200 mg L^{-1}) with stirring at 100 rpm at room temperature (20 ± 1 °C). Pollutants remaining in the filtrate were analyzed in a UV–vis spectrophotometer; the full spectrum before and after adsorption was shown in Figure S2, at a wavelength of 540 nm to test Cr(VI) based on the DPC method [10]. The Cr(VI) adsorption capacity of the adsorbent (q_e) was calculated using Equation (1):

$$q_e = \frac{(C_0 - C_e)V}{m} \quad (1)$$

where q_e is the Cr(VI) equilibrium adsorption capacity, and C_0 (mg L^{-1}) and C_e (mg L^{-1}) represent the initial and equilibrium concentrations of Cr(VI) in solution, respectively. V (L) and m (g) refer to the volume of the solution and the mass of samples, respectively.

The effects of initial Cr(VI) concentration (5–500 mg L^{-1}, the standard curve was shown in Figure S3), pH (1.0, 3.0, 5.0, 7.0, 9.0, 11.0, the optimum pH was shown in Figure S4), and temperature (20–60 °C) on the adsorption process were investigated. All the adsorption experiments were performed in triplicate (n = 3) at least with the mean taken. The flow of the experiment was shown in Scheme 1.

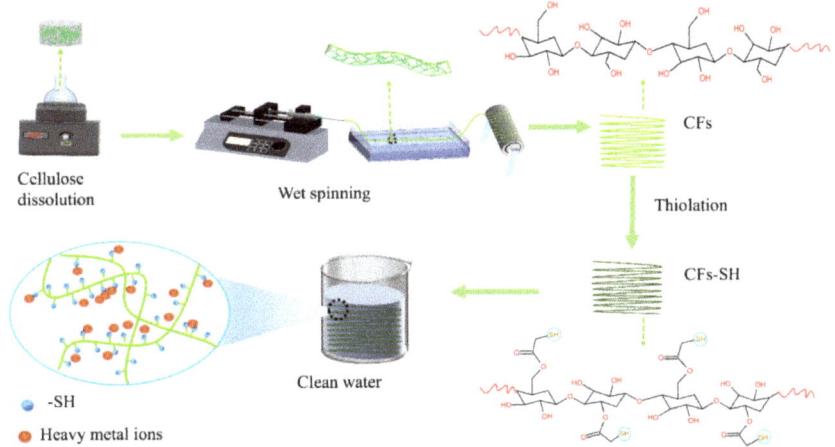

Scheme 1. Preparation of in-depth sulfhydryl-modified CFs–SH.

3. Test Results and Discussion

3.1. Characterization of Pure CFs and CFs–SH

3.1.1. FTIR Analysis

Figure 1 shows the FTIR spectra of pure CFs and CFs–SH. The FTIR spectra of both aerogels revealed bands at 3362 cm^{-1}, which can be attributed to –OH stretching vibration of hydrogen bonds. They play a pivotal role in the process of cellulose dissolution and regeneration, which indicate that the fibers have good hydrophilicity [23–25]. The characteristic peak of sulfhydryl (–SH) is 2250 cm^{-1}. The C=O stretching vibration peak is 1746 cm^{-1}. 1279 cm^{-1} and 1160 cm^{-1} are all the C–O–C stretching vibration peaks. These changes indicate that the esterification reaction between cellulose and thioglycolic acid occurred, resulting in the modification of pure CFs to CFs–SH.

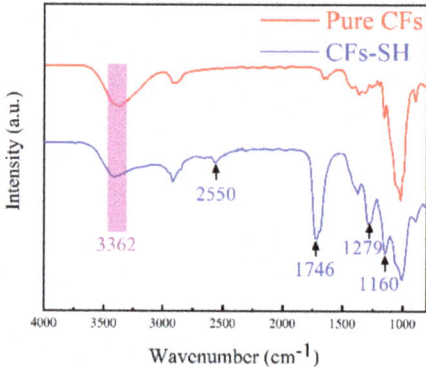

Figure 1. FTIR spectra of pure CFs and CFs–SH.

3.1.2. XRD Analysis

Figure 2 shows the X-ray diffraction patterns of pure CFs and CFs–SH. Note that the crystallinity of CFs–SH decreased compared to pure CFs, this is because the modification of the –SH breaks hydrogen bonds between parts of the cellulose, resulting in a change in crystallinity. At the same time, the mechanical properties of the fibers are reduced (from 155.64 MPa to 69.51 MPa, as shown in Figure S5), which indicates that the modification by –SH was successful.

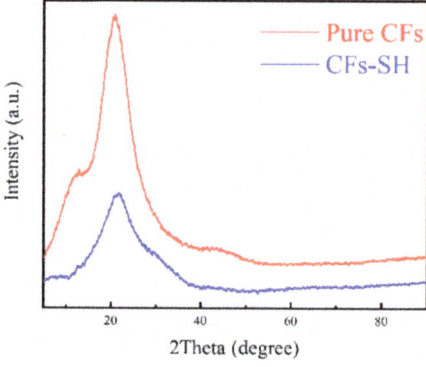

Figure 2. XRD patterns of pure CFs and CFs–SH.

3.1.3. SEM-EDS Analysis

The pure CFs obtained by wet spinning and freeze drying have smooth outer walls (Figure 3a) and a rich pore structure inside (Figure 3b), which makes the modification of –SH easier and more thorough. On the contrary, only the surface of the dried CFs can be modified by –SH, and the interior contains almost no sulfur element (Figure S6). The crystallinity of the modified cellulose fibers decreased by –SH, and the porous structure inside the fibers collapsed and stacked (Figure 3c,d). Fortunately, the interior of the CFs has been completely modified, the SEM-EDS cross–section of CFs–SH (Figure 3e–h), so that CFs–SH has a relatively large adsorption capacity.

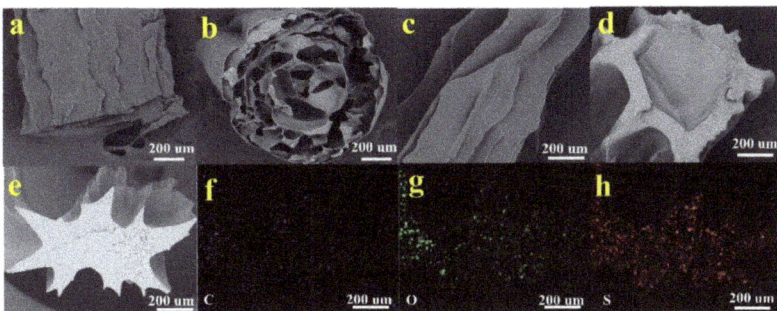

Figure 3. SEM-EDS images of pure CFs and CFs–SH, (**a**) surface of pure CFs, (**b**) cross section of pure CFs, (**c**) surface of CFs–SH, (**d,e**) cross section of CFs–SH, (**f–h**) EDS mapping of C, O, and S.

3.1.4. TG/DTG Analysis

The thermal stability of pure CFs and CFs–SH were displayed by TG and DTG curves (Figure 4a,b), respectively. Initially, all samples underwent a slight weight loss below 100 °C, belonging to the evaporation of adsorbed or bound water [26,27]. At 343 °C, pure CFs have the maximum decomposition rate, indicating that the decomposition temperature of pure CFs is at 343 °C. At 150 °C, in CFs–SH, the ester group contained in the cellulose successfully modified by thioglycolic acid was thermally decomposed, resulting in a large mass change occurring. The reason why the thermal decomposition temperature of other cellulose (300 °C) was lower than CFs (343 °C) was that the modification reduces the crystallinity of cellulose, which also proved the success of the modification.

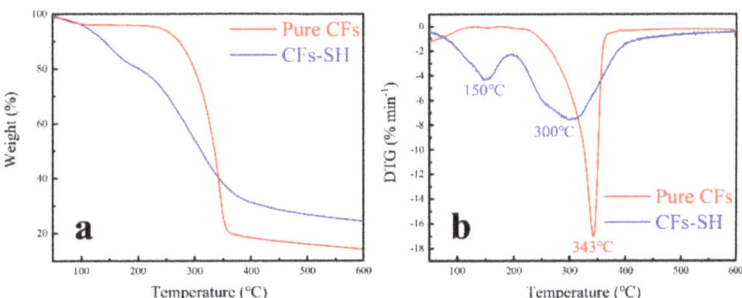

Figure 4. TG (**a**) and DTG (**b**) of pure CFs and CFs–SH.

3.2. Cr(VI) Adsorption

3.2.1. Effect of Contact Time and Adsorption Kinetics

Figure 5a shows the adsorption kinetics of the as-prepared pure CFs and CFs–SH. Cr(VI) was rapidly captured by –SH in the CFs–SH within the first two minutes, as shown in the Figure 5a, before achieving the q_e value at 250 s. The adsorption capacity of Cr(VI) is greatly increased by modifying CFs with –SH. Two kinetic models were used to explore the Cr(VI) adsorption behavior in CFs and CFs–SH, in the following equations:

$$\ln(q_e - q_t) = \ln q_e - k_1 t \tag{2}$$

$$\frac{t}{q_t} = \frac{1}{k_2 q_e^2} + \frac{t}{q_e} \tag{3}$$

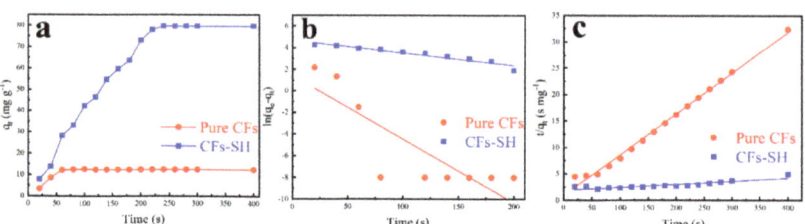

Figure 5. Cr(VI) adsorption behavior in CFs–SH: (**a**) Adsorption kinetics (pH = 3.0; V = 20.0 mL; m ≅ 50 mg; C_0 ≅ 200 mg L^{-1}), (**b**) pseudo-first-order kinetics, (**c**) pseudo-second-order kinetics.

As shown in Figure 5b,c, and Table 1, the fitting results revealed that a pseudo-second-order kinetic model (r^2 = 0.954) could better describe the adsorption behavior of CFs–SH than a pseudo-first-order one. This indicates that the Cr(VI) adsorption of the adsorbents is mainly dependent on the number of accessible active sites of CFs–SH [28–30].

Table 1. Pseudo first order and Pseudo second order.

Material	Pseudo First Order			Pseudo Second Order		
	q_e (mg g^{-1})	$k_1 \times 10^{-3}$ (min^{-1})	r^2	q_e (mg g^{-1})	$k_1 \times 10^{-3}$ (min^{-1})	r^2
Pure CFs	12.34	58.52	0.672	11.49	77.20	0.992
CFs–SH	79.73	11.55	0.906	78.32	5.86	0.954

For excellent adsorbents, high adsorption efficiency is critical for future industrial applications [31]. The effect of CFs–SH was confirmed by evaluating the initial Cr(VI) adsorption rate (s) of both adsorbents, following Equation (4).

$$h = k_2 \times q_e^2 \; t \to 0 \tag{4}$$

The adsorption efficiency of CFs–SH is calculated to be 0.319 mg g^{-1} s^{-1}, which is due to the large number of adsorption sites brought by the deeply –SH for CFs, which can reach the adsorption equilibrium in a very short time. On the contrary, fibers modified only on the surface layer did not have good adsorption effect (Figure S7). More importantly, the residual Cr(VI) in the treated solution was only 0.03 ± 0.01 mg L^{-1}, which met the World Health Organization (WHO) maximum allowable value of Cr(VI) in drinking water [4].

3.2.2. Cr(VI) Adsorption Isotherm on CFs–SH

With Cr(VI) concentrations of 200 mg L^{-1}, 0.05 g CFs–SH was added to various 20 mL Cr(VI) solutions. Each sample was left to stand at 20 °C, 30 °C, 40 °C, 50 °C, and 60 °C for 300 s to reach adsorption equilibrium. Subsequently, a preferential analysis was conducted to obtain the final Cr(VI) concentration in the solution. The results were then applied to the Langmuir and Freundlich models to investigate the Cr(VI) adsorption isotherm, using Equations (5) and (6).

$$\frac{C_e}{q_e} = \frac{C_e}{q_m} + \frac{1}{q_m \times K_L} \tag{5}$$

$$\ln q_e = \ln K_F + \frac{1}{n} \ln C_e \tag{6}$$

q_e (mg g^{-1}) and q_m (mg g^{-1}) are the Cr(VI) adsorption equilibrium capacity and theoretical maximum monolayer Cr(VI) adsorption capacity of CFs–SH, respectively. K_L (L mg^{-1}) represents the Langmuir adsorption equilibrium constant. k_F and $1/n$ are the adsorption equilibrium constant and intensity of the concentration effect on adsorption, respectively.

Figure 6a shows the results of the Cr(VI) adsorption equilibria of CFs–SH. The isotherm parameters were deduced by fitting the data in Figure 6b,c, and these parameters are summarized in Table 2. The adsorption capacity of the adsorbent increases with temperature, which is consistent with the findings reported in the literature. This illustrates that the adsorption process in the adsorbent is an endothermic reaction [32,33].

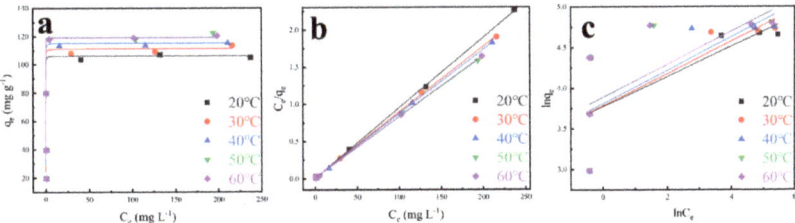

Figure 6. (a) Adsorption isotherms, (b) Langmuir and (c) Freundlich isotherms of Cr(VI) adsorption by CFs–SH at 20 °C, 30 °C, 40 °C, 50 °C and 60 °C (pH = 3.0; V = 20 mL; m ≅ 50 mg; C$_0$ ≅ 50–500 mg L^{-1}).

Table 2. Langmuir model and Freundlich model.

Temperature	Langmuir Isotherm			Freundlich Isotherm		
	q_m (mg g^{-1})	$k_1 \times 10^{-3}$	r^2	k_2	n	r^2
20 °C	105.08	9.42	0.999	0.185	3.77	0.471
30 °C	113.57	8.80	0.999	0.198	3.78	0.491
40 °C	115.58	8.61	0.999	0.205	3.81	0.470
50 °C	122.51	8.15	0.999	0.203	3.88	0.393
60 °C	120.60	8.24	0.999	0.200	3.89	0.376

A comparison of the correlation coefficients (r^2) indicated that the Langmuir model could better describe the adsorption behavior of the adsorbent, rather than the Freundlich model. This illustrated that similar energy was required during the Cr(VI) adsorption at all sites on CFs–SH, thus confirming the monolayer adsorption of Cr(VI) onto the external and pore surfaces of CFs–SH [34]. The maximum Cr(VI) adsorption capacity of CFs–SH was 120.60 mg g^{-1} (Table 2).

3.2.3. Adsorption Ability of CFs–SH

Cyclic adsorption/desorption performance was analyzed to investigate the durable performance in terms of performance consistency and the structural stability of CFs-SH and the importance of loading. Unfortunately, after one cycle, the adsorption performance of CFs–SH dropped sharply to 12.49% by the fifth cycle (shown in Figure S8), but it was sufficient as an inexpensive adsorption material.

A comparison of the Cr(VI) adsorption performance by the as-prepared adsorbent relative to that of other composites is summarized in Table 3. CFs–SH exhibited relatively high adsorption capacity; although the adsorption capacity does not have a great advantage, its adsorption rate reaches 0.319 mg g^{-1} s^{-1}, which is very important for applications.

Table 3. Comparison of the Cr(VI) adsorption performance of CFs–SH with that of other composites.

Composites	Adsorption Capacity (mg g^{-1})	Adsorption Rate (mg g^{-1} s^{-1})	Ref.
Peanut shell biochar	22.93	0.013	[4]
HI	353.87	0.098	[11]
SBA-15	66.50	0.004	[21]
PAIN/TiO$_2$	394.43	0.219	[35]
Brazilian-pine fruit coat	240.00	0.008	[36]
CFs–SH	120.60	0.319	This study

3.3. Other Heavy Metal Ions Adsorption

In previous studies, –SH also had an adsorption effect on other heavy metal ions [37]. To broaden the application of CFs–SH, six different heavy metal ions (Cd^{2+}, Pb^{2+}, Cu^{2+}, Co^{2+}, Zn^{2+}, and Ni^{2+}) were used to test their adsorption properties. As shown in Figure 7, CFs–SH adsorbs 20 mL, 200 mg L^{-1} different heavy metal ion solutions for 300 s under 20 °C. The adsorption efficiency of CFs–SH was relatively low. The adsorption and adsorption efficiency displayed by CFs–SH is relatively low, but it shows that it has an adsorption effect on various heavy metal ions, which is more promising.

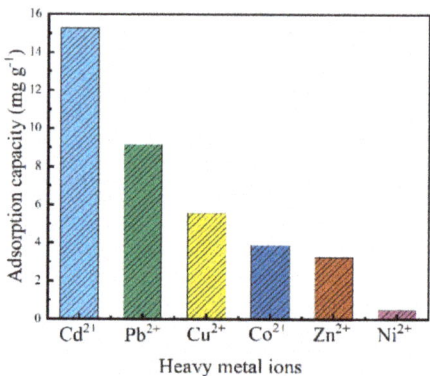

Figure 7. Other heavy metal ions' adsorption behavior in CFs–SH, including Cd^{2+}, Pb^{2+}, Cu^{2+}, Co^{2+}, Zn^{2+}, and Ni^{2+} (pH = 3.0; V = 20.0 mL; m $\cong$ 50 mg; $C_0 \cong$ 200 mg L^{-1}).

4. Conclusions

In this study, CFs–SH with porous structure CFs was deeply modified by sulfhydryl. Therefore, CFs-SH possessed a large number of adsorption sites, which made it become an adsorbent with extremely strong adsorption activity for Cr(VI) and other heavy metal ions. The maximum Cr(VI) equilibrium adsorption capacity of CFs–SH was 120.60 mg g^{-1} at 50 °C, and the Cr(VI) adsorption isotherm agreed well with the Langmuir model. More importantly, CFs–SH could reach the adsorption equilibrium within 300 s, the adsorption rate could reach 0.319 mg g^{-1} s^{-1} at 20 °C, and CFs–SH had certain reusability, which was very useful for practical applications. For future research, improving the reusability of CFs–SH and expanding its adsorption capacity will be the primary task, followed by the improvement of the adsorption efficiency of other heavy metal ions to more effectively remove heavy metal pollution in water.

Supplementary Materials: The following supporting information can be downloaded at: https://www.mdpi.com/article/10.3390/polym14071482/s1, Figure S1: Standard curve of Cr(VI), Figure S2: The full UV-Vis spectrum of different solution. Figure S3. Standard curve of Cr(VI). Figure S4. Adsorption rate under different pH conditions. Figure S5. The stress-strain curve of pure CFs and CFs–SH. Figure S6. SEM–EDS images of cellulose fibers after drying at 50 °C. Figure S7. The adsorption performance of cellulose fibers after drying at 50 °C. Figure S8. Cyclic adsorption/desorption performance of CFs–SH.

Author Contributions: W.W. and F.Y. contributed equally to this work with the experimental scheme; conceptualization, W.W. and F.Y.; data curation, Z.B., S.Z. and H.Q.; formal analysis, Z.B. and H.Q.; funding acquisition, D.L. and J.L. (Jian Li); methodology, W.W., F.Y. and W.J.; project administration J.L. (Jian Li) and D.L.; supervision, D.L.; writing—original draft, W.W. and Z.B.; writing—review & editing J.L. (Jie Liu) and D.L. All authors have read and agreed to the published version of the manuscript.

Funding: This research was funded by the Fundamental Research Funds for the Central Universities (2572021BB02); State Key Laboratory of Bio-Fibers and Eco-Textiles (Qingdao University) and the College Student Innovation and Entrepreneurship Training Program, Northeast Forestry University, China (DC2020101).

Institutional Review Board Statement: Not applicable.

Informed Consent Statement: Not applicable.

Data Availability Statement: The data presented in this study are available in the manuscript and Supplementary Material.

Acknowledgments: This research was supported by the Fundamental Research Funds for the Central Universities (2572021BB02); State Key Laboratory of Bio-Fibers and Eco-Textiles (Qingdao University) and the College Student Innovation and Entrepreneurship Training Program, Northeast Forestry University, China (DC2020101). The authors are grateful for providing infrastructural facilities and assistance.

Conflicts of Interest: The authors declare no conflict of interest.

References

1. Singh, H.P.; Mahajan, P.; Kaur, S.; Batish, D.R.; Kohli, R. Chromium toxicity and tolerance in plants. *Environ. Chem. Lett.* **2013**, *11*, 229–254. [CrossRef]
2. Ukhurebor, K.E.; Aigbe, U.O.; Onyancha, R.B.; Nwankwo, W.; Osibote, O.A.; Paumo, H.K.; Ama, O.M.; Adetunji, C.O.; Siloko, I.U. Effect of hexavalent chromium on the environment and removal techniques: A review. *J. Environ. Manag.* **2021**, *280*, 111809. [CrossRef] [PubMed]
3. Jiang, B.; Gong, Y.; Gao, J.; Sun, T.; Liu, Y.; Oturan, N.; Oturan, M.A. The reduction of Cr(VI) to Cr(III) mediated by environmentally relevant carboxylic acids: State-of-the-art and perspectives. *J. Hazard. Mater.* **2019**, *365*, 205–226. [CrossRef]
4. Murad, H.A.; Ahmad, M.; Bundschuh, J.; Hashimoto, Y.; Zhang, M.; Sarkar, B.; Ok, Y.S. A remediation approach to chromium-contaminated water and soil using engineered biochar derived from peanut shell. *Environ. Res.* **2021**, *204*, 112125. [CrossRef]
5. Huang, Z.; Liu, J.; Liu, Y.; Xu, Y.; Li, R.; Hong, H.; Shen, L.; Lin, H.; Liao, B.-Q. Enhanced permeability and antifouling performance of polyether sulfone (PES) membrane via elevating magnetic Ni@MXene nanoparticles to upper layer in phase inversion process. *J. Membr. Sci.* **2021**, *623*, 119080. [CrossRef]
6. Chen, F.; Guo, S.; Wang, Y.; Ma, L.; Li, B.; Song, Z.; Huang, L.; Zhang, W. Concurrent adsorption and reduction of chromium(VI) to chromium(III) using nitrogen-doped porous carbon adsorbent derived from loofah sponge. *Front. Environ. Sci. Eng.* **2021**, *16*, 57. [CrossRef]
7. Li, T.; Shen, J.; Huang, S.; Li, N.; Ye, M. Hydrothermal carbonization synthesis of a novel montmorillonite supported carbon nanosphere adsorbent for removal of Cr (VI) from waste water. *Appl. Clay Sci.* **2014**, *93–94*, 48–55. [CrossRef]
8. Dong, X.; Ma, L.Q.; Li, Y. Characteristics and mechanisms of hexavalent chromium removal by biochar from sugar beet tailing. *J. Hazard. Mater.* **2011**, *190*, 909–915. [CrossRef]
9. Zheng, C.; Yang, Z.; Si, M.; Zhu, F.; Yang, W.; Zhao, F.; Shi, Y. Application of biochars in the remediation of chromium contamination: Fabrication, mechanisms, and interfering species. *J. Hazard. Mater.* **2020**, *407*, 124376. [CrossRef]
10. Xia, S.; Song, Z.; Jeyakumar, P.; Shaheen, S.M.; Rinklebe, J.; Ok, Y.S.; Bolan, N.; Wang, H. A critical review on bioremediation technologies for Cr(VI)-contaminated soils and wastewater. *Crit. Rev. Environ. Sci. Technol.* **2019**, *49*, 1027–1078. [CrossRef]
11. Khan, A.R.; Awan, S.K.; Husnain, S.M.; Abbas, N.; Anjum, D.H.; Abbas, N.; Benaissa, M.; Mirza, C.R.; Mujtaba-ul-Hassan, S.; Shahzad, F. 3d flower like 8-Mno2/Mxene nano-hybrids for the removal of hexavalent Cr from wastewater. *Ceram. Int.* **2021**, *47*, 25951–25958. [CrossRef]
12. Daneshvar, E.; Zarrinmehr, M.J.; Kousha, M.; Hashtjin, A.M.; Saratale, G.D.; Maiti, A.; Vithanage, M.; Bhatnagar, A. Hexavalent chromium removal from water by microalgal-based materials: Adsorption, desorption and recovery studies. *Bioresour. Technol.* **2019**, *293*, 122064. [CrossRef] [PubMed]
13. Dilamian, M.; Noroozi, B. Rice straw agri-waste for water pollutant adsorption: Relevant mesoporous super hydrophobic cellulose aerogel. *Carbohydr. Polym.* **2021**, *251*, 117016. [CrossRef] [PubMed]

14. Zou, Y.; Zhao, J.; Zhu, J.; Guo, X.; Chen, P.; Duan, G.; Liu, X.; Li, Y. A mussel-inspired polydopamine-filled cellulose aerogel for solar-enabled water remediation. *ACS Appl. Mater. Interfaces* **2021**, *13*, 7617–7624. [CrossRef]
15. Ali, N.; Awais; Kamal, T.; Ul-Islam, M.; Khan, A.; Shah, S.J.; Zada, A. Chitosan-coated cotton cloth supported copper nanoparticles for toxic dye reduction. *Int. J. Biol. Macromol.* **2018**, *111*, 832–838. [CrossRef]
16. Nie, Y.; Wang, J.; Zhang, Z.; Liu, X.; Zhang, X. Trends and research progresses on the recycling of ionic liquids. *Chem. Ind. Eng. Prog.* **2019**, *38*, 100–110.
17. Choi, H.Y.; Bae, J.H.; Hasegawa, Y.; An, S.; Kim, I.S.; Lee, H.; Kim, M. Thiol-functionalized cellulose nanofiber membranes for the effective adsorption of heavy metal ions in water. *Carbohydr. Polym.* **2020**, *234*, 115881. [CrossRef]
18. Dickinson, E. Biopolymer-based particles as stabilizing agents for emulsions and foams. *Food Hydrocoll.* **2017**, *68*, 219–231. [CrossRef]
19. Lin, X.; Tran, D.T.; Song, M.H.; Yun, Y.S. Development of quaternized polyethylenimine-cellulose fibers for fast recovery of Au(Cn)2(-) in alkaline wastewater: Kinetics, isotherm, and thermodynamic study. *J. Hazard. Mater.* **2022**, *422*, 126940.
20. Orelma, H.; Virtanen, T.; Spoljaric, S.; Lehmonen, J.; Seppala, J.; Rojas, O.J.; Harlin, A. Cyclodextrin-functionalized fiber yarns spun from deep eutectic cellulose solutions for nonspecific hormone capture in aqueous matrices. *Biomacromolecules* **2018**, *19*, 652–661.
21. Liu, F.; Wang, A.; Xiang, M.; Hu, Q.; Hu, B. Effective adsorption and immobilization of Cr(VI) and U(VI) from aqueous solution by magnetic amine-functionalized SBA-15. *Sep. Purif. Technol.* **2021**, *282*, 120042. [CrossRef]
22. Gao, Q.; Wang, X.; Wang, H.; Liang, D.; Zhang, J.; Li, J. Sulfhydryl-modified sodium alginate film for lead-ion adsorption. *Mater. Lett.* **2019**, *254*, 149–153. [CrossRef]
23. Chen, M.; Zhang, X.; Zhang, A.; Liu, C.; Sun, R. Direct preparation of green and renewable aerogel materials from crude bagasse. *Cellulose* **2016**, *23*, 1325–1334. [CrossRef]
24. Tarchoun, A.F.; Trache, D.; Klapötke, T.M.; Derradji, M.; Bessa, W. Ecofriendly isolation and characterization of microcrystalline cellulose from giant reed using various acidic media. *Cellulose* **2019**, *26*, 7635–7651. [CrossRef]
25. Sun, L.; Miao, M. Dietary polyphenols modulate starch digestion and glycaemic level: A review. *Crit. Rev. Food Sci. Nutr.* **2020**, *60*, 541–555. [CrossRef]
26. Zhang, H.; Chen, Y.; Wang, S.; Ma, L.; Yu, Y.; Dai, H.; Zhang, Y. Extraction and comparison of cellulose nanocrystals from lemon (*Citrus limon*) seeds using sulfuric acid hydrolysis and oxidation methods. *Carbohydr. Polym.* **2020**, *238*, 116180. [CrossRef]
27. Dai, H.; Huang, Y.; Zhang, Y.; Zhang, H.; Huang, H. Green and facile fabrication of pineapple peel cellulose/magnetic diatomite hydrogels in ionic liquid for methylene blue adsorption. *Cellulose* **2019**, *26*, 3825–3844. [CrossRef]
28. Park, M.J.; Nisola, G.M.; Vivas, E.L.; Limjuco, L.; Lawagon, C.P.; Gil Seo, J.; Kim, H.; Shon, H.K.; Chung, W.-J. Mixed matrix nanofiber as a flow-through membrane adsorber for continuous Li+ recovery from seawater. *J. Membr. Sci.* **2016**, *510*, 141–154. [CrossRef]
29. Shi, X.-C.; Zhang, Z.-B.; Zhou, D.-F.; Zhang, L.-F.; Chen, B.-Z.; Yu, L.-L. Synthesis of Li+ adsorbent (H_2TiO_3) and its adsorption properties. *Trans. Nonferrous Met. Soc. China* **2013**, *23*, 253–259. [CrossRef]
30. Limjuco, L.; Nisola, G.M.; Lawagon, C.P.; Lee, S.-P.; Gil Seo, J.; Kim, H.; Chung, W.-J. H_2TiO_3 composite adsorbent foam for efficient and continuous recovery of Li+ from liquid resources. *Colloids Surfaces A Physicochem. Eng. Asp.* **2016**, *504*, 267–279. [CrossRef]
31. Ju, P.; Liu, Q.; Zhang, H.; Chen, R.; Liu, J.; Yu, J.; Liu, P.; Zhang, M.; Wang, J. Hyperbranched topological swollen-layer constructs of multi-active sites polyacrylonitrile (PAN) adsorbent for uranium(VI) extraction from seawater. *Chem. Eng. J.* **2019**, *374*, 1204–1213. [CrossRef]
32. Xiao, G.; Tong, K.; Zhou, L.; Xiao, J.; Sun, S.; Li, P.; Yu, J. Adsorption and desorption behavior of lithium ion in spherical PVC-MnO_2 ion sieve. *Ind. Eng. Chem. Res.* **2012**, *51*, 10921–10929. [CrossRef]
33. Luo, X.; Zhang, K.; Luo, J.; Luo, S.; Crittenden, J. Capturing lithium from wastewater using a fixed bed packed with 3-D MnO_2 ion cages. *Environ. Sci. Technol.* **2016**, *50*, 13002–13012. [CrossRef] [PubMed]
34. Santoso, S.P.; Kurniawan, A.; Soetaredjo, F.E.; Cheng, K.-C.; Putro, J.; Ismadji, S.; Ju, Y.-H. Eco-friendly cellulose–bentonite porous composite hydrogels for adsorptive removal of azo dye and soilless culture. *Cellulose* **2019**, *26*, 3339–3358. [CrossRef]
35. Wang, N.; Feng, J.; Yan, W.; Zhang, L.; Liu, Y.; Mu, R. Dual-functional sites for synergistic adsorption of Cr(Vi) and Sb(V) by polyaniline-Tio2 hydrate: Adsorption behaviors, sites and mechanisms. *Front. Environ. Sci. Eng.* **2022**, *16*, 105. [CrossRef]
36. Vaghetti, J.C.; Lima, E.C.; Royer, B.; Brasil, J.L.; da Cunha, B.M.; Simon, N.M.; Cardoso, N.F.; Noreña, C.P.Z. Application of Brazilian-pine fruit coat as a biosorbent to removal of Cr(VI) from aqueous solution—Kinetics and equilibrium study. *Biochem. Eng. J.* **2008**, *42*, 67–76. [CrossRef]
37. Zhang, J.; Wang, Y.; Liang, D.; Xiao, Z.; Xie, Y.; Li, J. Sulfhydryl-modified chitosan aerogel for the adsorption of heavy metal ions and organic dyes. *Ind. Eng. Chem. Res.* **2020**, *59*, 14531–14536. [CrossRef]

Article

Tung Oil Thermal Treatment Improves the Visual Effects of Moso Bamboo Materials

Tong Tang [1,2,*], Benhua Fei [2], Wei Song [2], Na Su [2] and Fengbo Sun [2,*]

[1] School of Art & Design, Qilu University of Technology (Shandong Academy of Sciences), Jinan 250353, China
[2] Key Laboratory of Bamboo and Rattan Science and Technology of the State Forestry Administration, Department of Biomaterials, International Center for Bamboo and Rattan, Beijing 100102, China; feibenhua@icbr.ac.cn (B.F.); j5international@163.com (W.S.); yuhesu122216@126.com (N.S.)
* Correspondence: tangtong@qlu.edu.cn (T.T.); sunfengbo@icbr.ac.cn (F.S.)

Abstract: Color is one of the most important characteristics of a material's appearance, which affects the additional value of bamboo and psychological feelings of users. Previous studies have shown that the dimensional stability, mildew resistance and durability of bamboo were improved after tung oil thermal treatment. In this study, the effects of tung oil thermal treatment on bamboo color at different temperatures and durations of time were investigated. The results show that the lightness (L^*) of bamboo decreased as the tung oil temperature or duration of time increased. The red–green coordinates (a^*) and color saturation (C^*) of bamboo were gradually increased as the tung oil temperature rose from 23 °C to 160 °C, while the a^* and C^* were gradually decreased when the temperature continued to rise from 160 °C to 200 °C. There was no significant difference in the yellow–blue coordinates (b^*) of bamboo when the duration was prolonged from 0.5 h to 3 h with tung oil thermal treatment at 140 °C. Eye movement data show that the popularity of bamboo furniture was significantly improved at 23–100 °C and slightly improved at 160–180 °C with tung oil treatment. Therefore, tung oil thermal treatment plays a positive role in improving visual effects and additional value of bamboo.

Keywords: tung oil; bamboo; thermal treatment; color

Citation: Tang, T.; Fei, B.; Song, W.; Su, N.; Sun, F. Tung Oil Thermal Treatment Improves the Visual Effects of Moso Bamboo Materials. *Polymers* 2022, 14, 1250. https://doi.org/10.3390/polym14061250

Academic Editor: Carlo Santulli

Received: 26 February 2022
Accepted: 17 March 2022
Published: 20 March 2022

Publisher's Note: MDPI stays neutral with regard to jurisdictional claims in published maps and institutional affiliations.

Copyright: © 2022 by the authors. Licensee MDPI, Basel, Switzerland. This article is an open access article distributed under the terms and conditions of the Creative Commons Attribution (CC BY) license (https://creativecommons.org/licenses/by/4.0/).

1. Introduction

Bamboo is an excellent biomaterial that improves the living environments of humans and has lightweight, good mechanical properties, processes properties and visual properties [1,2]. Bamboo has been widely used in architecture, decoration and furniture. Color is an important characteristic for the evaluation of the application value of bamboo [3,4]. The color of bamboo is perceived by human eyes at a light wavelength at 400–700 nm after absorption and reflection on the bamboo surface, relating to the color system composed of chromophores and auxochromes. However, bamboo is prone to deformation and mildew. Therefore, it is necessary to modify bamboo before it is used as an engineering material.

In order to improve the durability of bamboo, the properties of bamboo can be improved by mechanical compression, surface coating, thermal treatment, in situ polymerization and so on [5–9]. Oil thermal treatment is considered to be an effective industrial modification method for improving the dimensional stability, mildew resistance and durability of bamboo and wood [10–12]. Compared with thermal treatment in steam [13,14] or a nitrogen atmosphere [15], oil thermal treatment can protect wood or bamboo from mold and fungi decay as well as prevent moisture access to its cell walls in the long term, as oil is not only a heat transfer medium but also an excellent modifier. Oil had been used as protective surface coating for over thousands years. Additionally, oil treatment not only improves wood properties but also enhances the aesthetic of wood. In addition, tung oil, as a vegetable oil, is friendly for the environment and wood itself [16]. The effect of oil thermal treatment on wood color is mainly related to the changes in chromophores and auxochromes in lignin, extracts and hemicellulose [17,18]. The oxidative polymerization of oil influences the color of the material, and the more oil absorbed by the material, the deeper the color of the material. For example, after Dubey et al. thermally treated radiata pine with linseed oil at 180 °C, the lightness (L^*) of the radiata pine significantly decreased, whereas the red–green coordinates (a^*) and yellow–blue coordinates (b^*) increased, which may be related to the oil absorption of radiata pine and the move of pyrolysis products, including quinine, low-molecular-weight sugar and amino acid to the surface of radiata pine [19].

Previous studies have shown that the synergistic effect of tung oil and thermal treatment (100–200 °C) could improve the dimensional stability and mildew resistance of bamboo and maintain the excellent mechanical properties of bamboo [12,20]. Additionally, tung oil can separate oxygen from bamboo during the treatment process, with less degradation of the bamboo's chemical structure compared with thermal treatment in air, allowing the excellent mechanical properties of bamboo to be maintained [17]. Tung oil is a transparent oil with an orange color containing a mass of unsaturated conjugated groups compared with other natural vegetable oils, such as soybean oil and flaxseed oil; thus, the natural drying time of tung oil is significantly shorter than that of the other oils. For instance, the natural drying time of tung oil is approximately one-tenth of that of linseed oil [21–23]. Moreover, tung oil is an excellent water repellent, even after severe aging and weathering [24]. Therefore, tung oil is more advantageous in industrial applications. At present, no systematic study on the effect of tung oil thermal treatment on bamboo color has been carried out. Moreover, due to unique chemical property of tung oil, the effects and mechanisms of thermal treatment with tung oil on bamboo could be different from other oils. In view of the importance of bamboo color for its application in the living environment, this study examined the effect of different temperatures (23–200 °C) and durations (0.5–3 h) of tung oil thermal treatment on bamboo color. The visual–physical and visual–psychological data were measured and analyzed to provide a theoretical basis for the application of tung oil in bamboo modification.

2. Materials and Methods

2.1. Materials

Five-year-old moso bamboo (*Phyllostachys edulis* (Carr.)J. Houz) was obtained from Xuancheng, Anhui, China. Moso bamboo from 1.5 m (height from base) to 3.5 m in height was used in this study. Defect-free bamboo materials were dried at room temperature and cut from the center region to dimensions of 20 × 50 × 5 mm^3 (longitudinal × tangential × radial), as shown in Figure 1a–c. Samples were then kept in a climate-controlled room until the moisture content reached approximately 12% before use. Tung oil was purchased from Emperor's craftsman, Shanghai, China.

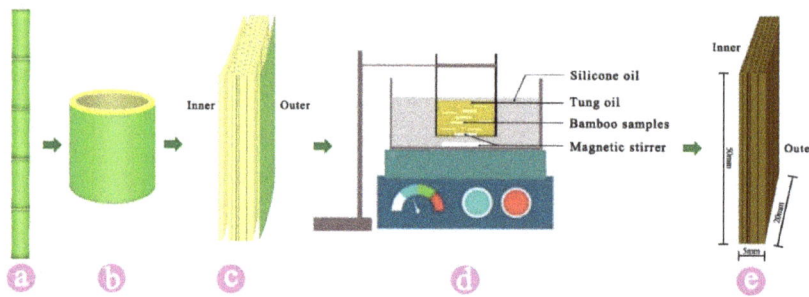

Figure 1. Schematic presentation of sample preparation. (**a**) Moso bamboo, (**b**) bamboo culm, (**c**) bamboo sample, (**d**) bamboo thermal treatment with tung oil and (**e**) bamboo sample.

2.2. Sample Preparation

Moso bamboo samples were treated with tung oil at 23 °C, 100 °C, 120 °C, 140 °C, 160 °C, 180 °C and 200 °C for 3 h (Figure 1d). Furthermore, in order to analyze the color as influenced by the duration of treatment, the bamboo samples were thermal treated with tung oil at 140 °C for 0.5 h, 1 h and 3 h. The 140 °C variable was chosen for a further analysis of how color is influenced by treatment duration, mainly because of the better mechanical performance at 140 °C than other temperatures (range from 23 to 200 °C) [12,20]. During thermal treatment of samples, the treatment temperature was maintained constantly within ±2 °C. After tung oil thermal treatment, the excess oil on samples' surface was wiped off, and the samples were then naturally dried. Following our conventions, 140 °C–3 h is the abbreviation of bamboo that underwent tung thermal treatment at 140 °C for 3 h, and likewise for the other treatments.

2.3. Visual–Physical Quantification

The color of bamboo before and after thermal treatment with tung oil was measured using a colorimeter (CC-6834, BYK-Gardner, Grazrid, Germany), referring to color space CIE $L^*a^*b^*$. Measurements were carried out by a D65 illuminant and 10° standard observer. To reduce errors, color parameters were measured at the center position of the sample surface. Furthermore, the measurement was performed more than 20 times to verify the results.

The CIE $L^*a^*b^*$ is one of the most frequently used methods to quantify surface color, as it describes the color as human eyes would recognize it. The relative color changes ΔL^*, Δa^* and Δb^* were calculated by L^*, a^* and b^* before and after the same sample was treated (for example: $\Delta a^* = a^*_{after\ treated} - a^*_{before\ treated}$).

From the relative color changes ΔL^*, Δa^* and Δb^*, the total color change ΔE^* was calculated by Equation (1):

$$\Delta E* = \sqrt{\Delta a*^2 + \Delta b*^2 + \Delta L*^2} \tag{1}$$

Based on the L^*, a^* and b^* color coordinates, the color saturation (C^*) was calculated according to Equation (2):

$$C* = \sqrt{a*^2 + b*^2} \tag{2}$$

The statistical analysis of color parameters was carried out with the use of IBM SPSS Statistics software and based on the least significant difference (LSD) or one-way ANOVA, with a significance level (p) of 0.05. The same superscript letters (a, b, c, d, e, f or g) marked in the same column indicate there were no significant differences between them.

2.4. Visual–Psychological Quanification

The effect of bamboo color changes on the psychological feelings of users was quantified using an eye tracker (Tobii X120, Tobii Technology AB, Danderyd, Sweden). A total of 40 participants were randomly selected from college students at Qilu University of Technology with normal vision, aged between 18 and 26 years old. Three different types of furniture, including two different styles of tables and one cabinet, were designed. The bamboo furniture models with corresponding colors after different tung oil treatments were created using graphic design software, which was used to quantify the changes in users' visual psychology. In the eye movement experiment, the same type of bamboo furniture that had undergone different tung oil treatments with varying temperature or treatment duration was put in a picture, and twelve groups of tests were designed in total. After the participants entered a laboratory with sound insulation and uniform light, a pre-experiment was conducted so participants could familiarize themselves with the testing process. Subsequently, the formal experiment began after the participants relaxed their eyes for one minute. The experiments involved observing bamboo furniture and making a choice about their favorite piece of furniture. To quantify the visual psychology of different types tung-oil-treated bamboo furniture, the eye tracker automatically recorded the eye fixation duration, fixation count and a hot-spot map of the participants as they observed different colors and different types of furniture during the experiment. The total fixation duration and fixation count of each participant was different, so it is more accurate to express them as a percentage.

3. Results and Discussion

The visual-physical quantity depends on color tone, lightness and saturation. The color tone is determined by the dominant wavelength of light reflected from the surface of the material. In the visible spectrum, different wavelengths of light cause different visual effects, resulting in red, orange, yellow, green, blue, purple or other corresponding tones. Lightness is the degree of color brightness, which refers to the reflection coefficient of light and depends on the intensity of the light. Saturation refers to the purity of color and depends on the range of the wavelength of light reflected from the surface.

Bamboo is usually processed into standard engineering materials, such as laminated timber and flattened timber after removing bamboo bark. Due to the differential radial gradient structure of bamboo, bamboo shows different colors in different radial gradient directions affected by the changes in chemical composition [25]. In this study, the color (tone, lightness and saturation) of bamboo (near the outer layer and near the inner layer) after different tung oil treatments was investigated in detail.

3.1. The Effect of Changes in Tung Oil Temperature on Bamboo Color

The color of bamboo after tung oil treatment for 3 h at different temperatures is shown in Table 1. The L^* of bamboo near the outer layer was decreased after tung oil treatment, which was decreased from 81.67 (untreated bamboo) to 78.41 (23 °C–3 h). The L^* of bamboo near the outer layer continued to decline as tung oil temperature increased, reaching only 37.77 after tung oil treatment at 200 °C. The a^* of bamboo after tung oil treatment was higher than that of the untreated bamboo, indicating that the redness of the bamboo increased after tung oil treatment. When the tung oil temperature increased from 23 °C to 160 °C, the a^* of bamboo near the outer layer gradually increased from 12.64 to 18.12, while the a^* of bamboo near the outer layer decreased to 14.94 as the tung oil temperature increased to 200 °C. No significant difference in the b^* of bamboo after the increase in tung oil temperature from 23 °C to 160 °C was observed, but the b^* within this temperature range was significantly higher than that of the bamboo that had undergone an increase in tung oil temperature from 180 °C to 200 °C. The variation trend of C^* of bamboo was similar to that of a^*, and the only difference was that the C^* of bamboo that had undergone tung oil thermal treatment at 200 °C was lower than that of untreated bamboo.

Table 1. The color of bamboo after tung oil treatment at different temperatures.

Treatment	Lightness L^*		Red–Green Coordinates a^*		Yellow–Blue Coordinates b^*		Color Saturation C^*	
	Outer	Inner	Outer	Inner	Outer	Inner	Outer	Inner
Untreated	81.67 [g] (0.80)	84.11 [g] (0.66)	10.26 [a] (0.33)	8.75 [a] (0.40)	25.56 [b] (0.68)	23.63 [b] (0.69)	27.55 [a] (0.73)	25.20 [b] (0.75)
23 °C–3 h	78.41 [fg] (1.90)	79.10 [ef] (2.50)	12.64 [ab] (1.03)	11.43 [b] (1.03)	36.08 [d] (1.96)	32.71 [d] (1.74)	38.23 [bc] (2.19)	34.66 [cd] (1.91)
100 °C–3 h	75.23 [ef] (1.11)	80.14 [fg] (0.76)	14.56 [bc] (0.38)	11.71 [b] (0.34)	39.67 [d] (0.56)	36.34 [e] (0.46)	42.26 [cd] (0.55)	38.18 [def] (0.46)
120 °C–3 h	74.72 [ef] (0.84)	78.87 [ef] (1.93)	14.97 [bcd] (0.17)	12.26 [bc] (1.04)	39.47 [d] (1.10)	35.28 [de] (1.31)	42.22 [cd] (1.05)	37.35 [de] (1.56)

Table 1. Cont.

Treatment	Lightness L^*		Red–Green Coordinates a^*		Yellow–Blue Coordinates b^*		Color Saturation C^*	
	Outer	Inner	Outer	Inner	Outer	Inner	Outer	Inner
140 °C–3 h	69.49 [cd] (2.03)	75.59 [cde] (1.99)	17.01 [de] (0.86)	13.89 [bcd] (1.07)	39.41 [d] (1.35)	37.31 [e] (1.20)	42.92 [d] (1.57)	39.82 [ef] (1.22)
160 °C–3 h	66.47 [c] (3.13)	71.75 [c] (2.95)	18.12 [e] (1.02)	16.07 [de] (1.18)	39.37 [d] (0.87)	38.53 [e] (1.13)	43.35 [d] (0.94)	41.76 [f] (1.30)
180 °C–3 h	51.46 [b] (1.98)	46.47 [b] (1.82)	17.80 [e] (1.04)	17.05 [e] (0.77)	30.39 [c] (1.94)	27.18 [b] (1.76)	35.22 [b] (2.11)	32.09 [c] (1.88)
200 °C–3 h	37.77 [a] (3.63)	32.39 [a] (2.39)	14.94 [bcd] (2.87)	12.60 [bc] (2.90)	19.58 [a] (4.77)	14.12 [a] (3.45)	24.65 [a] (5.45)	18.94 [a] (4.42)

* Note: Mean values followed by the same superscript letters (a, b, c, d, e, f or g) in the same column are not significantly different at <0.05. Values in parentheses are standard deviations. Outer: the surface near the outer layer of bamboo. Inner: the surface near the inner layer of bamboo.

The L^* of untreated bamboo from near the outer layer was lower than that from near the inner layer, whereas the a^* and b^* were higher than that from near the inner layer. The bamboo color is mainly related to the chemical structure of lignin and extractives [26,27]. Wei et al. measured the chemical composition of inner, middle and outer moso bamboo layers, which showed that the content of lignin in the outer layer was higher than in the inner layer [28]. The content of lignin in the outer layer was higher than in the inner layer of bamboo, which indicates a high bulk density increases the chromophore concentration in the outer layer, leading to more chromophores in the outer layer [29]. Hence, the color intensity of the bamboo's outer layer was increased, and a reduction in lightness could be observed. Although bamboo has a gradient structure, the color change rule of bamboo near the inner layer was similar to that of the outer layer (Table 1).

In order to further analyze the effect of tung oil temperature on bamboo color, statistical analysis was conducted on the changes in color, as shown in Figure 2. Both tung oil thermal treatment and tung oil treatment at room temperature changed bamboo color, and the ΔE^* increased with rising tung oil temperature. A significant color difference was observed between 23 °C–3 h and untreated bamboo, which was mainly related to the oxidation of tung oil on bamboo. Yoo et al. reported that the L^* of wood was decreased, whereas the a^* and the b^* were increased on the tung-oil-finished wood surface [22], which shows similar results with tung-oil-treated bamboo at 23 °C. The a^* and b^* of bamboo were gradually decreased as the tung oil treatment temperature increased from 160 °C to 200 °C. The same trend was also found in wood after thermal treatment [19,30–32].

With rising tung oil temperature, the ΔE^* was gradually increased (Table S1), which was closely related to the degradation of bamboo chemical components and the aging of tung oil on bamboo. The chemical structure of lignin and extractives changed after thermal treatment, resulting in the variation in bamboo color [33]. Lignin is an aromatic compound that produces a significant number of chromophores during thermal treatment [34]. Previous studies have suggested that the β-O-4 bond in the propyl structure of the benzene ring side chain is acidified and broken during thermal treatment, which reduces the esterification structure and increases the content of phenolic hydroxyl [35–38]. The phenolic hydroxyl is gradually oxidized to quinone compounds, leading to the increase in a^* and b^*. In addition, the dehydration reaction occurs in conjugated C=C in the aromatic ring and C=O during thermal treatment, resulting in the increase in ketones and conjugated carbonyl groups in lignin [39]. The increase in conjugated structure and extension of conjugated system leads to increased L^* [27,40–42]. Extractives mainly include phenolic, alcohol and aldehyde compounds [43], and they can be acid-catalyzed by acetic acid during thermal treatment, which causes the self-condensation and oxidation of compounds, forming a new chromogenic system composed of conjugated double bonds, carbonyl or a quinone structure. The spectral absorption is enhanced and extended to the range of visible light, resulting in the bamboo becoming darker and redder.

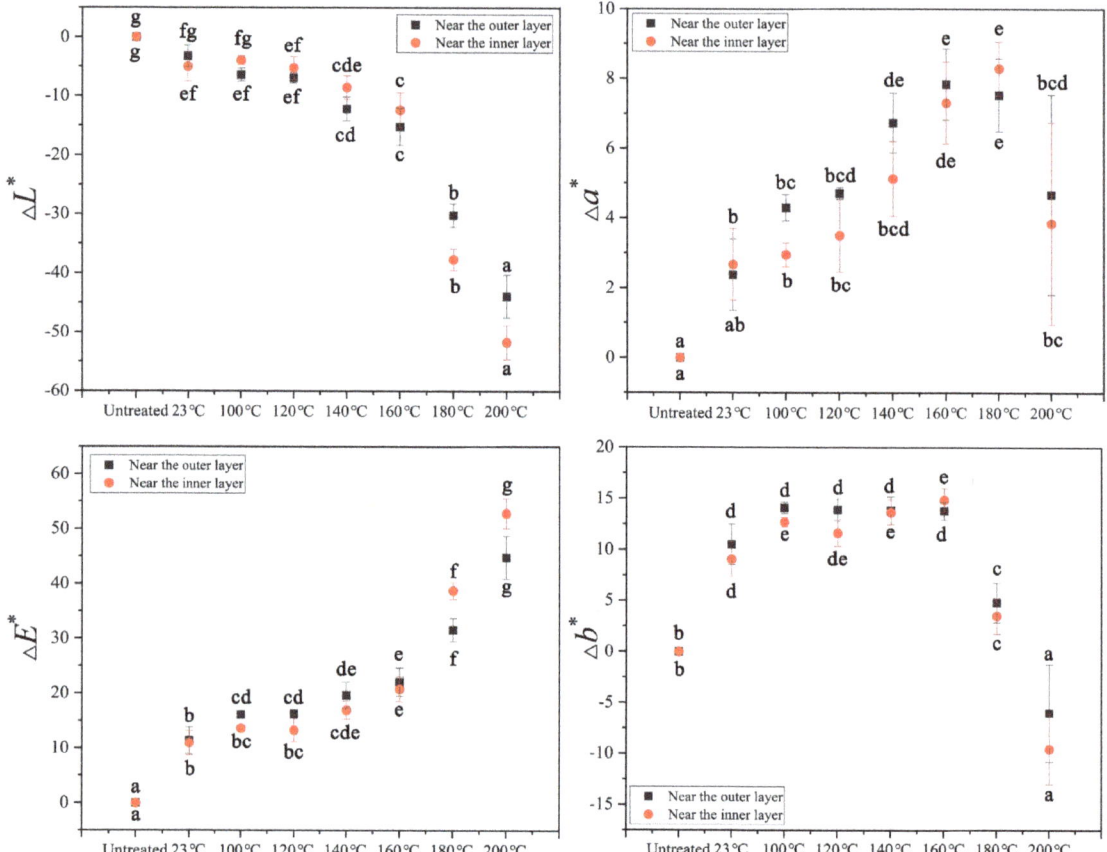

Figure 2. The variation in bamboo color after tung oil treatment at different temperatures. (* Note: Mean values followed by the same superscript letters (a, b, c, d, e, f or g) in the same column are not significantly different at <0.05. Values in parentheses are standard deviations. Outer: the surface near the outer layer of bamboo. Inner: the surface near the inner layer of bamboo.).

Tung oil has major components of α-eleostearic acid (77–82%) with three conjugated double bonds (at carbons 9 *cis*,11 *trans*, and 13 *trans*), oleic acid (3.5–12.7%) with one double bond, and linoleic acid (8–10%) with two nonconjugated double bonds [22]. Partially conjugated double bonds are altered to nonconjugated double bonds after thermal treatment by oxidation, thermal rearrangement or both [44]. Similarly to other vegetable oils [45], the color of tung oil deepens with thermal treatment. The changed chemical structure of tung oil also influences bamboo color. In addition, some studies have demonstrated that more oil absorption results in a deeper color of the material [20,46]. After the thermal treatment and drying processes, an oil layer is formed on the bamboo surface, enhancing surface color changes in bamboo [6]. The chemical structure of bamboo and tung oil change during thermal treatment, and oil absorption changes along with the thermal temperature [12], which comprehensively influence bamboo color.

3.2. The Effect of Changes in Tung Oil Treatment Duration on Bamboo Color

The color of bamboo after tung oil thermal treatment at 140 °C under different treatment durations of time is shown in Table 2. With the extension of tung oil treatment duration, the L^* of bamboo decreased and the a^* of bamboo increased, but the tung oil treatment duration had no significant impact on the b^* of bamboo. After thermal treatment with tung oil for over 1 h, the C^* of bamboo did not change significantly. Comparing Table 1 with Table 2, the ΔE^* of 140 °C–0.5 h was higher than that of 23 °C–3 h but lower than that of 100 °C–3 h. Furthermore, the ΔE^* of 140 °C–1 h was higher than that of 120 °C–3 h but lower than that of 140 °C–3 h. Significant differences among the color of bamboo were observed after thermal treatment for 0–1 h at 140 °C. However, the differences reduced with increasing treatment duration.

Table 2. The color of bamboo after tung oil treatment at different durations of time.

Treatment	Lightness L^*		Red–Green Coordinates a^*		Yellow–Blue Coordinates b^*		Color Saturation C^*	
	Outer	Inner	Outer	Inner	Outer	Inner	Outer	Inner
Untreated	81.67 [g] (0.80)	84.11 [g] (0.66)	10.26 [a] (0.33)	8.75 [a] (0.40)	25.56 [b] (0.68)	23.63 [b] (0.69)	27.55 [a] (0.73)	25.20 [b] (0.75)
140 °C–0.5 h	76.21 [f] (1.54)	78.15 [def] (2.16)	14.09 [bc] (0.55)	12.92 [bc] (0.89)	38.07 [d] (1.42)	36.06 [de] (0.76)	40.60 [cd] (1.46)	38.31 [def] (0.88)
140 °C–1 h	71.31 [de] (1.35)	73.81 [cd] (2.96)	16.48 [cde] (0.43)	14.55 [cde] (1.04)	40.05 [d] (0.78)	37.01 [e] (1.23)	43.31 [d] (0.83)	39.77 [ef] (1.44)
140 °C–3 h	69.49 [cd] (2.03)	75.59 [cde] (1.99)	17.01 [de] (0.86)	13.89 [bcd] (1.07)	39.41 [d] (1.35)	37.31 [e] (1.20)	42.92 [d] (1.57)	39.82 [ef] (1.22)

* Note: Mean values followed by the same superscript letters (a, b, c, d, e, f or g) in the same column are not significantly different at <0.05. Values in parentheses are standard deviations. Outer: the surface near the outer layer of bamboo. Inner: the surface near the inner layer of bamboo.

The changes in bamboo color after different tung oil treatment durations are shown in Figure 3. The L^* and a^* of bamboo significantly changed after tung oil thermal treatment at 140 °C within 0–3 h, which was mainly related to the gradual increase in quinone compounds, ketones and the conjugated system during thermal treatment. The b^* of bamboo had no significant changes after tung oil thermal treatment at 140 °C within 0.5–3 h (Figure 3), which indicates a rapid increase in b^* occurred in the early thermal treatment and an extended treatment duration had little impact on b^*.

The ΔE^* was affected by thermal treatment conditions (Table S2) and the type of material. The ΔE^* was gradually increased with the thermal temperature rising from 23 °C to 200 °C (Figure 2), which was also gradually increased over the thermal treatment duration, which was extended to 1 h at 140 °C (Figure 3). The ΔE^* had greater values after the oil thermal treatment than after air thermal treatment and nitrogen thermal treatment under the same conditions [6]. The ΔE^* was also related to the type of material, for example, moso bamboo had an obviously higher value of ΔE^* compared with Scots pine after thermally treated in air under the same conditions [6,47]. Nevertheless, bamboo scrimber thermally treated with methyl silicone oil had a similar value of ΔE^* to moso bamboo thermally treated with tung oil [11].

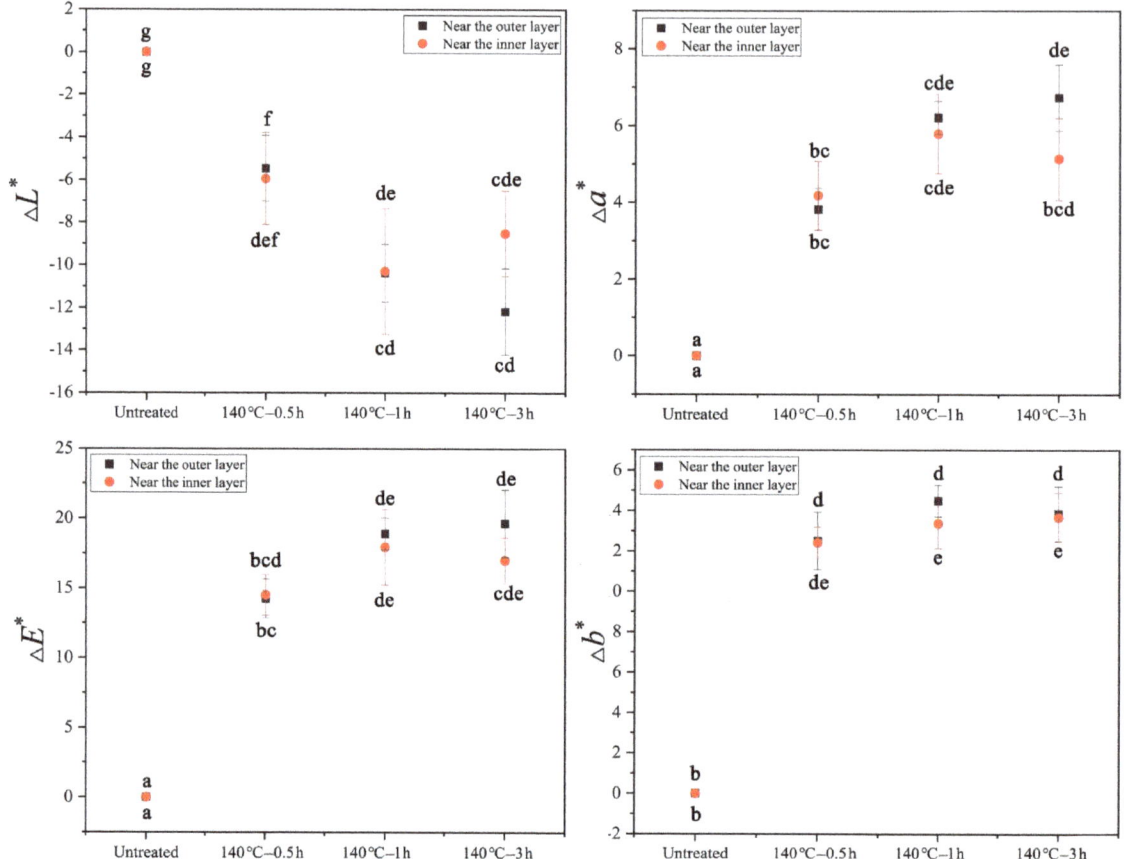

Figure 3. The variation in bamboo color after tung oil treatment at 140 °C for 0–3 h. (* Note: Mean values followed by the same superscript letters (a, b, c, d, e, f or g) in the same column are not significantly different at <0.05. Values in parentheses are standard deviations. Outer: the surface near the outer layer of bamboo. Inner: the surface near the inner layer of bamboo.).

3.3. The Effect of Tung Oil Treatment on Visual–Psychological Quantification

Eye trackers are used to precisely record the trajectory of human eye movement. Visual responses and characteristics can be obtained from eye movement data. The fixation points of observers are usually distributed across the most important, comprehensible or informative objects. The regions recognized as the interesting spots experience more fixation counts and a longer fixation duration [48–51]. The hot-spot map as a common representation of fixation duration that appears in the studies of eye movements, which reflects the attention of participants [52,53]. Together, the fixation count, fixation duration and hot-spot map were used to analyze the popularity of different bamboo furniture pieces after tung oil treatment.

To simulate bamboo used in the living environment, three different types of furniture were designed and matched to bamboo color after tung oil treatment. The eye movement data show that furniture was made of bamboo after tung oil treatment at 23 °C–3 h, 100 °C–3 h, 160 °C–3 h and 180 °C–3 h experience more fixation counts and a longer fixation duration compared with untreated bamboo (Figure 4), indicating the popularity of bamboo furniture was improved after tung oil treatment at 23 °C, 100 °C, 160 °C and 180 °C. The popularity of bamboo furniture was significantly improved after tung oil treatment at 23 °C and 100 °C, whereas a slightly improved popularity of bamboo furniture was observed after tung oil treatment at 160 °C and 180 °C (Figure 4). The popularity of bamboo furniture after tung oil thermal treated at 140 °C and 200 °C decreased according to the eye movement data. A longer fixation duration and greater fixation count of bamboo furniture made with untreated bamboo compared with that furniture made of bamboo treated at 140 °C–0.5 h, 140 °C–1 h and 140 °C–3 h (Figure 4b,d), which suggests the participants had a greater preference for untreated bamboo compared to the bamboo treated with tung oil at 140 °C.

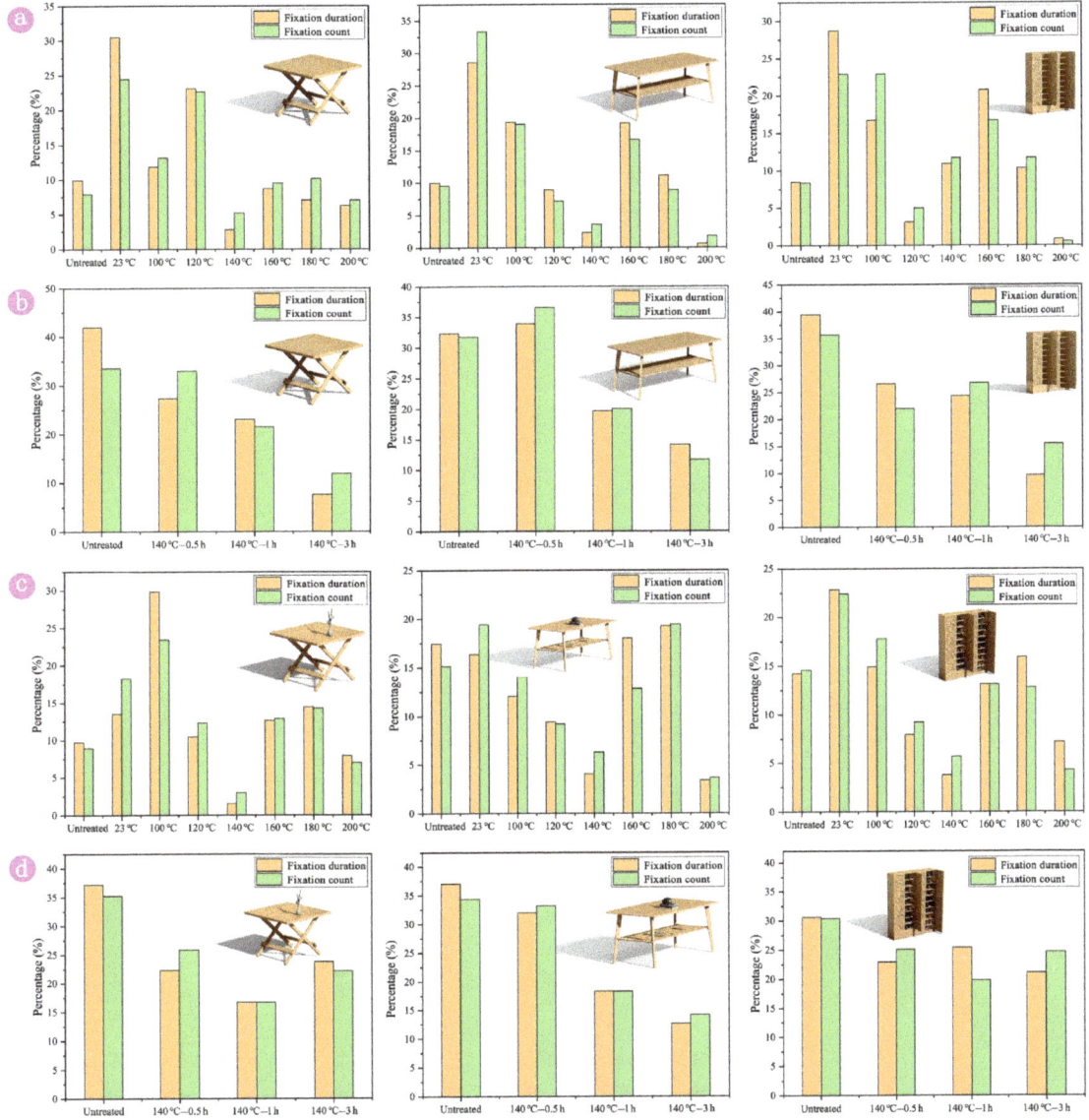

Figure 4. The fix duration and fix count of different bamboo furniture. (**a**,**b**) Furniture was made of the outer layer of bamboo, (**c**,**d**) furniture was made of the inner layer of bamboo.

The fix duration and fix count had some differences among the different types of furniture, and the hot-spot map for bamboo furniture is shown in Figure 5. In the hot-spot map, the area of the furniture was overlaid with red and yellow colors, indicating that this area received a longer total fixation duration. By contrast, the green color indicates a shorter fixation duration. The bamboo treated at 23 °C–3 h, 100 °C–3 h, 160 °C–3 h and 180 °C–3 h had a larger area of the red and yellow colors, which suggests that the participants had a greater preference for bamboo that had been treated at 23 °C–3 h, 100 °C–3 h, 160 °C–3 h and 180 °C–3 h. Interestingly, the area of the red and yellow colors for bamboo treated at 120 °C–3 h and 200 °C–3 h was gradually decreased as furniture volume increase. According to the hot-spot map, the table (small furniture) that had been treated with tung oil at 200 °C was more popular than the cabinet (large furniture).

Figure 5. The hot-spot map of different types of bamboo furniture. Bamboo (near the outer layer) after thermal treatment with tung oil for 3h: (a_1–c_1) untreated, (a_2–c_2) 23 °C, (a_3–c_3) 100 °C, (a_4–c_4) 120 °C, (a_5–c_5) 140 °C, (a_6–c_6) 160 °C, (a_7–c_7) 180 °C and (a_8–c_8) 200 °C.

Previous studies have shown that tung oil thermal treatment could improve the dimensional stability and mildew resistance of bamboo, and the durability of bamboo was gradually improved with the increase in tung oil thermal temperature [12]. According to a comprehensive analysis of bamboo properties and the popularity of bamboo furniture, the popularity of bamboo after tung oil thermal treatment at 23 °C and 100 °C was significantly improved and the durability was slightly improved, which may be more suitable for furniture or interior decorative materials with high appearance requirements. By contrast, the popularity of bamboo after tung oil thermal treatment at 160 °C and 180 °C was slightly improved and the durability was significantly improved, which is more suitable for furniture or engineering materials with high durability requirements. Although the durability of bamboo after tung oil thermal treatment at 200 °C had the greatest improvement, the popularity of bamboo was relatively low, suggesting that bamboo treated with tung oil at 200 °C may be more applicable to outdoor engineering materials, such as floorings and fencing or small-area decoration materials. Thus, the application of tung oil modification technology to bamboo not only improved the durability but also improved the visual effects of bamboo.

4. Conclusions

The oil absorption of bamboo and chemical structure of bamboo and tung oil comprehensively influenced bamboo color. Bamboo became darker with the increase in tung oil treatment temperature and duration of tung oil treatment time. The L^* of bamboo after tung oil thermal treatment at 200 °C was decreased by more than 50% compared with untreated bamboo. Bamboo became redder and had a higher saturation as the tung oil temperature rose from 23 °C to 160 °C. When the tung oil treatment temperature was higher than 180 °C, bamboo became greener and had a lower saturation with increasing tung oil temperature. The L^* and a^* of bamboo were closely related to tung oil treatment duration. However, the b^* of bamboo had no significant change over the duration of tung oil treatment from 0.5 h to 3 h at 140 °C. According to eye movement data, the popularity of furniture made by bamboo treated at 23 °C–3 h, 100 °C–3 h, 160 °C–3 h and 180 °C–3 h was increased compared with that made with untreated bamboo. Comprehensive analysis of bamboo properties and the popularity of bamboo furniture suggests that bamboo after tung oil thermal treatment at 23 °C and 100 °C is more suitable for engineering materials with high appearance requirements, whereas bamboo after tung oil thermal treatment at 160 °C and 180 °C is more suited to engineering materials with high durability requirements. The application of tung oil modification technology in the production of bamboo engineering materials not only improved the durability of bamboo but also improved the visual effects of bamboo, which would expand the application field of bamboo.

Supplementary Materials: The following supporting information can be downloaded at: https://www.mdpi.com/article/10.3390/polym14061250/s1, Table S1: Visualizations of bamboo after tung oil treatment at different temperature, Table S2: Visualizations of bamboo after tung oil treatment at different duration of time.

Author Contributions: Conceptualization, T.T. and B.F.; validation, F.S.; investigation, W.S.; writing, T.T.; visualization, N.S. All authors have read and agreed to the published version of the manuscript.

Funding: This research was funded by the Key Laboratory of National Forestry and Grassland Administration/Beijing for Bamboo & Rattan Science and Technology, grant number ICBR-2020-15 and the Science & Technology Research and Development Program of Guizhou Forestry Administration for Rural Industrial Revolution and Characteristic Forestry Industry, grant number GZMC-ZD20202112.

Data Availability Statement: All data or used during the study appear in the submitted article.

Acknowledgments: The authors appreciate the valuable comments and suggestions of the editors and reviewers.

Conflicts of Interest: The authors declare no conflict of interest. The funders had no role in the design of the study; in the collection, analyses, or interpretation of data; in the writing of the manuscript; or in the decision to publish the results.

References

1. Chen, Q.; Wei, P.L.; Tang, T.; Fang, C.H.; Fei, B.H. Quantitative visualization of weak layers in bamboo at the cellular and subcellular levels. *ACS Appl. Bio Mater.* **2020**, *3*, 7087–7094. [CrossRef] [PubMed]
2. Chen, C.J.; Li, Z.H.; Mi, R.Y.; Dai, J.Q.; Xie, H.; Pei, Y.; Li, J.G.; Qiao, H.Y.; Tang, H.; Yang, B.; et al. Rapid processing of whole bamboo with exposed, aligned nanofibrils toward a high-performance structural material. *ACS Nano* **2020**, *14*, 5194–5202. [CrossRef] [PubMed]
3. Wan, Q.; Hu, Q.W.; Chen, B.W.; Fang, H.; Ke, Q.; Song, S.S. Study on the visual cognition of laminated bamboo furniture. *For. Prod. J.* **2021**, *71*, 84–91. [CrossRef]
4. Li, R.R.; Chen, J.J.; Wang, X.D. Prediction of the color variation of moso bamboo during CO_2 laser thermal modification. *Bioresources* **2020**, *15*, 5049–5057. [CrossRef]
5. Chen, H.; Zhang, Y.T.; Yang, X.; Ji, H.; Zhong, T.H.; Wang, G. A comparative study of the microstructure and water permeability between flattened bamboo and bamboo culm. *J. Wood Sci.* **2019**, *65*, 64. [CrossRef]
6. Lee, C.H.; Yang, T.H.; Cheng, Y.W.; Lee, C.J. Effects of thermal modification on the surface and chemical properties of moso bamboo. *Constr. Build. Mater.* **2018**, *178*, 59–71. [CrossRef]
7. Kadivar, M.; Gauss, C.; Tomazello-Filho, M.; Ahrar, A.J.; Ghavami, K.; Savastano, H. Optimization of thermo-mechanical densification of bamboo. *Constr. Build. Mater.* **2021**, *298*, 123860. [CrossRef]
8. Huang, S.S.; Jiang, Q.F.; Yu, B.; Nie, Y.J.; Ma, Z.Q.; Ma, L.F. Combined chemical modification of bamboo material prepared using vinyl acetate and methyl methacrylate: Dimensional stability, chemical structure, and dynamic mechanical properties. *Polymers* **2019**, *11*, 1651. [CrossRef]
9. Azadeh, A.; Ghavami, K. The influence of heat on shrinkage and water absorption of *Dendrocalamus giganteus* bamboo as a functionally graded material. *Constr. Build. Mater.* **2018**, *186*, 145–154. [CrossRef]
10. Cheng, D.L.; Li, T.; Smith, G.D.; Xu, B.; Li, Y.J. The properties of moso bamboo heat-treated with silicon oil. *Eur. J. Wood Wood Prod.* **2018**, *76*, 1273–1278. [CrossRef]
11. Yuan, Z.R.; Wu, X.W.; Wang, X.Z.; Zhang, X.; Yuan, T.C.; Liu, X.M.; Li, Y.J. Effects of one-step hot oil treatment on the physical, mechanical, and surface properties of bamboo scrimber. *Molecules* **2020**, *25*, 4488. [CrossRef] [PubMed]
12. Tang, T.; Zhang, B.; Liu, X.M.; Wang, W.B.; Chen, X.F.; Fei, B.H. Synergistic effects of tung oil and heat treatment on physicochemical properties of bamboo materials. *Sci. Rep.* **2019**, *9*, 12824. [CrossRef] [PubMed]
13. Kekkonen, P.M.; Ylisassi, A.; Telkki, V.V. Absorption of water in thermally modified pine wood as studied by nuclear magnetic resonance. *J. Phys. Chem. C* **2014**, *118*, 2146–2153. [CrossRef]
14. Gawron, J.; Marchwicka, M. Color changes of ash wood (*Fraxinus excelsior* L.) caused by thermal modification in air and steam. *For. Wood Technol.* **2021**, *116*, 21–27. [CrossRef]
15. Bytner, O.; Laskowska, A.; Drozdzek, M.; Kozakiewicz, P.; Zawadzki, J. Evaluation of the dimensional stability of black poplar wood modified thermally in nitrogen atmosphere. *Materials* **2021**, *14*, 1491. [CrossRef]
16. Sedliačiková, M.; Moresová, M.; Kocianová, A. Mapping the supply of colour tones of wood and furniture products in slovakian small and medium-sized enterprises. *Forests* **2021**, *12*, 1775. [CrossRef]
17. Lee, S.H.; Ashaari, Z.; Lum, W.C.; Halip, J.A.; Ang, A.F.; Tan, L.P.; Chin, K.L.; Tahir, P.M. Thermal treatment of wood using vegetable oils: A review. *Constr. Build. Mater.* **2018**, *181*, 408–419. [CrossRef]
18. Bytner, O.; Drozdzek, M.; Laskowska, A.; Zawadzki, J. Temperature, Time, and Interactions between Them in Relation to Colour Parameters of Black Poplar (*Populus nigra* L.) Thermally Modified in Nitrogen Atmosphere. *Materials* **2022**, *15*, 824. [CrossRef]
19. Dubey, M.K.; Pang, S.; Walker, J. Changes in chemistry, color, dimensional stability and fungal resistance of Pinus radiata D. Don wood with oil heat-treatment. *Holzforschung* **2012**, *66*, 49–57. [CrossRef]
20. Tang, T.; Chen, X.F.; Zhang, B.; Liu, X.M.; Fei, B.H. Research on the physico-mechanical properties of moso bamboo with thermal treatment in tung oil and its influencing factors. *Materials* **2019**, *12*, 599. [CrossRef]
21. Žlahtič, M.; Mikac, U.; Serša, I.; Merela, M.; Humar, M. Distribution and penetration of tung oil in wood studied by magnetic resonance microscopy. *Ind. Crop. Prod.* **2017**, *96*, 149–157. [CrossRef]
22. Yoo, Y.; Youngblood, J.P. Tung oil wood finishes with improved weathering, durability, and scratch performance by addition of cellulose nanocrystals. *ACS Appl. Mater. Inter.* **2017**, *9*, 24936–24946. [CrossRef] [PubMed]
23. Arminger, B.; Jaxel, J.; Bacher, M.; Gindl-Altmutter, W.; Hansmann, C. On the drying behavior of natural oils used for solid wood finishing. *Prog. Org. Coat.* **2020**, *148*, 105831. [CrossRef]
24. He, Z.B.; Qian, J.; Qu, L.J.; Yan, N.; Yi, S.L. Effects of tung oil treatment on wood hygroscopicity, dimensional stability and thermostability. *Ind. Crop. Prod.* **2019**, *140*, 111647. [CrossRef]
25. Wu, J.Y.; Zhong, T.H.; Zhang, W.F.; Shi, J.J.; Fei, B.H.; Chen, H. Comparison of colors, microstructure, chemical composition and thermal properties of bamboo fibers and parenchyma cells with heat treatment. *J. Wood Sci.* **2021**, *67*, 56. [CrossRef]
26. Kurei, T.; Tsushima, R.; Okahisa, Y.; Nakaba, S.; Funada, R.; Horikawa, Y. Creation and structural evaluation of the three-dimensional cellulosic material "White-Colored Bamboo". *Holzforschung* **2021**, *75*, 180–186. [CrossRef]
27. Yu, H.X.; Zheng, H.L.; Zhan, M.Y.; Zhang, W.F.; Wang, J.; Pan, X.; Zhuang, X.W. Surface characterization and biodegradability of sodium hydroxide-treated moso bamboo substrates. *Eur. J. Wood Wood Prod.* **2020**, *79*, 443–451. [CrossRef]
28. Wei, X.; Wang, G.; Smith, L.M.; Jiang, H. The hygroscopicity of moso bamboo (*Phyllostachys edulis*) with a gradient fiber structure. *J. Mater. Res. Technol.* **2021**, *15*, 4309–4316. [CrossRef]

29. Zhang, Y.; Naebe, M. Lignin: A review on structure, properties, and applications as a light-colored UV absorber. *ACS Sustain. Chem. Eng.* **2021**, *9*, 1427–1442. [CrossRef]
30. Gašparík, M.; Gaff, M.; Kačík, F.; Sikora, A. Color and chemical changes in teak (*Tectona grandis* L. f.) and meranti (*Shorea* spp.) wood after thermal treatment. *Bioresources* **2019**, *14*, 2667–2683.
31. Cuccui, I.; Negro, F.; Zanuttini, R.; Espinoza, M.; Allegretti, O. Thermo-vacuum modification of teak wood from fast-growth plantation. *Bioresources* **2017**, *12*, 1903–1915. [CrossRef]
32. Aydemir, D.; Gunduz, G.; Ozden, S. The influence of thermal treatment on color response of wood materials. *Color Res. Appl.* **2012**, *37*, 148–153. [CrossRef]
33. Yu, H.X.; Pan, X.; Wang, Z.; Yang, W.M.; Zhang, W.F.; Zhuang, X.W. Effects of heat treatments on photoaging properties of moso bamboo (*Phyllostachys pubescens* Mazel). *Wood Sci. Technol.* **2018**, *52*, 1671–1683. [CrossRef]
34. Wang, X.D.; Hou, Q.X.; Zhang, X.; Zhang, Y.C.; Liu, W.; Xu, C.L.; Zhang, F.D. Color evolution of poplar wood chips and its response to lignin and extractives changes in autohydrolysis pretreatment. *Int. J. Biol. Macromol.* **2020**, *157*, 673–679. [CrossRef] [PubMed]
35. Sun, S.N.; Li, H.Y.; Cao, X.F.; Xu, F.; Sun, R.C. Structural variation of eucalyptus lignin in a combination of hydrothermal and alkali treatments. *Bioresource Technol.* **2015**, *176*, 296–299. [CrossRef] [PubMed]
36. Li, J.B.; Henriksson, G.; Gellerstedt, G. Lignin depolymerization/repolymerization and its critical role for delignification of aspen wood by steam explosion. *Bioresource Technol.* **2007**, *98*, 3061–3068. [CrossRef] [PubMed]
37. Mitsui, K.; Inagaki, T.; Tsuchikawa, S. Monitoring of hydroxyl groups in wood during heat treatment using NIR spectroscopy. *Biomacromolecules* **2008**, *9*, 286–288. [CrossRef]
38. Kim, J.Y.; Hwang, H.; Oh, S.; Kim, Y.S.; Kim, U.J.; Choi, J.W. Investigation of structural modification and thermal characteristics of lignin after heat treatment. *Int. J. Biol. Macromol.* **2014**, *66*, 57–65. [CrossRef]
39. Keating, J.; Johansson, C.I.; Saddler, J.N.; Beatson, R.P. The nature of chromophores in high-extractives mechanical pulps: Western red cedar (*Thuja plicata* Donn) chemithermomechanical pulp (CTMP). *Holzforschung* **2006**, *60*, 365–371. [CrossRef]
40. Zhang, Y.M.; Yu, W.J.; Zhang, Y.H. Effect of Steam Heating on the color and chemical properties of *Neosinocalamus affinis* bamboo. *J. Wood Chem. Technol.* **2013**, *33*, 235–246. [CrossRef]
41. Zhang, H.; Liu, X.X.; Fu, S.Y.; Chen, Y.C. Fabrication of light-colored lignin microspheres for developing natural sunscreens with favorable UV absorbability and staining resistance. *Ind. Eng. Chem. Res.* **2019**, *58*, 13858–13867. [CrossRef]
42. Wei, Y.X.; Wang, M.J.; Zhang, P.; Chen, Y.; Gao, J.M.; Fan, Y.M. The role of phenolic extractives in color changes of locust wood (*Robinia pseudoacacia*) during heat treatment. *Bioresources* **2017**, *12*, 7041–7055.
43. Wang, Y.P.; Wu, L.B.; Wang, C.; Yu, J.Y.; Yang, Z.Y. Investigating the influence of extractives on the oil yield and alkane production obtained from three kinds of biomass via deoxy-liquefaction. *Bioresour. Technol.* **2011**, *102*, 7190–7195. [CrossRef] [PubMed]
44. Zhuang, Y.W.; Ren, Z.Y.; Jiang, L.; Zhang, J.X.; Wang, H.F.; Zhang, G.B. Raman and FTIR spectroscopic studies on two hydroxylated tung oils (HTO) bearing conjugated double bonds. *Spectrochim. Acta, A.* **2018**, *199*, 146–152. [CrossRef]
45. Yang, D.; Wu, G.C.; Lu, Y.; Li, P.Y.; Qi, X.G.; Zhang, H.; Wang, X.G.; Jin, Q.Z. Comparative analysis of the effects of novel electric field frying and conventional frying on the quality of frying oil and oil absorption of fried shrimps. *Food Control* **2021**, *128*, 108195. [CrossRef]
46. Ohshima, K.; Sugimoto, H.; Sugimori, M.; Sawada, E. Effect of the internal structure on color changes in wood by painting transparent. *Color Res. Appl.* **2021**, *46*, 645–652. [CrossRef]
47. Kamperidou, V.; Barboutis, I.; Vasileiou, V. Response of colour and hygroscopic properties of Scots pine wood to thermal treatment. *J. Forestry Res.* **2013**, *24*, 571–575. [CrossRef]
48. Renshaw, J.A.; Finlay, J.E.; Tyfa, D.; Ward, R.D. Understanding visual influence in graph design through temporal and spatial eye movement characteristics. *Interact. Comput.* **2004**, *16*, 557–578. [CrossRef]
49. Gao, Y.; Zhang, T.; Zhang, W.K.; Meng, H.; Zhang, Z. Research on visual behavior characteristics and cognitive evaluation of different types of forest landscape spaces. *Urban For. Urban Gree.* **2020**, *54*, 126788. [CrossRef]
50. Yang, X.; Wang, R.H.; Tang, C.L.; Luo, L.H.; Mo, X.H. Emotional design for smart product-service system: A case study on smart beds. *J. Clean. Prod.* **2021**, *298*, 126823. [CrossRef]
51. Lin, C.J.; Chang, C.C.; Lee, Y.H. Evaluating camouflage design using eye movement data. *Appl. Ergon.* **2014**, *45*, 714–723. [CrossRef] [PubMed]
52. Ding, M.; Song, M.J.; Pei, H.N.; Cheng, Y. The emotional design of product color: An eye movement and event-related potentials study. *Color Res. Appl.* **2021**, *46*, 871–889. [CrossRef]
53. Wang, Y.H.; Yu, S.H.; Ma, N.; Wang, J.L.; Hu, Z.G.; Liu, Z.; He, J.B. Prediction of product design decision making: An investigation of eye movements and EEG features. *Adv. Eng. Inform.* **2020**, *45*, 101095. [CrossRef]

Article

Construction of Konjac Glucomannan/Oxidized Hyaluronic Acid Hydrogels for Controlled Drug Release

Hongyi Wu, Nitong Bu, Jie Chen, Yuanyuan Chen, Runzhi Sun, Chunhua Wu * and Jie Pang *

College of Food Science, Fujian Agriculture and Forestry University, Fuzhou 350002, China;
liccfas@163.com (H.W.); bnt15850446180@163.com (N.B.); 17373679172@163.com (J.C.);
chenyy6760@163.com (Y.C.); srz1206971425@163.com (R.S.)
* Correspondence: 000q818025@fafu.edu.cn (C.W.); pang3721941@fafu.edu.cn (J.P.)

Abstract: Konjac glucomannan (KGM) hydrogel has favorable gel-forming abilities, but its insufficient swelling capacity and poor control release characteristics limit its application. Therefore, in this study, oxidized hyaluronic acid (OHA) was used to improve the properties of KGM hydrogel. The influence of OHA on the structure and properties of KGM hydrogels was evaluated. The results show that the swelling capacity and rheological properties of the composite hydrogels increased with OHA concentration, which might be attributed to the hydrogen bond between the KGM and OHA, resulting in a compact three-dimensional gel network structure. Furthermore, epigallocatechin gallate (EGCG) was efficiently loaded into the KGM/OHA composite hydrogels and liberated in a sustained pattern. The cumulative EGCG release rate of the KGM/OHA hydrogels was enhanced by the increasing addition of OHA. The results show that the release rate of composite hydrogel can be controlled by the content of OHA. These results suggest that OHA has the potential to improve the properties and control release characteristics of KGM hydrogels.

Keywords: konjac glucomannan; oxidized hyaluronic acid; hydrogels

Citation: Wu, H.; Bu, N.; Chen, J.; Chen, Y.; Sun, R.; Wu, C.; Pang, J. Construction of Konjac Glucomannan/Oxidized Hyaluronic Acid Hydrogels for Controlled Drug Release. *Polymers* **2022**, *14*, 927. https://doi.org/10.3390/polym14050927

Academic Editor: Ki Hyun Bae

Received: 30 January 2022
Accepted: 22 February 2022
Published: 25 February 2022

Publisher's Note: MDPI stays neutral with regard to jurisdictional claims in published maps and institutional affiliations.

Copyright: © 2022 by the authors. Licensee MDPI, Basel, Switzerland. This article is an open access article distributed under the terms and conditions of the Creative Commons Attribution (CC BY) license (https://creativecommons.org/licenses/by/4.0/).

1. Introduction

Hydrogels are three-dimensional polymer networks connected by cross-linked covalent bonds and weak cohesive forces in the form of hydrogen or ionic bonds [1]. Due to their excellent biocompatibility, soft texture, and high permeability to small hydrophilic molecules, hydrogels have been attractive candidates for a wide range of applications, such as cartilage scaffolds, drug delivery, and sensors [2]. Furthermore, hydrogels have been used in a variety of forms, including scaffolds, injectable hydrogels, hydrogel nanoparticles, microgels, nanofibers, and hydrogel membranes [3]. However, most hydrogels have weak stability due to their high water content, which greatly limits their application [4]. This drawback can be overcome by fabricating a "composite or hybrid hydrogel" system.

Konjac glucomannan (KGM) is a kind of natural polysaccharide that is extracted from the tuber of the Amorphophallus konjac K. Koch plant. It primarily consists of a linear chain of β-1-4-linked D-glucose and D-mannose, with a glucose to mannose ratio of approximately 1:1.6, with a low degree of acetyl groups at the side-chain C-6 position [5]. Due to its excellent film-forming ability and good biodegradable, biocompatible, and gel-forming properties [6,7], KGM has been widely applied in the field of biomedicine, food, health care, and cosmetics, among others [8,9]. In particular, KGM has been used as a raw material for the preparation of hydrogels and films that show drug-loading and sustained-release properties [5,6,9]. The preparation methods of KGM gel mainly include alkaline treatment, borate cross-linking, polymer compositing, high-voltage electric field preparation, and metal ion cross-linking after modification [6]. However, as a drug carrier, pure KGM hydrogel has a low drug release amount due to the strong hydrogen bond between KGM and other molecules [7]. Therefore, other substances need to be added to KGM hydrogel to improve its release properties.

Hyaluronic acid (HA), widely present in all organisms [10], is a nonsulfated glycosaminoglycan consisting of alternating β-1,4-D-glucuronic acid and β-1,3-N-acetyl-D-glucosamine monosaccharides [11]. Due to its non-immunogenicity and excellent degradable and excellent biocompatible properties, HA has been widely used in tissue engineering, drug delivery, and immune regulation [12]. Furthermore, as a drug carrier, it has been shown that the controlled release from HA has many benefits, such as maintaining optimum drug concentration, enhancing therapeutic effects, improving treatment efficiency with less drug, lowering toxicity, and prolonging in vivo release rates [13]. To expand the application of HA, modification is usually required. Research shows that modified HA could be mixed with chitosan and gelatin to form a hydrogel with enhanced properties [14,15]. As a modified product of HA, oxidized hyaluronic acid (OHA) has similar characteristics to HA and can be used to improve the properties of the hydrogel. The preparation of a KGM/OHA hydrogel has not been reported. In this study, HA was oxidized by sodium periodate to prepare OHA, and then mixed with KGM to form a composite hydrogel. The physical and chemical properties of the KGM/OHA composite hydrogels were evaluated by FT-IR, rheometer, and SEM. Moreover, the epigallocatechin gallate (EGCG) release properties of KGM/OHA hydrogels were also discussed. At present, the study of KGM/OHA composite hydrogels has not been reported. Therefore, this study could provide feasible advice for the preparation of KGM/OHA hydrogels and a preliminary evaluation of the properties of KGM/OHA hydrogels.

2. Materials and Methods

2.1. Materials

HA (molecular weight: 200–400 kDa), sodium periodate, EGCG, and hydroxylamine hydrochloride were purchased from Shanghai Macklin Biochemical Technology Co., Ltd. (Shanghai, China). KGM (molecular weight: 200–2000 kDa) was purchased from Zhaotong Sanai Organic Konjac Development Co., Ltd. (Yunnan, China). Glycol and anhydrous sodium carbonate were purchased from Sinopharm Chemical Reagent Co., Ltd. (Shanghai, China). The dialysis bag (M_W 3.5 kDa cut-off) was purchased from Shanghai Yuanye Biotechnology Co., Ltd. (Shanghai, China).

2.2. Preparation of OHA

OHA was prepared by the previous method with some modifications [16]. HA (1 g) was added to 100 mL of deionized water and stirred at 500 r/min for 2 h until the HA was completely dissolved. Then, 0.535 g of sodium periodate was added to the HA solution; 1mL of glycol was added to stop the reaction after 24 h of light protection. The dialysis was performed with a dialysis bag (M_W 3.5 kDa cut-off) for 3 d. OHA was obtained after freeze drying the dialysate for 24 h.

2.3. Oxidation Rate Determination

The oxidation rate was calculated according to the previous method [17]. A total of 0.05 g of OHA was added to 25 mL (0.25 mol/L) of hydroxylamine hydrochloride solution into which 2–3 drops of methyl orange were dropped. The solution was shaken for 2 h at 90 r/min. The solution of methyl orange was titrated with 0.1 mol/L of sodium hydroxide until it turned yellow. The oxidation rate can be calculated by the following formula:

$$\eta = \frac{\text{mol of}(-CHO)}{2 \cdot \text{mol of OHA}} = \frac{V_{NaOH} \cdot N_{NaOH} \times 0.001}{2 \cdot W \div 400} \quad (1)$$

where η is the oxidation rate, V_{NaOH} is the volume of sodium hydroxide consumed by titration (mL), N_{NaOH} is the concentration of sodium hydroxide (mol/L), W is the mass of OHA (g), and 400 is the molar mass of the basic cyclic unit of OHA.

2.4. Preparation of Hydrogels

The fabrication process of the hydrogels was conducted as follows: First, a certain amount of OHA was weighed and dissolved in deionized water through stirring (400 r/min) for 1 h. Then, the KGM powder was added to the OHA solution under stirring at 400 r/min. The compositions of the hydrogel samples are detailed in Table 1. After adding the KGM, 0.1 mol/L of sodium carbonate solution was added to the KGM/OHA solution to adjust the pH = 11. Subsequently, the KGM/OHA solution was placed in the water baths at 40 °C under stirring at 400 r/min for 2 h. Then, the KGM/OHA solution was transferred to the 80 °C water baths for 1 h to form hydrogel. The steps of the hydrogels' preparation are shown in Figure 1. The obtained hydrogels were cooled to 25 °C, by running water, for further evaluation.

Table 1. Composition of KGM and OHA in hydrogels.

Sample	KO-0	KO-1	KO-2	KO-3
KGM% (w/v)	1	1	1	1
OHA% (w/v)	0	0.1	0.3	0.5

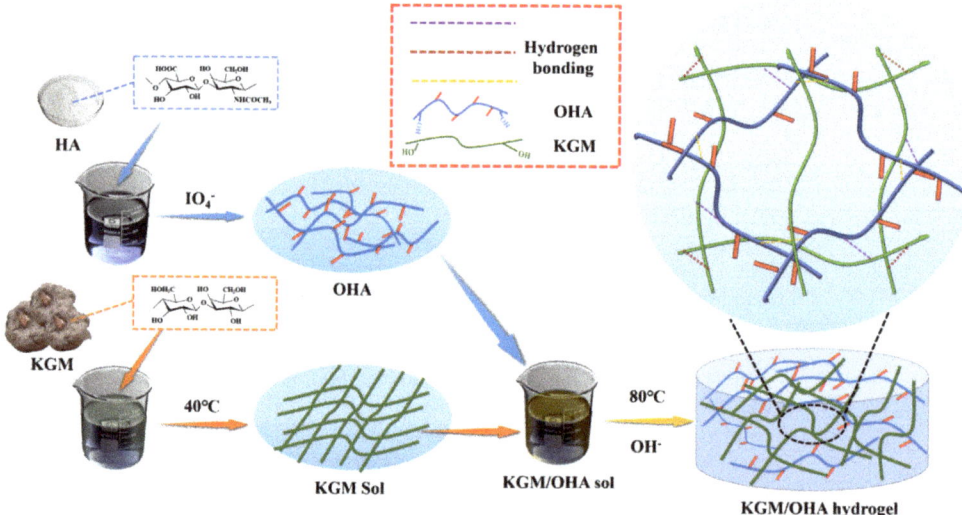

Figure 1. Synthesis diagram of KGM/OHA hydrogel.

2.5. Preparation of EGCG-Loaded Hydrogels

In order to load the EGCG into the OHA/KGM hydrogel, EGCG was directly added to deionized water at a concentration of 0.8 mg/mL, followed by a 10 min ultrasound. The freeze-dried hydrogel was then placed in the EGCG solution with 100 mL of EGCG solution for each gram of freeze-dried gel. The EGCG was loaded onto the hydrogel after standing at room temperature for 12 h [18].

2.6. FT-IR Characterization

The FT-IR of the OHA and KGM/OHA freeze-dried composite gels was measured by an FT-IR spectrometer (Bruker VERTEX 70, Thermo Fisher Scientific Co., Ltd., Waltham, MO, USA) using the KBr pressed pellet method. A small number of samples were mixed with potassium bromide, ground carefully for approximately 5 min, and then pressed. The FT-IR spectra were measured with wavelengths from 400 cm^{-1} to 4000 cm^{-1} with a resolution of 4 cm^{-1} and 64 scan acquisitions.

2.7. Rheological Test

The rheological properties of the KGM/OHA hydrogels were measured using a rotational rheometer (MCR301, Anton Parr, Austria) with a standard parallel-plate geometry (PP-50, 50 mm of diameter, and 1 mm gap). Steady shear flow behavior was measured at 25 °C and the scans were conducted at a shear rate range of 0.1 s^{-1} to 1000 s^{-1}. The measurement was conducted in oscillation amplitude mode at a constant temperature of 25 °C, an angular frequency of 1.0 rad/s, and a strain range of 0.1–100%. After the linear viscoelastic region of the samples was obtained, the measurement was carried out in oscillation frequency mode, in which the strain was fixed at 1% and the frequency sweep was in a range of 0.1–100 rad/s. The values of the frequency-dependent G' and G'' were recorded. The shear stress sweep was tested from 0.1 Pa to 100 Pa at a constant temperature of 25 °C.

2.8. Scanning Electron Microscopy (SEM)

The microstructures of the samples were observed through SEM (Nova NanoSEM 230, FEI CZECH REPUBLIC S.R.O., Brno, Czech Republic). The cross section of the freeze-dried gel was coated with gold and scanned at an accelerated voltage of 15 KV.

2.9. Swelling Properties of Hydrogels

The prepared hydrogel was freeze dried for 24 h and weighed immediately after it was taken out. The dry hydrogel was then immersed in 100 mL of PBS (0.01 M, pH = 7.4). After a certain interval of time, the hydrogel was weighed. The swelling rate was calculated as the formula:

$$SD = \frac{M_t - M_0}{M_0} \times 100\% \qquad (2)$$

where M_0 is the weight of dry hydrogel and M_t is the weight of hydrogel after a period of time.

2.10. In Vitro Degradation Rate (DR)

The freeze-dried OHA/KGM hydrogels were weighed and then immersed in the PBS (0.01 M, pH = 7.4) solution at 25 °C. The hydrogels were washed with deionized water and freeze dried for 24 h. The degradation rate of hydrogel was calculated as the formula:

$$DR = \frac{W_0 - W_t}{W_0} \times 100\% \qquad (3)$$

where W_0 is the initial weight of the dried hydrogel and W_t is the weight of the dried hydrogel weighed every 4 days.

2.11. EGCG Loading Determination

The concentration of EGCG was measured using the Folin-Ciocalteu phenol method [19]. The absorbance of the solution was recorded at a 717 nm wavelength using a UV–vis spectrophotometer (UV-1780, Shimadzu, Japan). The concentration and content of EGCG in the solution were calculated by the standard curve of EGCG absorbance. The entrapment efficiencies (EEs) of the KGM/OHA hydrogel for EGCG were determined using the following equation:

$$EEs = \frac{m_0 - m_1}{m_0} \times 100\% \qquad (4)$$

where m_0 is the total amount of EGCG contained in the suspension and m_1 is the amount of free EGCG.

2.12. In Vitro Release Studies

In order to evaluate the sustained release performance of KGM/OHA hydrogel, the KGM/OHA hydrogel, loaded with EGCG, was immersed in 50 mL of PBS (0.01 M, pH = 7.4) and continuously shaken at 60× g rpm. The temperature was maintained at 37 °C. Then, 1 mL of the solution was taken out every 1 h to measure the content of EGCG, and the release amount of EGCG was, also, calculated according to the Folin–Ciocalteu phenol method. The concentration and content of EGCG in the PBS were calculated by the standard curve of EGCG absorbance. The cumulative release rate of EGCG was calculated by the EGCG content in the hydrogel and the solution before the release experiment. The formula is as follows:

$$Cumulative\ release\ rate = \frac{m_t}{m_0 - m_1} \times 100\% \quad (5)$$

where m_t is the EGCG content in the PBS solution after a certain release time. The volume of PBS was kept at 50 mL. The experimental data were obtained from three parallel experiments.

3. Result and Discussion

3.1. Characterization of OHA

Sodium periodate can oxidize two adjacent hydroxyl groups in HA into two aldehyde groups to form OHA. Because the amino group in hydroxylamine hydrochloride can react with the aldehyde group in OHA, the reaction between hydroxylamine hydrochloride and the aldehyde group can be used to verify whether OHA can be successfully prepared [20]. The content of the aldehyde group in OHA was calculated according to the reaction formula, and the oxidation rate of OHA was 43.8 ± 2.26%, which was consistent with previous research [21]. To further confirm the successful formation of the dialdehyde groups in OHA, FT-IR spectra of HA and OHA were recorded. As shown in Figure 2, compared with HA, a new absorption peak at approximately 1725 cm^{-1}, corresponding to the dialdehyde groups, was observed in the spectrum of OHA [15], which confirmed the successful preparation of OHA. Similar results were also reported by Li et al. [22].

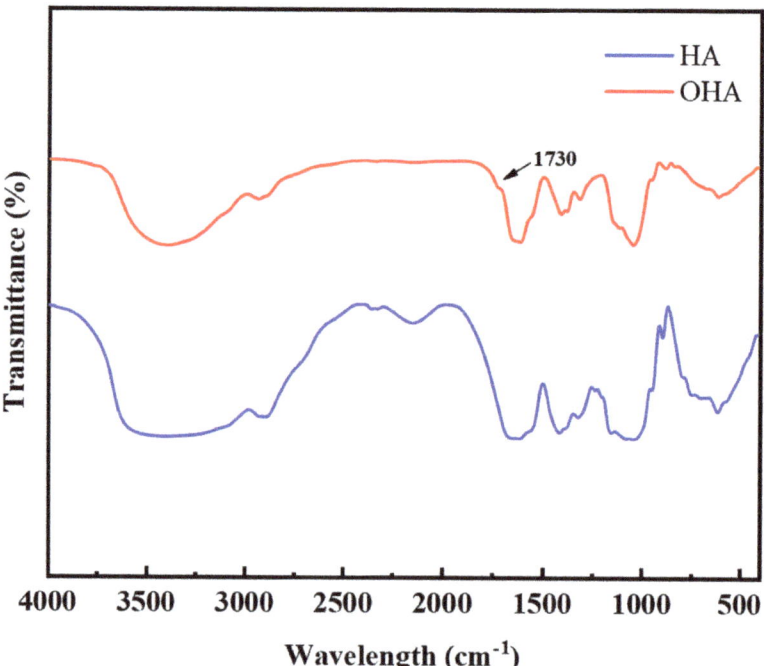

Figure 2. FT-IR spectra of HA and OHA.

3.2. Appearance Characteristics of Hydrogels

The photographic appearance of the hydrogels is shown in Figure 3. It was obvious that all of the hydrogels were pale yellow due to the deacetylation of KGM during the formation of the hydrogels [23]. The original slight yellow color of the OHA hydrogel might be responsible for this. With the increase in the OHA concentration, the composite hydrogels' color became deeper. Meanwhile, the structure of all the hydrogels was flat without collapse, indicating the good potential of KGM and OHA for the formulation of hydrogels.

Figure 3. An image of the hydrogels. From left to right, the hydrogels are KO-0, KO-1, KO-2 and KO-3, respectively.

3.3. FT-IR Spectra of Hydrogels

FT-IR was usually used to observe functional groups in polysaccharides. Changes occurring at a molecular level were often difficult to observe from a macroscopic perspective. The FT-IR spectra of the KO-0, KO-1, KO-2 and KO-3 hydrogels are shown in Figure 4a. The spectra of the KO-0, KO-1, KO-2 and KO-3 hydrogels were basically similar. However, the O-H bond absorption peak near 3406 cm^{-1} moved with the addition of OHA. The peaks of KO-0, KO-1, KO-2 and KO-3 appeared at 3396 cm^{-1}, 3406 cm^{-1}, 3420 cm^{-1}, and 3404 cm^{-1}, respectively. From KO-0 to KO-2, the site of the O-H bond absorption peak gradually deviated to the high wavelength due to the stretching of the O-H bond. This was attributed to the hydrogen bond interaction between the KGM/OHA hydrogels that gradually increased with the increase in the OHA content [24]. However, the hydrogen bond absorption peak site of the KO-3 hydrogel was close to that of KO-1, indicating that the hydrogen bond interaction of KGM/OHA hydrogel was weakened when the OHA concentration was 0.5%.

3.4. Rheological Analysis of Hydrogels

To investigate the effect of the OHA amount on the rheological properties of composite hydrogels, the steady rheological behavior of the composite hydrogels with different OHA contents was measured, as shown in Figure 4b. It was found that the apparent viscosity of all the hydrogels could be maintained at a relatively high level at a low shear rate. With the increase in shear rate, the apparent viscosity of the hydrogels gradually decreased and finally stabilized at an almost identical low viscosity level. It shows that the properties of pseudoplastic fluid were consistent with the rheological properties of KGM studied previously [25]. At a lower shear rate, the three-dimensional network structure of the hydrogels was not damaged, so the viscosity of the hydrogels could be maintained at a high level. However, with the increase in shear rate, the molecular chain gradually fractured, leading to an increased degree of damage to the three-dimensional network of the hydrogels and a decrease in the viscosity of the hydrogels [26]. In addition, adding

OHA can significantly improve the viscosity of the mixed system, which might be due to the hydrogen-bond interaction between the OHA and KGM that made the hydrogels' network structure more stable. Among the three hydrogels containing OHA, the hydrogels with an OHA concentration of 0.3% had the highest viscosity. This was because the main effect of the OHA addition on the KGM hydrogel was to reinforce the network structure at a concentration of 0–0.3%. However, the viscosity of the hydrogel decreased at an OHA concentration of 0.5%. This may be because the excessive OHA addition might have impacted the original structure of the KGM gel, allowing the molecules to slide.

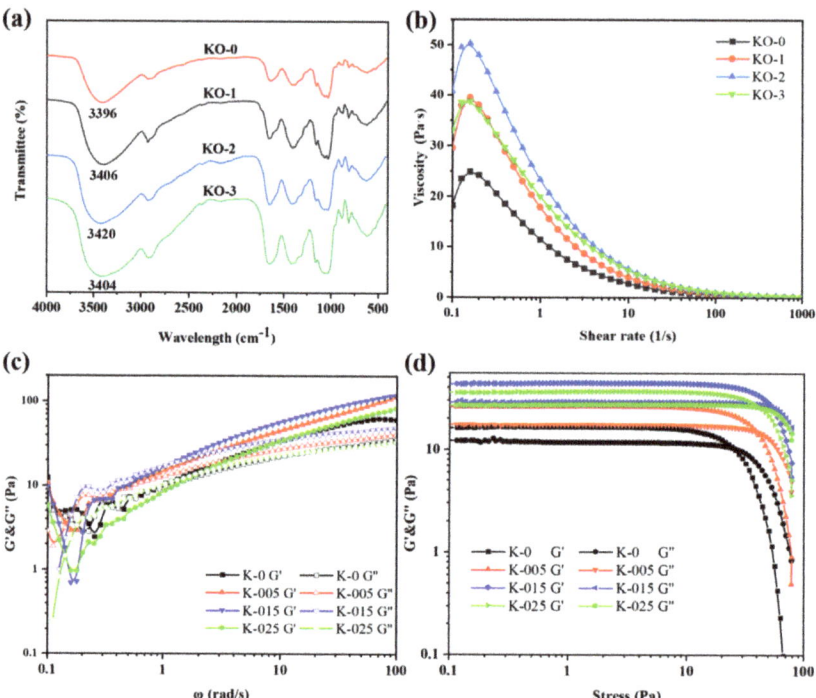

Figure 4. FT-IR spectra of KO-0 hydrogel, KO-1 hydrogel, KO-2 hydrogel and KO-3 hydrogel (**a**). The viscosity–shear rate curves (**b**), the viscoelasticity–angular frequency curves (**c**) and the viscoelasticity–shear stress curves (**d**) of KO-0, KO-1, KO-2 and KO-3 hydrogels.

Frequency scanning is one of the most common test modes in polymer dynamic rheological testing. Through frequency scanning, the frequency dependence of the storage modulus (G') and loss modulus (G'') can be obtained in order, so as to judge the relaxation time of the polymers. The relationship between the viscoelasticity and angular frequency of the hydrogels is shown in Figure 4c. G' represents the elasticity of hydrogel while G'' represents the viscosity of the hydrogel. It has been previously stated that when the G' value of a substance is greater than the G'' value, the substance acts as a viscoelastic solid, indicating that the substance behaves like a gel [25,27]. It can be seen that the G' values of all the hydrogels were higher than the G'' values in the whole angular frequency measurement range (0.1–100 rad/s), and all the hydrogels exhibited stable viscoelastic solid morphology. These results indicate that hydrogels can keep their original hydrogel network structure in a middle frequency band and have good structural stability. In addition, the viscoelasticity of all the hydrogels containing OHA was higher than that of the KGM hydrogels, indicating that OHA could improve the stability of the KGM hydrogel network. This result was consistent with the steady rheological study. The hydrogels with an OHA concentration of 0.3% still had the highest viscoelasticity, which was consistent with the performance of viscosity and the shear rate of the hydrogels.

The mechanical properties of the hydrogels reflected the homogeneity state and the interfacial interactions between its components, which was measured by oscillatory shear rheology [27]. The hydrogels behaved as viscoelastic fluids under a certain degree of shear stress, and the critical value of shear stress was also shown as the intersection of the G' and G'' curves [28]. As shown in Figure 4d, the viscoelasticity of the KGM hydrogel was lower than that of the hydrogel containing OHA. The intersection point of the G' and G'' curves of the KGM hydrogel was more advanced than those of the other hydrogels containing OHA. These results demonstrate that the mechanical properties of the KGM hydrogels were significantly lower than those of the hydrogels containing OHA [26], indicating that OHA could improve the mechanical properties of hydrogel and make the hydrogel structure more stable. Furthermore, the hydrogel structure at an OHA concentration of 0.5% was weaker, which was consistent with the above experiments.

3.5. SEM Analysis

Figure 5 shows the morphology of the KGM and KGM/OHA freeze-dried hydrogels with different OHA concentrations. It can be seen that all the hydrogels present continuous network structures, which was the structural characteristic of stable hydrogels. This was similar to the observation of the KGM hydrogel network structure under alkaline conditions by Mu et al. [29]. It indicated that the network structure of KGM occupied the majority in the gel system of KGM/OHA, which corresponded to the higher content of KGM relative to OHA in the experimental group design. In addition, the network structure of the KGM/OHA hydrogels was more compact than that of the KGM hydrogels after adding OHA. It can be speculated that the OHA molecules wound onto the KGM gel network, and its molecular chains were connected by the hydrogen bond interaction, which strengthened the network structure of the KGM hydrogel and thus enhanced the stability of the KGM hydrogel. However, when the content of OHA was 0.5%, due to the hydrophilicity of OHA, a large number of holes appeared in the network structure of the KGM/OHA hydrogel after freeze drying [30]. Adding too much OHA increased the water content within the hydrogel, which led to many water-filled holes in the structure of the hydrogel. Therefore, the excessive OHA made the water content of the hydrogel higher, thereby resulting in a decrease in the viscosity and stability of the KGM/OHA hydrogel at 0.5% OHA content.

3.6. Swelling Properties

Swelling is one of the main properties of hydrogels. The swelling properties of the hydrogel within 6h are shown in Figure 6a. It can be observed that the swelling rate of all the prepared hydrogels increased, with the increase in soaking time, and finally reached a relatively stable level. It indicated that the hydrogels could swell well in the PBS solution with a pH of 7.4, which was similar to the swelling trend of the KGM hydrogels observed previously [31]. In addition, it can be seen, from the broken line diagram of gel swelling, that the swelling rate of the KGM hydrogel was higher than that of the KGM/OHA hydrogel. The swelling rate of the hydrogels was usually related to the cross-linking density and the network structure of the hydrogels [32]. As the cross-linking density increased, the swelling rate of the hydrogels decreased, which was probably due to the denser network formed in the hydrogels with a higher cross-linking density, thus making it difficult for water molecules to enter the internal space of the hydrogels [33]. Therefore, with the addition of OHA, the swelling rate of the hydrogels became lower because the cross-linking density of the KGM hydrogel was lower than that of the KGM/OHA hydrogel. When the OHA concentration ranged from 0 to 0.3%, the decreasing swelling rate showed that the cross-linking density of the hydrogel network gradually increased. However, when the OHA concentration was 0.5%, excessive OHA made the microstructure of the hydrogel appear porous, which reduced the cross-linking density of the hydrogels and increased the swelling rate.

3.7. In Vitro Degradation Rate

In order to study the degradation performance of hydrogels, the hydrogels were soaked in PBS solution and then removed every 4 days for drying and weighing, followed by a recording of the mass loss of the hydrogels, which can be utilized for reflecting the degradation rate of the hydrogels. The degradation rate of the hydrogels in different proportions is shown in Figure 6b. The results show that the degradation rate of the hydrogel increased gradually within 16 d, which was a common phenomenon of hydrogel degradation in vitro. In addition, the degradation rate of hydrogels containing OHA was significantly higher than that of the KGM hydrogels within 16 d, and the degradation rate of hydrogels in vitro increased with the increase in the OHA supplemental level. This phenomenon was due to the high degradation capacity of the OHA [33]. The cyclic monomer of the OHA contained an aldehyde group, which was beneficial to degradation [34]. Therefore, the addition of OHA could make KGM hydrogel possess a higher degradation rate.

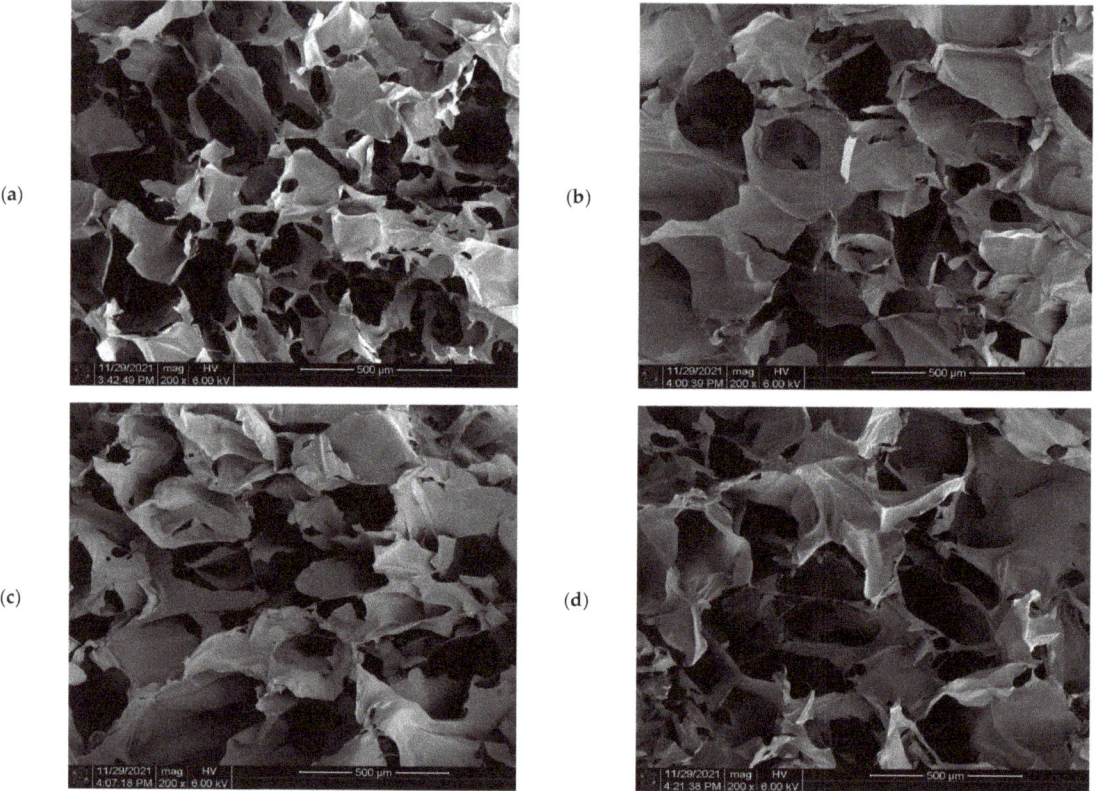

Figure 5. The SEM images of KO-0 hydrogels (**a**), KO-1 hydrogels (**b**), KO-2 hydrogels (**c**) and KO-3 hydrogels (**d**).

3.8. Load and Release Behavior of Hydrogels

The entrapment efficiencies (EEs) of KO-0, KO-1, KO-2, and KO-3 hydrogels were $18.43 \pm 1.99\%$, $19.13 \pm 1.94\%$, $22.40 \pm 1.24\%$, and $20.87 \pm 2.28\%$, respectively. The EEs of the hydrogels were usually related to the microstructure of the hydrogel network. When the network structure of the hydrogels was relatively loose, the area available for EGCG attachment was less [35]. The increase in the number of pores in the hydrogel could also improve EEs, which was based on the principle that the pores could increase the internal surface area of the hydrogel [36]. SEM showed that from KO-0 to KO-2, the network of hydrogels became more compact. KO-2 hydrogels had the densest network structure and contained many pores and, thus, had the highest EEs. Compared with the KO-2 hydrogel, the KO-3 hydrogel had slightly more pores, but the excessive OHA impacted the molecular structure of the KGM and made the structure of the hydrogel not compact. Moreover, the reduced load space of EGCG on the hydrogel resulted in the EEs of the KO-3 hydrogel being slightly lower than that of the KO-2 hydrogel.

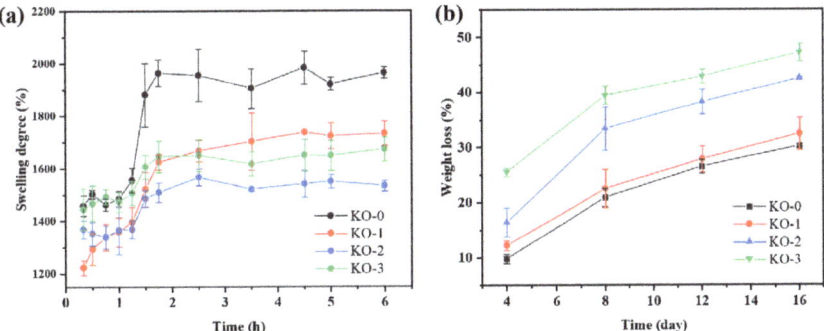

Figure 6. Swelling curve of KO-0 hydrogels, KO-1 hydrogels, KO-2 hydrogels, and KO-3 hydrogels (**a**). In vitro degradation curve of KO-0 hydrogels, KO-1 hydrogels, KO-2 hydrogels, and KO-3 hydrogels (**b**).

Figure 7 shows the EGCG release rates of KGM/OHA hydrogels in different proportions. The EGCG release amounts of KO-0, KO-1, KO-2, and KO-3 hydrogels after 10 h were $31.40 \pm 0.73\%$, $54.89 \pm 2.81\%$, $57.62 \pm 4.11\%$, and $62.44 \pm 1.97\%$, respectively. Figure 7 also shows that the amount of EGCG released in the first 6 h was significantly higher than that in the following 4 h. The phenomenon observed by Wang et al. [37] and Yuan et al. [38] did not appear in this experiment: hydrogels showed an explosive release of the loaded substance in the first 1 h of the release experiment and a sustained release in the following several hours. In this experiment, all hydrogels showed higher release rates in the first 5 h of the release experiment than in the subsequent hours. Although the burst release rate was lower, KGM/OHA hydrogels showed a longer burst release period of EGCG than other KGM-based hydrogels. This phenomenon was attributed to the fact that OHA could prolong the in vivo release [13]. The other reason was that the molecular structure of EGCG contained a large number of hydroxyl groups so that it was easy to have a strong hydrogen bond interaction with the hydroxyl groups on the KGM and OHA molecular chains [39]. The gradual increase in the EGCG release of the KO-1, KO-2, and KO-3 hydrogels was due to the hydrophilicity and degradation of the OHA [30,34]. The hydrophilic properties of the OHA allowed water to enter the internal structure of the KGM/OHA hydrogels more quickly, which allowed the EGCG in the hydrogel to be released into the buffer more easily. The rapid degradation of the OHA made the structure of the gel become gradually fluffy during the release experiment, which made EGCG easy to release. It could be concluded that the release rate and amount were related to the content of OHA and that they can be easily controlled by the addition of OHA. Therefore, KGM/OHA hydrogels, capable of the long-term, high-speed, and controlled release of EGCG, can be used for biomedical and food applications, as a promising gel material.

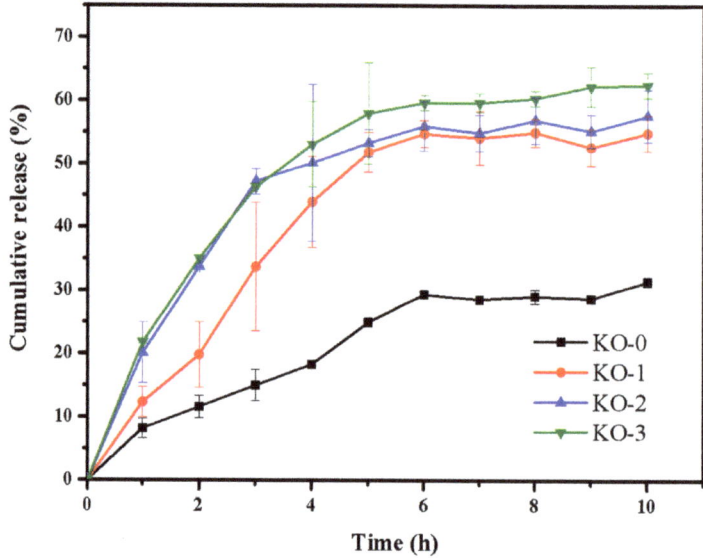

Figure 7. Cumulative release curves of KO-0, KO-1, KO-2 and KO-3 hydrogels.

4. Conclusions

In this study, a stable composite hydrogel was prepared by incorporating KGM with OHA, after which alkali processing and thermal treatment were conducted. The effect of OHA content on various gel properties was evaluated. The obtained hydrogel was pale yellow, smooth in surface, and had a favorable swelling capacity, which qualified the essential requirements for ideal drug-delivery applications. The OHA played an effective role in adjusting the swelling ratio and increasing the biodegradation rate. The rheological analysis shows that, when the concentration of OHA is 0.3%, KGM/OHA has the most stable network structure and the greatest mechanical properties. When observed through SEM, the firm and porous structure of the KGM/OHA hydrogel demonstrates a rich storage space for drug loading. Furthermore, both EGCG encapsulation efficiency and the release properties of the hydrogels were significantly raised with the presence of OHA. The overall results suggest that the KGM/OHA hydrogel, loaded with EGCG, exhibited potential applications in controlled release.

Author Contributions: Conceptualization, H.W. and C.W.; Methodology, H.W., J.C. and N.B.; Software, N.B.; Validation, H.W. and C.W.; Formal Analysis, H.W.; Investigation, H.W., J.C. and Y.C.; Resources, R.S.; Data Curation, H.W.; Writing—Original Draft Preparation, H.W.; Writing—Review and Editing, C.W.; Visualization, N.B.; Supervision, C.W.; Project Administration, J.P.; Funding Acquisition, J.P. All authors have read and agreed to the published version of the manuscript.

Funding: This work was supported by the National Natural Science Foundation of China (Grant No. 31772045), the Fuzhou Science and Technology Bureau (2019-G-50), and the Fujian Science and Technology Planning Project (2018N2002).

Institutional Review Board Statement: Not applicable.

Informed Consent Statement: Not applicable.

Data Availability Statement: Data are contained within the article.

Conflicts of Interest: The authors declare no conflict of interest.

References

1. Mitura, S.; Sionkowska, A.; Jaiswal, A. Biopolymers for hydrogels in cosmetics: Review. *J. Mater. Sci. Mater. Med.* **2020**, *31*, 50. [CrossRef] [PubMed]
2. Fan, C.; Liao, L.; Zhang, C.; Liu, L. A tough double network hydrogel for cartilage tissue engineering. *J. Mater. Chem. B* **2013**, *1*, 4251–4258. [CrossRef] [PubMed]
3. Yazdi, M.K.; Vatanpour, V.; Taghizadeh, A.; Taghizadeh, M.; Ganjali, M.R.; Munir, M.T.; Habibzadeh, S.; Saeb, M.R.; Ghaedi, M. Hydrogel membranes: A review. *Mater. Sci. Eng. C* **2020**, *114*, 111023. [CrossRef] [PubMed]
4. Sun, J.; Zhao, X.; Illeperuma, W.R.K.; Chaudhuri, O.; Oh, K.H.; Mooney, D.J.; Vlassak, J.J.; Suo, Z. Highly stretchable and tough hydrogels. *Nature* **2012**, *489*, 133–136. [CrossRef]
5. Zhou, L.; Xu, T.; Yan, J.; Li, X.; Xie, Y.; Chen, H. Fabrication and characterization of matrine-loaded konjac glucomannan/fish gelatin composite hydrogel as antimicrobial wound dressing. *Food Hydrocoll.* **2020**, *104*, 105702. [CrossRef]
6. Lin, W.; Ni, Y.; Liu, D.; Yao, Y.; Pang, J. Robust microfluidic construction of konjac glucomannan-based micro-films for active food packaging. *Int. J. Biol. Macromol.* **2019**, *137*, 982–991. [CrossRef]
7. Yang, D.; Yuan, Y.; Wang, L.; Wang, X.; Mu, R.; Pang, J.; Xiao, J.; Zheng, Y. A Review on Konjac Glucomannan Gels: Microstructure and Application. *Int. J. Mol. Sci.* **2017**, *18*, 2250. [CrossRef]
8. Chen, H.; Lan, G.; Ran, L.; Xiao, Y.; Yu, K.; Lu, B.; Dai, F.; Wu, D.; Lu, F. A novel wound dressing based on a Konjac glucomannan/silver nanoparticle composite sponge effectively kills bacteria and accelerates wound healing. *Carbohydr. Polym.* **2018**, *183*, 70–80. [CrossRef]
9. Chen, L.; Liu, Z.; Zhuo, R. Synthesis and properties of degradable hydrogels of konjac glucomannan grafted acrylic acid for colon-specific drug delivery. *Polymer* **2005**, *46*, 6274–6281. [CrossRef]
10. Necas, J.; Bartosikova, L.; Brauner, P.; Kolar, J. Hyaluronic acid (hyaluronan): A review. *Vet. Med.* **2008**, *53*, 397–411. [CrossRef]
11. Varela-Aramburu, S.; Su, L.; Mosquera, J.; Morgese, G.; Schoenmakers, S.M.C.; Cardinaels, R.; Palmans, A.R.A.; Meijer, E.W. Introducing Hyaluronic Acid into Supramolecular Polymers and Hydrogels. *Biomacromolecules* **2021**, *22*, 4633–4641. [CrossRef]
12. Zhang, S.; Hou, J.; Yuan, Q.; Xin, P.; Cheng, H.; Gu, Z.; Wu, J. Arginine derivatives assist dopamine-hyaluronic acid hybrid hydrogels to have enhanced antioxidant activity for wound healing. *Chem. Eng. J.* **2020**, *392*, 123775. [CrossRef]
13. Bayer, I.S. Hyaluronic Acid and Controlled Release: A Review. *Molecules* **2020**, *25*, 2649. [CrossRef]
14. Sanmartín-Masiá, E.; Poveda-Reyes, S.; Ferrer, G.G. Extracellular matrix-inspired gelatin/hyaluronic acid injectable hydrogels. *Int. J. Ploym. Mater.* **2017**, *66*, 280–288. [CrossRef]
15. Chen, Y.; Su, W.; Yang, S.; Gefen, A.; Lin, F. In situ forming hydrogels composed of oxidized high molecular weight hyaluronic acid and gelatin for nucleus pulposus regeneration. *Acta Biomater.* **2013**, *9*, 5181–5193. [CrossRef]
16. França, C.G.; Sacomani, D.P.; Villalva, D.G.; Nascimento, V.F.; Dávila, J.L.; Santana, M.H.A. Structural changes and crosslinking modulated functional properties of oxi-HA/ADH hydrogels useful for regenerative purposes. *Eur. Polym. J.* **2019**, *121*, 109288. [CrossRef]
17. Zhao, H.; Heindel, N.D. Determination of Degree of Substitution of Formyl Groups in Polyaldehyde Dextran by the Hydroxylamine Hydrochloride Method. *Pharm. Res.* **1991**, *8*, 400–402. [CrossRef]
18. Wang, L.; Li, Y.; Lin, L.; Mu, R.; Pang, J. Novel synthesis of mussel inspired and Fe^{3+} induced pH-sensitive hydrogels: Adhesion, injectable, shapeable, temperature properties, release behavior and rheological characterization. *Carbohydr. Polym.* **2020**, *236*, 116045. [CrossRef]
19. Donsì, F.; Voudouris, P.; Veen, S.J.; Velikov, K.P. Zein-based colloidal particles for encapsulation and delivery of EGCG. *Food Hydrocoll.* **2017**, *63*, 508–517. [CrossRef]
20. Pandit, A.H.; Mazumdar, N.; Ahmad, S. Periodate oxidized hyaluronic acid-based hydrogel scaffolds for tissue engineering applications. *Int. J. Biol. Macromol.* **2019**, *137*, 853–869. [CrossRef]
21. Zhao, Y.; Li, Y.; Peng, X.; Yu, X.; Cheng, C.; Yu, X. Feasibility study of oxidized hyaluronic acid cross-linking acellular bovine pericardium with potential application for abdominal wall repair. *Int. J. Biol. Macromol.* **2021**, *184*, 831–842. [CrossRef]
22. Li, S.; Pei, M.; Wan, T.; Yang, H.; Gu, S.; Tao, Y.; Liu, X.; Zhou, Y.; Xu, W.; Xiao, P. Self-healing Hyaluronic Acid Hydrogels Based on Dynamic Schiff Base Linkages as Biomaterials. *Carbohydr. Polym.* **2020**, *250*, 116922. [CrossRef]
23. Yang, X.; Li, A.; Li, D.; Li, X.; Li, P.; Sun, L.; Guo, Y. Improved physical properties of konjac glucomannan gels by co-incubating composite konjac glucomannan/xanthan systems under alkaline conditions. *Food Hydrocoll.* **2020**, *106*, 105870. [CrossRef]
24. Wang, J.; Liu, C.; Shuai, Y.; Cui, X.; Nie, L. Controlled release of anticancer drug using graphene oxide as a drug-binding effector in konjac glucomannan/sodium alginate hydrogels. *Colloid Surface B.* **2014**, *113*, 223–229. [CrossRef]
25. Yuan, C.; Zou, Y.; Cui, B.; Fang, Y.; Lu, L.; Xu, D. Influence of cyclodextrins on the gelation behavior of κ-carrageenan/konjac glucomannan composite gel. *Food Hydrocoll.* **2021**, *120*, 106927. [CrossRef]
26. Ma, X.; Xu, T.; Chen, W.; Wang, R.; Xu, Z.; Ye, Z.; Chi, B. Improvement of toughness for the hyaluronic acid and adipic acid dihydrazide hydrogel by PEG. *Fiber Polym.* **2017**, *18*, 817–824. [CrossRef]
27. Liu, Q.; Zhan, C.; Barhoumi, A.; Wang, W.; Santamaria, C.; McAlvin, J.B.; Kohane, D.S. A Supramolecular Shear-Thinning Anti-Inflammatory Steroid Hydrogel. *Adv. Mater.* **2016**, *28*, 6680–6686. [CrossRef]
28. Rad, E.R.; Vahabi, H.; Formela, K.; Saeb, M.R.; Thomas, S. Injectable poloxamer/graphene oxide hydrogels with well-controlled mechanical and rheological properties. *Polym. Advan. Technol.* **2019**, *30*, 2250–2260. [CrossRef]

29. Mu, R.; Wang, L.; Du, Y.; Yuan, Y.; Ni, Y.; Wu, C.; Pang, J. Synthesis of konjac glucomannan-silica hybrid materials with honeycomb structure and its application as activated carbon support for Cu(II) adsorption. *Mater. Lett.* **2018**, *226*, 75–78. [CrossRef]
30. França, C.G.; Plaza, T.; Naveas, N.; Santana, M.H.A.; Manso-Silván, M.; Recio, G.; Hernandez-Montelongo, J. Nanoporous silicon microparticles embedded into oxidized hyaluronic acid/adipic acid dihydrazide hydrogel for enhanced controlled drug delivery. *Micropor. Mesopor. Mater.* **2021**, *310*, 110634. [CrossRef]
31. Tang, J.; Chen, J.; Guo, J.; Wei, Q.; Fan, H. Construction and evaluation of fibrillar composite hydrogel of collagen/konjac glucomannan for potential biomedical applications. *Regen. Biomater.* **2018**, *5*, 239–250. [CrossRef]
32. Jiang, Y.; Reddy, C.K.; Huang, K.; Chen, L.; Xu, B. Hydrocolloidal properties of flaxseed gum/konjac glucomannan compound gel. *Int. J. Biol. Macromol.* **2019**, *133*, 1156–1163. [CrossRef]
33. Wang, S.; Chi, J.; Jiang, Z.; Hu, H.; Yang, C.; Liu, W.; Han, B. A self-healing and injectable hydrogel based on water-soluble chitosan and hyaluronic acid for vitreous substitute. *Carbohydr. Polym.* **2021**, *256*, 117519. [CrossRef]
34. Li, H.; Wu, B.; Mu, C.; Lin, W. Concomitant degradation in periodate oxidation of carboxymethyl cellulose. *Carbohydr. Polym.* **2011**, *84*, 881–886. [CrossRef]
35. Shoaib, T.; Espinosa-Marzal, R.M. Influence of Loading Conditions and Temperature on Static Friction and Contact Aging of Hydrogels with Modulated Microstructures. *ACS Appl. Mater. Inter.* **2019**, *11*, 2610–2616. [CrossRef]
36. Johnson, K.; Muzzin, N.; Toufanian, S.; Slick, R.A.; Lawlor, M.W.; Seifried, B.; Moquin, P.; Latulippe, D.; Hoare, T. Drug-impregnated, pressurized gas expanded liquid-processed alginate hydrogel scaffolds for accelerated burn wound healing. *Acta Biomater.* **2020**, *112*, 101–111. [CrossRef]
37. Yuan, Y.; Wang, L.; Mu, R.; Gong, J.; Wang, Y.; Li, Y.; Ma, J.; Pang, J.; Wu, C. Effects of konjac glucomannan on the structure, properties, and drug release characteristics of agarose hydrogels. *Carbohydr. Polym.* **2018**, *190*, 196–203. [CrossRef]
38. Wang, L.; Du, Y.; Yuan, Y.; Mu, R.; Gong, J.; Ni, Y.; Pang, J.; Wu, C. Mussel-inspired fabrication of konjac glucomannan/microcrystalline cellulose intelligent hydrogel with pH-responsive sustained release behavior. *Int. J. Biol. Macromol.* **2018**, *113*, 285–293. [CrossRef]
39. Sun, J.; Jiang, H.; Li, M.; Lu, Y.; Du, Y.; Tong, C.; Pang, J.; Wu, C. Preparation and characterization of multifunctional konjac glucomannan/carboxymethyl chitosan biocomposite films incorporated with epigallocatechin gallate. *Food Hydrocoll.* **2020**, *105*, 105756. [CrossRef]

Article

Determination of Moisture Content and Shrinkage Strain during Wood Water Loss with Electrochemical Method

Zongying Fu [1], Hui Wang [2], Jingpeng Li [3] and Yun Lu [1,*]

[1] Key Laboratory of Wood Science and Technology of National Forestry and Grassland Administration, Research Institute of Wood Industry, Chinese Academy of Forestry, Beijing 100091, China; zyfu@caf.ac.cn
[2] State Key Laboratory of Animal Nutrition, Institute of Animal Science, Chinese Academy of Agricultural Sciences, Beijing 100193, China; wanghui_lunwen@163.com
[3] Key Laboratory of High Efficient Processing of Bamboo of Zhejiang Province, China National Bamboo Research Center, Hangzhou 310012, China; lijp@caf.ac.cn
* Correspondence: y.lu@caf.ac.cn

Citation: Fu, Z.; Wang, H.; Li, J.; Lu, Y. Determination of Moisture Content and Shrinkage Strain during Wood Water Loss with Electrochemical Method. *Polymers* **2022**, *14*, 778. https://doi.org/10.3390/polym14040778

Academic Editor: Ganesh Dattatraya Saratale

Received: 15 January 2022
Accepted: 13 February 2022
Published: 16 February 2022

Publisher's Note: MDPI stays neutral with regard to jurisdictional claims in published maps and institutional affiliations.

Copyright: © 2022 by the authors. Licensee MDPI, Basel, Switzerland. This article is an open access article distributed under the terms and conditions of the Creative Commons Attribution (CC BY) license (https://creativecommons.org/licenses/by/4.0/).

Abstract: Moisture content and shrinkage strain are essential parameters during the wood drying process. The accurate detection of these parameters has very important significance for controlling the drying process and minimizing drying defects. The presented study describes an electrochemical method to determine wood moisture content and shrinkage strain during drying, and the accuracy of this method is also evaluated. According to the results, the electrical resistance of the samples increased with the decrease in wood moisture content. As the moisture content changed from 42% to 12%, the resistance increased from 1.0×10^7 Ω to 1.2×10^8 Ω. A polynomial fitting curve was fitted with a determination coefficient of 0.937 to describe the relationship between moisture content and electrical resistance. In addition, both the shrinkage strain and resistance change rate increased with the decrease in wood moisture content, especially for the moisture content range of 23% to 8%, where the shrinkage strain and resistance change rate increased by 4% and 30%, respectively. The shrinkage strain increased exponentially with the increase in the resistance change rate; thereby, an exponential regression equation was proposed with a determination coefficient of 0.985, expressing the correlation between the two. This demonstrates the feasibility of the electrochemical method for measuring wood moisture content and shrinkage strain.

Keywords: electrochemical method; moisture content; shrinkage strain; wood water loss

1. Introduction

Wood drying is an essential step in wood processing, and it is also the most energy- and time-demanding step. When the drying quality of wood is ensured, the reduction in time and energy consumption can result in economic benefits. This enhancement requires persistent improvements of the process to obtain the best trade-off between the drying rate and quality, which can also be facilitated by a greater understanding of the drying process, especially for moisture content (MC) and shrinkage strain.

Wood MC is one of the most important parameters that need to be rapidly and accurately measured during the drying process, because both the control and adjustment of drying environmental conditions are mainly dependent on the MC. The oven drying method is one of the most traditional testing methods of wood MC, which has high precision, but the speed of the testing is slow, and continuous online testing is unachievable. The probe test method is the most popular approach in the wood processing enterprise due to the advantages of quick measurement and data evaluation. However, the method has certain limitations, as the testing accuracy is mainly concentrated on an MC below 30% and is also influenced by the temperature, probe depth and probe location, among other factors. As the measurement methods become increasingly advanced, some non-destructive methods, such as the X-ray densitometry method [1,2], X-ray microscopy [3,4], computed tomography scan [5,6] and the nuclear magnetic resonance approach [7,8], are applied to measure wood moisture content. Although the above advanced measuring methods have the characteristics of a fast measurement speed, high precision and imaging capabilities, some limitations still exist such as their high cost, their complex operation and a demanding application environment in industrial production.

Shrinkage strain is another characteristic of significant interest that can be used as an indicator to evaluate the wood drying process. It can be determined by the traditional slicing and strain gauge methods. For the slicing method, a manual measurement error is inevitable because of the contact measurement by a vernier caliper, or micrometer, and it has a specific requirement for the size of test specimens [9]. In the strain gauge method, a perfect bond between the strain gauges and the sample is crucial. Additionally, the temperature and relative humidity have a significant impact on the test results [10]. To improve the precision of the measurement and to visualize the shrinkage of wood, some non-contact optical measurement methods such as digital image correlation and near-infrared spectroscopy have been applied for the determination of shrinkage strain [11–14]. However, the optical methods require real-time image acquisition for the wood drying process, which limits their application in wood industry.

Electrochemical methods are usually simple, rapid, accurate and sensitive characterization approaches, which have played an essential role in scientific research and industrial applications. The main principle is to accurately establish the relationship between the electrical parameters and the substance being measured, based on the electrochemical properties and changing rules of substances in the solution, and then conduct the qualitative and quantitative analyses for components. The electrochemical methods have been widely used in the fields of electrochemical power sources [15,16], chemical and biological sensors [17,18], corrosion of metals [19,20] and biotechnology [21]. However, the application of electrochemical methods in the field of wood science is scarce.

There is a vast amount of literature on the relationship between electrical properties and wood MC. The electrical properties of wood strongly depend on MC, exhibiting changes that span almost 10 orders of magnitude over the range of possible MCs [22]. As electrical resistance measurements provide information about wood MC, Brischke et al. designed a system for the long-term recording of wood MC with internal conductively glued electrodes [23]. Further, Brischke and Lampen determined resistance characteristics for a total of 27 wood-based materials and established a functional relation between electrical resistance and wood MC in a range between 15 and 50% MC [24]. In addition, the relationship between MC and material resistance is different in various MC ranges. It is affected by the wood species, experimental variables and calibration experiments [25–27]. Currently, industrial tests of commercial online MC meters have shown low accuracy of individual readings [28].

Taking into account the aspects mentioned above, some methods based on the electrical properties of wood have been used to determine wood MC. However, an electrochemical workstation is first used in the determination of wood MC, which has higher signal resolution and measuring stability. No previous research discusses the feasibility of electrochemical workstations in measuring shrinkage strain during wood drying. Thus, the presented study describes an electrochemical method to determine wood MC and shrinkage strain during drying, and the accuracy of this method is also evaluated.

2. Materials and Methods

2.1. Preparation of the Samples and Testing Equipment

Forty-five-year-old eucalyptus (*Eucalyptus exserta*) trees with a diameter of 40 cm at breast height were obtained from Tilestone town (110°42′ E, 22°09′ N), Gaozhou, China. Several flat-sawn lumbers with a dimension of 120 mm (tangential) × 25 mm (radial) × 900 mm (longitudinal) were sawn from one log, wrapped with a preservative film and stored in a freezer to keep them in the green condition. In this experiment, one flat-sawn lumber with an initial MC of 42% and no visible defects was chosen. Eight test specimens with the dimension of 120 mm (tangential) × 25 mm (radial) × 10 mm (longitudinal) were machined from the lumber and equally divided into two groups (Figure 1). One group was used to measure MC, and the other was for the determination of shrinkage strain.

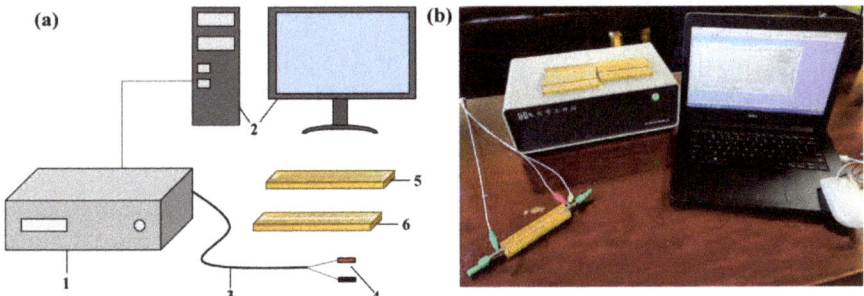

Figure 1. Diagram of the testing device and wood specimens (**a**): 1—electrochemical workstation; 2—computer host and monitor; 3—electric wire; 4—conductive clip; 5—test specimen for MC; 6—test specimen for shrinkage strain. Testing system of the electrochemical method (**b**).

A CHI760E electrochemical workstation test system (Shanghai Chenhua instrument equipment Co., LTD, Shanghai, China) was employed to measure the MC and shrinkage strain. The electrochemical workstation is an electronic instrument that controls the potential difference between the working electrodes and reference electrodes. In this study, one of the electrodes was connected to a working electrode, and the other was connected to a counter electrode and reference electrode together. The potentiostat controls the potential between the working electrode and the reference electrode and measures the current at the counter electrode so that a plot of potential vs. current can be created. The applied potential range was ±10 V, the current range was ±250 mA and the lower limit of the current measurement was below 50 pA. The electrochemical software of CHI760E (Shanghai Chenhua instrument equipment Co., LTD, Shanghai, China) was used for data acquisition, storage and processing. The diagram of the testing device and wood specimens is shown in Figure 1a, while the testing system can be observed in Figure 1b.

2.2. Determination of MC and Shrinkage Strain

The MC was measured by a current vs. time curve (i–t). The parameters were set as follows: the voltage was kept at a constant value of 1 V, the sampling interval was 0.1 s, the running time was 240 s and the sensitivity was set to 1×10^{-6} A/V. Before testing, both ends of the test specimens were sprayed with conductive coatings, which acted as contacts to enhance the conductivity between the wood and electrodes.

The shrinkage strain was measured by the current–time curve, combined with linear sweep voltammetry (LSV). The parameter settings of the current–time curve followed the measurement mentioned above regarding the moisture content. The parameter settings for LSV were as follows: the voltage ranged from −1 V to 1 V, the scan rate was 0.1 V/s, the sampling interval was 0.01 V and the sensitivity was 1×10^{-6} A/V. Before the electrochemical test, a conductive band with a thickness of 3 mm was sprayed on the surface of test specimens, as shown in Figure 1a.

Firstly, the initial weight for all test specimens and the initial length of the conductive band for shrinkage strain test specimens were obtained, and the electrochemical tests for all test specimens were performed. After that, the test specimens were dried at a constant temperature of 60 °C in a DKN611-type drying oven (Yamato Scientific Co., LTD, Tokyo Japan) to obtain different MC stages of wood specimens. The test specimens were taken out from the drying oven at drying times of 0.5 h, 1 h, 1.5 h, 2 h, 3 h and 4 h. The samples were weighed, followed by electrochemical measurements. For the shrinkage strain test specimens, the measurement of conductive band length was added. After finishing the tests, the specimens were oven dried to the absolute dry state and weighed. The moisture content, shrinkage strain and resistance change rate were calculated using Equations (1)–(3), respectively.

$$MC = \frac{M - M_0}{M_0} \times 100\% \tag{1}$$

where MC is the moisture content of test specimens, M is the weight of test specimens at different MCs and M_0 is the weight after oven drying treatment of the test specimens.

$$S = \frac{L_0 - L_i}{L_0} \times 100\% \tag{2}$$

where S is the shrinkage strain of specimens, L_0 is the initial length of the conductive band on the surface of test specimens and L_i is the length of the conductive band on the surface of test specimens at different moisture contents.

$$\Delta R = \frac{R_i - R}{R_i} \times 100\% \tag{3}$$

where ΔR is the resistance change rate for shrinkage strain measurement specimens, R_i is the initial resistance at the MC of 42% and R is the resistance at different MCs.

3. Results and Discussion

3.1. The MC of Test Specimens

The MC of test specimens obtained with various drying times is shown in Figure 2. The MC of test specimens at the initial stage was about 42% and decreased with drying time. The changing trend almost remained the same for the specimens of MC and shrinkage strain. The mean value of MC at drying times of 0.5 h, 1 h, 1.5 h, 2 h, 3 h and 4 h was 31%, 23%, 17%, 12%, 8% and 5%, respectively. Therefore, all discussions and analyses after this section are based on these MC stages.

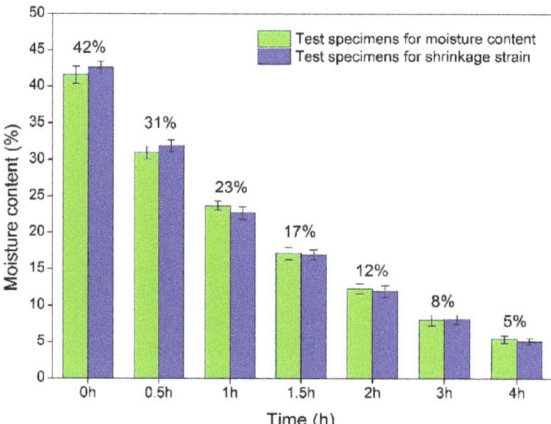

Figure 2. Changes in MC with the applied drying times.

3.2. Analysis of MC Test Results

The electrical properties of wood can strongly depend on the MC, exhibiting changes that span almost ten orders of magnitude in the range of possible MCs [22]. The changes in current with time at different wood MC stages are shown in Figure 3a. As shown, the current had a positive correlation with moisture content. However, the phenomenon disappeared as the MC became lower than 12%. When MC decreased from 42% to 12%, the current was reduced from 1.2×10^{-7} A to 1.0×10^{-8} A, whereas the current showed nearly constant values at smaller MCs of 8% and 5%. This is because when the wood MC is lower than 10%, wood is similar to an insulator [22]. Thereby, the effect of MC on the electrical signal disappeared. The current slightly decreased with time and gradually reached a steady state at around 240 s. The resistance increased with the decrease in wood MC under a constant voltage of 1 V (Figure 3b). As the MC changed from 42% to 12%, the resistance increased from 1.0×10^7 Ω to 1.2×10^8 Ω, while the increasing trend continued at lower concentrations. For MCs of 8% and 5%, there was a tiny difference in resistance.

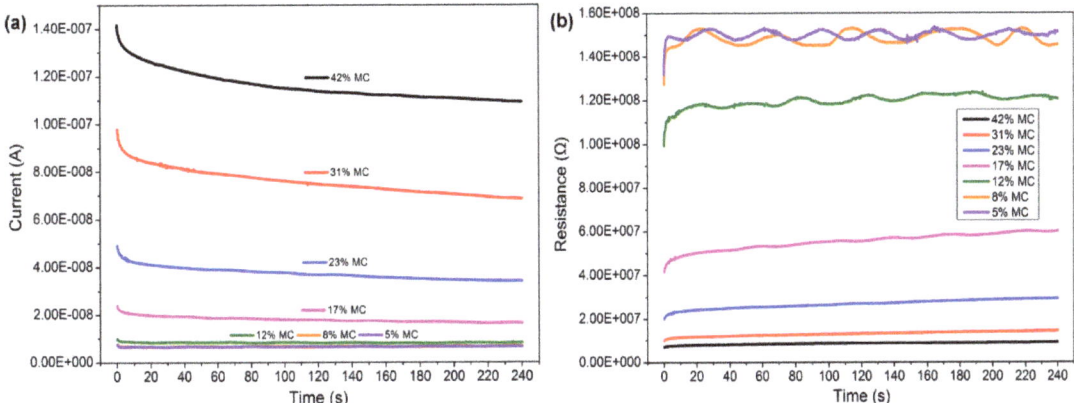

Figure 3. Changes in current (**a**) and resistance (**b**) over time at a potential of 1 V for the measurement of moisture content.

On the other hand, in the case where the MC decreased from 42% to 17%, the variation in electrical resistance with time was negligible, except for the first 20 s. For wood MC below 12%, the resistance fluctuated with time to a limited extent. This behavior may be related to the uneven distribution of the MC in wood specimens at the low-MC stage. According to the calculation formula of resistivity, the resistance was converted into resistivity, and the resistivity was about $10^6 \sim 10^7$ Ω m at an MC range of 42% to 5%, which agrees with the results of the literature. As reported, the resistivity could be about $10^{15} \sim 10^{16}$ Ω m for oven-dried wood and $10^3 \sim 10^4$ Ω m for wood at the fiber saturation point [22,29].

The relationship between MC and resistance can be observed in Figure 4. As it can be observed, the electrical resistance increased in a nonlinear way with the decrease in MC, and there was an apparent increasing trend as MC was reduced from 31% to 17%. Similar results reported by Barański et al. showed a nonlinear dependence of wood resistance on the moisture content [25]. In this study, a polynomial fitting curve was employed, and the determination coefficient was 0.937, which indicated that the resistance of wood specimens was capable of explaining more than 93.7% of the wood MC. The relationship between MC and resistance, according to the fitting curve, can be expressed as Equation (4). Thus, wood MC can be obtained directly by the changes in resistance using the electrochemical approach.

$$MC = 47.59 - 1.17R + 0.013R^2 - 4.8 \times 10^{-5}R^3 \quad (4)$$

where MC is the moisture content of wood test specimens, and R is 10^{-6} times the electrical resistance of wood test specimens.

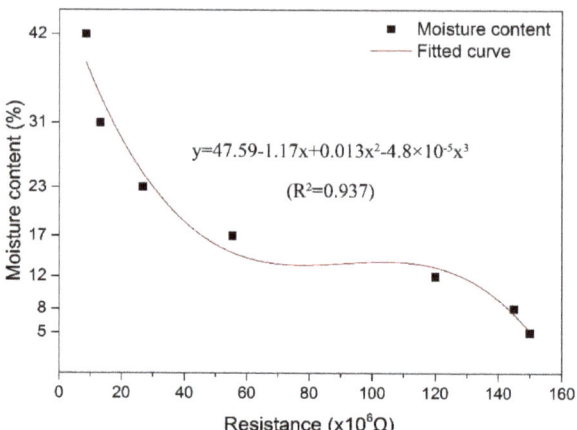

Figure 4. The fitting curve for wood MC to electrical resistance.

3.3. Analysis of Shrinkage Strain Test Results

The variation in the current and resistance at different MC stages for measuring shrinkage strain is shown in Figure 5. The current increased significantly with the decrease in moisture content, increasing from about 0.014 A to 0.028 A as the MC dropped from 42% to 5%. In contrast, the resistance decreased gradually with decreasing MC. Furthermore, the variation in electrical resistance ranged from 70 Ω to 35 Ω within the measured range of the moisture content. However, this phenomenon conflicts with the inverse relationship between electrical resistance and MC observed in Figure 3b. The reason can be explained as follows: comparing Figure 3b with Figure 5b, the electrical resistance in test specimens of MC was about 6~7 orders of magnitude higher than in the test specimens of shrinkage strain. This observation revealed that the conductivity of the conductive silver paint used in measuring shrinkage strain was much better than that of the wood itself, and the electrical current could move through the conductive band with low resistivity instead of wood tissue. Therefore, the resistance change with MC in measuring shrinkage strain was determined by the conductive band. Generally, the value of resistance decreased with the decreasing length of the electrical conductor. With the shrinkage of the wood, the length of the conductive band shortened, and in conjunction with the accumulation of the conductive silver paint, the electrical resistance gradually decreased with the wood shrinkage generated by the decrease in MC.

The LSV method was used to explore the changes in the current and resistance at different voltages. The current and resistance changes with the MC at a range of $-1\sim1$ V are presented in Figure 6. As observed in Figure 6a, the slope of the lines represents the inverse of the resistance, which varies with the changes in MC. The slope also confirms the variation in resistance at different MC stages. As seen in Figure 6b, the resistance remained constant when the voltage changed from -1 V to 1 V at each MC, but there was a great difference at different MCs. A particular situation can be seen at the MCs of 8% and 5%, where the curves of resistance vs. voltage overlap, indicating that the change in electrical resistance generated by shrinkage was very small at this MC stage. Moreover, Figure 5 also provides a piece of evidence as the voltage does not affect the testing results of electrical resistance. Thus, the activation voltage was free to choose at the coverage of -1 V to 1 V in this study.

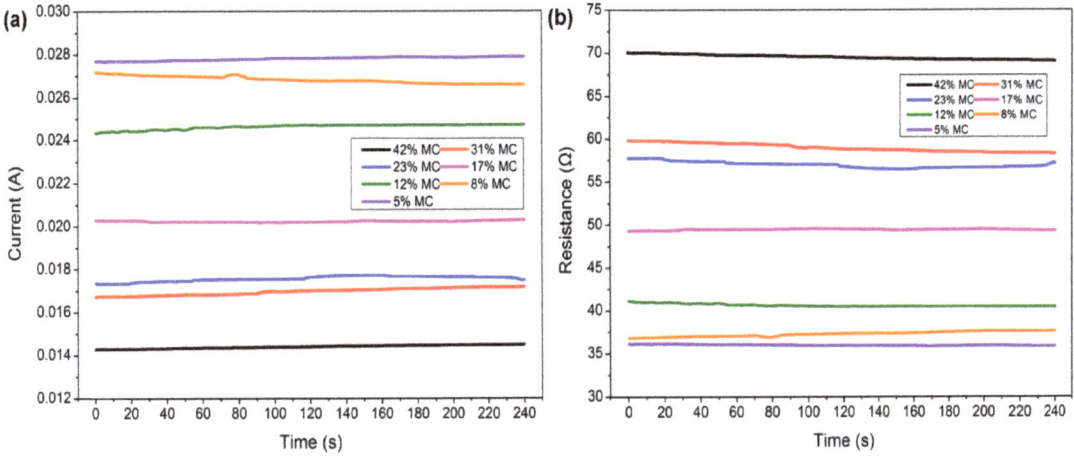

Figure 5. Current (**a**) and resistance (**b**) change over time under a constant voltage of 1 V for measuring shrinkage strain.

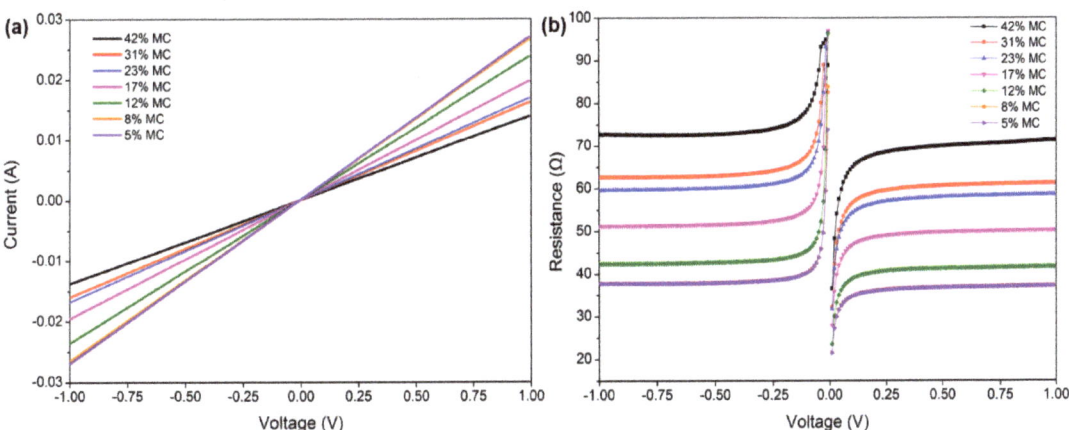

Figure 6. Plots of current (**a**) and resistance (**b**) with voltage for measuring shrinkage strain.

Figure 7 shows the variation in the resistance change rate and shrinkage strain at different MCs. Wood shrinkage was strongly connected with the water in the wood; once the wood MC dropped to the fiber saturation points, shrinkage occurred. It is widely accepted that the fiber saturation point is at an MC of approximately 30% for most wood species, which is a turning point for wood physics and mechanical properties [30,31]. Therefore, the shrinkage strain in this study was discussed from the MC of 31%, and no shrinkage was considered at the initial MC of 42%. As observed in Figure 7, both the shrinkage strain and resistance change rate increased with the decrease in wood MC, especially for the MC range of 23% to 8%, where the shrinkage strain and resistance change rate increased by 4% and 30%, respectively. As observed, the shrinkage of wood clearly increased with the decrease in MC below the FSP [9,32]. From Section 3.3, the electrical resistance gradually decreased with the wood shrinkage generated by the decrease in MC, and thus the resistance change rate increased with decreasing MC. In the case of shrinkage strain, its value increased from 1% at an MC of 23% to 5% at an MC of 8%, and the matching values for the resistance change rate changed from 17.5% to 47.5%. These results demonstrate a close correspondence between shrinkage strain and the resistance change rate.

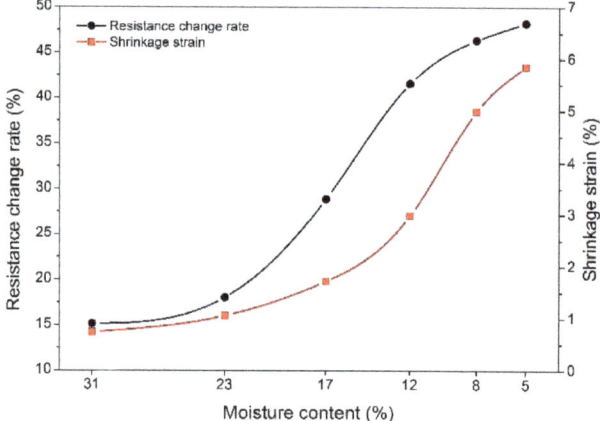

Figure 7. Resistance change rate and shrinkage strain at different MC stages.

In order to describe the correlations between shrinkage strain and the resistance change rate, the fitting curve between the two parameters is presented in Figure 8. As shown, the shrinkage strain increased exponentially with the increasing resistance change rate. The regression equation between shrinkage strain and the resistance change rate was described by Equation (5), and the determination coefficient reached 0.985, indicating that the prediction success rate of shrinkage strain was as high as 98.5% using this equation. This result provides a decent approach for the determination of wood shrinkage strain and also shows the feasibility of the electrochemical method in determining the wood shrinkage behavior.

$$S = 0.852 + 0.021 e^{0.113 \Delta R} \tag{5}$$

where S is the shrinkage strain, and ΔR is the resistance change rate.

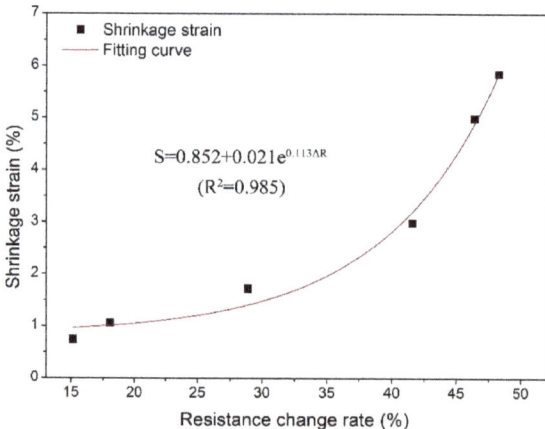

Figure 8. Fitting curve for the resistance change rate to shrinkage strain of wood.

4. Conclusions

The present research discussed the applicability of electrochemical methods for the determination of MC and shrinkage strain in wood drying, evaluating the precision as well.

According to the test results for MC, the electrical resistance clearly increased with the decreasing wood MC, especially in the MC range of 42% to 12%. A polynomial fitting curve with a determination coefficient of 0.937 was employed to describe the relationship between MC and electrical resistance.

In the case of the measurement of shrinkage strain, the electrical resistance gradually decreased with the decrease in MC, and the voltage did not affect the results of electrical resistance. The resistance change rate was further chosen as the correlation parameter to characterize shrinkage strain. The shrinkage strain increased exponentially with the increase in the resistance change rate. An exponential regression equation with the determination coefficient of 0.985 was determined to describe the correlation between shrinkage strain and the resistance change rate.

The findings of this article demonstrate the feasibility of the electrochemical approach to determine the MC and shrinkage strain in the wood drying process. Additional studies will be conducted on full-size specimens to achieve the applicability of this method in industrial production.

Author Contributions: Conceptualization, Z.F. and Y.L.; methodology, H.W. and J.L.; software, Y.L. and H.W.; validation, J.L.; formal analysis, J.L.; investigation, H.W.; resources, J.L.; data curation, Z.F.; writing—original draft preparation, Z.F.; writing—review and editing, Y.L. and H.W.; visualization, J.L.; supervision, Y.L.; project administration, Z.F.; funding acquisition, Z.F. All authors have read and agreed to the published version of the manuscript.

Funding: This study was financed by a Grant-in-aid for scientific research from the Youth Program of National Natural Science Foundation of China (Grant No. 31800478) and the National Natural Science Foundation of China (Grant No. 31870535 and 32122058).

Institutional Review Board Statement: Not applicable.

Informed Consent Statement: Not applicable.

Data Availability Statement: The data presented in this study are available upon request from the corresponding author.

Conflicts of Interest: The authors declare no conflict of interest.

References

1. Baettig, R.; Rémond, R.; Perré, P. Measuring moisture content profiles in a board during drying: A polychromatic X-ray system interfaced with a vacuum/pressure laboratory kiln. *Wood Sci. Technol.* **2006**, *40*, 261–274. [CrossRef]
2. Yu, L.; Hao, X.; Cai, L.; Shi, S.Q.; Jiang, J.; Lu, J. An Investigation of Moisture Gradient in Wood during Drying Using X-ray Radiation and Numeric Methods. *For. Prod. J.* **2014**, *64*, 199–205. [CrossRef]
3. Watanabe, K.; Saito, Y.; Avramidis, S.; Shida, S. Non-destructive Measurement of Moisture Distribution in Wood during Drying Using Digital X-ray Microscopy. *Dry. Technol.* **2008**, *26*, 590–595. [CrossRef]
4. Tanaka, T.; Avramidis, S.; Shida, S. Evaluation of moisture content distribution in wood by soft X-ray imaging. *J. Wood Sci.* **2009**, *55*, 69–73. [CrossRef]
5. Watanabe, K.; Lazarescu, C.; Shida, S.; Avramidis, S. A Novel Method of Measuring Moisture Content Distribution in Timber During Drying Using CT Scanning and Image Processing Techniques. *Dry. Technol.* **2012**, *30*, 256–262. [CrossRef]
6. Lindgren, O.; Seifert, T.; Du Plessis, A. Moisture content measurements in wood using dual-energy CT scanning—A feasibility study. *Wood Mater. Sci. Eng.* **2016**, *11*, 312–317. [CrossRef]
7. Rosenkilde, A.; Glover, P. High Resolution Measurement of the Surface Layer Moisture Content during Drying of Wood Using a Novel Magnetic Resonance Imaging Technique. *Holzforschung* **2002**, *56*, 312–317. [CrossRef]
8. Xu, K.; Lu, J.; Gao, Y.; Wu, Y.; Li, X. Determination of moisture content and moisture content profiles in wood during drying by low-field nuclear magnetic resonance. *Dry. Technol.* **2017**, *35*, 1909–1918. [CrossRef]
9. Fu, Z.; Zhao, J.; Yang, Y.; Cai, Y. Variation of Drying Strains between Tangential and Radial Directions in Asian White Birch. *Forests* **2016**, *7*, 59. [CrossRef]
10. Cheng, W.; Morooka, T.; Liu, Y.; Norimoto, M. Shrinkage stress of wood during drying under superheated steam above 100 °C. *Holzforschung* **2004**, *58*, 423–427. [CrossRef]
11. Peng, M.; Ho, Y.-C.; Wang, W.-C.; Chui, Y.H.; Gong, M. Measurement of wood shrinkage in jack pine using three dimensional digital image correlation (DIC). *Holzforschung* **2012**, *66*, 639–643. [CrossRef]
12. Khoo, S.-W.; Karuppanan, S.; Tan, C.-S. A Review of Surface Deformation and Strain Measurement Using Two-Dimensional Digital Image Correlation. *Metrol. Meas. Syst.* **2016**, *23*, 461–480. [CrossRef]

13. Mallet, J.; Kalyanasundaram, S.; Evans, P.D. Digital Image Correlation of Strains at Profiled Wood Surfaces Exposed to Wetting and Drying. *J. Imaging* **2018**, *4*, 38. [CrossRef]
14. Han, Y.; Park, Y.; Park, J.-H.; Yang, S.-Y.; Eom, C.-D.; Yeo, H. The shrinkage properties of red pine wood assessed by image analysis and near-infrared spectroscopy. *Dry. Technol.* **2016**, *34*, 1613–1620. [CrossRef]
15. Huang, Y.; Peng, L.; Liu, Y.; Zhao, G.; Chen, J.Y.; Yu, G. Biobased Nano Porous Active Carbon Fibers for High-Performance Supercapacitors. *ACS Appl. Mater. Interfaces* **2016**, *8*, 15205–15215. [CrossRef]
16. Pulido, Y.F.; Blanco, C.; Anseán, D.; García, V.M.; Ferrero, F.; Valledor, M. Determination of suitable parameters for battery analysis by Electrochemical Impedance Spectroscopy. *Measurement* **2017**, *106*, 1–11. [CrossRef]
17. Yang, J.; Chen, J.; Zhou, Y.; Wu, K. A nano-copper electrochemical sensor for sensitive detection of chemical oxygen demand. *Sens. Actuators B Chem.* **2011**, *153*, 78–82. [CrossRef]
18. Wang, Y.-C.; Su, M.; Xia, D.-H.; Wu, Z.; Qin, Z.; Xu, L.; Fan, H.-Q.; Hu, W. Development of an electrochemical sensor and measuring the shelf life of tinplate cans. *Measurement* **2018**, *134*, 500–508. [CrossRef]
19. Arellano-Pérez, J.; Negrón, O.R.; Jiménez, R.E.; Gómez-Aguilar, J.; Uruchurtu-Chavarín, J. Development of a portable device for measuring the corrosion rates of metals based on electrochemical noise signals. *Measurement* **2018**, *122*, 73–81. [CrossRef]
20. Ma, C.; Wang, Z.; Behnamian, Y.; Gao, Z.; Wu, Z.; Qin, Z.; Xia, D.-H. Measuring atmospheric corrosion with electrochemical noise: A review of contemporary methods. *Measurement* **2019**, *138*, 54–79. [CrossRef]
21. Salimi, A.; Kavosi, B.; Navaee, A. Amine-functionalized graphene as an effective electrochemical platform toward easily miRNA hybridization detection. *Measurement* **2019**, *143*, 191–198. [CrossRef]
22. Glass, S.V.; Zelinka, S.L. Moisture relations and physical properties of wood. In *Wood Handbook: Wood as an Engineering Material*; United States Department of Agriculture Forest Service, Forest Service, Forest Products Laboratory: Madison, WI, USA, 2021.
23. Brischke, C.; Rapp, A.O.; Bayerbach, R. Measurement system for long-term recording of wood moisture content with internal conductively glued electrodes. *Build. Environ.* **2008**, *43*, 1566–1574. [CrossRef]
24. Brischke, C.; Lampen, S.C. Resistance based moisture content measurements on native, modified and preservative treated wood. *Holz Als Roh- Werkst.* **2014**, *72*, 289–292. [CrossRef]
25. Barański, J.; Suchta, A.; Barańska, S.; Klement, I.; Vilkovská, T.; Vilkovský, P. Wood Moisture-Content Measurement Accuracy of Impregnated and Nonimpregnated Wood. *Sensors* **2021**, *21*, 7033. [CrossRef] [PubMed]
26. Davidson, R. The effect of temperature on the electrical resistance of wood. *For. Prod. J.* **1958**, *8*, 160–164.
27. Stamm, A.J. The Electrical Resistance of Wood as a Measure of Its Moisture Content. *Ind. Eng. Chem.* **1927**, *19*, 1021–1025. [CrossRef]
28. Nilsson, M. Evaluation of Three In-Line Wood Moisture Content Meters. Master's Thesis, Luleå University of Technology, Skellefteå, Sweden, 2010.
29. Stamm, A.J. *Wood and Cellulose Science*; Ronald Press: New York, NY, USA, 1964; 549p.
30. Pang, S.; Herritsch, A. Physical properties of earlywood and latewood of Pinus radiata D. Don: Anisotropic shrinkage, equilibrium moisture content and fibre saturation point. *Holzforschung* **2005**, *59*, 654–661. [CrossRef]
31. Gerhards, C.C. Effect of moisture content and temperature on the mechanical properties of wood: An analysis of immediate effects. *Wood Fiber Sci.* **2007**, *14*, 4–36.
32. Hernandez, R.E.; Pontin, M. Shrinkage of three tropical hardwoods below and above the fiber saturation point. *Wood Fiber Sci.* **2007**, *38*, 474–483.

Article

Lignin Distribution on Cell Wall Micro-Morphological Regions of Fibre in Developmental *Phyllostachys pubescens* Culms

Bo Liu, Lina Tang, Qian Chen, Liming Zhu, Xianwu Zou, Botao Li, Qin Zhou, Yuejin Fu * and Yun Lu *

Research Institute of Wood Industry, Chinese Academy of Forestry, Beijing 100091, China; liubo@criwi.org.cn (B.L.); 18211090798@163.com (L.T.); chenqian0610@126.com (Q.C.); lzhulm@caf.ac.cn (L.Z.); xwzou@caf.ac.cn (X.Z.); botaoLi@163.com (B.L.); zhouqin567@sina.com (Q.Z.)
* Correspondence: bj-fyj@163.com (Y.F.); y.lu@caf.ac.cn (Y.L.)

Citation: Liu, B.; Tang, L.; Chen, Q.; Zhu, L.; Zou, X.; Li, B.; Zhou, Q.; Fu, Y.; Lu, Y. Lignin Distribution on Cell Wall Micro-Morphological Regions of Fibre in Developmental *Phyllostachys pubescens* Culms. *Polymers* **2022**, *14*, 312. https://doi.org/10.3390/polym14020312

Academic Editor: Adriana Kovalcik

Received: 18 December 2021
Accepted: 11 January 2022
Published: 13 January 2022

Publisher's Note: MDPI stays neutral with regard to jurisdictional claims in published maps and institutional affiliations.

Copyright: © 2022 by the authors. Licensee MDPI, Basel, Switzerland. This article is an open access article distributed under the terms and conditions of the Creative Commons Attribution (CC BY) license (https://creativecommons.org/licenses/by/4.0/).

Abstract: Bamboo is a natural fibre reinforced composite with excellent performance which is, to a certain extent, an alternative to the shortage of wood resources. The heterogeneous distribution and molecular structure of lignin is one of the factors that determines its performance, and it is the key and most difficult component in the basic research into the chemistry of bamboo and in bamboo processing and utilization. In this study, the distribution of lignin components and lignin content in micro-morphological regions were measured in semi-quantitative level by age and radial location by means of visible-light microspectrophotometry (VLMS) coupled with the Wiesner and Maule reaction. There as guaiacyl lignin and syringyl lignin in the cell wall of the fibre. Lignin content of the secondary cell wall and cell corner increased at about 10 days, reached a maximum at 1 year, and then decreased gradually. From 17 days to 4 years, the lignin content of the secondary cell wall in the outer part of bamboo is higher than that in the middle part (which is, in turn, higher than that in the inner part of the bamboo). VLSM results of the micro-morphological regions showed that bamboo lignification developed by aging. Guaiacyl and syringl lignin units can be found in the cell wall of the fibre, parenchyma, and vessel. There was a difference in lignin content among different ages, different radial location, and different micro-morphological regions of the cell wall. The fibre walls were rich in guaiacyl lignin in the early stage of lignification and rich in syringyl units in the later stage of lignification. The guaiacyl and syringyl lignin deposition of bamboo green was earlier than that of the middle part of bamboo culm, and that of the middle part of bamboo culm was earlier than that of bamboo yellow. The single molecule lignin content of the thin layer is higher than that of thick layers, while the primary wall is higher than the secondary cell wall, showing that lignin deposition is consistent with the rules of cell wall formation. The obtained cytological information is helpful to understand the origin of the anisotropic, physical, mechanical, chemical, and machining properties of bamboo.

Keywords: *Phyllostachys pubescens*; fibre; lignification; micro-morphological regions of cell wall; syringyl lignin; guaiacyl lignin; visible-light spectrophotometry

1. Introduction

Bamboo is a profuse, long-lasting resource and the quickest developing and most adaptable plant on Earth. Bamboo has a wide range of applications. Due to its sustainability, extraordinary growth rate, accessibility, light weight, high mechanical strength, and good toupghness, it has been widely used as structural materials and bio-composites for industrial applications [1,2]. Bamboo is also considered to be an important source of biofuels and biochemical production [3]. Many excellent properties of bamboo stem are governed by the properties of the cell wall, which can be described in terms of the sub microstructure of cell wall and the localization of cell wall components of cellulose, hemicellulose, and lignin.

Being the most abundant biopolymer on earth, lignin shows beneficial structural properties [4]. Lignin is an amorphous polymer of benzene linked by the ether bond of propane and carbon-carbon (C-C) bond [5]. Lignin has various unique characteristics, including biocompatibility, antioxidant, antimicrobial, redox activity, etc. [6]. Lignin can be used in various industrial applications, including bio-fuels, chemicals, polymers, etc. Lignin can also be utilized in biomedical applications, such as drug delivery. However, these applications depend on the source, chemical modification, and physico-chemical properties of lignin [7]. An important role of lignin in the wood cell wall is to function as a cross-linking matrix between moisture sensitive cellulose and hemicelluloses. Thereby, lignin is considered to be a special compound closely related to the mechanical strength of cell wall [8,9]. Wood parenchyma cells do not generally contain lignin, but in bamboo culms, lignin widely exists in all kinds of lignified bamboo tissues [10]. It is the important component of the fibre cell wall, parenchyma cell wall, and vessel cell wall [11]. Therefore, understanding the structural distribution of bamboo lignin on the cell wall, especially "seeing" the micro structure of bamboo in the sense of chemical element distribution, is of great significance for making full use of bamboo lignin resources [12].

Studies on lignin in bamboos have been carried out in species of the genus *Phyllostachys* characteristically lacking free fibre strands [13,14], and possessing free fibre strands is associated with the vascular bundle [9,15]. The studies mainly focused on the lignification progress [16,17], various lignin content [10], peroxidase in lignification [18], lignin structure [4,15], and lignin industrial application [5], etc. However, the heterogeneous distribution and molecular structure of lignin are one of the factors that determine the performance of bamboo. Considering the heterogeneous distribution of lignin and the quantitative analysis of the micro morphological region of cell wall from the perspective of bamboo processing and utilization performance is of great significance for the basic research of bamboo chemistry and bamboo processing and utilization.

The vascular bundles of *Phyllostachys pubescens* are typically of the type III, which has a central vascular strand with four small fibre caps, one adjacent to the protoxylem, one to the phloem and one to each of the two large metaxylem vessels [19,20]. The in situ distribution and the content of lignin component within the fiber located at bamboo green, bamboo timber and bamboo yellow, and also located at different cell wall micromorphological regions within fibre cap on radial and longitudinal location were visualized by complementary microscopy techniques (visible-light microspectrophotometry) coupled with the biochemical method (Wiesner and Maule reactions), especially in different cell wall growth stages. The obtained cytological information is helpful to understand the origin of the anisotropic, physical, mechanical, chemical, and machining properties of bamboo.

2. Materials and Methods

2.1. Bamboo Samples

12- and 17-day-old immature bamboo shoots and 1- and 4-year-old mature bamboo culms of *Phyllostachys edulis* Carr. Lehaie were harvested on 23 April 2007 in Miaoshanwu Bamboo Garden (29°44′–30°12′ N, 119°25′–120°09′ E and 20 m altitude) of Semitropical Forestry Institute, Chinese Academy of Forestry, Zhejiang Province, China. Blocks of about 2 cm along the grain, including the green and yellow parts of bamboo culm, were cut from the middle part of the 13th internode above the ground and preserved in Formaldehyde-acetic acid-ethanol Fixative (FAA). 3 blocks for each bamboo age were taken as repetition.

2.2. Sectioning and Microscopy

Transverse sections of 20 μm in thickness, including the whole radial culm wall, were cut with a sliding microtome (Yamato Kohki TU-213, Saitama, Japan). All sections were observed directly under light microscope (Olympus BX50F4, Tokyo, Japan) for checking integrity of vascular bundles. Three sections from each sample block as repetition were used for detection of autofluorescence of cell walls by fluorescence microscopy before and after chemical treatments. The average value of nine sections from three sample blocks was as the value of lignin content.

2.3. Mäule Reaction, Wiesner Reaction and Visible-Light Spectrophotometry

Mäule and Wiesner reactions were classically applied to stain the syringylpropane and guaiacylpropane units of the cell wall lignin, respectively. After rinsing by distilled water, half of the transverse sections were treated in 1% $KMnO_4$ for 5 min followed by three washing in distilled water, then immersed in 3% HCl for 1 min and washed again in distilled water before being mounted in 29% ammonia and immediately observed by fluorescence microscope [21]. Other transverse sections were treated with 2% phloroglucinol in 95% ethanol for 5 min and mounted in 6 M HCl [17]. Absorption spectra from 450 nm to 650 nm at interval of 5 nm were measured with a visible-light microspectrophotometer (UNIVAR, Austria, spot size: 1.0 µm, band width: 10 µm) on the sections after Mäule and Wiesner reactions. Measurements were repeated for three times at every wavelength for the measuring points in each micro-morphological regions of cell wall. Visible-light absorption value of fibre was measured on four different levels: (1) different bamboo ages; (2) different radial positions of bamboo culms, including the green part, the middle part and the yellow part of bamboo culm; (3) different positions of vascular bundle fiber cap, from inner-side to outer-side of fibre cap adjacent to phloem in vascular bundles (Figure 1, four-year-old fibre in Mäule reaction); and (4) different positions of cell micro-morphological regions between two adjacent cells (Figure 2), including cell corner (CC), compound middle lamella (CML), the primary wall (PW), and the layers of secondary wall (SW).

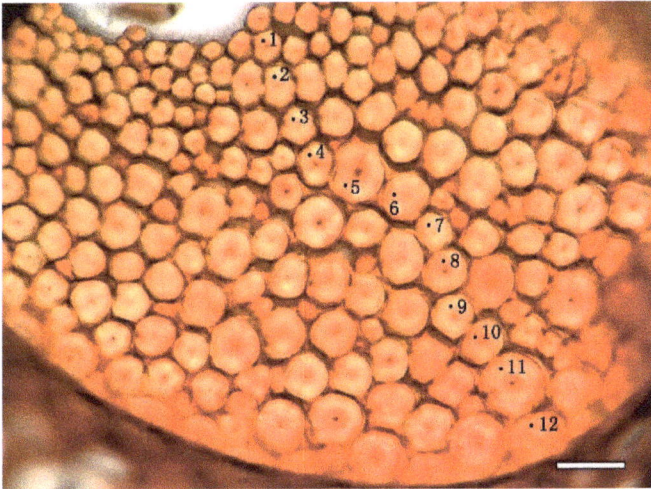

Figure 1. Testing points on different cell wall layers from inside to outside of the outer fibre cluster in the vascular bundle (four-year-old bamboo, the outer part of bamboo, Mäule reaction (Bar = 20 µm)).

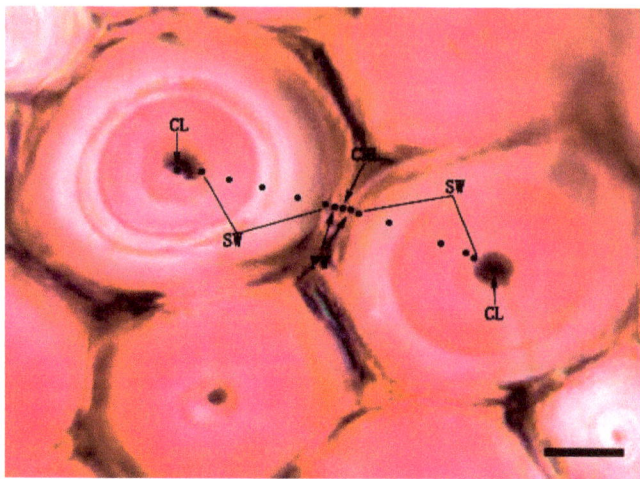

Figure 2. Testing points on different micro-morphological regions of fibre wall in lignin content test—four-year-old, the outer part of bamboo, Wiesner reaction (Bar = 5 μm). CML, cell middle lamella; PW, primary wall; SW, secondary wall; CL, cell lumen.

3. Results and Discussion

3.1. Histochemical Staining of the Cell Walls and Determination of Lignin Components

In histochemical staining, chemicals mainly reacted with syringyl lignin of cell wall in Mäule reaction. Then, the cell walls of dicotyledon wood fibres are stained into amaranth, but the cell walls of gymnosperm wood fibres are stained brown [22]. The Wiesner reaction has universal applicability to guaiacyl and syringyl lignin, which cell walls were stained into red in Wiesner reaction. In this study, fibres of *Phyllostachys pubescens* displayed light brown to dark brown in Mäule reaction, and pink or red in Wiesner reaction.

The layering of fibre generally alternate thick and thin layers with different fibrillar orientation [22]. To find out whether there is the difference of lignin content between thin layer and thick layer of the secondary wall, as an illustration, the secondary wall of four-year-old fibre in Wiesner reaction was divided into two micro-morphological regions for research, which were marked as B (the thin layer of the secondary wall) and H (the thick layer of the secondary wall). The detailed measurements were detected on micro-morphological regions of double cell walls between two fibre lumen. From one side to the other side, visible-light absorption value was measured gradually on B, PW, CML, PW and H in turn (Figure 2, four-year-old fibre in Mäule reaction). Due to the temporary nature of the colour reactions, all of the measurements were performed within 10 min. Then visible-light absorption spectra taken were averaged to give the mean spectrum, from which the mean absorbance at each developmental stage of the secondary wall was obtained.

Figures 3 and 4 showed the color reactions of fibre cell walls in Mäule and Wiesner reactions. Since the fibers in the vascular bundle of bamboo over one-year-old are extremely tough, when sliced with sliding microtome, the quality of slicing is not high, and the fiber cell wall is particularly easy to crack. The high strength and toughness of fibers often causes the edge of slicing blade to crack. The damage rate of slices increased with the increase of bamboo age. Cell wall of 12-day-old fibre did not show characteristic colours in both reactions (Figure 3A–C, Figure 4A–C). Cell wall of 17-day-old fibre showed back-brown in Mäule reaction (Figure 3D–F), especially in the fibres near the centre of the vascular bundle, according to the mature degree of fibre cell wall. However, cell wall of 17-day-old fibre in Weisner reaction showed pink colour on the area near the centre of the vascular bundle, indicating that fibre began lignification 17 days from shooting. Both one-year-old and four-year-old fibre showed brown in Mäule reaction (Figure 3G–I), and reddish pink in Wiesner reaction (Figure 4G–I), wherever in the green or yellow part of bamboo. This reflected the different developmental progress and distribution of guaiacyl and syringyl lignin. The positive reactions with Mäule and Weisner reagents demonstrated that both the guaiacyl (G) and syringyl (S) units were the components of fibre lignin. This accords with the general view which angiosperm plants contained abundant G and S units in lignin. Lybeer and Koch [23] found that the lignin of fibre in *Gigantochloa levis* and *Phyllostachys viridiglaucescens* (Carr.) Riv. & Riv. culms had G, S and p-hydroxyphenypropane (H) units. That was the three diferent phenyl propane monomers: coniferyl alcohol, syringyl alcohol, and coumaryl alcohol precursors. In softwoods, coniferyl alcohol is higher. In hardwoods, syringyl alcohol is abundant, and in crops and grasses, coumaryl alcohol is dominant [5,24,25]. The different proportions of these three monomers determine the important biological characteristics of plants, such as different rigidity and stiffness characteristics, water absorption, antifungal and insecticidal characteristics. Although there exists a little H lignin in monocotyledon plants, it has still no effective histochemical method for its detection and visual microscopic observation.

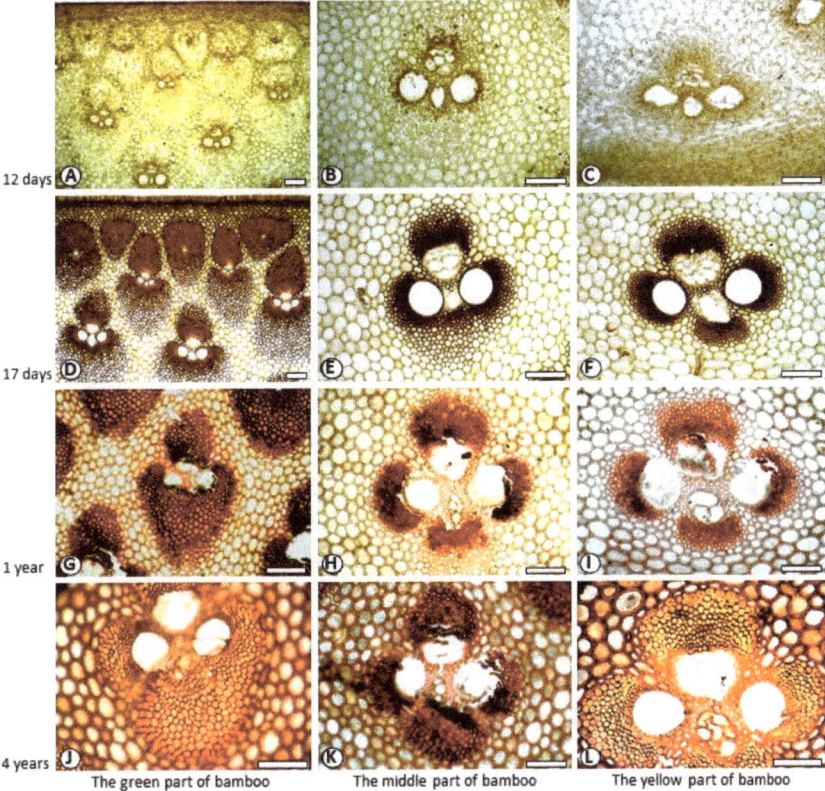

Figure 3. Fibre staining in Mäule reaction. Bar = 100 μm. (**A**) 12-day-old, the green part of bamboo; (**B**) 12-day-old, the middle part of bamboo, (**C**) 12-day-old, the yellow part of bamboo; (**D**) 17-day-old, the green part of bamboo; (**E**). 17-day-old, the middle part of bamboo, (**F**). 17-day-old, the yellow part of bamboo; (**G**). 1-year-old, the green part of bamboo; (**H**). 1-year-old, the middle part of bamboo; (**I**). 1-year-old, the middle part of bamboo; (**J**). 4-year-old, the green part of bamboo; (**K**). 4-year-old, the middle part of bamboo; (**L**). 4-year-old, the yellow part of bamboo.

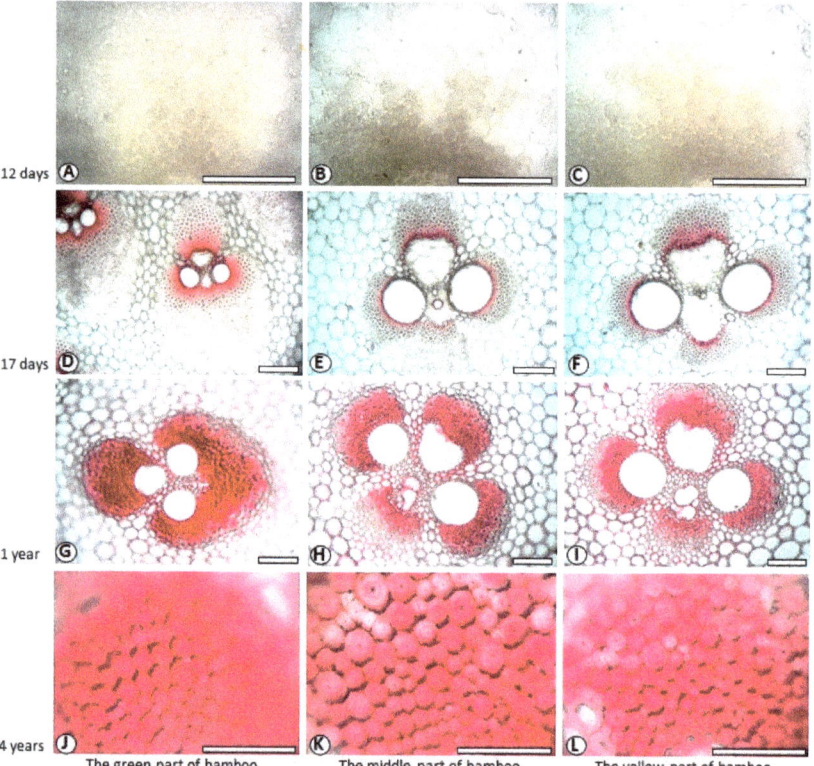

Figure 4. Fibre staining in Weisner reaction. Bar = 100 μm. (**A**) 12-day-old, the green part of bamboo; (**B**) 12-day-old, the middle part of bamboo, (**C**) 12-day-old, the yellow part of bamboo; (**D**) 17-day-old, the green part of bamboo; (**E**). 17-day-old, the middle part of bamboo, (**F**). 17-day-old, the yellow part of bamboo; (**G**). 1-year-old, the green part of bamboo; (**H**). 1-year-old, the middle part of bamboo; (**I**). 1-year-old, the middle part of bamboo; (**J**). 4-year-old, the green part of bamboo; (**K**). 4-year-old, the middle part of bamboo; (**L**). 4-year-old, the yellow part of bamboo.

3.2. Visible-Light Absorption Spectra

Visible light absorption spectra varied remarkably with culm age. Figure 5A,B showed the visible-light spectra taken on the thickest layer of fibre in different ages after Mäule and Weisner reactions. The spectrum exhibited respectively the absorption peak, summarized in Table 1. From 12 days to 4 years, absorption spectrum revealed clear peaks at 500 nm and 510 nm in the Mäule reaction. Absorption maxima of the 12-day-old fibre, 17-day-old fibre, and 4-year-old fibre were at 515 nm, 505 nm, and 520 nm, respectively, in the Weisner reactions. However, the spectra of four-year-old fibre had an unclear shoulder. Most of the absorption values were near the upper limit of the reference range, except a little peak at 500 nm. So, 500 nm was selected (arrowed in Figure 5B) for the detection of four-year-old fibre. Then, the absorption values on different micro-regions of fibre cell wall were measured at corresponding absorption spectra peak value.

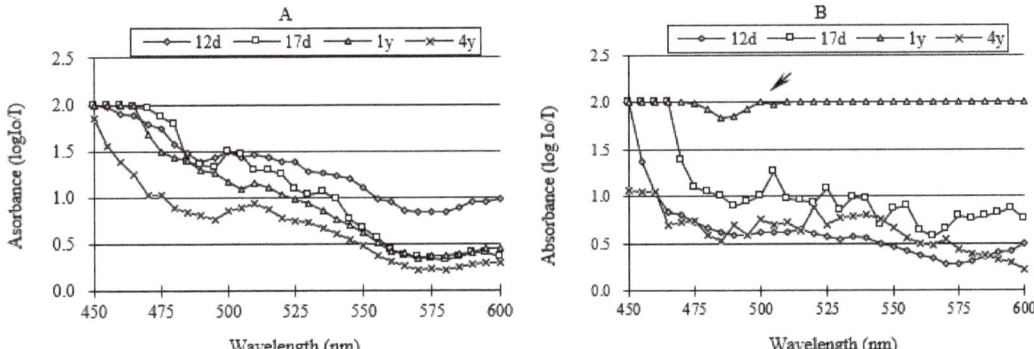

Figure 5. Visible-light absorption spectra for Mäule (**A**) and Wiesner (**B**) reactions of fibre in different ages. 12d, 12-day-old; 17d, 17-day-old; 1y, 1-year-old; 4y, 4-year-old.

Table 1. Absorption peaks with Mäule and Wiesner reactions of fibre.

Type of Reaction	Absorption Peak of Spectra (nm)			
	12-Day-Old	17-Day-Old	1-Year-Old	4-Year-Old
Mäule reaction	500	500	510	510
Wiesner reaction	515	505	500	520

3.3. Variation of Lignin Content of SW and CC in Different Ages

For SW, the absorbance values were very low in both of Mäule and Wiesner reactions, when fibre was 12-day-old (Figure 6A,B). This was consistent with the results of Figures 3 and 4, and indicated that fibre was in the swelling stage. It was the beginning of fibre lignification. When the fibre was 17-day-old, the curve showed the absorbance value of SW ascended to the top, before descending in the Mäule reaction. Since the Mäule reaction was the characteristic reaction to distinguish syringyl lignin [26], the content of syringyl lignin was the most abundance in SW of 17-day-old fibre and gradually decreased when growth was occurring (Figure 6A). The value was nearly equaled to the level of shooting age when fibre was four years old. In the Wiesner reaction, the content of lignin in SW increased at one period, and reached the absorption maximum at one year before decreasing gradually (Figure 6B). Since both guaiacyl and syringyl lignin react with Wiesner reagents, the absorbance value indicates the both lignin units from the stain colour by means of visible-light microspectrophotometry. It was found that the content of guaiacyl lignin increased from 17-day-old to 1-year-old. Comparing the results of Wiesner and Mäule reaction, it was concluded that guaiacyl lignin was increasing when syringyl lignin was decreasing. The content of both guaiacyl and syringyl lignin decreased a little after one year.

For cell corner, the variation trends of lignin components were generally similar to those of secondary wall in different ages. The difference was that the content of syringyl lignin on cell corner was always higher than that on secondary wall in the middle and yellow parts of bamboo culms in Mäule reaction (Figure 6C) when fibre was four years old. So, the rapid decreasing of guaiacyl lignin must be the reason that both content of lignin units decreased in the middle and yellow parts of bamboo culm from one-year-old to four-year-old samples in contrast with the results of Wiesner reaction.

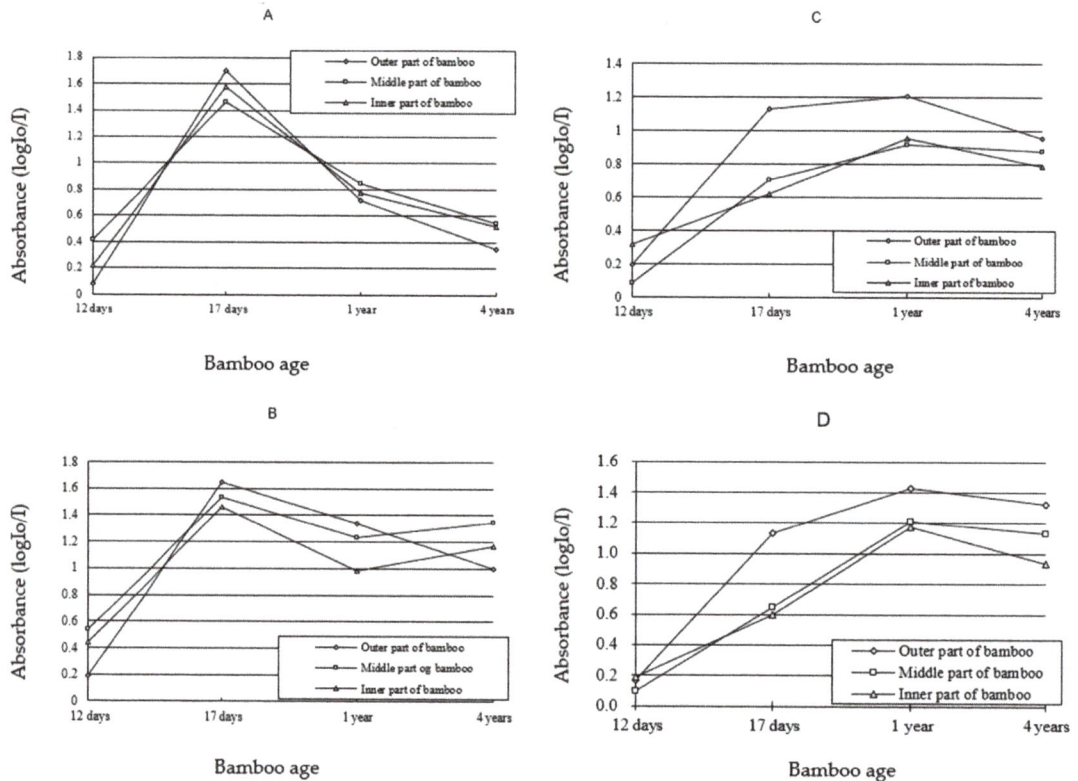

Figure 6. Change curve of absorbance of Mäule and Wiesner reaction in fibre secondary wall and cell corner during lignification process—(**A**) Secondary wall, Mäule reaction; (**B**) secondary wall, Wiesner reaction; (**C**): cell corner, Mäule reaction; (**D**): cell corner, Wiesner reaction.

In the initial period of lignification, the lignin content of fibre cell wall was low. However, it enhanced rapidly following cell development and fibre became one cell type containing the highest concentration of lignin. It owed to different functions of cells during bamboo growth process. As mechanical supporting tissue, fibre cells deposited lignin rapidly in order to improve the mechanics property and the resistance to outside attacks. This is consistent with the literature "the difference of lignin components reflects the difference of cell functions" [10]. In particular, the variation in the modulus was mainly due to the variation in the cell wall lignification level and its composition [12]. The observed difference in the modulus of elasticity between developing and fully lignified cell walls is due to the filling of the spaces with lignin and an increase in the packing density of the cell wall during lignification [27]. In addition, the results of lignin heterogeneity researched in *Quercus mongolica* latewood indicated that guaiacyl lignin was abundant in fibre secondary wall during the initial period of lignification [28]. However, the content and proportion of syringyl lignin would increase gradually with the development of lignification. Syringyl lignin would become the main unit in bamboo culm. Fukushima and Terashima [29] studied the lignin components of sugarcane and rice by means of UV microspectrophotometry. They suggested guaiacyl lignin is the main body in secondary wall of protoxylem vessel. In this study, syringyl lignin was abundant in fibre secondary wall of *Phyllostachys pubescens* during the initial stage of lignification, but decreased in the middle and last stages of lignification, accompanying with the increasing of guaiacyl lignin.

These results are different from the research results of *Quercus mongolica*, sugarcane and rice. This may be due to the difference of plant genus.

3.4. Variation of Lignin Content of SW and CC in Different Radial Location

The results of Mäule reaction showed that syringyl lignin of SW deposited more in the green and yellow parts of bamboo culm than in the middle part of bamboo culm when fibre was 17 day sold. The sequence of lignin content was the green part, yellow part and middle part of bamboo culm. However, syringyl lignin of SW deposited more in the middle part of bamboo culm than in the other part of bamboo culm, when the fibre was twelve days old, one year old or four year sold. The sequence of lignin content was the middle part, yellow part and green part of bamboo culm. The results of Wiesner reaction indicated that the whole lignin of syringyl and guaiacyl units was the most abundant in yellow part of bamboo culm, and the lesser abundant in green part of bamboo culm, when fibre was 12-day-old. While it varied to the most abundant in the green part of bamboo culm, and the lesser abundant in the middle part of bamboo culm, when fibre was after 12-day-old.

There was some difference of lignin content on CC with SW of fibre. When fibre was 12-day-old and 4-year-old, the sequence of syringyl lignin content on CC was the middle part, yellow part and green part of bamboo culm. But when fibre was 17-day-old and 1-year-old, the sequence of syringyl lignin content on CC was the green part, middle part and yellow part of bamboo culm. For the whole lignin of syringyl and guaiacyl units on CC, it was the totally same with the distribution of lignin on SW from 12-day-old to 4-year-old.

In different development period of *Phyllostachys pubescens*, the lignification progress of SW and CC experienced from the green part of bamboo culm to the yellow part of bamboo culm from 17-day-old to 4-year-old samples, except it varied greatly on radial location when the fibre was 12 days old. In the middle and last period of fibre lignification, the lignification progress and degree were similar to each other on radial location. These results supported the conclusion which the lignification of fibre and parenchyma cell developed from the outside to inside of bamboo culm of *Phyllostachys pubescens* [13].

Along the radius bamboo culm the modulus, hardness and carbohydrates concentration also had a gradient trend [30]. The change trend of elastic modulus and hardness was positive correlation with lignin content, and the carbohydrate concentration was negative correlation with lignin content [12]. This is the relationship between the lignin distribution in micro-morphological regions and the macro properties of bamboo.

3.5. Variation of Lignin Content of Different Cell Wall Micro-Morphological Regions

As shown in Figure 7A,B, the absorbance value showed syringyl lignin distribution on different micro-morphological regions of fibre cell wall in Mäule reaction when the fibre was four years old. The curve and histogram of absorbance showed the syringyl lignin content of CML was the most abundant, the syringyl lignin content of PW was the secondary, and the syringyl lignin content of SW was the least. Furthermore, the syringyl lignin content of thin layer of SW was always higher than that of thick layer of SW. Both single molecule lignin content of primary wall is higher than the secondary cell wall.

Figure 7C,D showed the both syringyl and guaiacyl lignin units distribution on different micro-morphological regions of fibre cell wall in Wiesner reaction when the fibre was four years old. As shown in the curve and histogram of absorbance, the trend of both syringyl and guaiacyl lignin distribution on different micro-morphological regions were similar to that of syringyl lignin distribution. The lignin content between two adjacent cells was the highest, followed by the cell lumen and, finally, the cell wall. It could be judged that the lignin monomer in the cell wall may come from the protoplast existing in the cell lumen, and the lignin monomer substances can permeate and flow between cells.

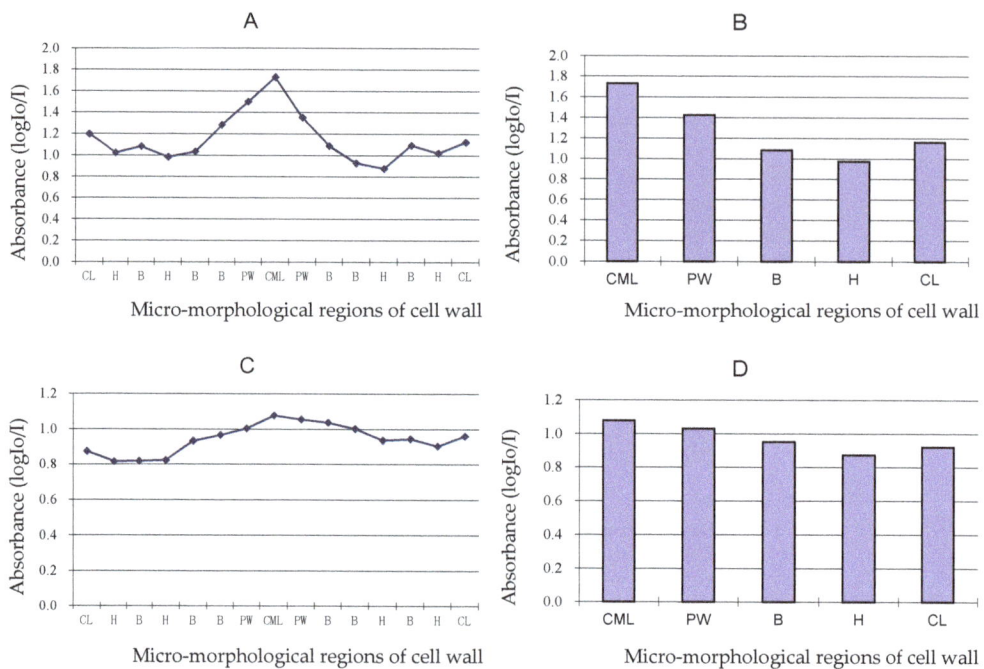

Figure 7. Curve and histogram of absorbance of Mäule (**A**,**B**) and Wiesner (**C**,**D**) reactions on different micro-morphological regions of four-year-old fibre wall in middle part of bamboo. CML, cell middle lamella, PW, primary wall; B, thin lamella of secondary wall; H, thick lamella of secondary wall; CL, cell lumen.

4. Conclusions

As one of the main components of the cell wall, lignin is distributed everywhere on different micro-morphological regions in the developmental fibre of *Phyllostachys pubescens*. Lignification develops by aging. Guaiacyl lignin units and syringyl lignin units can be found in the cell wall of the fibre, parenchyma, and vessel. The difference in lignin content among different ages, different radial location, and different micro-morphological regions of the cell wall was observed in this paper. The fibre walls were rich in guaiacyl lignin in the early stage of lignification, and lignin rich in syringyl units were deposited in the later stage. The guaiacyl and syringyl lignin deposition of bamboo green was earlier than that of the middle part of bamboo culm, and that of the middle part of bamboo culm was earlier than that of bamboo yellow. The multilayer of the fibre secondary cell wall has alternating thick and thin layers, while both single molecule lignin content of thin layer is higher than that of thick layers and the primary wall is higher than the secondary cell wall. This also shows that the rules of lignin deposition is consistent with the rules of cell wall formation. It is considered that lignin plays an important role in cell wall formation and cell wall mechanical properties. Lignin is related to the physical and mechanical properties of bamboo. Therefore, the study of the distribution and change of lignin in bamboo development is conducive to the mastery and prediction of various properties in the bamboo development, and has guiding significance for bamboo and lignin industrial utilization.

Author Contributions: Formal analysis, B.L. (Bo Liu), L.T. and X.Z.; investigation, L.Z., B.L. (Botao Li) and Q.Z.; methodology, Q.C.; project administration, B.L. (Bo Liu); resources, B.L. (Bo Liu) and Y.L.; supervision, Y.F.; writing—original draft, B.L. (Bo Liu); writing—review and editing, YL. All authors have read and agreed to the published version of the manuscript.

Funding: This research was funded by the "National Natural Science Foundation of China", grant Nos. 31870539, 31870535, 32122058.

Institutional Review Board Statement: Not applicable.

Informed Consent Statement: Not applicable.

Data Availability Statement: The data presented in this research are available on request from the corresponding author.

Acknowledgments: This study was financially supported by the National Natural Science Foundation 31870539, 31870535 and 32122058. The authors would like to express appreciation to International Center for Bamboo and Rattan, and Reasearch Institue of Subtropical Forestry, Chinese Academy of Forestry for the assistance with sample collecting.

Conflicts of Interest: The authors declare no conflict of interest.

References

1. Huang, Y.X.; Ji, Y.H.; Yu, W.J. Development of bamboo scrimber: A literature review. *J. Wood Sci.* **2019**, *65*, 25. [CrossRef]
2. Huang, S.S.; Jiang, Q.F.; Yu, B.; Nie, Y.J.; Ma, Z.Q.; Ma, L.F. Combined chemical modifichation of bamboo material prepared using vinyl acetate and methyl methacrylate: Dimensional stability, chemical structure, and dynamic mechanical properties. *Polymers* **2019**, *11*, 1651. [CrossRef]
3. Nirmala, C.; Bisht, M.S.; Bajwa, H.K.; Santosh, O. Bamboo: A rich source of natural antioxidants and its applications in the food and pharmaceutical industry. *Trends Food Sci. Technol.* **2018**, *77*, 91–99. [CrossRef]
4. Xu, G.F.; Shi, Z.J.; Zhao, Y.H.; Deng, J.; Dong, M.Y.; Liu, C.T.; Vignesh, M.; Mai, X.M.; Guo, Z. Structural characterization of lignin and its carbohydrate complexes isolated from bamboo (*Dendrocalamus sinicus*). *Int. J. Biol. Macromol.* **2019**, *126*, 376–384. [CrossRef]
5. Verma, S.; Hashmi, S.A.R.; Mili, M.; Hada, V.; Prashant, N.; Naik, A.; Rathore, S.K.S.; Srivastava, A.K. Extraction and applications of lignin from bamboo: A critical review. *Eur. J. Wood Wood Prod.* **2021**, *79*, 1341–1357. [CrossRef]
6. Kaur, P.J.; Kardam, V.; Pant, K.K.; Naik, S.N.; Satya, S. Characterization of commercially important Asian bamboo species. *Eur. J. Wood Wood Prod.* **2016**, *74*, 137–139. [CrossRef]
7. Figueiredo, P.; Lintinen, K.; Hirvonen, J.T.; Kostiainen, M.A.; Santos, H.A. Properties and chemical modifcations of lignin: Towards lignin-based nanomaterials for biomedical applications. *Prog. Mater. Sci.* **2018**, *93*, 233–269. [CrossRef]
8. Salmén, L.; Burgert, I. Cell wall features with regard to mechanical performance. A review COST Action E35 2004-2008: Wood machining micromechanics and fracture. *Holzforschung* **2009**, *63*, 121–129. [CrossRef]
9. Lybeer, B.; Koch, G. Lignin distribution in the tropical bamboo species *Gigantochloa levis*. *IAWA J.* **2005**, *26*, 443–456. [CrossRef]
10. Yang, S.M.; Liu, X.E.; Shang, L.L.; Ma, J.F.; Tian, G.L.; Jiang, Z.H. The Characteristics and Representation Methods of Lignin for Bamboo. *Polym. Polym. Matrix Compos.* **2020**, *34*, 7177–7182.
11. Igor, C. Unraveling the regulatory network of bamboo lignification. *Plant Physiol.* **2021**, *187*, 673–675.
12. Jin, K.X.; Ling, Z.; Jin, Z.; Ma, J.F.; Yang, S.M.; Liu, X.E.; Jiang, Z.H. Local Variations in Carbohydrates and Matrix Lignin in Mechanically Graded Bamboo Culms. *Polymers* **2022**, *14*, 143. [CrossRef] [PubMed]
13. Itoh, T. Lignification of bamboo (*Phyllostachys heterocycla* Mitf.) during its growth. *Holzforschung* **1990**, *44*, 191–200. [CrossRef]
14. Qu, C.; Shinjiro, O.; Takao, K. Characterization of immature bamboo (*Phyllostachys nigra*) component changes with its growth via heteronuclear single-quantum coherence nuclear magnetic resonance spectroscopy. *J. Agric. Food Chem.* **2020**, *68*, 9896–9905. [CrossRef] [PubMed]
15. Zhu, Y.K.; Huang, J.W.; Wang, K.L.; Wang, B.; Sun, S.L.; Lin, X.C.; Song, L.L.; Wu, A.M.; Li, H.L. Characterization of Lignin Structures in Phyllostachys edulis (Moso Bamboo) at Different Ages. *Polymers* **2020**, *12*, 187. [CrossRef] [PubMed]
16. Suzuki, K.; Itoh, T. The changes in cell wall architecture during lignification of bamboo, Phyllostachys aurea Carr. *Trees* **2001**, *15*, 137–147. [CrossRef]
17. Lin, J.X.; He, X.Q.; Hu, Y.X.; Kuang, T.Y.; Ceulemans, R. Lignification and lignin heterogeneity for various age classes of bamboo (*Phyllostachys pubescens*) stems. *Physiol. Plant.* **2002**, *144*, 296–302. [CrossRef]
18. Gan, X.H.; Ding, Y.L. Investigation on the Variation of Fiber Wall in Phyllostachys edulis Culms. *J. For. Res.* **2006**, *19*, 457–462.
19. Liu, B. Study on the Formation of Cell Wall during the Development of *Phyllostachys pubescens*. Ph.D. Thesis, Chinese Academy of Forestry, Beijing, China, 2008.
20. Zhang, X.X.; Yu, Z.X.; Yu, Y.; Wang, H.K.; Li, J.H. Axial compressive behavior of Moso Bamboo and its components with respect to fiber-reinforced composite structure. *J. For. Res.* **2019**, *30*, 2371–2377. [CrossRef]
21. Watanabe, Y.; Kojima, Y.; Ona, T.; Asada, T. Histochemical study on heterogeneity of lignin in Eucalyptus species II. The distribution of lignins and polyphenols in the walls of various cell types. *IAWA J.* **2004**, *25*, 283–295. [CrossRef]
22. Parameswaran, N.; Liese, W. On the fine structure of bamboo fibres. *Wood Sci. Technol.* **1976**, *10*, 231–246.
23. Lybeer, B.; Koch, G. A top chemical and semi quantitative study of the signification during ageing of bamboo culms (*phyllostachys viridiglaucescens*). *IAWA J.* **2005**, *26*, 99–109. [CrossRef]

24. He, X.Q. Histo- and Cytological Studies on the Lignification of Bamboo Stem (*Phyllostachys pubescens* Mazel). Ph.D. Thesis, The Chinese Academy of Sciences, Beijing, China, 1999.
25. Lu, F.; Ralph, J.; Agric, J. The DFRC method for lignin analysis. 2. Monomers from isolated lignins. *Food Chem.* **1998**, *46*, 547–552. [CrossRef] [PubMed]
26. Xiong, W.Y.; Ding, Z.F.; Li, Y.F. Mediacy growth and innernode growth of bamboo. *Sci. Silvae Sin.* **1980**, *16*, 81–89.
27. Huang, Y.H.; Fei, B.H. Comparison of the mechanical characteristics of fifibers and cell walls from moso bamboo and wood. *BioResources* **2017**, *12*, 8230–8239. [CrossRef]
28. Antonova, G.F.; Varaksina, T.N.; Zheleznichenko, T.V.; Stasova, V.V. Lignin deposition during earlywood and latewood formation in Scots pine stems. *Wood Sci. Technol.* **2014**, *48*, 919–936. [CrossRef]
29. Fukushima, K.; Terashima, N. Hererogeneity in formation of lignin. 15. Formation and structure of lignin in comparession wood of pinus-thundergii studied by microautoradiography. *Wood Sci. Technol.* **1991**, *25*, 371–381.
30. Wang, X.Q.; Ren, H.Q.; Zhang, B.; Fei, B.H.; Burgert, I. Cell wall structure and formation of maturing fifibres of moso bamboo (*Phyllostachys pubescens*) increase buckling resistance. *J. R. Soc. Interface* **2012**, *9*, 988–996. [CrossRef]

Article

Improving the Mould and Blue-Stain-Resistance of Bamboo through Acidic Hydrolysis

Zixuan Yu [1,2], Xiaofeng Zhang [1,2], Rong Zhang [1,2], Yan Yu [3] and Fengbo Sun [1,2,*]

1. Department of Biomaterials, International Center for Bamboo and Rattan, Beijing 100102, China; yuzixuan@icbr.ac.cn (Z.Y.); zhangxf@icbr.ac.cn (X.Z.); zhangrong@icbr.ac.cn (R.Z.)
2. SFA and Beijing Co-Built Key Laboratory of Bamboo and Rattan Science & Technology, State Forestry and Grassland Administration, Beijing 100102, China
3. College of Material Engineering, Fujian Agriculture and Forestry University, Fuzhou 350002, China; yuyan9812@outlook.com
* Correspondence: sunfengbo@icbr.ac.cn

Citation: Yu, Z.; Zhang, X.; Zhang, R.; Yu, Y.; Sun, F. Improving the Mould and Blue-Stain-Resistance of Bamboo through Acidic Hydrolysis. *Polymers* **2022**, *14*, 244. https://doi.org/10.3390/polym14020244

Academic Editor: Antonios N. Papadopoulos

Received: 8 November 2021
Accepted: 28 December 2021
Published: 7 January 2022

Publisher's Note: MDPI stays neutral with regard to jurisdictional claims in published maps and institutional affiliations.

Copyright: © 2022 by the authors. Licensee MDPI, Basel, Switzerland. This article is an open access article distributed under the terms and conditions of the Creative Commons Attribution (CC BY) license (https:// creativecommons.org/licenses/by/ 4.0/).

Abstract: Bamboo is much more easily attacked by fungus compared with wood, resulting in shorter service life and higher loss in storage and transportation. It has been long accepted that the high content of starch and sugars in bamboo is mainly responsible for its low mould resistance. In this paper, acetic acid, propionic acid, oxalic acid, citric acid, and hydrochloric acid were adopted to hydrothermally hydrolyze the starch in bamboo, with the aims to investigate their respective effect on the mould and blue-stain resistance of bamboo, and the optimized citric acid in different concentrations were studied. The starch content, glucose yields, weight loss, and colour changes of solid bamboo caused by the different acidic hydrolysis were also compared. The results indicated that weak acidic hydrolysis treatment was capable of improving mould-resistant of bamboo. The mould resistance increased with the increased concentration of citric acid. Bamboo treated with citric acid in the concentration of 10% could reduce the infected area ranging to 10–17%, the growth rating of which could reach 1 resistance. The content of soluble sugar and starch remained in bamboo decreased significantly from 43 mg/g to 31 mg/g and 46 mg/g to 23 mg/g, respectively, when the citric acid concentration varied from 4% to 10%. Citric acid treatments of 10% also caused a greatest surface colour change and weight loss. The results in this study demonstrated citric acid treatment can effectively reduce the starch grain and soluble sugars content and improve mould resistance of bamboo, which can be attributed to the reduction of starch grain and soluble carbohydrates (such as glucose, fructose, and sucrose, etc.) in bamboo.

Keywords: acidic treatment; bamboo; colour; mould-resistant; starch

1. Introduction

Bamboo is easily attacked by various fungi, among which the infection caused by mould fungi is the most common and serious. Mould fungi can infect bamboo during storage, transportation, processing, and utilizing, with notable appearance, and in some cases, it can cause illness to humans. This greatly limits the application of solid bamboo and bamboo-derived products. The low mould resistance of bamboo is mainly attributed to its high content of starch and free sugars, which can act as feed for fungus or insects [1,2].

Many inorganic and organic preservatives, such as CCA, ACQ, CCC, DDAC, and IPBC, demonstrate a certain degree of mould resistance. Nevertheless, few of them perform well as a bamboo-mould inhibitor [3]. This is because the majority of them were actually developed for wood materials. Furthermore, the anatomic characteristics and chemical compositions of bamboo are quite different from wood [4]. There is no transverse conductive tissues in bamboo, which makes bamboo inaccessible to many preservatives with high molecular weight. Furthermore, the wide applications of these preservatives might lead to major environmental concerns [5–7]. Recently, wood furfurylation was found to be a highly efficient approach to deal with the mould issue on wood [8], but the colour of the treated wood was typically turned into black or brown, losing its original natural appearance.

Since the existence of starch and free sugars in bamboo is directly related to its mould deterioration, a reasonable way to improve the durability of bamboo is to remove these nutritious components from bamboo. The starch in bamboo is mainly located in the cavities of its ground parenchyma cells with normally less than 6% of total bamboo mass. The plant starch is stored in crystalline form and is nearly insoluble in water at ambient temperatures [9,10]. However, it can be hydrolysed to sugars in an acidic solution or enzyme catalyst. Clausen [11] pointed that multifactorial fatty acid emulsifications incorporated as an appropriate adjuvant are effective in inhibiting mould on wood products. Organic acids, such as boric, citric, and sorbic acids, have long been used in the food industry as preservatives [12,13]. Tang [14,15] found that bamboo dipped in 7% propionic acid or 10% acetic acid were effective in inhibiting mould growth completely. However, their effectiveness was mainly attributed to their acidity. Other monocotyledons, like palm woods, which also contain much starch and soluble carbohydrates, were also prevented from fungal colonization by acetic and propionic acid [16]. Sun also found that 1% HCl was helpful to improve mould resistance of bamboo [17], but the supplementary tests with water immersing the treated specimens performed significant resistance as well. The results indicated that acidic treatments are promising against fungus-affected bamboo. However, the differences of acid types in inhibiting mould and their mechanisms are still not identified completely.

In this study, four kinds of low molecular weight organic acids, namely acetic acid, propionic acid, oxalic acid, and citric acid, as well as one inorganic acid, namely hydrochloric acid, were adopted to hydrothermally hydrolyze the starch in bamboo. The mould resistance, weight loss, and colour changes of solid bamboo as well as the glucose yields in the respective hydrolysates caused by the different acidic hydrolysis were compared. The objective of this study is to evaluate the significance of starch in bamboo for mould growth and whether acidic treatment can achieve satisfactory mould resistance for bamboo.

2. Material and Methods

2.1. Materials

Moso bamboo (*Phyllostachys pubescens* Mazel ex H. de Lehaie) strips with regular cross section were purchased from Hangzhou Dazhuang Flooring Co., Ltd. (Dazhuang Flooring Co., Ltd., Hangzhou, China). The samples for mould resistance test had dimensions of 50 mm (longitudinal) × 20 mm (tangential) × 5 mm (radial), whereas the ones for both weight loss and colour change measurement were 20 mm (longitudinal) × 20 mm (tangential) × 5 mm (radial). For each treatment, there were 12 specimens for mould resistance tests and 10, respectively, for weight loss and colour change tests. Five kinds of common acids, namely acetic acid, propionic acid, oxalic acid, citric acid, and hydrochloric acid, were purchased from Beijing Chemicals (Beijing, China). The mould fungus (*Aspergillus niger* van. Tieghem) and the blue-staining fungus (*Botryodiplodia theobromae* Pat.) were purchased from the Institute of Forest Ecology Environment and Protection, Chinese Academy of Forestry (Beijing, China). The mould fungus of *Penicillium citrinum* and *Trichoderma viride* were purchased from Institute of Microbiology, Chinese Academy of Sciences (Beijing, China).

2.2. Acidic Treatment

Bamboo specimens were soaked in the aqueous solutions of acetic acid, propionic acid, oxalic acid, citric acid, and hydrochloric acid with concentration (W/W) of 2%, 2%, 2%, 2%, and 0.7%, respectively, for 1 h at room temperature. Then, the samples together with the acidic solutions were transferred to a drying oven and heated at a temperature of 90 °C for 3 h. Afterwards, the residual acids in the treated samples were removed by repetitive washing with deionized water. All samples were then oven dried at 105 °C. Bamboo specimens treated with citric acid of 4%, 6%, 8%, and 10% concentration were soaked for 1 h at room temperature as well and then oven-dried at 90 °C for 2.5 h, then washed with deionized water and oven dried at 105 °C.

2.3. Mould Resistance Tests

The tests of laboratory mould resistance were carried out according to a Chinese national standard GB/T 18261-2000. The treated and control (untreated) blocks were placed in petri dishes containing agar (2% agar) with the selected fungus and incubated for 4 weeks at 25 °C and 85% relative humidity. The change of each sample was photographed and recorded every day. The samples were visually rated for the growth of fungi on the following scale: 0 = no growth, 1 = 25 percent, 2 = 50 percent, 3 = 75 percent, and 4 = 100 percent coverage with mould. To evaluate growth of fungi objectively, the infection area on bamboo surface after the test was estimated by Matlab (MathWorks, America, R2011b). The infection degree of bamboo was obtained from ratio of hypha pixels to the whole block pixels.

2.4. Soluble Sugar Content in the Hydrolysates

Ion chromatography with an ampere detector (850, Metrohm, Switzerland) was used to test the content of glucose in the hydrolysates. Solutions after acidic treatments at 90 °C were collected and centrifuged, and the supernatant fluid was recovered and injected into the centrifuge tube for analysis. The injection volume was 10 mL. The ion chromatographic analysis by using a Hamilton RCX-30 aminex was performed at 20 °C. The ultrapure water solution was 2.0 mM NaOH + 0.5 mM NaAc. The flow rate of the eluent was 1.0 mL/min and kept for 60 min.

High-performance liquid chromatography (Waters Alliance e2695, Waters Co., Milford, MA, USA) was also used to test soluble sugar content in the hydrolysates of citric acid treatments. The working condition of HPLC was Bio-Rad87-H column, column temperature of 65 °C, and sulfuric acid as mobile phase. The flow rate was 0.6 mL/min.

2.5. Soluble Sugars and Starch Content Remained in Bamboo

Residual soluble sugars content of bamboo was determined based on the method of anthrone colorimetry. The main procedure was as follows: 0.1 g of bamboo powders were ground by pestle by using 80% alcohol and then heated in water baths at 80 °C for 30 min. The supernatants collected by centrifugation were transferred in a 100-mL volumetric flask after three replications. Next, 4 mL of anthrone solution was added to the diluted solution in the ice water bath and shaken well. Next, we boiled the tube in a boiling water bath for 10 min. The absorbance at 630 nm was read.

Residual starch of bamboo was determined using the method of anthrone colorimetry too. The residues after extraction were dried at 80 °C and thereafter boiled with deionized water for 15 min. We next put the centrifugal tube into an ice water bath, then added 2 mL perchloric acid (9.2 mol/L) for 15 min, and centrifuged it for supernatant. Next, 4.6 mol/L perchloric acid were added to the sediment and extract for 15 min to obtain the supernatant. Then, we washed the sediments with 7 mL distilled water and performed a water bath for 20 min, and then, the supernatants were transferred to the volumetric bottle. We then took 2 mL solutions in the tube and added them into 4 mL anthrone and boiled them for 7.5 min. The absorbance at 630 nm was determined.

2.6. Color Change

The colour changes of specimens due to acidic treatments were measured by a portable chromatic aberration meter (BYK Gardner-6834, Geretsried, Germany). The value of CIE L*, a*, and b* colour parameters of the samples were obtained directly from the meter. Color change was evaluated according to the ISO 7724 standard. The overall colour change (ΔE^*) of the treated samples was calculated according to the following equation:

$$\Delta E^* = \sqrt{(\Delta L^*)^2 + (\Delta a^*)^2 + (\Delta b^*)^2} \tag{1}$$

where ΔL*, Δa*, and Δb* represent the changes in L*, a*, and b* of samples after acidic treatment, respectively. Ten replicates were tested, and the average value was calculated for colour analysis.

2.7. Weight Loss Ratio

Weight loss ratio of samples was calculated according to Equation (2):

$$WL = \frac{W_0 - W_1}{W_0} \times 100\% \qquad (2)$$

where WL is weight loss ratio of each sample (%); W_0 and W_1 represent the oven dried weight of samples before and after acidic treatment, respectively.

2.8. SEM Bbservation

An Environmental Scanning Electron Microscopy (ESEM XL 30, FEI, Hillsboro, OR, USA) was used to compare the microstructure of bamboo before and after acidic treatment. The acceleration voltage was set at 7–10 kV.

3. Results and Discussion

3.1. Treatments with Acids Concentration below 2%

3.1.1. Mould-Resistance Test

Bamboo infected by *A. niger* and *B. theobromae*, respectively, were observed periodically. The occurrence and propagation of both fungi on the treated samples were all much later and slower than those on the control samples. This phenomenon was especially obvious during the period of the first 3–5 days. For example, *A. niger* appeared on the control bamboo strips as early as the third day and then spread over the whole surface in less than five days. In contrast, the occurrence of mycelia on the treated bamboo was postponed by 2–7 days, dependent on the acid used. Furthermore, the treatments with oxalic acid, citric acid, and hydrochloric acid showed better fungus resistance performance than the other two acids, as the date of first occurrence was postponed, and the mycelia on the surfaces were much thinner during the first week. During the second week, the propagation of fungi started to speed up, but it did not cover the whole surface yet. Until the fourth week, all the treated samples were seriously infected by the fungi, with mycelia spreading over the surface. The growth rating of for treated groups ranged between 2 and 3 compared to 4 for the control group (Table 1). The results showed that although the distribution density and growth rate of mycelia and spores on the surface of treated bamboo were much lower than that of untreated samples, significant differences were observed for 2% organic acidic treatments in terms of the infected areas ratio at the final date of a standard mould-resistance experiment. For *B. theobromae*, the visible blue stain happened during the second week. The growth rating ranged between 2 and 3 for treated groups and up to 4 for control group.

Table 1. Fungus growth rating of bamboo treated with different acids.

Varieties of Treatment	A. niger	Penicillium citrinum	Trichoderma viride	B. theobromae
Control	4	4	4	4
Acetic acid	3	3	3	3
Propionic acid	3	3	3	3
Oxalic acid	3	3	2	2
Citric acid	3	2	2	2
Hydrochloric acid	2	2	2	2

The general improvement in mould and blue-stain resistance of bamboo after acidic treatments could be to a large extent attributed to the reduction of nutritious components in bamboo, including free sugars and starch particles. Previous studies [2,18,19] have indicated that the presence of considerable quantities of starch and sugars in bamboo make it more attractive to organisms, especially mould and stain fungi and borer beetles. Wingfield and Schmidt also pointed out that stain fungi can easily get nutrients from soluble sugars and starch [20,21]. Huang demonstrated that a combination of pressurized hydrothermal treatment and amylase hydrolysis could improve the mould resistance of Moso bamboo due to the reduction of free sugars and starch content [22]. The acidic treatments in the present study follow a similar mechanism. Figure 1 shows a weight loss of 5.04–6.39% of bamboo after various acidic treatments. This indicated that some starch and extractives were removed after hydrothermal acidic treatments, representing a significant reduction in the nutritious contents in bamboo. Another possible reason is that some water-solution-free sugars in bamboo were also washed out during the experiments. Schmit showed that the contents of starch and sugars of palm woods samples were considerably reduced by three days of watering [16]. However, although acidic treatments can improve the fungus resistance of bamboo to a certain extent, it seems that the earlier differences in mould time of the five groups may be more influenced by the remaining nutrients. In addition, the removal of starch and free glucose was incapable of totally solving the problem, as serious mould growth still occurred after one month of incubation. This contrasts to the results on palm woods, where the decrease of starch and sugar contents reduced mould and blue-stain growth [16]. A possible explanation might be related to the acidic decomposition of xylan in bamboo, producing additional monosaccharides or oligosaccharides that instead facilitate the appearance of mildews.

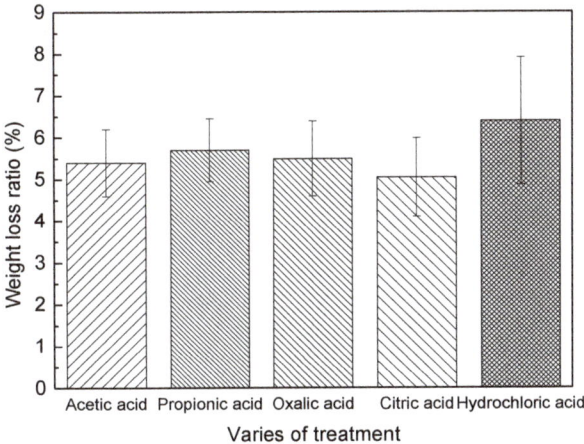

Figure 1. Weight loss of bamboo caused by five different acidic treatments.

3.1.2. Glucose Yields in Different Hydrolysates

The glucose yields in different hydrolysates are shown in Figure 2. The glucose yields in hydrolysates derived from acetic acid and propionic acid were 288 and 372 mg/L, respectively, significantly lower than the 500, 557, and 588 mg/L, respectively, from oxalic acid, citric acid, and hydrochloric acid. Among the five kinds of acids used in the present study, the inorganic hydrochloric acid showed the highest hydrolysis performances although its concentration in the solution was only 0.7%, significantly lower than 2% for the other four organic acids. In addition, acetic acid and propionic acid were much weaker than oxalic acid and citric acid in hydrolysis because they provided lower glucose yields in the hydrolysates. The acidic hydrolysis of starch is a highly complex process. This process is related to several factors, including hydrolysis time, temperature, acid types and concentrations, etc. During the process of treatment, organic acids may become involved in esterification reactions with starch. However, the key reason for the differences in glucose yields by different organic acids cannot be highly clear without sufficient investigation. In addition, mass loss was mainly influenced by the hydrolysis of bamboo starch and the dissolution of some extractives, but there was no significant difference among the four organic acids. The result indicated that the hydrolysis reaction and other chemical reactions take place when bamboo is treated with the organic acids selected.

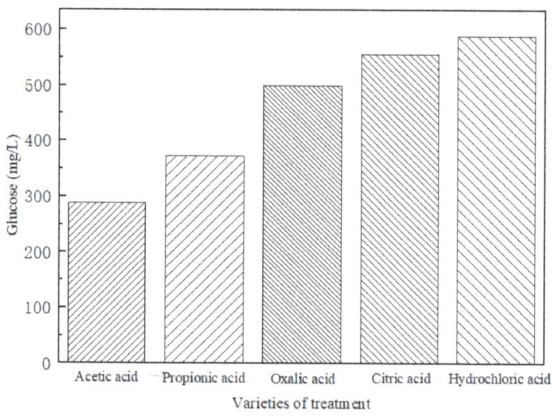

Figure 2. Glucose content in five different acid hydrolysates.

Several papers [14,23] showed that the prevention of fungal growth on bamboo samples was mainly inhibited by impregnating sample outer layers with acids of low pH value so that fungi could not start their growth. The main chemical components of bamboo culms include cellulose, hemicellulose, and lignin, which account for above 70–90% of total bamboo mass [24,25]. Moreover, there are about 2–6% starch, 2% deoxidized saccharide, 2–4% fat, and 0.8–6% protein in bamboo [26]. Cellulose is highly difficult to be hydrolysed under the present weak acidic solutions. Xylose contains more than 95% of the sugar units in the hemicellulose of Moso bamboo [27]. Therefore, the starch in bamboo was obviously the main glucose source in this study. Starch has been the primary source of various sugars in the industry (such as glucose, maltose, some oligosaccharide, etc.), with the action of acid, for a long time [28,29]. Additionally, the free glucoses in bamboo can be seen as another source of glucoses in the hydrolysate. Figure 3 illustrates the cross-section images of the untreated and acid-treated bamboo. It can be observed that the starch grains in the parenchyma cells disappeared mostly after being treated with propionic acid, oxalic acid, citric acid, and hydrochloric acid. However, this does not mean the whole starch had been hydrolysed into glucose. Some of it could still exist in bamboo but with the new structure or even smaller sizes that were invisible under SEM.

3.1.3. Color Changes Due to the Treatments

Table 2 shows the colour changes (ΔL^*, Δa^*, Δb^*, ΔE^*) of bamboo due to acidic treatments. The surface colour changed from light to dark, as indicated by a decrease in lightness (L^*). Besides, the treated surfaces became reddish and blue with increased a^* and reduced b^*, respectively. It can also be observed that the overall colour change value (ΔE^*) increased after acidic treatments. Similar to the tendency of glucose yields with different acidic treatments, the samples treated with hydrochloric acid showed highest ΔE^* (6.83), followed by those treated with oxalic acid (5.62), citric acid (5.98), acetic acid (5.15), and propionic acid (4.64). Color changing after different treatments indicates the existence of chemical reactions in the acidic conditions. The different discoloration behaviors of bamboo indicate that the surface colour of bamboo was influenced by acid types, concentration, and pH, etc. It has been reported that the lignin and extractives in bamboo (such as tannins and colour pigments, etc.) can lead to the colour changes of bamboo/wood under the reaction of light and oxygen [30–32]. In this study, the colour variations on bamboo surface may be caused by the dissolution or reaction of bamboo extractives with acids and heat as well as minor hydrolysis of hemicelluloses.

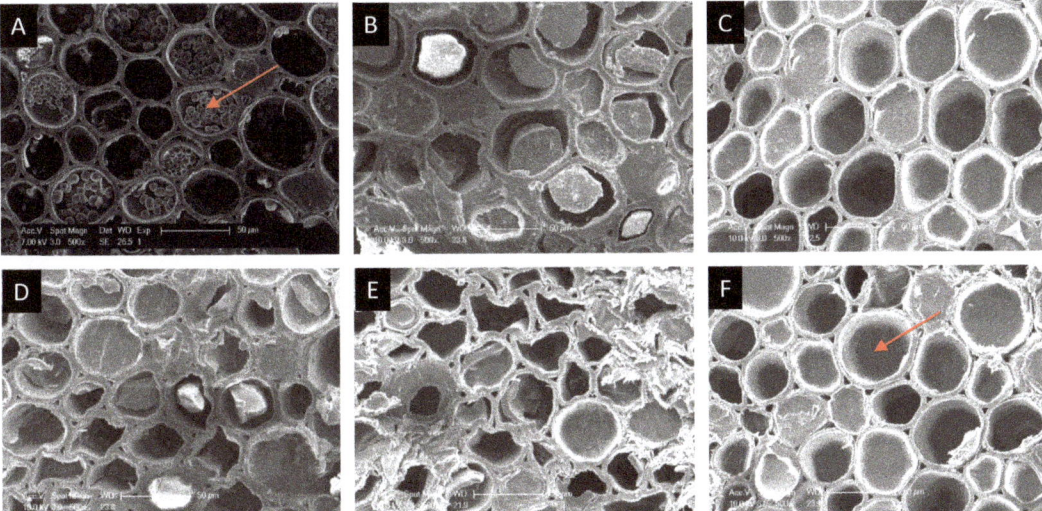

Figure 3. Distribution of starch in bamboo parenchyma cells ((**A**) Control sample (starch grains (red arrow) are rich in the parenchyma cells)): (**B**) acetic acid treatment sample, (**C**) propionic acid treatment sample, (**D**) oxalic acid treatment sample, (**E**) citric acid treatment sample, and (**F**) hydrochloric acid treatment sample (starch grains disappeared (red arrow) in the parenchyma cells).

Table 2. Color parameter changes of bamboo due to acid treatments.

Parameter	Acetic Acid	Propionic Acid	Oxalic Acid	Citric Acid	Hydrochloric Acid
ΔL^*	−4.329	−2.945	−4.897	−4.979	−6.127
Δa^*	0.46	0.249	1.557	0.22	2.214
Δb^*	−2.747	−3.578	−2.276	−3.304	−2.061
ΔE^*	5.15	4.64	5.62	5.98	6.83

3.2. Citric Acid Treatments with Different Concentration

Since the acidic treatments with oxalic acid, citric acid, and hydrochloric acid showed a better fungus resistance performance in the first week, further study should be focused on the mould resistance treated with higher acid concentration, which can make the hydrolysis of starch grains highly effective. Though the oxalic acid and hydrochloric acid are both effective agents for mould and blue-stain resistance, citric acid is superior to all of them considering their environmentally friendly and edibility. Therefore, bamboo treated with citric acid in different concentration was further investigated.

3.2.1. Mould Resistance Test

Bamboo strips that treated with citric acid from 4–10% concentration were tested in group for their ability against mould and blue stain fungi. The result shows that citric acid treatment was highly effective in refraining bamboo mildew, and the degree of refraining increased with the increased concentration.

Figure 4 shows that the mycelia and spores of *A. niger* appeared on the surface of bamboo strips on the third day when treated with 4% concentration level of citric acid, whereas the time can be postponed until the seventh day, as long as the concentration of the citric acid is changed to 10%. For the *Penicillium citrinum* and the *Trichoderma viride*, the time of the appearance of mycelia and spores can be postponed from the fourth day to the thirteenth and sixteenth day, respectively, by adjusting the concentration level of the citric acid as well. For *B. theobromae*, the visible blue-stain spots occurrence lasted 14 days. The chart illustrate the significant upward trend in the occurrence of mycelia on the treated bamboo.

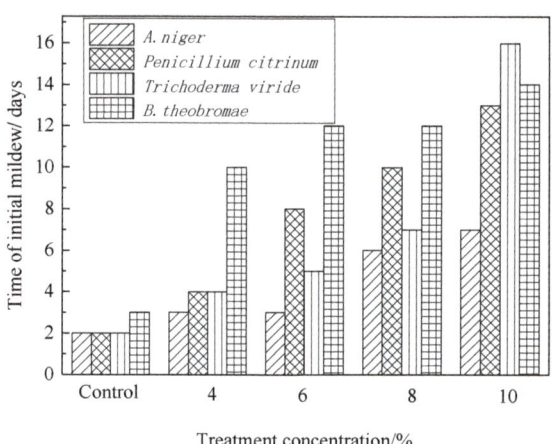

Figure 4. Time of initial mildew for bamboo treatment with different citric acid.

Obviously, the average infected areas for four treated groups decreased with increased concentration of citric acid. As an example, Figure 5 shows the bamboo samples infected by *A. niger* in different periods. As shown in Figure 6, for bamboo strips treated with citric acid concentration ranging from 4% to 10%, the average areas infected by *A. niger*, *Penicillium citrinum*, *Trichoderma viride*, and *B. theobromae* decreased from 35% to 13%, 50% to 17%, 74% to 10%, and 28% to 10%, respectively. Moreover, the mildew performance of the 10% concentration treatment was significantly higher than other three treatments, reaching a range of infected area of 10–17% that was three to eight times that treated with 4% concentration citric acid after four weeks when infected by four different moulds.

From the results of infected areas, it is known that the growth rating can be controlled within 1 resistance by treated with 8% or 10% citric acid. Additionally, for *A. niger*, *Penicillium citrinum*, and *B. theobromae*, the prevention efficiency of bamboo treated with 10% concentration was 75%, while *Trichoderma viride* was slightly lower.

3.2.2. Glucose Yields and Starch Remained

To clarify the effect of citric acid concentration on bamboo, soluble sugar and starch remaining in bamboo were tested separately. Both of them showed a downtrend with increased acid concentration (Figure 7). The soluble sugar remaining in bamboo was 43 mg/g, 36 mg/g, and even 31 mg/g when citric acid concentration was 4%, 8%, and 10%, respectively. The content of starch remaining in bamboo was 46 mg/g, 37 mg/g, and even 23 mg/g when citric acid concentration was 4%, 8%, and 10%, respectively.

To further clarify the hydrothermal citric acid treatments in different concentration, dissolved glycosyls were tested by analyzing conditioning fluid. The glucose and xylose yields in hydrolysates derived from citric acid are shown in Figure 8. A marked effect was observed in the citric acid of 10%. The increases in glucose and xylose ranged from 310 and 338 mg/L to 392 and 412 mg/L, respectively, indicating acidic hydrolysis of starch increased with increased citric acid concentration. Consistently, the weight loss of treated bamboo increased with increased concentration. It was measured that the weight loss of bamboo was from 6% to 7.6% as citric acid concentration increased from 4% to 10% (Figure 9).

Figure 5. The bamboo samples infected by *A. niger* in different periods (**A1–A5** show the mould growth on samples at Day 0, 7, 14, 21, and 30 when treated with 4% citric acid; **B1–B5** show the mould growth on samples at Day 0, 7, 14, 21, and 30 when treated with 6% citric acid; **C1–C5** show the mould growth on samples at Day 0, 7, 14, 21, and 30 when treated with 4% citric acid; **D1–D5** show the mould growth on samples at Day 0, 7, 14, 21, and 30 when treated with 4% citric acid.).

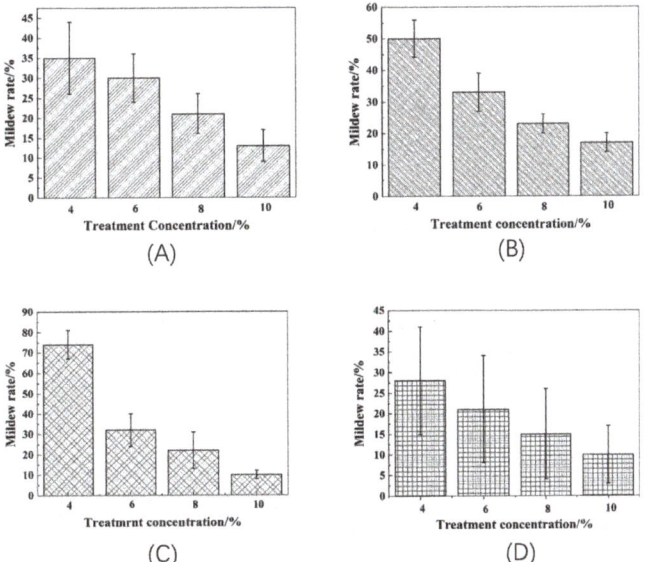

Figure 6. Mildew rate of bamboo in different treatment concentrations (**A–D** is for *A. niger*, *Penicillium citrinum*, *Trichoderma viride*, and *B. theobromae*, respectively).

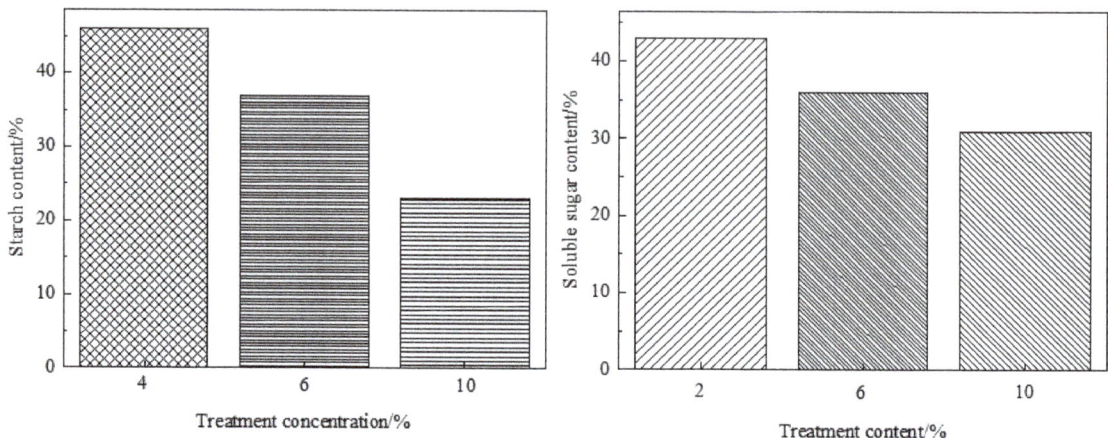

Figure 7. Starch and soluble sugar remaining in bamboo.

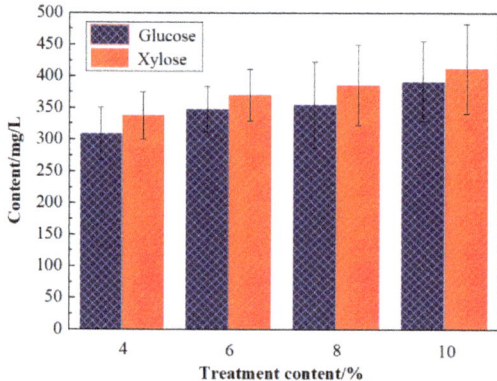

Figure 8. Glucose and xylose contents in different hydrolysates.

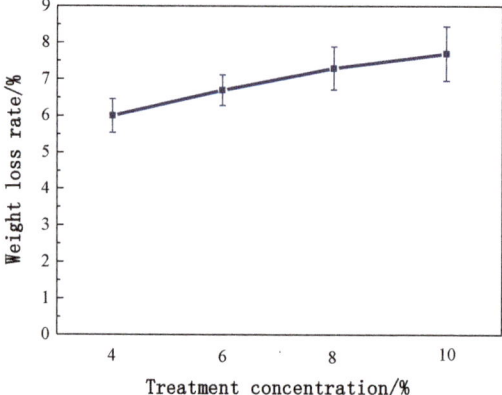

Figure 9. Weight loss of bamboo caused by different citric acidic treatments.

The weight loss of bamboo and contents of sugars in hydrolysate showed a similar trend. Based on the results of Figure 9, it is clear that citric acid concentration contributed to hydrolysis and dissolution of starch in bamboo. Furthermore, the sugars in the conditioning fluid were mainly produced by citric acid hydrolysis. On the other hand, citric acid has the potential to react with cell wall polymers. Feng [33] pointed out that citric-acid-treated wood exhibited considerable improvements in dimensional stability, as wood and citric acid can react by ester linkages. The improvement of bamboo mould resistance could also be due to esterification between citric acid and bamboo.

3.2.3. Color Changes Due to the Treatments

Considering the chemical changes brought about by citric acid, especially in its high concentrations, colour changes of treated bamboo in different concentrations were tested. The result shows that the parameter ΔE^* of 10% citric acid treatment had the highest value, 5.002. For the gradually increasing discoloration of bamboo, one of the reasons the colour changed on the bamboo surface was due to the dissolution of bamboo extractives; additionally, the high concentrated citric acid reacted readily under heating conditions. The colour change can be attributed to the decomposition of citric acid to unsaturated acids, such as acetone dicarboxylic acid and aconitic acid, when heated at high temperatures. Feng [33] reported that samples treated with citric acid at variable WPG levels caused colour change. For example, at 6% of WPG level, a slight yellowing was visible, and the wood was even darker at higher WPG level.

4. Conclusions

Mould resistance of bamboo strips treated with low-molecular-weight organic acids and inorganic acid were first tested, and then effect of citric acid with different concentrations were studied. Bamboo treated with acetic acid, propionic acid, oxalic acid, citric acid, and hydrochloric acid in a low concentration could improve their fungus growth rating from 4 in control samples to 2 or 3 resistance. However, there was no remarkable difference in final inflected areas of bamboo surface among different acidic treatments. Citric acid is effective to refrain mildew, and the mould resistance increased with the increased concentration of citric acid, and the fungus growth rating could reach 1 resistance when citric acid concentration was greater than 8%, while bamboo treated with citric acid in the concentration of 10% could control the infected area in the range of 10–17%. The improved mould and blue-stain resistance of treated bamboo could be attributed to the reduced nutrients in bamboo due to the hydrolysis of starch grains in parenchyma cells and the dissolution of soluble sugar. Compared with the treatments with other acid concentration, 10% citric acid treatments produced the greatest dissolved glycosyls content, colour change, and weight loss value.

Author Contributions: Writing—original draft preparation, Z.Y.; formal analysis, Z.Y.; investigation, X.Z., R.Z. and Y.Y.; methodology, F.S.; supervision, F.S.; writing—review and editing, F.S. All authors have read and agreed to the published version of the manuscript.

Funding: This research was funded by the Basic Scientific Research Funds of International Center for Bamboo and Rattan (Grant No. 1632020017).

Institutional Review Board Statement: Not applicable.

Informed Consent Statement: Not applicable.

Data Availability Statement: The data presented in this research are available on request from the corresponding author.

Conflicts of Interest: The authors declare no conflict of interest.

References

1. Okahisa, Y.; Yoshimura, T.; Imamura, Y. Imamura, An application of the alkaline extraction-glucoamylase hydrolysis method to analyze starch and sugar contents of bamboo. *J. Wood Sci.* **2005**, *51*, 542–545. [CrossRef]
2. Magel, E.; Kruse, S.; Lütje, G.; Liese, W. Soluble carbohydrates and acid invertases involved in the rapid growth of developing culms in Sasa palmata (Bean) Camus. *Bamboo Sci. Cult.* **2006**, *19*, 23–29.

3. Sun, F.; Bao, B.; Ma, L.; Chen, A.; Duan, X. Mould-resistance of bamboo treated with the compound of chitosan-copper complex and organic fungicides. *J. Wood Sci.* **2012**, *58*, 51–56. [CrossRef]
4. Fengel, D.; Shao, X. A chemical and ultrastructural study of the Bamboo species Phyllostachys makinoi Hay. *Wood Sci. Technol.* **1984**, *18*, 103–112.
5. Townsend, T.; Dubey, B.; Tolaymat, T.; Solo-Gabriele, H. Preservative leaching from weathered CCA-treated wood. *J. Environ. Manag.* **2005**, *75*, 105–113. [CrossRef]
6. Dubey, B.; Townsend, T.; Solo-Gabriele, H.; Bitton, G. Impact of surface water conditions on preservative leaching and aquatic toxicity from treated wood products. *Environ. Sci. Technol.* **2007**, *41*, 3781–3786. [CrossRef]
7. Nayebare, K.P.; Zhang, S.Y.; Wu, H.; Yang, S.; Li, S.; Sun, F.; Goodell, B. Enzymatic biocatalysis of bamboo chemical constituents to impart anyimold properties. *Wood Sci. Technol.* **2018**, *52*, 619–635. [CrossRef]
8. Li, W.; Wang, H.; Ren, D.; Yu, Y.; Yu, Y. Wood modification with furfuryl alcohol catalysed by a new composite acidic catalyst. *Wood Sci. Technol.* **2015**, *49*, 845–856. [CrossRef]
9. Kaper, T.; Van Der Maarel, M.; Euverink, G.; Dijkhuizen, L. Exploring and exploiting starch-modifying amylomaltases from thermophiles. *Biochem. Soc. Trans.* **2004**, *32*, 279–282. [CrossRef]
10. Tester, R.F.; Debon, S.J. Annealing of starch—A review. *Int. J. Biol. Macromol.* **2000**, *27*, 1–12. [CrossRef]
11. Clausen, C.A.; Coleman, R.D.; Yang, V.W. Fatty acid-based formulations for wood protection against mold and sapstain. *For. Prod. J.* **2010**, *60*, 301–304. [CrossRef]
12. Barbosa-Canovas, G.; Pothakamury, U.R.; Palou, E.; Swanson, B.G. *Chemicals and Biochemicals Used in Food Preservation, Nonthermal Preservation of Foods*; Marcel Dekker, Inc.: New York, NY, USA, 1998; Chapter 8; pp. 215–233.
13. Doores, S. *Organic Acids. Antimicrobials in Foods*; Davidson, P., Branen, A., Eds.; Marcel Dekker: New York, NY, USA, 1993; p. 95.
14. Tang, T.K.H.; Schmidt, O.; Liese, W. Environment-friendly short-term protection of bamboo against moulding. *J. Timber Deve. Assoc. India* **2009**, *55*, 8–17.
15. Tang, T.K.H.; Schmidt, O.; Liese, W. Protection of bamboo against mould using environment-friendly chemicals. *J. Trop. For. Sci.* **2012**, *24*, 285–290.
16. Schmidt, O.; Magel, E.; Frühwald, A.; Glukhykh, L.; Erdt, K.; Kaschuro, S. Influence of sugar and starch content of palm wood on fungal development and prevention of fungal colonization by acid treatment. *Holzforschung* **2016**, *70*, 783–791. [CrossRef]
17. Sun, F.; Zhou, Y.; Bao, B.; Chen, A.; Du, C. Influence of solvent treatment on mould resistance of bamboo. *Bioresources* **2011**, *6*, 2091–2100.
18. Liese, W. Protection of bamboo in service. *World Bamboo Ratt.* **2003**, *1*, 29–33. (In Chinese)
19. Kumar, S.; Shukla, K.S.; Dev, I.; Dobriyal, P.B. *Bamboo Preservation Techniques: A Review*; International Bamboo and Rattan and Indian Council of Forestry Research Educaton: Dehra Dun, India, 1994.
20. Wingfield, M.J.; Seifert, K.A.; Webber, J.F. *Webber, Ceratocystis and Ophiostoma: Taxonomy, Ecology and Pathogenicity*; APS Press: St. Paul, MN, USA, 1993; pp. 141–151, 269–287.
21. Schmidt, O. *Wood and Tree Fungi. Biology, Damage, Protect, and Use*; Springer: Berlin, Germany, 2006.
22. Huang, X.D.; Hse, C.Y.; Shupe, T.F. Study on the mould resistant properties of moso bamboo treated with high pressure and amylase. *Bioresources* **2014**, *9*, 497–509. [CrossRef]
23. Schmidt, O.; Wei, D.S.; Tang, T.K.H.; Liese, W. Bamboo and fungi. *J. Bamboo Ratt.* **2013**, *12*, 1–14.
24. Wei, D.S. Bamboo Inhabiting Fungi and Their Damage to the Substrate. Master's Dissertation, University of Hamburg, Hamburg, Germany, 2014.
25. Yusoff, R.B.; Takagi, H.; Nakagaito, A.N. Tensile and flexural properties of polylactic acid-based hybrid green composites reinforced by kenaf, bamboo and coir fibers. *Ind. Crops Prod.* **2016**, *94*, 562–573. [CrossRef]
26. Jiang, Z.H. *Bamboo and Rattan in the World*; China Forestry Publishing House: Beijing, China, 2008.
27. Li, Z.; Jiang, Z.; Fei, B.; Cai, Z.; Pan, X. Comparison of bamboo green, timber and yellow in sulfite, sulfuric acid and sodium hydroxide pretreatments for enzymatic saccharification. *Bioresour. Technol.* **2014**, *151*, 91–99. [CrossRef]
28. Pirt, S.J.; Whelan, W.J. The determination of starch by acid hydrolysis. *J. Sci. Food Agric.* **1951**, *2*, 224–228. [CrossRef]
29. Jacobs, H.; Eerlingen, R.C.; Rouseu, N.; Colonna, P.; Delcour, J.A. Acid hydrolysis of native and annealed wheat, potato and pea starches-DSC melting features and chain length distributions of lintnerised starches. *Carbohydr. Res.* **1998**, *308*, 359–371. [CrossRef]
30. Wang, X.Q.; Ren, H.Q. Comparative study of the photo-discoloration of moso bamboo (Phyllostachys pubescens Mazel) and two wood species. *Appl. Surf. Sci.* **2008**, *254*, 7029–7034. [CrossRef]
31. Pastore, T.C.M.; Santos, K.O.; Rubim, J.C. Aspectrocolorimetric study on the effect of ultraviolet irradiation of four tropical hardwoods. *Bioresour. Technol.* **2004**, *93*, 37–42. [CrossRef] [PubMed]
32. Jamalirad, L.; Doosthoseini, K.; Koch, G.; Mirshokraie, S.A.; Welling, J. Investigation on bonding quality of beech wood (*Fagus orientalis* L.) veneer during high temperature drying and aging. *Eur. J. Wood Prod.* **2012**, *70*, 497–506. [CrossRef]
33. Feng, X.; Xiao, Z.; Sui, S.; Wang, Q.; Xie, Y. Esterification of wood with citric acid: The catalytic effects of sodium hypophosphite (SHP). *Holzforschung* **2014**, *68*, 427–433. [CrossRef]

Article

Preparation of PVA–CS/SA–Ca^{2+} Hydrogel with Core–Shell Structure

Shuai Zhang [1,2], Yu Wan [1,3], Weijie Yuan [1,2,*], Yaoxiang Zhang [1,2], Ziyuan Zhou [1,2], Min Zhang [1,3], Luzhen Wang [4] and Ran Wang [1,2]

1. Experimental Center of Forestry in North China, Chinese Academy of Forestry, Beijing 102300, China; zhang.s@caf.ac.cn (S.Z.); wanyu96115a@163.com (Y.W.); zyx631022@263.net (Y.Z.); zhouziyuan@caf.ac.cn (Z.Z.); zhanglaoshi00@163.com (M.Z.); wangran@caf.ac.cn (R.W.)
2. National Permanent Scientific Research Base for Warm Temperate Zone Forestry of Jiulong Mountain in Beijing, Beijing 102300, China
3. School of Environmental Science and Engineering, Shaanxi University of Science and Technology, Xi'an 710021, China
4. Qinghai Provincial Investigation, Design &Research Institute of Water Conservancy & Hydropower Co., Ltd., Xining 810000, China; wlzroger@163.com
* Correspondence: yuanwj@caf.ac.cn

Citation: Zhang, S.; Wan, Y.; Yuan, W.; Zhang, Y.; Zhou, Z.; Zhang, M.; Wang, L.; Wang, R. Preparation of PVA-CS/SA-Ca^{2+} Hydrogel with Core–Shell Structure and Its Properties. *Polymers* 2022, 14, 212. https://doi.org/10.3390/polym14010212

Academic Editors: Jingpeng Li, Yun Lu, Huiqing Wang and Vijay Kumar Thakur

Received: 16 November 2021
Accepted: 1 January 2022
Published: 5 January 2022

Publisher's Note: MDPI stays neutral with regard to jurisdictional claims in published maps and institutional affiliations.

Copyright: © 2022 by the authors. Licensee MDPI, Basel, Switzerland. This article is an open access article distributed under the terms and conditions of the Creative Commons Attribution (CC BY) license (https://creativecommons.org/licenses/by/4.0/).

Abstract: Hydrogels are highly hydrophilic polymers that have been used in a wide range of applications. In this study, we prepared PVA–CS/SA–Ca^{2+} core–shell hydrogels with bilayer space by cross-linking PVA and CS to form a core structure and chelating SA and Ca^{2+} to form a shell structure to achieve multiple substance loading and multifunctional expression. The morphology and structure of core–shell hydrogels were characterized by scanning electron microscopy (SEM) and Fourier transform infrared spectroscopy (FTIR). The factors affecting the swelling properties of the hydrogel were studied. The results show that the PVA–CS/SA–Ca^{2+} hydrogel has obvious core and shell structures. The SA concentration and SA/Ca^{2+} cross-linking time show a positive correlation with the thickness of the shell structure; the PVA/CS mass ratio affects the structural characteristics of the core structure; and a higher CS content indicates the more obvious three-dimensional network structure of the hydrogel. The optimal experimental conditions for the swelling degree of the core–shell hydrogel were an SA concentration of 5%; an SA/Ca^{2+} cross-linking time of 90 min; a PVA/CS mass ratio of 1:0.7; and a maximum swelling degree of 50 g/g.

Keywords: core–shell structure; hydrogel; double load; response surface

1. Introduction

Hydrogels are a class of hydrophilic polymers with a three-dimensional network structure formed by cross-linking with the action of covalent and hydrogen bonds [1–3], which have a large specific surface area [4], high carrier strength [5], controllability [6] and a wide range of physicochemical adjustability [7,8]. Since the 1960s, when the Czech scholar Wichterle first produced poly (2-hydroxyethyl methacrylate) hydrogels [9], hydrogels have been widely used in medicine, engineering, agriculture, forestry, environmental protection and the information industry due to their good biocompatibility, sensitivity to environmental changes and superb molecular designability [10–15].

Hydrogel is a material similar to the tissue of living organisms, which has a wide range of medical applications because of its excellent biological properties. It can both encapsulate drugs for slow release into the body and load dressings for contact with tissue wounds [16]. High-strength composite hydrogels can also be used as scaffold materials [17]. The high hydrophilicity of hydrogel can hold a large amount of water, play the role of water retention and drought prevention and promote the growth of crops and forests [18]; hydrogel has a strong adsorption effect and can adsorb heavy metals and pollutants in the soil [19]; hydrogel can also remove a wide range of aqueous pollutants containing toxic dyes and organic pollutants [20,21]. With the development of artificial intelligence, hydrogel strain sensors can be produced, which have enabled the extensive development of flexible wearable devices [22]. Due to its own properties, hydrogel can take on important roles. In addition, it can respond to changes in the external environment, for example, when factors such as temperature and humidity, pH and light intensity change, the three-dimensional network structure of the hydrogel will also change in response, which is why hydrogels are also known as smart polymers [23]. In addition, the composite of hydrogel and different media will expand its application areas, for example, the introduction of hydrogel into conductive media can produce conductive hydrogel [24].

The properties of hydrogels are determined by the polymer network structure. Single-material hydrogels are isotropic in terms of microstructure and macroscopic properties, lacking an ordered structure, and have poor mechanical properties, which in turn limit their application in some fields. A double network structure is usually used to solve the mechanical property problem, and a biocompatible material is used to solve the problem of poor biocompatibility. Double network hydrogels usually contain two kinds of networks: one is rigid, which is easy to break and can increase the tensile stress of the hydrogel by dissipating energy; and the other is flexible, which can increase the tensile strain of the hydrogel [25,26]. The polyvinyl alcohol (PVA) hydrogel is a class of highly hydrophilic, non-toxic, degradable elastomeric materials with good film formation and chemical stability, but a single PVA hydrogel is prone to swelling and adhesion and poor stability. Chitosan (CS) has good biocompatibility and tunability, but the intermolecular hydrogen bonding force is strong, and when it is solely used to prepare hydrogels, the products suffer from disadvantages such as their brittleness and poor mechanical properties. The cross-linking of PVA material with CS can make up for the deficiencies of both materials [27]. Sodium alginate (SA) is a green biomass sodium salt with strong affinity to dyes and metal ions because of its abundant hydroxyl and carboxyl groups, but the mechanical strength of a single SA hydrogel is weak, while cross-linking with Ca^{2+} through chelation can significantly enhance the stability of the structure [28].

Previous studies of hydrogels have shown them to be versatile but only able to perform a specific function. In this study, we propose to use polyvinyl alcohol (PVA) and chitosan (CS) to build the core of the hydrogel, and sodium alginate (SA) and Ca^{2+} to construct the shell, forming a core–shell structure with a bilayer space that does not affect each part of the structure in order to achieve the purpose of loading two substances, so that one hydrogel can achieve two different functions.

2. Experimental Method
2.1. Raw Materials and Reagents

Polyvinyl alcohol (PVA) was purchased from Sinopharm Group Chemical Reagent Co., (Shanghai, China, AR); chitosan (CS) and sodium alginate (SA) were purchased from Chengdu Kelong Chemical Reagent Factory, (Chengdu, China, AR); calcium chloride was purchased from Tianjin Tianli Chemical Reagent Co., (Tianjin, China, AR); citric acid was purchased from Tianjin Comio Chemical Reagent Co., (Tianjin, China, AR); dipotassium hydrogen phosphate was purchased from Tianjin Chemical Reagent Factory, (Tianjin, China, AR); sodium dihydrogen phosphate was purchased from Tianjin Fuchen Chemical Reagent Factory, (Tianjin, China, AR); and D-(+)-gluconic acid delta-lactone was purchased from Shanghai Maclean Biochemical Co., (Shanghai, China, AR).

Centrifuge (80-1), Shanghai Pudong Physical Optical Instrument Factory (Shanghai, China); freeze dryer (LGJ-10), Beijing Songyuan Huaxing Technology Development Co. (Beijing, China); scanning electron microscope (FEI Q45), FEI Inc. (Hillsboro, OR, USA); X-ray photoelectron spectroscopy (Vario EL III), Kratos Analytical Ltd. (Manchester, UK); Fourier transform infrared spectroscopy (VECTOR-22), Bruker Co. (Karlsruhe, Germany).

2.2. Preparation of PVA–CS/SA–Ca^{2+} Core–Shell Hydrogels

A PVA solution with a mass fraction of 10% was prepared by dissolving 1 g of PVA in 10 mL of distilled water and mechanically stirring at 120 rpm for 2 h in a water bath at 90 °C. Subsequently, 0.4 g of CS and citric acid were dispersed in 10 mL of distilled water at 20 °C and mechanically stirred at 120 rpm for 4 h to obtain a CS suspension with a mass fraction of 4%. Then, different volumes of CS suspension were added to the PVA solution and mechanically stirred at 180 rpm for 1 h at 60 °C. Moreover, 1 g of CaCl$_2$ and 1 g of gluconolactone were added to the reaction system and stirred for 30 min before 1 mL of different concentrations of SA solution was pipetted into the above mixed solution and left for different times to form hydrogels. The obtained hydrogels were referred to as PVA–CS/SA–Ca^{2+} core–shell hydrogels. Finally, the dried core–shell hydrogels were obtained by cyclic freeze–thawing 3 times and then washed several times with distilled water and vacuum freeze-dried and stored in a drying tower for use, as shown in Figure 1.

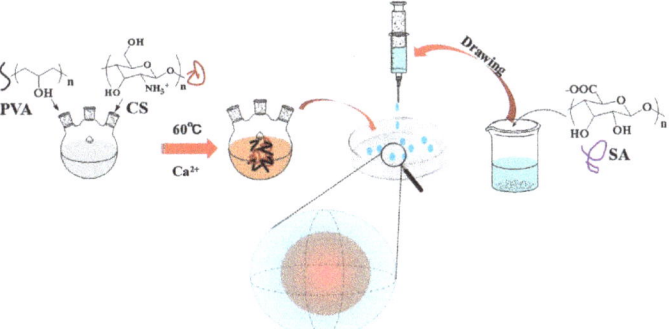

Figure 1. Preparation of PVA–CS/SA–Ca^{2+} core–shell hydrogel.

The amounts of CS, SA and SA/Ca^{2+} cross-linking time used to prepare the hydrogels are listed in Table 1.

Table 1. Reaction conditions of PVA–CS/SA–Ca^{2+} core–shell hydrogels.

Sample	PVA/mL	CS/mL	SA/%	SA/Ca^{2+} Cross-Linking Time/min
1	10	20	4	30
2	10	20	4	60
3	10	20	4	90
4	10	20	2	30
5	10	20	6	30
6	10	15	4	30
7	10	25	4	30

2.3. Structural Characterization and Performance Testing of PVA–CS/SA–Ca^{2+} Core–Shell Hydrogels

2.3.1. Microscopic Morphology of PVA–CS/SA–Ca^{2+} Core–Shell Hydrogels (SEM)

The core–shell structure of the freeze–thawed core–shell hydrogel was separated and freeze-dried under vacuum, after which it was cut into thin slices with a thickness of 0.5 mm and sprayed with gold at a test voltage of 25 kV to a thickness of 10 nm. The microscopic morphology of the dried core–shell samples was observed using an FEI Q45 scanning electron microscope.

2.3.2. Structure of PVA–CS/SA–Ca^{2+} Core–Shell Hydrogels (FTIR)

The core–shell structure of the core–shell hydrogel after vacuum drying for 24 h was characterized by using a VECTOR-22 Fourier transform infrared spectrometer with a set wavelength range of 500~4000 cm^{-1} and 32 scans.

2.3.3. Swelling Properties of PVA–CS/SA–Ca^{2+} Core–Shell Hydrogels

The effects of SA concentration, SA and Ca^{2+} cross-linking time and PVA/CS mass ratio on the swelling of PVA–CS/SA–Ca^{2+} core–shell hydrogels were investigated to determine the range of each parameter and to provide a basis for optimizing the process parameters for the swelling of core–shell hydrogels. The freeze-dried core–shell hydrogels were weighed (W_0), added to 7 mL of disodium hydrogen phosphate at a concentration of 1/15 mol/L and 3 mL of potassium dihydrogen phosphate buffer at a concentration of 1/15 mol/L, and left to absorb water at 25 °C, removed at intervals of 20 min, dried on moist filter paper and precisely weighed (W_t)—all of which was repeated several times until the swelling equilibrium was reached, and the experiments were conducted in parallel nine times. The swelling degree (SR) of the core–shell hydrogel was calculated according to the following equation [29]:

$$SR = \frac{W_t - W_0}{W_0}$$

where W_0 and W_t are the initial dry weight (g) and the mass (g) of the sample at t min of water absorption, respectively; SR (g/g) is the swelling degree of hydrogel.

2.4. Statistical Analysis

The statistical analyses were performed by the SPSS computer program (SPSS Statistic 20.0) software using one-way analysis of variance (ANOVA). Values were presented as means ± standard deviations (SD) of triplicate determinations. Statistical significance was set at $p < 0.05$.

3. Results and Analysis

3.1. Microscopic Morphology of PVA–CS/SA–Ca^{2+} Core–Shell Hydrogels

The SEM photographs of PVA–CS/SA–Ca^{2+} core–shell hydrogel is shown in Figure 2. The surface of the hydrogel is rough with many folds, which can provide a larger surface area. Here, A is a cross-sectional view of the core–shell structure, and B and C are the core and shell structures, respectively. The core structure of the hydrogel is a relatively loose and three-dimensional network structure with more pores, and the shell structure is relatively compact with fewer pores, which is mainly because the network structure formed by the cross-linking of PVA and CS molecular chains is looser, while the chelating force of SA and Ca^{2+} is strong and the chelating rings formed are tightly connected, so the shell structure formed is denser. As shown in Figure 2D–F, with the increase in cross-linking time, the shell wall of the core–shell hydrogel becomes thicker and the pore structure increases, which indicates that with the increase in cross-linking time, the reaction between SA and Ca^{2+} will be more complete and the three-dimensional structure will be more stereoscopic and stable. In Figure 2G–I, the hydrogel shell walls become thicker and denser with the increasing SA concentration, which indicates that the increase in SA concentration makes more SA coordination atoms form chelate rings with Ca^{2+} and occupy the pores of the 3D network structure. As shown in Figure 2J–L, the three-dimensional network structure of the hydrogel core structure becomes more and more significant as the PVA/CS mass ratio decreases. These results indicate that the SA/Ca^{2+} cross-linking time and SA concentration have a strong influence on the shell structure of hydrogels, and a longer cross-linking time means a larger SA concentration, a thicker hydrogel shell wall and a more stable structure; the PVA/CS mass ratio can regulate the spatial conformation of the hydrogel nuclear structure. The structure of the hydrogel determines its properties, and the properties determine its usage, so a different PVA/CS mass ratio, SA/Ca^{2+} cross-linking time and SA concentration can be set according to the different usage of core–shell hydrogel to achieve the purpose of controlling the core–shell structure of hydrogel.

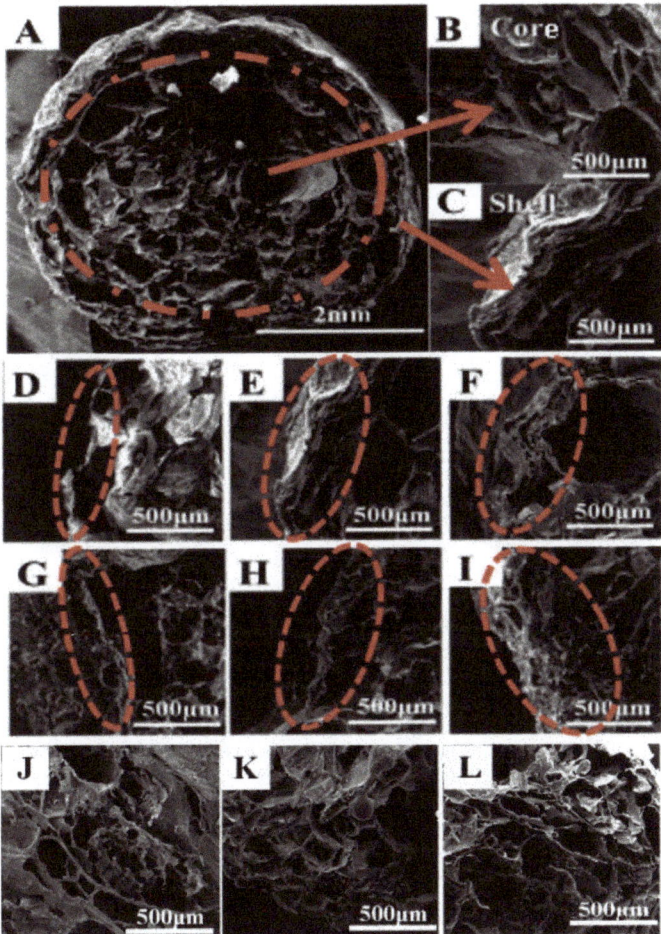

Figure 2. SEM photograph of PVA–CS/SA–Ca^{2+} core–shell hydrogel ((**A–C**) are core–shell hydrogels with 4% SA, PVA/CS = 1:0.8, and SA/Ca^{2+} cross-linking time of 30 min; (**D–F**) are core–shell hydrogels with 4% SA, PVA/CS = 1:0.8, and SA/Ca^{2+} cross-linking time of 30 min, 60 min and 90 min; (**G–I**) are core–shell hydrogels with PVA/CS = 1:0.8, SA/Ca^{2+} cross-linking time of 30 min, and SA concentrations of 2%, 4% and 6%, respectively; (**J–L**) are core–shell hydrogels with 4% SA, SA/Ca^{2+} cross-linking time of 30 min, and PVA/CS of 1:0.6, 1:0.8 and 1:1).

3.2. Structure of Core–Shell Hydrogel

Figure 3 shows the FTIR spectra of the PVA–CS/SA–Ca^{2+} core–shell hydrogel (4% SA, SA/Ca^{2+} cross-linking time 30 min, PVA/CS = 1:0.6) and its raw materials. In Figure 3a, 3500 cm^{-1} shows the superposition peaks of the amino (-NH) and hydroxyl (-OH) groups of CS, 1360 cm^{-1}, 1090 cm^{-1} and 650 cm^{-1} which show the out-of-plane bending vibration absorption peaks of carbon and nitrogen bonds (C–N), ether groups (C–O–C) and N–H of CS, respectively, and 1680 cm^{-1} and 1590 cm^{-1}, which show the characteristic absorption peaks of the undeacylated carbonyl group of CS (C=O) characteristic absorption peaks. Additionally, 3300 cm^{-1}, 1680 cm^{-1}, 1400 cm^{-1} and 1100 cm^{-1} are the absorption peaks of -OH, unpolymerized carbon–carbon double bond (–C=C–), alkyl and carbon–oxygen single bond (C–O) of PVA, respectively. Compared with the spectra of CS and PVA, no new characteristic peaks appear in PVA/CS, except that the peaks of -NH and –OH of CS at 3500 cm^{-1} are shifted to 3490 cm^{-1}, indicating that there is no chemical reaction between PVA and CS, but a hydrogen bond cross-linking between –NH or –OH of CS and –OH of PVA. In addition, the C=O absorption peaks of CS at 1680 cm^{-1} and 1540 cm^{-1} were shifted to 1700 cm^{-1} and 1620 cm^{-1}, respectively, and the intensity of the peaks also significantly increased, which might be due to the reaction between the core and shell structures of the hydrogel and the introduction of the carboxyl group of SA.

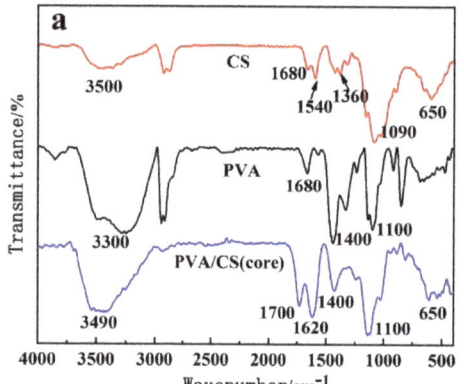 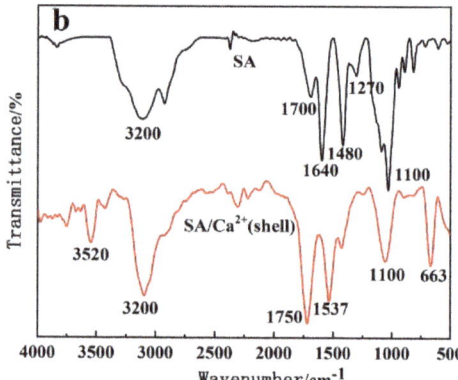

Figure 3. FTIR spectra of PVA–CS/SA–Ca^{2+} core–shell hydrogel and its raw materials: (**a**) shows the FTIR spectra of PVA/CS nuclear structure and its raw materials PVA and CS; (**b**) shows the FTIR spectra of SA/Ca^{2+} shell structure and its raw material SA.

In Figure 3b, the absorption peaks of SA are O–H, C–O, and C–O–C at 3200 cm^{-1}, 1270 cm^{-1} and 1100 cm^{-1}, respectively, and the absorption peaks of C=O and -OH are at 1700 cm^{-1}, 1640 cm^{-1} and 1480 cm^{-1}, respectively. The -OH peak of SA in the SA/Ca^{2+} shell structure has a narrowed peak shape and weakened intensity, while the C=O peaks changed and shifted, mainly due to the chelation of SA –OH and –COOH with Ca^{2+} to form the shell structure. The absorption peaks at 1520 cm^{-1} and 663 cm^{-1} were the same as those at 1360 cm^{-1} (C–N) and 650 cm^{-1} (N–H) in Figure 3a, respectively, indicating that the core and shell structures of the hydrogel were cross-linked through the hydrogen bonding between the CS amino group and the SA carboxyl group, which confirmed the above suspicion.

3.3. Swelling Properties of PVA–CS/SA–Ca^{2+} Core–Shell Hydrogels

3.3.1. Single Factor Screening for Swelling Performance

(1) The effect of SA concentration.

Figure 4 shows the variation graphs of the swelling degree of PVA–CS/SA–Ca^{2+} core–shell hydrogels at different SA concentrations. The maximum swelling of the core–shell hydrogels with different SA concentrations differed slightly, but the equilibrium swelling time was significantly different, showing a significant increase with the increase in SA concentration, indicating that the increase in SA concentration reduces the water absorption efficiency of hydrogels. Combined with the results of SEM observation (Figure 2G–I), the increase in SA concentration increases the cross-linkage with Ca^{2+}, which leads to a thicker shell wall and denser structure, and decreases the pore area, rendering it difficult for water to enter the hydrogel, and the equilibrium swelling time increases.

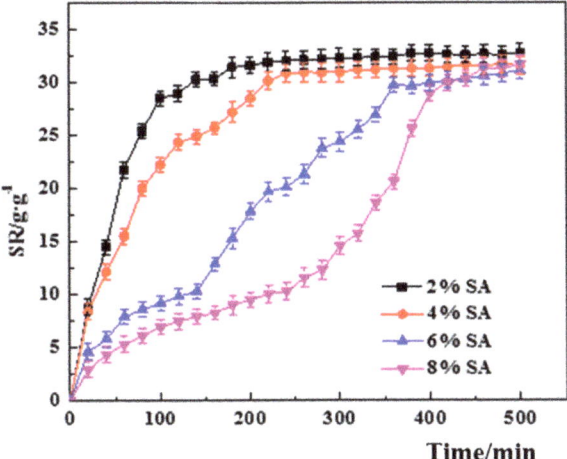

Figure 4. Effect of SA concentration on the swelling degree of PVA–CS/SA–Ca^{2+} core–shell hydrogel.

(2) Effect of SA/Ca^{2+} cross-linking time.

Figure 5 shows the variation curves of the swelling degree of the core–shell hydrogel under different cross-linking times of SA/Ca^{2+}. The final equilibrium swelling of the core–shell hydrogel was basically the same at approximately 30 ± 1.4 g/g for different cross-linking times in the figure, but the equilibrium swelling time became longer and longer with the increase in cross-linking time, which was similar to the effect of SA concentration on the swelling of the core–shell hydrogel. This is because the cross-linking time of SA/Ca^{2+} affects the wall thickness of the core–shell hydrogel, and a longer cross-linking time means a thicker hydrogel shell wall, leading to an increasingly longer swelling time.

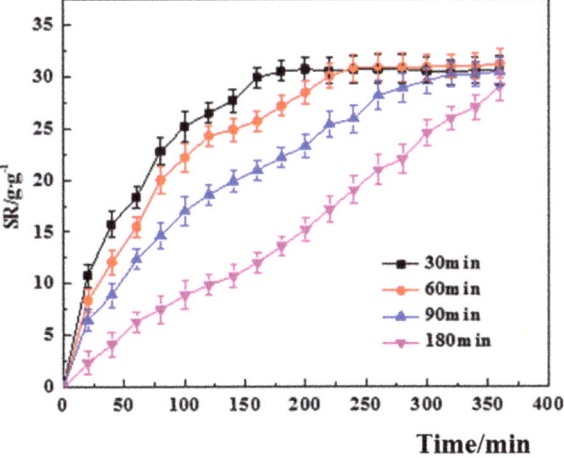

Figure 5. Effect of SA/Ca^{2+} cross-linking time on the swelling degree of PVA–CS/SA–Ca^{2+} core–shell hydrogel.

(3) Effect of PVA/CS ratio.

Figure 6 shows the variation curves of the swelling degree of PVA–CS/SA–Ca^{2+} core–shell hydrogels with different PVA/CS mass ratios. With the increase in CS content, the swelling degree of the core–shell hydrogel shows a trend of increasing and then decreasing.

When the PVA/CS mass ratio is 1:0.8, the swelling of the hydrogel is the largest, reaching 32 ± 1.4 g/g. This is due to the strong hydrogen bonding in CS, which can cross-link with PVA to form a more stable three-dimensional network skeleton when the content increases, which can also be reflected by the SEM results (Figure 2J–L). At the same time, CS contains many hydrophilic functional hydroxyl and carboxyl groups, which improve the water absorption of hydrogel. However, when the CS content is excessive, the three-dimensional network structure formed by cross-linking will be more compact, which reduces the surface pore space and affects the entry of water molecules.

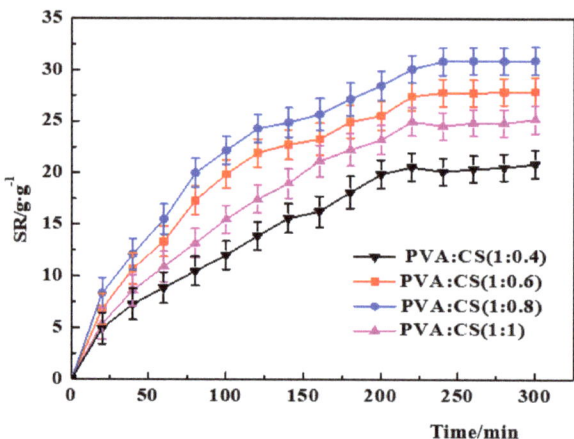

Figure 6. Effect of the PVA/CS ratio on the swelling degree of PVA–CS/SA–Ca^{2+} core–shell hydrogel.

3.3.2. Response Surface Analysis

From Section 3.3.1, it can be concluded that the SA concentration, SA/Ca^{2+} cross-linking time and the PVA/CS mass ratio all affect the maximum swelling of PVA–CS/SA–Ca^{2+} core–shell hydrogels as well as the equilibrium swelling time; so in order to further investigate the magnitude of the effect of the three factors on the swelling of the core–shell hydrogels and the optimal conditions for the swelling of core–shell hydrogels, a response surface analysis was performed. The response surface analysis was performed.

(1) Response modeling.

Based on the experiments with the PVA–CS/SA–Ca^{2+} core–shell hydrogel swelling degree as the one-way variable method combined with the Box–Behnken central combination test design principle, the SA concentration (A), SA/Ca^{2+} cross-linking time (B) and PVA/CS mass ratio (C) were selected as independent variables and the swelling degree of core–shell hydrogel as the response value, and Table 2 shows the experimental factor levels.

Table 2. Factors and levels for Box–Behnken design.

Level	SA Concentration	SA/Ca^{2+} Cross-Linking Time	PVA/CS
1	5	60	1:0.7
2	6	90	1:0.8
3	7	120	1:0.9

(2) Model significance and response surface analysis.

Using Design Expert 10 software to optimize the experimental design and processing of the swelling degree of the core–shell hydrogel, according to the optimized results, the influence of the constant term, linear term, interaction term and square term on the swelling degree of the hydrogel can be obtained, and the analysis of the experimental results is quadratically fitted to obtain the quadratic multinomial regression equation between the swelling degree of the core–shell hydrogel (response value SR) and each influencing factor:

$$SR = -301.834 - 54.867 \times A - 1.62499 \times B + 970.199 \times C + 0.313121 \times A \times B + 15.5803 \times A \times B - 0.130687 \times A \times C + 0.237412 \times A^2 - 0.00146488 \times B^2 - 425.35 \times C^2$$

Table 3 shows the ANOVA results of the quadratic regression model, from which it can be seen that the effect of the model on the swelling degree of the core–shell hydrogel is highly significant ($p < 0.01$), and the experimental and predicted values are highly correlated ($R^2 = 0.9810$, $R^2_{adj} = 0.9566$), indicating that the model fits well experimentally and can be used to predict the optimal conditions for the swelling degree of the core–shell hydrogel. The degree of influence of each factor on the swelling degree of the core–shell hydrogel was inferred from the magnitude of F value as: SA/Ca^{2+} cross-linking time (B) > PVA/CS mass ratio (C) > SA concentration (A). The primary terms A, B, C, interaction term AC and secondary term C2 of this model were significant ($p < 0.05$), and the interaction term BC reached a highly significant level ($p < 0.01$), indicating that the swelling of the core–shell hydrogels increased with the increase in SA concentration and SA/Ca^{2+} cross-linking time.

Table 3. Analysis of variance for quadric regression model.

Variation Source	Sum of Squares	Freedom	Mean Square	F Value	p Value	Significance
Model	1414.60	9	157.18	40.20	<0.0001	**
A	131.263	1	131.23	33.56	0.0007	**
B	222.81	1	222.81	56.98	0.0001	**
C	166.01	1	166.01	42.45	0.0003	**
AB	352.96	1	352.96	0.40	0.5460	
AC	24.86	1	24.86	6.36	0.0397	*
BC	1.57	1	1.57	90.26	<0.0001	**
A^2	0.24	1	0.24	0.061	0.8125	
B^2	7.32	1	7.32	1.87	0.2136	
C^2	499.24	1	499.24	127.67	<0.0001	**
Residual	27.37	7	3.91			
Spurious term	27.37	3	9.12			
Pure error	0.00	4	0.00			
Total value	1441.98	16				
Model determination coefficient			$R^2 = 0.9810$			
Model adjustment determination Coefficient			$R^2_{adj} = 0.9566$			

Note: ** indicates highly significant ($p < 0.01$); * indicates significant ($p < 0.05$).

Figure 7 shows the contour (a) and three-dimensional plot (b) of the interaction between SA concentration and SA/Ca^{2+} cross-linking time on the swelling degree of response values. From Figure 7a, it can be seen that the response surface is relatively flat and the contours are approximately circular, and the changes of hydrogel swelling are not obvious with the increase in cross-linking time when the SA concentration is at low and high values. This indicates that the interaction between SA concentration and SA/Ca^{2+} cross-linking time is not significant. Figure 8 shows the contour (a) and 3D plot (b) of the interaction between SA concentration and PVA/CS mass ratio on the swelling degree of the response value. The contour in Figure 8a is elliptical, and the swelling degree increases and then decreases with the increase in PVA/CS mass ratio when the SA concentration is at low and high values. This indicates that the interaction between SA concentration and PVA/CS mass ratio is more significant. Figure 9 shows the contour (a) and 3D plot (b) of the interaction between the SA/Ca^{2+} cross-linking time and the PVA/CS mass ratio on the swelling degree of the response value, the contour in Figure 9a is elliptical, and the swelling degree increases and then decreases with the increase in PVA/CS mass ratio when the cross-linking time is at low and high values. This indicates that the interaction of the cross-linking time and PVA/CS mass ratio has a significant effect on the swelling degree.

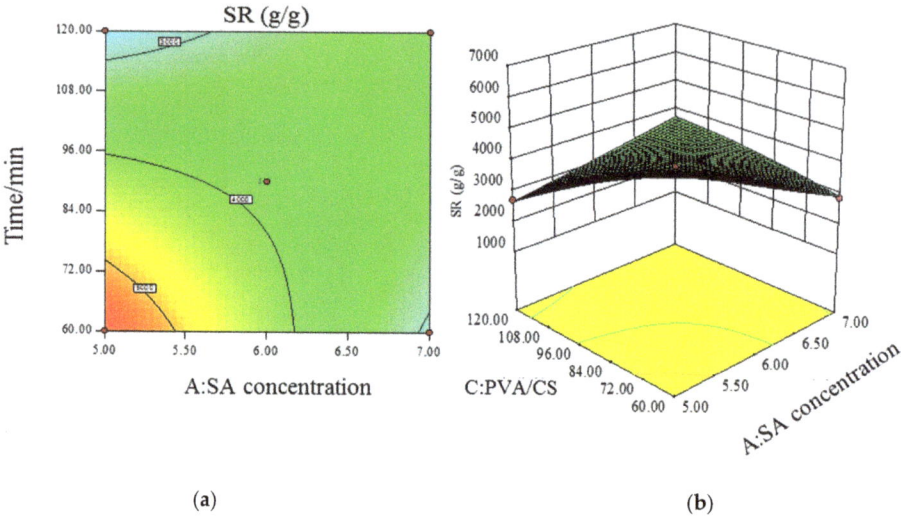

Figure 7. Contour (a) and 3D plot (b) of SA concentration versus SA/Ca^{2+} cross-linking time interaction versus response value SR.

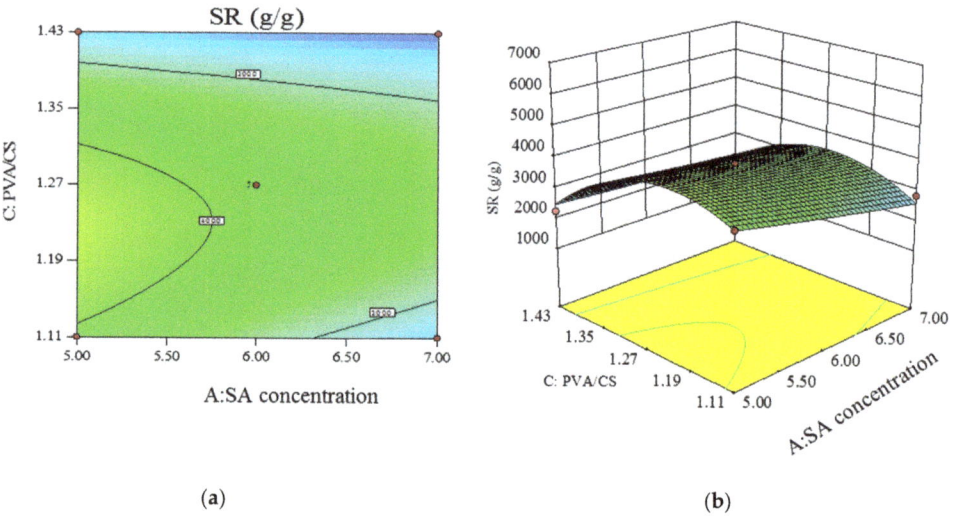

Figure 8. Contour (a) and 3D plot (b) of SA concentration versus PVA/CS addition ratio interaction versus response value SR.

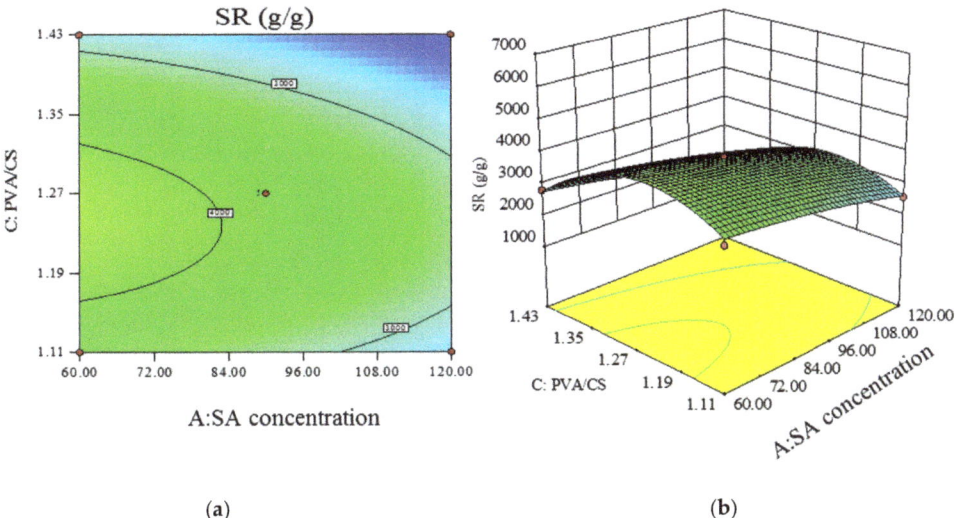

Figure 9. Contour (**a**) and 3D plot (**b**) of SA/Ca^{2+} cross-linking time versus PVA/CS addition ratio interaction versus response value SR.

(3) Prediction of optimal experimental conditions for swelling degree.

We can see from the predicted results that the optimal experimental conditions for the core–shell hydrogel are an SA concentration of 5%, SA/Ca^{2+} cross-linking time of 90 min, and PVA/CS mass ratio of 1:0.7, when the maximum swelling degree can be obtained, which is approximately 50 g/g. To verify this conclusion, we conducted experimental measurements of the swelling degree under these experimental conditions, and the maximum swelling degree results obtained were 49 ± 1.2 g/g, which is in general agreement with the predicted results.

4. Discussions

4.1. Swelling Properties of Hydrogels

PVA and SA are commonly used cross-linking materials for hydrogels, and a large number of different types of hydrogels have been explored by previous studies for the preparation and swelling performance testing. The maximum swelling of CS/PVA hydrogels prepared by Luo et al. was approximately 20 g/g [30]; the maximum swelling of sodium alginate/polyvinyl alcohol (SA/PVA) hydrogels prepared by Ji et al. using the freeze–thaw method was approximately 10 g/g, and it could be increased to approximately 20 g/g when calcium alginate/polyvinyl alcohol (CA/PVA) was prepared using the Ca^{2+} release method and the freeze–thaw cycling method [31]. Polymalic acid is the only kind of high molecular polymer that contains a fat bond and can be quickly hydrolyzed, and the hydrolysis product is malic acid monomer, which can be degraded in the tricarboxylic acid cycle. The hydrogel prepared with PVA reaches a swelling equilibrium after 24 h water absorption, and the maximum swelling degree is approximately 10.5 g/g [32]. In addition, environmental factors also affect the swelling properties of hydrogels. The maximum swelling rate of sodium alginate/graphene oxide hydrogels prepared by Zhuang et al. was approximately 15 g/g under neutral conditions, but the maximum swelling rate can grow to 35 g/g when adjusting the pH to 13 [33]; Hülya found in his study that temperature affects the swelling of hydrogels, and the swelling rate is at its highest at 40 °C [34].

The swelling property is one of the most important parameters for evaluating the structure of hydrogels, and a high swelling degree reduces the stability and mechanical strength of hydrogels. In general, hydrogels that can be used repeatedly tend to limit the maximum swelling degree, except for disposable products such as diapers and hemostats that can be used with high swelling degree hydrogels. The mechanical strength of hydrogels depends on the cross-linking density, so the appropriate cross-linking conditions can be selected according to specific needs.

4.2. Dual Loading Function of PVA–CS/SA–Ca^{2+} Core–Shell Hydrogels

The main objective of this study was to construct the core–shell structure of hydrogel for the purpose of loading two substances for multifunctional expression. After preparing the core–shell hydrogel, in order to verify the dual loading function of the hydrogel, we added degradation bacteria in the core structure and photocatalyst in the shell structure, taking methylene blue (MB) as the research object, and analyzed the degradation of MB. The results showed that MB was first degraded by the degradation bacteria after entering the hydrogel, and the color of the solution becomes lighter. When the light is irradiated through the solution to the hydrogel, MB was further degraded by the photocatalyst. The core–shell hydrogel showed a gradual degradation function, which indicates that the core–shell hydrogel can achieve multi-functional expression.

5. Conclusions

(1) The PVA–CS/SA–Ca^{2+} hydrogels have obvious nucleation and shell structures as observed by SEM. The SA concentration and SA/Ca^{2+} cross-linking time are positively correlated with the thickness of the shell structure; the PVA/CS mass ratio affects the structural characteristics of the nucleation structure, and a higher CS content means a more obvious three-dimensional network of the hydrogels' structure.

(2) The optimal experimental conditions for the swelling of the core–shell hydrogels were obtained by single-factor screening and surface response analysis, which were an SA concentration of 5%, SA/Ca^{2+} cross-linking time of 90 min, PVA/CS mass ratio of 1:0.7 and maximum swelling of 50 g/g.

Author Contributions: Conceptualization, W.Y. and M.Z.; methodology, Y.W.; validation, S.Z., Z.Z.; formal analysis, S.Z. and R.W.; investigation, S.Z. and L.W.; resources, L.W.; data curation, S.Z.; writing—original draft preparation, S.Z.; writing—review and editing, W.Y.; visualization, S.Z. and Y.W.; supervision, M.Z. and Y.Z.; project administration, Y.Z.; funding acquisition, W.Y. and S.Z. All authors have read and agreed to the published version of the manuscript.

Funding: This study was supported by the Science and Technology Program for Innovative Talent granted by the National Forestry and Grassland Administration of China (No.2019132604) and the National Natural Science Foundation of China (No. 32001375).

Conflicts of Interest: We have no conflicts of interest to declare.

References

1. Xu, N.; Ma, N.; Yang, X.; Ling, G.; Yu, J.; Zhang, P. Preparation of intelligent DNA hydrogel and its applications in biosensing. *Eur. Polym. J.* **2020**, *137*, 109951. [CrossRef]
2. Sennakesavan, G.; Mostakhdemin, M.; Dkhar, L.K.; Seyfoddin, A.; Fatihhi, S.J. Acrylic acid/acrylamide based hydrogels and its properties—A review. *Polym. Degrad. Stab.* **2020**, *180*, 109308. [CrossRef]
3. Syuuhei, K. Fabrication of thermoresponsive degradable hydrogel made by radical polymerization of 2-methylene-1,3-dioxepane: Unique thermal coacervation in hydrogel. *Polymer* **2019**, *179*, 121633.
4. Dixit, A.; Bag, D.S.; Kalra, S. Synthesis of strong and stretchable double network (DN) hydrogels of PVA-borax and P(AM-HEMA) and study of their swelling kinetics and mechanical properties. *Polymer* **2017**, *119*, 263–273. [CrossRef]
5. Tao, G.; Wang, Y.; Cai, R.; Chang, H.; Song, K.; Zuo, H.; Zhao, P.; Xia, Q.; He, H. Design and performance of sericin/poly(vinyl alcohol) hydrogel as a drug delivery carrier for potential wound dressing application. *Mater. Sci. Eng.* **2019**, *101*, 341–351. [CrossRef]
6. Qiao, Y.; Xu, S.; Zhu, T.; Tang, N.; Bai, X.; Zheng, C. Preparation of printable double-network hydrogels with rapid self-healing and high elasticity based on hyaluronic acid for controlled drug release. *Polymer* **2020**, *186*, 121994. [CrossRef]
7. Raghuwanshi, V.S.; Garnier, G. Characterisation of hydrogels: Linking the nano to the microscale. *Adv. Colloid Interface Sci.* **2019**, *274*, 102044. [CrossRef]
8. Qi, X.; Su, T.; Tong, X.; Xiong, W.; Zeng, Q.; Qian, Y.; Zhou, Z.; Wu, X.; Li, Z.; Shen, L.; et al. Facile formation of salecan/agarose hydrogels with tunable structural properties for cell culture. *Carbohydr. Polym.* **2019**, *224*, 115208. [CrossRef]

9. Wichterle, O.; Lím, D. Hydrophilic gels for biological use. *Nature* **1960**, *185*, 117–118. [CrossRef]
10. Polat, T.G.; Duman, O.; Tun, S. Agar/κ-carrageenan/montmorillonite nanocomposite hydrogels for wound dressing applications. *Int. J. Biol. Macromol.* **2020**, *164*, 4591–4602. [CrossRef]
11. Seo, J.W.; Moon, J.H.; Jang, G.; Jung, W.K.; Park, Y.H.; Park, K.T.; Bae, H. Cell-Laden Gelatin Methacryloyl Bioink for the Fabrication of Z-Stacked Hydrogel Scaffolds for Tissue Engineering. *Polymers* **2020**, *12*, 3027. [CrossRef]
12. Kaur, R.; Sharma, R.; Chahal, G.K. Synthesis of lignin-based hydrogels and their applications in agriculture: A review. *Chem. Pap.* **2021**, *75*, 4465–4478. [CrossRef]
13. Bojarczuk, K.; Karliński, L.; Hazubska-Przybył, T.; Kieliszewska-Rokicka, B. Influence of mycorrhizal inoculation on growth of micropropagated Populus×canescens lines in metal-contaminated soils. *New For.* **2015**, *46*, 195–215. [CrossRef]
14. Mu, R.; Liu, B.; Chen, X.; Wang, N.; Yang, J. Hydrogel adsorbent in industrial wastewater treatment and ecological environment protection. *Environ. Technol. Innov.* **2020**, *20*, 101107. [CrossRef]
15. Gajewski, P.; Lewandowska, A.; Szcześniak, K.; Przesławski, G.; Marcinkowska, A. Optimization of the Properties of Photocured Hydrogels for Use in Electrochemical Capacitors. *Polymers* **2021**, *13*, 3495. [CrossRef] [PubMed]
16. Hoffman, A.S. Hydrogels for biomedical applications. *Adv. Drug Deliv. Rev.* **2012**, *64*, 18–23. [CrossRef]
17. Zhang, Y.S.; Khademhosseini, A. Advances in engineering hydrogels. *Science* **2017**, *356*, 6337. [CrossRef] [PubMed]
18. Giweta, M.; Garedew, E. Hydrogel-A Promising Technology for Optimization of Nutrients and Water in Agricultural and Forest Ecosystems. *Int. J. Environ. Sci. Nat. Resour.* **2020**, *23*, 106–111.
19. Alizadehgiashi, M.; Khuu, N.; Khabibullin, A.; Henry, A.; Tebbe, M.; Suzuki, T.; Kumacheva, E. Nanocolloidal hydrogel for heavy metal scavenging. *ACS Nano* **2018**, *12*, 8160–8168. [CrossRef]
20. Verma, A.; Thakur, S.; Mamba, G.; Gupta, R.K.; Thakur, P.; Thakur, V.K. Graphite modified sodium alginate hydrogel composite for efficient removal of malachite green dye. *Int. J. Biol. Macromol.* **2020**, *148*, 1130–1139. [CrossRef]
21. Dai, L.; Cheng, T.; Xi, X.; Nie, S.; Ke, H.; Liu, Y.; Tong, S.; Chen, Z. A versatile TOCN/CGG self-assembling hydrogel for integrated wastewater treatment. *Cellulose* **2020**, *27*, 915–925. [CrossRef]
22. Xia, S.; Song, S.; Jia, F.; Gao, G. A flexible, adhesive and self-healable hydrogel-based wearable strain sensor for human motion and physiological signal monitoring. *J. Mater. Chem. B* **2019**, *7*, 4638–4648. [CrossRef] [PubMed]
23. Zhao, Q.; Liang, Y.; Ren, L.; Yu, Z.; Zhang, Z.; Ren, L. Bionic intelligent hydrogel actuators with multimodal deformation and locomotion. *Nano Energy* **2018**, *51*, 621–631. [CrossRef]
24. Hu, C.; Zhang, Y.; Wang, X.; Xing, L.; Shi, L.; Ran, R. Stable, strain-sensitive conductive hydrogel with antifreezing capability, remoldability, and reusability. *ACS Appl. Mater. Interfaces* **2018**, *10*, 44000–44010. [CrossRef]
25. Matsuda, T.; Kawakami, R.; Namba, R.; Nakajima, T.; Gong, J.P. Mechanoresponsive self-growing hydrogels inspired by muscle training. *Science* **2019**, *363*, 504–508. [CrossRef]
26. Sharma, B.; Thakur, S.; Mamba, G.; Gupta, R.K.; Gupta, V.K.; Thakur, V.K. Titania modified gum tragacanth based hydrogel nanocomposite for water remediation. *J. Environ. Chem. Eng.* **2020**, *9*, 104608. [CrossRef]
27. Thakur, S.; Sharma, B.; Verma, A.; Chaudhary, J.; Tamulevicius, S.; Thakur, V.K. Recent progress in sodium alginate based sustainable hydrogels for environmental applications. *J. Clean. Prod.* **2018**, *198*, 143–159. [CrossRef]
28. Gao, X.; Guo, C.; Hao, J.; Zhao, Z.; Long, H.; Li, M. Adsorption of heavy metal ions by sodium alginate based adsorbent-a review and new perspectives. *Int. J. Biol. Macromol.* **2020**, *164*, 4423–4434. [CrossRef] [PubMed]
29. Zhang, M.; Wan, Y.; Wen, Y.; Li, C.; Kanwal, A. A novel Poly (vinyl alcohol)/carboxymethyl cellulose/yeast double degradable hydrogel with yeast foaming and double degradable property. *Ecotoxicol. Environ. Saf.* **2020**, *187*, 109765. [CrossRef] [PubMed]
30. Luo, C.; Zhao, Y.; Sun, X.; Hu, B. Developing high strength, antiseptic and swelling-resistant polyvinyl alcohol/chitosan hydrogels for tissue engineering material. *Mater. Lett.* **2020**, *280*, 128499. [CrossRef]
31. Qin, C.; Ou, K.; Dong, X.; Ji, X.; He, J. Preparation and properties of calcium alginate/polyvinyl alcohol hydrogel. *J. Mater. Sci. Eng.* **2018**, *36*, 739–744.
32. Zhang, J.; Chen, D.; Liang, G.; Xu, W.; Tao, Z. Biosynthetic Polymalic Acid as a Delivery Nanoplatform for Translational Cancer Medicine. *Trends Biochem. Sci.* **2020**, *46*, 213–224. [CrossRef]
33. Zhuang, Y.; Yu, F.; Chen, H.; Zheng, J.; Ma, J.; Chen, J. Alginate/graphene double-network nanocomposite hydrogel beads with low-swelling, enhanced mechanical properties, and enhanced adsorption capacity. *J. Mater. Chem. A* **2016**, *4*, 10885–10892. [CrossRef]
34. Arslan, H.; Pfaff, A.; Lu, Y.; Stepanek, P.; Müller, A.H. Stimuli-R esponsive Spherical Brushes Based on d-G alactopyranose and 2-(Dimethylamino) ethyl Methacrylate. *Macromol. Biosci.* **2014**, *14*, 81–91. [CrossRef] [PubMed]

Article

Local Variations in Carbohydrates and Matrix Lignin in Mechanically Graded Bamboo Culms

Kexia Jin [1,3], Zhe Ling [2], Zhi Jin [4], Jianfeng Ma [1], Shumin Yang [1], Xinge Liu [1,*] and Zehui Jiang [1,*]

1. Key Lab of Bamboo and Rattan Science & Technology, International Center for Bamboo and Rattan, Beijing 100102, China; jinkexia@zju.edu.cn (K.J.); majf@icbr.ac.cn (J.M.); yangsm@icbr.ac.cn (S.Y.)
2. Co-innovation Center of Efficient Processing and Utilization of Forest Resources, College of Chemical Engineering, Nanjing Forestry University, Nanjing 210037, China; jjling19@njfu.edu.cn
3. State Key Lab of Chemical Engineering, College of Chemical and Biological Engineering, Zhejiang University, Hangzhou 310027, China
4. Research Institute of Wood Industry, Chinese Academy of Forestry, Beijing 100091, China; lucy870826@163.com
* Correspondence: liuxinge@icbr.ac.cn (X.L.); jiangzehui@icbr.ac.cn (Z.J.)

Abstract: The mechanical performance of bamboo is highly dependent on its structural arrangement and the properties of biomacromolecules within the cell wall. The relationship between carbohydrates topochemistry and gradient micromechanics of multilayered fiber along the diametric direction was visualized by combined microscopic techniques. Along the radius of bamboo culms, the concentration of xylan within the fiber sheath increased, while that of cellulose and lignin decreased gradually. At cellular level, although the consecutive broad layer (Bl) of fiber revealed a relatively uniform cellulose orientation and concentration, the outer Bl with higher lignification level has higher elastic modulus (19.59–20.31 GPa) than that of the inner Bl close to the lumen area (17.07–19.99 GPa). Comparatively, the cell corner displayed the highest lignification level, while its hardness and modulus were lower than that of fiber Bl, indicating the cellulose skeleton is the prerequisite of cell wall mechanics. The obtained cytological information is helpful to understand the origin of the anisotropic mechanical properties of bamboo.

Keywords: multilayered bamboo fiber; topochemistry; microscopic imaging; gradient micromechanics

Citation: Jin, K.; Ling, Z.; Jin, Z.; Ma, J.; Yang, S.; Liu, X.; Jiang, Z. Local Variations in Carbohydrates and Matrix Lignin in Mechanically Graded Bamboo Culms. *Polymers* **2022**, *14*, 143. https://doi.org/10.3390/polym14010143

Academic Editor: Antonios N. Papadopoulos

Received: 15 November 2021
Accepted: 27 December 2021
Published: 31 December 2021

Publisher's Note: MDPI stays neutral with regard to jurisdictional claims in published maps and institutional affiliations.

Copyright: © 2021 by the authors. Licensee MDPI, Basel, Switzerland. This article is an open access article distributed under the terms and conditions of the Creative Commons Attribution (CC BY) license (https://creativecommons.org/licenses/by/4.0/).

1. Introduction

Bamboo, with the advantage of sustainability, extraordinary growth rate, ready availability, low weight, excellent mechanical strength and superior toughness, has been widely used as a structural material and bio-composite for industrial application [1,2]. Compared with other common building materials, bamboo is stronger than most timbers, and its strength-to-weight ratio is higher than that of common wood, cast iron, aluminum alloy and structural steel [3–5]. In terms of tissue types, the bamboo culm consists of fibers, parenchyma and conducting elements (including the xylem vessels, sieve tubes, and companion cells). As the source of the superior mechanical properties of bamboo, the fibers' density and quality vary significantly along and across the bamboo culm. For example, along the diametric direction, the longitudinal tensile modulus of elasticity for the outermost layer was 3–4 times as high as that of the innermost layer, while the longitudinal tensile strength ranged from 115.94 to 328.15 MPa from the outermost layer to the innermost layer [6]. Numerous studies have focused on the mechanical properties of bamboo and bamboo composites [7–9]. However, the origin of the anisotropic mechanical properties across and along the culm are poorly understood.

Actually, much of the mechanical behavior of bamboo is governed by the properties of the cell wall, which, in turn, can be described in terms of the submicroscopic structure of the wall and the localization of cell wall components of cellulose, hemicelluloses, and lignin.

Cellulose is the main structural fiber in the plant kingdom and has remarkable mechanical properties for a polymer: its Young's modulus is roughly 130 GPa, and its tensile strength is close to and even more than 1 GPa [10]. The properties of hemicelluloses and lignin are similar to common engineering polymers. Lignin, for instance, has a modulus of roughly 3 GPa and a strength of about 50 MPa [11]. An important role of lignin in the wood cell wall is to function as a cross-linking matrix between moisture sensitive cellulose and hemicelluloses, thereby, lignin contributes to the mechanical rigidity [12]. The modulus of xylan varies from a value of 8 GPa at low moisture contents (0–10%) to 10 MPa at moisture contents near saturation (70%). Meanwhile, the incorporation of xylan in the cell wall, especially in the secondary wall, has a strength-enhancing effect on the joint strength individual fiber crossings [13]. The arrangement of the basic building blocks in bamboo cell walls and the variations in cellular structure give rise to a remarkably wide range of mechanical properties.

Unlike the typical three-layered structure of wood secondary wall, bamboo exhibits a polylamellate secondary wall with alternating broad and narrow lamellae that arise from the alternation in the orientation of cellulose microfibrils in a matrix of intertwined hemicelluloses and lignin. To date, macro- and micro-structural investigations have been reported comprehensively. For example, Huang et al. [14] and Wang et al. [15] have investigated the effect of different locations within the vascular bundle and non-multilayered fibers and the age of bamboo on mechanical properties. However, quite few works focus on the mechanical properties of multilayered fiber, to date. Since the multilayered fiber with alternating hierarchical structure was considered to be one of the factors that contribute to the high tensile strength of bamboo [16], insight of its structure, especially "seeing" bamboo macro- and micro-structure in a chemical sense, has great significance for full utilization of bamboo resources.

To the best of our knowledge, this work is the first report to reveal the fine mechanical details within the multilayered fiber of bamboo and correlate them with the chemical features of the cell wall. The in situ distribution of polysaccharides and lignin within the fiber located at bamboo green (Bg), bamboo timber (Bt) and bamboo yellow (By) were visualized by complementary microscopy techniques, including FT-IR microscopy and confocal Raman microscopy. The microscopes allowed correlative imaging of the same biomass sample under near-physiological conditions and at high chemical and spatial resolutions at the tissue, cellular, and molecular levels. In parallel, nanoindentation technique was applied to investigate the cell wall mechanical properties of fibers along the radius of bamboo culm. This information will contribute to greater fundamental understanding of the mechanical design of bamboo fiber cell walls and the mechanically graded structure of bamboo culms.

2. Experimental

2.1. Materials

A three-year-old moso bamboo (*Phyllostachys pubescens*) culm was collected from the local forest in Anhui Taiping experimental center, International Center for Bamboo and Rattan. Blocks of a length about 2–3 cm along the grain were cut from the middle part of the 10th internode (numbered from the ground level). Without any further preparation, a series of 15-μm-thick cross-sections, located at around the middle part of culm wall were cut from the freshly blocks on a rotary microtome (LEICA RM2165, Wetzlar, Germany) for the FT-IR and Raman imaging.

2.2. FT-IR Imaging Analysis

The FT-IR microspectroscopic imaging system (Spotlight400, PerkinElmer Ltd., Waltham, MA, USA), which combines a microscope with a FT-IR spectrometer, was used to obtain the FT-IR microspectroscopic spectra and image. Visible images were obtained using a charge-coupled device (CCD) camera, which made it possible to observe the specimen and select the area of interest for spectral analyses. FT-IR microspectroscopic image was obtained by a liquid nitrogen cooled mercury cadmium telluride (MCT) line array (16 × 1 element) detector. The air-dried bamboo section was placed over a KBr window supported on the Spotlight stage. All FT-IR images were taken using Spectrum IMAGE Software (Perkin Elmer) and collected in transmission mode in the region of 4000–750 cm^{-1} at 8 cm^{-1} spectral resolution with a 6.25 μm × 6.25 μm spatial resolution and eight scans co-added. The obtained IR spectra were then processed by using software Spectra IR developed by Perkin Elmer Inc. The functions of atmosphere correction, flat correction and baseline offset correction were applied in turn to create corrected spectra.

2.3. Confocal Raman Imaging Analysis

For Raman chemical imaging, the bamboo section was placed on a glass slide with a drop of distilled water, covered by a coverslip (0.17 mm thickness) and sealed with nail-polish to prevent evaporation during measurement. Raman spectra were acquired with a confocal Raman microscope (LabRam HR Evolution, Horiba Jobin Yvon, Paris, France). Measurements were conducted with an Olympus 100×Oil objective (PlanC N 100×, Oil, NA = 1.25) and a 532-nm laser. The Raman light was detected by an air-cooled, front-illuminated spectroscopic electron-multiplying charge-coupled device (EMCCD) behind a grating (300 grooves mm^{-1}) spectrometer with a spectral resolution of 2 cm^{-1}. For mapping, 0.3 μm steps were chosen and every pixel corresponds to one scan. The overview chemical images separated cell wall layers and marked the defined distinct cell wall areas to calculate the average spectra from the areas of interest. The Labspect6 software was used for spectral and image processing and analysis. Before a detailed analysis, the calculated average spectra were baseline corrected using the linear least squares algorithm.

2.4. Nanoindentation Test

A Hysitron TI 950 nanoindentation instrument (Triboindenter, Hysitron Inc., Minneapolis, MN, USA) was used to measure the nanoindentation modulus, nanoindentation hardness, yield strength, and creep of the materials by calculating the load-displacement curve derived from nano-indenter loading and unloading on the materials. The standard holding time was 50 s, while loading times or unloading times varied at 5 s, 15 s, and 25 s, respectively. A Berkovich diamond pyramid was used for indentation, with indentor tip radius of about 100 nm. Creep compliance tests were performed by using 300 μN, 400 μN, and 500 μN constant loads. Sample blocks for nanoindentation tests were vertically fixed on the cylindrical holder. The transverse (cross section) surface of the bamboo was polished by an ultra-microtome with a diamond knife to keep the cutting surface smooth. Before testing, the smoothed samples were conditioned for at least 24 h at 22 °C and 60% relative humidity in a room that housed the nano-indenter to ensure uniform moisture content. At least 35 valid points in cell corner (CC), outer layer of fiber secondary wall (SW-out) and inner layer of fiber secondary wall (SW-in) of each sample were tested, and the elastic modulus and hardness were obtained by averaging. All samples were treated equally to compare the mechanical properties on a relative basis.

3. Results and Discussion

3.1. FT-IR Chemical Image of Bamboo Culm

The FT-IR spectra of the bamboo fiber sheaths extracted from Bg, Bt and By are shown in Figure 1a. FT-IR spectra indicated peak changes in fingerprint regions at various tissues and cell locations. The main differences in the absorption spectra are visible at wavenumbers 1730, 1508, 1369, 1240, 1158, and 896 cm^{-1}. The IR band at 1508 cm^{-1} corresponds to a C=C stretching the vibration of the aromatic rings of lignin. The carbohydrates peaks at 1730, 1369, and 1158 cm^{-1} are assigned, respectively, for unconjugated C=O in xylan, C–H deformation in cellulose and hemicelluloses, and C–O–C vibration in cellulose and hemicelluloses [17,18]. Moreover, the band at 1240 and 896 cm^{-1} is a diagnostic peak for cellulose by the C–O–C vibration and C–H deformation in cellulose, respectively [19,20]. The relative absorbance of characteristic FT-IR absorbance bands is plotted as a function of the fiber sheaths (Figure 1b). It was noted that the bands assigned to cellulose (1240 and 896 cm^{-1}), and to lignin (1508 cm^{-1}) showed obvious decrease in the absorbance peaks from Bg to By for fiber sheath, while the xylan band at 1730 cm^{-1} displayed an increase in the corresponding regions. Notably, the intensity of hemicelluloses related band at 1369 and 1158 cm^{-1} kept constant, probably due to these two band areas containing partial contribution from cellulose which displayed a declining tendency in band intensity.

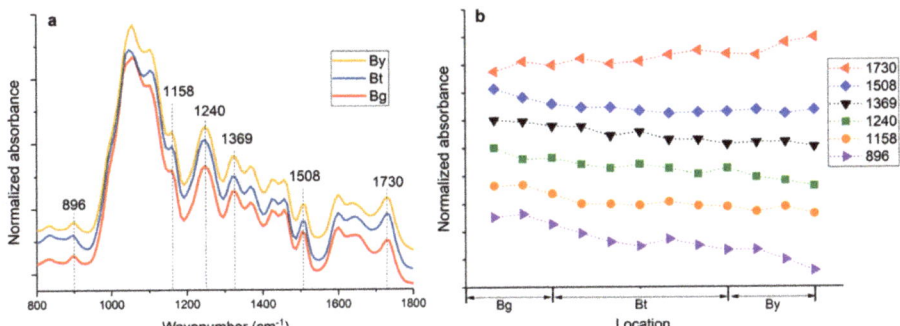

Figure 1. (**a**) FT-IR spectra of fiber sheath extracted from Bg to By; (**b**) Change in relative intensity of absorbance bands for fiber sheath from Bg to By.

Previously, the investigation of carbohydrate and lignin in the bamboo cell wall mainly focused on the total value of lignin in bamboo tissues, comprising all tissues, such as fibers, vessels and parenchyma, rather than specifying the lignin and values for individual cell types [21–23]. Actually, the extent of compositional variation was found to be influenced by the age of the culm, in certain ages by the position of the vascular bundle, and most strikingly, by the proportion of fibers within the vascular bundle and surrounding parenchyma. The FT-IR microspectroscopic data can be displayed as chemical images at specific wavelengths at the tissue level with a spatial resolution near 6.25 μm (Transmission mode). Figure 2 shows the functional group chemical images of bamboo culm transverse sections. The red color represents high intensity and the blue color stands for little or no intensity. Thus, chemical images at bands near 1730, 1508, and 1240 cm^{-1} can show the relative concentrations of xylan, lignin, and cellulose, respectively. The chemical images indicated that the xylan (Figure 2a) and lignin (Figure 2b) mostly accumulated in the fiber sheath within the vascular bundles and appeared much less in the ground parenchymatic regions. This is to be expected, because the vascular bundles are believed to be the main mechanical support for the whole bamboo culm. Interestingly, for a single vascular bundle, there is also a heterogeneity in compositional distribution. High lignin and xylan concentration was visualized in the outermost part of the fiber sheath. Comparatively, cellulose showed a more homogeneous distribution pattern for the fiber sheaths within the vascular bundles (Figure 2c). It has been revealed that the average tensile modulus for the bamboo fiber is three times higher than that of the parenchyma [24]. Cellulose has been proved to be the main structural component and has remarkable mechanical properties. Thus, the higher concentration of cellulose in the fiber sheaths area may partly explain the superior modulus and strength properties compared with parenchyma cells.

3.2. Confocal Raman Chemical Image of Bamboo Fiber

To further explore the chemical constituent distribution of bamboo fiber at the cellular and sub-cellular level in situ, confocal Raman microscopy with high spatial resolution (<0.5 μm) was employed. The strong band at 2897 cm^{-1} was assigned to the stretching of the C–H and C–H$_2$ groups of carbohydrates [25]. The spectral fingerprints for the characteristic bands of lignin were identified at 1598 and 1656 cm^{-1}, attributed to the stretching vibrations of the aromatic ring and ring-conjugated C=C bonds in coniferyl alcohol units, respectively [26].

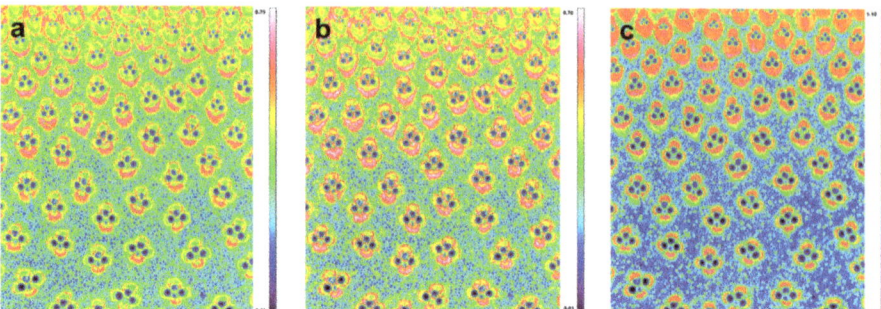

Figure 2. FT-IR images of the relative concentration and distribution of xylan (**a**), lignin (**b**), and cellulose (**c**).

By integrating over the CH and CH_2 stretching vibrations (2800 cm^{-1} to 2918 cm^{-1}), high intensity and thus high carbohydrates concentration were observed especially in the secondary wall (SW) of fiber (Figure 3a–c). Due to the high spatial resolution, the alternating broad (Bl) and narrow layers (Nl) can be easily differentiated in the polylamellate secondary walls, with the Bl having higher carbohydrates concentration than the Nl. Raman image by calculating the band ranges from 1540 cm^{-1} to 1660 cm^{-1} displayed the heterogeneity in lignin distribution. As shown in Figure 3d–f, lignin concentration was the highest in cell corner (CC) and compound middle lamella (CML). Within the SW, the lignin concentration decreased from the outer layer to the cell lumen, with relative higher concentration in the lumen edge as well as the interface between adjacent layers of the multilayered fiber.

To obtain better chemical composition changes among the Bg, Bt and By, the average Raman spectra in CC, CML, Bl, and Nl were extracted, respectively (Figure 4). It was noted that the Bg displayed higher lignin Raman intensity (1598 and 1656 cm^{-1}) followed by Bt and fewest in By especially in CC and CML. In fiber SW, the average lignin Raman intensity of Bl was higher than Nl indicating higher lignin concentration in Bl. Along the diametric direction, although the Nl showed a similar lignin Raman strength, the Bl of Bg showed a relative higher lignin intensity. Meanwhile, the carbohydrates' Raman intensity (2897 cm^{-1}) both in Bl and Nl of the Bg was higher than that of the Bt and By. This detailed examination is consistent with the data with respect to the chemical change at specific wavelengths at tissue level observed by FT-IR, as shown in Figure 1.

In plant tissues, lignin does not exist in an independent polymer, but it is associated with cellulose and hemicelluloses, forming complexes with them through physical admixture and covalent bonds [27]. Specifically, in herbaceous plants, hydroxycinnamic acids (HCA) are attached to lignin and hemicelluloses, via ester and ether bonds as bridges between them forming lignin/phenolics–carbohydrate complexes [28,29]. By integrating over the band regions from 1131–1190 cm^{-1}, assigned to cinnamoyl ester bond in HCA [30], it was demonstrated that HCA mainly deposited within the CC and CML of the corresponding cells (Figure 3g–i), showing a similar distribution of lignin.

To better observe the variability and distribution of carbohydrates (2897 cm^{-1}), lignin (1598 cm^{-1}) and HCA (1172 cm^{-1}) concentrations in various layers of fiber wall, a double cell wall line scan was carried out (Figure 5). As the scan moved from the left lumen-secondary wall interface to the right lumen-secondary wall interface, lignin and attached HCA concentration gradually increased, reached the maximum value in CML, and then declined to a low value in the SW region of the right cell wall (Figure 5b,c). The heterogeneity in lignin distribution at a cellular level has also been reported in a two-month old moso bamboo and other woody biomass [15,31,32]. In the case of carbohydrates, the CML area with the highest lignin intensity showed the lowest concentration of 5.5 k intensity units

(Figure 5a). Similarly, the carbohydrates distribution along the adjacent cell wall displayed a gradual increase from CML to lumen. Characteristically the Bl and Nl of fiber wall has the alternating compositional distribution pattern, with low carbohydrates concentration in the Nl.

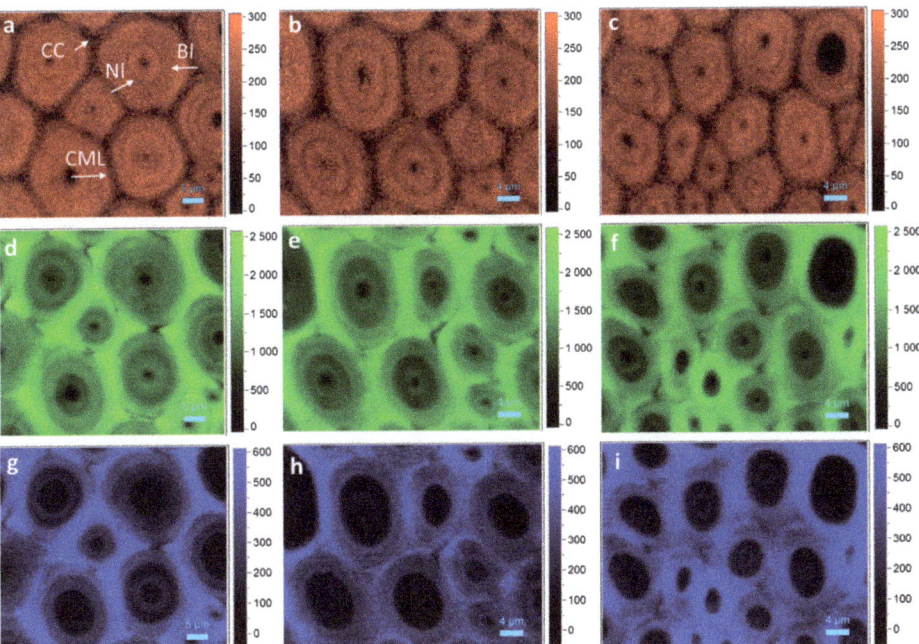

Figure 3. Raman images showing the distribution of fiber wall components from Bg (**a,d,g**), Bt (**b,e,h**) and By (**c,f,i**). (**a–c**) Carbohydrates, 2800–2918 cm^{-1}; (**d–f**) Lignin, 1540–1660 cm^{-1}; (**g–i**) HCA, 1120–1200 cm^{-1}.

To visualize the variation in carbohydrates to lignin ratio, the concentration of carbohydrates along the segment points was divided by lignin concentration. As shown in Figure 5d, the ratio decreased from 0.25 in SW of fiber to 0.075 in CML. Within the fiber SW, the ratio varied periodically, with the higher values in the Bl. By comparison, for several locations in the spruce and *Conus alba* L. fiber SW, the ratio was constant and reflected the fact that at these locations the concentrations of both lignin and carbohydrates increased or declined simultaneously [33,34]. The variation in distribution pattern largely stems from the mechanical roles of fibers in different species. As known, the ultrastructure of most of the bamboo fibers is characterized by thick polylamellate SW. This lamellation consists of alternating Bl and Nl with differing chemical compositions, which leads to an extremely high tensile strength, as demonstrated in engineering constructions with bamboo culms [35]. Comparatively, in the cell walls of xylem fibers or tracheids of normal wood that shares lower tensile strength, the polylamellate construction with different chemical composition does not exist. Additionally, it has been generally accepted and largely cited in the literature that the deposition of HCA is highly correlated with lignin [36,37]. However, it was found that the ratio of HCA to carbohydrates displayed a regular pattern that increased consecutively from 0.2 in the SW-lumen interface to 2.4 in CML (Figure 5e), while the ratio of HCA to lignin along the multi-layered fiber wall varied significantly (Figure 5f), indicating less dependence between these two polymers. During cell wall formation, the incorporation of lignin within the polysaccharide cell wall framework is generally regarded as the final stage of the typical differentiating process (Donaldson 2001). Thus, the accompanied distribution pattern between HCA and carbohydrates demonstrated that HCA probably participated in the cell wall biosynthesis prior to the lignin.

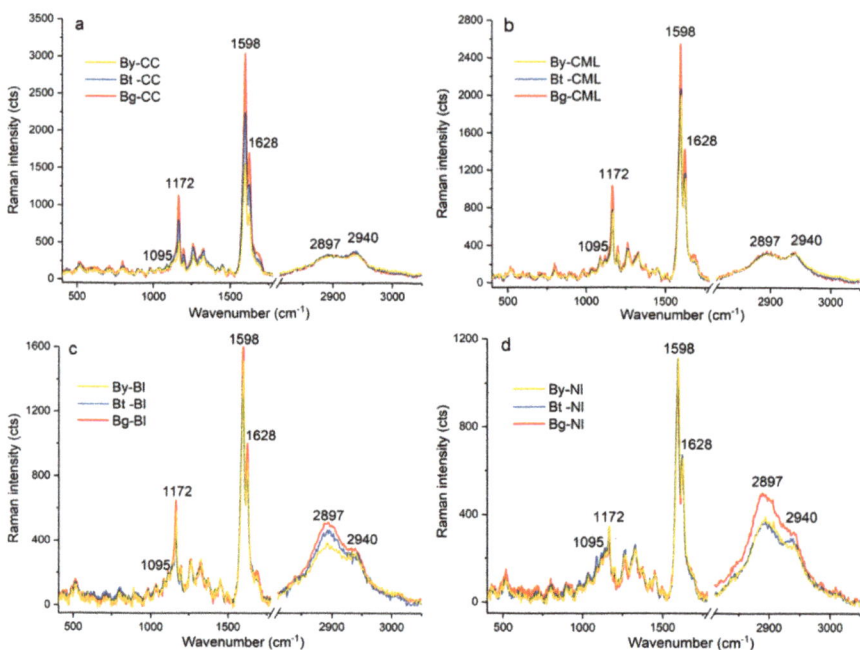

Figure 4. Average Raman spectra extracted from the CC (**a**), CML (**b**), Bl (**c**) and Nl (**d**) of fiber located at Bg, Bt and By.

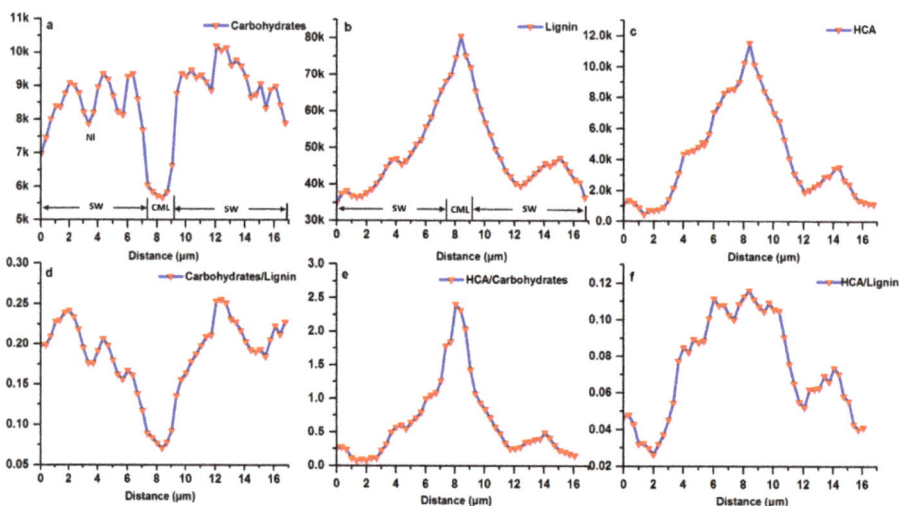

Figure 5. Raman line scan of Bg fiber wall, (**a**) Carbohydrates; (**b**) Lignin; (**c**) HCA; (**d**) Carbohydrates to lignin ratio; (**e**) HCA to carbohydrates ratio; (**f**) HCA to lignin ratio.

3.3. The Variation in the Mechanical Properties of Fiber Cell Walls

As shown in Figure 6a–c, the mean value of indentation modulus and hardness of bamboo fiber within a single fiber wall, displayed an obvious difference. The outer layer of fiber SW in all the Bg, Bt and By areas has higher modulus (19.59–20.31 GPa) and hardness (428–445 MPa) than the inner layer of fiber close to the lumen area with modulus of 17.07–19.99 GPa and hardness of 410–440 MPa. Although previous studies suggested that the microfibril angle is negatively correlated with the elastic modulus of fiber cell wall [38,39], the Raman spectral analysis result revealed a relatively uniform cellulose orientation distribution in the multilayered fiber by extracting the cellulose orientation sensitive band around 1097 cm^{-1} (Figure 6d). This result is consistent with the earlier typical microfibril angle model of multilayer fiber by Parameswaran & Liese [16]. Thus, the variation in the modulus was mainly due to the variation in the cell wall lignification level and its composition, which has been confirmed by lignin Raman line scan (Figure 5b). Similarly, in the spruce tracheid wall, the observed difference in the modulus of elasticity between developing and fully lignified cell walls is due to the filling of spaces with lignin and an increase in the packing density of the cell wall during lignification [40].

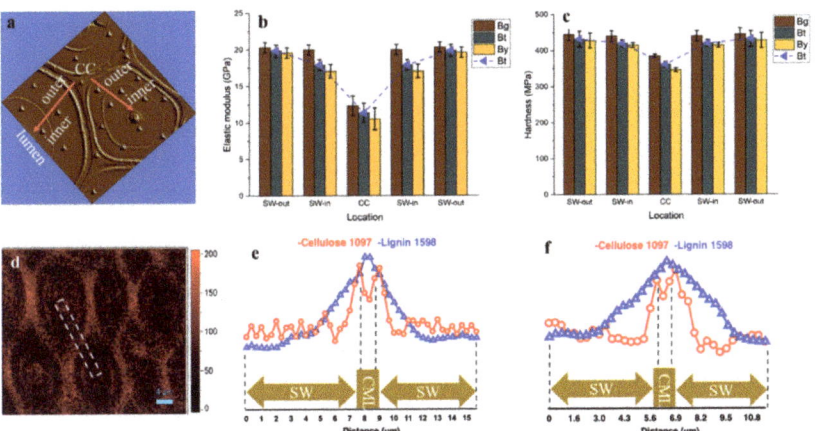

Figure 6. (**a**) The image obtained after indenting shows the actual position of indents; (**b**,**c**) The elastic modulus and hardness of bamboo fiber at different locations; (**d**) Raman image showing the cellulose microfibrils orientation distribution by integrating over 1097 cm^{-1}; (**e**,**f**) the Raman line scan of Bg (**e**) and Bt (**f**) fiber wall. The white dotted line indicates the typical region of the Raman spectra collected. SW-out: outer layer of fiber secondary wall; SW-in: inner layer of fiber secondary wall.

Additionally, along the radius bamboo culm the modulus and hardness also had a gradient trend. The Bg fiber presented a relatively higher modulus and hardness than that located at Bt and By due to a higher lignification degree as supported by the FT-IR and Raman results. Although the Bt and By displayed a relatively lower modulus and hardness than that of Bg, their values are still superior to those of other lignocellulosic biomass as listed in Table 1. This is probably because bamboo has a special cell wall structure with alternating broad and narrow layers, smaller microfibril angle and higher density than wood [35,41,42]. Earlier work has investigated anatomical, physical and mechanical properties of different wood and bamboo material, and implies that most properties were mainly governed by the fibers [4,42–46]. Moreover, this micromechanics data in individual fiber agreed well with the previous research that the longitudinal tensile modulus of elasticity of bamboo for the outermost layer was 3–4 times as high as that for the innermost layer [6], probably due to the density variations in the fiber sheath, which acts like a plant muscle within the bamboo culm. Thus, when the single bamboo fiber with various modulus values are assembled together to form a fiber sheath, it contributes to the superior macro-mechanical properties in Bg, though translating the extraordinary mechanical properties of micro-scale individual fiber to the macroscale fiber sheath will inevitably face the fundamentally non-ideal stress transfer.

Table 1. Fiber cell wall mechanical properties of different materials biomass.

Species	Modulus (GPa)	Hardness (GPa)	References
Spruce	17.1	0.38	[40]
Masson pine	19.18	0.53	[41]
Chinese fir	17.8	0.42	
Manchurian ash	17.5	0.48	[47]
Loblolly pine	14.2–17.6	0.43–0.53	[48]
Hemp	12.3	0.41	[49]
Cotton stalk	16.3	0.85	
Flax stalk	17.4	0.39	[50]
Bamboo (*Dendrocalamus farinosus*)	18.56	410.7	[51]
Bamboo (moso)	19.59–20.31	0.43–0.45	This work

Although the CC regions displayed the highest lignification level, its modulus (10.52–12.32 GPa) and hardness (346–385 MPa) were lower than that of fiber SW, which was probably due to the trace amount of the cell wall skeleton substance deposited in this region. Actually, it has been stated that the impact of lignin content on mechanics seems to depend on the specific structural configuration of the plant cell wall, with high microfibril angles likely to be a requirement for a visible impact of lignin on stiffness [52,53]. Since only fiber walls along the culm of a specific internode was examined, the results may not generally be valid for the whole tissue or plant. However, the findings pointed out that besides the structural variety and complexity, the heterogeneity in composition distribution also contributed to the mechanically graded bamboo culms.

4. Conclusions

Combined microscopic techniques have been used to non-destructively investigate the compositional heterogeneity and variation in cell wall mechanics in moso bamboo. At the tissue level, the fiber sheath has a higher concentration of carbohydrates and lignin than the ground parenchyma cells from Bg to By. For the multilayer bamboo fiber, the Bl revealed a higher carbohydrates and lignin concentration than the Nl, and the outer Bl showed a higher lignification degree than the inner Bl, yet the consecutive Bl revealed a relatively uniform cellulose orientation and concentration. Furthermore, the Bg displayed the highest elastic modulus and hardness followed by Bt and then by By, and the elastic modulus and hardness decreased from the outer Bl to the inner Bl. Comparatively, the CC displayed the highest lignification level but the lowest hardness and modulus.

Author Contributions: Z.J. (Zehui Jiang) and X.L. conceived and designed the experiments, K.J. and J.M. performed the experiments and analyzed the data, Z.J. (Zhi Jin), Z.L., X.L. and S.Y. provided insightful comments on an earlier version of this paper, K.J. and J.M. wrote the manuscript. All authors have read and agreed to the published version of the manuscript.

Funding: This work was funded by the Fundamental Research Funds of ICBR (NO. 1632021001), the National Natural Science Foundation of China (NO. 31670565), and the National Natural Science Foundation for Youth (NO. 32001270).

Institutional Review Board Statement: Not applicable.

Data Availability Statement: All data supporting the findings of this study are available within the article.

Acknowledgments: We would like to thank Dong Wang from the Research Institute of Wood Industry, Chinese Academy of Forestry for providing the nanoindentation measurements.

Conflicts of Interest: The authors declare no conflict of interest.

References

1. Huang, Y.; Ji, Y.; Yu, W.J. Development of bamboo scrimber: A literature review. *J. Wood Sci.* **2019**, *65*, 25. [CrossRef]
2. Huang, S.; Jiang, Q.; Yu, B.; Nie, Y.; Ma, Z.; Ma, L. Combined chemical modification of bamboo material prepared using vinyl acetate and methyl methacrylate: Dimensional stability, chemical structure, and dynamic mechanical properties. *Polymers* **2019**, *11*, 1651. [CrossRef]
3. Li, Z.; Chen, C.; Mi, R.; Gan, W.; Dai, J.; Jiao, M.; Xie, H.; Yao, Y.; Xiao, S.; Hu, L. A strong, tough, and scalable structural material from fast-growing bamboo. *Adv. Mater.* **2020**, *32*, 1906308. [CrossRef]
4. Anokye, R.; Bakar, E.S.; Ratnasingam, J.; Awang, K. Bamboo properties and suitability as a replacement for wood. *Pertanika J. Sch. Res. Rev.* **2016**, *2*, 64–80. [CrossRef]
5. Sharma, B.; van der Vegte, A. Engineered bamboo for structural applications. In *Nonconventional and Vernacular Construction Materials*; Woodhead Publishing: Sawston, UK, 2020; pp. 597–623. [CrossRef]
6. Yu, H.Q.; Fei, B.H.; Ren, H.Q.; Jiang, Z.H.; Liu, X.E. Variation in tensile properties and relationship between tensile properties and air-dried density for Moso bamboo. *Front. For. China* **2008**, *3*, 127–130. [CrossRef]
7. Dixon, P.G.; Gibson, L.J. The structure and mechanics of Moso bamboo material. *J. R. Soc. Interface* **2014**, *11*, 20140321. [CrossRef] [PubMed]
8. Verma, C.S.; Sharma, N.K.; Chariar, V.M.; Maheshwari, S.; Hada, M.K. Comparative study of mechanical properties of bamboo laminate and their laminates with woods and wood based composites. *Compos. Part B Eng.* **2014**, *60*, 523–530. [CrossRef]
9. Deng, J.C.; Wang, G. Axial tensile properties and flexibility characteristics of elementary units from multidimensional bamboo-based composites: Radial and tangential moso bamboo slivers. *Holzforschung* **2018**, *72*, 779–787. [CrossRef]
10. Mittal, N.; Ansari, F.; Gowda, V.K.; Brouzet, C.; Chen, P.; Larsson, P.T.; Toth, S.V.; Lundell, F.; Wågberg, L.; Kotov, N.A.; et al. Multiscale control of nanocellulose assembly: Transferring remarkable nanoscale fibril mechanics to macroscale fibers. *ACS Nano* **2018**, *12*, 6378–6388. [CrossRef] [PubMed]
11. Gibbson, L.J. The hierarchical structure and mechanics of plant materials. *J. R. Soc. Interface* **2012**, *9*, 2749–2766. [CrossRef]
12. Salmén, L.; Burgert, I. Cell wall features with regard to mechanical performance. A review COST Action E35 2004-2008: Wood machining micromechanics and fracture. *Holzforschung* **2009**, *63*, 121–129. [CrossRef]
13. Miletzky, A.; Fischer, W.J.; Czibula, C.; Teichert, C.; Bauer, W.; Schennach, R. How xylan effects the breaking load of individual fiber-fiber joints and the single fiber tensile strength. *Cellulose* **2015**, *22*, 849–859. [CrossRef]
14. Huang, Y.; Fei, B.; Wei, P.; Zhao, C. Mechanical properties of bamboo fiber cell walls during the culm development by nanoindentation. *Ind. Crops. Prod.* **2016**, *92*, 102–108. [CrossRef]
15. Wang, X.Q.; Ren, H.Q.; Zhang, B.; Fei, B.H.; Burgert, I. Cell wall structure and formation of maturing fibres of moso bamboo (*Phyllostachys pubescens*) increase buckling resistance. *J. R. Soc. Interface* **2012**, *9*, 988–996. [CrossRef] [PubMed]
16. Parameswaran, N.; Liese, W. On the fine structure of bamboo fibres. *Wood Sci. Technol.* **1976**, *10*, 231–264.
17. Kačuráková, M.; Wellner, N.; Ebringerová, A.; Hromádková, Z.; Wilson, R.H.; Belton, P.S. Characterisation of xylan type polysaccharides and associated cell wall components by FT-IR and FT-Raman spectroscopies. *Food Hydrocolloid* **1999**, *13*, 35–41. [CrossRef]
18. Chang, S.S.; Salmén, L.; Olsson, A.M.; Clair, B. Deposition and organisation of cell wall polymers during maturation of poplar tension wood by FT-IR microspectroscopy. *Planta* **2014**, *239*, 243–254. [CrossRef] [PubMed]
19. Michell, A.J. 2nd-derivative FT-IR spectra of native celluloses. *Carbohyd. Res.* **1990**, *197*, 53–60. [CrossRef]
20. Peng, H.; Salmén, L.; Stevanic, J.S.; Lu, J.X. Structural organization of the cell wall polymers in compression wood as revealed by FTIR microspectroscopy. *Planta* **2019**, *250*, 163–171. [CrossRef]
21. Domínguezrobles, J.; Sánchez, R.; Espinosa, E.; Savy, D.; Mazzei, P.; Piccolo, A.; Rodríguez, A. Isolation and characterization of gramineae and fabaceae soda lignins. *Int. J. Mol. Sci.* **2017**, *18*, 327. [CrossRef]
22. Yuan, Z.; Wen, Y. Evaluation of an integrated process to fully utilize bamboo biomass during the production of bioethanol. *Bioresour. Technol.* **2017**, *236*, 202–211. [CrossRef]
23. Li, Z.Q.; Jiang, Z.H.; Fei, B.H.; Cai, Z.Y.; Pan, X.J. Comparison of bamboo green, timber and yellow in sulfite, sulfuric acid and sodium hydroxide pretreatments for enzymatic saccharification. *Bioresour. Technol.* **2014**, *151*, 91–99. [CrossRef] [PubMed]
24. Zou, L.; Jin, H.; Lu, W.Y.; Li, X. Nanoscale structural and mechanical characterization of the cell wall of bamboo fibers. *Mat. Sci. Eng. C-Mater.* **2009**, *29*, 1375–1379. [CrossRef]
25. Gierlinger, N.; Schwanninger, M. The potential of Raman microscopy and Raman imaging in plant research. *Spectroscopy* **2012**, *21*, 69–89. [CrossRef]
26. Agarwal, U.P.; Ralph, S.A. FT-Raman spectroscopy of wood: Identifying contributions of lignin and carbohydrate polymers in the spectrum of black spruce (*Picea mariana*). *Appl. Spectrosc.* **1997**, *51*, 1648–1655. [CrossRef]
27. Azuma, J.I.; Tetsuo, K. Lignin-carbohydrate complexes from various sources. *Methods Enzymol.* **1998**, *161*, 12–18. [CrossRef]
28. Toscan, A.; Morais, A.R.C.; Paixão, S.M.; Alves, L.; Andreaus, J.; Camassola, M.; Dillon, A.J.P.; Lukasik, R.M. Effective extraction of lignin from elephant grass using imidazole and its effect on enzymatic saccharification to produce fermentable sugars. *Ind. Eng. Chem. Res.* **2017**, *56*, 5138–5145. [CrossRef]
29. Sun, R.C.; Tomkinson, J. Comparative study of lignins isolated by alkali and ultrasound-assisted alkali extractions from wheat straw. *Ultrason. Sonochem.* **2002**, *9*, 85–93. [CrossRef]

30. Ma, J.F.; Zhou, X.; Ma, J.; Ji, Z.; Zhang, X.; Xu, F. Raman microspectroscopy imaging study on topochemical correlation between lignin and hydroxycinnamic acids in *Miscanthus sinensis*. *Microsc. Microanal.* **2014**, *20*, 956–963. [CrossRef]
31. Decou, R.; Serk, H.; Menard, D.; Pesquet, E. Analysis of lignin composition and distribution using fluorescence laser confocal microspectroscopy. *Methods Mol. Biol.* **2017**, *1544*, 233–247. [CrossRef]
32. Wei, P.L.; Ma, J.F.; Jiang, Z.H.; Liu, R.; An, X.; Fei, B.H. Chemical constituent distribution within multilayered cell walls of moso bamboo fiber tested by confocal Raman microscopy. *Wood Fiber Sci.* **2017**, *49*, 12–21.
33. Agarwal, U.P. Raman imaging to investigate ultrastructure and composition of plant cell walls: Distribution of lignin and cellulose in black spruce wood (*Picea mariana*). *Planta* **2006**, *224*, 1141–1153. [CrossRef]
34. Ma, J.F.; Ji, Z.; Zhou, X.; Zhang, Z.H.; Xu, F. Transmission electron microscopy, fluorescence microscopy, and confocal Raman microscopic analysis of ultrastructural and compositional heterogeneity of *Cornus alba* L. wood cell wall. *Microsc. Microanal.* **2013**, *19*, 243–253. [CrossRef]
35. Parameswaran, N.; Liese, W. Structure of septate fibers in bamboo. *Holzforschung* **1977**, *31*, 55–57.
36. Humphreys, J.M.; Chapple, C. Rewriting the lignin roadmap. *Curr. Opin. Plant Biol.* **2002**, *5*, 224–229. [CrossRef]
37. Boerjan, W.; Ralph, J.; Baucher, M. Lignin biosynthesis. *Annu. Rev. Plant Biol.* **2003**, *54*, 519–546. [CrossRef] [PubMed]
38. Jäger, A.; Bader, T.; Hofstetter, K.; Eberhardsteiner, J. The relation between indentation modulus, microfibril angle, and elastic properties of wood cell walls. *Compos. Part A-Appl. Sci.* **2011**, *42*, 677–685. [CrossRef]
39. Yu, Y.; Fei, B.; Zhang, B.; Yu, X. Cell-wall mechanical properties of bamboo investigated by in-situ imaging nanoindentation. *Wood Fiber Sci.* **2007**, *39*, 527–535.
40. Gindl, W.; Gupta, H.S.; Grünwald, C. Lignification of spruce tracheid secondary cell walls related to longitudinal hardness and modulus of elasticity using nano-indentation. *Can. J. Bot.* **2002**, *80*, 1029–1033. [CrossRef]
41. Huang, Y.; Fei, B. Comparison of the mechanical characteristics of fibers and cell walls from moso bamboo and wood. *BioResources* **2017**, *12*, 8230–8239. [CrossRef]
42. Srivaro, S.; Rattanarat, J.; Noothong, P. Comparison of the anatomical characteristics and physical and mechanical properties of oil palm and bamboo trunks. *J. Wood Sci.* **2018**, *64*, 186–192. [CrossRef]
43. Park, S.; Jang, J.; Wistara, I.; Hidayat, W.; Lee, M.; Febrianto, F. Anatomical and physical properties of Indonesian bamboos carbonized at different temperatures. *J. Korean Wood Sci. Technol.* **2018**, *46*, 656–669.
44. Gołofit, T.; Zielenkiewicz, T.; Gawron, J. FTIR examination of preservative retention in beech wood (*Fagus sylvatica* L.). *Eur. J. Wood Wood Prod.* **2012**, *70*, 907–909. [CrossRef]
45. Roman, K.; Roman, M.; Szadkowska, D.; Szadkowski, J.; Grzegorzewska, E. Evaluation of physical and chemical parameters according to energetic eillow (*Salix viminalis* L.) cultivation. *Energies* **2021**, *14*, 2968. [CrossRef]
46. Trzciński, G.; Tymendorf, Ł.; Kozakiewicz, P. Parameters of trucks and loads in the transport of scots pine wood biomass depending on the season and moisture content of the load. *Forests* **2021**, *12*, 223. [CrossRef]
47. Wu, Y.; Wang, S.; Zhou, D.; Xing, C.; Zhang, Y. Use of nanoindentation and silviscan to determine the mechanical properties of 10 hardwood species. *Wood Fiber Sci.* **2009**, *41*, 64–73.
48. Tze, W.T.Y.; Wang, S.; Rials, T.G.; Pharr, G.M.; Kelley, S.S. Nanoindentation of wood cell walls: Continuous stiffness and hardness measurements. *Compos. Part A-Appl. Sci.* **2007**, *38*, 945–953. [CrossRef]
49. Li, X.; Wang, S.; Du, G.; Wu, Z.; Meng, Y. Variation in physical and mechanical properties of hemp stalk fibers along height of stem. *Ind. Crops Prod.* **2013**, *42*, 344–348. [CrossRef]
50. Keryvin, V.; Lan, M.; Bourmaud, A.; Parenteau, T.; Charleux, L.; Baley, C. Analysis of flax fibres viscoelastic behaviour at micro and nano scales. *Compos. Part A-Appl. Sci.* **2015**, *68*, 219–225. [CrossRef]
51. Yang, X.; Tian, G.; Shang, L.; Lv, H.; Yang, S. Variation in the cell wall mechanical properties of *Dendrocalamus farinosus* bamboo by nanoindentation. *BioResources* **2014**, *9*, 2289–2298. [CrossRef]
52. Morandim-Giannetti, A.A.; Agnelli, J.A.M.; Lanças, B.Z.; Magnabosco, R.; Casarin, S.A.; Bettini, S.H. Lignin as additive in polypropylene/coir composites: Thermal, mechanical and morphological properties. *Carbohyd. Polym.* **2012**, *87*, 2563–2568. [CrossRef]
53. Özparpucu, M.; Rüggeberg, M.; Gierlinger, N.; Cesarino, I.; Vanholme, R.; Boerjan, W.; Burgert, I. Unravelling the impact of lignin on cell wall mechanics-a comprehensive study on young poplar trees downregulated for cinnamyl alcohol dehydrogenase (cad). *Plant J.* **2017**, *91*, 480–490. [CrossRef] [PubMed]

Article

Tailoring Poly(lactic acid) (PLA) Properties: Effect of the Impact Modifiers EE-g-GMA and POE-g-GMA

Edson Antonio dos Santos Filho *, Carlos Bruno Barreto Luna, Danilo Diniz Siqueira, Eduardo da Silva Barbosa Ferreira and Edcleide Maria Araújo

Department of Materials Engineering, Federal University of Campina Grande, Campina Grande 58429-900, Brazil; brunobarretodemaufcg@hotmail.com (C.B.B.L.); danilodinizsiqueira@gmail.com (D.D.S.); eduardosbf95@gmail.com (E.d.S.B.F.); edcleidemaraujo@gmail.com (E.M.A.)
* Correspondence: edson.a.santos.f@gmail.com

Abstract: Poly(ethylene-octene) grafted with glycidyl methacrylate (POE-g-GMA) and ethylene elastomeric grafted with glycidyl methacrylate (EE-g-GMA) were used as impact modifiers, aiming for tailoring poly(lactic acid) (PLA) properties. POE-g-GMA and EE-g-GMA was used in a proportion of 5; 7.5 and 10%, considering a good balance of properties for PLA. The PLA/POE-g-GMA and PLA/EE-g-GMA blends were processed in a twin-screw extruder and injection molded. The FTIR spectra indicated interactions between the PLA and the modifiers. The 10% addition of EE-g-GMA and POE-g-GMA promoted significant increases in impact strength, with gains of 108% and 140%, respectively. These acted as heterogeneous nucleating agents in the PLA matrix, generating a higher crystallinity degree for the blends. This impacted to keep the thermal deflection temperature (HDT) and Shore D hardness at the same level as PLA. By thermogravimetry (TG), the blends showed increased thermal stability, suggesting a stabilizing effect of the modifiers POE-g-GMA and EE-g-GMA on the PLA matrix. Scanning electron microscopy (SEM) showed dispersed POE-g-GMA and EE-g-GMA particles, as well as the presence of ligand reinforcing the systems interaction. The PLA properties can be tailored and improved by adding small concentrations of POE-g-GMA and EE-g-GMA. In light of this, new environmentally friendly and semi-biodegradable materials can be manufactured for application in the packaging industry.

Keywords: poly(lactic acid); impact modifier; polymer blends; properties

Citation: dos Santos Filho, E.A.; Luna, C.B.B.; Siqueira, D.D.; Ferreira, E.d.S.B.; Araújo, E.M. Tailoring Poly(lactic acid) (PLA) Properties: Effect of the Impact Modifiers EE-g-GMA and POE-g-GMA. *Polymers* **2022**, *14*, 136. https://doi.org/10.3390/polym14010136

Academic Editors: Jingpeng Li, Yun Lu and Huiqing Wang

Received: 28 October 2021
Accepted: 13 November 2021
Published: 30 December 2021

Publisher's Note: MDPI stays neutral with regard to jurisdictional claims in published maps and institutional affiliations.

Copyright: © 2021 by the authors. Licensee MDPI, Basel, Switzerland. This article is an open access article distributed under the terms and conditions of the Creative Commons Attribution (CC BY) license (https:// creativecommons.org/licenses/by/ 4.0/).

1. Introduction

With new technologies and new product development, there is more and more concern about the environmental impacts that these materials can cause. In this context, polymeric commodity materials have drawn attention because they are derived from petroleum, since they have high resistance to degradation. As a consequence, they can contribute significantly to the accumulation of waste in natural ecosystems, and thus increase pollution [1–3].

Research has advanced toward the development of new eco-friendly materials using "green" technology, targeting materials that favor a closed life cycle, such as biopolymers and biodegradable polymers. Biopolymers are produced from raw materials derived from renewable sources, such as: cellulose, sugar cane, corn and others [4–6]. On the other hand, biodegradable polymers are those that undergo degradation from the action of microorganisms in environments that are considered bioactive [7,8].

Poly(lactic acid)—PLA is a biodegradable biopolymer derived from natural resources such as corn, sugar cane and rice. It is an aliphatic polyester thermoplastic and is currently widely used in the processing industry. Regarding the production method, it can be synthesized either by direct condensation polymerization of the lactic acid monomer, or by ring opening of the lactide [9–12]. PLA has properties similar to crystal polystyrene

(PS) and polyethylene terephthalate (PET). In view of this, PLA has aroused interest in the production industry and the scientific community [13–16]. However, PLA is a very rigid and brittle polymer, which makes it impractical for applications requiring high impact strength. To overcome this problem, the properties of PLA can be tailored by adding impact modifiers.

There are several studies in the literature [17–19] on the production of polymer blends by adding natural rubber, polyethylene, ethylene vinyl acetate (EVA) and others into the PLA matrix. Aiming for a higher degree of synergy of mechanical properties, in general, impact modifiers functionalized with glycidyl methacrylate (GMA) are added, aiming at a higher interaction with PLA [20,21]. This may improve the interfacial adhesion between the two phases, leading to higher impact strength and ductility in the system. Therefore, it is desirable to investigate the influence of various impact modifiers that are compatible with PLA, helping to expand the database of "green blends".

The present investigation aimed to evaluate the influence of elastomeric impact modifiers (POE-g-GMA and EE-g-GMA) addition to the PLA matrix, in order to tailor the thermal, mechanical, thermomechanical properties and the morphology.

2. Methodology

2.1. Materials

Poly(lactic acid) (PLA) manufactured by NatureWorks, supplied in pellet form by 3D LAB (Betim, Brazil), with a density of 1.24 g/cm^3 and a flow index (FI) = 6 g/10 min. Elastomeric ethylene graphted with glycidyl methacrylate (EE-g-GMA), supplied by Coace® Plastic (Xiamen, China), FI = 3–8 g/10 min (190 °C/2.16 kg), containing up to 0.8% of GMA. Poly(ethylene octene) grafted with glycidyl methacrylate (POE-g-GMA), supplied by Coace® Plastic (Xiamen, China), FI = 8–16 g/10 min (190 °C/2.16 kg), containing up to 0.8% of GMA. Styrene-(ethylene-butylene)-styrene copolymer (SEBS-MA), FI = 5 g/10 min (230 °C/5 kg), with 1.7% of maleic anhydride, marketed under code FG1901, containing 30% styrene, supplied by Kraton (Houston, Texas, USA). Ethylene methyl acrylate copolymer (EMA), with 24% of methyl acrylate and FI = 7 g/10 min (190 °C/2.16 kg), supplied by Arkema (Colombes Cedex, France), in granule form.

2.2. Materials Processing

Prior to the blends preparation, a study of the different impact modifiers influence on PLA was performed. All materials were dried in a vacuum oven at 60 °C for 24 h. Table 1 illustrates the mass proportions (%) of the compositions that were used in the development of the blends. These compositions were developed to analyze the degree of interaction between PLA and functionalized impact modifiers by means of torque rheometry and the Molau test.

Table 1. Obtained systems compositions.

Materials	PLA (%)	EE-g-GMA (%)	POE-g-GMA (%)	SEBS-MA (%)	EMA (%)
PLA	100	-	-	-	-
PLA/EE-g-GMA	70	30	-	-	-
PLA/POE-g-GMA	70	-	30	-	-
PLA/SEBS-MA	70	-	-	30	-
PLA/EMA	70	-	-	-	30

After preliminary tests, EE-g-GMA and POE-g-GMA were selected as impact modifiers for PLA, aiming to obtain the binary blends PLA/EE-g-GMA and PLA/POE-g-GMA, with the ratios of 95/5; 92.5/7.5 and 90/10%.

The PLA/EE-g-GMA and PLA/POE-g-GMA blends were dry mixed and later processed in a modular corotating twin-screw extruder, model ZSK (D = 18 mm and L/D = 40), from Coperion Werner & Pfleiderer, with temperature of 170 °C in zones 1 and 2, and 180 °C in the other zones, screw rotation speed of 250 rpm and controlled feed rate of 4 kg/h, with screw profile configured with distributive and dispersive mixing elements

(Figure 1). For comparison, the pure PLA was processed under the same conditions as the blends. After processing the systems by extrusion, the materials were granulated and dried in an oven without vacuum for 24 h at a temperature of 60 °C.

Figure 1. Extruder screw profile used with distributive and dispersive mixing elements [22].

The pure PLA and the blends obtained by extrusion were injection molded in an Arburg Allrounder 207C Golden Edition model to obtain specimens. The molding temperature was 180 °C in all zones and the mold temperature was 20 °C.

2.3. Characterizations

2.3.1. Torque Rheometry

The rheology curves were obtained in a Thermo Scientific Haake PolyLab QC mixer (Waltham, MA, USA), with roller-type rotors, at 180 °C and rotor speed of 60 rpm, under air atmosphere for 10 min.

2.3.2. Molau Test

The Molau test was performed by dissolving 1 g of PLA and the blends, in 50 mL of N-methyl-2-pyrrolidone (NMP), under magnetic stirring at 100 °C.

2.3.3. Fourier Transform Infrared Spectroscopy (FTIR)

Fourier transform infrared spectroscopy (FTIR) was performed on a BRUKER Vertex 70 Spectrometer (attenuated total reflectance—ATR) (Billerica, MA, USA), in the range 4000 to 400 cm^{-1}, with 32 scans and a resolution of 4 cm^{-1}.

2.3.4. Impact Strength Test

The Izod impact strength test was performed on notched specimens according to ASTM D256 [23] in a Ceast model Resil 5.5 J device (Turin, Italy), operating with a 2.75 J hammer, at room temperature. The results were analyzed with an average of 10 specimens.

2.3.5. Shore D Hardness Test

The penetration resistance measurement was carried out according to ASTM D2240 [24], in Shore-Durometer Hardness Type "D" (São Paulo, Brazil) equipment with a 50 N load controlled by springs calibrated using standard indentors for the durometer. The indenter was pressed into the sample for 15 s at five random points on the sample.

2.3.6. Heat Deflection Temperature (HDT)

The heat deflection temperature (HDT) was obtained according to ASTM D648, in a Ceast equipment, model HDT 6 VICAT (Turin, Italy) with a load of 1.82 MPa and heating rate of 120 °C/h (method A). The temperature was determined after the sample was deflected 0.25 mm. The results were analyzed with an average of 3 specimens.

2.3.7. Thermogravimetry (TG)

Thermogravimetry (TG) was obtained in a Shimadzu DTG 60H equipment (Kyoto, Japan), using about 5 mg of sample, heating rate of 10 °C/min and gas flow rate of 100 mL/min, from 30 to 500 °C under nitrogen atmosphere.

2.3.8. Differential Scanning Calorimetry (DSC)

Differential scanning calorimetry (DSC) analysis was performed in a TA Instruments DSC-Q20 equipment (New Castle, DE, USA). The scanning was performed from 30 to 200 °C under heating rate of 10 °C/min, gas flow rate of 50 mL/min and nitrogen atmosphere with a mass of approximately 6 mg. The blends degree of crystallinity was calculated based on the curves obtained in the DSC analyses, according to Equation (1) [25]:

$$\%Xc = \frac{\Delta Hm - \Delta Hcc}{\Delta H_{100\%PLA} \times W_{PLA}} \quad (1)$$

where: ΔH_m = melting enthalpy; ΔH_{cc} = cold crystallization enthalpy; $\Delta H_{100\%,PLA}$ = melting enthalpy for 100% crystalline PLA (93,7 J/g) [26]; W_{PLA} = mass fraction of PLA; X_c = degree of crystallinity.

2.3.9. Scanning Electron Microscopy (SEM)

Scanning electron microscopy (SEM) analyses were performed on the specimen's fracture surface submitted to the impact test. A scanning electron microscope—VEGAN 3 TESCAN (Brun, Tchéquia)—was used at a voltage of 30 kV under high vacuum. The samples fracture surfaces were coated with gold (sputtering—Shimadzu metallizer—IC 50, using a current of 4 mA for a period of 2 min).

3. Results and Discussion

3.1. Torque Rheometry

Figure 2 illustrates the torque vs. time curves, as well as the magnification with the stabilized torque for the pure PLA, the impact modifiers and the binary blends, with the fixed ratios of 70/30%.

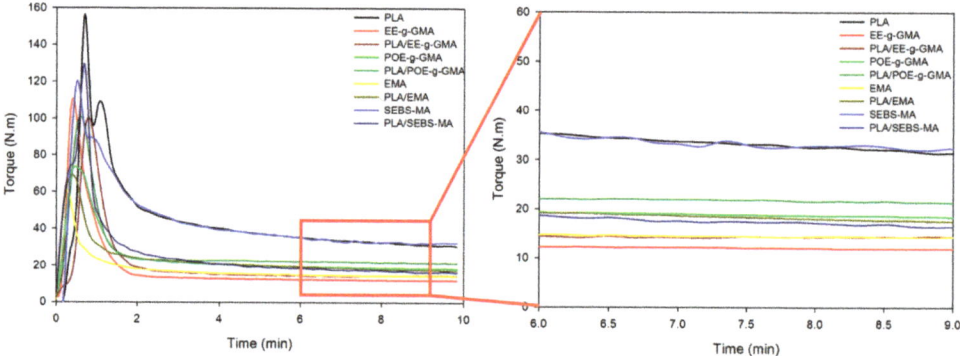

Figure 2. Torque versus time curves of pure PLA, impact modifiers and blends (70/30% by mass).

In Figure 2, it was observed in the first minutes of mixing an initial peak referring to the loading of the material—i.e., when the solid material enters the mixing chamber —that it promotes a resistance to the rotor's rotation and, as a consequence, there was an increase in torque. The torque then started to reduce, due to the materials plasticization. After approximately 4 min, the curves became constant with small oscillations. This behavior is related to the viscosity stability under the conditions used in this process, i.e., speed of 60 rpm and temperature of 180 °C. It was noted that after 6 min, PLA showed an average torque of 35 N·m, a value similar to SEBS-MA. The impact modifiers EE-g-GMA, POE-g-GMA and EMA, showed lower average torque, 12 N·m, 19 N·m and 15 N·m, respectively. As for the blends, it was observed that there was an increase in the average torque compared to the pure modifiers, with the exception of SEBS-MA, which had a significant reduction. This increase in torque is an indication that there was some interaction between the components. Brito et al. (2016) showed that PLA reacts with polymers functionalized with GMA, generating a reaction between the carboxyl or hydroxyl groups present in PLA and the epoxy group present in GMA.

3.2. Molau Test

The Molau test is a fractional dissolution experiment, which is widely used as a qualitative test to indicate whether there has been a reaction between the polymer and the functionalized impact modifiers [27,28]. Therefore, PLA, impact modifiers and blends were dissolved at a ratio of 70/30%.

From this test, it can be evaluated which impact modifiers reacted with PLA, considering that, if there is an interaction between both components, upon dissolution of PLA in the NMP solvent, the modifier is "dragged", thus changing the solvent color. Figure 3 shows the dissolution of the PLA, impact modifiers and blends in 50 mL volumetric flasks containing the NMP.

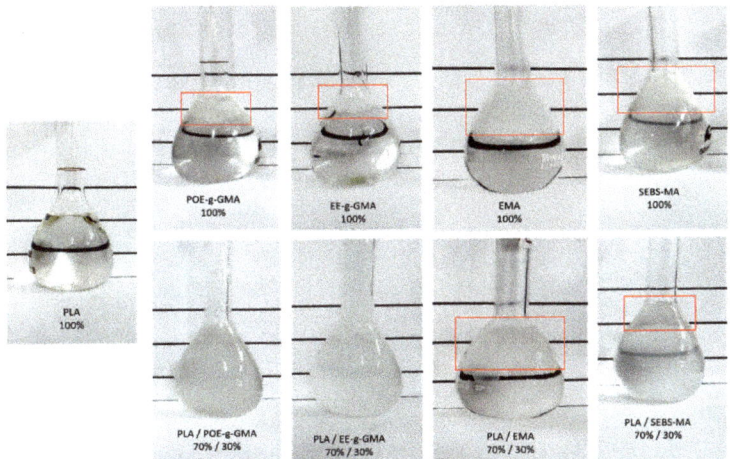

Figure 3. Dissolution of pure PLA, impact modifiers and blends (70/30%) in NMP solvent.

Pure PLA was found to dissolve completely in the solvent. The impact modifiers, on the other hand, do not dissolve, as it is possible to clearly observe two distinct phases, indicated in red. It was then possible to notice that the blends containing the modifiers POE-g-GMA and EE-g-GMA showed a single phase with a milky appearance, suggesting indications that there was an interaction between the materials. It can also be observed that the solutions of PLA/EMA and PLA/SEBS-MA blends presented an insoluble part in the solution, which is an indication that there probably were no interactions between the phases [28,29].

3.3. Fourier Transform Infrared Spectroscopy (FTIR)

The literature [20,30,31] showed that GMA has an interaction with PLA, confirming the trend verified in torque rheometry and Molau test. Regarding this, the modifiers EE-g-GMA and POE-g-GMA are probably good candidates for tailoring the PLA properties.

Figure 4 shows the infrared spectra of pure PLA, the modifiers EE-g-GMA and POE-g-GMA, and the blends PLA/EE-g-GMA and PLA/POE-g-GMA, in the 70/30% ratios.

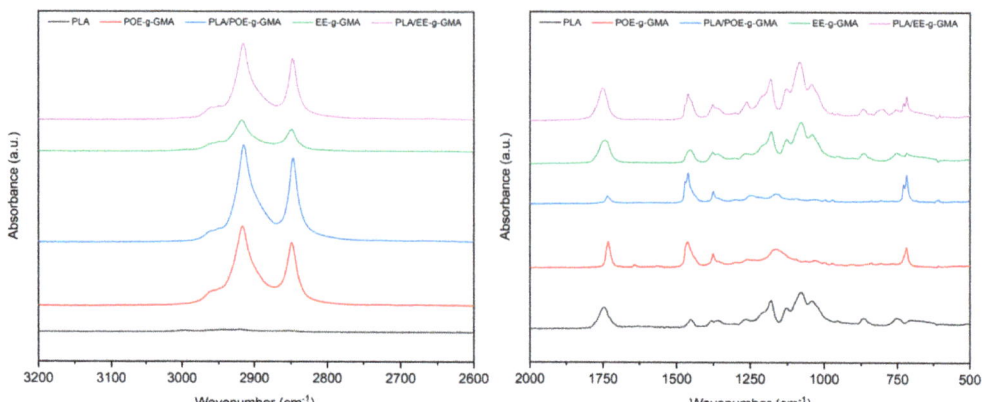

Figure 4. Infrared spectra of pure PLA, EE-g-GMA, POE-g-GMA and the blends (70/30%).

The main bands that can be visualized in the spectra are: 750 cm^{-1} for C=O stretching; 865 cm^{-1} for C-COO stretching; 1050 cm^{-1} for C-CH$_3$ stretching; 1080 cm^{-1} for C-CH$_3$ stretching and -CO stretching (ester); 1185 cm^{-1} for asymmetric CH$_3$ stretching and -CO stretching (ester); 1260 cm^{-1} for CH stretching and COC stretching; 1360 cm^{-1} for symmetric CH$_3$ stretching and CH bending; 1452 cm^{-1} for asymmetric CH$_3$ stretching; 1740 cm^{-1} for the C=O stretching [32–35]. It was also observed that there were characteristic bands present in EE-g-GMA, POE-g-GMA and the blends at 2860 cm^{-1} referring to the -CH stretching; 2915 cm^{-1} referring to the CH$_3$ symmetric stretching and the -CH stretching; 2995 cm^{-1} referring to the CH$_3$ asymmetric stretching. This is an indication that there was a chemical reaction between these impact modifiers and PLA [28,36].

Figure 5a,b show the FTIR spectra of the modifiers EE-g-GMA and POE-g-GMA, compared to pure PLA and their respective blends in the 70/30% ratio, in the spectrum between 500 cm^{-1} and 1500 cm^{-1}. It can be seen that some bands are similar for all three spectra. However, it is worth considering a low intensity band at 920 cm^{-1}, which is typical for GMA (grafted onto both modifiers). This band intensity is related to the GMA grafting degree in the modifiers, which, according to the supplier, is above or equal to 0.8%. It can be seen that there is a decrease in the intensity of this band, thus indicating that a reaction likely occurred during processing and an opening of the epoxy ring, justifying this reduction [33,37].

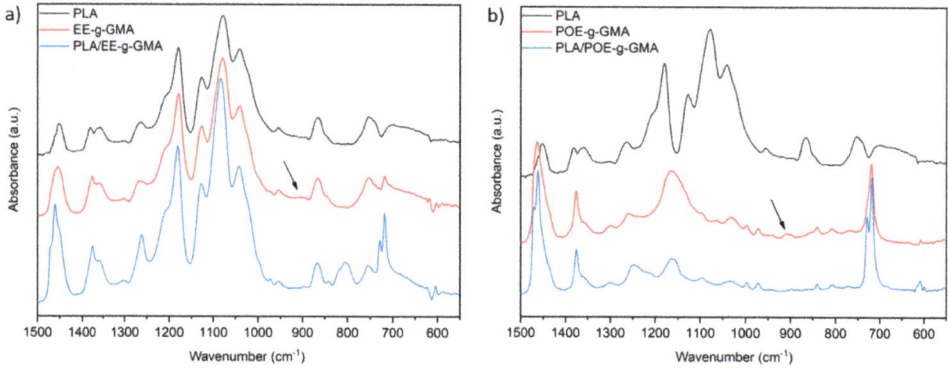

Figure 5. Infrared spectra of pure PLA, (**a**) EE-g-GMA and of the blend PLA/EE-g-GMA and (**b**) POE-g-GMA and of the blend PLA/POE-g-GMA.

3.4. Impact Strength Test

Figure 6 shows the impact strength of PLA and the blends containing 5; 7.5 and 10% of the impact modifiers EE-g-GMA and POE-g-GMA. This test is important to evaluate the energy dissipation in the material and whether they exhibit a brittle or ductile character.

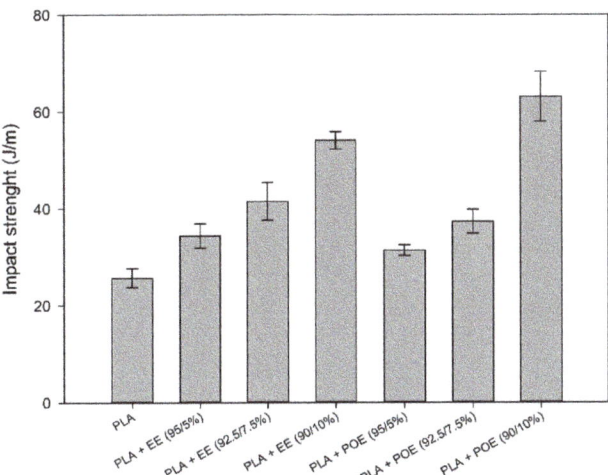

Figure 6. Impact strength of pure PLA and blends containing 5; 7.5 and 10% EE-g-GMA and POE-g-GMA.

Literature [38–40] showed that the impact strength of PLA can be increased by adding other materials, whether they are polymeric or not. It was observed that pure PLA showed an impact strength of approximately 27 J/m, a value close to the literature [41,42]. With the impact modifiers addition there was an increase in this property. As the content of EE-g-GMA and POE-g-GMA increased, the impact strength increased continuously. Thus, there are indications that there was an improvement in the PLA toughness with the addition of these modifiers, given that the blends show gains of 30, 60 and 108% with the addition of 5; 7.5 and 10% of EE-g-GMA, respectively, while 20, 42 and 140% increased with the addition of 5; 7.5 and 10% of POE-g-GMA, respectively. Apparently, the addition of 10% POE-g-GMA tended to maximize performance under impact, suggesting that there is a critical concentration to Improve the synergy of behavior under impact.

In view of the results presented, EE-g-GMA and POE-g-GMA possibly act by dissipating energy under impact and consequently delaying the propagation of cracks, which are quite common in brittle polymers such as PLA. This behavior was probably due to the impact modifiers elastomeric character used, contributing to an increase of the impact strength of PLA. In addition, it can take into consideration that there is probably an interaction between the GMA functional group and the PLA, generating good interactions between both phases, as verified in the SEM later on. These results are of great importance for green polymer technology, since the addition of a small percentage of the impact modifier already results in an increase in the impact strength of PLA.

3.5. Shore D Hardness Test

Figure 7 shows the Shore D hardness of the pure PLA and the blends, with 5; 7.5 and 10% contents of the impact modifiers EE-g-GMA and POE-g-GMA. In this test, the penetration resistance of the material is evaluated.

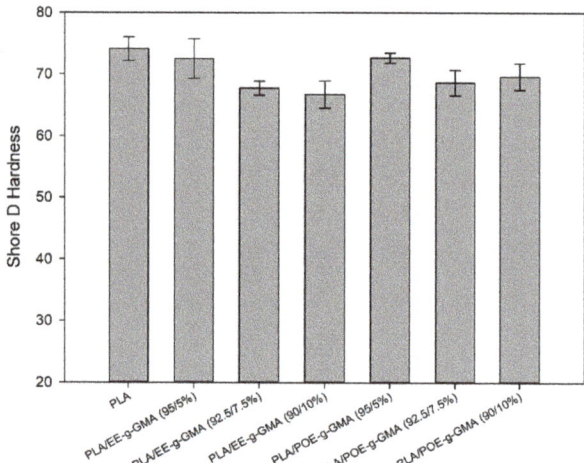

Figure 7. Shore D hardness of pure PLA and the blends containing 5; 7.5 and 10% 5; 7.5 and 10% EE-g-GMA and POE-g-GMA.

Pure PLA showed the highest value of hardness, with a value of 75 Shore D due to the high stiffness of this material [43–45]. The addition of 5% EE-g-GMA and POE-g-GMA subtly reduced the Shore D hardness compared to pure PLA. However, it was not a significant decrease, since they are within the experimental error. For higher concentrations of impact modifiers (7.5% and 10%), there was a more obvious reduction in Shore D hardness due to greater flexibility, as seen in impact strength.

It was also observed that, as the proportion of modifiers increased, there was also a proportional reduction of these values in Shore D hardness, in view of the increased flexibility and greater amount of material with elastomeric behavior. These results are important for the plastics processing industry, as it increased impact strength and did not significantly compromise penetration resistance. This suggests that new semi-biodegradable materials can be produced for practical applications, contributing to the expanded use of sustainable materials.

3.6. Differential Scanning Calorimetry (DSC)

The melting parameters, crystallization and the degree of crystallinity of the pure PLA and the blends are summarized in Table 2. The DSC curves obtained during the second heating and cooling can be seen in Figure 8a,b, respectively. In Figure 8a, it was noted that all compositions showed three main events during heating. The first event is a variation from baseline, which is related to the Tg of PLA, around 62.5 °C [46,47], a value above room temperature, confirming its brittle characteristic, observed in the impact test. The blends PLA/EE-g-GMA and PLA/POE-g-GMA also presented this Tg event, however, without significant variations.

Table 2. Data obtained in the DSC of the compositions.

	Tg (°C)	Tcc (°C)	ΔHcc	Tm (°C)	ΔHm	Tc (°C)	Xc (%)
PLA	63.3	120.6	21.75	154.3	22.61	-	0.92
PLA/EE-g-GMA (95/5%)	62.7	113.3	10.80	152.0	15.12	70.4	4.85
PLA/EE-g-GMA (92.5/7.5%)	63.2	113.9	13.77	152.2	18.25	70.3	5.17
PLA/EE-g-GMA (90/10%)	62.4	113.1	14.76	150.8	18.58	71.5	4.53
PLA/POE-g-GMA (95/5%)	61.8	118.1	20.53	152.0	22.10	-	1.76
PLA/POE-g-GMA (92.5/7.5%)	62.7	119.7	22.56	153.5	23.76	-	1.38
PLA/POE-g-GMA (90/10%)	62.5	120.3	20.06	152.7	22.53	-	2.93

Tg: Glass temperature transition; Tcc: Cold crystallization temperature; ΔHcc: Cold crystallization enthalpy; Tm: Melting temperature; ΔHm: Melting enthalpy; Tc: Crystallization temperature; Xc: Crystallization degree.

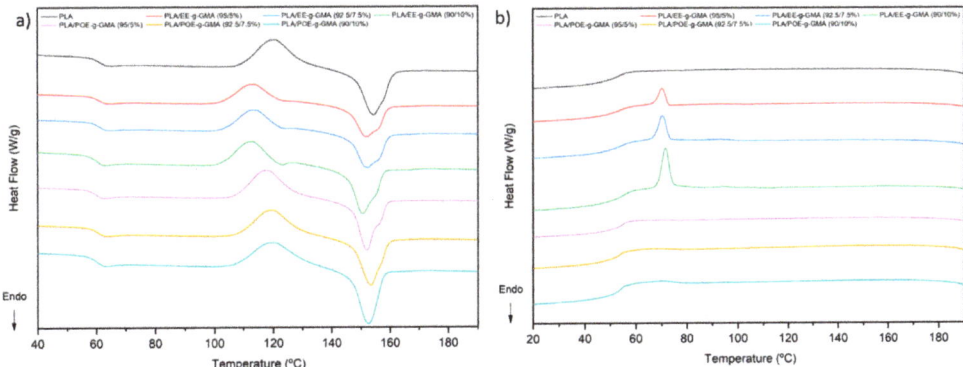

Figure 8. DSC curves of pure PLA and the blends containing 5; 7.5 and 10% EE-g-GMA and POE-g-GMA, (a) heating and (b) cooling.

Subsequently, between 110 °C and 120 °C, the samples presented an exothermic peak, referring to cold crystallization (Tcc), an intrinsic phenomenon of PLA. The cold crystallization process originates from the rearrangement of the amorphous region into a crystalline phase. Note that the compositions containing EE-g-GMA show a 7 °C reduction in Tcc, indicating that this impact modifier inhibits this process. Finally, there is an endothermic peak referring to the crystalline fusion, with peak temperature at approximately 152 °C for all compositions. However, it is noted that some compositions showed evidence of double peaks, especially those containing EE-g-GMA. Double peaks form in the presence of crystals of distinct sizes and shapes, melting at different temperatures [48,49].

It can also be observed that during cooling, the composition containing EE-g-GMA showed an exothermic peak related to crystallization at around 70 °C. This can probably be explained by the fact that ethylene crystallizes rapidly, unlike ethylene octene, in which the polymer chain is larger and more difficult to reorganize and crystallize.

From Table 2, it can be observed that there was an increase in the crystallinity of the blends PLA/EE-g-GMA and PLA/POE-g-GMA, compared to pure PLA. Such behavior suggests that the incorporation of low concentration of EE-g-GMA and POE-g-GMA promoted a nucleating effect, contributing to an increase of the crystallinity. Apparently, EE-g-GMA was more effective in enhancing the degree of crystallinity, since it increased by an average of 400% over PLA.

Although POE-g-GMA acted as a nucleating agent, the performance was not comparable to EE-g-GMA. This may be explained by the larger polymer chain size of POE-g-GMA, which hinders crystallization. There was also the presence of the exothermic peak during cooling, as ethylene crystallizes more easily; the higher crystallinity can be attributed to this factor.

3.7. Thermogravimetry (TG)

Figure 9 illustrates the TG curves of pure PLA and the blends containing 5; 7.5 and 10% of the impact modifiers EE-g-GMA and POE-g-GMA, and Table 3 presents the data obtained from this analysis.

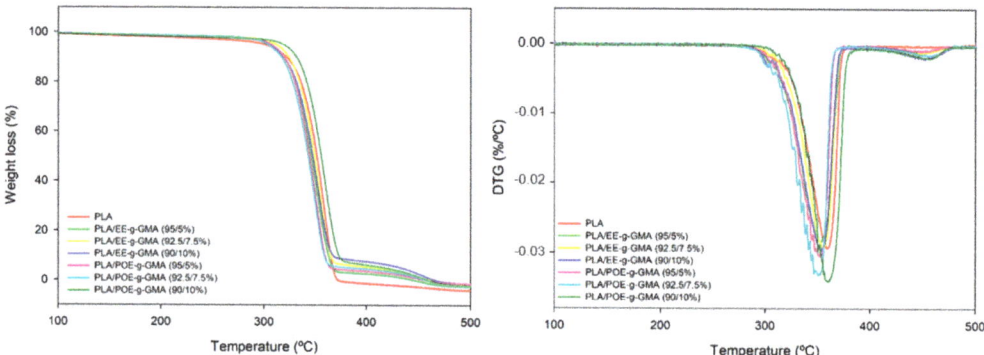

Figure 9. TG and DTG curves of pure PLA and the blends containing 5; 7.5 and 10% 5; 7.5 and 10% EE-g-GMA and POE-g-GMA.

Table 3. Degradation data obtained from TG of the compositions.

	$T_{5\%}$ (°C)	$T_{50\%}$ (°C)	T_{max} (°C)	Residue at 500 °C (%)
PLA	295.9	348.5	370.2	0.00
PLA/EE-g-GMA (95/5%)	306.4	345.5	435.9	2.24
PLA/EE-g-GMA (92.5/7.5%)	312.0	348.8	452.4	3.07
PLA/EE-g-GMA (95/10%)	310.0	346.8	459.8	3.27
PLA/POE-g-GMA (95/5%)	303.9	344.1	447.2	3.25
PLA/POE-g-GMA (92.5/7.5%)	306.7	342.2	451.0	2.56
PLA/POE-g-GMA (95/10%)	318.8	355.0	453.1	2.46

It is possible to verify that PLA presents a single decomposition step initiated at approximately 300 °C. Further, there was the process of the primary bonds rupture due to the thermal energy, causing the material degradation without presenting residues from 385 °C on. As for the blends, it was observed in the initial degradation temperature Tonset ($T_{5\%}$) that there was an increase from 10 to 20 °C. This increase indicates an improvement in the thermal stability of PLA, especially with a higher proportion of impact modifiers. This suggests that there is good interaction between the PLA and the impact modifiers, generating a stabilizing effect, and acting as additives while shifting the stability to a higher temperature. Furthermore, the increase in the degree of crystallinity as verified in the DSC, in general, contributes to improving the thermal stability [50].

According to Table 3, it can also be seen that the blends present a second degradation step around 420 °C, referring to the impact modifiers EE-g-GMA and POE-g-GMA degradation. Thus, it can be inferred that both are more thermally stable than pure PLA. It is also observed that the T_{max}, presents a considerable increase of about 70 °C.

Finally, it can be noted that the pure PLA showed no residue at 500 °C, while all the blends showed an average of 3% of the initial mass. This phenomenon indicates that there is an improvement in the thermal stability of PLA due to the chemical and physical interactions between the materials [51]. It is worthwhile to note that in the blends containing EE-g-GMA, there is an increase of the residue as the proportion introduced in the PLA increases. On the other hand, with the introduction of POE-g-GMA, there is a proportional decrease of the residue at 500 °C.

3.8. Heat Deflection Temperature (HDT)

Figure 10 shows the HDT behavior of pure PLA and the blends containing 5; 7.5 and 10% of the impact modifiers EE-g-GMA and POE-g-GMA.

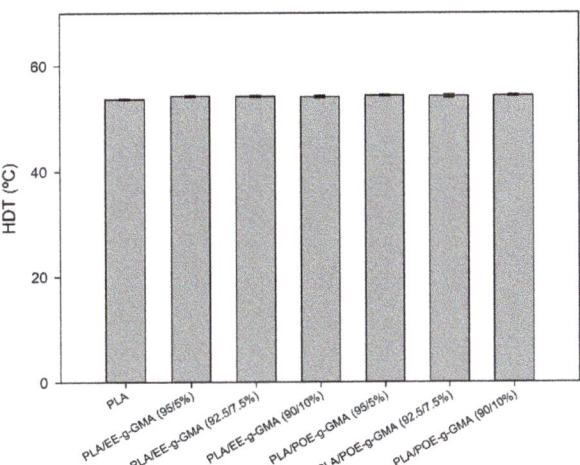

Figure 10. Heat deflection temperature of pure PLA and blends containing 5; 7.5 and 10% of EE-g-GMA and POE-g-GMA.

It can be observed that pure PLA presented a value of HDT on the order of 53.6 °C, a value close to that reported in the literature [52,53]. The PLA/EE-g-GMA and PLA/POE-g-GMA blends, regardless of the impact modifier content, showed no significant variations compared to pure PLA. Such behavior suggests that the addition of a GMA functionalized impact modifier in low concentration did not promote a deleterious effect on the thermomechanical strength in the PLA matrix.

From a technological point of view, the results are important for new eco-friendly materials production and development, with a good balance of properties, especially impact strength and HDT. Regarding the blends, EE-g-GMA and POE-g-GMA provided comparable values in HDT, considering that the values are within the experimental error.

3.9. Scanning Electron Microscopy (SEM)

Figure 11 shows the SEM images of the pure PLA and the blends samples containing 5; 7.5 and 10% of the impact modifiers EE-g-GMA and POE-g-GMA, at a magnification of 5000×, after being subjected to the impact strength test.

From the micrographs, the phase dispersion and interfacial adhesion can be analyzed. The pure PLA exhibits a morphology with smooth lines and without plastic deformation, characteristic of amorphous and exhibiting brittle fracture polymers [46,54,55], as can be seen in Figure 11a. In Figure 11b.1–b.3, are the blends containing 5; 7.5 and 10% EE-g-GMA, respectively, and Figure 11c.1–c.3 the blends containing 5; 7.5 and 10% POE-g-GMA, respectively.

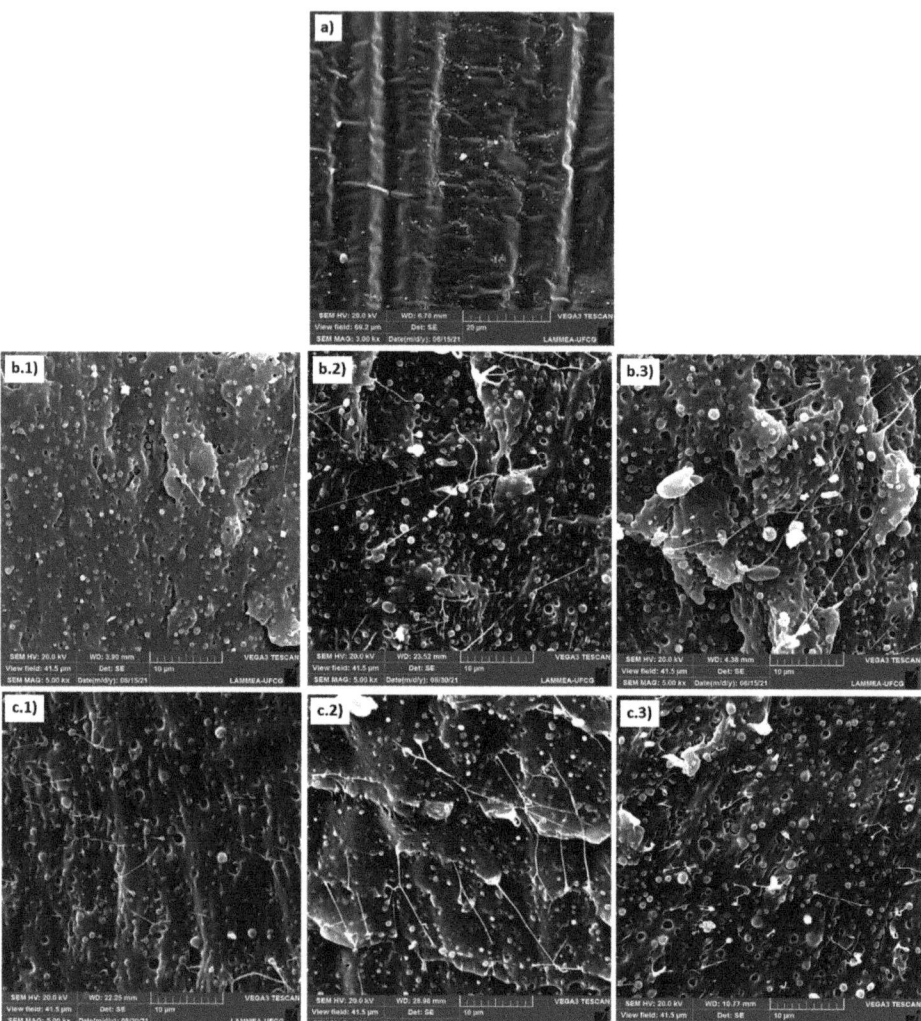

Figure 11. Micrographs obtained by SEM after impact strength test of: (**a**) pure PLA, (**b**) PLA/EE-g-GMA and (**c**) PLA/POE-g-GMA, in the proportions of: (1) 5%; (2) 7.5% and (3) 10%.

Figure 11 shows that the PLA/EE-g-GMA and PLA/POE-g-GMA blends showed phase separation, where EE-g-GMA and POE-g-GMA particles are dispersed in the PLA matrix. Clearly, there was the presence of well adhered EE-g-GMA and POE-g-GMA particles in the PLA matrix, justifying the increases in impact strength. However, it was also noted that the morphology showed some voids, indicating that some particles were pulled out during the impact test. At the same time, it can be seen that increasing the content of the modifiers EE-g-GMA and POE-g-GMA caused an increase in the average particle size, especially at 10%, which can be attributed to the coalescence phenomenon [56]. However, it can be observed that the dispersion of both modifiers, EE-g-GMA and POE-g-GMA, was uniform in the PLA matrix, contributing to increase the energy dissipation level. Furthermore, some fibrils were noted in the PLA/EE-g-GMA and PLA/POE-g-GMA blends, as well as whitish zones, characteristic of a plastic deformation, and corroborating with the impact strength results.

4. Conclusions

The mechanical, thermal, thermomechanical properties and morphology of semi-biodegradable blends of PLA with EE-g-GMA and POE-g-GMA were investigated. It was found that the properties of PLA can be tailored by adding small concentrations of EE-g-GMA and POE-g-GMA, generating promising eco-friendly materials. The blends PLA/EE-g-GMA and PLA/POE-g-GMA showed better impact properties and thermal stability, compared to pure PLA. At the same time, HDT was not affected, due to the nucleating effect of the impact modifiers EE-g-GMA and POE-g-GMA in the PLA matrix. As a consequence, the increase in crystallinity contributed to maintaining the thermomechanical strength (HDT), Shore D hardness, and shifting the thermal stability to higher temperature of the PLA/EE-g-GMA and PLA/POE-g-GMA blends. The morphology obtained by SEM suggested a good interaction between PLA and the EE-g-GMA and POE-g-GMA systems, due to the glycidyl methacrylate functional group. In general, new ecoproducts can be manufactured with the PLA/EE-g-GMA and PLA/POE-g-GMA blends, aiming at applications in the packaging industry. These blends are less polluting to the environment and non-toxic, contributing to sustainable development.

Author Contributions: All authors participated in the drafting the article or revising it critically content, approving the final version submitted. Conceptualization, E.A.d.S.F., C.B.B.L., D.D.S. and E.M.A.; methodology, C.B.B.L., D.D.S. and E.d.S.B.F.; formal analysis, E.d.S.B.F. and C.B.B.L.; investigation, E.A.d.S.F., E.d.S.B.F. and C.B.B.L.; resources, E.M.A.; writing—original draft preparation, E.A.d.S.F. and C.B.B.L.; writing—review and editing, E.A.d.S.F., C.B.B.L. and E.M.A.; visualization, E.d.S.B.F. and C.B.B.L.; supervision, E.M.A.; project administration, E.M.A.; funding acquisition, E.M.A. All authors have read and agreed to the published version of the manuscript.

Funding: This work was carried out with support from the CAPES (Coordination for the Improvement of Higher Education Personnel, Brasilia/DF, Brazil)—Funding code 001 and National Council for Scientific and Technological Development—CNPq—Process number: 312014/2020-1. Prof Edcleide Araújo is CNPq fellows.

Institutional Review Board Statement: Not applicable.

Informed Consent Statement: Not applicable.

Acknowledgments: The authors would like to thank UFCG for the laboratory infrastructure, MCTI/CNPq and Coace® Plastic for the donation of the impact modifiers.

Conflicts of Interest: There is no conflict of interest and all authors have agreed with this sub-mission and they are aware of the content.

References

1. Manfra, L.; Marengo, V.; Libralato, G.; Costantini, M.; De Falco, F.; Cocca, M. Biodegradable polymers: A real opportunity to solve marine plastic pollution? *J. Hazard. Mater.* **2021**, *416*, 125763. [CrossRef] [PubMed]
2. Silva, R.V.; de Brito, J. Plastic wastes. In *Waste and Supplementary Cementitious Materials in Concrete*; Siddique, R., Cachim, P., Eds.; Woodhead Publishing: Sawston, UK, 2018; pp. 199–227.
3. dos Santos Filho, E.A.; Siqueira, D.D.; Araújo, E.M.; Luna, C.B.B.; de Medeiros, E.P. The Impact of the Macaíba Components Addition on the Biodegradation Acceleration of Poly (ϵ-Caprolactone) (PCL). *J. Polym. Environ.* **2021**, *1*, 1–18. [CrossRef]
4. Morales, A.; Labidi, J.; Gullón, P.; Astray, G. Synthesis of advanced biobased green materials from renewable biopolymers. *Curr. Opin. Green Sustain. Chem.* **2021**, *29*, 100436. [CrossRef]
5. Pinto, L.; Bonifacio, M.A.; De Giglio, E.; Santovito, E.; Cometa, S.; Bevilacqua, A.; Baruzzi, F. Biopolymer hybrid materials: Development, characterization, and food packaging applications. *Food Packag. Shelf Life* **2021**, *28*, 100676. [CrossRef]
6. Delamarche, E.; Massardier, V.; Bayard, R.; dos Santos, E. A Review to Guide Eco-Design of Reactive Polymer-Based Materials. In *Reactive and Functional Polymers Volume Four: Surface, Interface, Biodegradability, Compostability and Recycling*; Gutiérrez, T.J., Ed.; Springer International Publishing: Cham, Germany, 2020; pp. 207–241.
7. Janczak, K.; Dąbrowska, G.B.; Raszkowska-Kaczor, A.; Kaczor, D.; Hrynkiewicz, K.; Richert, A. Biodegradation of the plastics PLA and PET in cultivated soil with the participation of microorganisms and plants. *Int. Biodeterior. Biodegrad.* **2020**, *155*, 105087. [CrossRef]
8. Richert, A.; Dąbrowska, G.B. Enzymatic degradation and biofilm formation during biodegradation of polylactide and polycaprolactone polymers in various environments. *Int. J. Biol. Macromol.* **2021**, *176*, 226–232. [CrossRef]

9. Jin, F.L.; Hu, R.R.; Park, S.J. Improvement of thermal behaviors of biodegradable poly(lactic acid) polymer: A review. *Compos. Part B Eng.* **2019**, *164*, 287–296. [CrossRef]
10. Nofar, M.; Sacligil, D.; Carreau, P.J.; Kamal, M.R.; Heuzey, M.C. Poly (lactic acid) blends: Processing, properties and applications. *Int. J. Biol. Macromol.* **2019**, *125*, 307–360. [CrossRef] [PubMed]
11. Vert, M.; Schwarch, G.; Coudane, J. Present and future of PLA polymers. *J. Macromol. Sci. Part A* **1995**, *32*, 787–796. [CrossRef]
12. Murariu, M.; Dubois, P. PLA composites: From production to properties. *Adv. Drug Deliv. Rev.* **2016**, *107*, 17–46. [CrossRef] [PubMed]
13. Cheroennet, N.; Pongpinyopap, S.; Leejarkpai, T.; Suwanmanee, U. A trade-off between carbon and water impacts in bio-based box production chains in Thailand: A case study of PS, PLAS, PLAS/starch, and PBS. *J. Clean. Prod.* **2017**, *167*, 987–1001. [CrossRef]
14. Leejarkpai, T.; Mungcharoen, T.; Suwanmanee, U. Comparative assessment of global warming impact and eco-efficiency of PS (polystyrene), PET (polyethylene terephthalate) and PLA (polylactic acid) boxes. *J. Clean. Prod.* **2016**, *125*, 95–107. [CrossRef]
15. Madival, S.; Auras, R.; Singh, S.P.; Narayan, R. Assessment of the environmental profile of PLA, PET and PS clamshell containers using LCA methodology. *J. Clean. Prod.* **2009**, *17*, 1183–1194. [CrossRef]
16. Wojtyła, S.; Klama, P.; Baran, T. Is 3D printing safe? Analysis of the thermal treatment of thermoplastics: ABS, PLA, PET, and nylon. *J. Occup. Environ. Hyg.* **2017**, *14*, D80–D85. [CrossRef] [PubMed]
17. Pongtanayut, K.; Thongpin, C.; Santawitee, O. The Effect of Rubber on Morphology, Thermal Properties and Mechanical Properties of PLA/NR and PLA/ENR Blends. *Energy Procedia* **2013**, *34*, 888–897. [CrossRef]
18. Lovinčić Milovanović, V.; Hajdinjak, I.; Lovriša, I.; Vrsaljko, D. The influence of the dispersed phase on the morphology, mechanical and thermal properties of PLA/PE-LD and PLA/PE-HD polymer blends and their nanocomposites with TiO_2 and $CaCO_3$. *Polym. Eng. Sci.* **2019**, *59*, 1395–1408. [CrossRef]
19. Aghjeh, M.R.; Nazari, M.; Khonakdar, H.A.; Jafari, S.H.; Wagenknecht, U.; Heinrich, G. In depth analysis of micro-mechanism of mechanical property alternations in PLA/EVA/clay nanocomposites: A combined theoretical and experimental approach. *Mater. Des.* **2015**, *88*, 1277–1289. [CrossRef]
20. Kumar, M.; Mohanty, S.; Nayak, S.K.; Rahail Parvaiz, M. Effect of glycidyl methacrylate (GMA) on the thermal, mechanical and morphological property of biodegradable PLA/PBAT blend and its nanocomposites. *Bioresour. Technol.* **2010**, *101*, 8406–8415. [CrossRef] [PubMed]
21. Wang, Y.-N.; Weng, Y.-X.; Wang, L. Characterization of interfacial compatibility of polylactic acid and bamboo flour (PLA/BF) in biocomposites. *Polym. Test.* **2014**, *36*, 119–125. [CrossRef]
22. Lima, J.C.C.; Araújo, J.P.; Agrawal, P.; Mélo, T.J.A. Efeito do teor do copolímero SEBS no comportamento reológico da blenda PLA/SEBS. *Rev. Eletrônica De Mater. E Process.* **2016**, *11*, 10–17.
23. ASTM D256-10(2018). *Standard Test Methods for Determining the Izod Pendulum Impact Resistance of Plastics*; ASTM International: West Conshohocken, PA, USA, 2018. [CrossRef]
24. ASTM D2240-15(2021). *Standard Test Method for Rubber Property—Durometer Hardness*; ASTM International: West Conshohocken, PA, USA, 2021. [CrossRef]
25. Xiao, H.; Lu, W.; Yeh, J.-T. Crystallization behavior of fully biodegradable poly(lactic acid)/poly(butylene adipate-co-terephthalate) blends. *J. Appl. Polym. Sci.* **2009**, *112*, 3754–3763. [CrossRef]
26. Auras, R.; Harte, B.; Selke, S. An Overview of Polylactides as Packaging Materials. *Macromol. Biosci.* **2004**, *4*, 835–864. [CrossRef] [PubMed]
27. Al-Malaika, S.; Kong, W. Reactive processing of polymers: Effect of in situ compatibilisation on characteristics of blends of polyethylene terephthalate and ethylene-propylene rubber. *Polymer* **2005**, *46*, 209–228. [CrossRef]
28. Feng, Y.; Hu, Y.; Yin, J.; Zhao, G.; Jiang, W. High impact poly(lactic acid)/poly(ethylene octene) blends prepared by reactive blending. *Polym. Eng. Sci.* **2013**, *53*, 389–396. [CrossRef]
29. Sookprasert, P.; Hinchiranan, N. Preparation of Natural Rubber-graft-Poly (lactic acid) Used as a Compatibilizer for Poly (lactic acid)/NR Blends. In Proceedings of the Macromolecular Symposia, Bordeaux, France, 5–10 July 2015; pp. 125–130.
30. Yang, W.; Dominici, F.; Fortunati, E.; Kenny, J.M.; Puglia, D. Melt free radical grafting of glycidyl methacrylate (GMA) onto fully biodegradable poly (lactic) acid films: Effect of cellulose nanocrystals and a masterbatch process. *RSC Adv.* **2015**, *5*, 32350–32357. [CrossRef]
31. Wu, N.; Zhang, H. Mechanical properties and phase morphology of super-tough PLA/PBAT/EMA-GMA multicomponent blends. *Mater. Lett.* **2017**, *192*, 17–20. [CrossRef]
32. Leroy, A.; Ribeiro, S.; Grossiord, C.; Alves, A.; Vestberg, R.H.; Salles, V.; Brunon, C.; Gritsch, K.; Grosgogeat, B.; Bayon, Y. FTIR microscopy contribution for comprehension of degradation mechanisms in PLA-based implantable medical devices. *J. Mater. Sci. Mater. Med.* **2017**, *28*, 87. [CrossRef]
33. Djellali, S.; Haddaoui, N.; Sadoun, T.; Bergeret, A.; Grohens, Y. Structural, morphological and mechanical characteristics of polyethylene, poly (lactic acid) and poly (ethylene-co-glycidyl methacrylate) blends. *Iran. Polym. J.* **2013**, *22*, 245–257. [CrossRef]
34. Jiang, X.; Luo, Y.; Tian, X.; Huang, D.; Reddy, N.; Yang, Y. Chemical Structure of Poly(Lactic Acid). In *Poly(Lactic Acid): Synthesis, Structures, Properties, Processing, and Applications*; John Wiley & Sons, Inc.: Hoboken, NJ, USA, 2010; pp. 67–82.

35. Herrera-Kao, W.A.; Loría-Bastarrachea, M.I.; Pérez-Padilla, Y.; Cauich-Rodríguez, J.V.; Vázquez-Torres, H.; Cervantes-Uc, J.M. Thermal degradation of poly(caprolactone), poly(lactic acid), and poly(hydroxybutyrate) studied by TGA/FTIR and other analytical techniques. *Polym. Bull.* **2018**, *75*, 4191–4205. [CrossRef]
36. Shao, H.; Qin, S.; Yu, J.; Guo, J. Influence of Grafting Degree on the Morphology and Mechanical Properties of PA6/POE-g-GMA Blends. *Polym. Plast. Technol. Eng.* **2012**, *51*, 28–34. [CrossRef]
37. Brito, G.F.; Agrawal, P.; Mélo, T.J.A. Mechanical and Morphological Properties of PLA/BioPE Blend Compatibilized with E-GMA and EMA-GMA Copolymers. *Macromol. Symp.* **2016**, *367*, 176–182. [CrossRef]
38. Boruvka, M.; Behalek, L.; Lenfeld, P.; Ngaowthong, C.; Pechociakova, M. Structure-related properties of bionanocomposites based on poly(lactic acid), cellulose nanocrystals and organic impact modifier. *Mater. Technol.* **2019**, *34*, 143–156. [CrossRef]
39. Jaratrotkamjorn, R.; Khaokong, C.; Tanrattanakul, V. Toughness enhancement of poly(lactic acid) by melt blending with natural rubber. *J. Appl. Polym. Sci.* **2012**, *124*, 5027–5036. [CrossRef]
40. Petchwattana, N.; Naknaen, P.; Narupai, B. Combination effects of reinforcing filler and impact modifier on the crystallization and toughening performances of poly (lactic acid). *Express Polym. Lett.* **2020**, *14*, 848–859. [CrossRef]
41. Agrawal, P.; Araújo, A.P.M.; Lima, J.C.C.; Cavalcanti, S.N.; Freitas, D.M.G.; Farias, G.M.G.; Ueki, M.M.; Mélo, T.J.A. Rheology, Mechanical Properties and Morphology of Poly(lactic acid)/Ethylene Vinyl Acetate Blends. *J. Polym. Environ.* **2019**, *27*, 1439–1448. [CrossRef]
42. Luna, C.B.B.; Siqueira, D.D.; Araújo, E.M.; Wellen, R.M.R. Annealing efficacy on PLA. Insights on mechanical, thermomechanical and crystallinity characters. *Momento* **2021**, *62*, 1–17. [CrossRef]
43. Teymoorzadeh, H.; Rodrigue, D. Biocomposites of Wood Flour and Polylactic Acid: Processing and Properties. *J. Biobased Mater. Bioenergy* **2015**, *9*, 252–257. [CrossRef]
44. Ferreira, E.d.S.B.; Luna, C.B.B.; Siqueira, D.D.; Araújo, E.M.; de França, D.C.; Wellen, R.M.R. Annealing Effect on Pla/Eva Blends Performance. *J. Polym. Environ.* **2021**, *1*, 1–14. [CrossRef]
45. Ma, P.; Hristova-Bogaerds, D.G.; Goossens, J.G.P.; Spoelstra, A.B.; Zhang, Y.; Lemstra, P.J. Toughening of poly(lactic acid) by ethylene-co-vinyl acetate copolymer with different vinyl acetate contents. *Eur. Polym. J.* **2012**, *48*, 146–154. [CrossRef]
46. da Silva, W.A.; Luna, C.B.B.; de Melo, J.B.d.C.A.; Araújo, E.M.; Filho, E.A.d.S.; Duarte, R.N.C. Feasibility of Manufacturing Disposable Cups using PLA/PCL Composites Reinforced with Wood Powder. *J. Polym. Environ.* **2021**, *29*, 2932–2951. [CrossRef]
47. Tejada-Oliveros, R.; Balart, R.; Ivorra-Martinez, J.; Gomez-Caturla, J.; Montanes, N.; Quiles-Carrillo, L. Improvement of Impact Strength of Polylactide Blends with a Thermoplastic Elastomer Compatibilized with Biobased Maleinized Linseed Oil for Applications in Rigid Packaging. *Molecules* **2021**, *26*, 240. [CrossRef] [PubMed]
48. Wellen, R.M.R.; Canedo, E.L. Complex cold crystallisation peaks in PET/PS blends. *Polym. Test.* **2015**, *41*, 26–32. [CrossRef]
49. Wellen, R.M.R.; Canedo, E.L.; Rabello, M.S. Melting and crystallization of poly(3-hydroxybutyrate)/carbon black compounds. Effect of heating and cooling cycles on phase transition. *J. Mater. Res.* **2015**, *30*, 3211–3226. [CrossRef]
50. Luna, C.B.B.; Ferreira, E.d.S.B.; da Silva, L.J.M.D.; da Silva, W.A.; Araújo, E.M. Blends with technological potential of copolymer polypropylene with polypropylene from post-consumer industrial containers. *Mater. Res. Express* **2019**, *6*, 125319. [CrossRef]
51. Zhou, Y.; Wang, J.; Cai, S.-Y.; Wang, Z.-G.; Zhang, N.-W.; Ren, J. Effect of POE-g-GMA on mechanical, rheological and thermal properties of poly(lactic acid)/poly(propylene carbonate) blends. *Polym. Bull.* **2018**, *75*, 5437–5454. [CrossRef]
52. Bubeck, R.A.; Merrington, A.; Dumitrascu, A.; Smith, P.B. Thermal analyses of poly(lactic acid) PLA and micro-ground paper blends. *J. Therm. Anal. Calorim.* **2018**, *131*, 309–316. [CrossRef]
53. Ghasemi, S.; Behrooz, R.; Ghasemi, I.; Yassar, R.S.; Long, F. Development of nanocellulose-reinforced PLA nanocomposite by using maleated PLA (PLA-g-MA). *J. Thermoplast. Compos. Mater.* **2017**, *31*, 1090–1101. [CrossRef]
54. Wang, X.; Mi, J.; Wang, J.; Zhou, H.; Wang, X. Multiple actions of poly(ethylene octene) grafted with glycidyl methacrylate on the performance of poly(lactic acid). *RSC Adv.* **2018**, *8*, 34418–34427. [CrossRef]
55. Gigante, V.; Canesi, I.; Cinelli, P.; Coltelli, M.B.; Lazzeri, A. Rubber Toughening of Polylactic Acid (PLA) with Poly(butylene adipate-co-terephthalate) (PBAT): Mechanical Properties, Fracture Mechanics and Analysis of Ductile-to-Brittle Behavior while Varying Temperature and Test Speed. *Eur. Polym. J.* **2019**, *115*, 125–137. [CrossRef]
56. Sundararaj, U.; Macosko, C.W. Drop Breakup and Coalescence in Polymer Blends: The Effects of Concentration and Compatibilization. *Macromolecules* **1995**, *28*, 2647–2657. [CrossRef]

Article

Sustainability Evaluation of Polyhydroxyalkanoate Production from Slaughterhouse Residues Utilising Emergy Accounting

Khurram Shahzad *, Mohammad Rehan , Muhammad Imtiaz Rashid , Nadeem Ali, Ahmed Saleh Summan and Iqbal Muhammad Ibrahim Ismail

Centre of Excellence in Environmental Studies, King Abdulaziz University, Jeddah 21589, Saudi Arabia; dr.mohammad_rehan@yahoo.co.uk (M.R.); irmaliks@gmail.com (M.I.R.); nbahadar@kau.edu.sa (N.A.); asumman@kau.edu.sa (A.S.S.); iqbal30@hotmail.com (I.M.I.I.)
* Correspondence: ksramzan@kau.edu.sa or shahzadkhu@gmail.com

Citation: Shahzad, K.; Rehan, M.; Rashid, M.I.; Ali, N.; Summan, A.S.; Ismail, I.M.I. Sustainability Evaluation of Polyhydroxyalkanoate Production from Slaughterhouse Residues Utilising Emergy Accounting. *Polymers* **2022**, *14*, 118. https://doi.org/10.3390/polym 14010118

Academic Editors: Jingpeng Li, Yun Lu and Huiqing Wang

Received: 20 October 2021
Accepted: 24 December 2021
Published: 29 December 2021

Publisher's Note: MDPI stays neutral with regard to jurisdictional claims in published maps and institutional affiliations.

Copyright: © 2021 by the authors. Licensee MDPI, Basel, Switzerland. This article is an open access article distributed under the terms and conditions of the Creative Commons Attribution (CC BY) license (https:// creativecommons.org/licenses/by/ 4.0/).

Abstract: High raw material prices and rivalry from the food industry have hampered the adoption of renewable resource-based goods. It has necessitated the investigation of cost-cutting strategies such as locating low-cost raw material supplies and adopting cleaner manufacturing processes. Exploiting waste streams as substitute resources for the operations is one low-cost option. The present study evaluates the environmental burden of biopolymer (polyhydroxyalkanoate) production from slaughtering residues. The sustainability of the PHA production process will be assessed utilising the Emergy Accounting methodology. The effect of changing energy resources from business as usual (i.e., electricity mix from the grid and heat provision utilising natural gas) to different renewable energy resources is also evaluated. The emergy intensity for PHA production (seJ/g) shows a minor improvement ranging from 1.5% to 2% by changing only the electricity provision resources. This impact reaches up to 17% when electricity and heat provision resources are replaced with biomass resources. Similarly, the emergy intensity for PHA production using electricity EU27 mix, coal, hydropower, wind power, and biomass is about 5% to 7% lower than the emergy intensity of polyethylene high density (PE-HD). In comparison, its value is up to 21% lower for electricity and heat provision from biomass.

Keywords: biopolymers; sustainability; emergy accounting; polyhydoxyalkanoates; slaughterhouse residues

1. Introduction

Plastics are used in almost every aspect of daily life. These materials have opened new horizons for research and development in academia and the industry [1]. They are used for a variety of services such as food packaging [2], low-density supplies [3], resilient chemical commodities [4], specialised niche structures in electronic fabrication, as well as their use in the medical field as implants and scaffolds [5,6]. However, the collection of huge amounts of waste has resulted from indiscriminate usage and disposal. Because they are non-biodegradable, managing and disposing of them is becoming increasingly complex [7]. In 2018, the global plastic output hit 359 million tons, with 67.5 percent of plastics going unrecycled and contaminating ecosystems [8]. Moreover, plastic degrades pure habitats and damages humans and other organisms, necessitating a quick micromanagement. The World Health Organisation (WHO) has asked scientists to develop alternative ways and materials to replace traditional plastics to limit environmental and human exposures [9]. Biodegradable polymers, particularly polyhydroxy-alkanoates (PHAs) [10], have been proposed as a promising alternative to non-biodegradable polyethylene plastic polymers [11] and can help to reduce plastic overgrowth [12]. Aside from their useful properties and biodegradability, PHAs are one of the most promising materials of the 21st century. They are used for various applications in the agricultural sector, municipal engineering, pharmacology, and biomedicine [13].

Bioplastic production currently accounts for only 0.6 percent (2.05 million tons) of the total plastic output, but it is expected to increase to 2.44 million tons by 2023 [14]. The valorisation of various types of trash as a carbon source for a more cost-effective microbial PHA synthesis and environmental friendliness is the subject of exceptional research efforts worldwide [15]. However, several factors must be considered for commercial PHA production to be sustainable and beneficial. The economic cost, stable feedstock properties and composition, simplicity of collection, transport, storage, and, most importantly, no competition with food and feed streams are among these aspects [16].

The natural capital storage is being depleted because of direct and indirect resource demand (e.g., oil, chemicals, minerals, and the treatment of human waste) [17]. In this light, material circularity is a significant issue in the quest for fossil-based raw materials and energy alternatives. As an analytical and quantitative technique, eco-efficiency is used to examine sustainable development [18].

The food, drug, cosmetics, and leather sectors, among others, generate a massive amount of organic waste, which can be a source of harmful diseases [19]. Almost half of the livestock is turned into byproducts, which retain a meaningful amount of energy (tallow: 3.98E 04 J/g average, meat, and bone meal: 1.85E 04 J/g average) [20]. From a bio-refinery perspective, a wide range of products and commodities may be obtained by adequately managing animal slaughtering residues and byproducts (i.e., biochemicals and pharmaceutical products from the blood and gelatinous material, clothes from skins, etc.) through various processes [21,22]. A conventional refinery produces a variety of petrol-based fuels and products. In contrast, a bio-refinery uses residual biomass from agriculture, forests, or industries as a resource to produce fuels, power, heat, and value-added chemicals. Implementing a bio-refinery plan should utilise growing and dynamic existing systems rather than building brand new complexes [23].

Waste management systems are incredibly complicated, necessitating extra caution when making assumptions and methodological decisions. Because garbage is on the verge of consumption and creation (via recycling), the outcomes of this study are very much reliant on the assessment approach chosen. Furthermore, emergy accounting (EMA) may be viewed as a new and more complete measure of eco-efficiency. It calculates the environmental system effort required to support a product or service delivered by a specific process [20]. This methodology has never been reported for the ecological evaluation of PHA production as per the authors' best knowledge. The effect of replacing energy provision resources from the business-as-usual scenario, i.e., the electricity mix from the grid and heat provision utilising natural gas, to various renewable energy ones will also be evaluated.

2. Methodology

As a methodological framework, EMA was employed in this study. Emergy is the energy (of a certain type) utilised in a system for transformations [24]. The limits in EMA are set at the level of the biosphere. The whole process demand and supply, including resource production, processing, and throwing away and the environmental inputs for the creation of storage and natural income flows (both renewable and non-renewable), which support the specified system directly or indirectly, are considered [25].

Ulgiati and Brown [26] indicate that the emergy input to a process is the result of all the resources and energy input traced back into all the processes or the chain of processes used in the process and expressed in the form of solar energy. Emergy is the measure of value for both energy and material resources at a standard basis, equivalent to the biosphere processes required to produce something [27]. The environmental services that are usually free of cost and outside the monetary economy and benefits of humans in the form of labour for processing the resource are embedded in the emergy value of the resource [28]. The emergy definition by Odum [29] is: "Emergy is the available energy (exergy) of one kind that is used up in transformations directly and indirectly to make a product or service".

The value is a unit of solar equivalent energy articulated in solar emjoules, abbreviated as (seJ). The solar emjoule (sej) is the emergy unit, which refers to the consumption of available energy in the transformations to deliver a product or service or functioning of a process. Solar energy is undoubtedly the most abundant form of energy accessible for Earth's activities. Thus, it is appropriate to utilise it as the reference kind of energy [30]. Resource production considers both evolutionary "trial and error" patterns and the diverse excellence of input flows, including solar, gravitational, and geothermal, each one of which is measured regarding its equivalency to solar radiation flow. Resource generation encapsulates the accessible energy flows invested inside the biosphere processes. The gradual development and the diverse qualities of input flows contributed by solar, geothermal, and gravitational, each computed concerning its equivalence to the solar radiation flow. Instead of merely "joule," the word "em-joule" is noticeably more related to biosphere dynamics than the energy content indicated by a simple joule. As a result, the total emergy (U) operating a process includes the overall "ecological production cost" of goods or services, calculated by Equation (1) by adding all inputs going into the process [31].

$$U = \sum_i f_i \times tr_i \qquad i = 1, 2, \ldots, n \tag{1}$$

where U stands for total energy, f_i for the i-th input of energy or matter, UEV for the i-th inflow's Unit Emergy Value, and n for the number of supporting inflows. The emergy investment per unit output may also be computed by using U and the process yield (Y), i.e., Unit Emergy Value, UEV, typically represented as sej/J or sej/g, as shown in Equation (2).

$$UEV = U/Y \tag{2}$$

where Y is termed process yield, which is the output as energy (joules), mass (grams), or other suitable service units. When Y is expressed in joules, as is the case for all the energy flows, the UEV is referred to as "transformity" described as sej/J. The solar radiation's transformity is set to 1 sej/J. All emergy values are calculated using a Global Emergy Baseline (GEB) as a reference, which is the total emergy accessible yearly to all biosphere activities. All UEVs in this work are connected to Brown et al., (2016)'s GEB [31], 12.0 E+24 sej/yr. For emergy calculations, renewable inputs (solar radiations, wind, rain potential, geothermal) and non-renewable inputs (soil erosion) for a process, system, or region are calculated by utilising the area within the boundary conditions of the system under study. The boundary conditions of the system can be selected at the plant area, regional area, national level, and global level. The renewable as well as local non-renewable (soil erosion) depend on the spatial location of the system. Normally, system boundaries are set at the national level considering the national area for renewable and local non-renewable flow calculations [32].

EMA is a supply-side approach, since it considers direct and indirect contributions to systems and labour and service contributions. Indeed, evaluating an investment entails calculating the inescapable cost of resource replacement [32]. Since its inception, the concept of EMA has been widely applied in a variety of fields [33], including agriculture [34,35]), waste management [20], petroleum production [36], and coal mining [37].

The conversion of animal slaughtering waste or animal residues involves several sub-processes. These sub-processes are the hydrolysis, rendering, biodiesel production, and fermentation process. The detailed description of the process design [38], development [39], as well as economic [40] and ecological evaluation using SPI methodology has been published earlier [41,42]. The material and energy flows for the current study have been considered from the published literature [16]. The renewable inputs (sun, wind, and rain potential) are calculated and used for Austria. For emergy accounting, material and energy flows for a production capacity of 10,000 t/yr PHA production have been considered, and the following assumptions has been made:

- Animal slaughtering residues are considered as waste having zero emergy content. Although it does have emergy content, it can only be available when animal production (farming) is produced to produce meat. Therefore, this study assumes that all emergy content is assigned to the main product "meat".

- Only mass and energy flows have been considered for analysis while infrastructure is out of the system boundary.
- Out of the renewable inputs (sun, wind, and rainfall), only the significant input has been accounted for to avoid double counting, considering that solar radiations are the sole energy source for wind movement and rainfall.
- Water input is supposed to be renewable, considering it is provided from the rain collected in a water reservoir or a lake or river.
- Water input has been converted to energy units by multiplying by the Gibbs free energy content of water.
- For wastewater treatment, only electricity consumption has been considered in the evaluation.
- The electrical demand is fulfilled using the European electricity mix (EU_27), while natural gas has been considered a source of heat provision.
- For wastewater treatment, electricity consumption has been considered as an input to the system.

The assumptions for emergy accounting have been that renewable energy flows and water are used as a renewable input. Therefore, a system diagram is required to organise the evaluation and adequately account for inputs and outflows from the processes or systems. Figure 1 is the system diagram of the PHA production process utilising slaughtering waste as starting material input. It helps to construct evaluation tables of the actual flow of material, energy, labour, and services. All information (energy, material, labour, and environmental services) shown in the diagram are evaluated in their standard units (J, kg, m^3, $, etc.). The information available in the form of evaluation table data is further processed to calculate different indices for PHA production.

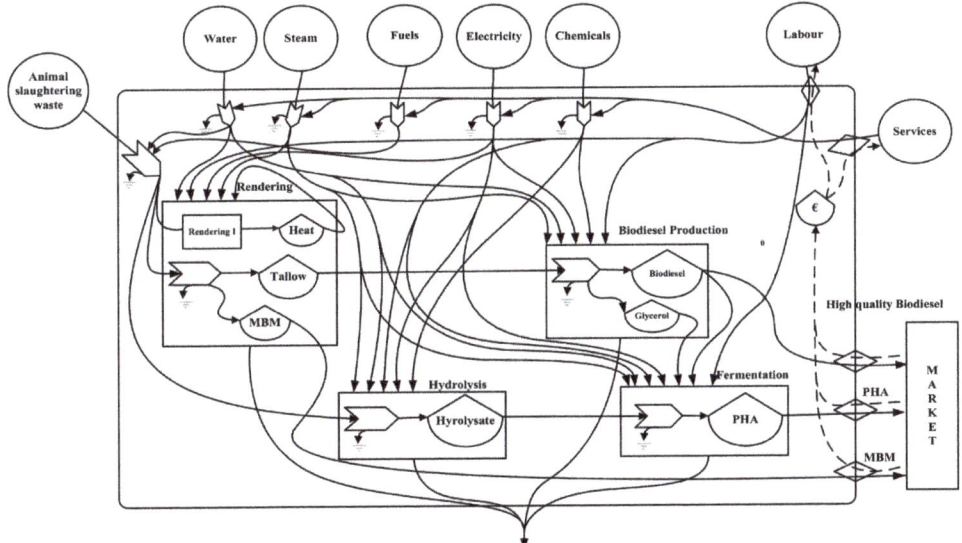

Figure 1. Emergy system diagram of PHA production process utilising slaughtering waste as starting material (the emergy symbol nomenclature is given Appendix B).

Specific emergy (seJ/unit) for the material flows, and transformity (seJ/J) for the energy flows is the amount of emergy necessary to create one unit of each input. It is a quality factor measuring the ecological assistance intensity supplied by the planet during the development of each input [43]. The emergy intensity values for different input flows have been shown in Table 1. The reference sources for the unit emergy intensity values have been written as a footnote at the end of the table.

Table 1. Transformities and other Emergy intensities of material and energy flows.

Item	Value	Unit	Variations	References
	All flows are evaluated on a yearly basis. The numbers in the first column refer to the calculation procedures. UEV values refer to the 12E+24 baseline [31].			
Renewable input (locally available)				
Sun	1.00E+00	seJ/J	0%	[a]
Wind (kinetic wnergy of wind used at the surface)	2.51E+03	seJ/J	0%	[b]
Rainfall (chemical potential)	3.05E+04	seJ/J	0%	[b]
Imported Input				
Diesel for transport	1.81E+05	seJ/J	0%	[c]
Electricity	1.20E+05	seJ/J	0%	[d]
Heat (natural gas)	2.76E+05	seJ/J	0%	[c]
Heat (rendering I products combustion)	1.21E+05	seJ/J	0%	[e]
Biodiesel	2.86E+09	sej/g	0%	[e]
Glycerol	2.86E+09	sej/g	0%	[e]
Ammonium hydroxide (NH_4OH)	6.38E+08	sej/g	0%	[f]
Chemicals	6.38E+08	sej/g	0%	[f]
Hydrochloric acid (HCl)	6.38E+08	sej/g	0%	[f]
Sodium hydroxide (NaOH)	6.38E+08	seJ/g	0%	[f]
Methanol CH_3OH	6.38E+08	sej/g	0%	[f]
Acid (sulfuric acid) H_2SO_4	8.86E+08	seJ/g	0%	[g]
Potassium hydroxide (KOH)	6.38E+08	sej/g	0%	[f]
Hydrolysate	6.74E+08	sej/g	0%	[e]
Fresh water (assumed from natural reservoir or collected rain)	3.05E+04	seJ/J	0%	[b]
Wastewater treatment electricity consumption	1.49E+05	sej/g	0%	[d]
Emergy to money ratio for Austria, 2012	3.38E+11	seJ/€	0%	[h]
Labour	2.00E+17	seJ/capita	0%	[h]
Services	3.38E+11	seJ/€	0%	[h]

References for transformities: [a] By definition; [b] After Odum et al., 2000 [44]; [c] Brown et al., 2011. [45]; [d] own calculation after Brown & Ulgiati, 2001, 2002, 2004 [46–48] and Buonocore et al., 2012 [49]; [e] own calculation in this study; [f] After Odum et al., 2000 [50]; [g] Fahd and Fiorentino 2012 [51]; [h] our calculation after NEAD, 2014 [52].

3. Results and Discussion

The results of the emergy accounting analysis have been shown in the emergy evaluation table for each sub-process, one by one, as follows:

3.1. Transportation Phase

The emergy content calculated for the transportation phase is 1.78E+19 seJ/yr and 1.19E+19 seJ/yr, with (L&S) and without labour and services (L&S), respectively. The emergy accounting for the transportation phase has been shown in Table 2.

Table 2. Emergy table of transportation phase.

Items	Units	Raw Amounts	Transformity (seJ/Unit)	Ref.	Emergy (seJ/yr)
Slaughtering residues	g/yr	2.43E+11	0.00E+00	[i]	0
Diesel for transport	J/yr	6.56E+13	1.81E+05	[c]	1.19E+19
Labour	working years	2.66E+01	2.00E+17	[h]	5.32E+18
Services	€/yr	1.74E+06	3.38E+11	[h]	5.89E+17
TOTAL EMERGY with L&S					1.78E+19
TOTAL EMERGY without L&S					1.19E+19

3.2. Rendering I

It is assumed that the "rendering facility processes 100,000 t/yr slaughtering residue, and it operates 250 day/yr". This facility requires 3.28E+02 h/yr to process 4.87E+03 t/yr of condemned material in the rendering I facility. This time is equivalent to the labour force of 2.05E-01person/yr. Therefore, the sum of share of the transportation service share and utility cost constitutes the service input for rendering I, which is 2.44E+05 €/yr. The inventory input and emergy content calculated for rendering I is given in Table 3.

Table 3. Emergy table for rendering I.

Items	Units	Raw Amounts	Transformity (seJ/Unit)	Ref.	Emergy (seJ/yr)
Slaughtering residues at plant	g/yr	4.87E+09	4.89E+07	[f]	2.38E+17
Electricity EU_27 mix	J/yr	1.07E+12	2.58E+05	[d]	2.35E+17
Heat_natural gas	J/yr	1.61E+13	2.76E+05	[c]	4.45E+18
Fresh water (assumed from natural reservoir or collected rain)	J/yr	6.25E+09	3.05E+04	[b]	1.91E+14
Electricity consumption for waste water treatment	J/yr	2.09E+09	2.58E+05	[d]	5.10E+14
Labour	working years	2.05E-01	2.00E+17	[h]	4.11E+16
Services	€/yr	2.44E+05	3.38E+11	[h]	8.25E+16
TOTAL EMERGY with L&S					5.09E+18
TOTAL EMERGY without L&S					4.97E+18

The total emergy content calculated for rendering I is 5.09E+18 seJ/yr and 4.97E+18 seJ/yr with L&S and without L&S, respectively. The emergy content contribution from each input to the total emergy content of the process has been shown in Figure 2. In the entire emergy content with L&S, the heat provision from natural gas has a maximum share contributing 87%, while services, electricity, and slaughtering residues have 2%, 5%, and 5% shares. Similarly, for the total emergy content, 90% share is contributed by heat from natural gas, the other 5% from electricity input, and 4% from slaughtering residues at the plant. This means that the raw material input has almost no relative contribution to the overall emergy content of the process. The utilities provide most of the emergy content to transform the raw material (waste residue) into a usable product.

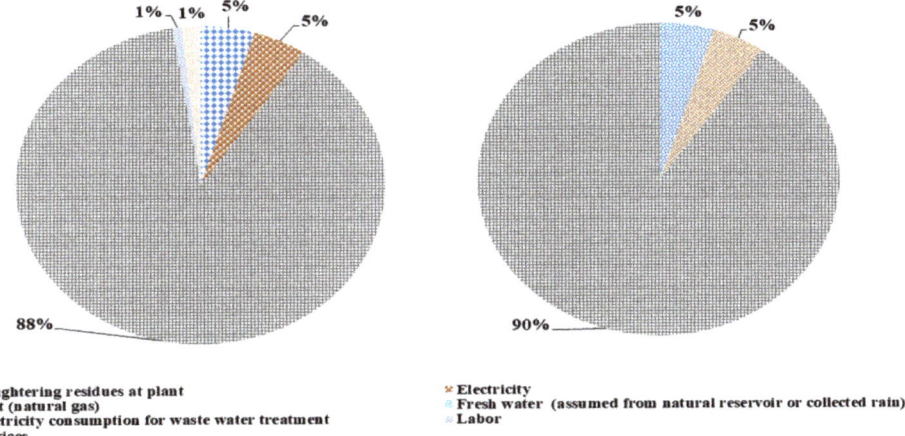

Figure 2. Emergy content, the contribution of different input flows for rendering I with and without L&S.

3.3. Rendering II

The processing time of 1.60E+04 h/yr is required to process 2.36E+05 t/yr slaughtering residue following the same assumption described in rendering I. It is equivalent to a

9.97 person/yr labour force. Therefore, the sum of transportation service share and utility cost constitutes the service input for rendering I, 1.21E+07 €/yr. The inventory input and emergy accounting data are shown in Table 4.

Table 4. Emergy table of rendering II.

Items	Units	Raw Amounts	Transformity (seJ/Unit)	Ref.	Emergy (seJ/yr)
Slaughtering residues	g/yr	2.36E+11	4.89E+07	[f]	1.16E+19
Electricity	J/yr	5.19E+13	2.58E+05	[d]	1.14E+19
Heat (natural gas)	J/yr	7.83E+14	2.76E+05	[c]	2.16E+20
Heat from rendering I products burning	J/yr	4.07E+13	1.22E+05	[f]	4.93E+18
Fresh water (assumed from natural reservoir or collected rain)	J/yr	3.04E+11	3.05E+04	[b]	9.26E+15
Electricity consumption for waste water treatment	J/yr	4.76E+05	2.58E+05	[d]	1.05E+11
Labour	working years	9.97E+00	2.00E+17	[h]	1.99E+18
Services	€/yr	1.21E+07	3.38E+11	[h]	4.09E+18
TOTAL EMERGY with L&S					2.50E+20
TOTAL EMERGY without L&S					2.44E+20

The total emergy content for rendering II with L&S is 2.50E+20 seJ/yr, while without L&S it is 2.44E+20 seJ/yr. The emergy content contribution of different input flows for rendering II has been shown in Figure 3. In the total emergy content with L&S, heat provision by natural gas and rendering I have a cumulative share of 88%. The rest 12% is contributed by services (2%), electricity (5%), slaughtering residues (4%), and labour (1%), respectively. The total emergy content without L&S is also contributed mainly by heat provision from natural gas and rendering I. It contributes up to 90%, while the rest, 10%, is equally shared between electricity and slaughtering residues. It also shows a similar trend, as shown in Figure 3, for transforming waste raw material input into useable products tallow and MBM. The emergy content of the process is allocated to the products following the mass allocation method.

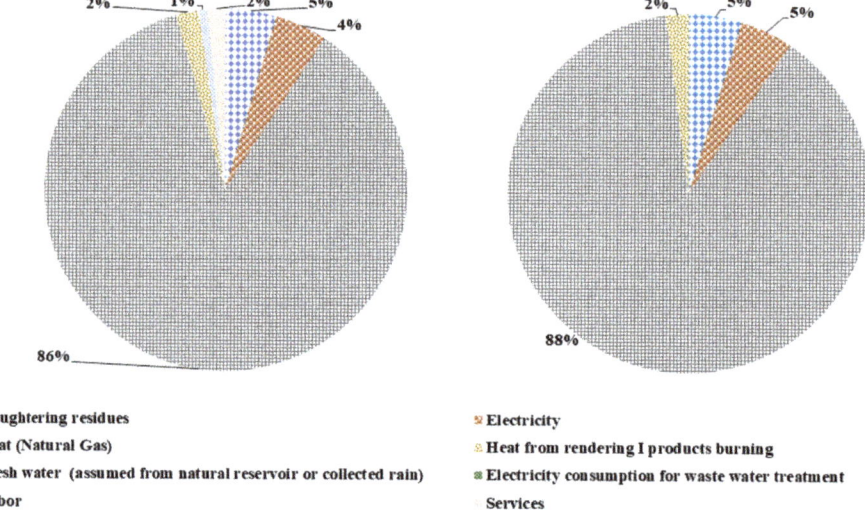

Figure 3. Emergy content share contribution of different input flows for rendering II with and without L&S.

3.4. Biodiesel Production

The labour input for the biodiesel production facility is calculated by assuming that it has a capacity of 100,000 t/yr and it is operating 250 days/yr. Producing 3.31 t of biodiesel requires 1.95E+3 h/yr, equivalent to 1.22 person/yr labour force. The cost of raw material, chemicals, and utilities constitutes services equal to 7.03E+06 €/yr. The inventory data obtained from a project partner and emergy accounting calculations for the biodiesel production process with and without L&S has been presented in Table 5.

Table 5. Emergy table of biodiesel production.

Items	Units	Raw Amounts	Transformity (seJ/Unit)	Ref.	Emergy (seJ/yr)
Tallow	g/yr	3.31E+10	2.57E+09	[e]	8.44E+19
Electricity	J/yr	8.46E+12	2.58E+05	[d]	1.86E+18
Heat (natural gas)	J/yr	4.77E+13	2.76E+05	[c]	1.32E+19
Methanol CH_3OH	g/yr	3.61E+09	6.38E+08	[f]	2.30E+18
Acid (sulfuric acid) H_2SO_4	g/yr	4.63E+08	8.86E+08	[g]	4.10E+17
Potassium hydroxide (KOH)	g/yr	5.96E+05	6.38E+08	[f]	3.80E+14
Fresh water (assumed from natural reservoir or collected rain)	J/yr	1.63E+10	3.05E+04	[b]	4.99E+14
Electricity consumption for waste water treatment	J/yr	2.64E+09	2.58E+05	[d]	5.83E+14
Labour	working years	1.22E+00	2.00E+17	[h]	2.43E+17
Services	€/yr	7.03E+06	3.38E+11	[h]	2.38E+18
TOTAL EMERGY with L&S					1.05E+20
TOTAL EMERGY without L&S					1.02E+20

The graphical representation of the biodiesel production emergy content relevant to the material and service inputs is shown in Figure 4. The emergy content contributed by the tallow input is 81% and 83% of emergy content of biodiesel production with and without L&S. The provision of energy in the form of heat from natural gas has 13% and the electricity provision and methanol consumption has 2% of each emergy content contribution for both scenarios. The L&S contribute 2% emergy content in the total emergy value of biodiesel production with labour and services.

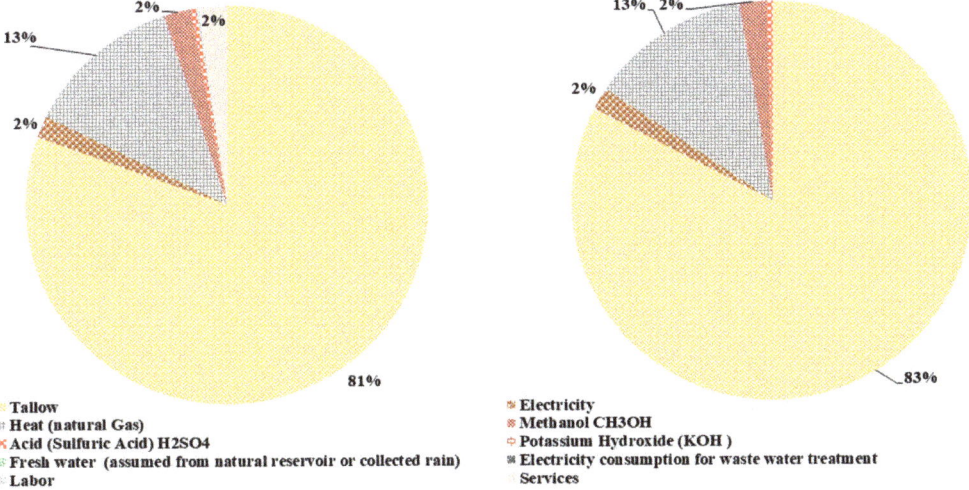

Figure 4. Contribution of different input flows for the emergy content of biodiesel production with and without L&S.

3.5. Hydrolysis

For a 10,000 t/yr PHA production, 150 batch/yr are required, requiring 1.2E+02 h/yr working time. It is equivalent to a 7.5E-01 person/yr labour force. Therefore, the sum of the cost for chemicals and utilities is equal to the services for this process, i.e., 4.22E+05 €/yr. The inventory input for hydrolysis and emergy content evaluation is shown in Table 6.

Table 6. Emergy table of hydrolysis.

Items	Units	Raw Amounts	Transformity (seJ/Unit)	Ref.	Emergy (seJ/yr)
Slaughtering residues	g/yr	1.73E+09	4.89E+07	[f]	8.44E+16
Electricity	J/yr	1.50E+11	2.58E+05	[d]	3.30E+16
Heat (natural gas)	J/yr	1.11E+12	2.76E+05	[c]	3.07E+17
Hydrochloric acid (HCl)	g/yr	2.04E+09	6.38E+08	[e]	1.30E+18
Sodium hydroxide (NaOH)	g/yr	7.36E+08	6.38E+08	[e]	4.70E+17
Fresh water (assumed from natural reservoir or collected rain)	J/yr	3.42E+09	3.05E+04	[b]	1.04E+14
Labour	working years	7.50E-01	2.00E+17	[h]	1.50E+17
Services	€/yr	4.22E+05	3.38E+11	[h]	1.43E+17
TOTAL EMERGY with L&S					2.49E+18
TOTAL EMERGY without L&S					2.20E+18

The total emergy content of hydrolysis for both scenarios, with and without L&S, is 2.49E+18 seJ/yr and 2.20E+18 seJ/yr, respectively. The emergy content contribution shown in Figure 5 reveals that the mineral acid and base are the key contributors in both scenarios.

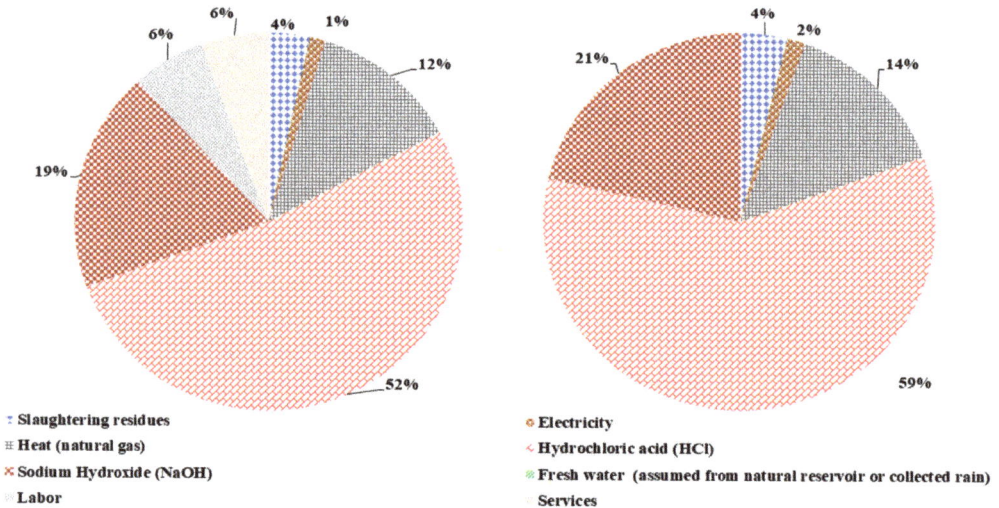

Figure 5. Contribution of different input flows for the emergy content of hydrolysis with and without L&S.

Correspondingly, the mineral acid and base contribute 71% and 80% for the emergy content with L&S and without L&S. The heat provision input with 12% and 14% shares is the next prominent contributor in both scenarios. Finally, slaughtering waste has a small contribution of 3% and 4%, while the electricity input has an equal share of 2% in both systems. Thus, L&S have a considerable cumulative percentage of 12% in the total emergy content for hydrolysis with L&S.

3.6. Fermentation Process

The inventory data for the fermentation process and emergy evaluation for the fermentation process are given in Table 7. The published literature reveals that the fermentation batch and the handling of the downstream processing of the fermentation broth requires about 55 h/batch. There would be 150 batch/yr for 10,000 t/yr PHA production [38]. It results in 8.25E+03 h/yr, which is equivalent to 5.16 person/yr labour force. The sum of the cost for chemicals and utilities for the fermentation process is 6.31E+08 €/yr, equal to services input.

Table 7. Emergy table of fermentation (PHA production) process.

Items	Units	Raw Amounts	Transformity (seJ/Unit)	Ref.	Emergy (seJ/yr)
Electricity	g/yr	1.81E+13	2.20E+05	[d]	3.98E+18
Heat (natural gas)	J/yr	1.05E+13	2.76E+05	[c]	2.91E+18
Glycerol	J/yr	3.31E+09	2.93E+09	[f]	9.70E+18
Biodiesel	g/yr	1.78E+10	2.93E+09	[f]	5.23E+19
Hydrolysate	g/yr	3.67E+09	5.98E+08	[f]	2.19E+18
Ammonium hydroxide (NH_4OH)	J/yr	7.67E+08	6.38E+08	[e]	4.89E+17
Chemicals	working years	7.82E+08	6.38E+08	[e]	4.99E+17
Fresh water (assumed from natural reservoir or collected rain)	€/yr	4.17E+11	3.05E+04	[b]	1.27E+16
Electricity consumption for waste water treatment	g/yr	6.74E+10	2.20E+05	[d]	1.48E+16
Labour	J/yr	5.16E+00	2.00E+17	[h]	1.03E+18
Services	€/yr	6.32E+08	3.38E+11	[h]	2.13E+20
TOTAL EMERGY with L&S					2.88E+20
TOTAL EMERGY without L&S					7.34E+19

The emergy content for a 10,000 t/yr PHA production with L&S is 2.88E+20 seJ/yr, while without L&S it is 7.34E+19 seJ/yr. The distribution of the total emergy content for the fermentation process with and without L&S into the emergy content of input shares is shown in Figure 6. The raw material's (biodiesel and glycerol) emergy content cumulative contribution to the with L&S scenario is 22%, while in the without L&S scenario it is the primary input provider, contributing almost 85% of the content. For the total emergy content with the L&S scenario, services are the leading donor, having 74% of the overall emergy content, while electricity and hydrolysate contribute about 3%. Similarly, the emergy content contribution share for hydrolysate is 3%, for electricity its 6%, and for heat it is 4%. At the same time, ammonium hydroxide, chemicals, and freshwater have a cumulative 2% share in the total emergy content of fermentation without L&S. This shows that the raw material input has a significantly high emergy content contribution to the fermentation process. In contrast, the services' input dominates the overall emergy content input.

3.7. Emergy-Based Performance Indicators

The emergy-based performance indicators calculated for the PHA production process, including the transportation phase and sub-processes like rendering I, rendering II, biodiesel production, and fermentation, have been shown in Table 8 using the system boundary at the biosphere level. The based performance indicators considering the PHA production facility area as the system boundary are given in Appendix A Table A1.

The emergy intensities of animal residue transportation, rendering I, rendering II, biodiesel production, hydrolysate, and PHA are 7.32E+07 seJ/$g_{animal\ residues\ transportation}$, 1.25E+05 seJ/$J_{Heat}$, 2.64E+09 seJ/$g_{(tallow,\ MBM)}$, 2.96E+09 seJ/$g_{(biodiesel,\ glycerol)}$, 6.80E+08 seJ/$g_{(hydrolysate)}$, and 2.88E+10 seJ/$g_{(PHA)}$, respectively. The ratio between the emergy content of imported resources (F) and the total emergy (U), i.e., "Emergy Yield Ratio", is 1.0 for all sub-processes. This value shows that almost all resources are imported or transported from outside of the system under examination. The ELR is a ratio between the sum of emergy content of imported input "F", labour "L", and services "S" and renewable input "R". The values of ELR for the sub-processes are 78.12 for the transportation phase, 12.58 and 12.58 for rendering I and rendering II, 13.87 for biodiesel production, 52.89 for hydrolysis, and 17.43 for the fermentation process. The values for ESI (ratio between EYR and ELR) are 0.01 for transportation, 0.08 for both rendering I and rendering II, 0.07 for biodiesel production, 0.02 for hydrolysis, and 0.06 for PHA production. The renewable fractions for different sub-processes are 1.26% for the transportation phase, 7.39% for rendering I, 7.36% for rendering II, 6.73% for biodiesel production, and 1.86% for hydrolysis 5.43% for the fermentation process.

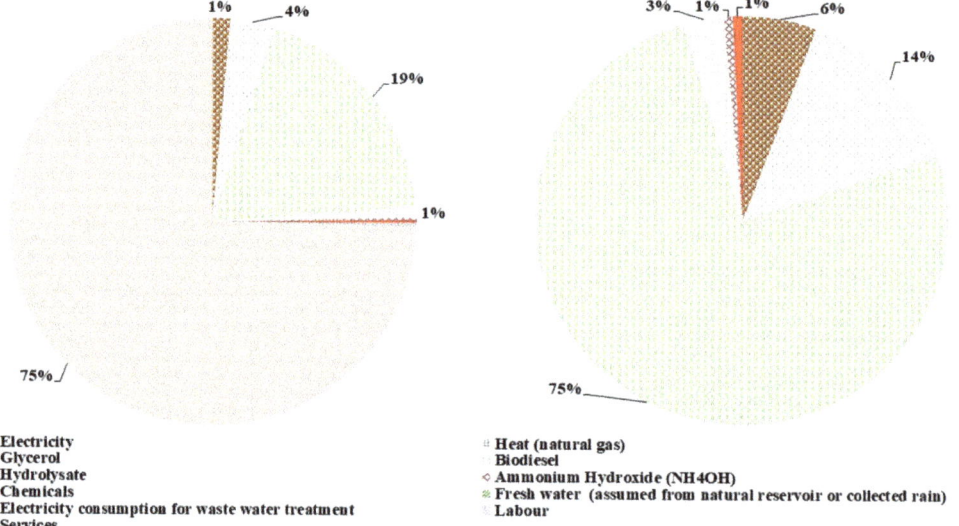

Figure 6. Contribution of input flows to the emergy content of the fermentation process with and without L&S.

3.8. Effect of the Change of Energy Provision Resource

A study about electricity production from different resources by Brown and Ulgiati (2002) [47] reveals % renewables from these energy systems. These % renewable values are also integrated as renewable input to the system under this study. The reported % renewable content for different energy resources are wind 86.61%, geothermal 69.67%, hydro 68.84%, natural gas (methane) 7.83%, and coal 8.79%.

3.8.1. Comparative Analysis of Biodiesel Production

The performance indicators calculated for the biodiesel production process fulfilling the energy demand from different energy resources are shown in Figure 7. The relevant values of this graph are given in Appendix A, Table A2.

Table 8. Emergy-based indicators calculated for the biobased PHA production overall process.

Emergy Accounting	Value	Unit
Transportation phase		
Emergy from local renewable resources, R	2.96E+17	seJ/yr
Emergy from imported resources, F	1.19E+19	seJ/yr
Total emergy, U = R + F + L + S	1.78E+19	seJ/yr
Emergy intensity	7.32E+07	seJ/$g_{animal\ residues\ transportation}$
Environmental yield ratio, EYR = U/(F + L + S)	1.00	
Environmental Loading Ratio, (ELR) = (F + L + S)/R	78.12	
Emergy Sustainability Index, EYR/ELR	0.01	
Renewable fraction, REN% = 1/(1 + ELR) or =R/U × 100	1.26%	
Rendering I		
Emergy from local renewable resources, R	4.05E+17	seJ/yr
Emergy from imported resources, F	4.97E+18	seJ/yr
Total emergy, U = R + F + L + S	5.09E+18	seJ/yr
Transformity of heat	1.25E+05	seJ/J_{Heat}
Environmental yield ratio, EYR = U/(F + L + S)	1.00	
Environmental Loading Ratio, (ELR) = (F + L + S)/R	12.54	
Emergy Sustainability Index, EYR/ELR	0.08	
Renewable fraction, REN% = 1/(1 + ELR) or =R/U × 100	7.39%	
Rendering II		
Emergy from local renewable resources, R	2.00E+19	seJ/yr
Emergy from imported resources, F	2.46E+20	seJ/yr
Total emergy, U = R + F + L + S	2.52E+20	seJ/yr
Emergy intensity	2.64E+09	seJ/$g_{(tallow,\ MBM)}$
Environmental yield ratio, EYR = U/(F + L + S)	1.00	
Environmental Loading Ratio, (ELR) = (F + L + S)/R	12.58	
Emergy Sustainability Index, EYR/ELR	0.08	
Renewable fraction, REN% = 1/(1 + ELR) or =R/U × 100	7.36%	
Biodiesel production		
Emergy from local renewable resources, R	7.62E+18	seJ/yr
Emergy from imported resources, F	1.03E+20	seJ/yr
Total emergy, U = R + F + L + S	1.058E+20	seJ/yr
Emergy intensity of biodiesel	2.96E+09	seJ/$g_{(biodiesel,\ glycerol)}$
Environmental yield ratio, EYR = U/(F + L + S)	1.00	
Environmental Loading Ratio, (ELR) = (F + L + S)/R	13.87	
Emergy Sustainability Index, EYR/ELR	0.07	
Renewable fraction, REN% = 1/(1 + ELR) or =R/U × 100	6.73%	
Hydrolysis		
Emergy from local renewable resources, R	4.69E+16	seJ/yr
Emergy from imported resources, F	7.05E+19	seJ/yr
Total emergy, U = R + F + L + S	2.49E+18	seJ/yr
Emergy intensity of hydrolysate	6.80E+08	seJ/$g_{(hydrolysate)}$
Environmental yield ratio, EYR = U/(F + L + S)	1.00	
Environmental Loading Ratio, (ELR) = (F + L + S)/R	52.89	
Emergy Sustainability Index, EYR/ELR	0.02	
Renewable fraction, REN% = 1/(1 + ELR) or =R/U × 100	1.86%	
Fermentation (PHA production) process		
Emergy from local renewable resources, R	1.59E+19	seJ/yr
Emergy from imported resources, F	7.34E+19	seJ/yr
Total emergy, U = R + F + L + S	2.88E+20	seJ/yr
Emergy intensity of PHA	2.88E+10	seJ/g_{PHA}
Environmental yield ratio, EYR = U/(F + L + S)	1.00	
Environmental Loading Ratio, (ELR) = (F + L + S)/R	17.43	
Emergy Sustainability Index, EYR/ELR	0.06	
Renewable fraction, REN% = 1/(1+ELR) or =R/U × 100	5.43%	

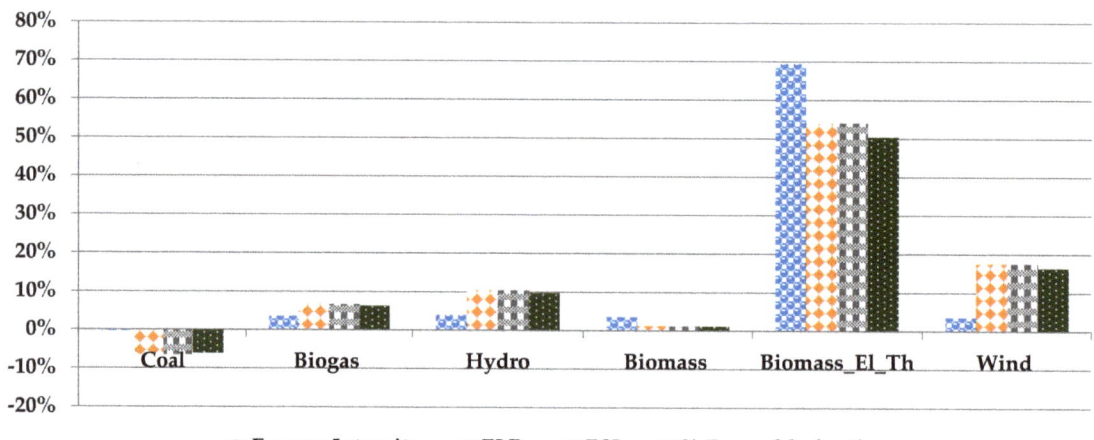

Figure 7. Comparative analysis of emergy flows and emergy-based indicators for the biodiesel production process.

The change of electricity provision resource shows an impact on different performance indicators. In the current study, the electricity provision form EU_27 mix is considered as the basic scenario, and the results calculated using other electricity provision systems are compared with it. The effect of the change of resource on the emergy intensity of biodiesel production (seJ/g) shows a negative impact on the electricity provision from coal while fulfilling the electricity demand using the electricity from biogas, hydro, biomass, and wind has a slightly positive effect, ranging between 3% to 4%. The provision of electricity and heat demand by biomass burning shows about a 69% improvement compared to the basic scenario of energy provision from the EU electricity mix and the natural gas for electricity and heat, respectively. The effect of changing energy resources on biodiesel transformity (seJ/J) compared to diesel transformity 1.81E+5 [45] has a positive impact with a 55% to 57% decrease in the transformity value for the electricity supply from the EU27 mix, coal, hydro power, wind, and biomass. In comparison, the electricity and heat provision from biomass has a maximum 87% decrease in the transformity value. The ELR value also shows a positive impact of −6% to 17% by replacing the electricity provision from more renewable resources. The change of electricity and heat provision source with biomass shows an improvement of the process of around 54%. Similarly, the ESI and % renewable fractions show an analogous effect, as both are dependent on ELR values.

3.8.2. Comparative Analysis of PHA Production

The performance indicators calculated for the PHA production process providing the required energy demand from different energy resources have been shown in Figure 8. The corresponding computational values are given in Appendix A, Table A3.

The replacement of electricity provision with electricity from renewable resources shows a positive impact. The emergy intensity for PHA production (seJ/g) shows a minor improvement ranging from 1.5% to 2% by changing only the electricity provision resources. This impact reaches up to 17% when electricity and heat provision resources are replaced with biomass resources. The comparison of the emergy intensity for PHA production utilising the electricity provision from the EU27 mix, coal, hydropower, wind power, and biomass is about 5% to 7% lower than the emergy intensity of fossil-fuel-derived polyethylene high density (PE-HD) [53]. This decrease in the emergy intensity value is 21% lower when biomass is utilised to fulfil the PHA production process's electric and heat provision. The exchange of electricity resources with electricity from coal negatively impacts the overall system, making it even less sustainable. The provision of electricity from wind farms represents the best available option, having the most significant improvements in ELR 11% and ESI 11%, and % a renewable fraction of 10.40%.

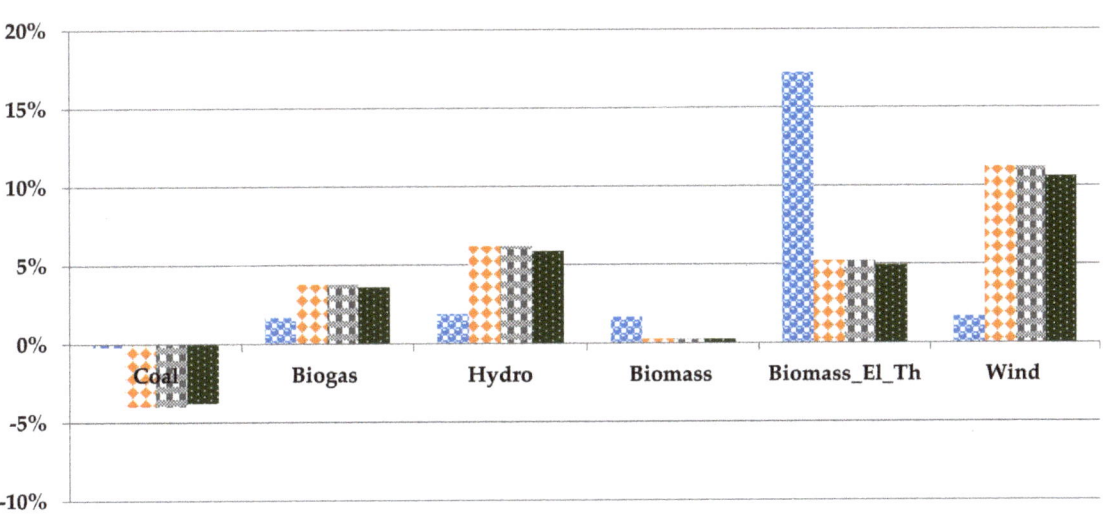

Figure 8. Comparative analysis of emergy flows and emergy-based indicators for the PHA production process.

The ELR must be calculated regarding the local environment (R only calculated within the hectare of land occupied by the plant) compared to the regional scale (where the inputs come from, with imported flows divided into renewable non-renewable fractions). The ELR calculated in a second way compares all the renewable input flows at the larger scale (as the renewable fractions of all flows) with the non-renewable flows at the same scale. In the case of a system expansion from the local to the regional scale, the concept of imported flows F becomes irrelevant. The non-renewable fraction of F overlaps to non-renewable N. The calculated ELR has a meaning at the larger scale. The values of ELR calculated at the local scale become huge and indicate that the process has too much loading on the local resources. This shows that the process is not sustainable and becomes sustainable only if buffered by the environmental forces available at a large scale (therefore, power plants and industries need a forest, a protected area, around them. The ELR is linked to the %REN, which is higher in the large-scale perspective.

These choices do not affect the EYR values because F at the numerator and denominator has the same value, whether F is split into renewable or non-renewable fractions. EYR is an indicator that informs about the extent to which the process is exploiting the local resources (N and R). The value of the EYR indicator rises in case the exploitation of local resources increases. Otherwise, its value is 1, as in the current case study, and indicates a process that converts the imported resources into other resources for export. Similarly, ESI is affected because of its definition as a ratio of EYR and ELR.

3.8.3. Evaluation of Waste

The emergy evaluation utilises local, regional, national, and biospheric boundary conditions that define the available renewable inputs to the system under consideration. Emergy dealing from a biosphere perspective has different meanings for anthropogenic waste terminology. It argues that no flow is a waste because every flow or residue from a process becomes an input to another process. In principle, any flow coming out as residue from a process is either a nutrient or a toxin, impacting the surrounding environment and leading to system evolution by favouring some processes and negatively affecting others. Similarly, Brown and Ulgiati (2010) [27] argued that the *"effect can be both positive and negative: Transformity does not suggest the outcome that might result from the interaction of a stressor with in an ecosystem, only that with high transformity the effect is greater. Where empower density of a stressor is significantly higher than the average empower density of the ecosystem it is released into, one can expect significant changes in ecosystem function"*. Similarly, the transformity of a product or service only explains the efficiencies of the current practices and compares optimisation potentials.

4. Conclusions

EMA evaluation results reveal that the main driving factors for hydrolysis are acid and base inputs. At the same time, heat provision also significantly affects the total emergy content of the process. Similarly, labour and services also have minor contributions to the entire emergy content of the sub-processes. The main contributing factor for rendering operations is heat provision, while residue transportation and electricity are minor contributors to the total emergy content of the process. For biodiesel production, the main contributor of emergy content is the tallow content, while heat provision is the minor contributor to the overall emergy content of the process. Tallow, a product of a highly energy-intensive rendering process and one of the main contributors of biodiesel production, indicates the possible potential of ecological optimisation by utilising energy from different resources. Similarly, the emergy accounting evaluation for the fermentation process shows that the main contributor to the total emergy content of this process is services, while biodiesel and glycerol also have significant contributions. EMA has the potential to rate energy technology. In this study, the exchange of energy from business as usual (EU mix grid electricity and natural gas for heating) to biomass for both heat and electricity represents the most environment-friendly option. The sustainability of biomass consumption as an energy source is in line with completing the short-term carbon cycle, including the fixation of CO_2 by photosynthesis and releasing the same amount of CO_2 in the combustion process. Similarly, in emergy accounting, wood or biomass are also accounted as renewable sources, which indicates that their replacement time is as fast as their use rate. The EMA, with real pluses and minuses, is a crucial methodology to attain global sustainability. There are a lot of methods available to assess the sustainable development of the process or products. It is up to the decision-makers (engineers in technology development especially) to look at their system boundary and select the suitable measurement methodology according to their needs. On the other hand, there are numerous uncertainties associated with renewable resource-based energy technologies, such as the location of the plant and the availability of renewable resources as a raw material for the plant and the energy production system. From the perspective of a chemical engineer, heat and energy integration in the design of new and existing plants may be different and more complex than state-of-the-art fossil-based energy systems.

Author Contributions: K.S.: conceptualisation, methodology, writing—original draft preparation. M.R.: validation, formal analysis, writing—original draft preparation. M.I.R.: investigation, resources, writing—original draft preparation. N.A.: data curation, writing—review and editing. A.S.S.: writing—review and editing, visualisation. I.M.I.I.: supervision, project administration, funding acquisition. All authors have read and agreed to the published version of the manuscript.

Funding: This research was funded by the Ministry of Education and King Abdulaziz University, Jeddah, Saudi Arabia, grant number (IFPHI-307-188-2020).

Institutional Review Board Statement: Not applicable.

Informed Consent Statement: Not applicable.

Data Availability Statement: Not applicable.

Acknowledgments: This research work was funded by the Institutional Fund Project under the grant number (IFPHI-307-188-2020). Therefore, the authors gratefully acknowledge the technical and financial support from the Ministry of Education and King Abdulaziz University, DSR, Jeddah, Saudi Arabia.

Conflicts of Interest: The authors declare no conflict of interest.

Appendix A

Table A1. Emergy-based indicators calculated for the biobased PHA production using facility area as the system boundary.

Emergy Accounting	Value	Unit
Transportation Phase		
Emergy from local renewable resources, R	4.34E+14	seJ/yr
Emergy from imported resources, F	1.19E+19	seJ/yr
Total emergy, U = R + F + L + S	1.78E+19	seJ/yr
Emergy intensity	7.32E+07	seJ/$g_{animal\ residues}$
Environmental yield ratio, EYR = U/(F + L + S)	1.00	
Environmental Loading Ratio, (ELR) = (F + L + S)/R	40956.33	
Emergy Sustainability Index, EYR/ELR	0.00002	
Renewable fraction, REN% = 1/(1 + ELR) or =R/U × 100	0.0024%	
Rendering I		
Emergy from local renewable resources, R	4.34E+14	seJ/yr
Emergy from imported resources, F	4.82E+18	seJ/yr
Total emergy, U = R + F + L + S	4.94E+18	seJ/yr
Transformity of heat	1.21E+05	seJ/J_{Heat}
Environmental yield ratio, EYR = U/(F + L + S)	1.00	
Environmental Loading Ratio, (ELR) = (F + L + S)/R	11382.85	
Emergy Sustainability Index, EYR/ELR	0.0001	
Renewable fraction, REN% = 1/(1 + ELR) or =R/U × 100	0.0088%	
Rendering II		
Emergy from local renewable resources, R	4.34E+14	seJ/yr
Emergy from imported resources, F	2.39E+20	seJ/yr
Total emergy, U = R + F + L + S	2.45E+20	seJ/yr
Emergy intensity	2.56E+09	seJ/$g_{(tallow,\ MBM)}$
Environmental yield ratio, EYR = U/(F + L + S)	1.00	
Environmental Loading Ratio, (ELR) = (F + L + S)/R	564420.33	
Emergy Sustainability Index, EYR/ELR	0.000002	
Renewable fraction, REN% = 1/(1 + ELR) or =R/U × 100	0.0002%	
Biodiesel production		
Emergy from local renewable resources, R	4.34E+14	seJ/yr
Emergy from imported resources, F	9.95E+19	seJ/yr
Total emergy, U = R + F + L + S	1.021E+20	seJ/yr
Emergy intensity	2.86E+09	seJ/$g_{(biodiesel,\ glycerol)}$
Environmental yield ratio, EYR = U/(F + L + S)	1.00	
Environmental Loading Ratio, (ELR) = (F + L + S)/R	235228.24	
Emergy Sustainability Index, EYR/ELR	0.0000043	
Renewable fraction, REN% = 1/(1 + ELR) or =R/U × 100	0.00043%	
Hydrolysis		
Emergy from local renewable resources, R	4.34E+14	seJ/yr
Emergy from imported resources, F	2.18E+18	seJ/yr
Total emergy, U = R + F + L + S	2.47E+18	seJ/yr
Emergy intensity of hydrolysate	6.74E+08	seJ/$g_{(hydrolysate)}$
Environmental yield ratio, EYR = U/(F + L + S)	1.00	

Table A1. Cont.

Emergy Accounting	Value	Unit
Environmental Loading Ratio, (ELR) = (F + L + S)/R	5695.91	
Emergy Sustainability Index, EYR/ELR	0.00018	
Renewable fraction, REN% = 1/(1+ELR) or =R/U × 100	0.018%	
Fermentation (PHA production) process		
Emergy from local renewable resources, R	4.34E+14	seJ/yr
Emergy from imported resources, F	6.58E+19	seJ/yr
Total emergy, U = R + F + L + S	2.74E+20	seJ/yr
Emergy intensity of PHA	2.74E+10	seJ/g_{PHA}
Environmental yield ratio, EYR = U/(F + L + S)	1.00	
Environmental Loading Ratio, (ELR) = (F + L + S)/R	630188.11	
Emergy Sustainability Index, EYR/ELR	0.0000016	
Renewable fraction, REN% = 1/(1 + ELR) or =R/U × 100	0.00016%	

Table A2. Comparative analytic values of emergy flows and emergy-based indicators for the biodiesel production process using alternate energy resources.

Indicators	EU_27	Coal	Biogas	Hydro	Biomass	Biomass_El_Th	Wind
Emergy from local renewable resources, R	7.62E+18	7.18E+18	7.88E+18	8.17E+18	7.44E+18	5.06E+18	8.92E+18
Emergy from imported resources, F	1.03E+20	1.04E+20	9.95E+19	9.90E+19	9.95E+19	2.98E+19	9.95E+19
Total emergy, U = R + F + L + S	1.06E+20	1.06E+20	1.02E+20	1.02E+20	1.02E+20	3.24E+19	1.02E+20
Emergy intensity	2.96E+09	2.97E+09	2.86E+09	2.84E+09	2.86E+09	9.06E+08	2.86E+09
Environmental yield ratio, EYR = U/(F + L + S)	1.00	1.00	1.00	1.00	1.00	1.00	1.00
Environmental Loading Ratio, (ELR) = (F + L + S)/R	13.87	14.77	12.94	12.42	13.71	6.37	11.43
Emergy Sustainability Index, EYR/ELR	0.072	0.068	0.077	0.081	0.073	0.157	0.087
Renewable fraction, REN% = 1/(1 + ELR) or =R/U × 100	6.73%	6.34%	7.17%	7.45%	6.80%	13.56%	8.05%
Percentage deviations:							
Emergy intensity		−0.39%	3.50%	3.95%	3.50%	69.40%	3.50%
ELR		−6.51%	6.68%	10.45%	1.17%	54.03%	17.58%
ESI = EYR/ELR		−6.51%	6.68%	10.45%	1.17%	54.03%	17.58%
% Renewable fraction		−6.08%	6.23%	9.75%	1.09%	50.40%	16.40%

Reference: Brown and Ulgiati 2002.

Table A3. Comparative analytic values of emergy flows and emergy-based indicators for the PHA production process using alternate energy resources.

Indicators	EU_27	Coal	Biogas	Hydro	Biomass	Biomass_El_Th	Wind
Emergy from local renewable resources, R	1.59E+19	1.53E+19	1.62E+19	1.66E+19	1.57E+19	1.38E+19	1.76E+19
Emergy from imported resources, F	7.34E+19	7.39E+19	6.87E+19	6.81E+19	6.87E+19	2.49E+19	6.87E+19
Total emergy, U = R + F + L + S	2.88E+20	2.88E+20	2.83E+20	2.83E+20	2.83E+20	2.39E+20	2.83E+20
Emergy intensity	2.88E+10	2.88E+10	2.83E+10	2.83E+10	2.83E+10	2.39E+10	2.83E+10
Environmental yield ratio, EYR = U/(F + L + S)	1.00	1.00	1.00	1.00	1.00	1.00	1.00
Environmental Loading Ratio, (ELR) = (F + L + S)/R	17.43	18.11	16.78	16.37	17.38	16.55	15.51
Emergy Sustainability Index, EYR/ELR	0.057	0.055	0.060	0.061	0.058	0.060	0.064
Renewable fraction, REN% = 1/(1 + ELR) or =R/U × 100	5.43%	5.23%	5.62%	5.76%	5.44%	5.70%	6.06%
Percentage deviation:							
Emergy intensity		−0.18%	1.64%	1.85%	1.64%	16.84%	1.64%
ELR		−3.89%	3.70%	6.08%	0.27%	5.05%	11.00%
ESI = EYR/ELR		−3.89%	3.70%	6.08%	0.27%	5.05%	11.00%
% Renewable fraction		−3.68%	3.50%	5.75%	0.25%	4.78%	10.40%

Reference: Brown and Ulgiati 2002.

Appendix B

Appendix B.1. Emergy Nomenclature

(From: Odum, H. T., 1996. "Environmental Accounting". J. Willey.)

Symbol	Description
	Energy circuit: A pathway whose flow is proportional to the quantity in the storage or source upstream.
	Source: Outside source of energy-delivering forces according to a programme controlled from the outside: a forcing function.
	Tank: A compartment of energy storage within the system storing a quantity as the balance of inflows and outflows. It is a state variable.
	Heat sink: Dispersion of potential energy into heat that accompanies all fundamental transformation processes and storages, loss of potential energy from further use by the system.
	Interaction: Interactive intersection of two pathways coupled to produce an outflow in proportion to a function of both control actions of one flow on another limiting factor action work gate.
	Consumer unit that transforms energy quality, stores it, and feeds it back autocatalytically to improve the inflow.
	Switching action: A symbol that indicates one or more switching actions
	Producer unit that collects and transforms low-quality energy under control interactions of high-quality flows
	Self-limiting energy receiver: A unit that has a self-limiting output when input drives are high because there is a limiting constant quality of material reacting on a circular pathway within.
	Box: Miscellaneous symbol to use for whatever unit or function is labelled.
	Constant-gain amplifier. A unit that delivers an output in proportion to the input I but is changed by a constant factor as long as the energy source S is sufficient
	Transaction: A unit that indicates a sale of goods or services (solid line) in exchange for the payment of money (dashed line). The price is shown as an external source.

Appendix B.2. Definitions of Terms

Available energy	Potential energy capable of doing work and being degraded in the process (units: kilocalories, Joules, etc.)
Useful energy	Available energy used to increase system production and efficiency. Power- Useful energy flow per unit time
Emergy	Available energy (exergy) of one kind previously required directly and indirectly to make a product or service (units: emjoules, emkilocalories, etc.)
Empower	Emergy flow per unit time (units: emjoules per unit time) Transformity-emergy per unit of available energy (units: emjoule per joule)
Solar emergy	Solar energy required directly and indirectly to make a product or service (units: solar emjoules)
Solar empower	Solar emergy flow per unit time (units: solar emjoules per unit time)
Emergy intensity	Emergy of one kind required to produce a product or service per unit of output of the product or service. There are two types of EIs: transformity and specific emergy
Solar transformity:	Solar emergy per unit of available energy (units: solar emjoules per Joule)
Specific emergy (solar)	Solar emergy per mass of a product (units: solar emjoules per gram)
Emdollars, (Em$)	Dollars of gross economic product due to an emergy contribution's proportion of the national empower

(After Odum, 1996).

References

1. Kourmentza, C.; Plácido, J.; Venetsaneas, N.; Burniol-Figols, A.; Varrone, C.; Gavala, H.N.; Reis, M.A. Recent Advances and Challenges towards Sustainable Polyhydroxyalkanoate (PHA) Production. *Bioengineering* **2017**, *4*, 55. [CrossRef] [PubMed]
2. Chen, G.G.Q. (Ed.) *Plastics from Bacteria: Natural Functions and Applications*; Springer Science & Business Media: Berlin/Heidelberg, Germany, 2009; Volume 14.
3. Koller, M. Poly(hydroxyalkanoates) for food packaging: Application and attempts towards implementation. *Appl. Food Biotechnol.* **2014**, *1*, 3–15.
4. Zinn, M.; Witholt, B.; Egli, T. Occurrence, synthesis and medical application of bacterial polyhydroxyalkanoate. *Adv. Drug Deliv. Rev.* **2001**, *53*, 5–21. [CrossRef]
5. Luef, K.P.; Stelzer, F.; Wiesbrock, F. Poly(hydroxy alkanoate)s in Medical Applications. *Chem. Biochem. Eng. Q.* **2015**, *29*, 287–297. [CrossRef]
6. Khosravi-Darani, K.; Bucci, D.Z. Application of poly(hydroxyalkanoate) in food packaging: Improvements by nanotechnology. *Chem. Biochem. Eng. Q.* **2015**, *29*, 275–285. [CrossRef]
7. Huang, L.; Chen, Z.; Wen, Q.; Ji, Y.; Wu, Z.; Lee, D.J. Toward flexible regulation of polyhydroxyalkanoate composition based on substrate feeding strategy: Insights into microbial community and metabolic features. *Bioresour. Technol.* **2020**, *296*, 122369. [CrossRef]
8. Alvarez-Santullano, N.; Villegas, P.; Mardones, M.S.; Durán, R.E.; Donoso, R.; González, A.; Sanhueza, C.; Navia, R.; Acevedo, F.; Pérez-Pantoja, D.; et al. Genome-Wide Metabolic Reconstruction of the Synthesis of Polyhydroxyalkanoates from Sugars and Fatty Acids by *Burkholderia* Sensu Lato Species. *Microorganisms* **2021**, *9*, 1290. [CrossRef]
9. WHO. WHO Calls for More Research into Microplastics and a Crackdown on Plastic Pollution. 2019. Available online: https://www.who.int/news/item/22-08-2019-who-calls-for-more-research-into-microplastics-and-a-crackdown-on-plastic-pollution#:~{}:text=The%20World%20Health%20Organization%20(WHO,to%20microplastics%20in%20drinking%2Dwater (accessed on 15 September 2021).
10. Gupta, J.; Rathour, R.; Maheshwari, N.; Thakur, I.S. Integrated analysis of Whole genome sequencing and life cycle assessment for polyhydroxyalkanoates production by *Cupriavidus* sp. ISTL7. *Bioresour. Technol.* **2021**, *337*, 125418. [CrossRef]
11. Sabbagh, F.; Muhamad, I.I. Production of poly-hydroxyalkanoate as secondary metabolite with main focus on sustainable energy. *Renew. Sustain. Energy Rev.* **2017**, *72*, 95–104. [CrossRef]
12. Kumar, V.; Darnal, S.; Kumar, S.; Kumar, S.; Singh, D. Bioprocess for co-production of polyhydroxybutyrate and violacein using Himalayan bacterium *Iodobacter* sp. PCH194. *Bioresour. Technol.* **2021**, *319*, 124235. [CrossRef]
13. Volova, T.; Kiselev, E.; Nemtsev, I.; Lukyanenko, A.; Sukovatyi, A.; Kuzmin, A.; Ryltseva, G.; Shishatskaya, E. Properties of degradable polyhydroxyalkanoates with different monomer compositions. *Int. J. Biol. Macromol.* **2021**, *182*, 98–114. [CrossRef]
14. Changwichan, K.; Silalertruksa, T.; Gheewala, S.H. Eco-Efficiency Assessment of Bioplastics Production Systems and End-of-Life Options. *Sustainability* **2018**, *10*, 952. [CrossRef]

15. Saratale, R.G.; Cho, S.K.; Ghodake, G.S.; Shin, H.S.; Saratale, G.D.; Park, Y.; Lee, H.S.; Bharagava, R.N.; Kim, D.S. Utilization of noxious weed water hyacinth biomass as a potential feedstock for biopolymers production: A novel approach. *Polymers* **2020**, *12*, 1704. [CrossRef]
16. Shahzad, K.; Ismail, I.M.I.; Ali, N.; Rashid, M.I.; Summan, A.S.A.; Kabli, M.R.; Narodoslawsky, M.; Koller, M. LCA, Sustainability and Techno-Economic Studies for PHA Production. In *The Handbook of Polyhydroxyalkanoates*; CRC Press: Boca Raton, FL, USA, 2020; Volume 2, pp. 455–485.
17. Ferronato, N.; Torretta, V. Waste Mismanagement in Developing Countries: A Review of Global Issues. *Int. J. Environ. Res. Public Health* **2019**, *16*, 1060. [CrossRef]
18. Caiado, R.G.G.; de Freitas Dias, R.; Mattos, L.V.; Quelhas, O.L.G.; Leal Filho, W. Towards sustainable development through the perspective of eco-efficiency—A systematic literature review. *J. Clean. Prod.* **2017**, *165*, 890–904. [CrossRef]
19. Nizami, A.S.; Rehan, M.; Waqas, M.; Naqvi, M.; Ouda, O.K.M.; Shahzad, K.; Miandad, R.; Khan, M.Z.; Syamsiro, M.; Ismail, I.M.I.; et al. Waste biorefineries: Enabling circular economies in developing countries. *Bioresour. Technol.* **2017**, *241*, 1101–1117. [CrossRef]
20. Santagata, R.; Viglia, S.; Fiorentino, G.; Liu, G.; Ripa, M. Power generation from slaughterhouse waste materials. An emergy accounting assessment. *J. Clean. Prod.* **2019**, *223*, 536–552. [CrossRef]
21. Jayathilakan, K.; Sultana, K.; Radhakrishna, K.; Bawa, A.S. Utilization of byproducts and waste materials from meat, poultry and fish processing industries: A review. *J. Food Sci. Technol.* **2012**, *49*, 278–293. [CrossRef]
22. Ali, A.M.; Nawaz, A.M.; Al-Turaif, H.A.; Shahzad, K. The economic and environmental analysis of energy production from slaughterhouse waste in Saudi Arabia. *Environ. Dev. Sustain.* **2020**, *23*, 4252–4269. [CrossRef]
23. Ben-Iwo, J.; Manovic, V.; Longhurst, P. Biomass resources and biofuels potential for the production of transportation fuels in Nigeria. *Renew. Sustain. Energy Rev.* **2016**, *63*, 172–192. [CrossRef]
24. Odum, H.T. *Environmental Accounting: Emergy and Environmental Decision Making*; John Wiley & Sons, Inc.: New York, NY, USA, 1996.
25. Brown, M.T.; Ulgiati, S. Emergy analysis and environmental accounting. In *Encyclopedia of Energy*; Elsevier: Amsterdam, The Netherlands, 2004; pp. 329–354. [CrossRef]
26. Brown, M.; Ulgiati, S. Emergy-based indices and ratios to evaluate sustainability: Monitoring economies and technology toward environmentally sound innovation. *Ecol. Eng.* **1997**, *9*, 51–69. [CrossRef]
27. Brown, M.T.; Ulgiati, S. Updated evaluation of exergy and emergy driving the geobiosphere: A review and refinement of the emergy baseline. *Ecol. Model.* **2010**, *221*, 2501–2508. [CrossRef]
28. Brown, M.T.; Buranakarn, V. Emergy indices and ratios for sustainable material cycles and recycle options. *Resour. Conserv. Recycl.* **2003**, *38*, 1–22. [CrossRef]
29. Odum, H.T. *Ecological and General Systems*; University of Colorado Press: Niwot, CO, USA, 1994; 644p.
30. Campbell, D.E. Emergy baseline for the Earth: A historical review of the science and a new calculation. *Ecol. Model.* **2016**, *339*, 96–125. [CrossRef]
31. Brown, M.T.; Campbell, D.E.; De Vilbiss, C.; Ulgiati, S. The geobiosphere emergy baseline: A synthesis. *Ecol. Model.* **2016**, *339*, 92–95. [CrossRef]
32. Chen, G.Q.; Jiang, M.M.; Chen, B.; Yang, Z.F.; Lin, C. Emergy analysis of Chinese agriculture. *Agric. Ecosyst. Environ.* **2006**, *115*, 161–173. [CrossRef]
33. Spagnolo, S.; Chinellato, G.; Cristiano, S.; Zucaro, A.; Gonella, F. Sustainability Assessment of Bioenergy at Different Scales: An Emergy Analysis of Biogas Power Production. *J. Clean. Prod.* **2020**, *277*, 124038. [CrossRef]
34. Moonilall, N.I.; Homenauth, O.; Lal, R. Emergy Analysis for Maize Fields under Different Amendment Applications in Guyana. *J. Clean. Prod.* **2020**, *258*, 120761. [CrossRef]
35. Zhao, C.; Chen, W. A Review for Tannery Wastewater Treatment: Some Thoughts under Stricter Discharge Requirements. *Environ. Sci. Pollut. Res.* **2019**, *26*, 26102–26111. [CrossRef]
36. Liu, J.; Lin, B.-L.; Sagisaka, M. Sustainability Assessment of Bioethanol and Petroleum Fuel Production in Japan Based on Emergy Analysis. *Energy Policy* **2012**, *44*, 23–33. [CrossRef]
37. Karuppiah, K.; Sankaranarayanan, B.; Ali, S.M. Evaluation of suppliers in the tannery industry based on emergy accounting analysis: Implications for resource conservation in emerging economies. *Int. J. Sustain. Eng.* **2021**. [CrossRef]
38. Titz, M.; Kettl, K.H.; Shahzad, K.; Koller, M.; Schnitzer, H.; Narodoslawsky, M. Process optimization for efficient biomediated PHA production from animal-based waste streams. *Clean Technol. Environ. Policy* **2012**, *14*, 495–503. [CrossRef]
39. Kettl, K.H.; Titz, M.; Koller, M.; Shahzad, K.; Schnitzer, H.; Narodoslawsky, M. Process design and evaluation of biobased polyhydroxyalkanoates (PHA) Production. *Chem. Eng. Trans.* **2011**, *25*, 983–988.
40. Shahzad, K.; Nizami, A.S.; Sagir, M.; Rehan, M.; Maier, S.; Khan, M.Z.; Ouda, O.K.M.; Ismail, I.M.I.; BaFail, A.O. Biodiesel production potential from fat fraction of municipal Waste in Makkah. *PLoS ONE* **2017**, *12*, e0171297. [CrossRef]
41. Kettl, K.-H.; Shahzad, K.; Narodoslawsky, M. Ecological footprint comparison of biobased PHA production from animal residues. *Chem. Eng. Trans.* **2012**, *29*, 439–444. [CrossRef]
42. Shahzad, K.; Kettl, K.-H.; Titz, M.; Koller, M.; Schnitzer, H.; Narodoslawsky, M. Comparison of ecological footprint for biobased PHA production from animal residues utilizing different energy resources. *Clean Technol. Environ. Policy* **2013**, *15*, 525–536. [CrossRef]

43. Ulgiati, S.; Raugei, M.; Bargigli, S. Overcoming the inadequacy of single-criterion approaches to life cycle assessment. *Ecol. Model.* **2006**, *190*, 432–442. [CrossRef]
44. Odum, H.T.; Odum, E.P. The energetic basis for valuation of ecosystem services. *Ecosystems* **2000**, *3*, 21–23. [CrossRef]
45. Brown, M.T.; Protano, G.; Ulgiati, S. Assessing geobiosphere work of generating global reserves of coal, crude oil, and natural gas. *Ecol. Model.* **2011**, *222*, 879–887. [CrossRef]
46. Brown, M.T.; Ulgiati, S. The role of environmental services in electricity production processes. *J. Clean. Prod.* **2001**, *10*, 321–334. [CrossRef]
47. Brown, M.T.; Ulgiati, S. Emergy evaluations and environmental loading of electricity production systems. *J. Clean. Prod.* **2002**, *10*, 321–334. [CrossRef]
48. Brown, M.T.; Ulgiati, S. Energy quality, emergy, and transformity: H.T. Odum's contributions to quantifying and understanding systems. *Ecol. Model.* **2004**, *178*, 201–213. [CrossRef]
49. Buonocore, E.; Franzese, P.P.; Ulgiati, S. Assessing the environmental performance and sustainability of bioenergy production in Sweden: A life cycle assessment perspective. *Energy* **2012**, *37*, 69–78. [CrossRef]
50. Odum, H.T. An energy hierarchy law for biogeochemical cycles. In *Emergy Synthesis: Theory and Applications of the Emergy Methodology, Proceedings of the First Biennial Emergy Analysis Research Conference, Gainesville, FL, USA, September 1999*; Brown, M.T., Brandt-Williams, S., Tilley, D., Ulgiati, S., Eds.; Center for Environmental Policy: Gainesville, FL, USA, 2000; pp. 235–248.
51. Fahd, S.; Fiorentino, G.; Mellino, S.; Ulgiati, S. Cropping bioenergy and biomaterials in marginal land: The added value of the biorefinery concept. *Energy* **2012**, *37*, 79–93. [CrossRef]
52. National Environmental Accounting Database (NEAD). 2014. Available online: www.cep.ees.ufl.edu/emergy/nead.shtml (accessed on 15 September 2021).
53. Buranakarn, V. Evaluation of Recycling and Reuse of Building Materials Using the Emergy Analysis Method. Ph.D. Thesis, University of Florida, Gainesville, FL, USA, 1998.

Article

Facile and Scalable Synthesis and Self-Assembly of Chitosan Tartaric Sodium

Sixuan Wei, Rujie Peng, Shilong Bian, Wei Han, Biao Xiao * and Xianghong Peng *

Key Laboratory of Optoelectronic Chemical Materials and Devices (Ministry of Education), School of Optoelectronic Materials & Technology, Jianghan University, Wuhan 430056, China; WSX15927474220@163.com (S.W.); 18007161071@163.com (R.P.); bianshilong5200@163.com (S.B.); e781206379@gmail.com (W.H.)
* Correspondence: biaoxiao@jhun.edu.cn (B.X.); pxh@jhun.edu.cn (X.P.)

Citation: Wei, S.; Peng, R.; Bian, S.; Han, W.; Xiao, B.; Peng, X. Facile and Scalable Synthesis and Self-Assembly of Chitosan Tartaric Sodium. *Polymers* **2022**, *14*, 69. https://doi.org/10.3390/polym14010069

Academic Editors: Jingpeng Li, Yun Lu and Huiqing Wang

Received: 12 November 2021
Accepted: 22 December 2021
Published: 25 December 2021

Publisher's Note: MDPI stays neutral with regard to jurisdictional claims in published maps and institutional affiliations.

Copyright: © 2021 by the authors. Licensee MDPI, Basel, Switzerland. This article is an open access article distributed under the terms and conditions of the Creative Commons Attribution (CC BY) license (https:// creativecommons.org/licenses/by/ 4.0/).

Abstract: Chitosan-based nanostructures have been widely applied in biomineralization and biosensors owing to its polycationic properties. The creation of chitosan nanostructures with controllable morphology is highly desirable, but has met with limited success yet. Here, we report that nanostructured chitosan tartaric sodium (CS-TA-Na) is simply synthesized in large amounts from chitosan tartaric ester (CS-TA) hydrolyzed by NaOH solution, while the CS-TA is obtained by dehydration-caused crystallization. The structures and self-assembly properties of CS-TA-Na are carefully characterized by Fourier-transform infrared spectroscopy (FTIR), nuclear magnetic resonance spectroscopy (^{1}H-NMR), X-ray diffraction (XRD), differential scanning calorimeter (DSC), transmission electron microscopy (TEM), a scanning electron microscope (SEM) and a polarizing optical microscope (POM). As a result, the acquired nanostructured CS-TA-Na, which is dispersed in an aqueous solution 20–50 nm in length and 10–15 nm in width, shows both the features of carboxyl and amino functional groups. Moreover, morphology regulation of the CS-TA-Na nanostructures can be easily achieved by adjusting the solvent evaporation temperature. When the evaporation temperature is increased from 4 °C to 60 °C, CS-TA-Na nanorods and nanosheets are obtained on the substrates, respectively. As far as we know, this is the first report on using a simple solvent evaporation method to prepare CS-TA-Na nanocrystals with controllable morphologies.

Keywords: chitosan; tartaric acid; nanocrystal; self-assembly; evaporation; dehydration of crystallization

1. Introduction

Chitosan, a cationic polysaccharide composed of β-(1–4) linked 2-acetamido-2-deoxy-β-D-glucopyranose and 2-amino-2-deoxy-β-D-glycopyranose, is an alkaline deacetylation product of chitin, the second most abundant polysaccharide, which mainly comes from the exoskeletons of crustaceans, insects, beetles, as well as the cell walls of fungi [1]. Many of the applications of chitosan in several fields are based on its biological and excellent cationic properties [2,3], including biocompatibility [4], low immunogenicity, low or no toxicity, and antibacterial and moisture retentive properties [5,6]. Chitosan-based nanomaterials (such as nanogels, nanofibers, and nanocrystals) have been paid increasing attention due to their size-specific and free amine properties. In previous studies, chitosan nanogels were usually prepared (1) by non-covalent cross-linking with sodium tripolyphosphate, (2) by chemical cross-linking with glutaraldehyde, genipin and dicarboxylic acid, (3) by electrostatic interactions through changing the pH of the medium [7]. As demonstrated in a recent study, chitosan tartaric acid based-nanogels can be also prepared in a reverse microemulsion system through a condensation reaction between carboxylic groups of dicarboxylic acids and amino groups of chitosan, in which 1-ethyl-3-(3-dimethylaminopropyl)-carbodiimide (EDC) and N-hydroxysuccinimide (NHS) were used as coupling agents [8,9]. Although, they could effectively deliver vitamin B$_{12}$ and blue dextran, the obtained nanogels can only be dispersed in an acid solution, limiting their scope of application. According to the research, chitosan nanostructures (e.g., nanowires, nanotubes, and nanorods) could be obtained during the electrochemical synthesis, through tuning the reaction conditions to adjust the hydrogen-bonding interactions of chitosan. The chitosan was degraded by ultrasound in an acidified propylene carbonate solution to obtain chitosan embryo, then the chitosan nanostructures were produced under an electric field [10]. Although this synthetic process was simple and could be performed under mild and usual experimental laboratory conditions, the resulting chitosan nanostructures could only exist on the surface of the electrode. Other chitosan nanofibers were prepared using electrospinning method under high-voltage conditions (15–25 kV); nevertheless, they were all amorphous with only thread-like morphology. Other recent studies indicated that chitin nanostructures with amine groups were excellent scaffolds for bone regeneration and cell adhesion, and in particular, chitosan with acetamide groups was the main subject [11,12]. Currently, carboxylated chitosan has been prepared from carboxylated chitin by alkaline deacetylation. The carboxylated chitin was produced with ammonium persulfate as a mild oxidant [13]. Due to the high crystallinity of natural chitin structure, this carboxylated chitosan showed nanorod morphology in the aqueous solution. Furthermore, studies showed that the nanostructures of chitin and chitosan were mostly dispersed in water, and the accumulation of nanoparticles could affect the cationic properties of chitosan, limiting its further applications in the solid state. Meanwhile, the structured chitosan-based nanomaterial pattern will be beneficial to the applications of nanoscale carriers and sensors.

In nature, based on the regular nanostructures as building blocks, many remarkable biological materials are formed by self-assembling the repetitive discrete components into higher-order structures [14], such as squid pens and crab shells [15], butterfly wings [16,17], and pearls [18]. Notably, the biomaterials mentioned above have excellent mechanical properties, anisotropy and even structural colors due to their non-covalently linked hierarchical architectures [14]. In particular, photonic hydrogels could be prepared by using twisted mesoporous chitosan nanofibrils as a precursor for acetylation and a platform for templating poly(methylmethacrylate) [19]. As reported in previous studies, chitosan-based photonic crystals with a layered structure and color tunability could be also fabricated using surface binding and polymerization method. During the synthesis, polymethacrylic acid was deposited on a morpho butterfly wing template [16]. To date, inspired by nature, the specifically designed photonic crystals based on chitosan nanofibrils can respond to organic solvent or moisture [20–22]. Similar to chitosan-based nanostructures, self-assembled peptide nanostructures possess unique physical and biological properties and have broad application prospects in the field of electronic devices and functional molecular recognition [23]. Polypeptides with unique helical structures have been self-assembled due to the strong hydrogen bonds between amino and carboxyl groups in polypeptide molecules and other non-chemical bonds [24]. To date, controllable morphologies and patterns of polypeptides have been realized by evaporating dehumidification solution on template substrates [23]. Based on this method, we intended to synthesize carboxylated derivatives of chitosan while retaining their semi-crystalline structure, and to investigate their controllable morphologies.

Owing to their special properties, the creation of chitosan nanostructures with controllable morphology is highly desirable, but has had limited success so far. Therefore, we intended to synthesize chitosan tartaric sodium (CS-TA-Na) which has –NH$_2$ and –COO– groups and a semi-crystalline structure by a simple synthesis method. The controllable morphologies of CS-TA-Na nanostructures were prepared using a solvent evaporation method. To the best of our knowledge, this is the first report using a facile and scalable method to prepare a mass of CS-TA-Na nanostructures. The study on chitosan-based nanostructures will promote the development of biomineralization and biosensors.

2. Experimental Section

2.1. Materials

Chitosan (95% deacetylation, the viscosity of 1 wt% chitosan solution was 100–200 mpa·s and the source was shrimp shell as reported by the supplier) was purchased from Macklin Reagent Biochemical Co. Ltd., Shanghai, China. The other agents were purchased from Shanghai Sinopharm Reagent Co., Shanghai, China. All reagents were analytical grade and used without further purification.

2.2. Sample Preparation

The synthetic method was prepared according to previous work [8,9] with some modifications. 0.5 wt% chitosan tartaric acid aqueous solution was obtained by dissolving the chitosan and 7.5 wt% tartaric acid in water, then the solution was poured into an open container and placed in the 90 °C air drying for 12–24 h to realize dehydration caused crystallization. After that, the sample was washed with ethanol by 6.5 wt% concentration of the dispersion to obtain the powder, named chitosan tartaric ester (CS-TA). We mixed 2 wt% NaOH aqueous solution and the ethanol dispersion of CS-TA at same volumes accompanied with stirring, then the solution was centrifuged and the centrifugal precipitate was washed with ethanol, then dried to obtain the CS-TA-Na powder.

Glass and polytetrafluoroethylene (PTFE) were washed with deionized water and ethanol, and then soaked in anhydrous ethanol for later use. CS-TA-Na powder was dispersed in pH = 7–8 solution to obtain the CS-TA-Na dispersed aqueous solution, which was then drop-cast on the substrate surface. The CS-TA-Na substrate was dried at different temperatures, and soaked in anhydrous ethanol for 30 min, and then dried for observation.

2.3. Instrumentation

The ATR-FT-IR spectra of the samples were recorded on FT-IR spectrometer (TENSOR 27, BRUKER, Billerica, MA, USA) with KBr powder as background, resolution set as 4 cm^{-1}, average number of scans set as 16 and ranging from 4000 cm^{-1} to 600 cm^{-1}. ^{1}H-NMR spectra for the samples in a mixed solvent of CF_3COOD and D_2O (2:98 /w_1:w_2) were acquired on NMR spectrometer (AVANCE NEO 400M and AVANCE NEO 500M, BRUKER, Billerica, MA, USA). X-ray diffraction (XRD) patterns were obtained using an XRD diffractometer (D8-Advance, BRUKER, Billerica, MA, USA) with Cu Kα radiation (λ = 0.15406 nm). The sample morphologies were characterized by transmission electron microscopy operated at 200 kV (HT7700 EXALENS, HITACHI, Tokyo, Japan), scanning electron microscope (SU8010, HITACHI, Tokyo, Japan) with sample coated with gold, and polarizing microscope (BX41-LED, OLYMPUS, Tokyo, Japan).

3. Results and Discussion

It is well known that functional groups are the reason why chitosan has many attractive properties. Therefore, the ATR-FTIR test was used to explore the changes of functional group during the synthesis of CS-TA and CS-TA-Na. Figure 1 shows the FTIR spectra of chitosan, CS-TA and CS-TA-Na. The spectrum of pure chitosan exhibited characteristic peaks at 1650 cm^{-1} and 1599 cm^{-1} originating from amide I (C=O stretching) and amide II (–NH$_2$ stretching) of N-acetylglucosamine and N-glucosamine units, respectively (Figure 1A). Compared to chitosan, CS-TA spectrum presented a new peak at 1729 cm^{-1}, which is attributed to the C=O stretching band [8,25]. It has been confirmed in previous studies that C6 alcohols and the –NH$_2$ groups of chitosan react with the carboxylic groups of tartaric acid during the dehydration caused crystallization (Figure 1B) [8,26]. What is noteworthy is that the CS-TA-Na spectrum demonstrated a strong absorption peak at 1600 cm^{-1} derived from the symmetric and asymmetric –COONa and –NH$_2$ bonds, indicating that CS-TA was hydrolyzed in NaOH aqueous solution to form salt bonds with –COONa groups (Figure 1C) [27,28]. In short, the ATR-FTIR results revealed that CS-TA-Na was featured in both carboxyl and amino functional groups.

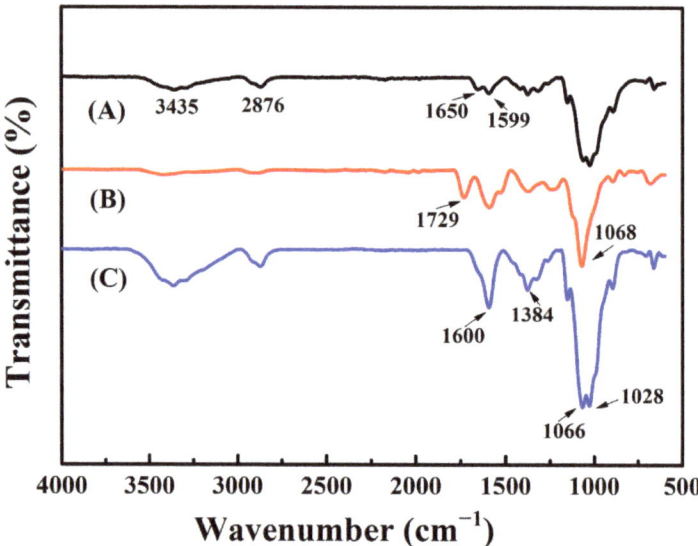

Figure 1. Attenuated total reflectance Fourier transform infrared (ATR-FTIR) spectra of (**A**) chitosan (CS), (**B**) chitosan tartaric ester (CS-TA) and (**C**) chitosan tartaric sodium (CS-TA-Na).

In order to accurately determine the chemical structure of the reactants and products, ^{1}H-NMR measurement was carried out. Figure 2 demonstrates typical ^{1}H-NMR spectra of chitosan and CS-TA-Na in CD_2COOD/D_2O = 2:98 ($w_1:w_2$) solvent. The ^{1}H-NMR spectra of CS-TA-Na exhibited an isolated peak at 4.15–4.21 ppm and 1.0–1.1 ppm, corresponding to the resonance of α-H protons of CS-TA-Na sample (Figure 2B). The presence of this resonance indicates the successful reaction between chitosan and tartaric acid. Meanwhile, the peak at 2.65 ppm was assigned to the three protons of N-acetylglucosamine (GlcNAc) units and $-CH_2COONa$ of glucosamine (GlcN) [26], while the peak at 3.1–3.5 ppm corresponds to H-2 proton of GlcN of chitosan and CS-TA-Na. The substitution degree (DA) of chitosan amino acid substituted by tartaric acid was calculated using H area and $-NH_2$ area on the pyranose ring, and the resulting DA value was 10.32%. The DA was calculated based on the ^{1}H-NMR spectrum considering the peak areas of the α-H of tartaric acid (4.15–4.2 ppm) and the GlcNAc (3.15–3.5 ppm) using the following equation [29]:

$$DA\ (\%) = [I_{\alpha\text{-H}}/(I_{GlcNAc}/5)] \times 100 \qquad (1)$$

$I_{\alpha\text{-H}}$ and I_{GlcNAc} are peak areas of the α-H of tartaric acid (4.15–4.2 ppm) and GlcNAc (3.15–3.5 ppm). The assignments and chemical shifts of the ^{1}H-NMR signals are given as follows: chitosan ^{1}H-NMN (D_2O/CD_3COOD, 500 MHz, 20 °C): d = 4.3–4.4 (1-H of GlcN), 3.1–3.4 (3-H, 4-H, 5-H, 6-H, 2-H of GlcNAc), 2.65 (2-H of GlcN), 1.65 (HN-COCH$_3$). CS-TA-Na ^{1}H-NMN (D_2O/CD_3COOD): d = 4.4–4.5 (1-H of GlcN), 3.15–3.5 (3-H, 4-H, 5-H, 6-H of GlcNAc), 2.75 (2-H of GlcN), 1.65 (HN-COCH$_3$), 4.15–4.2 (α-H of tartaric acid), 2.9–3.11 (α'-H of tartaric acid) [28,29].

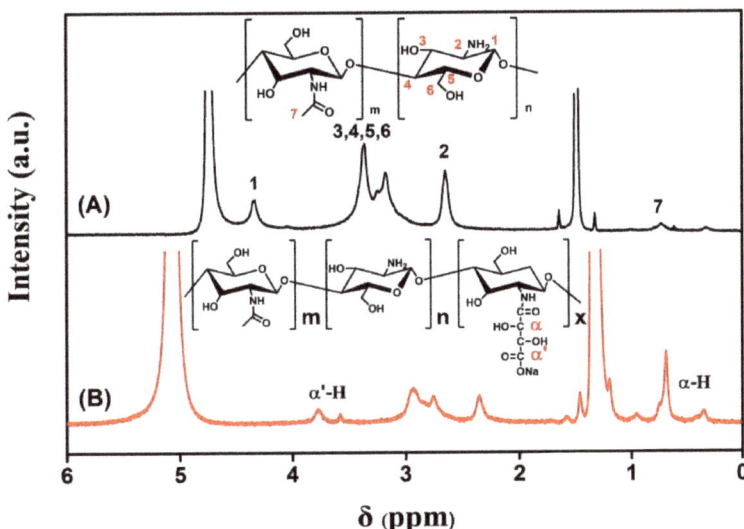

Figure 2. ^{1}H-NMR (nuclear magnetic resonance) spectra of (**A**) chitosan and (**B**) CS-TA-Na in CD$_2$COOD/D$_2$O = 2:98 (w_1:w_2) solvent.

The XRD test can further obtain the crystallization information of the materials. XRD patterns of chitosan, CS-TA and CS-TA-Na are shown in Figure 3. Compared to that of chitosan, the intensities of the crystalline diffraction peak at 2θ = 19.9° for 110 reflection increased, indicating that there were relatively regular lattices in CS-TA and CS-TA-Na (Figure 3A,B). Interestingly, a new peak at 2θ = 37.4° was observed in the pattern of CS-TA-Na, which is attributed to the stronger interaction between Na$^+$ and NH···O=C groups (Figure 3B) [30]. As previously reported, the nanoparticles of chitosan crosslinked with tartaric acid were essentially amorphous, because the crystal domain of chitosan was disturbed by the carboxyethyl groups [26]. In this work, the easy crystallization properties of tartaric acid would cause CS-TA to form relatively regular lattices.

Figure 4 shows the schematic of the synthesis process of CS-TA-Na. Firstly, chitosan was dissolved in tartaric acid solution, and its amine group protonated in the acid aqueous solution. Chitosan tartrate acid solution is a clear solution and stable for several months. Secondly, the chitosan tartaric acid aqueous solution was cast on the substrate, and then, dehydrated to cause crystallization and form CS-TA. As far as we know, the tartaric acid would crystallize during the dehydration and evaporation process, the crosslinking reaction and entanglement effects between chitosan and tartaric acid would cause the CS-TA to form a tightly stacked and long-range crystal structure. Such a result was proved by XRD results. Thirdly, CS-TA was hydrolyzed by NaOH solution to form salt linkages in between CS-TA-Na when CS-TA was added to NaOH solution, in which CS-TA-Na processes –NH$_2$ and –COONa groups. It is noteworthy that the regular crystal lattice of CS-TA-Na remained during the hydrolysis process at room temperature. To the best of our knowledge, this is the first report that the CS-TA-Na is synthesized using water evaporation and crystallization method without a complex purification process, and with this method, gram-grade CS-TA-Na can be obtained under the usual laboratory conditions.

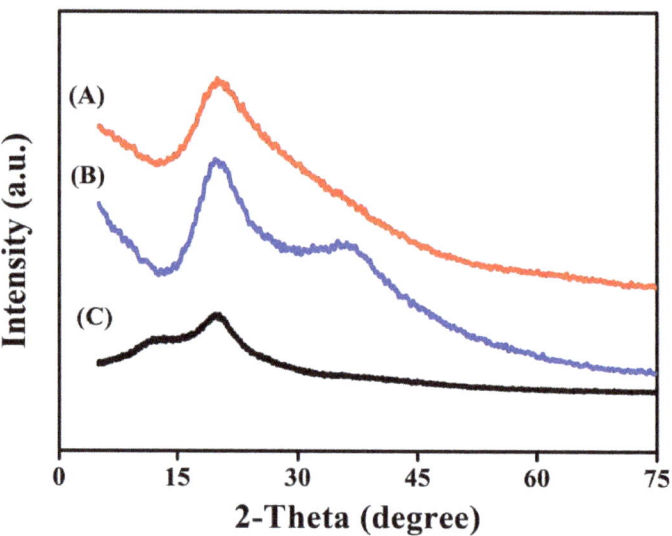

Figure 3. X-ray diffraction (XRD) patterns of (**A**) CS-TA, (**B**) CS-TA-Na, (**C**) chitosan.

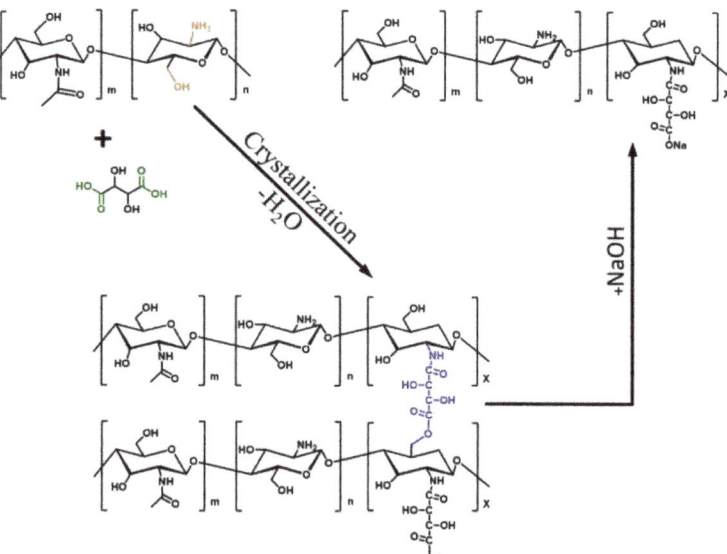

Figure 4. Schematic of the synthesis process of CS-TA-Na.

Figure 5 presents the photos of CS-TA-Na powder, CS-TA-Na water dispersion, and the TEM image of CS-TA-Na nanocrystals. It is worth noting that CS-TA-Na demonstrated a high yield and favorable water dispersibility. Meanwhile, the TEM image shows that the CS-TA-Na were about 20–50 nm in length and 10–15 nm in width. Interestingly, due to the hydrogen bonding interactions within and in between CS-TA-Na molecules, CS-TA-Na was prone to form a multilayer structure of polycrystals, leading to the parallel arrangement of CS-TA-Na nanocrystals and nanorods [31]. Such parallel structures are beneficial to self-assembly, resulting in ordered CS-TA-Na structures.

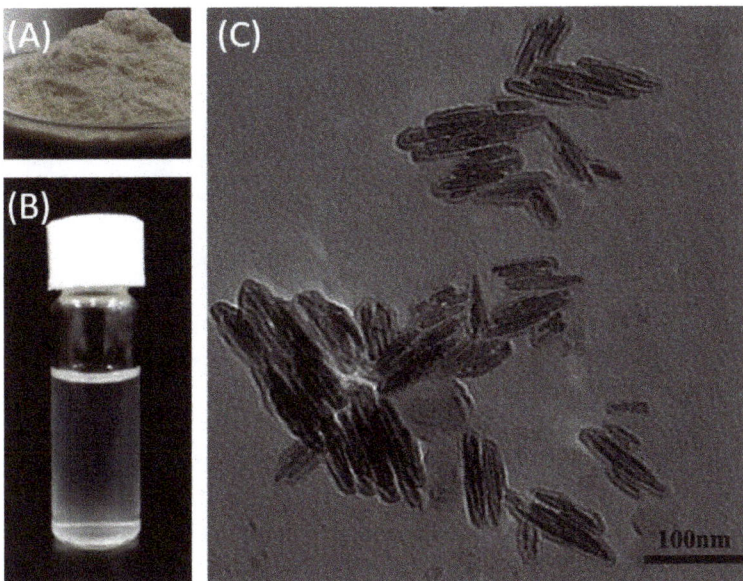

Figure 5. Photo of (**A**) CS-TA-Na powder and (**B**) its aqueous solution. (**C**) Transmission electron microscope (TEM) image of CS-TA-Na nanocrystals.

Figure 6 demonstrates the morphologies of self-assembled CS-TA-Na on glass and PTFE substrate at 4 °C, 25 °C and 60 °C, respectively. As shown in Figure 6G, the morphologies presented a dependence on the water evaporation rate and the surface characteristics of the substrate. When the water evaporation was slow (at 4 °C), CS-TA-Na had ample time to adjust and self-assemble, resulting in a more ordered structure. The assembled nanorods were more than 1 μm in length and 50 nm in width (Figure 6A,D) [31]. When the air-drying temperature increased, those dispersed CS-TA-Na nanocrystals tended to assemble quickly and interlace. At 25 °C, interlaced nanorods and nanosheets were observed (Figure 6B,E). Further increasing the drying temperature to 60 °C could lead to the flower-like pattern composed of vertical nanocrystals and nanosheets, as shown in Figure 6C,F. Obviously, there existed the vertical CS-TA-Na assembly on both the glass and PTFE substrate at 60 °C, suggesting that the morphologies of the patterns were mostly affected by the strong vertical convection originating from the higher rate of solvent evaporation. Therefore, the three-dimensional orders of the CS-TA-Na patterns could be controlled by tuning the rate of water evaporation and the surficial conditions of the substrate [23,32].

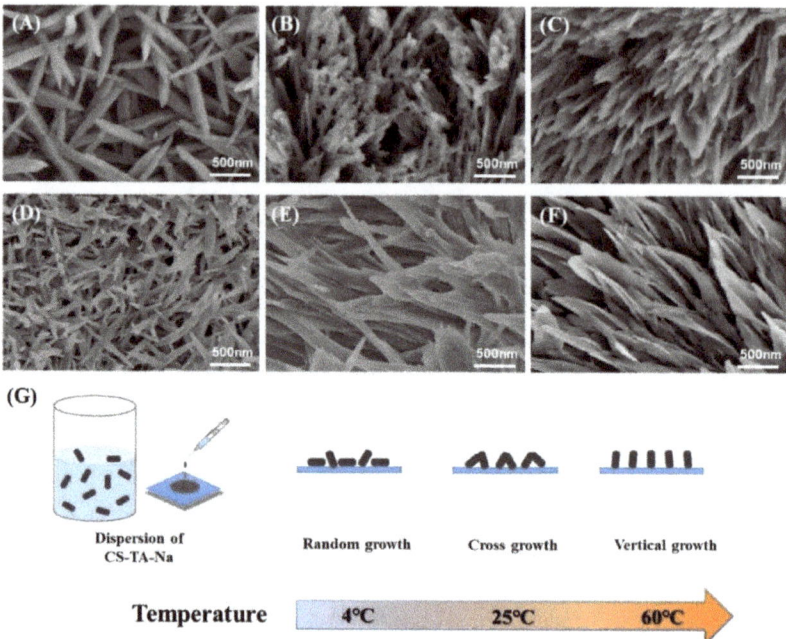

Figure 6. Scanning electron microscope (SEM) images: 5 μL 0.05 wt% CS-TA-Na dispersion drop-cast on glass substrate at (**A**) 4 °C, (**B**) 25 °C and (**C**) 60 °C, on PTFE substrate at (**D**) 4 °C, (**E**) 25 °C and (**F**) 60 °C. (**G**) Schematic of self-assembly of CS-TA-Na at different temperatures.

The self-assembly process of CS-TA-Na follows the nucleation and growth mechanism, as shown in Figure 7. First, CS-TA-Na nanocrystals dispersion was dripped onto the substrate to form a uniform liquid layer (Figure 7A). Then, nanocrystals moved towards the edge to form the seeds for nucleus growth on the substrate edge during the solvent evaporation. Finally, nanocrystals continued to grow around the initial nuclei to form the nanocrystals' domain and dendrimer. The bright birefringence pattern of the nanocrystals with the main dendritic morphology was observed after the dispersion droplets were cast and evaporated at 60 °C for 60 s (Figure 7B,C). Meanwhile, an obvious coffee ring was more proof that the self-assembly of CS-TA-Na nanocrystals was influenced by fluid mechanical effects [33].

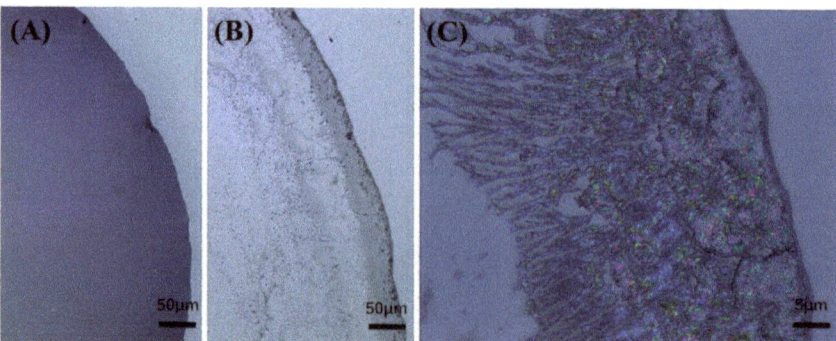

Figure 7. Polarizing optical microscope (POM) images of 0.05 wt% CS-TA-Na water dispersion self-assembled at 60 °C with the lapse of time: (**A**) 0 s, (**B**) 60 s, (**C**) enlarged view of (**B**).

4. Conclusions

In summary, we presented, for the first time, a facile and scalable method by which we can readily produce a mass of CS-TA by solvent-evaporation caused crystallization, in which tartaric acid was used as the crystallization and the crosslinking agent. Subsequently, CS-TA-Na featured in surface carboxyl, amine functionalities and regular crystal lattice structure was prepared by hydrolysis with NaOH aqueous solution. The resulting CS-TA-Na nanocrystals were 20–50 nm in length and 10–15 nm in width in solution. The nanostructure patterns of CS-TA-Na (including nanosheets, nanorods) were then obtained by dripping its aqueous dispersions on various substrates and subsequently evaporating the solvent. Meanwhile, the exact shape and morphology of the nanostructures can be controlled by the air-drying temperature. As a result, the favorite CS-TA-Na nanosheets standing on the substrate were acquired at 60 °C on various substrates. It is the first time that self-assembly nanostructures with controllable morphology have been reported for chitosan. Importantly, this work proves a simple route to prepare chitosan-based nanostructure patterns.

Author Contributions: Author Contributions: Conceptualization, writing-original draft preparation, supervision, S.W., R.P., S.B., W.H.; review and editing, B.X. and X.P. All authors have read and agreed to the published version of the manuscript.

Funding: This work was financially supported by the Opening Project of key Laboratory of Optoelectronic Chemical Materials and Devices of Ministry of Education, Jianghan University (JDGD-202021), The Discipline Cultivation Plan of Material Science and Engineering, Jianghan University (03100023). The National Natural Science Foundation of China (52103213), and the project of state guiding regional development for Hubei province (2019ZYYD005).

Institutional Review Board Statement: Not applicable.

Informed Consent Statement: Not applicable.

Data Availability Statement: The data that support the findings of this study are available from the corresponding author, X.P., upon reasonable request.

Conflicts of Interest: The authors declare no conflict of interest.

References

1. Federer, C.; Kurpiers, M.; Bernkop-Schnürch, A. Thiolated chitosans: A multi-talented class of polymers for various applications. *Biomacromolecules* **2021**, *22*, 24–56. [CrossRef]
2. Feng, Y.; Gao, H.L.; Wu, D.; Weng, Y.T.; Wang, Z.Y.; Yu, S.H.; Wang, Z. Biomimetic lamellar chitosan scaffold for soft gingival tissue regeneration. *Adv. Funct. Mater.* **2021**, *31*, 2105348. [CrossRef]
3. Zhou, Y.; Fang, Y.; Li, P.; Yan, L.; Fan, X.; Wang, Z.; Zhang, W.; Liu, H. Ampholytic Chitosan/Alginate Composite Nanofibrous Membranes with Super Anti-Crude Oil-Fouling Behavior and Multifunctional Oil/Water Separation Properties. *ACS Sustain. Chem. Eng.* **2019**, *7*, 15463–15470. [CrossRef]
4. Xia, J.; Zhang, H.; Yu, F.; Pei, Y.; Luo, X. Superclear, Porous Cellulose Membranes with Chitosan-Coated Nanofibers for Visualized Cutaneous Wound Healing Dressing. *ACS Appl. Mater. Interfaces* **2020**, *12*, 24370–24379. [CrossRef] [PubMed]
5. Sahariah, P.; Másson, M. Antimicrobial Chitosan and Chitosan Derivatives: A Review of the Structure-Activity Relationship. *Biomacromolecules* **2017**, *18*, 3846–3868. [CrossRef]
6. Cai, Y.; Zhong, Z.; He, C.; Xia, H.; Hu, Q.; Wang, Y.; Ye, Q.; Zhou, J. Homogeneously Synthesized Hydroxybutyl Chitosans in Alkali/Urea Aqueous Solutions as Potential Wound Dressings. *ACS Appl. Bio Mater.* **2019**, *2*, 4291–4302. [CrossRef]
7. Du, Z.; Liu, J.; Zhang, T.; Yu, Y.; Zhang, Y.; Zhai, J.; Huang, H.; Wei, S.; Ding, L.; Liu, B. A study on the preparation of chitosan-tripolyphosphate nanoparticles and its entrapment mechanism for egg white derived peptides. *Food Chem.* **2019**, *286*, 530–536. [CrossRef] [PubMed]
8. Pujana, M.A.; Pérez-Álvarez, L.; Iturbe, L.C.C.; Katime, I. Water dispersible pH-responsive chitosan nanogels modified with biocompatible crosslinking-agents. *Polymer* **2012**, *53*, 3107–3116. [CrossRef]
9. Ruiz-Caro, R.; Veiga, M.D.; Meo, C.D.; Cencetti, C.; Coviello, T.; Matricardi, P.; Alhaique, F. Mechanical and Drug Delivery Properties of a Chitosan–Tartaric Acid Hydrogel Suitable for Biomedical Applications. *J. Appl. Polym. Sci.* **2011**, *123*, 842–849. [CrossRef]
10. Gong, J.; Hu, X.; Wong, K.W.; Zheng, Z.; Yang, L.; Lau, W.M.; Du, R. Chitosan Nanostructures with Controllable Morphology Produced by a Nonaqueous Electrochemical Approach. *Adv. Mater.* **2008**, *20*, 2111–2115. [CrossRef]

11. Duan, B.; Shou, K.; Su, X.; Niu, Y.; Zheng, G.; Huang, Y.; Yu, A.X.; Zhang, Y.; Xia, H.; Zhang, L. Hierarchical Microspheres Constructed from Chitin Nanofibers Penetrated Hydroxyapatite Crystals for Bone Regeneration. *Biomacromolecules* **2017**, *18*, 2080–2089. [CrossRef] [PubMed]
12. Suzuki, S.; Teramoto, Y. A simple inkjet process to fabricate microstructures of chitinous nanocrystals for cell patterning. *Biomacromolecules* **2017**, *18*, 1993–1999. [CrossRef] [PubMed]
13. Jin, T.; Kurdyla, D.; Hrapovic, S.; Leung, A.C.W.; Régnier, S.; Liu, Y.; Moores, A.; Lam, E. Carboxylated Chitosan Nanocrystals: A Synthetic Route and Application as Superior Support for Gold-Catalyzed Reactions. *Biomacromolecules* **2020**, *21*, 2236–2245. [CrossRef] [PubMed]
14. Ye, D.; Lei, X.; Li, T.; Cheng, Q.; Chang, C.; Hu, L.; Zhang, L. Ultrahigh Tough, Super Clear and Highly Anisotropic Nanofibers-Structured Regenerated Cellulose Films. *ACS Nano* **2019**, *13*, 4843–4853. [CrossRef]
15. Ling, S.; Chen, W.; Fan, Y.; Zheng, K.; Jin, K.; Yu, H.; Buehler, M.J.; Kaplan, D.L. Biopolymer nanofibrils: Structure, modeling, preparation, and applications. *Prog. Polym. Sci.* **2018**, *85*, 1–56. [CrossRef]
16. Yang, Q.; Zhu, S.; Peng, W.; Yin, C.; Wang, W.; Gu, J.; Zhang, W.; Ma, J.; Deng, T.; Feng, C.; et al. Bioinspired Fabrication of Hierarchically Structured, pH-Tunable Photonic Crystals with Unique Transition. *ACS Nano* **2013**, *7*, 4911–4918. [CrossRef]
17. Li, Q.; Zeng, Q.; Shi, L.; Zhang, X.; Zhang, K.Q. Bio-inspired sensors based on photonic structures of Morpho butterfly wings: A review. *J. Mater. Chem. C* **2016**, *4*, 1752–1763. [CrossRef]
18. Liu, C.; Xu, G.; Du, J.; Sun, J.; Wan, X.; Liu, X.; Su, J.; Liang, J.; Zheng, G.; Xie, L.; et al. Mineralization of Nacre-like Structures Mediated by Extrapallial Fluid on Pearl Nucleus. *Cryst. Growth Des.* **2017**, *18*, 32–36. [CrossRef]
19. Nguyen, T.D.; Peres, B.U.; Carvalho, R.M.; MacLachlan, M.J. Photonic Hydrogels from Chiral Nematic Mesoporous Chitosan Nanofibril Assemblies. *Adv. Funct. Mater.* **2016**, *26*, 2875–2881. [CrossRef]
20. Huang, G.; Yin, Y.; Pan, Z.; Chen, M.; Zhang, L.; Liu, Y.; Zhang, Y.; Gao, J. Fabrication of 3D Photonic Crystals from Chitosan That Are Responsive to Organic Solvents. *Biomacromolecules* **2014**, *15*, 4396–4402. [CrossRef]
21. Retamal, M.J.; Corrales, T.P.; Cisternas, M.; Moraga, N.; Diaz, D.; Catalan, R.; Seifert, B.; Huber, P.; Volkmann, U.G. Surface morphology of vapor-deposited chitosan: Evidence of solid-state dewetting during the formation of biopolymer films. *Biomacromolecules* **2016**, *17*, 1142–1149. [CrossRef] [PubMed]
22. Zhu, K.; Duan, J.; Guo, J.; Wu, S.; Lu, A.; Zhang, L. High-Strength Films Consisted of Oriented Chitosan Nanofibers for Guiding Cell Growth. *Biomacromolecules* **2017**, *18*, 3904–3912. [CrossRef]
23. Chen, J.; Qin, S.; Wu, X.; Chu, P.K. Morphology and Pattern Control of Diphenylalanine Self-Assembly via Evaporative Dewetting. *ACS Nano* **2015**, *10*, 832–838. [CrossRef]
24. Qin, J.; Luo, T.; Kiick, K.L. Self-assembly of stable nanoscale platelets from designed elastin-like peptide—Collagen-like peptide bioconjugates. *Biomacromolecules* **2019**, *20*, 1514–1521. [CrossRef] [PubMed]
25. Demarger-Andre, S.; Domard, A. Chitosan carboxylic acid salts in solution and in the solid state. *Carbohydr. Polym.* **1994**, *23*, 211–219. [CrossRef]
26. Huang, J.; Xie, H.H.; Hu, S.; Gong, J.Y.; Jiang, C.J.; Xie, T.; Ge, Q.; Wu, Y.; Cui, Y.; Liu, S.W.; et al. Preparation, Characterization and Biochemical Activities of N-(2-Carboxyethyl)chitosan from Squid Pens. *J. Agric. Food Chem.* **2015**, *63*, 2464–2471. [CrossRef]
27. Huang, W.; Wang, Y.; Huang, Z.; Wang, X.; Chen, L.; Zhang, Y.; Zhang, L. On-Demand Dissolvable Self-healing Hydrogels Based on Carboxymethyl Chitosan and Cellulose Nanocrystal for Deep Partial Thickness Burn Wound Healing. *ACS Appl. Mater. Interfaces* **2018**, *10*, 41076–41088. [CrossRef] [PubMed]
28. Kurita, K.; Akao, H.; Yang, J.; Shimojoh, M. Nonnatural Branched Polysaccharides: Synthesis and Properties of Chitin and Chitosan Having Disaccharide Maltose Branches. *Biomacromolecules* **2003**, *4*, 1264–1268. [CrossRef] [PubMed]
29. Phongying, S.; Aiba, S.I.; Chirachanchai, S. Direct chitosan nanoscaffold formation via chitin whiskers. *Polymer* **2007**, *48*, 393–400. [CrossRef]
30. Fang, Y.; Zhang, R.; Duan, B.; Liu, M.; Lu, A.; Zhang, L. Recyclable Universal Solvents for Chitin to Chitosan with Various Degrees of Acetylation and Construction of Robust Hydrogels. *ACS Sustain. Chem. Eng.* **2017**, *5*, 2725–2733. [CrossRef]
31. Okuyama, K.; Noguchi, K.; Miyazawa, T.; Yui, T.; Ogawa, K. Molecular and Crystal Structure of Hydrated Chitosan. *Macromolecules* **1997**, *30*, 5849–5855. [CrossRef]
32. Zhou, J.; Man, X.; Jiang, Y.; Doi, M. Structure Formation in Soft-Matter Solutions Induced by Solvent Evaporation. *Adv. Mater.* **2017**, *29*, 1703769. [CrossRef] [PubMed]
33. Lohani, D.; Basavaraj, M.G.; Satapathy, D.K.; Sarkar, S. Coupled effect of concentration, particle size and substrate morphology on the formation of coffee rings. *Colloid Surf. A-Physicochem. Eng. Asp.* **2019**, *589*, 124387. [CrossRef]

Article

Effects of Raw Material Source on the Properties of CMC Composite Films

Yao Yao [1,†], Zhenbing Sun [1,†], Xiaobao Li [1,†], Zhengjie Tang [1], Xiaoping Li [1,2,*], Jeffrey J. Morrell [3,*], Yang Liu [4], Chunli Li [1] and Zhinan Luo [1]

1. Yunnan Key Laboratory of Wood Adhesives and Glue Products, Southwest Forestry University, Kunming 650224, China; yaoy1012@163.com (Y.Y.); sunzhenbing66@163.com (Z.S.); lxb15925024878@163.com (X.L.); zhengjietang@163.com (Z.T.); LCL2106448925@163.com (C.L.); lzn5960929@163.com (Z.L.)
2. International Joint Research Center for Biomass Materials, Southwest Forestry University, Kunming 650224, China
3. National Centre for Timber Durability and Design Life, University of the Sunshine Coast, Brisbane, QLD 4102, Australia
4. Qingdao Huicheng Adhesive Co., Ltd., Qingdao 266021, China; liuyang801129@126.com
* Correspondence: lxp810525@163.com (X.L.); jmorrell@usc.edu.au (J.J.M.)
† These authors contributed equally to this work.

Citation: Yao, Y.; Sun, Z.; Li, X.; Tang, Z.; Li, X.; Morrell, J.J.; Liu, Y.; Li, C.; Luo, Z. Effects of Raw Material Source on the Properties of CMC Composite Films. *Polymers* 2022, 14, 32. https://doi.org/10.3390/polym14010032

Academic Editors: Jingpeng Li, Yun Lu and Huiqing Wang

Received: 15 November 2021
Accepted: 20 December 2021
Published: 22 December 2021

Publisher's Note: MDPI stays neutral with regard to jurisdictional claims in published maps and institutional affiliations.

Copyright: © 2021 by the authors. Licensee MDPI, Basel, Switzerland. This article is an open access article distributed under the terms and conditions of the Creative Commons Attribution (CC BY) license (https://creativecommons.org/licenses/by/4.0/).

Abstract: Sodium carboxymethyl cellulose (CMC) can be derived from a variety of cellulosic materials and is widely used in petroleum mining, construction, paper making, and packaging. CMCs can be derived from many sources with the final properties reflecting the characteristics of the original lignocellulosic matrix as well as the subsequent separation steps that affect the degree of carboxy methyl substitution on the cellulose hydroxyls. While a large percentage of CMCs is derived from wood pulp, many other plant sources may produce more attractive properties for specific applications. The effects of five plant sources on the resulting properties of CMC and CMC/sodium alginate/glycerol composite films were studied. The degree of substitution and resulting tensile strength in leaf-derived CMC was from 0.87 to 0.89 and from 15.81 to 16.35 MPa, respectively, while the degree of substitution and resulting tensile strength in wooden materials-derived CMC were from 1.08 to 1.17 and from 26.08 to 28.97 MPa, respectively. Thus, the degree of substitution and resulting tensile strength tended to be 20% lower in leaf-derived CMCs compared to those prepared from wood or bamboo. Microstructures of bamboo cellulose, bamboo CMC powder, and bamboo leaf CMC composites' films all differed from pine-derived material, but plant source had no noticeable effect on the X-ray diffraction characteristics, Fourier transform infrared spectroscopy spectra, or pyrolysis properties of CMC or composites films. The results highlighted the potential for using plant source as a tool for varying CMC properties for specific applications.

Keywords: Chinese pine wood; pine needles; bamboo culms; bamboo leaves; industrial hemp hurd; CMC; DS; mechanical properties; TG; XRD; FTIR

1. Introduction

Sodium carboxymethyl cellulose (CMC) derived from the cellulose-containing materials is widely used in petroleum extraction, cement modification, textile production, paper making, soil improvement, and water pollution treatments [1–3]. While preparation of CMCs via acid-catalyzed reactions of cellulose with chloroacetic acid is relatively straightforward, the properties of the resulting product can vary widely depending on the plant source as well as the method of cellulose separation, which produces different degrees of substitution of carboxyl methyl groups on the cellulose hydroxyls. These differences have stimulated research to identify alternative sources for CMC including plant foliage and bark [4–8]; for example, the DS (degree of substitution) of CMC derived from corn straw was between 0.6 to 0.7 [4] and the DS of CMC derived from rice straw and reed were lower than 1.0 [5]. Foliage has the advantage of being continually harvested, while bark is often a low-value by-product of timber processing. The resulting materials can then be used to create biodegradable plastics to replace petroleum-derived materials or be combined with other materials to produce antimicrobial systems [9–11]. Compounded CMC, glycerol, dioscorea mucus, and Ag nanoparticles to prepare a material with strong antibacterial property and its maximum tensile strength was 12.21 MPa [10]. A composite film was prepared by adding okra mucus and nano zinc oxide in the CMC solution, and the inhibition diameters of the optimal composite sample against *Glucose aureus* and *Escherichia coli* were 14.35 ± 0.21 mm and 10.31 ± 0.21 mm, respectively, but its tensile strength was 10.26 ± 0.66 MPa [11].

Film properties can be further amended with extenders and thickeners such as starch, gelatin, pectin, and sodium alginate [2], while chitosans can be used to improve tensile strength or limit microbial attack [12,13]. The tensile strength and elongation at the break of the CMC composites prepared with 1.5% sodium alginate, 0.5% CMC, and 1.5% chitosan were 65.32 MPa and 17.85%, respectively [12]. Sodium alginate, CMC, and pyrogallic acid were used to make an antibacterial material for food packaging, and its tensile strength and elongation at the break were 24.22 ± 0.58 MPa and 39.60 ± 0.28%, respectively, and the inhibition diameters of glucose aureus and escherichia coli were 34.0 ± 1.1 mm and 18.0 ± 1.0 mm, respectively [13]. Glycerol, sorbitol, xylitol, and fructose are commonly used as plasticizers in CMC composites, but they can also enhance oxygen resistance [14,15]. Glycerol has the best plasticizing effect and the elongation at the break increases from 68.1% to 69.6% as the addition of glycerol in starch/sodium alginate/CMC composites increases from 0% to 7% [14]. Adding glycerol to sodium alginate/CMC composites can improve its oxygen resistance, and the optimum addition amount is 3% [15]. There is a diverse array of potential CMC additives, but there are relatively few studies comparing the properties of CMC films from different plant sources.

The objective of this study was to assess the impact of five different cellulose sources on the resulting properties of CMC alone or in a composite film.

2. Materials and Methods

2.1. Materials

The materials were obtained from Kunming, Yunnan Province (Table 1). All materials were air-dried and ground to powder to pass through a 40- to 60-mesh screen prior to processing. The materials differed in terms of % of lignin, hemicellulose, and cellulose with hemp hurd having the lowest lignin content and bamboo, pine wood, and hemp hurd having the most cellulose.

Table 1. Lignin, cellulose, and hemicelluloses content of materials used to produce CMCs.

Source	Lignin (%)	Holo-Cellulose (%)	Cellulose (%)	Hemicellulose (%)	Source
Pine wood	23.0	-	39.9	14.9	[16]
Bamboo	26.4 (0.04)	66.7 (0.02)	43.9 (0.02)	22.8 (0.03)	[17]
Pine needles	29.3 (0.3)	40.8 (0.3)	20.5 (0.07)	20.3 (0.4)	[18]
Bamboo leaves	25.2 (0.8)	57.3 (0.5)	19.5 (0.4)	37.7 (0.5)	[19]
Hemp Hurd	20.9 (0.1)	70.8 (0.1)	42.9 (0.1)	27.9 (0.2)	[19]

2.2. Cellulose Preparation

The ground powder of each material was soaked in an excess of 95% ethanol for 6 h at 70 °C to remove fatty acids. The 5-g samples were then collected by filtration, transferred to a 500-mL flask along with a mixture of 5 mL of glacial acetic acid, 2.5 g of sodium chlorite, and 375 mL of distilled water, and heated at 75 °C for 1 h. The solution was decanted, and 5 mL of glacial acetic acid and 2.5 g sodium chlorite were added and heated for an additional hour. The process was repeated until the material was white, indicating lignin digestion. The solids were collected by filtration and treated with 75 mL of 17.5% aqueous

NaOH at room temperature for 45 min to remove the hemicelluloses. Then, the samples were neutralized by repeated washing with distilled water before being dried at 104 °C and stored in a dessicator until needed.

2.3. CMC and CMC Composition Films' Preparation

CMC preparation: Four g of cellulose, 80 mL of 95% ethanol, and 20 mL of 30% NaOH solution were mixed and stirred for 60 min at 30 °C. Then, 5 g of sodium chloroacetate were added and the temperature was increased to 65 °C and held for 3 more hours with stirring. Glacial acetic acid (90%) solution was added to reduce the pH of the mixture and then the samples were washed with alcohol until the pH was 7. The neutralized samples were oven-dried at 65 °C and stored for later use (Figure 1).

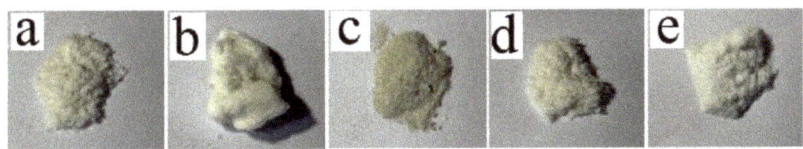

Figure 1. Examples of CMC powders from (**a**) pine wood, (**b**) bamboo, (**c**) pine needles, (**d**) bamboo leaves, and (**e**) hemp hurd.

CMC film preparation: Two g of CMC powder were dispersed in 98 mL of distilled water and stirred at 900–1000 rpm at 70 °C until the CMC was completely dissolved. The solution was amended with 1.4 g of sodium alginate and 0.25 g of glycerol and thoroughly mixed before being sonicated to remove air bubbles. The resulting liquid was placed in a polytetrafluoroethylene (PTFE) mould and allowed to solidify (Figure 2).

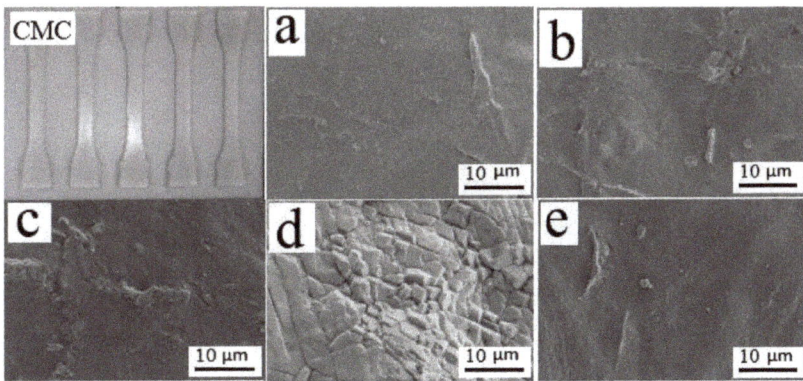

Figure 2. Example of CMC composite dog-bones used to assess tensile strength and scanning electron micrographs of CMC composites derived from (**a**) pine wood, (**b**) bamboo culm, (**c**) pine needles, (**d**) bamboo leaves, and (**e**) hemp hurd, showing slightly different topographies based upon parent material.

2.4. Tensile Properties and Opacity of CMC Composites' Films

Tensile strength (MPa) and elongation at break (%) were measured on 10 0.089- to 0.098-mm by 150-mm-long dog-bone samples of each material on a Universal Testing Machine according to procedures described in GB/T 1040.1-2006 (Plastics Determination of tensile properties). A load was applied to failure at a rate of 1 mm/min.

The opacity of the CMC composite films was tested by cutting 10- by 40-mm-long samples and placing them on the inner surface on one side of a cuvette and then measuring absorbance at 600 nm on a 752# ultraviolet spectrophotometer (XP-Spectrum Company, Shanghai, China). Five tests were performed for each material.

2.5. Material Charaterization

Microstructure: Samples of the parent materials, the extracted cellulose, reacted CMC powder, and the resulting CMC composite film were placed on an aluminium grid and examined by field emission scanning electron microscopy on a Nova NanoSEM 450 microscope (FEI, Hillsboro, OR, USA). A minimum of five fields were examined per material.

Degree of CMC Substitution: The degree of substitution on the hydroxyls had important effects on the resulting CMC properties. The degree of substitution was determined by the ash alkali method [4], wherein 1.5 g of CMC powder were placed into a crucible and washed four to five times with 80% ethanol at 50 °C to 70 °C to remove any residual soluble salt and then washed once with 100% ethanol. The samples were dried at 104 °C. Then, 1 g of the oven-dried sample was placed in a crucible and then heated in a muffle furnace of 300 °C until no smoke was observed. The temperature was then increased to 700 °C for 15 min. The sample was removed from the oven and allowed to cool to 200 °C before being transferred to a beaker with 100 mL of distilled water and 50 mL of 0.1 mol/L sulfuric acid standard titration solution. The beaker was heated to boiling for 10 min. Then 2–3 drops of methyl red indicator solution were added. A 0.1 mol/L sodium hydroxide standard solution was added dropwise until the solution turned from red to white. The amount required to reach the end point was then used to calculate the degree of substitution (DS) based on Equations (1) and (2), as follows.

$$B = \frac{c_1 V_1 - c_2 V_2}{m} \quad (1)$$

$$DS = \frac{162B}{1000 - 80B} = \frac{0.162B}{1 - 0.08B} \quad (2)$$

where B is the Amount of carboxymethyl substance contained in the sample, mmol/g; m is the Quality of the sample, g; c_1 is the Concentration of sulfuric acid standard titration solution, mol/L; V_1 is the Volume of sulfuric acid standard titration solution, mL; c_2 is the Concentration of sodium hydroxide standard titration solution, mol/L; and V_2 is the Volume of sodium hydroxide standard titration solution, mL.

X-ray Diffractometer analysis: The relative degree of crystallinity of the raw materials, the extracted cellulose, and the reacted CMC was examined by X-ray diffractometry on a Rigaku Ultima IV X-ray diffractometer (Rigaku Corp, Tokyo, Japan) (XRD, Ulti,) using a scanning angle from 5° to 80°, a step size of 0.026° (accelerating current = 30 mA and voltage = 40 kV), and Cu-Kα radiation of $\lambda = 0.154$ nm.

Fourier Transform Infrared Spectroscopy (FTIR): Extracted cellulose, CMC, and CMC composite powder from each plant source were mixed with KBr, pressed into a pellet, and analyzed on a Nicolet i50 FTIR Analyzer (Thermo Scientific, Waltham, MA, USA). Samples were subjected to 64 scans and the resulting spectra were baseline corrected and then analyzed for differences in spectra for different raw materials.

Thermogravimetric (TG) analysis: Approximately 5.0 to 6.0 mg of the original air-dried raw materials as well as the extracted cellulose, the CMC, and the CMC composites film were ground to pass an 80-mesh to 120-mesh and placed into sample holders for analysis on a TGA92 thermo gravimetric analyzer (KEP Technologies EMEA, Caluire, France). N_2 was used as the shielding gas and Al_2O_3 as the reference compound. The temperature was increased from room temperature (approx. 20–23 °C) to 800 °C at a rate of 10 °C/min to produce thermogravimetric curves.

3. Results and Discussion

3.1. The Physical and Mechanical Properties of CMC Composites' Films

The Color of the CMC powders varied slightly with source and the resulting films contained small particles or protrusion (Figures 1 and 2). CMCs prepared from bamboo leaves contained slender particles, which were not present in the other materials (Figure 3).

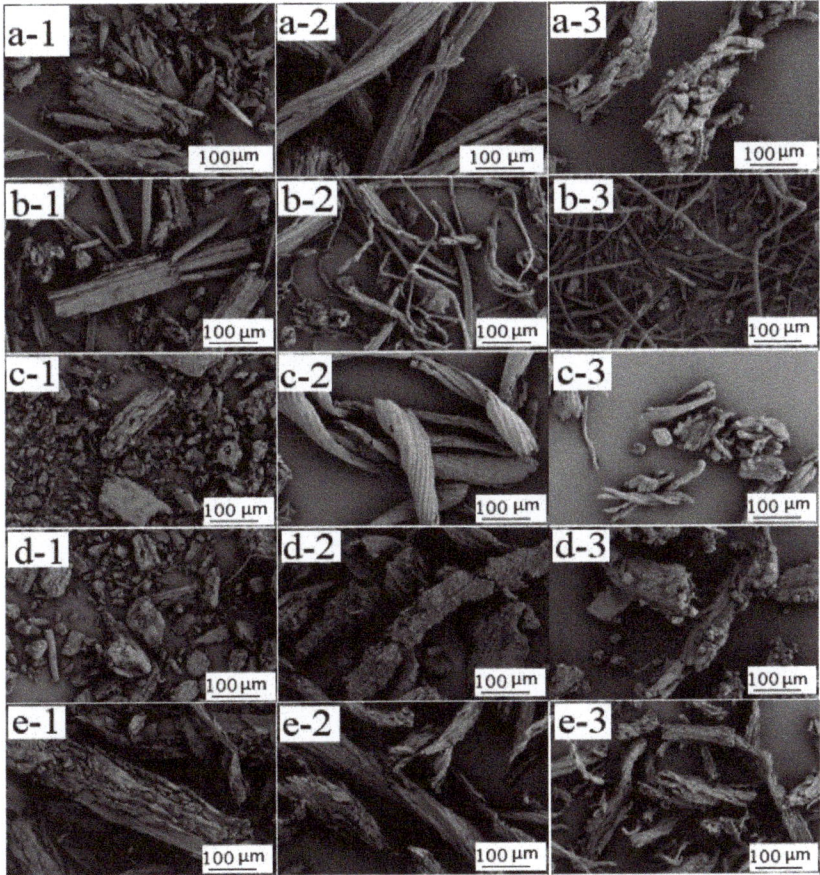

Figure 3. SEM images showing microstructure differences for the (**1**) the raw material, (**2**) the extracted cellulose, and (**3**) the resulting CMC derived from (**a**) pine wood, (**b**) bamboo culm, (**c**) pine needles, (**d**) bamboo leaves, and (**e**) hemp hurd.

The degrees of substitution of hydroxyls were similar for CMCs from bamboo foliage and pine needles but were nearly 20% lower than those for CMCs from pine wood, bamboo culm, or hemp hurd (Table 2). The lower levels of substitution on leaves or needles were surprising since the cellulose in these materials should be less heavily lignified and, therefore, more accessible to substitution. Tensile strength was also lower for the foliage-derived CMCs. The tensile strength of CMC composite films ranged from 15.81 MPa to 28.97 MPa, which were much higher than results from previous studies [9–11,13], while elongation at break (%) ranged from 3.37% to 6.60%, which were lower than previous reports [12]. Chitosan addition might help improve tensile strength [12] and will be explored in future studies.

Table 2. Characteristics of CMCs and CMC films produced from different cellulosic sources.

Source	Degree of Substitution	Tensile Strength (MPa)	Elongation at Break (%)	Opacity (A/mm)
Pine wood	1.17 (0.026)	27.50 (1.93)	3.50 (1.23)	4.34 (0.27)
Bamboo culm	1.08 (0.046)	28.97 (3.17)	3.50 (1.42)	3.67 (0.26)
Pine needles	0.87 (0.025)	15.81 (2.19)	6.60 (0.79)	6.60 (0.18)
Bamboo leaves	0.89 (0.071)	16.35 (1.27)	3.67 (0.34)	4.34 (0.31)
Hemp hurd	1.12 (0.088)	26.08 (2.69)	3.37 (0.39)	5.22 (0.14)

Values represents means of three samples for degree of substitution, 10 samples for tensile strength and elongation, and five samples for opacity, while figures in parentheses represent one standard deviation.

Opacity of the resulting films was similar for the pine wood, bamboo culm, and foliage, while films from pine needles were slightly higher.

Increased degree of substitution was highly correlated with increased tensile strength ($r = 0.94$), which is in line with previous research [7]. Elongation at break and opacity were both negatively correlated with degree of substitution ($r = -0.68$ and -0.49, respectively), suggesting that increased substitution disrupted the integrity of the cellulose chain. CMCs can be classified by degree of substitution as low (0.4 to 1.0), high (1.0 to 1.6), or super high (1.7 to 3.0), respectively [5,8,9]. The foliage-derived CMCs had low degrees of substitution and could be used for petroleum extraction or cement modification. CMCs made from pine wood, bamboo culm, or hemp hurd had high degrees of substitution and could be used for making food package [5]. The results illustrated the potential for obtaining CMCs from specific plant materials based upon ultimate application.

XRD spectra from the parent materials contained two featured cellulose peaks with $2\theta = 15.8°$ and $2\theta = 20.8°$, respectively (Figure 4). The cellulose peak at $2\theta = 15.8°$ decreased in the isolated cellulose (Figure 5), although it was still present. The parent materials or the extracted cellulose all contained Type I cellulose, which is one of the six cellulose isomers [20,21]. No peaks were present at $2\theta = 15.8°$ for any material after extraction and reaction with chloroacetic acid, indicating successful conversion into CMC [22], but four new peaks were observed between 32° to 75° (Figure 6). These peaks have not been noted in previous studies and merit further study.

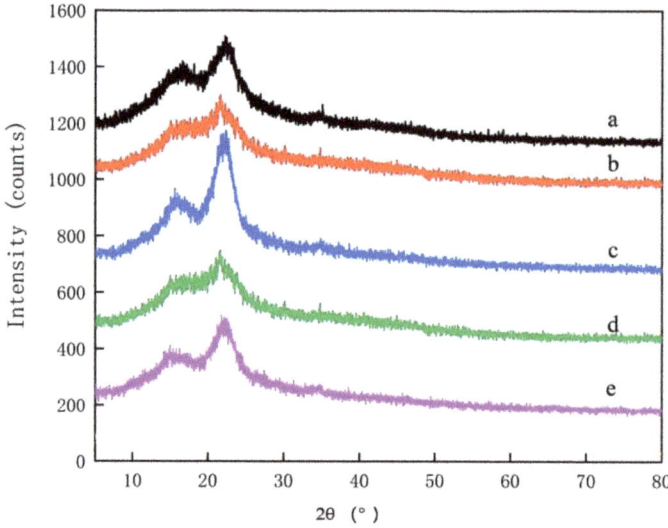

Figure 4. XRD spectra of parent materials used to produce CMCs: (**a**) pine wood, (**b**) bamboo culm, (**c**) pine needles, (**d**), bamboo leaves, or (**e**) hemp hurd.

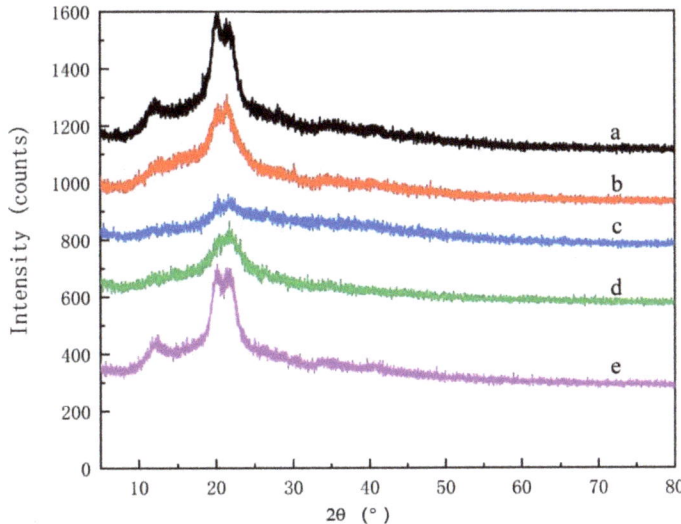

Figure 5. XRD spectra of cellulose derived from (**a**) pine wood, (**b**) bamboo culm, (**c**) pine needles, (**d**) bamboo leaves, or (**e**) hemp hurd.

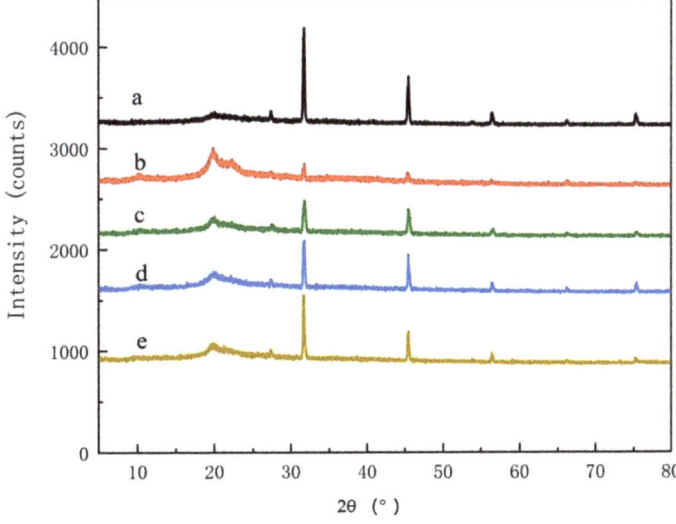

Figure 6. XRD spectra of CMCs derived from (**a**) pine wood, (**b**) bamboo culm, (**c**) pine needles, (**d**) bamboo leaves, or (**e**) hemp hurd.

3.2. The FTIR Characterization of CMC Composites

Peaks typically found for materials containing lignin and hemicellulose were absent form the cellulose regardless of raw material source [23,24] (Figure 7). These results indicated that the extraction process successfully removed the polymers, leaving a pure cellulose residue for CMC production. Two peaks at 1592 cm^{-1} and 1421 cm^{-1} corresponding to COO$^-$ stretching of the carboxylic group and O-H stretching of CMC, respectively, were found in CMC for all five materials, indicating that the materials were successfully reacted (Figure 8) [5,25]. These same peaks were also detected in spectra from the CMC composite films, suggesting that reactions did not alter the hydroxyl interactions. However, three new peaks were observed at 1035 cm^{-1}, 2889 cm^{-1}, and 2945 cm^{-1} in the composites (Figure 9), which means that the CMC composites were different from the CMC powder and this new peak need be researched more in the future.

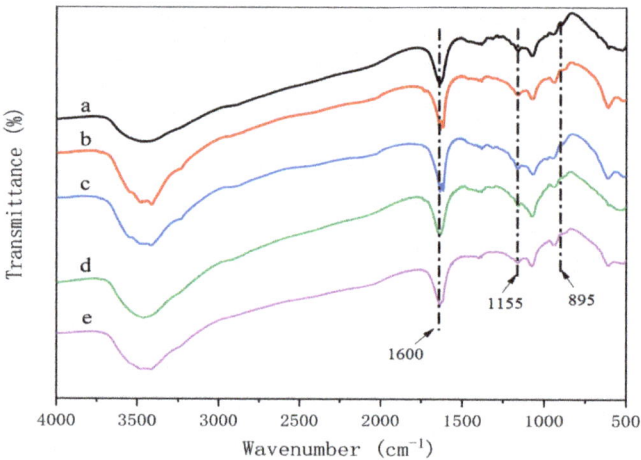

Figure 7. FTIR spectra of cellulose derived from (**a**) pine wood, (**b**) bamboo culm, (**c**) pine needles, (**d**), bamboo leaves, or (**e**) hemp hurd.

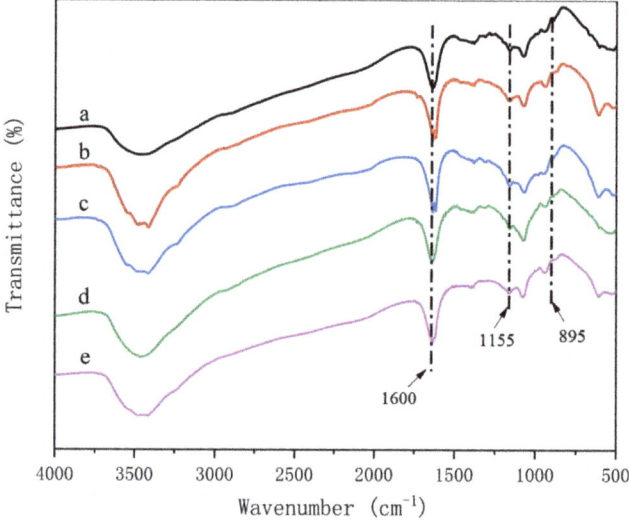

Figure 8. FTIR spectra of CMCs derived from (**a**) pine wood, (**b**) bamboo culm, (**c**) pine needles, (**d**) bamboo leaves, or (**e**) hemp hurd.

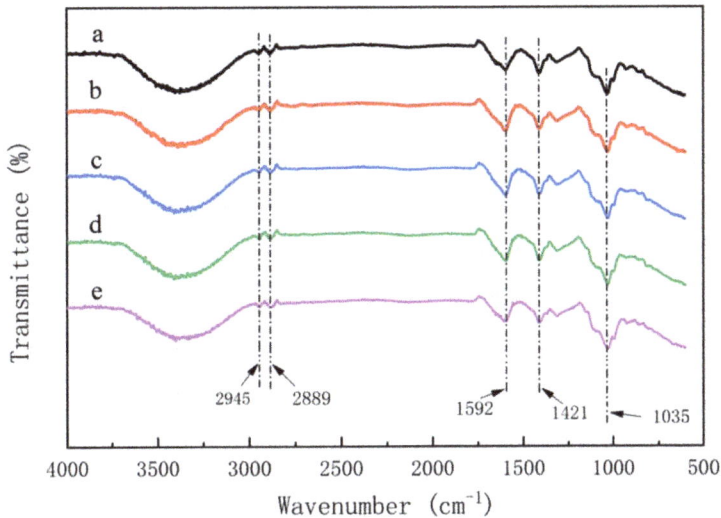

Figure 9. FTIR spectra of CMC composite films derived from (**a**) pine wood, (**b**) bamboo culm, (**c**) pine needles, (**d**) bamboo leaves, or (**e**) hemp hurd.

3.3. The Pyrolysis Characteristics of CMC Composites

The thermogravimetric (TG) curves for the parent materials, isolated cellulose, CMC, and CMC composite films are shown in Figures 10–13, respectively.

The amount of residue remaining after pyrolysis varied widely between raw materials, reflecting the inorganic elements present (Table 3). The highest residues in parent materials were found with bamboo culms and leaves as well as pine needles, while residues were lower in pine wood and hemp hurd. Bamboo contained elevated levels of silica, while pine timber contained a number of minerals. Bamboo leaves were associated with the highest residues in purified cellulose and the resulting CMC, while hemp hurd was associated with the lowest residue levels for both of these materials.

Table 3. Residue remaining (as a % of the original material) after pyrolysis.

Materials	Residual Material (%)				
	Pine Wood	Pine Needle	Bamboo Culm	Bamboo Leaves	Hemp Hurd
Parent material	19.50	26.42	27.82	25.12	14.32
Cellulose	14.46	14.12	12.98	26.66	11.84
CMC	33.01	33.92	35.84	48.08	26.60
CMC-composite film	19.72	26.20	27.82	29.44	22.42

Two pyrolysis peaks were noted in the resulting TG curves (Figures 10–13), supporting previous studies [25]. The first peak occurred near 100 °C, reflecting the removal of residual moisture. The second pyrolysis peak for pine, bamboo, pine needles, bamboo leaves, and hemp hurd parent materials occurred between 315 and 360 °C (Table 4), while the peak occurred between 340 and 352 °C for the extracted cellulose and 290 and 310 °C for the CMC. The peak occurred between 320 and 360 °C for the CMC film. The wider range in peak temperatures for the parent materials was consistent with their varying levels of cell wall polymers. Cellulose extraction followed by CMC production resulted in more uniform peak temperatures, while film formation resulted in an increase in the peak temperature and more variability among parent materials. The results, however, were fairly similar and suggest that the parent materials produced uniform cellulose, CMC, and CMC film derivatives.

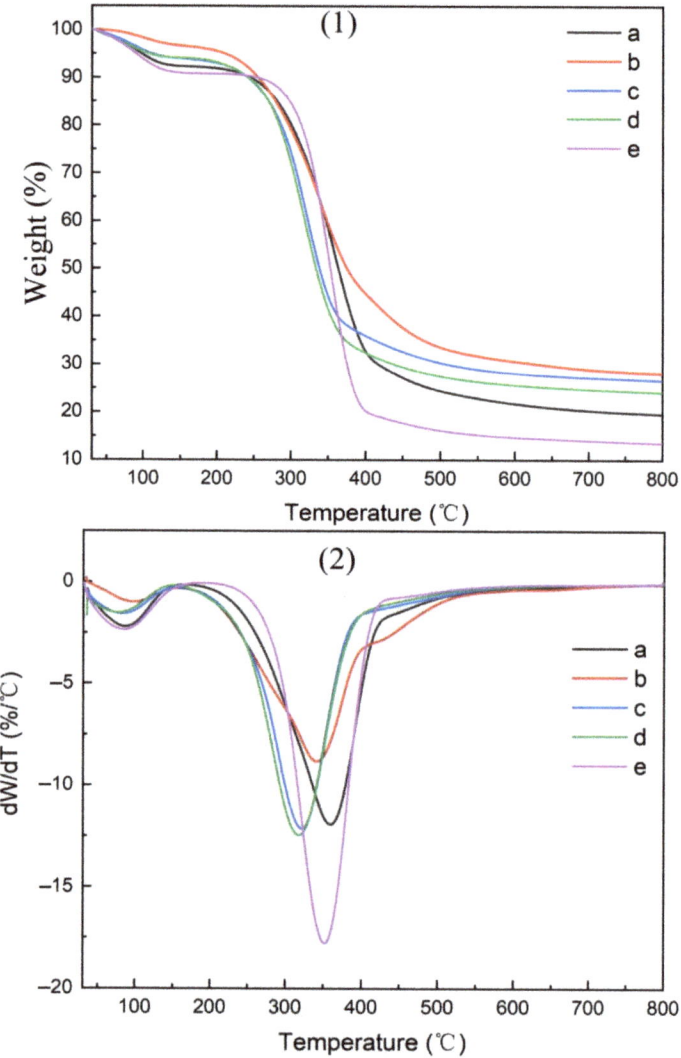

Figure 10. TG curves of materials used to produce CMCs derived from (a) pine wood, (b) bamboo culm, (c) pine needles, (d), bamboo leaves, or (e) hemp hurd, shown as thermogravimetric analysis (**1**) or differential thermogravimetric analysis (**2**).

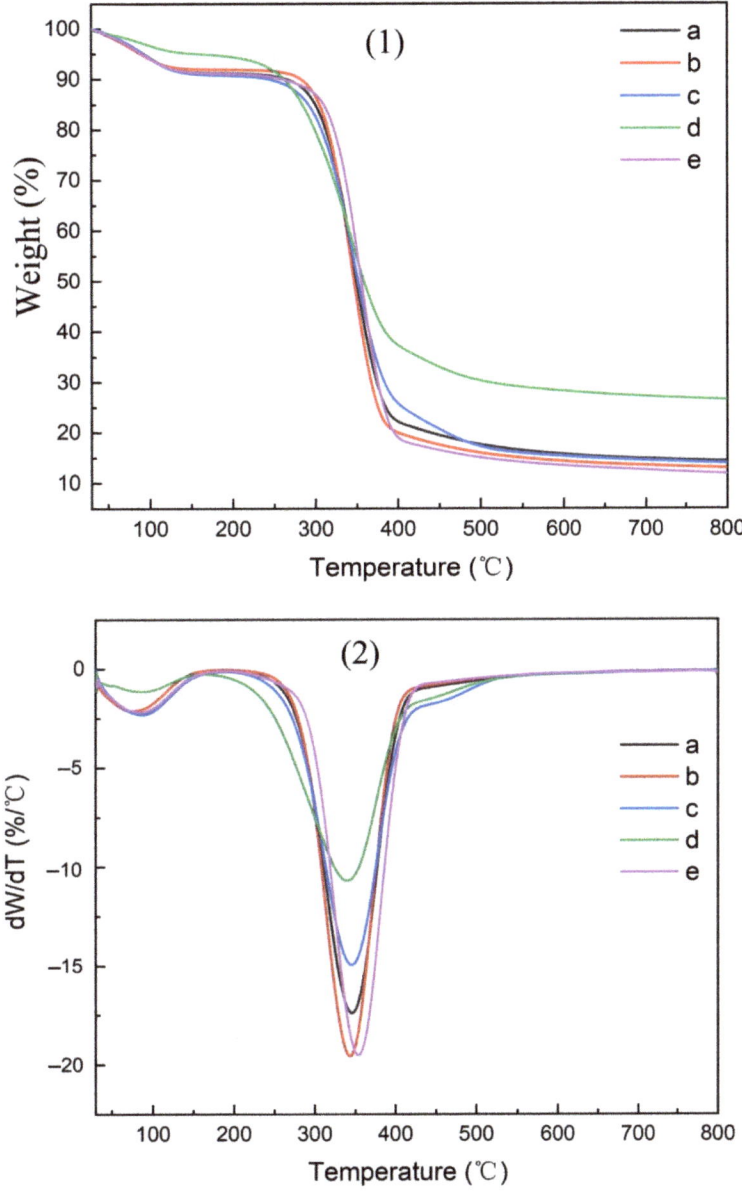

Figure 11. TG curves of cellulose derived from (a) pine wood, (b) bamboo culm, (c) pine needles, (d), bamboo leaves, or (e) hemp hurd, shown as thermogravimetric analysis (**1**) or differential thermogravimetric analysis (**2**).

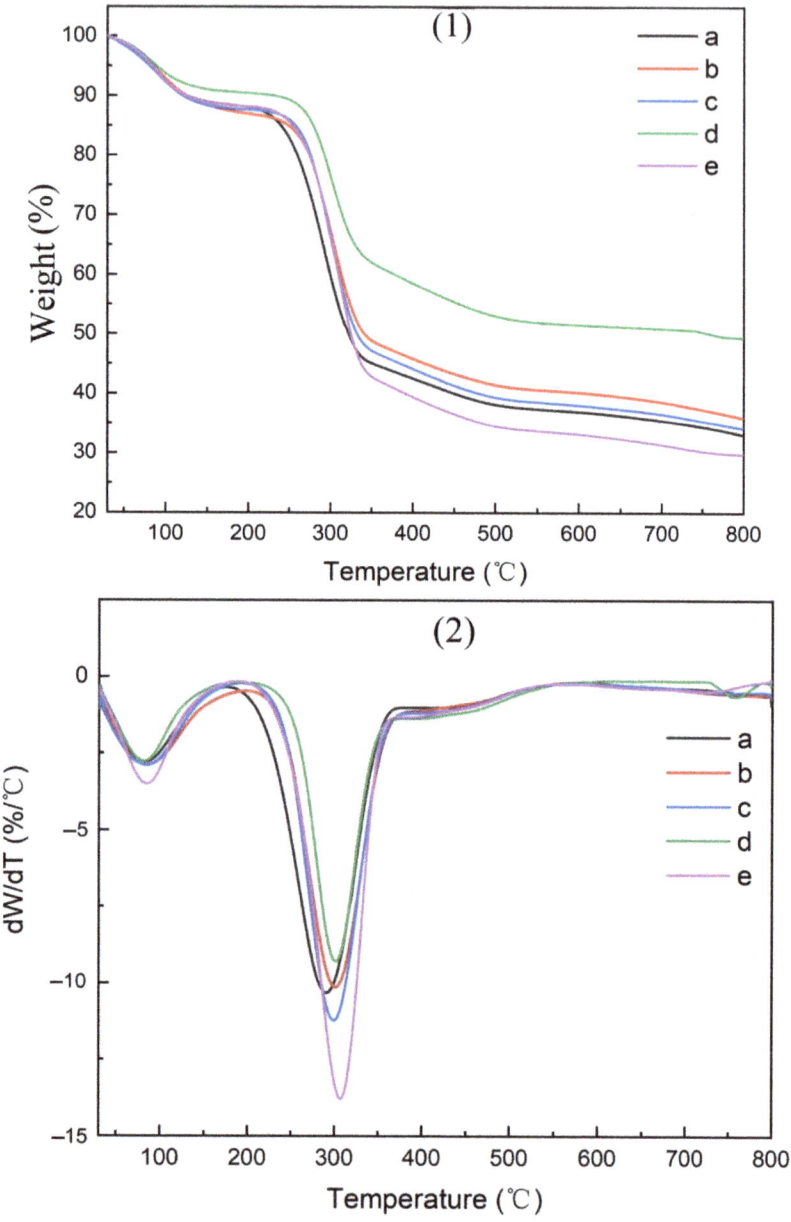

Figure 12. TG curves of CMCs derived from (a) pine wood, (b) bamboo culm, (c) pine needles, (d), bamboo leaves, or (e) hemp hurd, shown as thermogravimetric analysis (**1**) or differential thermogravimetric analysis (**2**).

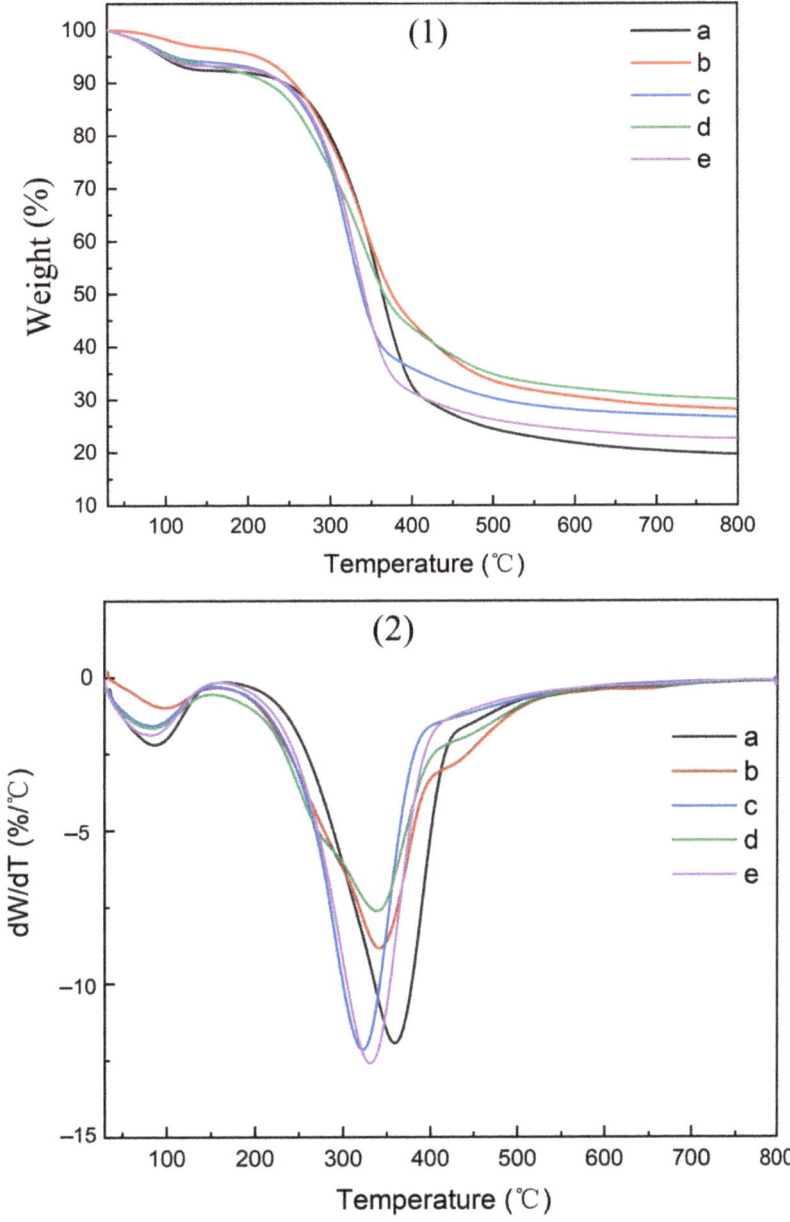

Figure 13. TG curves of CMC-composite films derived from (a) pine wood, (b) bamboo culm, (c) pine needles, (d), bamboo leaves, or (e) hemp hurd, shown as thermogravimetric analysis (**1**) or differential thermogravimetric analysis (**2**).

Table 4. Temperature at which the second peak was observed during pyrolysis.

Materials	Peak Temperature (°C)				
	Pine Wood	Pine Needle	Bamboo Culm	Bamboo Leaves	Hemp Hurd
Parent materials	360	320	340	314	351
Cellulose	342	342	340	342	352
CMC	290	298	298	290	310
CMC-composite film	360	320	340	340	325

4. Conclusions

Raw material source had marked effects on the degree of substitution of CMCs and, ultimately, the tensile strength of the resulting CMC films. The results highlight the potential for creating CMCs for selective applications by careful selection of parent materials. Leaves appear to be better suited for fillers for petroleum extraction or cement, while wood tissues from hemp hurd, bamboo, and pine were more suited for creating films for packing. Further studies are planned with chitosan as an additive.

Author Contributions: Conceptualization, X.L. (Xiaoping Li) and J.J.M.; methodology, X.L. (Xiaoping Li) and J.J.M.; validation, Y.Y., Z.S., X.L. (Xiaobao Li), Z.T.,C.L. and Z.L.; formal analysis, X.L. (Xiaoping Li) and J.J.M.; investigation, Y.Y., Z.S., X.L. (Xiaobao Li), Z.T., C.L. and Z.L.; resources, X.L. (Xiaoping Li) and Y.L.; data curation, Z.T.; writing—original draft preparation, X.L. (Xiaoping Li); writing—review and editing, J.J.M., X.L. (Xiaoping Li); supervision, X.L. (Xiaoping Li); project administration, X.L. (Xiaoping Li) and J.J.M.; funding acquisition, X.L. (Xiaoping Li). All authors have read and agreed to the published version of the manuscript.

Funding: This study was supported by the National Nature Science Foundation (31870551) Top Young Talents in Yunnan Province (YNWR-QNBJ-2018-120) and 111 Project (D21027).

Institutional Review Board Statement: Not applicable.

Informed Consent Statement: Not applicable.

Data Availability Statement: The data presented in this study are available from the listed authors.

Conflicts of Interest: The authors declare no conflict of interest.

References

1. Zhao, X.; Clifford, A.; Poon, R.; Mathews, R.; Zhitomirsky, I. Carboxymethyl cellulose and composite films prepared by electrophoretic deposition and liquid-liquid particle extraction. *Colloid Polym. Sci.* **2018**, *296*, 927–934. [CrossRef]
2. Yao, Y.; Sun, Z.; Li, X.; Tang, Z.; Li, X. Research status of carboxymethyl cellulose composite film. *Packag. Eng.* **2021**, *9*, 1–10.
3. Nguyen-Sy, T.; Tran-Le, A.D.; Nguyen-Thoi, T.; Langlet, T. A multi-scale homogenization approach for the effective thermal conductivity of dry lime–hemp concrete. *J. Build. Perform. Simul.* **2018**, *11*, 179–189. [CrossRef]
4. Wang, W. Preparation of carboxymethyl cellulose from straw of crop. *Tianjin Chem. Ind.* **2004**, *18*, 10–12.
5. Dai, X.; Xue, Y.; Peng, W.; Liu, Z.; Ren, H. Synthesis and characterization of carboxymethylated cellulose from straw and reed. *Guangzhou Chem. Ind.* **2014**, *42*, 36–39.
6. Ge, B.; Wu, M.; Li, L.; He, F. The research of high viscosity carboxymethyl cellulose sodium by apple marc. *Food Sci. Technol.* **2008**, *33*, 45–48.
7. Yu, X.; Liu, J.; Gan, L.; Li, H.; Long, M. Optimization and characterization of sodium carboxymethyl cellulose with a high degree of substitution prepared from bamboo shavings. *Mod. Chem. Ind.* **2015**, *35*, 109–114.
8. Li, R.; Yang, L.; Wang, Q.; Kong, Z.; Liang, L.; An, Y. Research on preparation of sodium carboxymethyl cellulose with high degree of substitution from bagasse regenerated cellulose. *Chem. Bioeng.* **2014**, *31*, 51–54.
9. Yaradoddi, J.S.; Banapurmath, N.R.; Ganachari, S.V.; Soudagar, M.E.M.; Mubarak, N.M.; Hallad, S.; Hugar, S.; Fayaz, H. Biodegradable carboxymethyl cellulose-based material for sustainable packaging application. *Sci. Rep.* **2020**, *10*, 21960. [CrossRef]
10. Wang, R.; Li, X.; Ren, Z.; Xie, S.; Wu, Y.; Chen, W.; Ma, F.; Liu, X. Characterization and antibacterial properties of biodegradable films based on CMC, mucilage from *Dioscorea opposita* Thunb. and Ag nanoparticles. *Int. J. Biol. Macromol.* **2020**, *163*, 2189–2198. [CrossRef] [PubMed]
11. Mohammadi, H.; Kamkar, A.; Misaghi, A. Nanocomposite films based on CMC, okra mucilage and ZnO nanoparticles: Physico mechanical and antibacterial properties. *Carbohydr. Polym.* **2018**, *181*, 351–357. [CrossRef]

12. Lan, W.; Zhang, R.; Wang, Y.; Zou, Q.; Liu, Y. Preparation and characterization of carboxymethyl cellulose/sodium alginate/chitosan composite membrane. *China Plast. Ind.* **2017**, *45*, 144–149.
13. Han, Y.; Wang, L. Sodium alginate/carboxymethyl cellulose films containing pyrogallic acid: Physical and antibacterial properties. *J. Sci. Food Agric.* **2017**, *97*, 1295–1301. [CrossRef] [PubMed]
14. Tang, Z.; He, Y. Effects of temperature and glycerol on the barrier properties of sodium alginate/sodium carboxymethyl cellulose membrane. *Light Ind. Sci. Technol.* **2015**, *31*, 25–27.
15. Tong, Q.; Xiao, Q.; Lim, L.T. Effects of glycerol, sorbitol, xylitol and fructose plasticisers on mechanical and moisture barrier properties of pullulan-alginate-carboxymethylcellulose blend films. *J. Food Sci. Technol.* **2013**, *48*, 870–878. [CrossRef]
16. Zhong, C.; Zhou, Z. Optimization of treatment process in wood extractives of pine. *Guangxi For. Sci.* **2016**, *45*, 266–270.
17. Li, X.; Xiao, R.; Morrell, J.J.; Wu, Z.; Du, G.; Wang, S.; Zhou, C.; Cappellazzi, J. Improving the performance of bamboo and Eucalyptus wood fiber/polypropylene composites using pectinase pre-treatments. *J. Wood Chem. Technol.* **2017**, *38*, 44–50. [CrossRef]
18. Tang, Z.; Yang, M.; Qiang, M.; Li, X.; Morrell, J.J.; Yao, Y.; Su, Y. Preparation of cellulose nanoparticles from foliage by bio-enzyme methods. *Materials* **2021**, *14*, 4557. [CrossRef]
19. Li, X.; Xiao, R.; Morrell, J.J.; Zhou, X.; Du, G. Improving the performance of hemp hurd/polypropylene composites using pectinase pre-treatments. *Ind. Crops Prod.* **2017**, *97*, 465–468. [CrossRef]
20. Peter, Z. Order in cellulosics: Historical review of crystal structure research on cellulose. *Carbohydr. Polym.* **2021**, *254*, 117417. [CrossRef]
21. Lu, Q.L.; Wu, J.; Li, Y.; Huang, B. Isolation of thermostable cellulose II nanocrystals and their molecular bridging for electroresponsive and pH-sensitive bio-nanocomposite. *Ind. Crops Prod.* **2021**, *173*, 114127. [CrossRef]
22. Jawad, Y.M.; Al-Kadhemy, M.F.H.; Salman, J.A.S. Synthesis, Structural and Optical Properties of CMC/MgO Nanocomposites. In Proceedings of the 3rd International Scientific Conference of Alkafeel University (ISCKU 2021), Najaf, Iraq, 23 July 2021; Trans Tech Publications Ltd.: Bäch, Switzerland, 2021; Volume 1039, pp. 104–114.
23. Naumann, A.; Gonzales, M.N.; Peddireddi, S.; Kues, U.; Polle, A. Fourier transform infrared microscopy and imaging: Detection of fungi in wood. *Fungal Genet. Biol.* **2005**, *42*, 829–835. [CrossRef] [PubMed]
24. Wu, Y.; Zhou, J.; Huang, Q.; Yang, F.; Wang, Y.; Liang, X.; Li, J. Study on the colorimetry properties of transparent wood prepared from six wood species. *ACS Omega* **2020**, *5*, 1782–1788. [CrossRef] [PubMed]
25. Badry, R.; Ezzat, H.A.; El-Khodary, S.; Morsy, M.; Elhaes, H.; Nada, N.; Ibrahim, M. Spectroscopic and thermal analyses for the effect of acetic acid on the plasticized sodium carboxymethyl cellulose. *J. Mol. Struct.* **2021**, *1224*, 129013. [CrossRef]

Article

Analysis and Calculation of Stability Coefficients of Cross-Laminated Timber Axial Compression Member

Qi Ye [1,2], Yingchun Gong [1,*], Haiqing Ren [1], Cheng Guan [2], Guofang Wu [1] and Xu Chen [1]

[1] Institute of Wood Industry, Chinese Academy of Forestry, Beijing 100091, China; yeqi97@bjfu.edu.cn (Q.Y.); renhq@caf.ac.cn (H.R.); gfwu@caf.ac.cn (G.W.); 18751856915@163.com (X.C.)
[2] School of Technology, Beijing Forestry University, Beijing 100083, China; cguan6@bjfu.edu.cn
* Correspondence: gongyingchun@caf.ac.cn

Abstract: Cross-laminated timber (CLT) elements are becoming increasingly popular in multi-storey timber-based structures, which have long been built in many different countries. Various challenges are connected with constructions of this type. One such challenge is that of stabilizing the structure against vertical loads. However, the calculations of the stability bearing capacity of the CLT members in axial compression in the structural design remains unsolved in China. This study aims to determine the stability bearing capacity of the CLT members in axial compression and to propose the calculation method of the stability coefficient. First, the stability coefficient calculation theories in different national standards were analyzed, and then the stability bearing capacity of CLT elements with four slenderness ratios was investigated. Finally, based on the stability coefficient calculation formulae in the GB 50005-2017 standard and the regression method, the calculation method of the stability coefficient for CLT elements was proposed, and the values of the material parameters were determined. The result shows that the average deviation between fitting curve and calculated results of European and American standard is 5.43% and 3.73%, respectively, and the average deviation between the fitting curve and the actual test results was 8.15%. The stability coefficients calculation formulae could be used to predict the stability coefficients of CLT specimens with different slenderness ratios well.

Keywords: cross-laminated timber; stability coefficient; axial compression; timber structures; slenderness ratio

Citation: Ye, Q.; Gong, Y.; Ren, H.; Guan, C.; Wu, G.; Chen, X. Analysis and Calculation of Stability Coefficients of Cross-Laminated Timber Axial Compression Member. *Polymers* **2021**, *13*, 4267. https://doi.org/10.3390/polym13234267

Academic Editor: Ľuboš Krišťák

Received: 14 November 2021
Accepted: 2 December 2021
Published: 6 December 2021

Publisher's Note: MDPI stays neutral with regard to jurisdictional claims in published maps and institutional affiliations.

Copyright: © 2021 by the authors. Licensee MDPI, Basel, Switzerland. This article is an open access article distributed under the terms and conditions of the Creative Commons Attribution (CC BY) license (https://creativecommons.org/licenses/by/4.0/).

1. Introduction

In recent years, cross-laminated timber (CLT) has been widely used as a prefabricated engineered wood product for wall panels, roof panels, and floor panels in mass timber construction and is composed of orthogonally oriented multi-layers of solid-sawn or engineered lumber glued with structural adhesive or mechanical fasteners such as dowels or nails [1,2]. As a kind of new structural material, there are many advantages for CLT. When compared with light wood-based construction, CLT constructions have better performance in fire including higher duration of fire resistance and greater ultimate load [3,4]. When compared with concrete constructions, CLT constructions have a lower thermal conductivity coefficient with the same thickness [5,6]. In addition, there are some other advantages of CLT construction including faster on-site construction times, lighter weight materials, use of a sustainable natural resource, carbon sequestration, lower embodied energy, and lower greenhouse gas emissions [7–9]. For these reasons, CLT elements have been successfully used in multi-storey timber-based structures.

Previous studies on CLT have mainly focused on the properties of the material itself such as bending, tensile, shear, and compressive strength [10–16]. Effects of thickness, wood species, lamina combinations, and so on [10–12] on the mechanical properties and properties of the CLT composites [13,14] were investigated. Our group also conducted some studies on CLT such as the size effect and prediction of compressive strength [15,16].

However, when using CLT in multi-storey timber-based structures, new challenges in stability bearing capacity need to be handled. Wei et al. [17] compared the axial compression properties between the CLT column and GLT (glued-laminated timber) column and the result showed that the CLT column specimens had worse stability bearing capacity but better ductility and energy absorption than GLT. Pina et al. [18] studied the effect of the number of layers, the size and location of openings, and angle between loading direction and longer edge of openings on CLT walls, but there are few formulae to calculate the stability bearing capacity of CLT. As a newly engineered wood product, the calculation of the stability bearing capacity of CLT members in axial compression in the structural design are not fully resolved. CLT stability coefficient calculation formulae from Canadian [19] and American [20] standards followed the calculation formulae of GLT, and formulae from the European [21] standard used the γ coefficient method to calculate the bending stiffness and then the stability coefficient. The structure of CLT is different from GLT, and the shear deformation of the CLT transverse layer should be considered. Therefore, this study aims to determine the stability bearing capacity of the CLT members in axial compression by experiments and to propose a calculation method of the CLT stability coefficient.

2. Stability Coefficient Calculation Formulae in Selected Countries

2.1. Stability Coefficient Formulae in Canada

Canadian wood structure product design code, CSA O86-2019 [19] uses a simplified method to calculate the stability coefficient of CLT products. The stability coefficient calculation followed the formulae of GLT, but some parameters such as the calculation of slenderness ratio were adjusted. The calculation formulae are shown in Equations (1)–(7).

$$\varphi = \left[1.0 + \frac{F_C K_{ZC} C_C^3}{35 E_{05}(K_{SE} K_T)}\right]^{-1} \quad (1)$$

$$F_C = f_c(K_D K_H K_{SC} K_T) \quad (2)$$

$$K_{ZC} = 6.3\left(\sqrt{12} r_{eff} l\right)^{-0.13} \leq 1.3 \quad (3)$$

$$C_C = L_e / \sqrt{12} r_{eff} \quad (4)$$

$$L_e = K_e l \quad (5)$$

$$r_{eff} = \sqrt{I_{eff} / A_{eff}} \quad (6)$$

$$I_{eff} = \sum_{i=1}^{n} \left(I_i + A_i a_i^2\right) \quad (7)$$

where φ is the stability coefficient under the axial compression of CLT; f_c is the standard strength in compression parallel to the grain of the laminae; F_c is the design value of compressive strength; K_D, K_H, K_{SC}, K_T are the load influence coefficient, system influence coefficient, environmental coefficient, and protective treatment coefficient, which are taken as 0.65, 1.0, 1.0, 1.0, respectively; K_{ZC} is the size-adjustment coefficient of compression resistance; C_c is the slenderness ratio of the specimen; E_{05} is the 5% quantile of elastic modulus of laminae, which was taken as 0.82 times the standard value of the modulus of elasticity E_{50} (also regarded as the mean value); K_{SE} is the environmental coefficient of elastic modulus, taken as 1.0; L_e is the calculated length of the CLT component; K_e is the calculated length adjustment coefficient, taken as 1.0 for simple support at both ends; l is the length of the laminate; r_{eff} is the effective turning radius of CLT section; I_{eff} is the effective moment of inertia of CLT; A_{eff} is the effective cross-sectional area of CLT; I_i is the moment of inertia of the i-th laminate; A_i is the cross-sectional area of the i-th laminate; and a_i is the distance between the centroid of the i-th laminate and the intermediate laminate.

2.2. Stability Coefficient Formulae in Europe

The European code for the design of timber structure products, Eurocode 5: Design of timber structures [21] does not have a separate description of the stability of CLT, so the relevant content of GLT is used to calculate the effective moment of inertia and the stability coefficient of CLT by the γ coefficient method. The formulae in the European standard are shown in Equations (8)–(15).

$$\varphi = \frac{1}{k + \sqrt{k^2 - \lambda_{rel}^2}} \tag{8}$$

$$k = 0.5\left[1 + \beta_c(\lambda_{rel} - 0.3) + \lambda_{rel}^2\right] \tag{9}$$

$$\lambda_{rel} = \frac{\lambda}{\pi}\sqrt{\frac{f_c}{E_{05}}} \tag{10}$$

$$\lambda = L_e\sqrt{\frac{A_{eff}}{I_{eff}}} \tag{11}$$

$$L_e = K_e l \tag{12}$$

$$I_{eff} = \frac{(EI)_{eff}}{E_{50}} \tag{13}$$

$$(EI)_{eff} = \sum_{i=1}^{n}\left(E_{50,i}I_i + \gamma_i E_{50,i} A_i a_i^2\right) \tag{14}$$

$$\gamma_i = \frac{1}{1 + \frac{\pi^2 E_{50,i} h_i}{L_e^2} \cdot \frac{h_0}{G_{90,0}}} \tag{15}$$

where k is instability coefficient; λ_{rel} is the relative slenderness ratio of CLT; β_c is taken as 0.1 (as same as glued laminated timber); λ is the slenderness ratio of CLT; $E_{50,i}$ is the standard value of the modulus of elasticity of the i-th laminate; γ_i is the connection effect coefficient of the i-th laminate; h_i is the thickness of the i-th laminate; h_0 is the thickness of the intermediate laminate; $G_{90,0}$ is the rolling shear modulus of the intermediate laminate, in the empirical formula, the rolling shear modulus of the laminate is taken as 1/10 of the shear modulus, the shear modulus of the laminate is taken to be 1/16 of the modulus of elasticity; the rest of the notations have the same meaning as before.

2.3. Stability Coefficient Formulae in the USA

American wood structure design code, NDS-2018 [20] adopts the shear analogy method for the calculation of the stability coefficient, following the formulae of calculating the stability coefficient of the GLT, but introducing the apparent bending stiffness in the calculation of the parameters to reflect the transverse shear effect of the middle layer, as shown in Equations (16)–(21).

$$\varphi = \frac{1+(F_{cE}/F_c)}{2c} - \sqrt{\left[\frac{1+(F_{cE}/F_c)}{2c}\right]^2 - \frac{F_{cE}/F_c}{c}} \tag{16}$$

$$F_{cE} = \frac{\pi^2 (EI)_{app-min}}{A_{eff} L_e^2} \tag{17}$$

$$(EI)_{app-min} = \frac{0.5184(EI)_{eff}}{1 + K_s \frac{(EI)_{eff}}{(GA)_{eff} l^2}} \tag{18}$$

$$L_e = K_e l \tag{19}$$

$$(EI)_{eff} = \sum_{i=1}^{n} \left(E_i I_i + E_i A_i a_i^2 \right) \tag{20}$$

$$(GA)_{eff} = \frac{a^2}{\frac{h_1}{2G_1 b} + \sum_{i=2}^{n-1} \frac{h_i}{G_i b} + \frac{h_n}{2G_n b}} \tag{21}$$

where F_{cE} is the critical buckling design value for compression members; c is the material-related parameter, taken as 0.9 (the same as for glued laminated timber); $(EI)_{app-min}$ is the reference apparent bending stiffness of CLT for panel buckling stability calculation; $(EI)_{eff}$ is the effective bending stiffness of the CLT section; K_s is the shear deformation adjustment coefficient, taken as 11.8; $(GA)_{eff}$ is the effective shear stiffness of the CLT section; E_i is the modulus of elasticity of the i-th laminate, the elastic modulus of the laminate perpendicular to the grain is about 1/30 of the elastic modulus along the grain; G_i is the shear modulus of the i-th laminate, with the longitudinal laminate taking the shear modulus and the transverse laminate taking the rolling shear modulus; and a is the centroid distance of the outermost two laminates.

2.4. Modification Based on Formulae in Selected Countries

The stability coefficient formulae of CLT in axial compression are stipulated based on the existing calculation methods in the timber structure design codes and related technical manuals of other countries. The stability coefficient is related to strength and the bending stiffness of the member. The values of compressive strength of CLT with different grades can be obtained from a certain standard, but the calculation methods of the effective bending stiffness are different. The Canadian standard uses the simplified method to calculate the effective bending stiffness, and the American standard uses the shear analogy method, while the European standard uses the γ coefficient.

Based on the CLT laminate parameters of the E1 grade set in the North American standard: ANSI/APA PRG 320-2019 [22], setting the width of the laminate as 1000 mm, three different methods were used to calculate the effective bending stiffness of the CLT component. The results are shown in Figure 1. It found that there exists a significant difference between the values by these three different methods when the slenderness ratio is small. However, with the increase in slenderness ratio, the values of effective bending stiffness using a different method approached the same number. The Chinese standard GB50005-2017, Standard for design of timber structure [23], stipulates that the calculation of CLT effective bending stiffness should have a simple design method, but this method ignored the interlaminar shear impact. The shear analogy method takes the shear deformation of each layer into account, but the method may only be used for relatively high slenderness ratios. The γ coefficient method originated from joggle beam theory, where the transverse laminates are used as the connecting parts between the longitudinal laminates, which takes the shear deformation of the longitudinal and transverse layers into account. It satisfies the action mechanism of CLT under axial load. Therefore, the γ coefficient method was selected to calculate the effective bending stiffness of CLT.

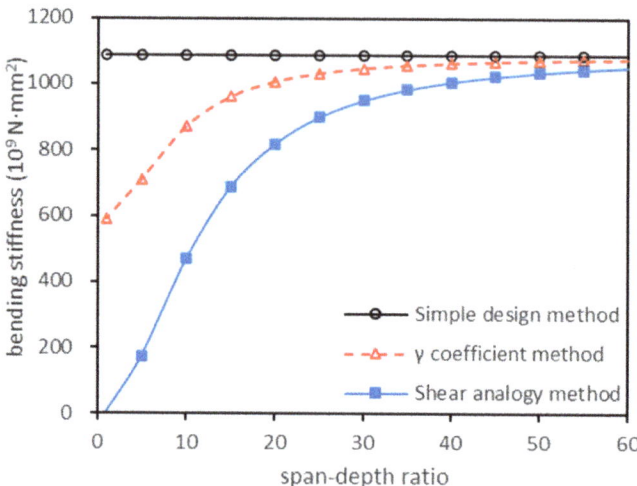

Figure 1. Effective bending stiffness curves with different methods.

The different process mode of the effect on the loading duration and the treatment methods of the resistance partial coefficient can also produce different calculation results of the stability coefficient. The differences between countries are shown in Table 1.

Table 1. The process modes of parameters in the stability coefficient formulae in different countries.

Countries	Loading Duration	Resistance Partial Coefficient	Form of Formula
China	$K_{cr,DOL} = K_{DOL}$	$\gamma_{cr,R} = \gamma_R$	segmented
Europe	$K_{cr,DOL} = K_{DOL}$	$\gamma_{cr,R} = \gamma_R$	continuous
Canada	$K_{cr,DOL} = 1.0$	$\gamma_{cr,R} \neq \gamma_R$	continuous
America	$K_{cr,DOL} = 1.0$	$\gamma_{cr,R} \neq \gamma_R$	continuous

Notes. $K_{cr,DOL}$ is the coefficient of load duration's influence on stable bearing capacity; K_{DOL} is the coefficient of load duration's influence on the strength of lumber or wood products; $\gamma_{cr,R}$ is the resistance partial coefficient of the stability bearing capacity meeting the reliability requirements; and γ_R is the resistance partial coefficient to meet reliability requirements.

Based on the parameters of the E1 grade laminates, the stability coefficient of CLT was calculated according to the methods of different national standards, while the results are shown in Figure 2. In addition, the effective bending stiffness of CLT was calculated by the simple design method, γ coefficient method, and shear analogy method, and then the effective bending stiffness was brought into the European formulae (Equations (8)–(15)) to calculate the stability coefficient, the results of which are shown in Figure 3.

Figure 2 shows that there were significant differences between the stability coefficient curves of different countries. In contrast to Figures 2 and 3, it was found that when the treatment method of the loading duration and the resistance partial coefficient were unified, the calculated stability coefficients showed little difference, even if the effective bending stiffness calculation method was different.

Zhu et al. [24] proposed a method for calculating the stability coefficient of components that was used in timbers and logs, imported dimension lumber, and glued laminated timber. This method determines the material coefficient values of different engineering wood products through regression analysis and the research results were adopted by GB 50005-2017. However, the calculation method of the stability coefficient for CLT products has not been given in the standard. Because CLTs are constructed differently to GLT and logs, the calculation method of the CLT stability coefficient cannot simply adopt the calculation method of the laminate and glued laminated timber.

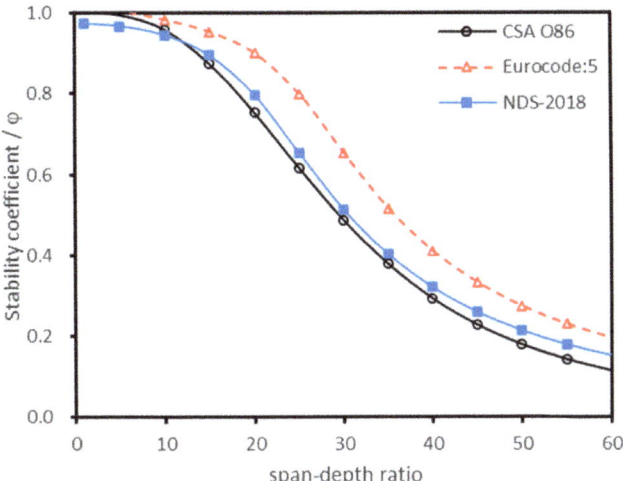

Figure 2. Stability coefficient curves of different countries.

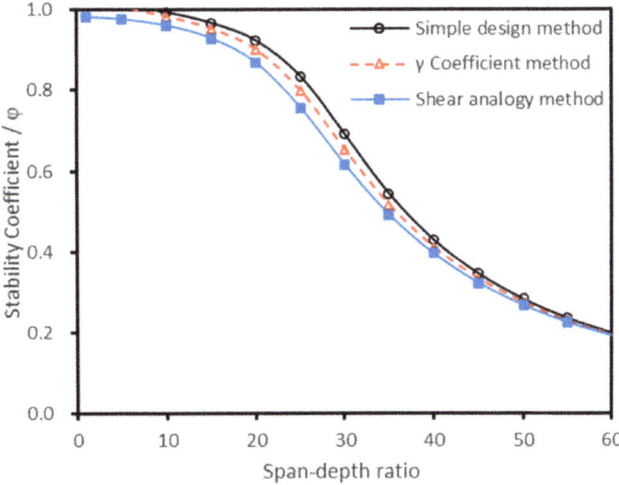

Figure 3. Stability coefficient curves of different calculation methods under the same treatment.

Based on the compression stability coefficient calculation method of wood components in Chinese standard GB 50005-2017, Standard for design of timber structure and theoretical regression analysis, the stability coefficient formulae for CLT in this paper were simply proposed and the values of the material coefficients in the formulae were obtained. Then, the stability bearing capacity and strength bearing capacity of CLT members were tested through experiments to verify the correctness of the formulae and regression results.

3. Establishment of Stability Coefficient Calculation Formulae in China
3.1. Slenderness Ratio Formula

Compared with GLT and lumber, the shear deformation of the CLT transverse layer was relatively larger, and its influence on the stability coefficient should be considered. CLT can be regarded as a split limb, with the longitudinal lamina as the column limb and the transverse lamina as the connection between the component limbs. The influence of shear deformation on the stability coefficient of the transverse laminate can be considered by the slenderness ratio conversion theory. Therefore, according to the bar buckling theory, the formula proposed by Dr. Wu was introduced to calculate the slenderness ratio as Equation (22).

$$\lambda_{eff} = \sqrt{\lambda + \frac{n\pi^2 E_{50} A_{eff}}{(GA)_{eff}}} \tag{22}$$

where λ_{eff} is the slenderness ratio after taking the shearing effect into account; n is 1.2 when the cross section is a rectangle.

3.2. Stability Coefficient Formulae

Based on the stability coefficient calculation formulae in GB 50005-2017, the shear deformation of the longitudinal and transverse layers was taken into account and the slenderness ratio was used; the calculation formulae are shown in Equations (23)–(26).

$$\varphi = \frac{a_c \pi^2 E_{05}}{\lambda_{eff}^2 f_c} (\lambda > \lambda_p) \tag{23}$$

$$\varphi = \left(1 + \frac{f_c \lambda_{eff}^2}{b_c \pi^2 E_{05}}\right)^{-1} (\lambda < \lambda_p) \tag{24}$$

$$\lambda_p = c_c \sqrt{\frac{E_{05}}{f_c}} \tag{25}$$

$$c_c = \pi \sqrt{\frac{a_c b_c}{b_c - a_c}} \tag{26}$$

where λ_p is the critical slenderness ratio of the specimens; I_{eff} should be obtained by the γ coefficient method; a_c, b_c, c_c are the correlation coefficients of CLT components. The correlation coefficients of logs, dimension lumber, and glued laminate have been determined in the Chinese standard, but there is still no final conclusion about the stability coefficient of CLT components.

3.3. Stability Coefficient Parameters

The correlation coefficients of a_c, b_c, c_c were obtained by the least square method. In the process of fitting, the corresponding relationship between slenderness ratio and stability coefficient should be used as the source data for fitting on the basis of the formulae in line with the processing method in China. Modify Equations (23) and (24) and define $y = \frac{1}{\varphi}$ and $x = \frac{f_c \lambda^2}{\pi^2 E_{05}}$ to obtain Equations (27) and (28).

$$y = \frac{1}{a_c} x (\lambda > \lambda_p) \tag{27}$$

$$y = \frac{1}{b_c} x + 1 \tag{28}$$

It can be found that the slenderness ratio conversion formula and the stability coefficient conversion formula are essentially a continuous fitting curve. On this basis, the curve can be divided by looking for an appropriate value of λ_p. The characteristic values of the dimension lumber were substituted into the formulae and the stability coefficient values of different grades were calculated in the range of 1 to 200 to slenderness ratio. With $\lambda_p = 1$ as the step, the source data on both sides of the cut-off point were fitted by the least squares method in the slenderness ratio range of 50~150. According to Equations (23) and (24), the correlation coefficients a_c, b_c and the corresponding error sum of squares for the specimens under different λ_p were obtained. The error sums of squares of both sides are summed, and the λ_p value with the minimum error sum of squares was taken as a value of this grade to the correlation coefficients a_c, b_c, and c_c.

The North American Product Standard ANSI/APA PRG 320-2019 classifies laminate materials used in CLT into four mechanical grades and two visual grades. After unifying the treatment method, the elastic modulus and compressive strength standard values in the product standard were substituted into the European standard and the American standard for fitting calculation. The CLT material parameters obtained by fitting are shown in Table 2. After unifying the treatment methods, the CLT material coefficients obtained were not significantly different. In this paper, we took the average values calculated by the European standard and the American standard, and the material coefficients were $a_c = 1.00$, $b_c = 2.92$, and $c_c = 3.89$.

Table 2. CLT material coefficients based on different standards.

Coefficient	PRG 320		Average
	Eurocode: 5	NDS-2018	
a_c	0.99	1.02	1.00
b_c	3.06	2.77	2.92
c_c	3.81	3.98	3.89

4. Materials and Methods

4.1. Materials

Japanese larch (*Larix kaempferi*) was harvested from the Dagujia Forest Farm of Liaoning Province, China. The diameters of logs were from 250 to 320 mm. A total of 351 logs with a 4.5 m length was cut into dimension lumber using four-faced sawing; the lumber sizes were 35 mm in thickness and 140 mm in width. The average density and moisture content of the lumber were 0.58 ± 0.07 g/cm^3 and $12 \pm 0.96\%$, respectively.

After cutting, each lumber was E-rated by nondestructive testing of stress waves. Lumbers with elastic modulus within 11~13 GPa were used as longitudinal laminates and counterparts with 8~10 GPa were used as transverse laminates. The one-component polyurethane was selected as an adhesive, and the pressure was 1.2 MPa. The cross section size of CLT was 300 mm × 105 mm and four different lengths were chosen for the research. The lengths of CLT specimens included 3950 mm, 3200 mm, 1950 mm, and 1200 mm and each kind was repeated three times. Relevant information is shown in Table 3.

Table 3. CLT specimen materials.

Number	Repetition Times	Size/mm	Slenderness Ratio
C1	3	3950 × 300 × 105	118.07
C2	3	3200 × 300 × 105	99.49
C3	3	1950 × 300 × 105	70.86
C4	3	1200 × 300 × 105	56.27

4.2. Compression Test

The test process was carried out with reference to the standard GB/T 50329-2012, Standard for test methods of timber structures [25]. The JSF-III high-precision static servo-hydraulic control testing machine produced by Chengdu Servo Hydraulic Equipment Co. Ltd. in Chengdu, China was used in the test with the power of 3 KW, dynamic accuracy of ±2%, and static accuracy of ±0.5%. The specimen was supported by a two-way knife hinge, and the specimen was fixed at both ends with the aid of a fixture to avoid horizontal displacement of the end. The diagram of test machine was shown in Figure 4. The test support device meets the five requirements of a two-way knife hinge in GB/T 50329-2012 as an axial compression bar support device. During the test, one pull wire displacement transducer (LVDT) was placed on each side of the span area of the specimen to measure the middle deflection of the specimen.

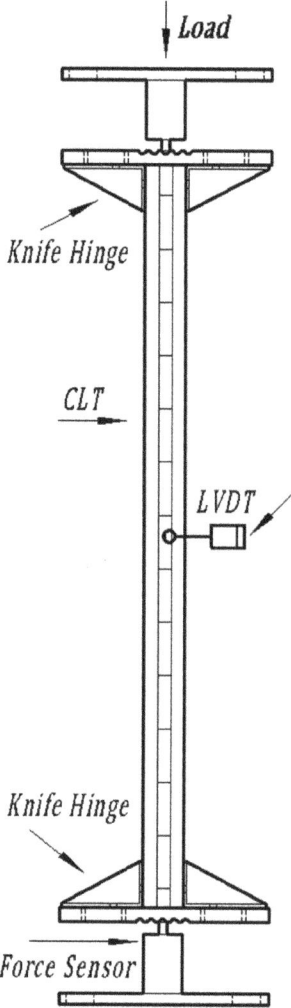

Figure 4. Specimen testing diagram.

In the process of specimen loading, it is necessary to first apply a preload F_0 to the specimen to eliminate the gap between the ends of the specimen and the fixture. In this study, the preloaded load value F_0 was taken as 10% of the estimated ultimate load value of the specimen. Then, we started to load at 1 kN/s evenly. Since specimens in group C1, C2 were relatively long and thin, the loading later changed to displacement control loading, the speed was 10 mm/min, and when the lateral deflection reached 60 mm, the loading was stopped; C3, C4 group specimens were loaded with 1 kN/s evenly, loading later changed to displacement control loading with the speed of 10 mm/min. The load continued increasing after the specimen load reached the maximum, and stopped loading after the load showed a downward trend with the deflection increasing and dropped to about 80% of the limit load.

After the test, a small specimen of length of 280 mm was cut from each end of each specimen in the C1~C4 group without surface damage to measure the compressive strength along the grain of CLT. The ratio between the stable bearing capacity of the specimen and the compressive strength along the grain of the small specimen is regarded as the stability coefficient of the CLT.

5. Results and Discussion

5.1. Results and Analysis of Stability Test

The axial load–lateral deflection curve showed that all specimens in the test were destabilized (Figure 5). C1 and C2 groups of specimens were elastically destabilized, and the surface of the specimen was restored to its original state after the end of axial loading; C3 and C4 groups of specimens were elastoplastically destabilized, and the surface of the specimen was damaged after the end of loading.

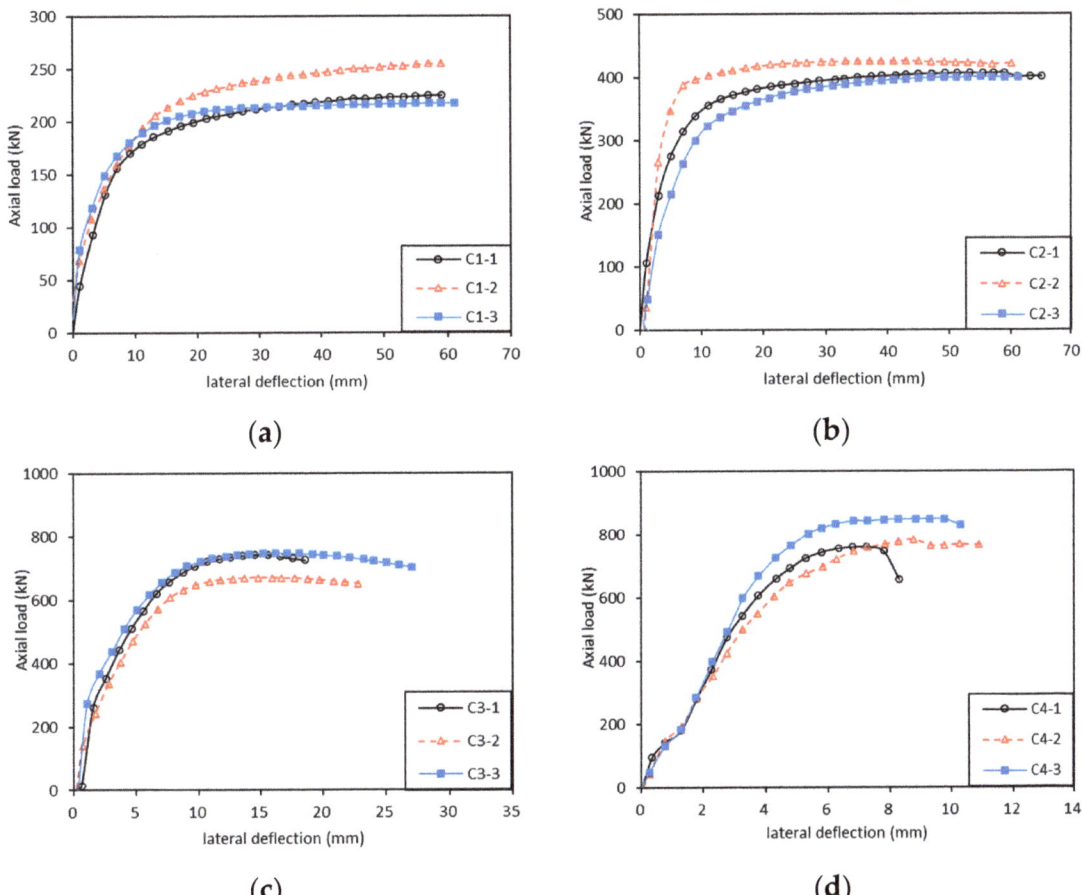

Figure 5. Axial load–lateral deflection curves of CLT specimens in axial compression. (**a**) C1 (3950 mm), (**b**) C2 (3200 mm), (**c**) C3 (1950 mm), (**d**) C4 (1200 mm).

The growth trend of lateral deflection is relatively stable with the increase in loads in the C1~C2 groups (Figure 5). In group C1, the load and deflection were approximately proportional when the specimen was in the load range of 0~200 kN. Without significant cracking or wrinkling, the lateral deflection increased slowly with load. After the lateral deflection exceeded 20 mm, the axial load almost stopped growing and the specimen started to bend rapidly. The central part of the axial compression specimen arched obviously, until the lateral deflection reached 60 mm, and the specimen stopped loading. Regarding group C2, the phenomenon was similar to C1 at first, but the proportional range extended to 0~300 kN. When the load reached approximately 60% of the ultimate load (245 kN), specimens started to bend. The instability process and the instability mode of specimens in group C2 were similar to specimens in group C1 and the destructive process of other specimens in groups were similar to the content described.

Observation of groups C3~C4 showed that plastic deformation begins to occur when compared with C1~C2 groups and the ultimate loads were obviously larger than them. In group C3, the first proportional stage occurred in the range of 0~250 kN. The lateral deflection started to grow rapidly from 250 to 600 kN in which the sound of splitting was heard. After the rapid growth, the load reached the limit of about 700 kN and the control mode changed to displacement control loading. With the increase in bending deformation, the loading value gradually decreased. Elastic instability is still the main kind of instability mode that took place in C3 group. Regarding group C4, plastic instability significantly occurred and bending did not occur significantly during the whole process, and other phenomena were similar to C3 group. At the same time, the loads produced shear forces on the cross section, making the transverse layer subjected to rolling shear. The rolling shear strength of a wood product is very low, so the rolling shear failure occurs mainly in the transverse layer. When the slenderness ratio is relatively small (C3~C4), members are closer to strength failure, mainly influenced by shear strength. The fitting curve between axial load and lateral deflection and the trend of stability coefficient with slenderness ratio was similar to the research results of Fu et al. [26] and Chen et al. [27] on the stability performance of slender concrete-filled steel tube columns and bamboo columns.

5.2. Stability Coefficient Comparison between Theoretical Curve and Actual Test Results

The test results of the CLT axial compression test are shown in Table 4.

Table 4. Results of CLT axial compression experiment.

Groups	Slenderness Ratio	F_s/kN	V_s/%	F_l/kN	V_l/%	φ_t	φ_c
C1	118.07	252.00	6.33	1311	4.42	0.192	0.227
C2	99.49	408.67	3.00	1219	7.34	0.335	0.320
C3	70.86	719.33	4.87	1286	3.90	0.559	0.630
C4	56.27	796.70	4.84	1143	9.00	0.697	0.745

Notes. F_s is the stable bearing capacity of the CLT specimen; V_s is the coefficient of variation of the test stable bearing capacity; F_l is the CLT ultimate bearing capacity, taking the minimum value of bearing capacity of the same group of specimens; V_l is the coefficient of variation of the bearing capacity of the specimen; φ_t is the actual test stability coefficient of the specimen, taking the ratio of the stability bearing capacity to the ultimate bearing capacity; and φ_c is the theoretical calculation stability coefficient.

As shown in Table 4, the compressive strength decreased with the increase in slenderness ratio, which was similar to the research conclusions of Wei et al. [28] and Pan et al. [29] on flattened bamboo–wood composite cross-laminated timber and the slender reinforced concrete columns wrapped with FRP, respectively.

By substituting the characteristic values of compressive strength along the grain, the standard values of elastic modulus and the CLT material parameters into the Chinese standard suggested formulae, the fitting curve of the theoretical stability coefficient were obtained. At the same time, the standard values of compressive strength and elastic modulus along the grain were substituted into the European standard and the American standard and the stability coefficient fitting curves were obtained. The measured scattering points of the stability coefficient of each CLT specimen were substituted into the fitting curve, which is shown in Figure 6. The stability coefficient curves fitted with the scattering points well. The overall average deviation was 8.15%, and the maximum average deviation for each group was 15.42% (C1). The maximum average deviation occurred at the maximum slenderness ratio. With an increase in the slenderness ratio, the stability coefficient gradually decreased, and the relative deviation became larger. In addition, the stability bearing capacity of the members with large slenderness ratio was easily affected by the initial geometric defects of the members.

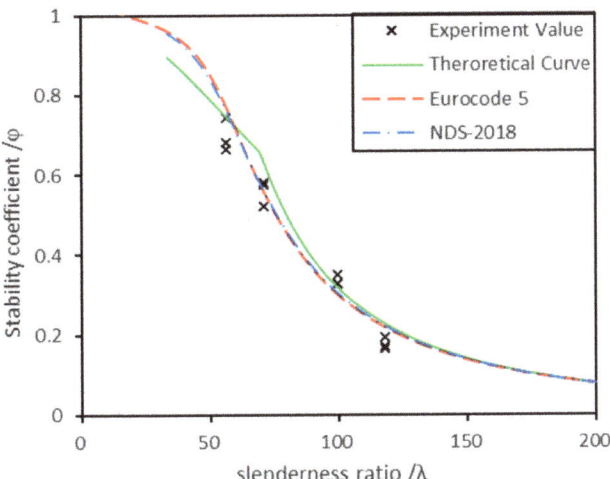

Figure 6. Comparison diagram of the theoretical calculation and experimental results.

The average deviation of the stability coefficient calculated following the suggested formulae from the European standard is 5.43% and that from the American standard is 3.73%. The maximal deviation with the European standard and American standard is 12.32% and 11.70%, respectively. They both occurred at the critical slenderness ratio ($\lambda = 70$), which should be on account of the discontinuity of the Chinese suggested formulae. The average deviation between the Chinese standard and American standard was lower than that of the European standard. This may be because the American standard slenderness ratio calculation method used in this paper takes the shear deformation factor into account, which is similar to the treatment method in the Chinese standard suggested formulae.

The resulting critical slenderness ratio of this paper was calculated to be about 70. Theoretically, the C1 ($\lambda = 118.07$) and C2 ($\lambda = 99.49$) groups showed elastic instability while the C3 ($\lambda = 70.86$) and C4 ($\lambda = 56.27$) groups showed elastic–plastic instability. In the actual test process, groups C1 and C2 had complete elastic instability, and group C4 had complete elastoplastic instability. However, group C3 still had a small amount of elastoplastic instability, which may because the slenderness ratio of group C3 was close to the critical slenderness ratio. The experimental results basically meet the theoretical inference.

6. Conclusions

In this study, CLT compression members with four slenderness ratios were used to test the stability coefficients and CLT stability coefficient formulae modified from the GLT stability coefficient formulae in the Chinese standard were used to calculate the theoretical values of these members and then compared with test results. The main conclusions are as follows:

(1) The experiment showed that with the increase in slenderness ratio of the CLT axial compression member, the stability bearing capacity gradually decreased. The C1 ($\lambda = 118.07$) and C2 ($\lambda = 99.49$) specimens were mainly characterized by elastic instability, and there was almost no damage to the specimens during the test; the C4 ($\lambda = 56.27$) specimens mainly showed elastic–plastic instability, and folding and shear failure occurred on the surface of the specimen. Both instability phenomena occurred in group C3 ($\lambda = 70.86$).

(2) By considering the CLT as a split limb, the longitudinal lamina as the column limb, and the transverse lamina as the connection between the component limbs, the slenderness ratio conversion theory, was proposed. Combined with the γ coefficient method, the calculation method of the stability coefficient for CLT elements were proposed.

(3) The material parameters a_c, b_c, c_c of the stability coefficient calculation formulae were 1.00, 2.92, and 3.89, respectively. The curve fit well with the calculated results of the European standard and American standard, and the average deviation was less than 6%. The average deviation between the fitting curve and the actual test results was 8.15%.

Author Contributions: Data curation, investigation, and writing—original draft, Q.Y.; Conceptualization, supervision, and writing—review and editing, Y.G.; Validation, H.R.; Writing—review and editing, C.G.; Methodology, G.W.; Data curation, X.C. All authors have read and agreed to the published version of the manuscript.

Funding: This research was funded by the Forestry Science and Technology Cooperation Project between Zhejiang Province and Chinese Academy of Forestry "Key technology for manufacturing engineering wood products of Chinese fir in high strength plantation" (No. 2021SY09) and the Natural Science Foundation of China "Study on Instability Mechanism and Calculation Method of Cross-laminated Timber under Compression" (No. 51808546).

Institutional Review Board Statement: Not applicable.

Informed Consent Statement: Not applicable.

Data Availability Statement: Data used in this study are included in this article.

Acknowledgments: The authors acknowledge Liangliang Huo (School of Civil Engineering, Harbin Institute of Technology) for the theoretical guidance and Guowei Zhang (School of Civil and Transportation Engineering, Beijing University of Civil Engineering and Architecture) for the experimental guidance.

Conflicts of Interest: The authors declare no conflict of interest.

References

1. He, M.; Tao, D.; Li, Z. State-of-the-art of research advances on multi-story timber and timber-hybrid structures. *J. Build. Struct.* **2016**, *37*, 1–9.
2. Nehdi, M.L.; Zhang, Y.; Gao, X.; Zhang, L.V.; Suleiman, A.R. Experimental investigation on axial compression of resilient nail-cross-laminated timber panels. *Sustainability* **2021**, *13*, 11257. [CrossRef]
3. Wang, Y.; Zhang, J.; Mei, F.; Liao, J.; Li, W. Experimental and numerical analysis on fire behaviour of loaded cross-laminated timber panels. *Adv. Struct. Eng.* **2020**, *23*, 22–36. [CrossRef]
4. Xing, Z.; Wang, Y.; Zhang, J.; Ma, H. Comparative study on fire resistance and zero strength layer thickness of CLT floor under natural fire and standard fire. *Constr. Build. Mater.* **2021**, *302*, 124368. [CrossRef]
5. Kosny, J.; Asiz, A.; Smith, I.; Shrestha, S.; Fallahi, A. A review of high R-value wood framed and composite wood wall technologies using advanced insulation techniques. *Energy Build.* **2014**, *72*, 441–456. [CrossRef]
6. Suthon, S.; Zoltán, P.; Hung Anh, L.D.; Hyungsuk, L.; Sataporn, J.; Jaipet, T. Physical, mechanical and thermal properties of cross laminated timber made with coconut wood. *Eur. J. Wood Wood Prod.* **2021**, *79*, 1519–1529.
7. Vanova, R.; Stompf, P.; Stefko, J.; Stefkova, J. Environmental impact of a mass timber building—A case study. *Forests* **2021**, *12*, 1571. [CrossRef]
8. Yao, T.; Zhu, E.; Niu, S. *Structure Technique Handbook for Cross-laminated Timber*; China Architecture & Building Press: Beijing, China, 2019.
9. Gagnon, S.; Pirvu, C. *CLT Handbook: Cross-Laminated Timber*; FPInnovations: Quebec, QC, Canada, 2011.
10. Sikora, K.S.; McPolin, D.; Harte, A. Effects of the thickness of cross-laminated timber (CLT) panels made from Irish Sitka spruce on mechanical performance in bending and shear. *Constr. Build. Mater.* **2016**, *116*, 141–150. [CrossRef]
11. Pang, S.-J.; Jeong, G.Y. Effects of combinations of lamina grade and thickness, and span-to-depth ratios on bending properties of cross-laminated timber (CLT) floor. *Constr. Build. Mater.* **2019**, *222*, 142–151. [CrossRef]
12. Wang, F.; Wang, X.; Yang, S.; Jiang, G.; Que, Z.; Zhou, H. Effect of different laminate thickness on mechanical properties of cross-laminated timber made from Chinese fir. *Sci. Silvae Sin.* **2020**, *56*, 168–175.
13. Li, Q.; Wang, Z.; Liang, Z.; Li, L.; Gong, M.; Zhou, J. Shear properties of hybrid CLT fabricated with lumber and OSB. *Constr. Build. Mater.* **2020**, *261*, 120504. [CrossRef]
14. Wang, Z.; Fu, H.; Gong, M.; Luo, J.; Dong, W.; Wang, T.; Chui, Y.H. Planar shear and bending properties of hybrid CLT fabricated with lumber and LVL. *Constr. Build. Mater.* **2017**, *151*, 172–177. [CrossRef]
15. Gong, Y.; Ye, Q.; Wu, G.; Ren, H.; Guan, C. Effect of size on compressive strength parallel to the grain of cross-laminated timber made with domestic larch. *Chin. J. Wood Sci. Technol.* **2020**, *35*, 42–46.

16. Gong, Y.; Ye, Q.; Wu, G.; Ren, H.; Guan, C. Prediction of the compressive strength of cross-laminated timber (CLT) made by domestic Larix Kaempferi. *J. Northwest For. Univ.* **2020**, *35*, 234–237.
17. Wei, P.; Wang, B.J.; Li, H.; Wang, L.; Peng, S.; Zhang, L. A comparative study of compression behaviors of cross-laminated timber and glued-laminated timber columns. *Constr. Build. Mater.* **2019**, *222*, 86–95. [CrossRef]
18. Pina, J.C.; Saavedra flores, E.I.; Saavedra, K. Numerical study on the elastic buckling of cross-laminated timber walls subject to compression. *Constr. Build. Mater.* **2019**, *199*, 82–91. [CrossRef]
19. CSA O86-2019. *Engineering Design in Wood*; Canadian Standards Association: Toronto, ON, Canada, 2019.
20. NDS-2018. *National Design Specification (NDS) for Wood Construction*, 2018 ed.; American Forest & Paper Association, American Wood Council: Washington, DC, USA, 2018.
21. EN 1995-1-1: 2004+A1. *Eurocode 5: Design of Timber Structures-Part 1-1: General-Common Rules and Rules for Buildings*; European Committee for Standardization: Brussels, Belgium, 2004.
22. ANSI/APA PRG 320-2019. *Standard for Performance-Rated Cross-Laminated Timber*; APA-The Engineered Wood Association: Tacoma, DC, USA, 2019.
23. GB 50005-2017. *Standard for Design of Timber Structure*; China Architecture & Building Press: Beijing, China, 2017.
24. Zhu, E.; Wu, G.; Zhang, D.; Pan, J. Unified method for calculation of stability coefficient of wood members in axial compression. *J. Build. Struct.* **2016**, *37*, 10–17.
25. GB/T 50329-2012. *Standard for Test Methods of Timber Structures*; China Architecture & Building Press: Beijing, China, 2012.
26. Fu, Z.-Q.; Ji, B.-H.; Lv, L.; Zhou, W.-J. Behavior of lightweight aggregate concrete filled steel tubular slender columns under axial compression. *Adv. Steel Constr.* **2011**, *7*, 144–156. [CrossRef]
27. Chen, S.; Wei, Y.; Hu, Y.; Zhai, Z.; Wang, L. Behavior and strength of rectangular bamboo scrimber columns with shape and slenderness effects. *Mater. Today Commun.* **2020**, *25*, 101392. [CrossRef]
28. Wei, P.; Wang, B.J.; Li, H.; Wang, L.; Gong, Y.; Huang, S. Performance evaluation of a novel cross-laminated timber made from flattened bamboo and wood lumber. *BioResources* **2021**, *16*, 5187–5202. [CrossRef]
29. Pan, J.; Xu, T.; Hu, Z. Experimental investigation of load carrying capacity of the slender reinforced concrete columns wrapped with FRP. *Constr. Build. Mater.* **2007**, *21*, 1991–1996. [CrossRef]

Article

Effects of Coupling Agent and Thermoplastic on the Interfacial Bond Strength and the Mechanical Properties of Oriented Wood Strand–Thermoplastic Composites

Ziling Shen, Zhi Ye, Kailin Li and Chusheng Qi *

MOE Key Laboratory of Wood Material Science and Utilization, Beijing Forestry University, Beijing 100083, China; shenziling@bjfu.edu.cn (Z.S.); ye.zhi037@gmail.com (Z.Y.); likailin2019@gmail.com (K.L.)
* Correspondence: qichusheng@bjfu.edu.cn

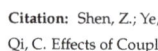

Citation: Shen, Z.; Ye, Z.; Li, K.; Qi, C. Effects of Coupling Agent and Thermoplastic on the Interfacial Bond Strength and the Mechanical Properties of Oriented Wood Strand–Thermoplastic Composites. *Polymers* **2021**, *13*, 4260. https://doi.org/10.3390/polym13234260

Academic Editors: Jingpeng Li, Yun Lu and Huiqing Wang

Received: 15 November 2021
Accepted: 2 December 2021
Published: 5 December 2021

Publisher's Note: MDPI stays neutral with regard to jurisdictional claims in published maps and institutional affiliations.

Copyright: © 2021 by the authors. Licensee MDPI, Basel, Switzerland. This article is an open access article distributed under the terms and conditions of the Creative Commons Attribution (CC BY) license (https://creativecommons.org/licenses/by/4.0/).

Abstract: Wood–plastic composites (WPC) with good mechanical and physical properties are desirable products for manufacturers and customers, and interfacial bond strength is one of the most critical factors affecting WPC performance. To verify that a higher interfacial bond strength between wood and thermoplastics improves WPC performance, wood veneer–thermoplastic composites (VPC) and oriented strand–thermoplastic composites (OSPC) were fabricated using hot pressing. The effects of the coupling agent (KH550 or MDI) and the thermoplastic (LDPE, HDPE, PP, or PVC) on the interfacial bond strength of VPC, and the mechanical and physical properties of OSPC, were investigated. The results showed that coupling agents KH550 and MDI improved the interfacial bond strength between wood and thermoplastics under dry conditions. MDI was better than KH550 at improving the interfacial bond strength and the mechanical properties of OSPC. Better interfacial bonding between plastic and wood improved the OSPC performance. The OSPC fabricated using PVC film as the thermoplastic and MDI as the coupling agent displayed the highest mechanical properties, with a modulus of rupture of 91.9 MPa, a modulus of elasticity of 10.9 GPa, and a thickness swelling of 2.4%. PVC and MDI are recommended to fabricate WPCs with desirable performance for general applications.

Keywords: wood–plastic composites; coupling agent; interfacial bonding; oriented strand–thermoplastic composites; wood veneer–thermoplastic composites

1. Introduction

Wood–plastic composites (WPCs) are innovative wood-based composites manufactured using wood, thermoplastic, and additives through injection [1], extrusion [2], and hot-pressing molding processes [3,4]. Due to their excellent weatherability, dimensional stability, and mechanical properties, WPCs are widely applied in many fields [5]. The performance of a WPC is determined by the properties of its raw materials [6], wood content [7], and the interfacial bonding between the wood and the plastic [8].

Recent studies have demonstrated the effect of thermoplastic type on interfacial bonding [9]. Since wood is a porous material, molten plastic can penetrate its pores and form mechanical bonds. Plastics with a higher melt flow rate have better permeability, leading to tighter interfacial bonding between the wood and the plastic [10]. Bekhta et al. [11] concluded that samples prepared from low-density polyethylene (LDPE) showed the lowest bonding strength, and samples prepared using polyamide showed the highest bonding strength. Stadlmann et al. [12] reported tensile shear strengths of 8.7 MPa and 3.0 MPa for birch bonded with PA6 and polypropylene (PP), respectively. Gaugler et al. [13] investigated a new methodology for rapidly assessing interfacial bonding between fibers and thermoplastics, and analyzed the interfacial bonding between wood fibers and high-density polyethylene (HDPE), PP, polyurethane (TPU), and polylactic acid (PLA). The PLA composites display the highest shear strengths (8.0–9.0 MPa), far higher than the other three plastics. Cavdar et al. [14] observed that the composite prepared using smaller plastic polymer molecules had a higher tensile modulus. LDPE, high-density polyethylene (HDPE), PP, and polyvinyl chloride (PVC) are among the most used thermoplastics in the WPC industry. However, the interaction between thermoplastics and coupling agents is lacking, so the effects of the plastic type on the interfacial bond strength and mechanical properties of WPC with coupling agents require further investigation.

Wood contains abundant hydroxyl and phenolic hydroxyl groups, making it highly popular; however, thermoplastics are nonpolar or weakly polar materials. The low compatibility between wood and plastic significantly impacts the interfacial bonding, leading to the low mechanical properties in the resulting composites [15]. Many studies have demonstrated that thermal modification [16,17], plasma treatment [18], and the addition of coupling agents [19,20] can improve interfacial bonding.

Coupling agents are easy to handle and have lower energy consumption than thermal and plasma treatment methods used to improve the interfacial bonding between wood and thermoplastics. Furthermore, the chemical reactions between the coupling agent and raw materials enhance the compatibility and interfacial bonding between WPC components [21]. Liu et al. [22] demonstrated that silane coupling agents significantly improved the interfacial bonding between wood and HDPE, because the formation of Si-O-C bonds reduced the content of hydrophilic hydroxyl groups on the wood surface. Moreover, the A171-treated samples had a higher bonding strength than KH550-treated samples. Previous studies have demonstrated that methylenediphenyl-4,4′-diisocyanate (MDI) as a WPC coupling agent enhanced the mechanical properties due to the formation of stable urethane bonds between the isocyanate groups of MDI and the hydroxyl groups of wood [23,24]. Maleic anhydride is also an effective coupling agent for WPCs [25]. Despite these results, there is no unified conclusion about which combination of plastics and coupling agents has the best effect on the interfacial bonding in WPCs. Therefore, evaluating the effect of both coupling agent and plastic on interfacial bonding has great significance for preparing high-performance WPC.

The interfacial bonding between wood and plastics is a key factor affecting the performance of WPC, including their mechanical strength, dimensional stability, and thermal properties. The samples with good bonding quality display a high bending strength, modulus of elasticity, and dimensional stability [11]. Preparing WPCs with thermoplastic films and veneers by hot pressing is an efficient method for interfacial evaluation [11,16]. However, it should be further verified whether this method is suitable for a WPC made of large strands using hot pressing.

This research aimed to establish an optimized combination of coupling agent and thermoplastic to produce oriented strand–thermoplastic composites (OSPCs) with desirable performances using hot pressing, and to verify that a higher interfacial bond strength between wood and thermoplastic improves the performance of an OSPC.

2. Materials and Methods

2.1. Materials

Poplar (*Populus tomentosa* Carr) veneers with dimensions of $400 \times 400 \times 1.2$ mm^3 and strands with lengths of 60–100 mm, widths of 15–20 mm, and thicknesses of 0.2–0.5 mm were obtained from Shandong province, China. All poplar veneers and strands were oven-dried until the final moisture content was below 3%. Thermoplastic films (HDPE, LDPE, PP, and PVC) with thicknesses of 0.1 mm were purchased from Wuhan Kaidi Plastic Products Company, and the plastic films were cut into 400×400 mm^2 pieces. The coupling agent MDI (PM200), with 30.0–32.0% cyanate (–NCO) groups, was purchased from Yantai Wanhua Polyurethane Company. MDI was diluted with acetone at a mass ratio of 1:2 before use. KH550 was purchased from Guangzhou Yong Zheng Chemical Industry Co., Ltd., Guangzhou, China.

2.2. Methods

2.2.1. Preparation of Veneer–Plastic Film Composites

The structure of the veneer–plastic film composite (VPC) is shown in Figure 1a. Roughly 60.0 g/m^2 coupling agent was coated onto the top surface of the first veneer, and one layer of plastic film was placed on the top surface of the veneer. Then, a second veneer, with both surfaces coated with a 60.0 g/m^2 coupling agent and grain perpendicular to the bottom veneer, was placed on the first layer of the plastic film. Repeating the above steps, the third veneer was placed on the top, with its grain parallel to the grain of the first veneer.

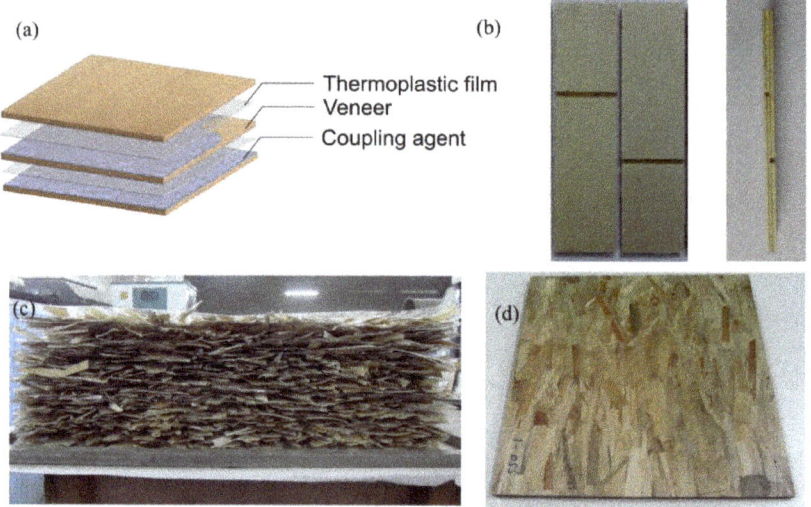

Figure 1. (**a**) Schematic diagram of the VPC; (**b**) interfacial bonding strength test samples; (**c**) forming mat of OSPC; (**d**) final OSPC product.

The formed mat was hot-pressed under 1.0 MPa pressure at 180 °C for 6 min, and then cold-pressed for 20 min until the temperature dropped below 40 °C. The effects of coupling agent and thermoplastic type on interfacial bonding were investigated, and Table 1 shows the experimental design for fabricating the VPC and OSPC.

Table 1. Experimental design for fabrication of the VPC and OSPC.

Products	Factors	Materials	Coupling Agent Content
VPC	Coupling agent	None MDI KH550	60.0 g/m^2
	Thermoplastic type	HDPE LDPE PP PVC	
OSPC	Coupling agent	None MDI KH550	2.0 wt%
	Thermoplastic type	HDPE PVC	

2.2.2. Preparation of Oriented Strand–Thermoplastic Composites

The poplar strands were sprayed with 2.0 wt% coupling agent in a drum blender and then divided into 18 equal parts. A sheet of thermoplastic film was first placed on release paper. One portion of the strands was placed on the film with an orientation angle of $\pm 20°$, and the above steps were repeated until a sheet of thermoplastic film covered the surface (Figure 1c). The thermoplastic films accounted for 20.0 wt% of the final panel. The mat was hot-pressed at 180 °C, and its thickness was controlled using a maximum pressure of 5.0 MPa. After 10 min of hot pressing, the mat was rapidly transferred into a cold press and pressed for 20 min under 1.0 MPa until the temperature dropped below 40 °C. At least 15 mm was trimmed on each side to obtain a final board with a dimension of $400 \times 400 \times 12$ mm^3 and a target density of 0.8 g/cm^3 (Figure 1d).

2.3. Characterization

2.3.1. Mechanical Strength

The VPC interfacial bond strength was evaluated according to GB/T 9846. The test samples (Figure 1b) were stretched at a cross-head speed of 10 mm/min using a universal material testing system. Thirty specimens were tested for each composite. The dry bond strength was measured after keeping the temperature and relative humidity under 20 °C and 65% for 7 days, until the weight was consistent. The wet bond strength was measured after submerging the samples in hot water (63 ± 3 °C) for 4 h.

The modulus of rupture (MOR) and modulus of elasticity (MOE) were evaluated according to ASTM D1037 using the three-point bending method. The cross-head speed was 5 mm/min during the static bending test, and 2 mm/min for the tensile test. Five specimens were tested for each sample.

2.3.2. Dynamic Mechanical Analysis (DMA)

The dynamic mechanical properties were analyzed using a TA Instruments Q800 at a frequency of 1 Hz, an amplitude of 20 μm, and a temperature range from room temperature to 180 °C. The storage modulus (E') and loss tangent ($tan\delta$) of VPCs with different coupling agents and thermoplastics were evaluated.

2.3.3. Morphological Examination

The morphology of the VPC was examined using an S3400 scanning electron microscope (SEM). SEM micrographs were used to investigate the interface between the wood and thermoplastics. The specimens were sputter-coated with gold for 1 min and dried in a vacuum at 100 °C for 1.5 h prior to the study.

2.3.4. Dimensional Stability

The thickness swelling (TS) and water absorption (WA) were evaluated according to ASTM D1037, and all samples were soaked in distilled water for 24 h. Five specimens were tested for each sample.

3. Results and Discussion

3.1. Interfacial Bond Strength of VPC

The effects of coupling agent and thermoplastic type on the interfacial bond strength of VPCs are shown in Figure 2. The coupling agent and plastic type had an evident influence on the interfacial bond strength of the VPC. Under dry conditions, the interfacial bond strength of the VPC both without a coupling agent and with KH550 varied with the thermoplastic type in the order PVC > PP > HDPE > LDPE (Figure 2a). The mechanical strength of thermoplastics and wood veneer, and the mechanical interlocking and chemical bonds formed at the interface, affected the interfacial bond strength of the VPC. PP had higher strength than HDPE and LDPE. In addition, PVC is a polar and amorphous polymer, in contrast to the nonpolar PP and HDPE, which provide better interfacial interactions with wood. Thus, the VPC made with PVC had the largest interfacial bond strength under both dry and wet conditions. Previous research showed that low-melt-viscosity polymers could penetrate deeper into the pores and gaps of wood and form better mechanical interlocking, thus giving a high interfacial bond strength [26]. Compared with dry conditions, the VPC under wet conditions had a lower interfacial bond strength (Figure 2c), which may have been caused by poorer wood mechanical properties and the breakage of chemical bonds produced by coupling agents.

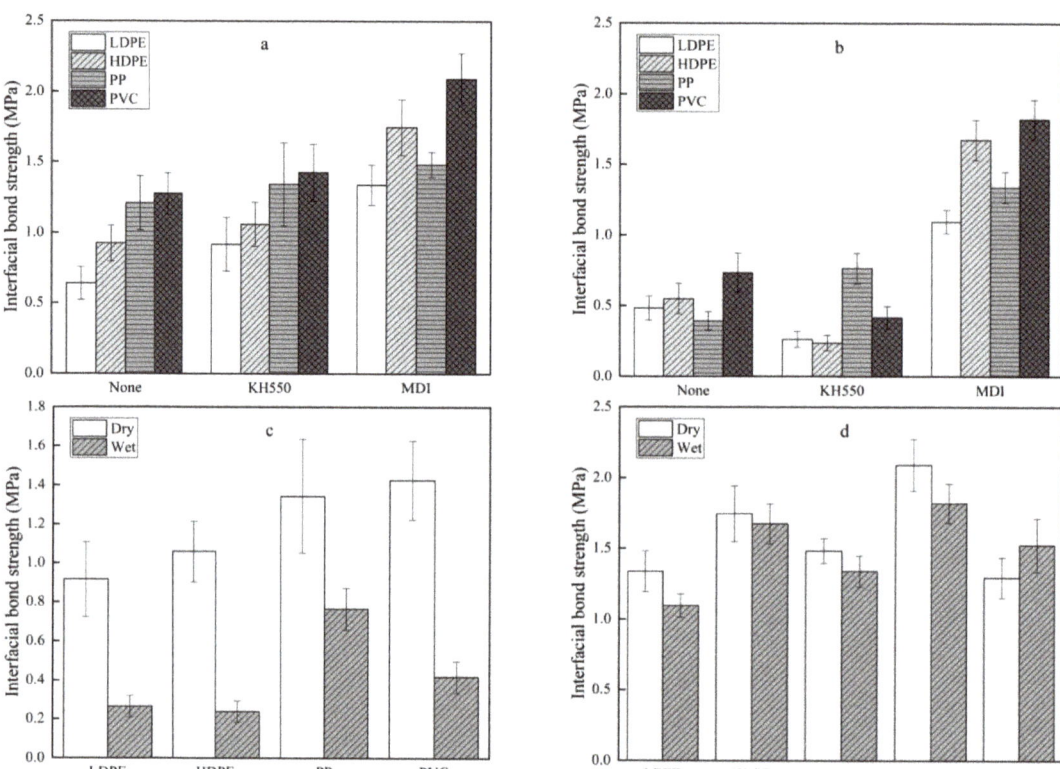

Figure 2. Interfacial bond strength of VPC under: (**a**) dry conditions; (**b**) wet conditions. The interfacial bond strength of VPC with the coupling agent: (**c**) KH550; (**d**) MDI.

Both MDI and KH550 improved the interfacial bond strength of the VPC, consistent with previous studies [27,28]. The strength enhancement was due to the chemical reaction between coupling agents and raw materials, which decreased the content of hydrophilic hydroxyl groups on the wood surface, due to the grafting reaction with the silane of KH550, generating Si-O-C bonds [22]. Urethane bonds were formed by the reaction between the isocyanate groups of MDI and the hydroxyl groups of wood and plastics [24].

The use of MDI as a coupling agent produced a higher interfacial bond strength than KH550. The average interfacial bond strength values for PVC, HDPE, PP, and LDPE under dry conditions were 2.1, 1.7, 1.5, and 1.3 MPa, respectively, and 1.8, 1.7, 1.3, and 1.1 MPa under wet conditions, respectively. Thus, MDI produced a better bonding interface between wood and plastics than KH550. The use of KH550 as a coupling agent weakened the interfacial bond of all VPCs under wet conditions, except for the VPC made with PP (Figure 2b). MDI as a coupling agent greatly strengthened the interfacial bond of the VPC under both dry and wet conditions (Figure 2a,b). The interfacial bond strength of the VPC with KH550 was 43.1–77.6% lower under wet conditions than under dry conditions

(Figure 2c), whilst with MDI as the coupling agent, the bond strength was 4.1–18.1% lower under wet conditions than under dry conditions (Figure 2d). Compared with MDI as the only adhesive, the addition of a thermoplastic increased the interfacial bond strength between wood veneers under dry conditions (Figure 2d), indicating the penetration of thermoplastics inside the wood may have increased the mechanical interlocking at the interface. The highest interfacial bond strength of the VPCs was obtained when PVC was used as the thermoplastic, and MDI as the coupling agent, though the VPC made from HDPE with MDI as the coupling agent also had a desirable interfacial bond strength.

3.2. Dynamic Mechanical Analysis of VPC

The normalized storage modulus (E') and loss tangent ($tan\delta$) were used to examine the dynamic thermomechanical properties of the VPC, as shown in Figure 3. When the temperature increased, E' first gradually decreased for all samples, and then rapidly dropped when the temperature reached softening temperature at 80 °C and 120 °C for PVC and HDPE. Previous studies support these results. For instance, Qi et al. [4] found that the melting temperature range of HDPE is 121.2–151.3 °C, and Li et al. [29] confirmed that PVC exhibited an exothermic peak at 82.8 °C. The VPC could not support a high load when the thermoplastics were completely melted. Both KH550 and MDI greatly increased the storage modulus of the VPC made from PVC and HDPE, indicating that the coupling agents greatly improved the interfacial bond strength, consistent with the results in Figure 2. The interface between the veneer and plastic changed from mechanical interlocking to chemical bonding, which improved the bond between wood and plastics and increased the sample's rigidity [16]. Figure 3 shows that both MDI and KH550 improved the stiffness of VPC made from PVC and HDPE, with a higher $tan\delta$, and MDI displayed better performance than KH550. Figure 3 also shows that PVC as the adhesive increased the plywood's stiffness compared with using MDI only; however, this advantage became a weakness at temperatures above the PVC softening temperature. Thus, the utilization temperature of wood–thermoplastic composites is limited, but they are suitable for general applications such as furniture, floor, and interior decoration.

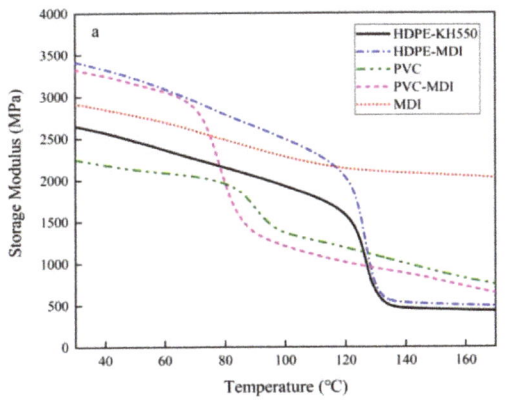
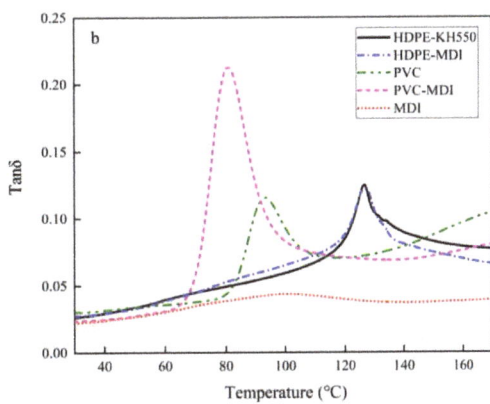

Figure 3. (a) Storage modulus and (b) loss tangent of VPC with: HDPE as the thermoplastic and KH550 as the coupling agent (HDPE-KH550); HDPE as the thermoplastic and MDI as the coupling agent (HDPE-MDI); PVC as the thermoplastic and no coupling agent (PVC); PVC as the thermoplastic and MDI as the coupling agent (PVC-MDI); and plywood with MDI as the adhesive (MDI).

3.3. Morphological Structure of VPC

The morphologies of VPCs with different thermoplastic types and coupling agents are shown in Figures 4 and 5. HDPE and PVC entered adjacent tracheids of a larger size, but they barely penetrated the small lumen (Figures 4b and 5d). Cracks were observed between the interfaces of longitudinal and transverse veneers in the VPC without coupling agents (Figures 4a and 5a). On the other hand, VPCs showed a tight interaction, without cracks, when MDI or KH550 was used as the coupling agent (Figures 4b and 5d). Altogether, these results prove that MDI and KH550 improve the interfacial bonds between the wood and thermoplastics, supporting the results for interfacial bond strength shown in Figure 2, and storage modulus in Figure 3.

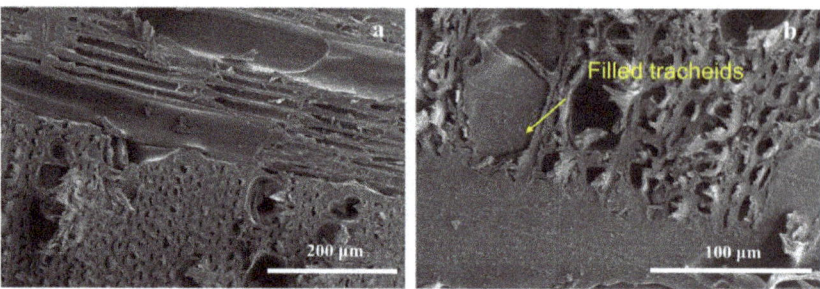

Figure 4. Interfacial SEM micrograph of the VPC made from HDPE: (**a**) without coupling agent; (**b**) with MDI as the coupling agent.

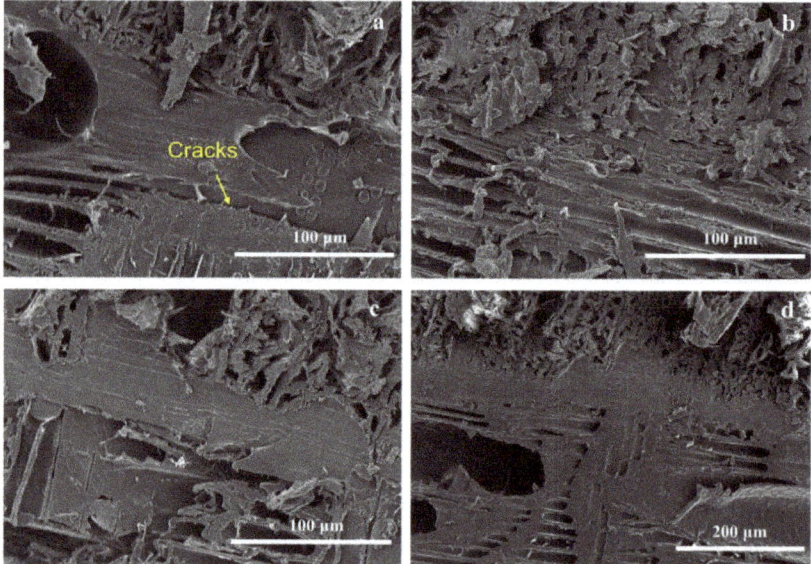

Figure 5. SEM micrograph of the interface between the veneer and: (**a**) LDPE; (**b**) LDPE with KH550 as the coupling agent; (**c**) PVC; (**d**) PVC with KH550 as the coupling agent.

3.4. Mechanical Properties of OSPC

Figure 6 shows the modulus of rupture and modulus of elasticity of OSPC. MDI and KH550 improved the MOR and MOE of OSPC in the parallel and vertical directions. MDI performed better than KH550, and the strength showed a greater improvement than the stiffness. The average MOR of OSPC made from PVC or HDPE, with MDI as the coupling agent, was 91.9 MPa and 86.1 MPa in the parallel direction, respectively, which was an increase of 235.9% and 134.5%, respectively, compared with OSPC without a coupling agent. Similarly, their average MOE was 10.9 GPa and 9.6 GPa in the parallel direction, which increased by 186.8% and 90.8%, respectively. A similar trend was obtained for the MOR and the MOE in the vertical direction.

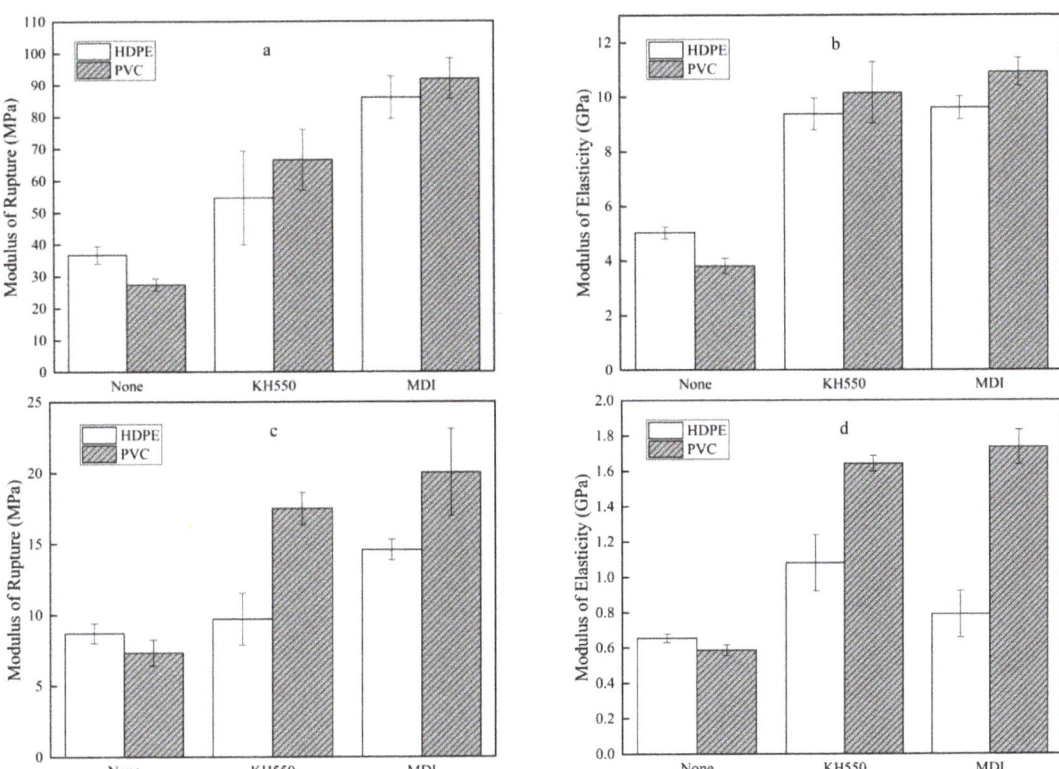

Figure 6. (**a**) MOR in the parallel direction, (**b**) MOE in the parallel direction, (**c**) MOR in the vertical direction, and (**d**) MOE in the vertical direction of OSPC.

The OSPC made from PVC without a coupling agent had a lower MOR and MOE in parallel and vertical directions than HDPE-OSPC because PVC is a flexible polymeric matrix with worse mechanical properties than HDPE [30]. However, the mechanical properties improved when MDI and KH550 were applied to strengthen the interface. This indicates that the interfacial bonding between PVC and wood was greatly enhanced by adding MDI and was higher than KH550, corresponding to the findings of Englund et al. [31] and verified by the interfacial bond strengths shown in Figure 2. These results reveal that better interfacial bond strength between wood and thermoplastics improves the mechanical properties of WPC.

3.5. OSPC Dimensional Stability

Figure 7 shows the effects of the coupling agent and thermoplastic type on the water absorption and thickness swelling after 24 h immersion in water. The average water absorption of the HDPE-OSPC made without coupling agent, with KH550 or MDI was 61.3%, 30.2%, and 12.7%, respectively, and 86.5%, 55.4%, and 15.2% for the PVC-OSPC. The moisture absorption in wood mainly occurs at pores, cracks, and hydrogen-bonding sites [32]. Both KH550 and MDI improved dimensional stability, especially MDI. When MDI was added, the average thickness swelling of the HDPE- and PVC-OSPC decreased by 27.9% and 53.6%, respectively. Coupling agents improved the interfacial bond between wood and plastic by reducing interfacial gaps and water penetration. The PVC-OSPC had a higher water absorption than HDPE-OSPC since PVC is more compatible with poplar water molecules than nonpolar HDPE.

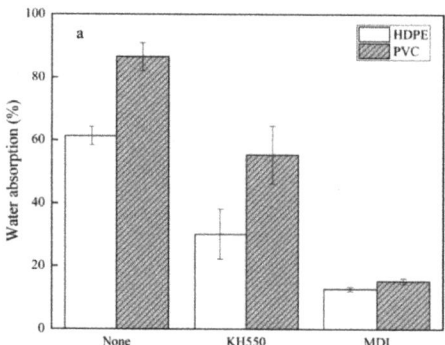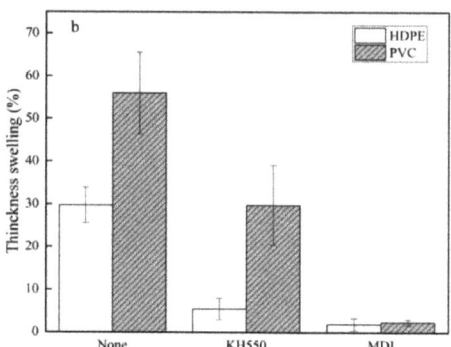

Figure 7. (a) Water absorption and (b) thickness swelling of OSPC.

Similarly, the average thickness swelling of the HDPE-OSPC without a coupling agent and with KH550 and MDI as the coupling agent was 29.7%, 5.4%, and 1.9%, respectively, and it was 56.0%, 29.7%, and 2.4%, respectively, for the PVC-OSPC. The addition of KH550 and MDI improved the thickness swelling of the OSPC, and MDI had a larger effect than KH550. The thickness swelling of the OSPC with MDI as the coupling agent was below 2.5%, indicating that these OSPC have potential applications in high-humidity and outdoor environments. The stronger interfacial bonding between wood and thermoplastic produced an OSPC with better dimensional stability for the specific thermoplastic.

4. Conclusions

Poplar veneer–thermoplastic composites and oriented strand–thermoplastic composites were fabricated using hot-pressing in this study. It was found that the use of both KH550 and MDI as coupling agents improved the interfacial bond strength between wood and thermoplastics under dry conditions. The use of MDI resulted in a much greater increase in the interfacial bond strength than KH550 under both dry and wet conditions, while KH550 had a negative effect under wet conditions. The better interfacial bond strength between wood and thermoplastics gave OSPC better mechanical properties and dimensional stability. The OSPC fabricated using PVC film and MDI gave the highest mechanical properties, with a MOR of 91.9 MPa and MOE of 10.9 GPa. PVC and MDI are recommended to fabricate WPC with a desirable performance for general use. However, WPC made from PVC was not suitable for high-temperature purposes, while HDPE could withstand higher temperatures. The results will guide the industry to produce high-performance WPC using hot pressing for general applications.

Author Contributions: Writing—original draft preparation, Z.S; formal analysis, Z.S.; investigation, Z.Y. and K.L.; conceptualization, C.Q.; methodology, C.Q.; supervision, C.Q.; writing—review and editing, C.Q. All authors have read and agreed to the published version of the manuscript.

Funding: This research was funded by the Fundamental Research Funds for the Central Universities (NO. 2021ZY30) and the National Natural Science Foundation of China (No. 31870536).

Institutional Review Board Statement: Not applicable.

Informed Consent Statement: Not applicable.

Data Availability Statement: The data presented in this research are available on request from the corresponding author.

Conflicts of Interest: The authors declare no conflict of interest.

References

1. Sohn, J.S.; Cha, S.W. Effect of Chemical Modification on Mechanical Properties of Wood-Plastic Composite Injection-Molded Parts. *Polymers* **2018**, *10*, 1391. [CrossRef]
2. Feng, C.X.; Li, Z.W.; Wang, Z.Y.; Wang, B.R.; Wang, Z. Optimizing torque rheometry parameters for assessing the rheological characteristics and extrusion processability of wood plastic composites. *J. Thermoplast. Compos. Mater.* **2019**, *32*, 123–140. [CrossRef]
3. Qi, C.S.; Yadama, V.; Guo, K.Q.; Wolcott, M.P. Preparation and properties of oriented sorghum-thermoplastic composites using flat hot-pressing technology. *J. Reinf. Plast. Compos.* **2015**, *34*, 1241–1252. [CrossRef]
4. Qi, C.; Wang, J.; Yadama, V. Heat Transfer Modeling of Oriented Sorghum Fibers Reinforced High-Density Polyethylene Film Composites during Hot-Pressing. *Polymers* **2021**, *13*, 3631. [CrossRef]
5. Ashori, A. Wood-plastic composites as promising green-composites for automotive industries! *Bioresour. Technol.* **2008**, *99*, 4661–4667. [CrossRef] [PubMed]
6. Gao, X.; Li, Q.; Cheng, W.; Han, G.; Xuan, L. Effects of moisture content, wood species, and form of raw materials on fiber morphology and mechanical properties of wood fiber-HDPE composites. *Polym. Compos.* **2018**, *39*, 3236–3246. [CrossRef]
7. Youssef, P.; Zahran, K.; Nassar, K.; Darwish, M.; El Haggar, S. Manufacturing of Wood-Plastic Composite Boards and Their Mechanical and Structural Characteristics. *J. Mater. Civ. Eng.* **2019**, *31*, 04019232. [CrossRef]
8. Luedtke, J.; Gaugler, M.; Grigsby, W.J.; Krause, A. Understanding the development of interfacial bonding within PLA/wood-based thermoplastic sandwich composites. *Ind. Crop. Prod.* **2019**, *127*, 129–134. [CrossRef]
9. Umar, K.; Yaqoob, A.; Ibrahim, M.; Parveen, T.; Safian, M. Environmental applications of smart polymer composites. *Smart Polym. Nanocompos. Biomed. Environ. Appl.* **2020**, *15*, 295–320.
10. Liu, Y.A.; Li, X.M.; Wang, W.H.; Sun, Y.A.; Wang, H.G. Decorated wood fiber/high density polyethylene composites with thermoplastic film as adhesives. *Int. J. Adhes. Adhes.* **2019**, *95*, 102391. [CrossRef]
11. Bekhta, P.; Müller, M.; Hunko, I. Properties of Thermoplastic-Bonded Plywood: Effects of the Wood Species and Types of the Thermoplastic Films. *Polymers* **2020**, *12*, 2582. [CrossRef] [PubMed]
12. Stadlmann, A.; Mautner, A.; Pramreiter, M.; Bismarck, A.; Muller, U. Interfacial Adhesion and Mechanical Properties of Wood-Polymer Hybrid Composites Prepared by Injection Molding. *Polymers* **2021**, *13*, 2849. [CrossRef]
13. Gaugler, M.; Luedtke, J.; Grigsby, W.J.; Krause, A. A new methodology for rapidly assessing interfacial bonding within fibre-reinforced thermoplastic composites. *Int. J. Adhes. Adhes.* **2019**, *89*, 66–71. [CrossRef]
14. Cavdar, A.D.; Kalaycioglu, H.; Mengeloglu, F. Tea mill waste fibers filled thermoplastic composites: The effects of plastic type and fiber loading. *J. Reinf. Plast. Compos.* **2011**, *30*, 833–844. [CrossRef]
15. Yanez-Pacios, A.J.; Martin-Martinez, J.M. Surface modification and improved adhesion of wood-plastic composites (WPCs) made with different polymers by treatment with atmospheric pressure rotating plasma jet. *Int. J. Adhes. Adhes.* **2017**, *77*, 204–213. [CrossRef]
16. Fang, L.; Xiong, X.Q.; Wang, X.H.; Chen, H.; Mo, X.F. Effects of surface modification methods on mechanical and interfacial properties of high-density polyethylene-bonded wood veneer composites. *J. Wood Sci.* **2017**, *63*, 65–73. [CrossRef]
17. Qi, C.; Guo, K. Effects of thermal treatment and maleic anhydride-grafted polyethylene on the interfacial shear stress between cotton stalk and high-density polyethylene. In Proceedings of the 2012 International Conference on Biobase Material Science and Engineering, Changsha, China, 21–23 October 2012; pp. 91–95.
18. Yanez-Pacios, A.J.; Martin-Martinez, J.M. Improved Surface and Adhesion Properties of Wood-Polyethylene Composite by Treatment with Argon-Oxygen Low Pressure Plasma. *Plasma Chem. Plasma Process.* **2018**, *38*, 871–886. [CrossRef]
19. Xiao, F.; Zhu, L.Z.; Yu, L.L. Evaluation of interfacial compatibility in wood flour/polypropylene composites by using dynamic thermomechanical analysis. *Polym. Compos.* **2020**, *41*, 3606–3614. [CrossRef]
20. Qi, C.S.; Guo, K.Q.; Liu, Y.Y. Preparation and properties of cotton stalk bundles and high-density polyethylene composites using hot-press molding. *J. Reinf. Plast. Compos.* **2012**, *31*, 1017–1024. [CrossRef]
21. Rao, J.; Zhou, Y.; Fan, M. Revealing the Interface Structure and Bonding Mechanism of Coupling Agent Treated WPC. *Polymers* **2018**, *10*, 266. [CrossRef]
22. Liu, Y.; Guo, L.; Wang, W.; Sun, Y.; Wang, H. Modifying wood veneer with silane coupling agent for decorating wood fiber/high-density polyethylene composite. *Constr. Build. Mater.* **2019**, *224*, 691–699. [CrossRef]
23. Yu, T.; Hu, C.Q.; Chen, X.J.; Li, Y. Effect of diisocyanates as compatibilizer on the properties of ramie/poly(lactic acid) (PLA) composites. *Compos. Part A Appl. Sci. Manuf.* **2015**, *76*, 20–27. [CrossRef]
24. Seo, Y.R.; Bae, S.U.; Gwon, J.; Wu, Q.L.; Kim, B.J. Effects of Methylenediphenyl 4,4′-Diisocyanate and Maleic Anhydride as Coupling Agents on the Properties of Polylactic Acid/Polybutylene Succinate/Wood Flour Biocomposites by Reactive Extrusion. *Materials* **2020**, *13*, 1660. [CrossRef] [PubMed]
25. Khamedi, R.; Hajikhani, M.; Ahmaditabar, K. Investigation of maleic anhydride effect on wood plastic composites behavior. *J. Compos. Mater.* **2019**, *53*, 1955–1962. [CrossRef]
26. Yuan, Q.; Wu, D.Y.; Gotama, J.; Bateman, S. Wood fiber reinforced polyethylene and polypropylene composites with high modulus and impact strength. *J. Thermoplast. Compos. Mater.* **2008**, *21*, 195–208. [CrossRef]
27. Cao, J.-X.; Liu, W.; Zhang, T.; Zhang, L. Preparation and characterization of eucalyptus wood flour/polypropylene composites by subcritical ethanol and 3-Aminopropyltriethoxysilane as cosolvent. *Polym. Compos.* **2018**, *39*, 3594–3604. [CrossRef]

28. Gregorova, A.; Hrabalova, M.; Wimmer, R.; Saake, B.; Altaner, C. Poly(lactide acid) composites reinforced with fibers obtained from different tissue types ofPicea sitchensis. *J. Appl. Polym. Sci.* **2009**, *114*, 2616–2623. [CrossRef]
29. Li, Q.; Shen, F.; Ji, J.; Zhang, Y.; Muhammad, Y.; Huang, Z.; Hu, H.; Zhu, Y.; Qin, Y. Fabrication of graphite/MgO-reinforced poly(vinyl chloride) composites by mechanical activation with enhanced thermal properties. *RSC Adv.* **2019**, *9*, 2116–2124. [CrossRef]
30. Ratanawilai, T.; Taneerat, K. Alternative polymeric matrices for wood-plastic composites: Effects on mechanical properties and resistance to natural weathering. *Constr. Build. Mater.* **2018**, *172*, 349–357. [CrossRef]
31. Englund, K.; Villechevrolle, V. Flexure and Water Sorption Properties of Wood Thermoplastic Composites Made with Polymer Blends. *J. Appl. Polym. Sci.* **2011**, *120*, 1034–1039. [CrossRef]
32. Wang, H.; Chang, R.; Sheng, K.C.; Adl, M.; Qian, X.Q. Impact Response of Bamboo-Plastic Composites with the Properties of Bamboo and Polyvinylchloride (PVC). *J. Bionic Eng.* **2008**, *5*, 28–33. [CrossRef]

Article

Compressive Failure Mechanism of Structural Bamboo Scrimber

Xueyu Wang, Yong Zhong, Xiangya Luo and Haiqing Ren *

Research Institute of Wood Industry, Chinese Academy of Forestry, Beijing 100091, China; wangxueyu1784@163.com (X.W.); zhongyong108@163.com (Y.Z.); luoxiangya0423@sina.cn (X.L.)
* Correspondence: renhq@caf.ac.cn

Abstract: Bamboo scrimber is one of the most popular engineering bamboo composites, owing to its excellent physical and mechanical properties. In order to investigate the influence of grain direction on the compression properties and failure mechanism of bamboo scrimber, the longitudinal, radial and tangential directions were selected. The results showed that the compressive load–displacement curves of bamboo scrimber in the longitudinal, tangential and radial directions contained elastic, yield and failure stages. The compressive strength and elastic modulus of the bamboo scrimber in the longitudinal direction were greater than those in the radial and tangential directions, and there were no significant differences between the radial and tangential specimens. The microfracture morphology shows that the parenchyma cells underwent brittle shear failure in all three directions, while the fiber failure of the longitudinal compressive specimens consisted of ductile fracture, and the tangential and radial compressive specimens exhibited brittle fracture. This is one of the reasons that the deformation of the specimens under longitudinal compression was greater than those under tangential and radial compression. The main failure mode of bamboo scrimber under longitudinal and radial compression was shear failure, and the main failure mode under tangential compression was interlayer separation failure. The reason for this difference was that during longitudinal and radial compression, the maximum strain occurred at the diagonal of the specimen, while during tangential compression, the maximum strain occurred at the bonding interface. This study can provide benefits for the rational design and safe application of bamboo scrimber in practical engineering.

Keywords: compression performance; load–displacement curves; digital image correlation (DIC); failure mode

Citation: Wang, X.; Zhong, Y.; Luo, X.; Ren, H. Compressive Failure Mechanism of Structural Bamboo Scrimber. *Polymers* **2021**, *13*, 4223. https://doi.org/10.3390/polym13234223

Academic Editors: Jingpeng Li, Yun Lu and Huiqing Wang

Received: 1 November 2021
Accepted: 29 November 2021
Published: 2 December 2021

Publisher's Note: MDPI stays neutral with regard to jurisdictional claims in published maps and institutional affiliations.

Copyright: © 2021 by the authors. Licensee MDPI, Basel, Switzerland. This article is an open access article distributed under the terms and conditions of the Creative Commons Attribution (CC BY) license (https://creativecommons.org/licenses/by/4.0/).

1. Introduction

Bamboo is an environmentally friendly biological material with superior physical and mechanical properties, which is widely distributed in Asia, America and other places [1,2]. Bamboo has a short growth cycle and can be harvested every 3–5 years. It has been used in the construction field for a long time. However, due to its small diameter and hollow structure, the further application of bamboo in the construction field is limited [3]. Therefore, some bamboo engineering products, such as bamboo scrimber [4], bamboo laminated timber [5] and bamboo particleboard [6], were explored. Among these, bamboo scrimber has become one of the most well-known bamboo engineering products, due to its high raw material utilization rate (above 90%) and good mechanical properties, and it has been widely used in the construction field [7].

When bamboo scrimber is used as a compression component in a building, its compression properties are important to evaluate its reliability and safety [8]. According to the direction of the bamboo bundle, the bamboo scrimber can be divided into a longitudinal direction (L) and a transverse direction. The direction parallel to the length of the bamboo bundle is the longitudinal direction, and the direction perpendicular to the length of the bamboo bundle is the transverse direction. The transverse direction can be divided into the radial direction (R) and the tangential direction (T), according to the assembly direction of the bamboo bundles, in which the height direction of assembly is the radial direction and the width direction of assembly is the tangential direction, as shown in Figure 1e. Previous studies showed that the compressive properties of bamboo scrimber were anisotropic [9], such that the stress–strain curves of the bamboo scrimber could be divided into elasticity, yield and softening stages under longitudinal compression, while that of bamboo scrimber under tangential compression had only the following two stages: elasticity and yield [10]. The compressive strength of bamboo scrimber under longitudinal compression, meanwhile, was greater than that compressed in a tangential direction (102.1 MPa vs. 54.4 MPa) [9].

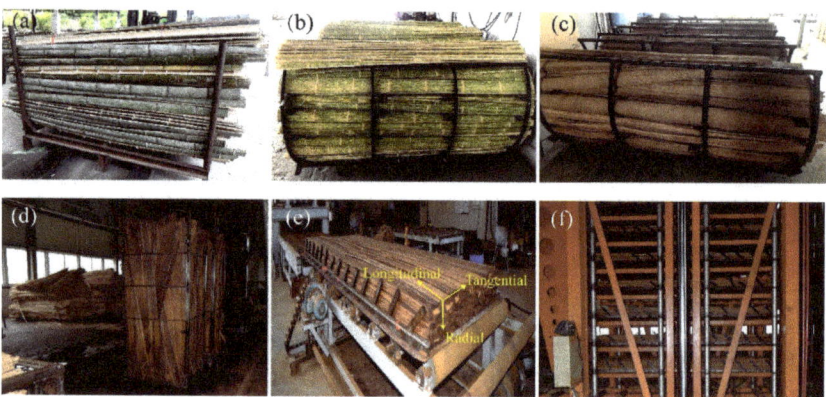

Figure 1. The manufacturing process of bamboo scrimber: (**a**) splitting; (**b**) fluffing; (**c**) heat treatment; (**d**) impregnating; (**e**) assembly; (**f**) hot pressing.

Previous studies have mainly focused on the macro-mechanical properties of bamboo scrimber under longitudinal and tangential compression directions. However, there are few studies on the radial compression performance of bamboo scrimber. Furthermore, the failure mechanism of bamboo scrimber under different compression directions has not been systematically studied.

Generally speaking, methods used to study the failure mechanism of materials include failure morphology analysis, stress state analysis, surface strain analysis and numerical methods. Stress state analysis can provide an explanation for the macro-failure mode of materials, while surface strain analysis can visualize the stress concentration phenomenon of materials more intuitively, so as to predict the failure location [11]. Micro-failure morphology can be used to speculate the failure behavior of materials [12], judge the failure mode of materials, and even reveal some details that cannot be observed at the macro scale [13]. Combining the results of stress state analysis with surface strain analysis explains the failure mechanism of a material. The numerical method consists of reproducing the force process of a material using mathematical models, which can directly reflect the stress state and failure process of materials. It is helpful to understand the macroscopic failure mode and predict the location of a failure [14].

In the past, the strain on the surface of materials was difficult to observe. With the application of the digital image correlation (DIC) method, the strain change of a material during the force loading process can be monitored without contact. The DIC method can convert the full-field strain information into an image, which intuitively reflects the stress concentration state of the material, so as to predict the damage location [15] and reveal the cause of the damage [16]. At present, the DIC method has been widely used to study the failure mechanism of wood [17] and bamboo [18].

In this paper, the compressive properties of bamboo scrimber in three directions (longitudinal direction, radial direction and tangential direction) were studied using a mechanical testing machine. During the compression process, the changes in strain field on the surface of the bamboo scrimber were observed by the DIC method. Combined with fracture morphology analysis of bamboo scrimber at macro and micro scales, the failure mechanism of bamboo scrimber under different directions was summarized. The results in this paper can provide a reference for the safe and rational application of bamboo scrimber in the construction field.

2. Materials and Methods

2.1. Materials

Bamboo scrimber was prepared from moso bamboo (*Phyllostachys pubescens*), 3–5 years old, harvested from Anhui Province, China. The manufacturing process of bamboo scrimber included the following six steps: splitting, fluffing, heat treatment, impregnating, assembly, and hot pressing. Firstly, the bamboo culms were longitudinally split into two semicircular bamboo tubes with a length of 2.4 m (Figure 1a). Without removing the inner and outer layers, the bamboo tubes were then pushed into a fluffing machine along the longitudinal direction to obtain the bamboo fiber bundles (Figure 1b). The bamboo fiber bundles were heated using steam at 180 °C for 2 h (Figure 1c), and were then immersed into phenol formaldehyde resin (PF, type: PF162510 with 45% of solid content, 36 CPs of viscosity, 10~11 pH, Beijing Dynea chemical industry Co., Ltd., Beijing, China) for 10~15 min at room temperature (Figure 1d), before being dried for 8 h at 45 °C. Finally, the bamboo fiber bundles were assembled along the longitudinal direction (Figure 1e) and hot pressed with a pressure of 4.0 MPa at 140 °C for 2 h (Figure 1f).

The specimens (Figure 2a) were divided into three groups according to grain direction, including longitudinal specimens L1~L54 (Figure 2b), tangential specimens T1~T54 (Figure 2c) and radial specimens R1~R54 (Figure 2d), with 54 specimens in each group. The dimensions of each sample were 20 mm × 20 mm × 30 mm, as shown in Figure 2. Before the test, all specimens were conditioned at 25 °C and 65% RH to arrive at an equilibrium moisture content. After equilibrium, the average air-dry density of bamboo scrimber specimens was 1.15 g/cm³.

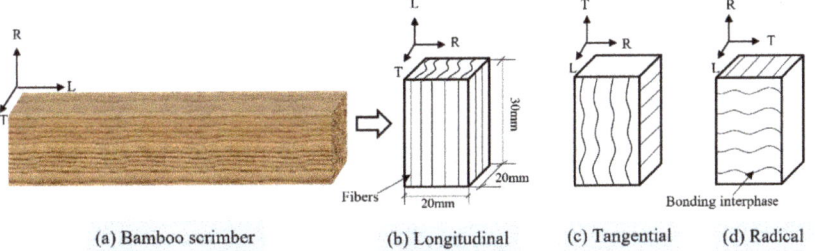

Figure 2. Schematic diagram of specimen preparation.

2.2. Test Methods

A universal test machine with a load capacity of 100 KN (Model: 5582, Instron Co., Ltd., Norwood, MA, USA) was used to evaluate the compression properties of each specimen (Figure 3b). The DIC method was used to collect the strain on the specimen surface during compression (Figure 3c). In order to facilitate observation, speckles were drawn on the surface of each bamboo scrimber specimen (Figure 3a). Before speckle drawing, the observation surface was ground with 100 mesh sandpaper, and then sprayed with a layer of white primer less than 0.5 mm thick to reduce the impact of surface roughness on the experimental results. During the test, the loading rate was 1 mm/min, the DIC image acquisition speed was 2 frames per second, the image pixels of each DIC image were 4896 × 3264 pixels, and the displacement accuracy of the specimen surface obtained by DIC was 0.01 pixel. When the load dropped to 70% of the maximum load or obvious cracks appeared on the specimen surface, the experiment was stopped [19].

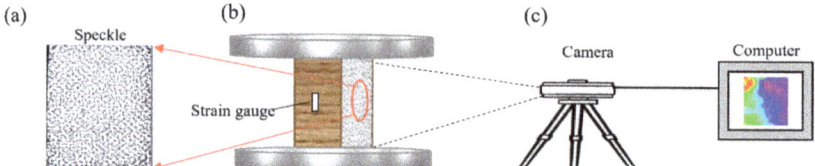

Figure 3. Schematic drawing of compression testing: (**a**) speckle on the specimen surface; (**b**) test schematic; (**c**) DIC.

Two strain gauges (Model: BX120-3AA, Xingtai City, Hebei Province, China) were attached at the middle positions of two opposite sides of the specimen to measure the strain. The date was collected by a multi-channel data acquisition instrument (TDS-530, Tokyo, Japan). The final result was the average of experimental tests of two strain gauges.

The compressive strength (σ) and elastic modulus (E) of the specimen can be calculated by Equations (1) and (2) [19,20], as follow:

$$\sigma = \frac{P_{max}}{bt} \quad (1)$$

$$E = \frac{\Delta P}{bt\Delta\zeta} \quad (2)$$

where σ is the compressive strength (MPa), P_{max} is the maximum load (kN), b is width of specimens (mm), t is the thickness of specimens (mm), E is the elastic modulus (GPa), ΔP is the load increment at the elastic stage of the stress–strain curve (N), $\Delta\zeta$ is the corresponding strain increment at the mid-span cross section (mm).

2.3. Microscopic Fracture Morphology

The micro-fracture morphology of longitudinal, tangential and radial specimens was observed by a scanning electron microscope (SEM, Hitachi S-4800, Tokyo, Japan) with low vacuum and an acceleration voltage of 10 kV, to investigate the compression mechanism of bamboo scrimber.

3. Results and Discussion

3.1. Compression Performance

Figure 4 shows the load–displacement curves of the bamboo scrimber compressed in the longitudinal direction (L), tangential direction (T) and radial direction (R). The curves of L, T and R all contained the following three stages: elastic stage (I), yield stage (II) and failure stage (III). In the elastic stage, the load increases linearly with the increase in displacement. In the yield stage, the load increases nonlinearly with the increase in displacement. In the failure stage, the load decreases with the increase in displacement. The critical point of the elastic stage and the yield stage is located by the second derivative of the curve, while the critical point of the yield stage and the failure stage is the maximum load points. In addition, the maximum displacements of L, T and R were 2.85 mm, 0.58 mm and 0.67 mm, respectively. The compression deformation of L was larger than those of T and R. However, for natural bamboo, the compression deformation in the radial direction was much larger than that in the longitudinal direction [21]. This difference was due to the changes in the micro-structure and chemical components of the bamboo scrimber [22].

Figure 5 shows the average compressive strength and elastic modulus of bamboo scrimber in three grain directions, with each value being the average of 54 values. Different letters represent significant differences, and the same letters represent no significant differences. The compressive strength of L, T and R was 94.67 MPa, 45.86 MPa and 48.80 MPa, respectively, and the elastic modulus of L, T and R was 14.61 GPa, 2.81 GPa and 3.18 GPa, respectively. The compressive strength of L was 2.06 and 1.94 times that of T and R, respectively, and the elastic modulus of L was 5.20 and 4.59 times that of T and R, respectively.

There were no significant differences in compressive strength and elastic modulus between T and R. The reason for this was that the compressive strength and elastic modulus in the longitudinal direction of the bamboo scrimber mainly depended on the characteristics of the bamboo fibers, while that of the bamboo scrimber in the radial and tangential directions was mainly determined by the bonding strength of the adhesive [9].

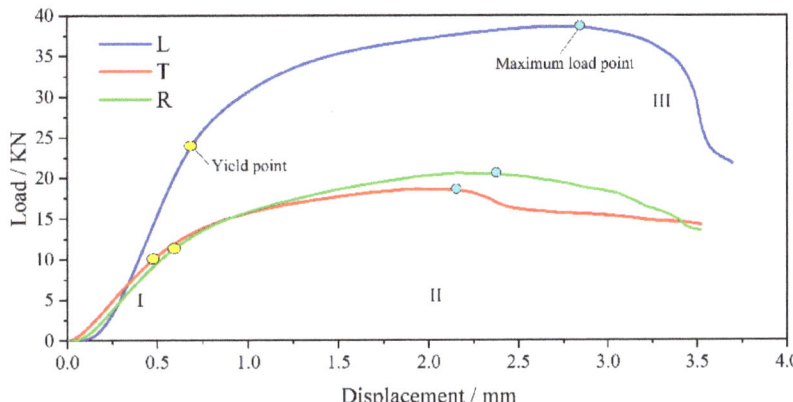

Figure 4. Load–displacement curves of bamboo scrimber in different grain directions. I elastic stage; II yield stage; III failure stage.

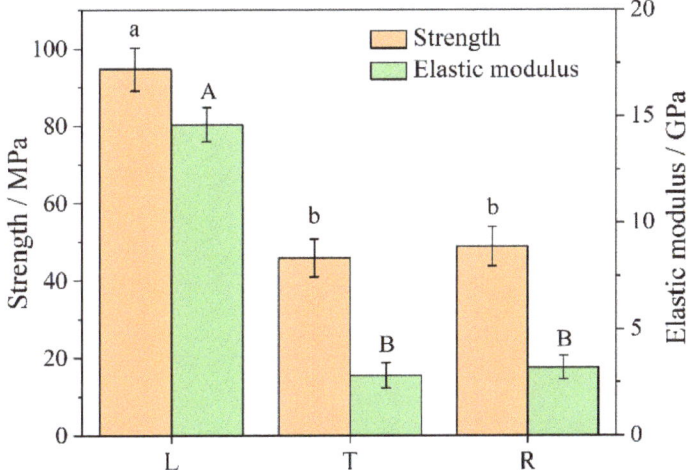

Figure 5. Compressive strength and elastic modulus of bamboo scrimber in three grain directions. Note: Lowercase letters indicate the difference in compressive strength, and uppercase letters indicate the difference in compressive elastic modulus. The same letter indicates insignificant difference ($p > 0.05$), and different letters indicate significant difference ($p < 0.05$).

3.2. Macroscopic Failure Morphology

In order to study the failure modes and reveal the failure mechanism of bamboo scrimber in three grain directions, the macroscopic failure morphology of each of the specimens was observed and divided into I~V modes (Figures 6–8). The proportion of each failure mode in the different grain direction groups is shown in Table 1.

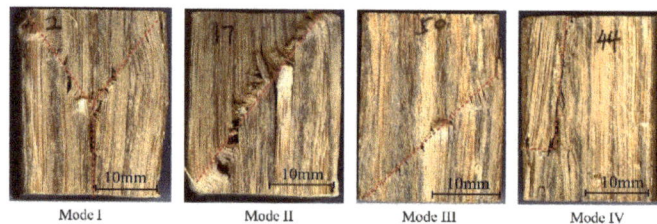

Figure 6. Failure modes of longitudinal compression specimens.

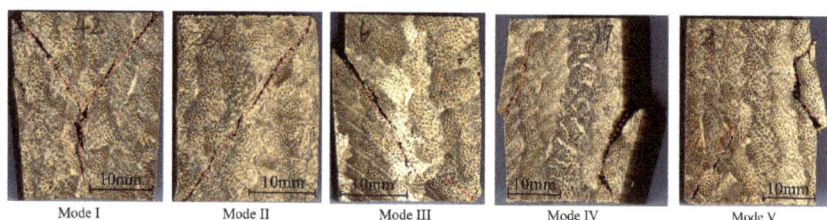

Figure 7. Failure modes of tangential compression specimens.

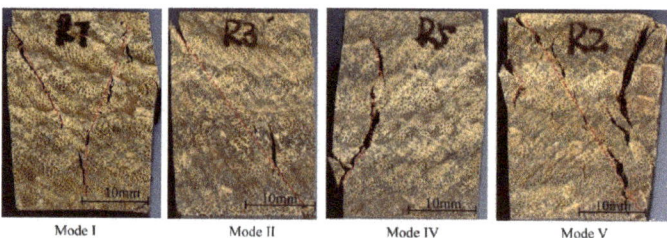

Figure 8. Failure modes of radial compression specimens.

Table 1. Proportion of failure modes for each group.

Specimen Group	Mode I	Mode II	Mode III	Mode IV	Mode V	Total
L	50%	26%	22%	2%	-	54
T	11%	18%	15%	22%	34%	54
R	50%	9%	-	15%	26%	54

Figure 6 shows the failure modes of the longitudinal compression specimens. The shape of the crack in mode I looks similar to the letter "Y". This crack first appeared in the middle of the specimen. As the load increased, the crack expanded along the diagonal direction parallel to the 45° angle, then the middle crack spread downwards, in a pattern similar to the letter "Y". The number of mode I failure specimens accounts for 50% of the total number. In addition, obvious fiber wrinkles and bulges could be observed on the failure surface. This failure mode is shear failure [11].

The crack in mode II was an oblique line along the diagonal direction, and the angle between the oblique line and the horizontal direction was about 60°. In this failure mode, the crack first appeared at the two diagonal corners of the specimen, and then developed along the diagonal direction. The number of mode II failure specimens accounts for 26% of the 54 specimens. This failure mode is also shear failure [9].

Similarly to mode II, mode III was also an oblique crack, but the angle of the crack in mode III was different from that in mode II. The angle between the oblique crack and the horizontal direction of mode III was about 45°. The number of specimens with this failure mode accounts for 22% of the 54 specimens. This failure mode is also shear failure.

Mode IV was a vertical crack. In this failure mode, the crack propagated along the bonding interface. With the increase in pressure, the sample divided into two parts. This failure mode may be caused by uneven impregnation, assembly and low bonding strength during the processing of bamboo scrimber, and it has the greatest impact on the properties of bamboo scrimber. This failure mode is associated with interlaminar separation failure. In this study, only one specimen was in this failure mode.

In summary, the failure modes I, II and III were shear failure, and failure mode IV was interlayer separation failure. Thus, shear failure was the main failure mode of bamboo scrimber under compression in the longitudinal direction.

Figure 7 shows the failure modes of bamboo scrimber under tangential compression. Similarly to longitudinal compression, bamboo scrimber under tangential compression showed "Y" cracks (mode I), diagonal cracks (mode II), 45° oblique cracks (mode III) and vertical cracks (mode IV). The proportion of each failure mode was 11%, 19%, 15% and 22%, respectively. The number of specimens in mode IV, under tangential compression, was greater than that under longitudinal compression. Differently from longitudinal compression, there was a mixed failure mode V under tangential compression, which consisted of the combination of shear failure and interlayer separation. This failure mode made up the largest proportion of the tangential compression specimens, which was 34%. Thus, interlayer separation failure was the main failure mode of the specimens under compression in the tangential direction. The reason for this was that the bamboo bundles were only connected by gluing in the transverse direction, so the glue layer and the bamboo bundle were easier to separate under tangential compression [23].

Figure 8 shows the failure modes of the bamboo scrimber under radial compression. Similarly to longitudinal compression and tangential compression, the failure mode also showed "Y" cracks (mode I), diagonal cracks (mode II), vertical cracks (mode IV) and mixed cracks (mode V). Among these, the proportion of mode I was the highest, which was 50%, and the proportions of mode II, mode IV and mode V were 10%, 15% and 25%, respectively. It is worth noting that the vertical crack (Figure 8, mode IV) in radial compression failure was different from that in longitudinal compression and tangential compression. This crack did not appear at the bonding interface, but was formed by the intersection of two oblique cracks. Thus, shear failure was the dominant failure mode for specimens that were compressed along the radial direction.

3.3. Microscopic Failure Morphology

In order to reveal the reason for the differences in macro-compression properties in three grain directions of bamboo scrimber, the microscopic damage morphologies of the longitudinal, tangential and radial compression specimens were analyzed. A representative sample was selected in each group; they were L17, T2 and R42, respectively (Figure 9).

When compressed in the longitudinal direction, the bamboo fibers were distorted (Figure 9(a2)), and a few bamboo fibers showed characteristics of shear failure and fiber debonding (Figure 9(a2)). It can be judged that the failure of the bamboo fibers during longitudinal compression was ductile [24]. However, when compressed in the tangential and radial directions, the fibers were broken neatly and without deformation (Figure 9(b2,c2)), which means that the failure of the fibers was brittle. This is one of the reasons why the deformation of specimens during longitudinal compression was greater than that of the specimens under tangential and radial compression (Figure 4).

On the other hand, the parenchyma cells showed brittle shear failure in all three groups (Figure 9(a3,b2,c3)). This was correlated with their short and thin-walled structures [25]. In addition, as shown in Figure 9(b3), the micro-failure morphology of the specimens at Position 2 was smooth and without fiber breakage, which further confirmed that interlayer separation failure occurred here.

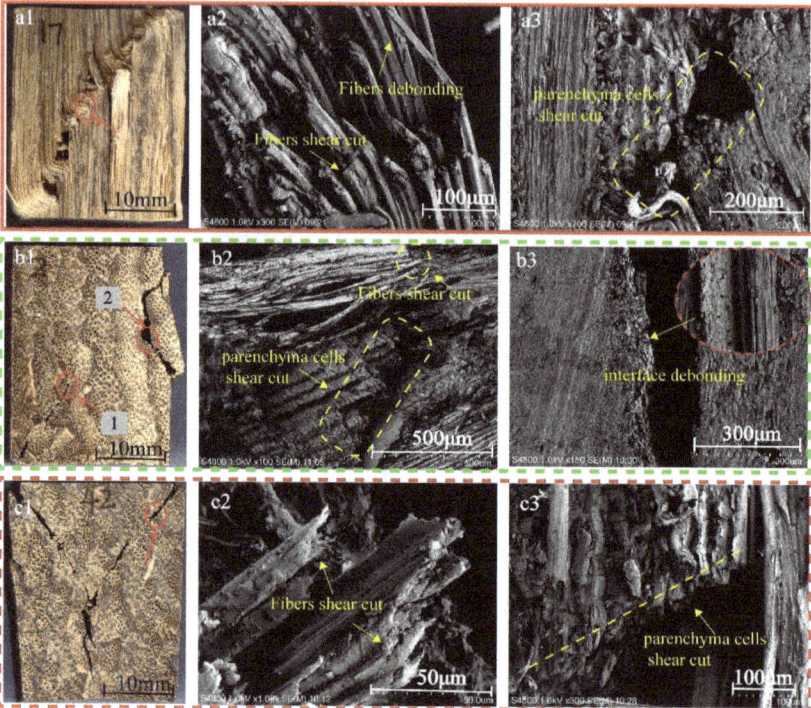

Figure 9. Micro-fracture morphology of bamboo scrimber in three grain directions: (**a1**) sampling location of longitudinal compression specimen; (**a2**) micro-fracture morphology of fibers; (**a3**) micro-fracture morphology of parenchyma cells; (**b1**) sampling location of tangential compression specimen; (**b2**) micro-fracture morphology of fibers and parenchyma cells of position 1; (**b3**) micro-fracture morphology of position 2; (**c1**) sampling location of radial compression specimen; (**c2**) micro-fracture morphology of fibers; (**c3**) micro-fracture morphology of parenchyma cells.

3.4. Failure Mechanism

3.4.1. Force Analysis

Through the analysis of the macroscopic and microscopic failure morphology of the bamboo scrimber under different grain directions, it was found that shear failure was the main failure mode. This was related to the stress distribution inside the specimen during the compression process. As shown in Figure 10, the stress (p_α) in any section of the specimen could be decomposed into normal stress(σ_α) and shear stress(τ_α). Shear stress dominates the compressive failure of bamboo scrimber [26]. According to the law of force composition, the stress p_α was as follows:

$$p_a = \sigma_0 \cos \alpha \tag{3}$$

The shear stress was [11] as follows:

$$\tau_\alpha = \sigma_0 \sin \alpha \cos \alpha \tag{4}$$

where α is the angle between the loading force and the normal direction of the section; F is the compression load; σ_0 is the normal stress of the cross section.

It can be observed from Equation (3) that when the α is 45°, the τ_α is the maximum. Thus, the crack expanded in the direction of the maximum force ($\alpha = 45°$) [27]. However, in the actual compression process, the specimen was easily twisted, leading to deflection of the crack angle, as with mode II (Figure 6).

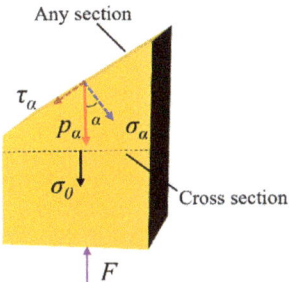

Figure 10. Schematic diagram of stress state during compression.

3.4.2. Strain Field Distribution

According to the above analysis, the main failure mode of the bamboo scrimber under longitudinal compression and radial compression was shear failure, while the main failure mode under tangential compression was layer separation failure. In order to further analyze the reason for this difference, the DIC method was used to monitor the surface strain field of the specimens. During the compression process, the crack morphology can only be observed on a specific surface, which is the surface shown in Figures 6–8. In order to analyze the failure mode and surface strain field together, this surface was selected for strain distribution observation. The strain distribution of the observation surface is the same as that of the opposite surface, which is different from the adjacent surface. L17 and T5 were selected as representative samples for shear failure and layer separation failure analysis, respectively.

Figure 11 shows the strain field distribution of the shear failure specimen L17 under the maximum load; its failure mode was mode II with diagonal cracks. Exx represents the strain in the X direction, eyy represents the strain in the Y direction, and exy represents the shear strain. When the load reached the maximum value, the maximum values of exx, eyy and exy all occurred at the diagonal of the specimen, indicating that a significant stress concentration occurred here, which is the location of the initial crack. This result is consistent with the above analysis.

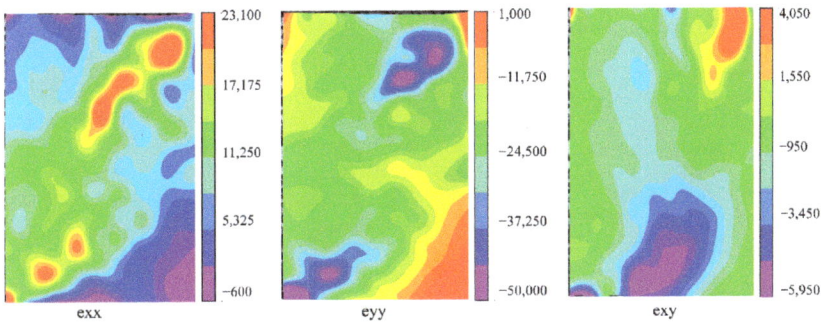

Figure 11. Strain field distribution of specimen L17 under the maximum load (shear failure).

There was a significant difference in the strain field of the tangential compression specimen T5. It can be observed from Figure 12 that the maximum exx, eyy and exy all appeared in the upper left corner of the specimens, and that the initial crack occurred here. On the other hand, the strain distributions of exx and eyy were stratified and parallel to the bonding layer of the specimen, resulting in interlayer separation failure at the bonding interphase. This indicates that the bonding interface affected the stress distribution and was prone to produce stress concentration, resulting in failure [19].

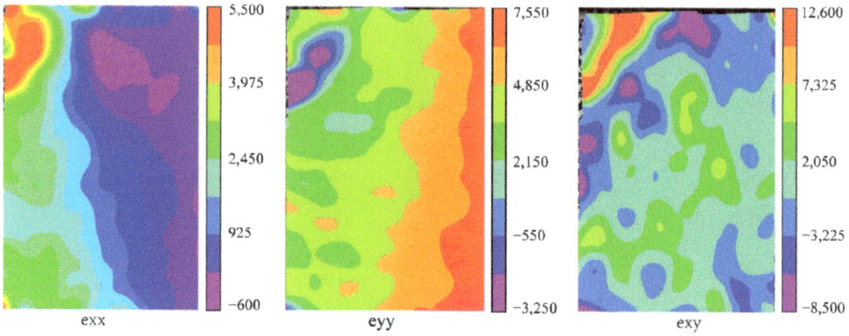

Figure 12. Strain field distribution of specimen T5 under the maximum load (layer separation failure).

4. Conclusions

In order to reveal the influence of grain direction on the compression properties and failure mechanisms of bamboo scrimber, the longitudinal, radial and tangential directions were selected. Moreover, the load–displacement curves, compression properties, macro and micro-failure morphology, stress state and strain field were analyzed. The main conclusions are summarized as follows:

(1) The compressive load–displacement curves of bamboo scrimber in longitudinal, tangential and radial directions all contained an elastic, yield and failure stage.

(2) The compressive strength of bamboo scrimber along the longitudinal direction was 94.67 MPa, which was 2.06 times that of the tangential specimens and 1.94 times that of the radial specimens. The elastic modulus of the longitudinal specimens was 14.61 GPa, which was 5.20 times that of the tangential specimens and 4.59 times that of the radial specimens.

(3) The micro-failure morphology shows that the parenchyma cells showed brittle shear failure in all three directions, and the fiber failure of the longitudinal compressive specimens was ductile fracture, while that of the tangential and radial compressive specimens was brittle fracture. This is one of the reasons that the deformation of the specimens under longitudinal compression was greater than in the tangential and radial directions.

(4) Under the compression of three grain directions, the macro-failure modes of bamboo scrimber include five modes, which are the "Y" crack (mode I), diagonal crack at 60° (mode II), diagonal crack at 45° (mode III), vertical crack (mode IV) and mixed crack (mode V). Among these, the longitudinal specimens include the modes I, II, III, and IV, the tangential specimens include the modes I, II, III, IV, and V, and the radial specimens include the modes I, II, IV, and V. The main failure mode of the longitudinal and radial compressive specimens was shear failure, while that of the tangential compressive specimens was interlayer separation failure. The reason for the difference was that the maximum strain occurred at the diagonal of the specimen during longitudinal and radial compression, while the maximum strain occurred at the bonding interface under tangential compression.

(5) Bamboo scrimber is a new kind of structural engineering material with broad application prospects. There has been little research on the bonding interface performance, long-term performance and seismic performance of bamboo scrimber when used as a structural material, something to which more attention needs to be paid.

Author Contributions: Data curation, investigation and writing original draft, X.W.; validation, Y.Z.; SEM tests, X.L.; writing–review and editing, H.R. All authors have read and agreed to the published version of the manuscript.

Funding: This research was funded by Chinese Academy of Forestry (CAFYBB2021ZX001).

Data Availability Statement: The date presented in this study are available on request from the corresponding author.

Acknowledgments: We would like to thank the Zhuji Anhui Co., Ltd. for its equipment support.

Conflicts of Interest: The authors declare that they have no known competing financial interests or personal relationships that could have appeared to influence the work reported in this paper.

References

1. Liese, W. Research on bamboo. *Wood Sci. Technol.* **1987**, *21*, 189–209. [CrossRef]
2. Scurlock, J.; Dayton, D.; Hames, B. Bamboo: An overlooked biomass resource? *Biomass Bioenergy* **2000**, *19*, 229–244. [CrossRef]
3. Elejoste, A.; Arevalillo, A.; Gabilondo, N.; Butron, A.; Peña-Rodriguez, C. Morphological Analysis of Several Bamboo Species with Potential Structural Applications. *Polymers* **2021**, *13*, 2126. [CrossRef]
4. Sun, Y.; Zhang, Y.; Huang, Y.; Wei, X.; Yu, W. Influence of Board Density on the Physical and Mechanical Properties of Bamboo Oriented Strand Lumber. *Forests* **2020**, *11*, 567. [CrossRef]
5. Mahdavi, M.; Clouston, P.L.; Arwade, S.R. A low-technology approach toward fabrication of Laminated Bamboo Lumber. *Constr. Build. Mater.* **2012**, *29*, 257–262. [CrossRef]
6. Wang, Y.-Y.; Peng, W.-J.; Chai, L.-Y.; Peng, B.; Min, X.-B.; He, D.-W. Preparation of adhesive for bamboo plywood using concentrated papermaking black liquor directly. *J. Cent. South Univ. Technol.* **2006**, *13*, 53–57. [CrossRef]
7. Huang, Y.; Ji, Y.; Yu, W. Development of bamboo scrimber: A literature review. *J. Wood Sci.* **2019**, *65*, 25. [CrossRef]
8. Li, H.-T.; Chen, G.; Zhang, Q.; Ashraf, M.; Xu, B.; Li, Y. Mechanical properties of laminated bamboo lumber column under radial eccentric compression. *Constr. Build. Mater.* **2016**, *121*, 644–652. [CrossRef]
9. Li, H.; Qiu, Z.; Wu, G.; Wei, D.; Lorenzo, R.; Yuan, C.; Zhang, H.; Liu, R. Compression Behaviors of Parallel Bamboo Strand Lumber Under Static Loading. *J. Renew. Mater.* **2019**, *7*, 583–600. [CrossRef]
10. Sheng, B.; Zhou, A.; Huang, D. Stress-strain relationship of parallel strand bamboo under uniaxial or pure shear load. *J. Civ. Archit. Environ. Eng.* **2015**, *37*, 24–31. [CrossRef]
11. Deng, C.; Zhang, H.; Yin, J.; Xiong, X.; Wang, P.; Sun, M.; Li, W.Q. Compression performance and failure mechanism of C/C-Cu composites. *Chin. J. Mater. Res.* **2017**, *31*, 182–186. [CrossRef]
12. Liu, H.; Jiang, Z.; Fei, B.; Hse, C.; Sun, Z. Tensile behaviour and fracture mechanism of moso bamboo (*Phyllostachys pubescens*). *Holzforschung* **2015**, *69*, 47–52. [CrossRef]
13. Habibi, M.K.; Lu, Y. Crack Propagation in Bamboo's Hierarchical Cellular Structure. *Sci. Rep.* **2014**, *4*, srep05598. [CrossRef]
14. Zhou, W.; Zhang, R.; Fang, D. Design and analysis of the porous $ZrO_2/(ZrO_2+Ni)$ ceramic joint with load bearing–heat insulation integration. *Ceram. Int.* **2016**, *42*, 1416–1424. [CrossRef]
15. Navaratnam, S.; Ngo, T.; Christopher, P.; Linforth, S. The use of digital image correlation for identifying failure characteristics of cross-laminated timber under transverse loading. *Measurement* **2020**, *154*, 107502. [CrossRef]
16. Wang, D.; Lin, L.; Fu, F. Fracture mechanisms of moso bamboo (*Phyllostachys pubescens*) under longitudinal tensile loading. *Ind. Crop. Prod.* **2020**, *153*, 112574. [CrossRef]
17. Bakir, K.; Aydemir, D.; Bardak, T. Dimensional stability and deformation analysis under mechanical loading of recycled PET-wood laminated composites with digital image correlation. *J. Clean. Prod.* **2021**, *280*, 124472. [CrossRef]
18. Krause, J.Q.; Silva, F.d.A.; Ghavami, K.; Gomes, O.d.F.M.; Filho, R.D.T. On the influence of Dendrocalamus giganteus bamboo microstructure on its mechanical behavior. *Constr. Build. Mater.* **2016**, *127*, 199–209. [CrossRef]
19. Standard Administration of China. *Method of Testing Compressive Strength Parallel to Grain of Wood*; GB/T 1935-2009; Standard Administration of China: Beijing, China, 2009; pp. 1–2.
20. Standard Administration of China. *Method for Determination of the Modulus of Elasticity in Compression Parallel to Grain of Wood*; GB/T 15777-2017; Standard Administration of China: Beijing, China, 2009; pp. 1–4.
21. Dixon, P.G.; Gibson, L.J. The structure and mechanics of Moso bamboo material. *J. R. Soc. Interface* **2014**, *11*, 20140321. [CrossRef] [PubMed]
22. Zhang, Y.; Huang, X.; Yu, Y.; Yu, W. Effects of internal structure and chemical compositions on the hygroscopic property of bamboo fiber reinforced composites. *Appl. Surf. Sci.* **2019**, *492*, 936–943. [CrossRef]
23. Yu, Y.; Liu, R.; Huang, Y.; Meng, F.; Yu, W. Preparation, physical, mechanical, and interfacial morphological properties of engineered bamboo scrimber. *Constr. Build. Mater.* **2017**, *157*, 1032–1039. [CrossRef]
24. Liu, H.; Peng, G.; Chai, Y.; Huang, A.; Jiang, Z.; Zhang, X. Analysis of tension and bending fracture behavior in moso bamboo (Phyllostachys pubescens) using synchrotron radiation micro-computed tomography (SRμCT). *Holzforschung* **2019**, *73*, 1051–1058. [CrossRef]

25. Chen, Q.; Dai, C.; Fang, C.; Chen, M.; Zhang, S.; Liu, R.; Liu, X.; Fei, B. Mode I interlaminar fracture toughness behavior and mechanisms of bamboo. *Mater. Des.* **2019**, *183*, 108132. [CrossRef]
26. Li, H.; Zhang, H.; Qiu, Z.; Su, J.; Wei, D.; Lorenzo, R.; Yuan, C.; Liu, H.; Zhou, C. Mechanical Properties and Stress Strain Relationship Models for Bamboo Scrimber. *J. Renew. Mater.* **2020**, *8*, 13–27. [CrossRef]
27. Cui, C.; Dong, J.; Mao, X. Effect of braiding angle on progressive failure and fracture mechanism of 3-D five-directional carbon/epoxy braided composites under impact compression. *Compos. Struct.* **2019**, *229*, 111412. [CrossRef]

Article

Study of *Aquilaria crassna* Wood as an Antifungal Additive to Improve the Properties of Natural Rubber as Air-Dried Sheets

Phattarawadee Nun-Anan, Sunisa Suchat, Narissara Mahathaninwong, Narong Chueangchayaphan, Seppo Karrila and Suphatchakorn Limhengha *

Faculty of Science and Industrial Technology, Prince of Songkla University, Surat Thani Campus, Surat Thani 84000, Thailand; phattarawadee.Anan@gmail.com (P.N.-A.); sunisa.su@psu.ac.th (S.S.); narissara.s@psu.ac.th (N.M.); narong.c@psu.ac.th (N.C.); seppo.karrila@gmail.com (S.K.)
* Correspondence: suphatchakorn.l@psu.ac.th

Abstract: Fungal growth on rubber sheets confers inferior properties and an unpleasant odor to raw natural rubber (NR) and products made from it, and it causes environmental concerns. The purpose of the present work was to investigate the effects of *Aquilaria crassna* wood (ACW) on the antifungal, physical and mechanical properties of NR as air-dried sheets (ADS) and ADS filled with ACW. The results show that the ACW-filled ADS had an increased Mooney viscosity, initial plasticity (P_O), and high thermo-oxidation plasticity (i.e., high plasticity retention index PRI). Additionally, superior green strength was observed for the ACW-filled ADS over the ADS without additive because of chemical interactions between lignin and proteins in NR molecules eliciting greater gel formation. A significant inhibition of fungal growth on the NR products during storage over a long period (5 months) was observed for ACW-filled ADS. Thus, it can be concluded that ACW could be applied as an antifungal additive that reduces fungal growth. This is a practically important aspect for the rubber industry, as fungal growth tends to spoil and cause the loss of NR sheets during storage. Moreover, the ACW is active as an incense agent, reducing negative impacts from odors that fungi, on rubber surfaces, release. Therefore, these filled intermediate NR products provide added value through, an environmentally friendly approach, this is pleasant to customers.

Keywords: *Aquilaria crassna* wood; natural rubber; antifungal; lignin; proteins

1. Introduction

Thailand is among the world's largest producers of natural rubber. Dry natural rubber (NR) is obtained from fresh latex suspension by processing it to block rubber, air-dried sheets, or ribbed smoked sheets. These intermediate rubber products can subsequently be processed into numerous consumer products like gloves, coatings, and tires. NR has excellent flexibility and tensile strength, and a good tear strength, together with low heat buildup [1]. Currently, NR products are rapidly being adopted by the food industry, and are used in contact with food or water, for example in foodstuff conveyor belts that need high flexibility [2]. However, the high moisture content of rubber sheets can enable fungal growth. This is a serious problem in rubber processing that can affect both the processing as well as the eventual properties of raw NR, and further on can also impact the final consumer products.

To overcome such problems, antifungal agents need to be added to the rubber during processing to reduce the growth of fungi, both in raw NR and its vulcanizates. Many studies have assessed effects of alternative natural additives, such as wood vinegar, as coagulating and antifungal agents for the rubber-sheet production process. For instance, the performance of wood vinegar from crude coconut shell as an additive for NR products has been studied [3]. It was found that the composition of wood vinegar (i.e., dominantly acetic acid and phenolic compounds) and its acidity improved the physical properties of raw NR, including plasticity and Mooney viscosity, while also inhibiting the growth of fungi on surfaces of the NR products [3]. Furthermore, using wood vinegar (produced from rubber wood) to inhibit fungi and malodors in NR products was also investigated [4]. The results showed that vinegar from rubber wood reduced the fungal colony counts on rubber sheets and decreased malodors while drying the rubber [4]. Recently, wood vinegars from para rubber wood, bamboo, and coconut shell have been tested as substitutes for a commercial acid (i.e., acetic acid) in coagulating rubber [5]. Regarding the physical and mechanical properties of the NR product, the results showed that the type of wood vinegar coagulant (i.e., para wood, bamboo, or coconut shell vinegar) did not cause significant differences from using the commercial coagulant. In addition, the rubber sheets coagulated with wood vinegar had comparatively less fungal growth than those using acetic acid [5].

Aquilaria crassna wood (ACW) is a local natural resource in Thailand, also grown in Vietnam, Laos, and Cambodia for the agar wood market [6]. At present, the fragrant *Aquilaria crassna* is popular in perfumes, decorations, and incenses [7]. However, the ACW residue from perfumery extraction is a waste material that can be considered for use in relatively low-cost rubber products. Moreover, ACW is a lignocellulosic material (i.e., composed mainly of cellulose, hemicellulose, and lignin) with only small quantities of other substances, such as phenols and acetyl groups [6]. Proteins account for approximately 0.5%wt of plant cell walls, where they are crosslinked with lignin [8,9]. It has also been reported that these components in wood contribute to its stiffness, strength and resistance against insects and pathogens (both in the plant and in the eventual wood products) [10]. *Aquilaria crassna* wood has many applications in food, cosmetics, and pharmaceutical products, depending on the purity and the wood source [10,11]. The ACW has been applied in agriculture as a peptizer, an anti-fungal, and an anti-termite substance, improving the longevity of other wood materials; it enables value-added wood products [12]. Therefore, it has potential to serve as an antifungal, also, in raw natural rubber.

In the present work, extracted *Aquilaria crassna* wood (ACW) was used as an environmentally friendly alternative additive for natural rubber in the form of air-dried sheets (ADS). The purpose of the study was to assess the effects of using ACW at various contents (0, 20, 40 and 60 phr) in NR products, under the hypothesis that ACW could enhance the key properties and antifungal characteristics of NR products (or specifically ADS). The ADS with and without ACW was analyzed for dirt, volatiles, nitrogen, and gel content, as well as for plasticity, Mooney viscosity, and plasticity retention index (PRI) together with their tensile properties. Moreover, the antifungal efficiencies of both ADS and ACW-filled ADS were evaluated by recording the area fractions of fungal growth on the rubber surfaces.

2. Materials and Methods
2.1. Materials

Field natural rubber latex was obtained from the RRIM 600 clonal variety of rubber trees, in a rubber plantation located in Surat Thani province, Thailand. The chemicals for filler dispersion and bentonite were supplied by the BASF Company (Rhineland-Palatinate, Ludwigshafen, Germany). Vultamol, used as dispersant, was produced by S&B minerals GmbH (Kifissia, Athens, Greece). Chemicals for the determination of dirt content, 2-mercaptobenzothiazole and turpentine oil, were produced by Merck KGaA (Darmstadt, Germany) and Vidhyasom CO., Ltd (Phra Nakhon, Bangkok, Thailand), respectively. In addition, boric acid, sulfuric acid (H_2SO_4), and potassium sulfate (K_2SO_4), used to analyze nitrogen content, were manufactured by Merck KGaA (Darmstadt, Germany). The copper (II) sulfate ($CuSO_4$) was obtained from VWR (Fontenay-sous-Bois, France).

2.2. Aquilaria crassna Wood (ACW) Powder Preparation

Extracted *Aquilaria crassna* wood was obtained from the perfumery process, in which fragrant substances are extracted. The extracted wood residue was sawn to small pieces and then ground in a ball mill for about 48 h. After that, the powder was passed through a 120-mesh screen before the preparation of an ACW dispersion. The functional groups in the ACW particles were analyzed by Fourier transform infrared spectroscopy (Perkin Elmer Inc., Waltham, MA, USA).

2.3. Preparation of Aquilaria crassna Wood Dispersion

Aquilaria crassna wood, as a filler, was first mixed with distilled water using a dispersing agent (Vultamol). The filler dispersion was prepared by ball milling, using ceramic balls for 72 h, to ensure the breakdown of filler aggregates (Table 1).

Table 1. Formulation of *Aquilaria crassna* wood dispersion.

Ingredient	Parts per Hundred of Rubber (phr)
Aquilaria crassna wood powder	20
bentonite	1
vultamol	1
water	78

2.4. Air-Dried Sheet Preparation

Field latex was tapped and collected from RRIM 600 clones in Surat Thani, Thailand. The latex was first diluted with clean water to 20% dry rubber content (%DRC). Then, the diluted latex was coagulated by using 5% w/v formic acid and left for 3 h to solidify. After that, the soft coagulum formed was passed through a nip between two steel rolls and washed with clean water before drying in a hot air oven at 50 °C until a constant weight was reached. Finally, air-dried sheets (ADS) were obtained.

2.5. Preparation of Aquilaria crassna Wood-Filled Rubber Sheet

The rubber latex of 20% DRC was prepared by diluting the initial field natural rubber latex with about 35% DRC using clean water. Then, 30% ACW dispersion was added into the latex to achieve the targeted contents (0, 20, 40 and 60 phr) in an aqueous solution/dispersion, and the mix was stirred well using a mechanical stirrer (Onilab, San Francisco, CA, USA) at 50 rpm for 10 min to homogenize the dispersion. The mixture was then added with 10% w/v formic acid to coagulate it in the coagulation tank and was allowed to solidify for 3 h. The coagulum was pressed by rollers to make a rubber sheet, which was further dried at 50 °C in a hot-air oven until constant weight. Then, *Aquilaria crassna* wood (ACW)-filled dry rubber sheets were obtained.

2.6. Preparation of NR Thin Films

NR thin films were prepared from NR sheet with a compression-molding machine (PR1D-W400L450 PM, Chareon Tut Co., Ltd., Bang Phli, Samut Prakarn, Thailand). First, the NR sheets were compressed at 160 °C for 10 min. After that, the films were immediately cooled down to room temperature, under pressure for 10 min by a cooling system. The rubber films were left at room temperature for one week before use, to allow the residual strains from molding to relax.

2.7. Characterization

2.7.1. Fourier Transform Infrared Spectroscopy (FTIR)

FTIR spectra of the rubber sheets were recorded with a Perkin-Elmer Spectrum spectrometer (Perkin Elmer Inc., Waltham, MA, USA). The analysis was carried out over the wavelength range of 4000–400 cm^{-1} with 32 scans acquired per recorded spectrum, at a resolution of 4 cm^{-1}.

2.7.2. Analysis of ACW

The chemical composition of ACW powder was characterized by X-ray fluorescence spectrometry (XRF) (PW2400, Philips & Co., Eindhoven, North Brabant, The Netherlands). First, ACW powder was dried to a constant weight in a hot-air oven at 100 °C for 14 h.

After that, the sample was treated in a muffle furnace by heating from 50 °C to 1000 °C. It was then mixed with a binder (WAX C) at a weight ratio of 2:1 and further compressed to a thin sheet. Finally, the sample sheets were characterized by XRF.

2.7.3. Characterization of Air-Dried Sheet (ADS)

The dirt content of air-dried sheets (ADS) was measured according to ISO 247-2, as described elsewhere [13]. The homogenized rubber sample (about 10 g) was first cut into small pieces and then immersed in a volumetric flask containing 200 mL of a mixture of turpentine oil (boiling point 154–198 °C) and 0.5 g of 2-mercaptobenzothiazole, as a peptizer. After that, the solution was left at room temperature for 48 h and then heated at 130 ± 5 °C to complete dissolution. The obtained solution was then separated through 45-mm sieves, after which the dirt retained was washed, dried and its weight was recorded. Finally, the weight percent of dirt content was calculated as follows:

$$\text{Dirt content (\% wt)} = \frac{B - A}{W} \times 100 \quad (1)$$

where A is the mass of the sieve (g), B is the mass of the sieve containing dirt (g), and W is mass of the rubber sample (g).

In addition, determination of the ash contents of the ADS samples was performed according to ISO 247-2, as described elsewhere [14]. First, the rubber sample was weighed and placed inside a crucible, and this was placed in a heat-treatment oven, an Optic Ivymen System (SNOL 3/1100 LHM01, Utena, Lithuania), at 550 °C, in order to achieve total oxidation of the sample. After that, the remaining ashes corresponded to the inorganic components in the ADS sample. The ash content was calculated as follows:

$$\text{Ash content (\% wt)} = \frac{m_2}{m_1} \times 100 \quad (2)$$

where m_1 and m_2 are the initial mass of rubber sample (g) and the final mass (g) of rubber sample after the oven treatment (ash), respectively.

Moreover, the volatile matter content (VM) of the ADS was also investigated according to ISO 248, as described elsewhere [15]. About 10 g of homogenized rubber was passed through a two-roll mill to control the thickness at less than 2 mm. The sample pieces were placed in a conventional circulating-air oven at 100 ± 5 °C for 4 h. After that, the rubber sample was cooled in a desiccator at room temperature for 30 min before weighing the rubber sample, and the volatile matter content was calculated as follows:

$$\text{Volatile matter content (\% wt)} = \left(\frac{W_1 - W_2}{W_1}\right) \times 100 \quad (3)$$

where W_1 is the initial mass of the sample and W_2 is the mass of the sample after heating.

2.7.4. Gel Content

The gel content in ADS was investigated in accordance with ISO 1166, as described elsewhere [16]. First, an ADS sample of about 0.1 g was cut into small pieces and suspended in 30 mL of toluene, followed by shaking for a few minutes, and was left at 25 °C for 20 h without stirring. After that, the insoluble rubber was separated by centrifuging at 22,000 g and the liquid fraction was removed from the tube. The remaining gel fraction was washed with acetone and then dried at 110 °C for 1 h. Then, the gel content of NR was calculated as follows:

$$\text{Gel content (\%wt)} = \left(\frac{m_0 - m_1}{m_0}\right) * 100 \quad (4)$$

where m_1 and m_0 are the dry weights after extraction and before extraction, respectively.

2.7.5. Measurement of Initial Plasticity

The plasticity of ADS was determined using a Plastimeter H-01 (CG Engineering Ltd., Part, Sam Khok, Pathum Thani, Thailand), in accordance with ISO 2007. First, the ADS was masticated and homogenized on a two-roll mill (YFCR 6″, Yong Fong machinery CO., LTD, Mueang Samut Sakhon, Samut Sakorn, Thailand) at ambient temperature. The sample was sheeted-out to 3.2–3.8 mm of thickness. After that, test pieces were cut from the thin sheet with a specimen-cutting press. Two sample sets were prepared for the determination of initial plasticity (P_O) and the plasticity retention index (PRI), using three replicates in each test.

2.7.6. Plasticity Retention Index (PRI) Determination

PRI is a measure of resistance of the NR product to thermal oxidation. The rubber pieces were inserted into an ageing chamber for 30 min at 140 °C, removed, and allowed to cool at room temperature. After that, the test pieces were immediately compressed by a plastimeter under a constant compressive force of 100 N at 100 °C for 15 s. The plasticity of the not-aged sample (P_O) and plasticity of the aged sample (P_{30}) were obtained by measuring their thickness changes. Finally, the median of three replicate samples was taken as the plasticity value. The plasticity retention index (PRI) was calculated as follows:

$$\mathrm{PRI} = \left(\frac{P_O}{P_{30}}\right) \times 100 \tag{5}$$

where P_O is the plasticity before aging in the oven and P_{30} is the plasticity after aging in the oven.

2.7.7. Mooney Viscosity

Mooney viscosity indicates the processability of an elastomer and it was measured by using a MV-2020 Mooney viscometer (Montech, Shinjuku, Tokyo, Japan) with a large rotor size, at 100 °C, and according to ISO 289.

2.7.8. Tensile Properties

The thin NR sheets were cut to dumbbell shapes with die type 5A, in accordance to ISO 527. Then, their tensile properties were measured using a tensile testing machine (Zwick Roell, Ulm, Germany) at room temperature with a crosshead speed of 500 mm/min and a 100-N load cell. The measurement was repeated five times for each type of rubber sample.

2.7.9. Antifungal Performance

The antifungal performance of the rubber sheets was investigated by measuring fungal growth area on the pre-dried rubber sheet surfaces after leaving them for 5 months. Area percentage of the fungal growth was calculated, as described elsewhere [17], by:

$$\% \text{ Fungal growth area} = \frac{A_{fungi}}{A_{NR}} \times 100 \tag{6}$$

where A_{fungi} is the fungal growth area on the rubber sheet surface (cm^2) and A_{NR} is the total area of rubber sheet surface (cm^2).

3. Results and Discussion

3.1. Characterization of the ACW Powder

The functional groups present in the *Aquilaria crassna* wood (ACW) powder are shown in Figure 1. It is clearly seen that ACW showed a broad peak at the wave number 3340 cm^{-1}, assigned to –OH stretching vibrations from the phenol, alcohol, and acid groups present in lignin of ACW (Table 2) [18,19]. These also caused the ACW absorption bands at 1730, 1620, and 1029 cm^{-1}, seen in Figure 1 and Table 2. The absorption peak at the wavelength range of 1740–1710 cm^{-1} is attributed to the carbonyl group (C–O) stretching vibrations in the esters and acids [19,20]. In addition, the broad peak around 1620 cm^{-1} indicates aromatic rings. Furthermore, the absorption peak at 1029 cm^{-1} is from the stretching vibrations of C–O [19].

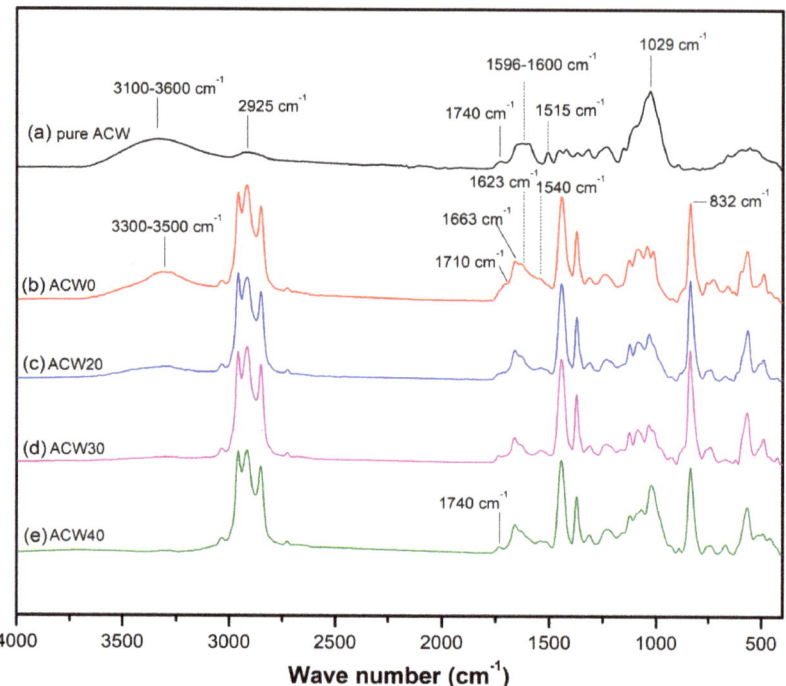

Figure 1. FTIR spectra of pure *Aquilaria crassna* wood (ACW) (**a**), and for air-dried sheets with various ACW contents: 0 phr (**b**), 20 phr (**c**), 40 phr (**d**), and 60 phr (**e**).

Table 2. The FTIR band assignment of *Aquilaria. crassna* wood and natural rubber.

Wavelength (cm^{-1})	Assignment
1240	Amide III band (in-phase combination of C–N stretch and N–H bend modes) [21]
1308	CH$_2$ wagging [19]
1375	CH$_3$ asymmetric deformation [19]
1448	CH$_2$ deformation [19]
1509	C–H stretching vibration of wood [13]
1540	Amide II: N–H and C–N stretching vibration of proteins in NR molecules [19]
1620	Aromatic ring stretching vibration [13]
1630	Amide I: R$_1$–(C=O)–NH–R$_2$ stretching vibration of proteins in NR molecules [19]
1660	C=C stretching vibration of *cis*-1,4-isoprene units [19]
1710	–(C=O)–OH stretching vibration of lipids in natural rubber [19]
1740–1720	C=O stretching vibration of aldehydes [13]
3280	N–H stretching vibration of proteins [19]

The chemical composition of *Aquilaria crassna* wood was analyzed by X-ray fluorescence spectrometer (XRF), with the results shown in Table 3. It is seen that *Aquilaria crassna* had CHNO content of about 98%. These major components correlated well to lignin's structure and a small quantity of proteins present in cell walls of plants (or in ACW) [10,19], which may affect in the properties of ADS products.

Table 3. Quantification of major elements in *Aquilaria crassna* wood.

Element	Concentration (%)
C	27.359
O	36.444
N	31.905
H	2.296
Na	0.034
Mg	0.076
Al	0.104
Si	0.195
P	0.043
S	0.145
Cl	0.136
K	0.243
Ca	0.874
Ti	0.011
Mn	0.028
Fe	0.090
Cu	0.011
Zn	0.006
Sr	0.002

3.2. FT–IR Spectra

Figure 1 and Table 2 show the FT-IR spectra for ADSs with various ACW contents (0, 20, 40, and 60 phr). It can be seen that the typical absorption peaks of natural rubber appeared at 1663 and 832 cm^{-1}, attributed to C=C stretching vibrations and =CH out-of-plane bending of cis-1,4-isoprene units in the natural rubber, respectively [21,22]. Reductions and shifts of the broad, strong –OH stretching of alcohol, phenol, and acid in lignin, at about 3100–3600 cm^{-1}, was observed [19]. The decrease of the band at 3100–3600 cm^{-1} can be attributed to intramolecular hydrogen bonding between lignin structures [23]. Furthermore, the overlapped absorption peaks at 1630 and 1540 cm^{-1} could possibly be due to proteins present in ACW [9,10], because the intensities of these bands increased with ACW content. However, the absorption peaks at 3280, 1630, and 1540 cm^{-1} were partly affected by mono- and di-peptides of proteins present in *Hevea* NR [24,25]. This relates to the elemental composition in ACW, as given in Table 3, attributed to the –NH group in the amines and amide II (N–H stretching vibrations) of proteins, respectively [24]. The peak ratios between –NH group (3280+1540 cm^{-1}) and the symmetric CH$_2$ stretching vibrations (2963 cm^{-1}) were used to estimate the nitrogen content in the NR samples, with the results shown in Table 4. It is clear that the ACW60 showed the highest peak ratio for proteins in NR, while the ACW40, ACW20 and ACW0 gave lower intensity ratios for proteins in the NR sample. It is notable that nitrogen content was directly used to estimate the level of proteins in raw NR products [25]. This implies that ADS with varying ACW content contains detectable proteins (i.e., protein in ACW) even though these are of different types than those present in *Hevea* NR. This result matches well the percentage of N atoms in *Aquilaria crassna* wood, as seen in Table 3. Also, it matches well the level of protein content calculated from the percentage of N atoms, as analyzed by the Kjeldahl method (Figure 2). Thus, it is interesting that it is possibly the amino acids of proteins in ACW and proteins in *Hevea* NR that react in ways that impact the various properties of raw NR.

Table 4. Physical properties of natural rubber with various *Aquilaria crassna* wood-filler contents (i.e., 0, 20, 40 and 60 phr).

Property\Sample	ACW0	ACW20	ACW40	ACW60
dirt content (% wt)	0.03 ± 0.06	6.88 ± 0.06	6.97 ± 0.06	8.23 ± 0.06
ash content (% wt)	0.43 ± 0.04	1.36 ± 20.04	1.86 ± 40.04	3.04 ± 60.04
volatile matter (VM) content (% wt)	0.94 ± 0.2	01.93 ± 0.40	2.99 ± 0.90	5.08 ± 1.00
nitrogen content (% wt)	0.410 ± 0.020	0.447 ± 0.122	0.487 ± 0.070	0.513 ± 0.115

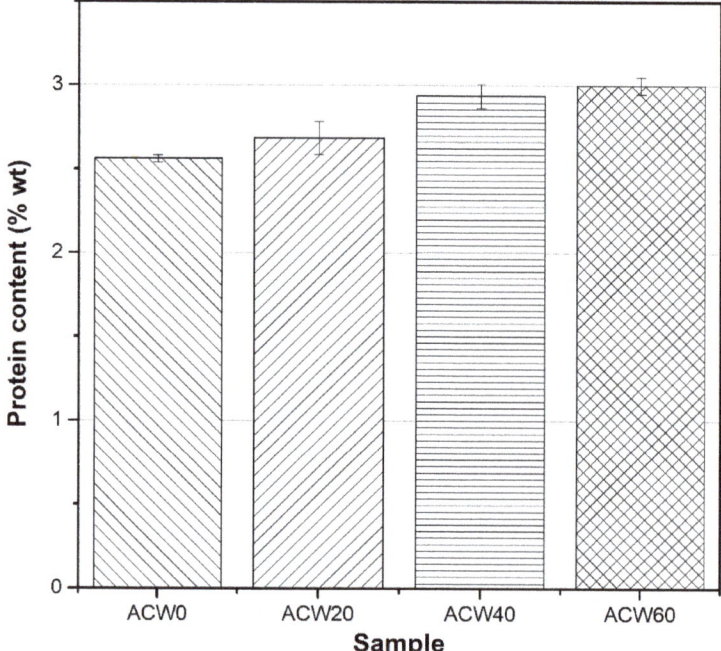

Figure 2. Protein content of NR as a function of *Aquilaria crassna* wood content.

Moreover, increasing *Aquilaria crassna* content caused changes in the absorbance ratio 1240 cm^{-1}/2963 cm^{-1} (where the band at 1240 cm^{-1} is assigned to C–O stretching in the phenolic compounds of lignin, and the band at 2963 cm^{-1} is assigned to symmetric CH_2 stretching vibrations) [19,26]. It is seen that the ratio 0.25 for ACW60 case is higher than those of the other samples, while the ratio is 0.15 for ACW0 (without any *Aquilaria crassna* in the ADS) (Figure 1). This intensity ratio could be attributed to crosslinking between ACW and protein molecules in NR, or to the filler–filler interactions of ACW. These possible interactions are reflected by an increase in hydrogen bonding in the filled NR [27], which may improve the physical properties of rubber sheets.

3.3. Physical Properties of Raw NR

The effects of ACW content on physical properties, including dirt, ash and volatile matter (VM) contents in ADS, are summarized in Table 4. The dirt content of ADS is a measure of the contamination level of the rubber sheets during the processing of latex to dry NR. It is seen that the dirt content of the rubber sheets increased with ACW content. The ACW60 sample showed the highest dirt content, of about 8.23 %wt, followed by the ACW40 (6.97 %wt) and ACW20 samples (6.88 %wt), in rank order (Table 4). In addition, ACW0 had the lowest dirt content, of about 0.03 %wt (Table 4). This is because it is a major component in ACW, as an additive in NR, and may have caused increased ash content in the NR sheets [10,28]. Therefore, ACW was the main contributor of the ash content in the filled rubber sheets. Furthermore, in most cases, a high ash content in NR is directly correlated with a high dirt content, as was also seen in the present work. This relation between dirt and ash contents in rubber sheets is shown in Figure 3. As is clearly seen, both the dirt and ash contents of the ADS increased with ACW content, due to *Aquilaria crassna's* composition and some soluble impurities present in the *Hevea* NR [10,29].

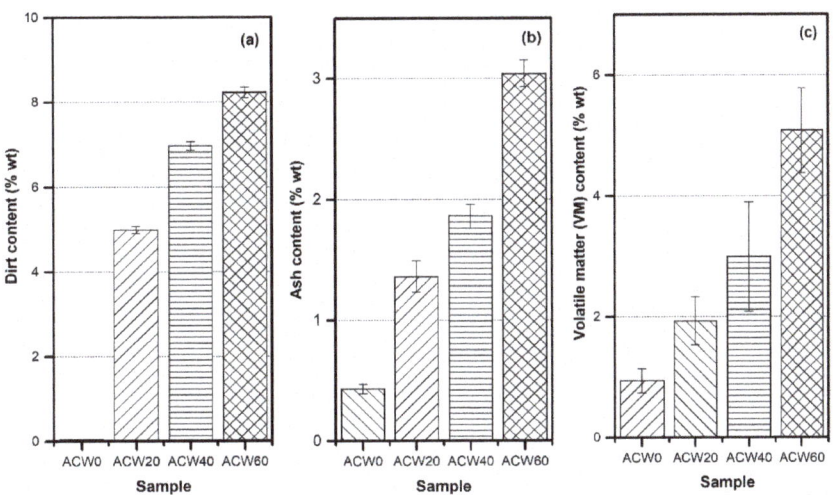

Figure 3. Dirt (**a**), ash (**b**) and volatile matter (VM) contents (**c**) of natural rubber as function Table 4. (i.e., 0, 20, 40 and 60 phr).

Table 4 and Figure 3c show the volatile matter (VM) contents in rubber sheets with different ACW contents. The volatile content of dry rubber is related to its moisture, as well as to its contamination with volatile matter during rubber processing (i.e., during coagulation of rubber latex) and showed the same trend as ash and dirt contents, as seen in Figure 3. The VM content increased with ACW content. The VM in ACW60 sample was about 5.08 %wt, which was higher than in the other samples. This behavior of the ADS product was caused by hydroxyl groups (–OH), as water absorption sites [30]. This may be associated with antifungal activity on the NR surface; due to their high moisture content, NR products are susceptible to fungal growth [17].

3.4. Plasticity of Rubber

Table 5 summarizes the initial plasticities (P_O) and Mooney viscosities of the ADSs. P_O and Mooney viscosity represent the ability of rubber to deform [3]. It was observed that the ADS without ACW (or ACW0) gave the lowest P_O, of about 40 (Table 5). On the other hand, P_O increased with ACW content. The ACW60 sample gave the highest P_O of about 61, followed by ACW40 and ACW20, at about 50 and 40, in rank order (Table 5). Increasing ACW content increased P_O in the NR products, due to phenolic compounds in ACW that can interact with the amine groups of NR molecules [31].

Furthermore, Mooney viscosity increased with ACW content in the rubber sheets, showing a trend similar to that for initial plasticity, as shown in Figure 4. The Mooney viscosity had the rank order ACW60 > ACW40 > ACW20 > ACW0. When the ACW content was 60 phr, the Mooney viscosity was the largest (96, Table 5). The increases in both P_O and Mooney viscosity were caused by (I) the intramolecular hydrogen bonding of lignin, and (II) the covalent and hydrogen bonds between lignin and protein in ACW [32,33]. Possibly, there were non-covalent bonds between NR and lignin [34] and chemical interactions between NR proteins via the active functional groups (or –OH groups) of lignin in ACW, as illustrated in Figure 5. These results match well the gel contents observed in ADS, which given in Table 5. It is notable that the gel content in NR represents the interactions between proteins and other active functional groups via hydrogen or other bonding [35,36]. In Table 5, it can be observed that the ACW60 gave the highest gel content, of about 50%wt followed by the ACW40, ACW20, and ACW0 samples (31, 20, and 8 %wt). This confirms the increased intramolecular hydrogen bonding of lignin in ACW with ACW content, by the interactions of lignin with proteins in rubber that contributed to the gel observed in NR. Furthermore, this result matches well the small shift in wavelengths at 1443–1445 cm^{-1} for WAC20, ACW40, and ACW60, as compared to the not-filled case (ACW0), associated with the vibrations of rubber molecules [34]. This shift indicates non-covalent interactions between the rubber molecules and lignin, and adsorption of NR onto the lignin in ACW [34] that increased with ACW content.

Table 5. Initial plasticities (P_O), plasticity retention indexes (PRI), and Mooney viscosities of the rubber sheets, along with their gel contents.

Sample	Initial Plasticity (P_O)	Plasticity Retention Index (PRI)	Mooney Viscosity [ML 1+4 (100 °C)]	Gel Content (%wt)
ACW0	38.2 ± 0.7	100.6 ± 2.0	64.9 ± 1.0	8.78 ± 0.14
ACW20	40.1 ± 1.1	91.2 ± 3.0	66.6 ± 0.1	20.53 ± 0.75
ACW40	50.1 ± 0.8	64.0 ± 1.0	75.1 ± 0.1	31.57 ± 1.85
ACW60	61.2 ± 1.0	45.1 ± 1.5	96.4 ± 1.2	50.22 ± 1.19

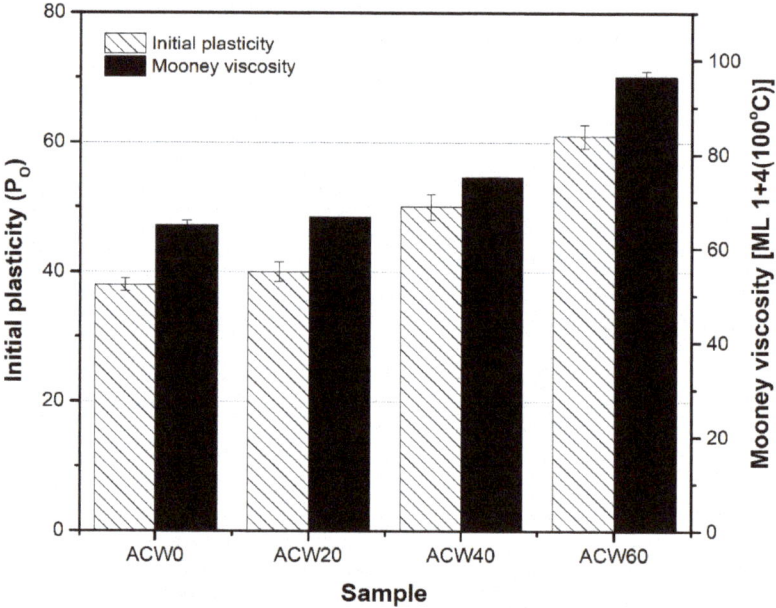

Figure 4. Initial plasticity and Mooney viscosity of natural rubber as a function of *Aquilaria crassna* wood content (i.e., 0, 20, 40, and 60 phr).

The oxidation resistance of raw NR is indicated by the plasticity retention index (PRI), summarized in Table 5. If the value is high, it shows that the rubber resisted thermo-oxidation. The results show that increasing ACW content improved the thermo-oxidation resistance of natural rubber. ACW60 had the highest PRI at about 110, indicating the best oxidation resistance, while ACW40 and ACW20 also showed higher PRI values (about 95 and 101) than that of the rubber sheets without ACW content (90). The highest PRI was found for the rubber sheet with the most ACW, so these rubber sheets resisted thermal oxidation comparatively well. The PRI for rubber sheets improved with ACW content, possibly, because of phenolic compounds and proteins in ACW [3,37].

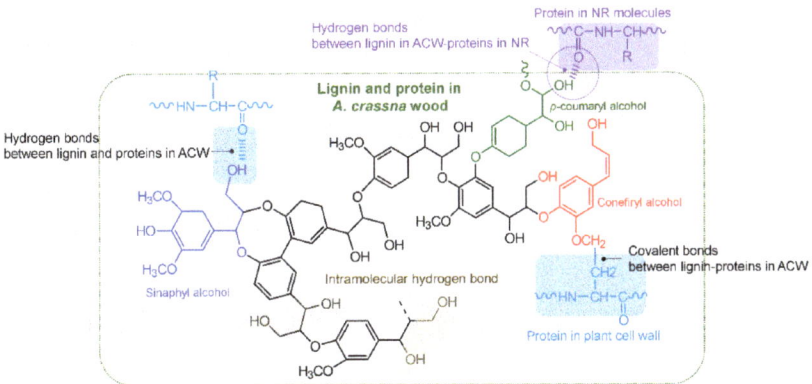

Figure 5. A proposed model for the interactions of lignin and plant proteins in the cell walls of *Aquilaria crassna* wood lignin with proteins in *Hevea* NR.

3.5. Tensile Properties

Figure 6 shows the stress–strain behavior of the NR films. It was observed that the ACW0 sample (without ACW) gave the lowest green strength, of about 0.27 MPa, while it exhibited an excellent elongation at break of approximately 669%. Furthermore, it should be noted that increased ACW content enhanced the green strength of ADS and conferred reduced elongation at break relative to the control cases without ACW. ACW60 showed the highest green strength, of about 2.19 MPa, while the ACW40 and ACW20 samples had lower green strengths (1.84 and 1.29 MPa, respectively, Table 6) than ACW60 (Figure 6). This is due to a high gel content (51%wt, Table 5), which encouraged the strength and stiffness of NR product [38]. On the other hand, the elongation at break of the rubber sheets decreased with ACW content (Figure 6). The ACW60 sample showed the lowest elongation at break (about 158%, Table 6) followed by the ACW40, ACW20, and ACW0 samples. This may be attributed to ACW, as a filler, inhibiting the orientation-induced crystallization of NR molecules [39], which decreased the elongation at break of the ADS.

Table 6. Mechanical properties of the NR films.

Property\Sample	ACW0	ACW20	ACW40	ACW60
100% modulus (MPa)	0.26 ± 0.01	0.56 ± 0.02	0.95 ± 0.01	1.94 ± 0.03
300% modulus (MPa)	0.24 ± 0.02	0.99 ± 0.04	1.61 ± 0.03	-
Green strength (MPa)	0.27 ± 0.01	1.29 ± 0.02	1.84 ± 0.01	2.19 ± 0.03
Elongation at break (%)	669.52 ± 29.17	434.22 ± 17.08	386.92 ± 11.16	158.36 ± 15.77

Table 6 summarizes the 100% and 300% moduli and toughnesses of the NR sheets as a function of ACW content (i.e., 0, 20, 40, and 60 phr). It is observed that ACW0 showed the lowest 100% and 300% moduli, due to its having the lowest gel content (or low intramolecular, physical, and chemical interactions) of about 8%wt (Table 5). Apparently, high lignin and protein in ACW60 acted as reinforcing fillers and enhanced the rigidity of the filled ADS. Possibly, the physical and chemical interactions between lignin and proteins in NR produced strong natural crosslink formations with high 100% and 300% moduli, along with greater stiffness, as shown in Figure 7. Therefore, it can be concluded that the use of ACW as an environmentally friendly filler increased the 100% and 300% moduli and the green strengths of the filled ADS products.

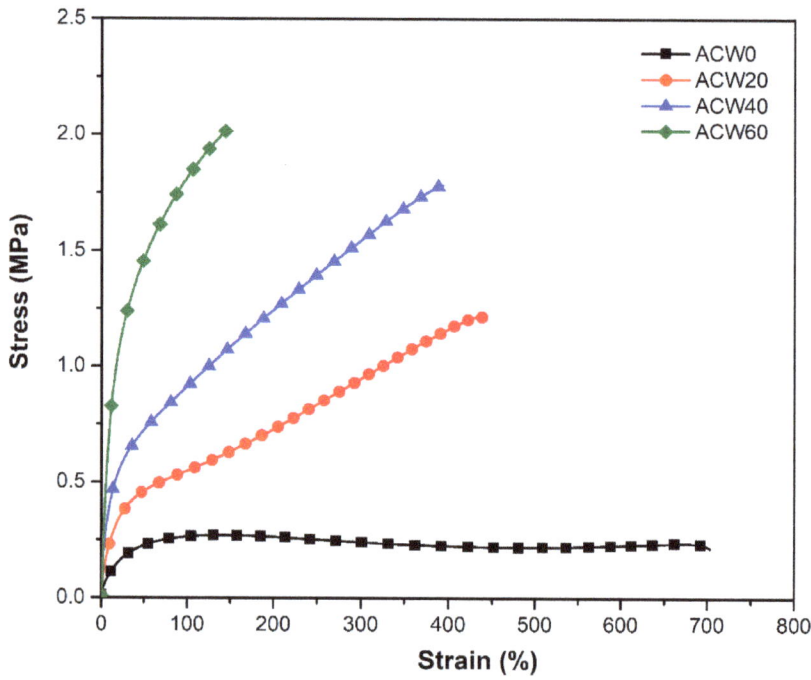

Figure 6. Stress–strain behavior of the ADS films as functions of *Aquilaria crassna* wood content (i.e., 0, 20, 40, and 60 phr).

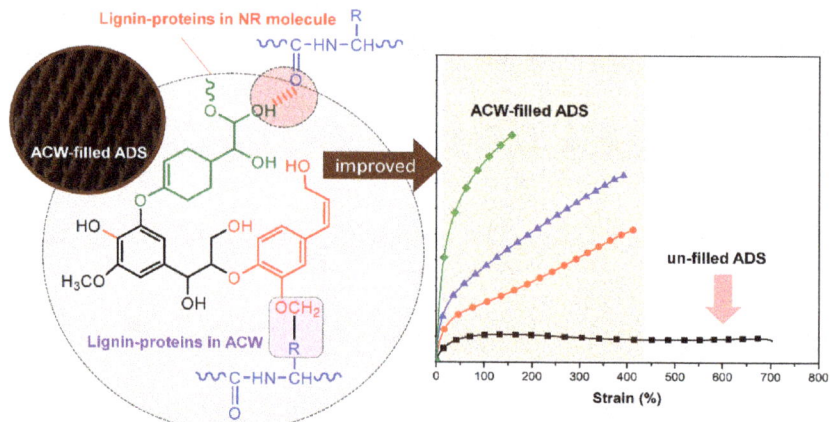

Figure 7. Possible physical and chemical interactions between lignin and proteins in NR.

3.6. Antifungal Performance of the Rubber Sheet

The antifungal performance of the ADS, as a function of ACW content, was measured by the area fractions of fungal growth on the ADS, when the rubber sheets were left at ambient conditions for 5 months. The percentage of fungal growth area was calculated according to Equation (5). It was observed that the dry rubber sheet had a yellow–brown color and was without fungi (white spots on the NR surface) before the storage for 5 months (Figure 8a). After storage for 5 months, the color of the rubber sheet had changed to black-and-brown and showed some white spots (i.e., fungi or bacteria) on the surfaces. The fungi grew well on the ACW0 sample, as shown in Figure 8b. In contrast, the rubber sheets with ACW at 20-phr loading had quite similar appearances (i.e., surface and color) before and after storage for 5 months. This suggests that ACW could inhibit fungal growth on surfaces of rubber sheets, as shown in Figure 8d, having high antifungal activity. However, the fungal growth area fractions were 65, 69, and 73% for ACW60, ACW40, and ACW20, respectively, while the highest fungal growth area fraction was 97%, for the ACW0 sample (without *Aquilaria crassna*, see Table 7). This indicates that ACW filler in the ADS had antifungal activity in the rubber product due to its composition, including, especially, phenolic compounds [4], which are antimicrobial [40]. Moreover, it had a strong antifungal effect on microorganisms [41].

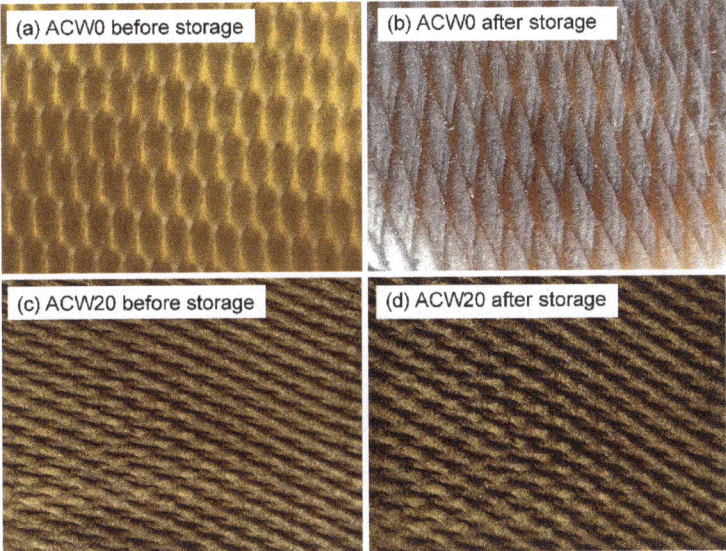

Figure 8. ACW0 and ACW20 filled ADS samples before and after storage (**a**–**d**) for 5 months at ambient temperature.

Table 7. Fungal growth area fractions at different *Aquilaria crassna* wood (ACW) contents in air-dried sheets (ADS).

Aquilaria crassna Wood Content (phr)	% Area of Fungal Growth
0	97.62 ± 2.45
20	73.70 ± 1.65
40	69.11 ± 0.56
60	65.23 ± 1.54

The growth of fungi on the rubber sheet surfaces is shown in Figure 9. It can be seen that the fungi were well distributed on ACW0 (without ACW filler) in Figure 9a. In contrast, the rubber sheets with ACW at 20, 40, and 60 phr did not have growth of fungi on their sheets' surfaces, as shown in Figure 9b–d, respectively. This confirms that the phenolic compounds in ACW inhibited fungal growth on the dry rubber products [14]. Using ACW as a filler in rubber sheets could improve their antifungal performance.

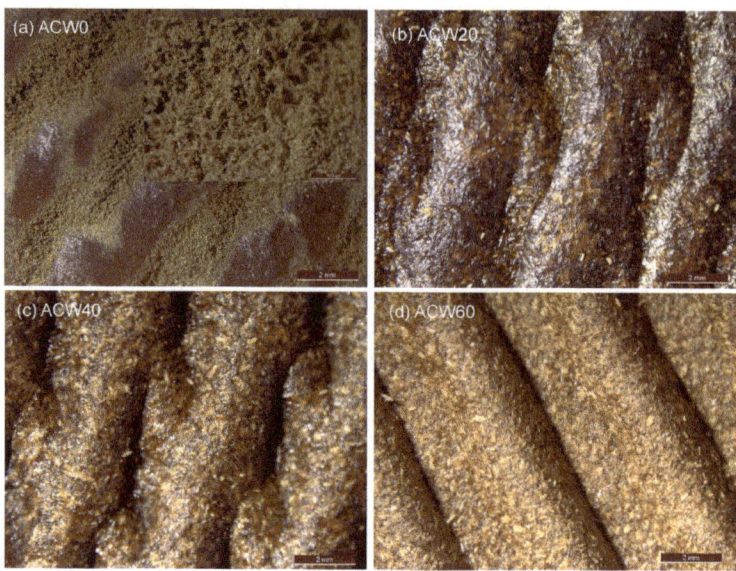

Figure 9. Morphological properties of the ACW0 (**a**), ACW20 (**b**), ACW40 (**c**), and ACW60 (**d**) samples after storage for 5 months at ambient temperature.

In addition, the changes in initial plasticity (P_O) and Mooney viscosity (Figure 10) of all ADS samples were tested after storage for 5 months (Figure 10a,b). The results correlated well with the ACW content, the quantity of proteins (Figure 2), and gel content (Table 5). Thus, it is concluded that *Aquilaria crassna* wood is suitable for use as a filler, having not only contributed to improving the plasticities and viscosities of the NR products by the provided lignin and proteins, but also in having improved the antifungal performance of the filled ADS product.

As is well known, in block rubber processing, unpleasant odors are released during the storage of fresh or dried cup-lump rubber, and from thermal degradation during the drying of shredded rubber. These odors from rubber factories have mainly been attributed to noxious volatile components, which are discharged into the atmosphere through a chimney during the drying stage in block rubber processing [42]. The odors have often led to complaints from the public, who are also irritated by other environmental problems associated with the rubber industry. Therefore, ACW powder, as an additive against fungi, could also serve as an incense, providing a pleasant scent to the NR product, especially in the case of cup lump, which will be studied in further work.

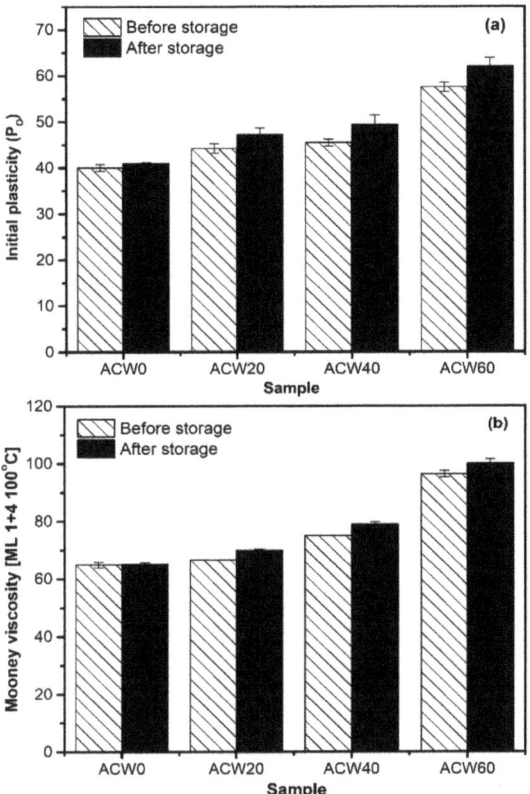

Figure 10. Initial plasticities (P_O) (**a**) and Mooney viscosities (**b**) of the ADS samples as a function Figure 4. (i.e., 0, 20 40, and 60 phr) after storage for 5 months.

4. Conclusions

The present work studied the effects of *Aquilaria crassna* wood (ACW) as an alternative antifungal additive on the properties of a raw NR product (i.e., air-dried sheets, ADS). Various properties of raw NR in ADS, including initial plasticity (P_O), plasticity retention index (PRI), Mooney viscosity and tensile properties (i.e., 100% and 300% moduli and green strength) were also considered. The results showed that ACW-filled ADS had improved P_O, Mooney viscosity, and tensile moduli, as well as green strength, because of the hydrogen bonding of lignin in ACW with proteins in NR. Improved resistance to thermo-oxidation and better retention of plasticity were also observed for the filled ADSs, because of phenolic compounds in the ACW. Moreover, ACW's antifungal activity inhibited or retarded fungal growth on the rubber sheets, as compared with the control without ACW, during storage for a long period (5 months), in a dose-dependent manner. Thus, ACW is an alternative antifungal additive, improving ADS properties and preventing fungal growth on NR products, while also acting as an incense to provide a pleasant scent. These characteristics represent important advantages to the rubber industry in its relationships with neighbors and customers. Additionally, rubber tree plantations and end users of the NR products stand to benefits therefrom.

Author Contributions: Conceptualization, P.N.-A., S.L., S.S., N.M., N.C. and S.K.; writing—original draft preparation, P.N.-A.; writing—review and editing, S.L. and S.K.; visualization, P.N.-A. and S.L.; supervision, S.L.; project administration, S.L.; funding acquisition, S.L. All authors have read and agreed to the published version of the manuscript.

Funding: This research was funded by the Natural Rubber Innovation Research Institute, Prince of Songkla University (Grant No. SIT6201100S).

Institutional Review Board Statement: Not applicable.

Informed Consent Statement: Not applicable.

Data Availability Statement: The data presented in this study are available on request from the corresponding author.

Acknowledgments: The authors would like to thank the Natural Rubber Innovation Research Institute, Prince of Songkla University (Grant No. SIT6201100S) and Faculty of Science and Industrial Technology Prince of Songkla University, Suratthani Campus for providing the facilities.

Conflicts of Interest: The authors declare no conflict of interest.

References

1. Amnuaypornsri, S.; Sakdapipanich, J.; Toki, S.; Hsiao, B.S.; Ichikawa, N.; Tanaka, Y. Strain-induced crystallization of natural rubber: Effect of proteins and phospholipids. *Rubber Chem. Technol.* **2008**, *81*, 753–766. [CrossRef]
2. Limhengha, S.; Limnararat, S.; Sriseubsai, W. Effect of ENR50/STR5L Blends on Properties of Foodstuff Conveyor Belts Compound. *Key Eng. Mater.* **2016**, *701*, 243–249. [CrossRef]
3. Prasertsit, K.; Rattanawan, N.; Ratanapisit, J. Effects of wood vinegar as an additive for natural rubber products. *Songklanakarin J. Sci. Technol.* **2011**, *33*, 425–430.
4. Chungsiriporn, J.; Pongyeela, P.; Iewkittayakorn, J. Use of wood vinegar as fungus and malodor retarding agent for natural rubber products. *Songklanakarin J. Sci. Technol.* **2018**, *40*, 87–92.
5. Kalasee, W.; Dangwilailux, P. Effect of Wood Vinegar Substitutes on Acetic Acid for Coagulating Natural Para Rubber Sheets during the Drying Process. *Appl. Sci.* **2021**, *11*, 7891. [CrossRef]
6. Cademartori, P.H.G.; dos Santos, P.S.; Serrano, L.; Labidi, J.; Gatto, D.A. Effect of thermal treatment on physicochemical properties of Gympie messmate wood. *Ind. Crops Prod.* **2013**, *45*, 360–366. [CrossRef]
7. Li, W.; Cai, C.-H.; Guo, Z.-K.; Wang, H.; Zuo, W.-J.; Dong, W.-H.; Mei, W.-L.; Dai, H.-F. Five new eudesmane-type sesquiterpenoids from Chinese agarwood induced by artificial holing. *Fitoterapia* **2015**, *100*, 44–49. [CrossRef] [PubMed]
8. Keller, B.; Templeton, M.D.; Lamb, C.J. Specific localization of a plant cell wall glycine-rich protein in protoxylem cells of the vascular system. *Proc. Natl. Acad. Sci. USA* **1989**, *86*, 1529–1533. [CrossRef]
9. Martius, C. Density, humidity, and nitrogen content of dominant wood species of floodplain forests (várzea) in Amazonia. *Holz Als Roh-Und Werkst.* **1992**, *50*, 300–303. [CrossRef]
10. Isikgor, F.H.; Becer, C.R. Lignocellulosic biomass: A sustainable platform for the production of bio-based chemicals and polymers. *Polym. Chem.* **2015**, *6*, 4497–4559. [CrossRef]
11. Paoli, G.D.; Peart, D.R.; Leighton, M.; Samsoedin, I. An ecological and economic assessment of the nontimber forest product gaharu wood in Gunung Palung National Park, West Kalimantan, Indonesia. *Conserv. Biol.* **2001**, *15*, 1721–1732. [CrossRef]
12. Datta, J.; Parcheta, P.; Surówka, J. Softwood-lignin/natural rubber composites containing novel plasticizing agent: Preparation and characterization. *Ind. Crops Prod.* **2017**, *95*, 675–685. [CrossRef]
13. Ehabe, E.; Ngolemasango, F.; Bonfils, F.; Sainte-Beuve, J. Precision associated with determination of dirt content of natural rubber. *J. Appl. Polym. Sci.* **2001**, *81*, 957–962. [CrossRef]
14. Giraldo-Vásquez, D.H.; Velásquez-Restrepo, S.M. Variation of technological properties of field natural rubber lattices from Hevea brasiliensis clones and natural rubber-based compounds. *Dyna* **2017**, *84*, 80–87. [CrossRef]
15. Vimalasiri, P.; Tillekeratne, L.; Weeraman, S.; Dekumpitiya, A. A rapid and accurate method for determining the volatile matter content of raw natural rubber. *Polym. Test.* **1987**, *7*, 317–323. [CrossRef]
16. Nun-anan, P.; Wisunthorn, S.; Pichaiyut, S.; Vennemann, N.; Kummerlöwe, C.; Nakason, C. Influence of alkaline treatment and acetone extraction of natural rubber matrix on properties of carbon black filled natural rubber vulcanizates. *Polym. Test.* **2020**, *89*, 106623. [CrossRef]
17. Baimark, Y.; Niamsa, N. Study on wood vinegars for use as coagulating and antifungal agents on the production of natural rubber sheets. *Biomass Bioenergy* **2009**, *33*, 994–998. [CrossRef]
18. Dahham, S.S.; Tabana, Y.M.; Ahmed Hassan, L.E.; Khadeer Ahamed, M.B.; Abdul Majid, A.S.; Abdul Majid, A.M.S. In vitro antimetastatic activity of Agarwood (*Aquilaria crassna*) essential oils against pancreatic cancer cells. *Alex. J. Med.* **2016**, *52*, 141–150. [CrossRef]
19. Esteves, B.; Velez Marques, A.; Domingos, I.; Pereira, H. Chemical changes of heat treated pine and eucalypt wood monitored by FTIR. *Maderas Cienc. Tecnol.* **2013**, *15*, 245–258. [CrossRef]

20. Rodrigues, J.; Faix, O.; Pereira, H. Determination of lignin content of Eucalyptus globulus wood using FTIR spectroscopy. *Holzforschung* **1998**, *52*, 46–50. [CrossRef]
21. Nallasamy, P.; Mohan, S. Vibrational spectra of cis-1, 4-polyisoprene. *Arab. J. Sci. Eng.* **2004**, *29*, 17–26.
22. Thuong, N.T.; Yamamoto, O.; Nghia, P.T.; Cornish, K.; Kawahara, S. Effect of naturally occurring crosslinking junctions on green strength of natural rubber. *Polym. Adv. Technol.* **2017**, *28*, 303–311. [CrossRef]
23. Kubo, S.; Kadla, J.F. Kraft lignin/poly (ethylene oxide) blends: Effect of lignin structure on miscibility and hydrogen bonding. *J. Appl. Polym. Sci.* **2005**, *98*, 1437–1444. [CrossRef]
24. Aik-Hwee, E.; Tanaka, Y.; Seng-Neon, G. FTIR studies on amino groups in purified Hevea rubber (short communication). *J. Nat. Rubber Res.* **1992**, *7*, 152–155.
25. Chaikumpollert, O.; Yamamoto, Y.; Suchiva, K.; Kawahara, S. Protein-free natural rubber. *Colloid Polym. Sci.* **2012**, *290*, 331–338. [CrossRef]
26. Tengroth, C.; Gasslander, U.; Andersson, F.O.; Jacobsson, S.P. Cross-linking of gelatin capsules with formaldehyde and other aldehydes: An FTIR spectroscopy study. *Pharm. Dev. Technol.* **2005**, *10*, 405–412. [CrossRef] [PubMed]
27. Colthup, N.B.; Daly, L.H.; Wiberley, S.E. *Introduction to Infrared and Raman Spectroscopy*, 2nd ed.; Academic Press, Inc.: New York, NY, USA; San Francisco, CA, USA; London, UK, 1975.
28. Eng, A.; Tanaka, Y.; Gan, S. Some properties of epoxidised deproteinised natural rubber. *J. Nat. Rubber Res.* **1997**, *12*, 82–89.
29. Tuampoemsab, S.; Sakdapipanich, J.; Tanaka, Y. Influence of some non-rubber components on aging behavior of purified natural rubber. *Rubber Chem. Technol.* **2007**, *80*, 159–168. [CrossRef]
30. Hill, C.A. *Wood Modification: Chemical, Thermal and Other Processes*; John Wiley & Sons: New York, NY, USA, 2007.
31. Manteghi, A.; Ahmadi, S.; Arabi, H. Enhanced thermo-oxidative stability through covalent attachment of hindered phenolic antioxidant on surface functionalized polypropylene. *Polymer* **2018**, *138*, 41–48. [CrossRef]
32. Cong, F.; Diehl, B.G.; Hill, J.L.; Brown, N.R.; Tien, M. Covalent bond formation between amino acids and lignin: Cross-coupling between proteins and lignin. *Phytochemistry* **2013**, *96*, 449–456. [CrossRef]
33. Košíková, B.; Gregorová, A. Sulfur-free lignin as reinforcing component of styrene–butadiene rubber. *J. Appl. Polym. Sci.* **2005**, *97*, 924–929. [CrossRef]
34. Pillai, K.V.; Renneckar, S. Cation−π interactions as a mechanism in technical lignin adsorption to cationic surfaces. *Biomacromolecules* **2009**, *10*, 798–804. [CrossRef] [PubMed]
35. Nimpaiboon, A.; Amnuaypornsri, S.; Sakdapipanich, J. Role of gel content on the structural changes of masticated natural rubber. *Adv. Mater. Res.* **2014**, *844*, 101–104. [CrossRef]
36. Yangthong, H.; Nun-anan, P.; Faibunchan, P.; Karrila, S.; Limhengha, S. The enhancement of cure and mechanical properties of natural rubber vulcanizates with waste Aquilaria crassna wood. *Ind. Crops Prod.* **2021**, *171*, 113922. [CrossRef]
37. McMahan, C.; Lhamo, D. Study of Amino Acid Modifiers in Guayule Natural Rubber. *Rubber Chem. Technol.* **2015**, *88*, 310–323. [CrossRef]
38. Ratanapisit, J.; Apiraksakul, S.; Rerngnarong, A.; Chungsiriporn, J.; Bunyakarn, C. Preliminary evaluation of production and characterization of wood vinegar from rubberwood. *Songklanakarin J. Sci. Technol.* **2009**, *31*, 343–349.
39. Jong, L. Modulus enhancement of natural rubber through the dispersion size reduction of protein/fiber aggregates. *Ind. Crops Prod.* **2014**, *55*, 25–32. [CrossRef]
40. Novriyanti, E.; Santosa, E. The role of phenolics in agarwood formation of Aquilaria crassna Pierre ex Lecomte and Aquilaria microcarpa Baill Trees. *Indones. J. For. Res.* **2011**, *8*, 101–113. [CrossRef]
41. Hwang, Y.-H.; Matsushita, Y.-I.; Sugamoto, K.; Matsui, T. Antimicrobial effect of the wood vinegar from Cryptomeria japonica sapwood on plant pathogenic microorganisms. *J. Microbiol. Biotechnol.* **2005**, *15*, 1106–1109.
42. Kamarulzaman, N.; Idris, N.; Nor, Z.M. Characteristics of odour concentration from rubber processing factories via olfactometry technique. *Chem. Eng. Trans.* **2012**, *30*, 121–126.

Article

Performance of Citric Acid-Bonded Oriented Board from Modified Fibrovascular Bundle of Salacca (*Salacca zalacca* (Gaertn.) Voss) Frond

Luthfi Hakim [1,2,*], Ragil Widyorini [3], Widyanto Dwi Nugroho [3] and Tibertius Agus Prayitno [3]

1. Department of Forest Product Technology, Faculty of Forestry, Universitas Sumatera Utara, Jl. Tri Dharma Ujung No. 1, Medan 20155, Indonesia
2. JATI-Sumatran Forestry Study Analysis Center, Universitas Sumatera Utara, Jl Tri Dharma Ujung No. 1, Medan 20155, Indonesia
3. Department of Forest Product Technology, Faculty of Forestry, Universitas Gadjah Mada, Jl. Agro No. 1, Yogyakarta 55281, Indonesia; rwidyorini@ugm.ac.id (R.W.); wdnugroho@ugm.ac.id (W.D.N.); ta_prayitno@ugm.ac.id (T.A.P.)
* Correspondence: luthfi@usu.ac.id

Citation: Hakim, L.; Widyorini, R.; Nugroho, W.D.; Prayitno, T.A. Performance of Citric Acid-Bonded Oriented Board from Modified Fibrovascular Bundle of Salacca (*Salacca zalacca* (Gaertn.) Voss) Frond. *Polymers* **2021**, *13*, 4090. https://doi.org/10.3390/polym13234090

Academic Editors: Jingpeng Li, Yun Lu and Huiqing Wang

Received: 30 October 2021
Accepted: 19 November 2021
Published: 24 November 2021

Publisher's Note: MDPI stays neutral with regard to jurisdictional claims in published maps and institutional affiliations.

Copyright: © 2021 by the authors. Licensee MDPI, Basel, Switzerland. This article is an open access article distributed under the terms and conditions of the Creative Commons Attribution (CC BY) license (https://creativecommons.org/licenses/by/4.0/).

Abstract: The fibrovascular bundle (FVB) in palm plants consists of fiber and vascular tissue. Geometrically, it is a long fiber that can be used as an oriented board raw material. This research aimed to examine the performance of citric acid-bonded orientation boards from modified FVB salacca frond under NaOH + Na_2SO_3 treatment and the bonding mechanism between the modified FVB frond and citric acid. The results showed that changes in the chemical composition of FVB have a positive effect on the contact angle and increase the cellulose crystallinity index. Furthermore, the mechanical properties of the oriented board showed that 1% NaOH + 0.2% Na_2SO_3 with 60 min immersion has a higher value compared to other treatments. The best dimension stability was on a board with the modified FVB of 1% NaOH + 0.2% Na_2SO_3 with 30 and 60 min immersion. The bonding mechanism evaluated by FTIR spectra also showed that there is a reaction between the hydroxyl group in the modified FVB and the carboxyl group in citric acid. This showed that the modified combination treatment of NaOH+Na_2SO_3 succeeded in increasing the mechanical properties and dimensional stability of the orientation board from the FVB salacca frond.

Keywords: fibrovascular bundle; salacca frond; modified orientation board; citric acid

1. Introduction

The fibrovascular bundle (FVB) is a tissue in monocot plants, specifically palm plants, that consists of xylem, phloem, sclerenchyma fibers, and parenchyma tissue. According to Hakim et al. [1], the FVB has good enough mechanical properties to be used as a raw material for making composite boards. Its advantages as a new natural fiber include being biodegradable, renewable, low in cost, environmentally friendly, light in weight, and abundant in nature. However, the FVB also has disadvantages such as high water absorption and anisotropic properties, low durability, and compatibility with some conventional matrices as well as adhesives [2,3].

Several types of research studies have been carried out on natural fibers using a chemical treatment to modify the surface structure and improve the compatibility between adhesive and raw material. This includes the use of alkali modification in natural fibers due to the convenience and simplicity of its application [4,5]. Previous research studies have used sodium hydroxide (NaOH); however, the combination of NaOH+Na_2SO_3 as a chemical for modification has not been widely reported. Therefore, the use of NaOH+Na_2SO_3 as a chemical modification exposes more –OH groups on the fiber surface and increases compatibility with adhesives since the surface structure facilitates higher interface bonding as raw material for composite boards [6,7]. Alkali and sodium sulfite modification with low concentration aims to expand cellulose and dissolve amorphous components in hemicellulose and lignin so that it can expose more –OH groups, increase the accessibility of crystalline cellulose, increase hydrophilic properties, and reduce sugar content so as to increase the adhesion bond between cellulose and adhesive. The bond between the modified FVB and the adhesive improves with the presence of more –OH groups due to the combined modification of $NaOH+Na_2SO_3$.

The use of modified FVB as a raw material for composite boards is more suitable when it is developed into an orientation board product [8]. Geometrically, it is a long fiber; therefore, compactness in orienting the raw materials is one of the advantages for improving mechanical properties and dimensional stability. Oriented board or oriented strand board (OSB) is defined as a three-layer structural board produced by strand material bonded with a thermosetting type of exterior adhesive and additives to increase dimensional stability such as water-resistant resin [9]. A previous study on the mechanical and physical properties of cement-bonded OSB by Papadopoulos et al. [10] showed that the geometry of raw materials affects OSB quality. Hassani et al. [11] improved the mechanical and physical properties of oriented strand lumber (OSL) by using fortification of nano-wollastonite on urea formaldehyde resin and stated that this treatment could improve dimensional stability and activate the bonds with cellulose hydroxyl groups. However, in this study, orientation boards made from modified FVB were manufactured using citric acid as a natural adhesive. This study focuses on the use of citric acid as a binder to FVB modified by alkali and sodium citrate in the manufacture of orientation boards.

Natural adhesives such as citric acid are used as an alternative in the manufacture of environmentally friendly composite boards [12–19]. However, research on the manufacture of oriented boards from modified FVB frond combined with citric acid as adhesive and the bonding mechanism between the modified FVB by $NaOH+Na_2SO_3$ combination and the citric acid adhesive has not been reported. Therefore, this research aims to evaluate the performance of citric acid-bonded oriented boards from modified FVB *S. zalacca* under $NaOH+Na_2SO_3$ treatment.

2. Materials and Methods

2.1. Materials

The fibrovascular bundle extracted from the *S. zalacca* frond was used as a raw material for the manufacture of oriented board. The extraction was based on Hakim et al.'s study [20], and the bundles were cut 25 cm in length before chemical treatment (Figure 1). The chemical solution was modified using sodium hydroxide (NaOH) anhydride and sodium sulfite (Na_2SO_3). Meanwhile, the citric acid (anhydrous) used as adhesive without further purification was supplied by Rudang Chemical Company (Indonesia).

Figure 1. The fibrovascular bundle extracted from *S. zalacca*.

2.2. Chemical Treatments and Chemical Content Changes

Before the board was manufactured, the fibrovascular bundle was chemically modified with NaOH and Na_2SO_3 and their combination, as shown in Table 1. The chemical characteristics analyzed after treatment were α-cellulose and hemicellulose determined based on ASTM D1103-60, Klason lignin, as well as ash content determined based on ASTM D110-84 and ASTM D1102-84, respectively.

Table 1. The chemical treatment of the fibrovascular bundle.

Treatment	NaOH+Na_2SO_3 Combination	Immersion Time (Minutes)
A0-30	Untreated, water immersion	30
A0-60	Untreated, water immersion	60
C1-30	1% NaOH	30
C1-60	1% NaOH	60
C12-30	1% NaOH + 0.2% Na_2SO_3	30
C12-60	1% NaOH + 0.2% Na_2SO_3	60
C14-30	1% NaOH + 0.4% Na_2SO_3	30
C14-60	1% NaOH + 0.4% Na_2SO_3	60

2.3. Contact Angle and Mechanical Properties of Modified Fibrovascular Bundles

The contact angle was measured using a method by Schellbach et al. [21]; meanwhile, two parallel fibrovascular bundles separated by 1–2 mm were attached to the sample holder and observed under the microscope. Furthermore, water was dripped between the two fibrovascular bundles to obtain a liquid that hangs, while the image of the liquid was photographed with a stereo optical camera and analyzed using IC-Measure version 2.0.0.245 to measure the contact angle.

Meanwhile, the mechanical properties of the modified fibrovascular bundle were evaluated based on the ASTM D-3379-75 standard. The FVB with 8–12% moisture content was cut to a length of about 90 mm + 0.1 mm and fixed on a 30 mm long paper frame using epoxy adhesive (ALF Epoxy Adhesive, P.T. Alfaglos, Semarang, Indonesia). The measurements were conducted in 50 replicates for each chemical treatment. The mechanical properties were determined using a universal testing machine (UTM Tensilon RTF 1350, Tokyo, Japan), with a 1 mm/min crosshead speed. Before the test, the supporting paper's middle part was cut out, and the test mounting was carried out according to the method proposed by Hakim et al. [1,20] (Figure 2). In addition, the density of the modified FVB was measured based on the method described by Munawar et al. [22].

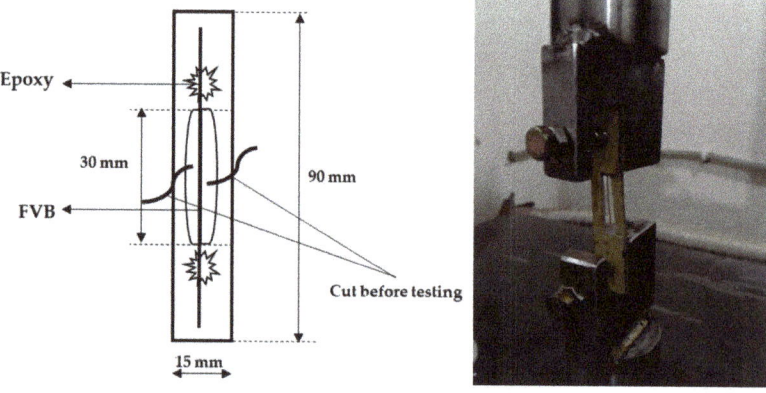

Figure 2. An illustration of FVB mechanical properties testing [20].

2.4. Index Crystallinity of Modified FVB

The crystallinity index (CrI) was determined by considering the regions of crystalline and amorphous cellulose. Crystalline cellulose is determined at a 2θ peak in the reflection plane position (I_{002}) (maximum intensity between 22.5° and 23°), while amorphous cellulose (I_{am}) is the minimum-intensity position (between 18° and 19°). Furthermore, the diffraction spectra were obtained at room temperature (20–22 °C) from radiation generated by Maximax X-ray Diffractometer-7000 (Shimadzu, Japan). The measurements were carried out at 40 kV and 20 mA with a detector placed on the range of 2θ from 5° to 70° at a scan speed of 2°/min. Subsequently, the percentage crystallinity index (%CI) of the cellulose was calculated using the formula provided by Segal et al. [23].

$$\%CI = \frac{(I_{002} - I_{am})}{I_{002}} \times 100$$

where I_{002} is the maximal peak intensity at a 2θ angle of approximately 22°–23° and I_{am} is the minimum peak intensity (amorphous region) at a 2θ angle of about 18°–19°.

2.5. The Manufacture of Oriented Board

After modification treatment, FVB was air-dried (10% moisture content) before being used as a raw material for orientation boards. The board was to be manufactured with dimensions of 250 × 250 × 8 mm^3 with a target density of 0.8 g/cm^3. The citric acid solution used as the adhesive was dissolved in water to a concentration of 60%. The resin content of citric acid is 30%. The adhesive solution was sprayed onto the raw materials and oven-dried at 75 °C for more than 8 h to reduce the moisture content to 4–6%. The fibrovascular bundle was manually hand-formed into a three-layer oriented mat using a forming box. Three different orientations (0°, 45°, and 90°) of the board were prepared for each fibrovascular bundle, each in three layers, namely face:core:back = 30%:40%:30% weight ratio (Figure 3). The three-layer oriented mat was hot-pressed at 180 °C for 10 min and a pressure of 3 MPa at three replications.

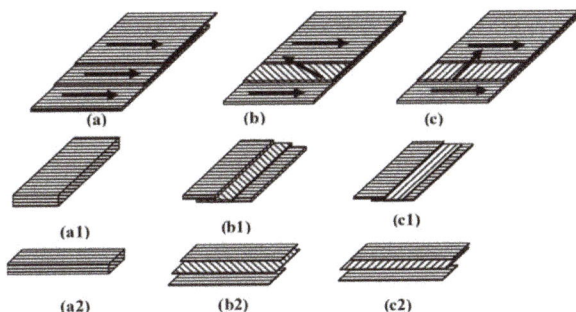

Figure 3. Layout for manufacturing fibrovascular bundle-oriented board. (**a**–**c**) are 0°, 90°, and 45° oriented, respectively. (**a1**–**c1**) are parallel (//) samples for the modulus of rupture (MOR) and modulus of elasticity (MOE). (**a2**–**c2**) are perpendicular (⊥) samples for MOR and MOE.

2.6. Evaluation of Oriented Board Properties

All of the samples were conditioned at environmental conditions for approximately a week to reach a moisture content of ±6% before the measurement of the oriented board properties according to the Japanese Industrial Standard for particleboard (JIS A 5908). The mechanical properties of the oriented board were measured for modulus of rupture (MOR), modulus of elasticity (MOE), the strength of the internal bond (IB), and screw holding power (SHP) using a universal testing machine (Tensilon RTF 1350, Tokyo, Japan), while the physical properties were tested for thickness swelling (TS) and water absorption

(WA). The static bending samples (MOR and MOE) of 200 × 50 × 8 mm³ were treated by the three-point bending method under dry conditions with a span distance of 150 mm and a crosshead speed of 10 mm/min. The IB specimens of 50 × 50 × 8 mm³ were randomly measured from both surfaces of each specimen. Furthermore, the SHP specimen of 75 × 50 × 8 mm³ was drawn two positions from the board of each sample with a speed of 2 mm/min, while the TS and WA of 50 × 50 × 8 mm³ were evaluated by water immersion for 24 h at 20 °C.

2.7. Fourier Transform Infrared (FTIR) Spectroscopy

The fibrovascular bundle-oriented board samples were previously soaked in boiling water for 2 h to remove unreacted citric acid, conditioned in room temperature water for 1 h, dried at 35 ± 5 °C for 12 h, and powdered through a 100-mesh screen. FTIR spectroscopy was performed at room temperature (approximately 25 °C) using the FTIR-4200 spectrophotometer (8201PC-Shimadzu, Tokyo, Japan) and the KBr disk method with 12 cm^{-1} of resolution to determine the assignment of absorbance bands to specific functional groups.

3. Results and Discussion

3.1. Properties Change of Modified FVB

Table 2 shows the contact angles, densities, crystallinity index, and chemical composition of modified FVB. The modified FVBs' α-cellulose content increased from 43.11% (treatment A0-30) to 53.38% (treatment C12-30) and slightly decreased again from treatments C12-60, C14-30, and C14-60 (50.71%, 39.62%, and 35.15%, respectively).

Table 2. Contact angle, density, crystallinity index, and chemical composition of modified FVB.

Treatment	Contact Angle (°)	Density (g/cm³)	CrI (%)	α-Cellulose (% wt)	Hemicellulose (% wt)	Lignin (% wt)
A0-30	92.30 ± 7.66	0.40 ± 0.09	–	43.11 ± 0.71	32.24 ± 0.87	28.18 ± 1.13
A0-60	84.47 ± 3.25	0.40 ± 0.08	53.04	43.13 ± 0.90	32.17 ± 0.79	28.09 ± 0.96
C1-30	55.83 ± 3.41	0.39 ± 0.12	–	46.45 ± 0.52	30.73 ± 0.94	27.11 ± 1.03
C1-60	46.20 ± 5.04	0.39 ± 0.17	56.25	48.15 ± 0.77	31.00 ± 1.57	26.33 ± 2.31
C12-30	44.38 ± 7.65	0.38 ± 0.06	–	53.38 ± 1.67	27.92 ± 0.76	22.05 ± 0.89
C12-60	34.88 ± 3.88	0.37 ± 0.18	58.33	50.71 ± 1.46	24.72 ± 0.66	21.33 ± 0.85
C14-30	31.98 ± 4.60	0.37 ± 0.09	–	39.62 ± 1.55	20.85 ± 0.69	21.22 ± 0.92
C14-60	24.02 ± 2.80	0.35 ± 0.12	47.43	35.15 ± 0.72	21.91 ± 0.98	18.78 ± 1.21

The increase in cellulose from A0-30 to C12-30 treatment does not imply an increase in cellulose content, but rather that the other components, hemicellulose and lignin, decreased. Meanwhile, the α-cellulose fraction reduces during treatments C12-60, C14-30, and C14-60 because the α-cellulose component dissolves. This reduction was caused by the NaOH + Na$_2$SO$_3$ concentration, which has the capacity to destroy amorphous cellulose at high concentrations and longtime immersion. Similar to the previous research, 3% NaOH treatment + steaming on *S. zalacca* FVB frond yielded 54.53% cellulose [24], and a 15% NaOH treatment on areca palm frond fiber (*Dypsis lutescens*) raised the cellulose content to 63.45% [25].

The hemicellulose content of modified FVB decreased from 32.24% with the 30 min water immersion treatment to 21.91% with the treatment modified with 1%NaOH + 0.4% Na$_2$SO$_3$ and 60 min immersion. In contrast to previous studies, the alkaline treatment of 3% on the FVB of the frond of *S. zalacca* became 74.09% [24]; 5% NaOH treatment on banyan tree root fibers decreases hemicellulose to 10.74% [26]. Hemicellulose is one of the components of lignocellulosic materials that is easily degraded under alkaline treatment because the structure of the hemicellulose components is nonlinear and more amorphous than crystalline [27,28].

The lignin component of the modified *S. zalacca* frond FVB decreased from 28.1% (treatment A0-30) to 18.7% (treatment C14-60). This pattern of decline is relatively the same as that in previous studies using modified alkali [24–26,29,30]. Furthermore, the NaOH + Na$_2$SO$_3$ treatment was more effective in maintaining α-cellulose, hemicellulose, and lignin content compared to previous studies by Darmanto et al. [31] using *S. zalacca* as raw material with higher concentration and high-energy steam.

The modified FVB had the highest density value of 0.40 g/cm^3 under the water treatment with 30 min immersion. Meanwhile, the lowest was 0.35 g/cm^3 under 1% NaOH + 0.4% Na$_2$SO$_3$ with 60 min immersion. This decreased density value is due to the degradation factor of the chemical components during the modification treatment, which causes the weight of FVB decrease.

The contact angle was measured to determine when the modified FVB is easily wetted by a liquid; meanwhile, the value obtained is 24.02° with treatment H (Table 2), which is much lower than the contact angle of bamboo, jute, ramie, and kenaf [4,32]. The modification of the combination of NaOH+Na$_2$SO$_3$ decreased the contact angle compared to the water immersion treatment. Furthermore, it was affected by the surface roughness due to changes in the structure of the chemical composition of the surface [33]. Surface changes due to the degradation of chemical composition and accessibility of hydroxyl groups on the surface of FVB are the two factors affecting the level of wettability. The NaOH+Na$_2$SO$_3$ treatment led to the removal of low molecular weight material such as hemicellulose and lignin, while the cellulose content increased. Based on previous research, the alkali modification reduced some impurities and wax in the FVB surface [5]. The results were almost similar to those of 3% NaOH with steaming treatment on the FVB of *S. zalacca* [24], 5.5% NaOH treatment of fiber-empty fruit bunches of oil palm [34], 15% NaOH treatment of palm fiber areca [25], 5% NaOH treatment of *Tridax procumbens* fiber [29], 5% NaOH treatment of oil palm mesocarp fiber [35], and 5% NaOH treatment of banyan tree root fiber [26].

Table 3 shows the mechanical properties of modified FVB. The higher maximum load of modified FVB was 122.57 ± 9.94 N after A0-60 treatment, while the lower maximum load was 57.15 ± 6.86 N after C14-60 treatment. Interestingly, the tensile strength increased from A0-30 to C12-60 treatment but decreased after C14-30 and C12-60 treatment. Conversely, the increase in tensile strength was not accompanied by the increase in maximum load. This phenomenon is explainable because modified FVB's tensile strength increases due to reduction in the FVB's transversal area. Furthermore, this reduction was due to the surface degradation during the modification treatment.

Table 3. Mechanical properties of modified FVB.

Treatment	Maximum Load (N)	Tensile Strength (MPa)	Young's Modulus (GPa)
A0-30	121.01 ± 13.53	217.19 ± 28.11	2.23 ± 0.44
A0-60	117.28 ± 8.82	222.10 ± 18.63	2.23 ± 0.42
C1-30	117.68 ± 8.77	253.32 ± 25.81	2.52 ± 0.40
C1-60	122.57 ± 9.94	275.41 ± 21.28	2.29 ± 0.41
C12-30	119.92 ± 9.33	313.93 ± 41.43	2.55 ± 0.24
C12-60	110.30 ± 6.67	331.66 ± 33.38	2.28 ± 0.36
C14-30	77.83 ± 6.06	267.72 ± 39.52	2.18 ± 0.43
C14-60	57.15 ± 6.86	237.99 ± 43.45	1.91 ± 0.38

Darmanto et al. [24] reported that a combination of 1% NaOH and high temperature with 2 bars pressure treatment resulted in a 355 MPa increase in the tensile strength of FVB *S. zalacca* frond. Compared to the result of Darmanto's research, this study was more effective and efficient because high temperatures and pressures were not used. Young's modulus has the same pattern as the tensile strength of the modified FVB. The higher Young's modulus of modified FVB increased from water treatment to C12-30 treatment but slightly decreased after C12-60, C14-30, and C14-60 treatment, respectively.

The mechanical properties of modified FVB are influenced by the material's crystallinity index. Figure 4 shows the X-ray diffraction spectra of modified FVB. The crystallinity index (CrI) of modified FVB increased from A0-60 treatment to C12-60 treatment before decreasing again after C14-60 treatment. The crystallinity index of modified FVB immersed in water for 60 min was 53.04%, whereas for the 1% NaOH treatment counterpart, it increased to 56.25%. In addition, the crystallinity index of FVBs subjected to 1% NaOH + 0.2% Na_2SO_3 treatment with 60 min immersion increased to 58.33%, whereas for the counterparts treated with 1% NaOH + 0.4% Na_2SO_3 with 60 min immersion, it decreased to 47.43%. The mechanical properties of fiber are affected by the crystallinity index. The higher the crystallinity index of a fiber, the higher is the predicted fiber strength [36]. The treatment of 1% NaOH + 0.2% Na_2SO_3 with 60 min immersion resulted in an optimal increase in the crystallinity index of modified FVB. This occurs due to the degradation of amorphous components, which include hemicellulose, lignin, and wax. These chemical components are sensitive to alkali, so they are easily removed by alkaline modification treatment [37]. The 1% NaOH + 0.4% Na_2SO_3 treatment with 60 min immersion decreased the crystallinity index because cellulose polymer is penetrated by sulfite ions (SO_3^-), so some of the cellulose and lignin components begin to dissolve [38].

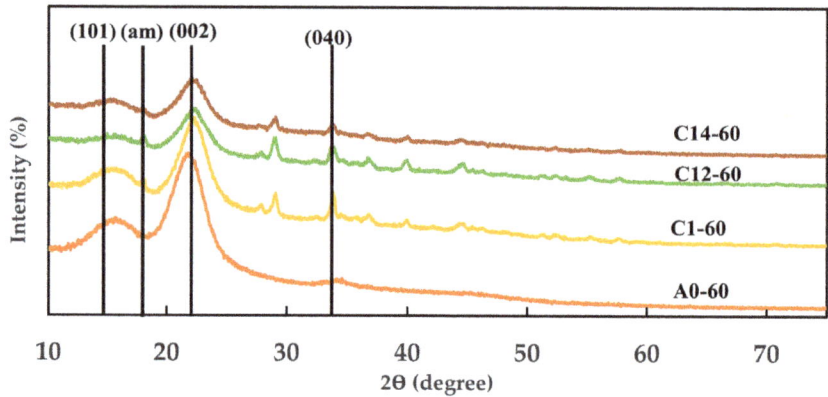

Figure 4. Diffractogram of X-ray diffraction of modified FVB. A0-60: water (60 min); C1-60: 1% NaOH (60 min); C12-60: 1% NaOH + 0.2% Na_2SO_3 (60 min); and C14-60: 1% NaOH + 0.4% Na_2SO_3 (60 min).

3.2. Modulus of Rupture (MOR) and Modulus of Elasticity (MOE) of a Modified Orientation Board

The MOR and MOE values of the oriented board were observed in two positions, namely perpendicular ($\perp$) and parallel (//). Generally, their values for an oriented board with 0° orientation angle were higher than those of the board with orientation angle 45° and 90°. The MOR value in the perpendicular position was higher than the MOR value in the parallel position for all orientation angles (0°, 45°, and 90°), which is similar to the previous research conducted by Munawar et al. and Baharin et al. [39,40]. Therefore, the MOR value increased with the combined treatment with 1% NaOH + 0.2% Na_2SO_3 and decreased with the combined treatment with 1% NaOH + 0.4% Na_2SO_3.

As shown in Figure 5, the highest MOR and MOE values of the orientation board are 23.4 MPa ($\perp$) and 11.4 MPa (//) with orientation angle 0°, 15.5 MPa ($\perp$) and 13.5 MPa (//) with orientation angle 45°, and 17.7 MPa ($\perp$) and 16.9 MPa (//) with orientation angle 90°, respectively. All of these highest values were achieved in the combination treatment of 1% NaOH + 0.2% Na_2SO_3 with 60 min immersion, whereas the lowest values were achieved in the 30 min water immersion treatment, and they were 11.8 MPa ($\perp$) and 4.9 MPa (//) with orientation angle 0°, 7.9 MPa ($\perp$) and 7.3 MPa (//) with orientation angle 45°, and 7.6 MPa ($\perp$) and 7.5 MPa (//) with orientation angle 90°, respectively.

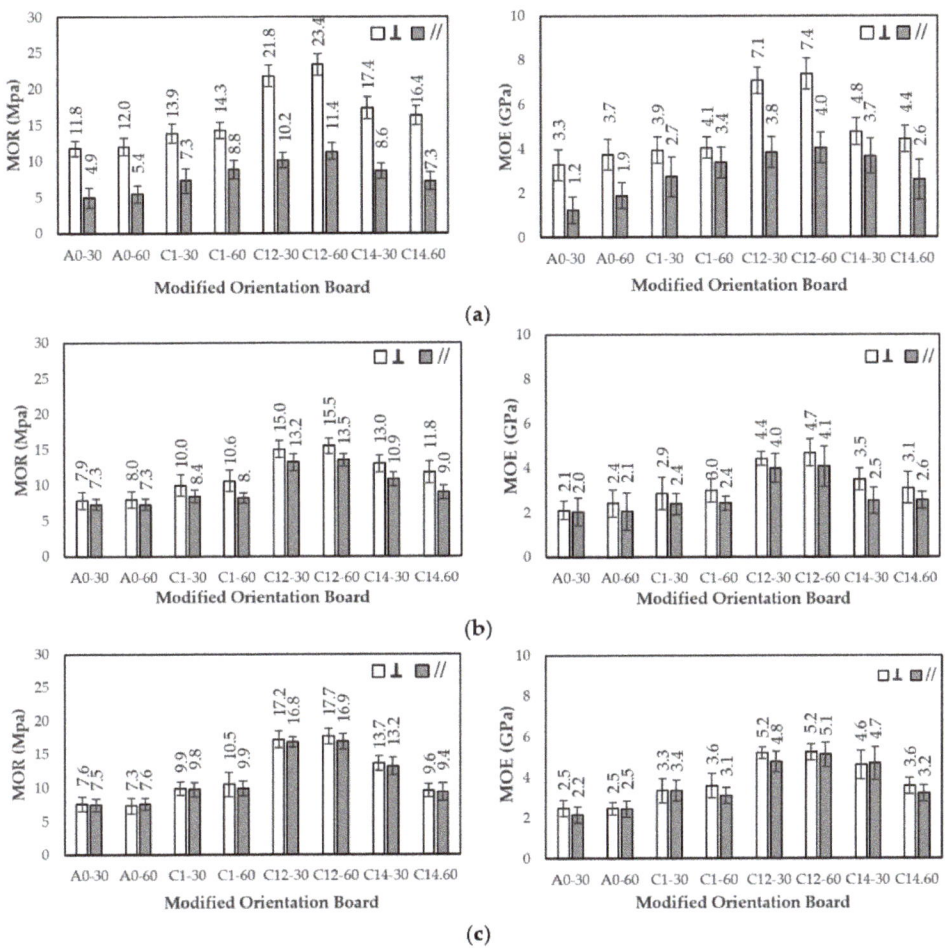

Figure 5. Histogram of MOR and MOE. (**a**–**c**) are 0°, 45°, 90°, respectively; ⊥: perpendicular, //: parallel; A0-30: water (30 min); A0-60: water (60 min); C1-30: 1% NaOH (30 min); C1-60: 1% NaOH (60 min); C12-30: 1% NaOH + 0.2% Na_2SO_3 (30 min); C12-60: 1% NaOH + 0.2% Na_2SO_3 (60 min); C14-30: 1% NaOH + 0.4% Na_2SO_3 (30 min); and C14-60: 1% NaOH + 0.4% Na_2SO_3 (60 min). Error bar represents the standard deviation.

The highest MOE values were obtained at 7.4 GPa (⊥) and 4.0 GPa (//) for 0° orientation angle, 4.7 GPa (⊥) and 4.1 GPa (//) for 45°, and 5.2 GPa (⊥) and 5.1 GPa (//) for 90°, respectively, in the combination treatment of 1% NaOH + 0.2% Na_2SO_3 for 60 min. Meanwhile, the lowest MOE values were obtained with the water immersion treatment for 30 min at 3.3 GPa (⊥) and 1.2 GPa (//) for 0°, 2.1 GPa (⊥) and 2.0 GPa (//) for 45°, and 2.5 GPa (⊥) and 2.2 GPa (//) for 90°, respectively.

Umemura et al. [41] reported that wood waste particleboard with citric acid as an adhesive has an MOR value of 10.7 MPa. Another study found that pressing at a temperature of 180 °C during the manufacture of particleboard using citric acid as an adhesive increases the MOR value by 19.6 MPa [42]. Furthermore, Liao et al. [17] stated that the addition of sucrose to the citric acid adhesive during the manufacture of low-density particleboard gives an MOR value of more than 6.0 MPa. The manufacture of orientation boards from natural fibers such as *Sansevieria* using phenol–formaldehyde (PF) adhesive also increases the MOR value by 403 MPa [39]. In this research, a board made using 30% citric acid adhesive and a compression temperature of 180 °C successfully produced an oriented board with an MOR value of 23.4 MPa. This occurred because the raw material in the form of long fibers subjected to alkali modification treatment can increase the mechanical value of the composite board obtained [43]. This shows that raw materials in the form of long natural fibers for the use of composite boards will result in good mechanical properties for the boards [44,45]. Meanwhile, alkali modification of long fibers such as FVB can increase the accessibility of cellulose, hemicellulose, and lignin to increase the FVB interface bond and citric acid adhesive, which will affect the strength of the composite board [46,47].

The research by Santoso et al. [48] on the manufacture of particleboard using particles of *Nypa fruticans* frond that bonded citric acid and maltodextrin adhesives give an MOE of 0.5 GPa. Meanwhile, Widyorini et al. [42] found that the addition of maltodextrin and the compression step to the particleboard made from the particles of salacca frond gives an MOE of 4.1 GPa. The research conducted by Kusumah et al. [18] using sorghum particles (bagasse) as a raw material for particleboard with citric acid adhesive gave an MOE of 5.27 MPa. Meanwhile, this research gave the highest MOE value of 8.2 MPa, which is higher than the previous results. The dimensions of the raw material in the form of long fibers influenced the mechanical properties of the composite board obtained. The perpendicular position test with the regular orientation of the FVB arrangement can increase compactness when subjected to a load.

3.3. Strength of Internal Bond (IB)

The IB increased from water treatment for 30 min to the combination treatment of 1% NaOH + 0.2% Na_2SO_3 and then decreased again at the 1% NaOH + 0.4% Na_2SO_3 treatment for 30 min. The orientation board with an angle of 0° has a greater IB value than boards with angles of 45° and 90°. The IB value of the orientation board made from the modified FVB frond of *S. zalacca* is shown in Figure 6.

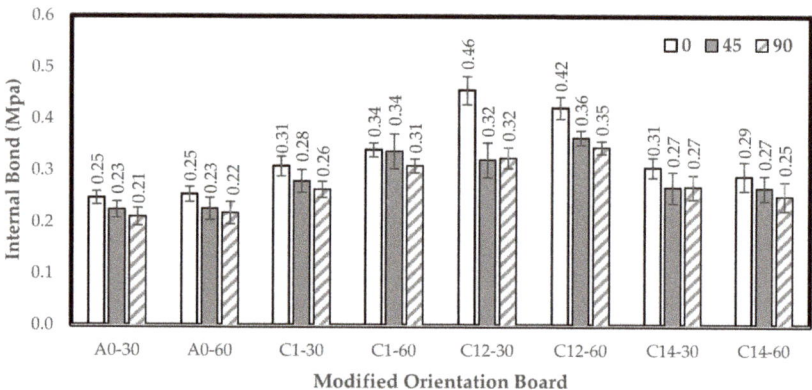

Figure 6. Histogram of the internal bond. A0-30: water (30 min); A0-60: water (60 min); C1-30: 1% NaOH (30 min); C1-60: 1% NaOH (60 min); C12-30: 1% NaOH + 0.2% Na_2SO_3 (30 min); C12-60: 1% NaOH + 0.2% Na_2SO_3 (60 min); C14-30: 1% NaOH + 0.4% Na_2SO_3 (30 min); and C14-60: 1% NaOH + 0.4% Na_2SO_3 (60 min). Error bar represents the standard deviation.

The highest IB value of the board was found in the combination treatment of 1% NaOH + 0.2% Na_2SO_3 with 30 min immersion and 1% NaOH + 0.2% Na_2SO_3 for 60 min immersion, and it was 0.46 MPa (0°) and 0.35 MPa (90°), respectively, while the lowest was 0.29 MPa (0°) and 0.25 MPa (90°), respectively, with 1% NaOH + 0.4% Na_2SO_3 treatment with 60 min immersion.

Moreover, the IB values of 90° and 45° orientation boards are lower than 0° because the 90° orientation boards are arranged crosswise, consisting of three layers of the face, core, and back for the bonding area and contact between the fibers on the surface of one layer with another layer to reduce. Some of the values in this research were still higher—0.37 MPa [49] and 0.15 MPa [50,51]—than the previous ones that used waste wood particles and citric acid as adhesives.

According to Munawar et al. [39], the orientation board of *Sansevieria* fiber bonded using phenol–formaldehyde (PF) had an IB value of 1.33 MPa, which is much higher compared to that of this study. Walther et al. [52] also stated that the IB value of the orientation board made from kenaf glued with PF is 0.44 MPa (with orientation angle 0°), which is lower compared to that of this study.

3.4. Screw Holding Power (SHP)

The value of SHP increased from control to the combination treatment of 1% NaOH + 0.2% Na_2SO_3 with 30 min immersion; however, it decreased again on the combined treatment of 1% NaOH + 0.4% Na_2SO_3 with 30 min immersion. The results showed that the oriented board with 0° and 90° angles of orientation have relatively similar SHP values. The highest SHP values were 519 N (0°), achieved in the combined treatment of 1% NaOH + 0.2% Na_2SO_3 with 60 min immersion, and 504 N (45°) and 495 N (90°), achieved in the combined treatment of 1% NaOH + 0.2% Na_2SO_3 with 30 min immersion, respectively. The lowest values of HPS were 318 N (0°), 307 N (45°), and 293.5 N (90°), achieved in the combined treatment of 1% NaOH + 0.4% Na_2SO_3 with 60 min immersion. The histogram of the SHP of the oriented board is shown in Figure 7.

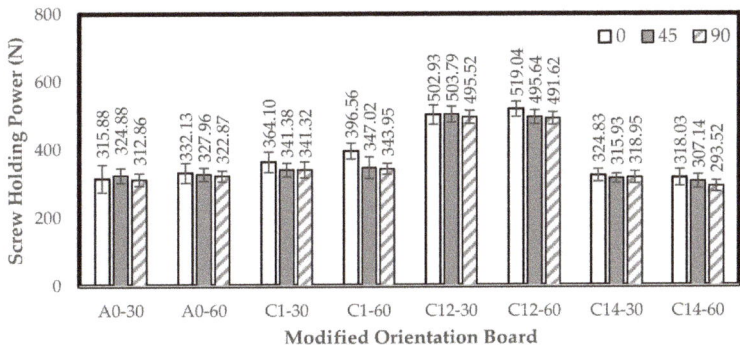

Figure 7. Histogram of screw holding power. A0-30: water (30 min); A0-60: water (60 min); C1-30: 1% NaOH (30 min); C1-60: 1% NaOH (60 min); C12-30: 1% NaOH + 0.2% Na_2SO_3 (30 min); C12-60: 1% NaOH + 0.2% Na_2SO_3 (60 min); C14-30: 1% NaOH + 0.4% Na_2SO_3 (30 min); and C14-60: 1% NaOH + 0.4% Na_2SO_3 (60 min). Error bar represents the standard deviation.

This research showed that the modified 1% NaOH + 0.2% Na_2SO_3 treatment with 60 min immersion was the best in influencing the screw holding strength. According to Hung et al. [53], the adhesion between the modification raw material and the adhesive increases with the SHP. Meanwhile, another factor that affects the SHP is the size and dimensions of the raw material. This is because the longer and thicker the raw material, the more is the SHP value of the composite board [54]. Compared to the use of particle raw material, the SHP value of the oriented board is higher than the particleboard with the same adhesive (citric acid) [19]. Finer particles are less effective in transferring the stress from particle to particle compared to FVB. Similarly, damage to the board due to screw threads occurs faster in particle raw materials compared to long-fiber raw materials.

The single-layer orientation board has a higher SHP value than the three-layer orientation board, while the 0° orientation angle is slightly higher than 45° and 90°. This showed that the orientation angle of 0° is a single-layer board, while 45° and 90° are a three-layer board.

3.5. Thickness Swelling (TS) and Water Absorption (WA)

The orientation board has the highest WA value in the water immersion treatment with 30 min immersion, which is 51.18% (0°), 54.87% (45°), and 55.26% (90°). By contrast, the lowest WA value is 43.50% (0°) in the 1% NaOH treatment with 30 min immersion, and 16.20% (45°) and 47.80% (90°) in the combination treatment of 1% NaOH + 0.2% Na_2SO_3 for 60 min. In addition, the orientation board has the highest TS value of 19.64% (0°), 21.65% (45°), and 22.65% (90°) in the 30 min water immersion treatment, while the lowest value is 14.99% (0°) and 16.20% (45°) in the 1% NaOH + 0.2% Na_2SO_3 with 60 min immersion, and 17% (90°) in the 1% NaOH + 0.2% Na_2SO_3 with 30 min immersion. Generally, the WA and TS values decreased from treatment A (30 min water immersion) to treatment F (1% NaOH + 0.2% Na_2SO_3 with 60 min immersion) and increased again in treatment G (1% NaOH + 0.4% Na_2SO_3 with 30 min immersion). The WA and TS values of the modified orientation board are shown in Figure 8.

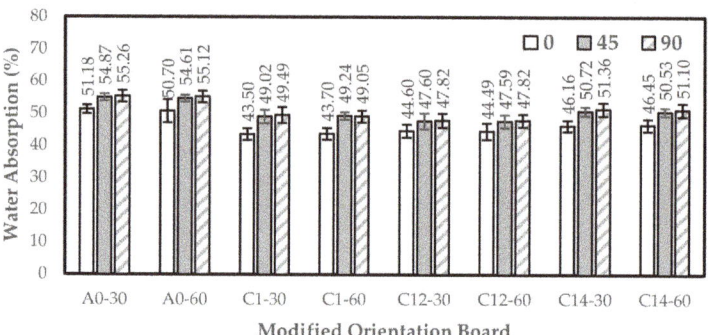

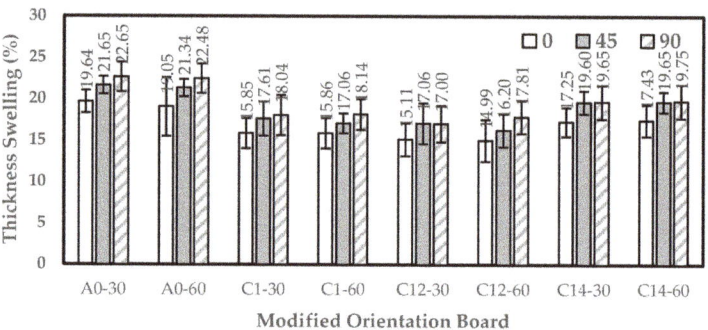

Figure 8. Histogram of water absorption and thickness swelling. A0-30: water (30 min); A0-60: water (60 min); C1-30: 1% NaOH (30 min); C1-60: 1% NaOH (60 min); C12-30: 1% NaOH + 0.2% Na_2SO_3 (30 min); C12-60: 1% NaOH + 0.2% Na_2SO_3 (60 min); C14-30: 1% NaOH + 0.4% Na_2SO_3 (30 min); and C14-60: 1% NaOH + 0.4% Na_2SO_3 (60 min). Error bar represents the standard deviation.

The results have lower WA and TS values compared to the previous research conducted by Kemalasari and Widyorini [55], which stated that the WA and TS of particleboard with the raw material of salacca frond particles with citric acid adhesive ranged from 36–

62.26% to 6.95–22.42%, respectively. Widyorini et al. [42] also researched the manufacture of particleboard using salacca frond particles and citric acid as adhesives and found a WA of 54.2%. It was also reported that the extractive substances in the salacca frond particles do not affect the water absorption of the particleboard bound to the citric acid [19]. Meanwhile, Kusumah et al. [56], who used citric acid as an adhesive, reported that the use of 30% citric acid is effective in reducing the water absorption of composite boards. Based on the alkali modification in this research, the combined treatment of 1% NaOH + 0.2% Na_2SO_3 with 60 min immersion succeeded in reducing water absorption to 43.0%.

3.6. Fourier Transform Infrared (FTIR) Spectroscopy

The effect of modified NaOH and the combination of NaOH + Na_2SO_3 on the chemical structure of the FVB orientation board of the salacca frond bonded with citric acid is shown in Figure 9. Based on the results, the ester functional group was detected at the peak of the wavelength around 1733 cm^{-1}, which shows the C=O stretching group, indicating the formation of carbonyl group or C=O group [16,19,42]. This showed a reaction/bond between the carboxyl group in citric acid and the hydroxyl group in the modified FVB. The peak with the highest intensity was discovered in treatment C12-30 (1% NaOH + 0.2% Na_2SO_3 for 30 min) and C12-60 (1% NaOH + 0.2% Na_2SO_3 for 60 min), while the modified treatment with a higher concentration of Na_2SO_3 (treatment C14-30 and C14-60) showed a decrease in intensity. This showed the mechanical properties of the orientation board, which has the best mechanical value in treatment C12-30 and C12-60.

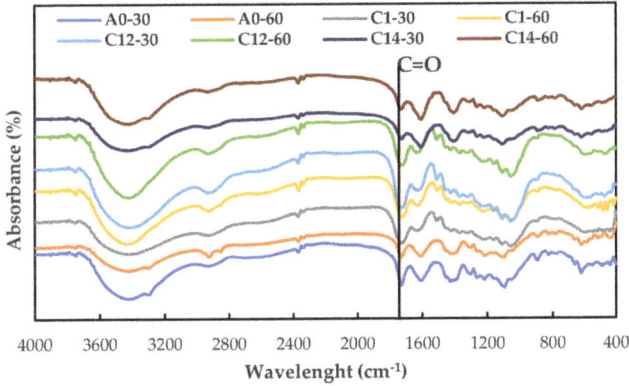

Figure 9. FTIR spectra of oriented board. A0-30: water (30 min); A0-60: water (60 min); C1-30: 1% NaOH (30 min); C1-60: 1% NaOH (60 min); C12-30: 1% NaOH + 0.2% Na_2SO_3 (30 min); C12-60: 1% NaOH + 0.2% Na_2SO_3 (60 min); C14-30: 1% NaOH + 0.4% Na_2SO_3 (30 min; and C14-60: 1% NaOH + 0.4% Na_2SO_3 (60 min).

Figure 10 shows the FTIR spectra of the modified FVB (FVB+C12-30 and FVB+C12-60) and the modified orientation board (board+C12-30 and board+C12-60). The modified FVB (FVB+C12-30 and FVB+C12-60) in the FTIR spectra did not show a peak at a wavelength of 1733 cm^{-1}, but there was a significant peak on the modified orientation board (board+C12-30 and board+C12-60) at a wavelength of 1733 cm^{-1}, which indicates the formation of a carbonyl group (C=O stretching) and an ester group C=O). This shows that there is a reaction between the hydroxyl group in the modified FVB and the carboxyl group in citric acid [42]. Furthermore, the formation of an ester bond on the modified orientation board made from modified FVB can increase the adhesive bond with the modified raw material. In addition, Menezzi et al. [57] described that esterification occurs between citric acid's carboxylic acid activities and the many aromatic and aliphatic hydroxyl groups of cellulose and lignin.

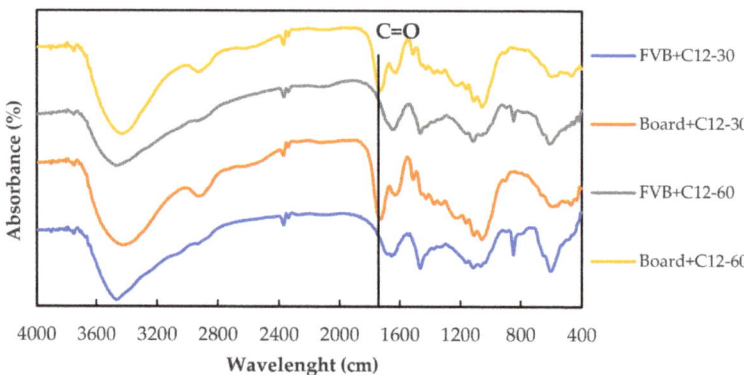

Figure 10. FTIR spectra of modified FVB (FVB+C12-30 and FVB+C12-60) and modified orientation board (board+C12-30 and board+C12-60). FVB+C12-30: modified fibrovascular bundle with 1% NaOH + 0.2% Na$_2$SO$_3$ (30 min); FVB+C12-60: modified fibrovascular bundle with 1% NaOH + 0.2% Na$_2$SO$_3$ (60 min); board+C12-30: oriented board with modified raw material (1% NaOH + 0.2% Na$_2$SO$_3$ (30 min); board+C12-60: oriented board with modified raw material (1% NaOH + 0.2% Na$_2$SO$_3$ (60 min).

As expected, orientation boards were manufactured from FVB modified with citric acid adhesive and their mechanical properties were tested. It was found that the combination of NaOH+Na$_2$SO$_3$ treatment influences the modulus of rupture, modulus of elasticity, internal bond, as well as screw holding power of the orientation board. The 1% NaOH + 0.2% Na$_2$SO$_3$ treatment with 30 and 60 min immersion increased the mechanical properties. As a structural panel, orientation boards made of modified FVB bonded with citric acid as a natural adhesive still rank below the wood-based composite with conventional adhesives. These orientation boards can be recommended as nonstructural panels such as partitions and insulation boards. In terms of the use of structural panels, FVB can be improved as a potential new raw material for the development of composite boards by combining it with other raw materials such as wood and veneer.

4. Conclusions

The research was conducted using the alkali modification of the fibrovascular bundle extracted from an *S. zalacca* frond for raw material of the orientation board. The results showed that the combined treatment of 1% NaOH + 0.2% Na$_2$SO$_3$ with 30 and 60 min immersion has a positive effect on increasing the mechanical properties of the orientation board. The 0° orientation angle also had a better effect on the MOR and MOE tests for the position perpendicular to the fiber. Furthermore, the bonding mechanism of FVB modified by the combination of 1% NaOH + 0.2% Na$_2$SO$_3$ by immersion for 30 and 60 min bonded citric acid adhesive showed good bonding quality. The FTIR spectra also showed the peak intensity of the wavelength at 1733 cm^{-1} with the detection of C–O stretching the carbonyl and ester groups, which indicated that there is a reaction between the hydroxyl group in the modified FVB and the carboxyl group in citric acid. Based on these results, the combination of 1% NaOH + 0.2% Na$_2$SO$_3$ treatment for 30 and 60 min immersion is successful in reducing water absorption and thickness swelling of the orientation board.

Author Contributions: Conceptualization, L.H. and R.W.; methodology, L.H., R.W., W.D.N. and T.A.P.; software, L.H. and W.D.N.; validation, L.H., R.W., W.D.N. and T.A.P.; formal analysis, L.H.; investigation, L.H.; resources, L.H.; data curation, L.H.; writing—original draft preparation, L.H.; writing—review and editing, R.W., W.D.N. and T.A.P.; visualization, L.H.; supervision, R.W., W.D.N.

and T.A.P.; project administration, L.H.; funding acquisition, L.H. All authors have read and agreed to the published version of the manuscript.

Funding: This research was funded by the Ministry of Education, Culture, Research, and Technology of Republic of Indonesia and the Indonesian Endowment Fund for Education (LPDP), Ministry of Finance, Republic of Indonesia [Awardee No.: 20161141030090].

Institutional Review Board Statement: Ethical review and approval were waived for this study, due to reason the study did not involve humans or animals.

Informed Consent Statement: The study did not involve humans.

Data Availability Statement: The data presented in this study are available on request from the corresponding author.

Acknowledgments: The authors are grateful to the Ministry of Education, Culture, Research, and Technology of the Republic of Indonesia and the Indonesian Endowment Fund for Education (LPDP), Ministry of Finance, Republic of Indonesia for supporting this research. Furthermore, the authors are also grateful to Rizqi Putri Winanti, Astri Winda Siregar, Yunida Syafriani Lubis, and Suri Fadilla for their assistance in the preparation of the FVB and fabrication of the oriented board.

Conflicts of Interest: The authors declare no conflict of interest.

References

1. Hakim, L.; Widyorini, R.; Nugroho, W.D.; Prayitno, T.A. Anatomical, chemical, and mechanical properties of fibrovascular bundles of salacca (snake fruit) frond. *BioResources* **2019**, *14*, 7943–7957. [CrossRef]
2. Gurunathan, T.; Mohanty, S.; Nayak, S.K. A review of the recent developments in biocomposites based on natural fibres and their application perspectives. *Compos. Part A Appl. Sci. Manuf.* **2015**, *77*, 1–25. [CrossRef]
3. Tamanna, T.A.; Belal, S.A.; Shibly, M.A.H.; Khan, A.N. Characterization of a new natural fiber extracted from *Corypha taliera* fruit. *Sci. Rep.* **2021**, *11*, 1–13. [CrossRef]
4. Sinebe, J.; Chukwuneke, J.L.; Omenyi, S.N. Surface energetics effects on mechanical strength of fibre reinforced polymer matrix. *J. Phys. Conf. Ser.* **2019**, *1378*, 042016. [CrossRef]
5. Ariawan, D.; Rivai, T.S.; Surojo, E.; Hidayatulloh, S.; Akbar, H.I.; Prabowo, A.R. Effect of alkali treatment of *Salacca Zalacca* fiber (SZF) on mechanical properties of HDPE composite reinforced with SZF. *Alex. Eng. J.* **2020**, *59*, 3981–3989. [CrossRef]
6. Deesoruth, A.; Ramasawmy, H.; Chummun, J. Investigation into the use of alkali treated screwpine (*Pandanus Utilis*) fibres as reinforcement in epoxy matrix. *Int. J. Plast. Technol.* **2014**, *18*, 263–279. [CrossRef]
7. Gholampour, A.; Ozbakkaloglu, T. A review of natural fiber composites: Properties, modification and processing techniques, characterization, applications. *J. Mater. Sci.* **2019**, *55*, 829–892. [CrossRef]
8. Boumediri, H.; Bezazi, A.; Del Pino, G.G.; Haddad, A.; Scarpa, F.; Dufresne, A. Extraction and characterization of vascular bundle and fiber strand from date palm rachis as potential bio-reinforcement in composite. *Carbohydr. Polym.* **2019**, *222*, 114997. [CrossRef]
9. Pizzi, A.; Papadopoulos, A.N.; Policardi, F. Wood Composites and Their Polymer Binders. *Polymers* **2020**, *12*, 1115. [CrossRef]
10. Papadopoulos, A.N.; Ntalos, G.A.; Kakaras, I. Mechanical and physical properties of cement-bonded OSB. *Holz Roh Werkst.* **2006**, *64*, 517–518. [CrossRef]
11. Hassani, V.; Papadopoulos, A.N.; Schmidt, O.; Maleki, S.; Papadopoulos, A.N. Mechanical and physical properties of oriented strand lumber (OSL): The effect of fortification level of nanowollastonite on UF resin. *Polymers* **2019**, *11*, 1884. [CrossRef] [PubMed]
12. Ciannamea, E.M.; Martucci, J.F.; Stefani, P.M.; Ruseckaite, R.A. Bonding quality of chemically-modified soybean protein concentrate-based adhesives in particleboards from rice husks. *J. Am. Oil Chem. Soc.* **2012**, *4302*, 1733–1741. [CrossRef]
13. Zhao, Z.; Umemura, K. Investigation of a new natural particleboard adhesive composed of tannin and sucrose. *J. Wood Sci.* **2014**, *60*, 269–277. [CrossRef]
14. Olivato, J.B.; Müller, C.M.O.; Carvalho, G.M.; Yamashita, F.; Grossmann, M.V.E. Physical and structural characterisation of starch/polyester blends with tartaric acid. *Mater. Sci. Eng. C* **2014**, *39*, 35–39. [CrossRef] [PubMed]
15. Yu, H.; Cao, Y.; Fang, Q.; Liu, Z. Effects of treatment temperature on properties of starch-based adhesives. *BioResources* **2015**, *10*, 3520–3530. [CrossRef]
16. Umemura, K.; Ueda, T.; Kawai, S. Characterization of wood-based molding bonded with citric acid. *J. Wood Sci.* **2011**, *58*, 38–45. [CrossRef]
17. Liao, R.; Xu, J.; Umemura, K. Low density sugarcane bagasse particleboard bonded with citric acid and sucrose: Effect of board density and additive content. *BioResources* **2015**, *11*, 2174–2185. [CrossRef]
18. Kusumah, S.S.; Umemura, K.; Yoshioka, K.; Miyafuji, H.; Kanayama, K. Utilization of sweet sorghum bagasse and citric acid for manufacturing of particleboard I: Effects of pre-drying treatment and citric acid content on the board properties. *Ind. Crop. Prod.* **2016**, *84*, 34–42. [CrossRef]

19. Widyorini, R.; Umemura, K.; Soraya, D.K.; Dewi, G.K.; Nugroho, W.D. Effect of citric acid content and extractives treatment on the manufacturing process and properties of citric acid-bonded salacca Frond. *BioResources* **2019**, *14*, 4171–4180.
20. Hakim, L.; Widyorini, R.; Nugroho, W.D.; Prayitno3, T.A. Radial variability of fibrovascular bundle properties of salacca (*Salacca zalacca*) fronds cultivated on Turi Agrotourism in Yogyakarta, Indonesia. *Biodiversitas J. Biol. Divers.* **2021**, *22*, 861. [CrossRef]
21. Schellbach, S.L.; Monteiro, S.N.; Drelich, J.W. A novel method for contact angle measurements on natural fibers. *Mater. Lett.* **2016**, *164*, 599–604. [CrossRef]
22. Munawar, S.S.; Umemura, K.; Kawai, S. Characterization of the morphological, physical, and mechanical properties of seven nonwood plant fiber bundles. *J. Wood Sci.* **2007**, *53*, 108–113. [CrossRef]
23. Segal, L.; Creely, J.J.; Martin, A.E., Jr.; Conrad, C.M. An Empirical method for estimating the degree of crystallinity of native cellulose using the X-ray diffractometer. *Text. Res. J.* **1959**, *29*, 786–794. [CrossRef]
24. Darmanto, S.; Rochardjo, H.S.B.; Widyorini, R. Effects of alkali and steaming on mechanical properties of snake fruit (Salacca) fiber. In Proceedings of the International Conference on Engineering Science and Nanotechnology 2016 (ICESNANO), Solo, Indonesia, 3–5 August 2017; Volume 1. [CrossRef]
25. Shanmugasundaram, N.; Rajendran, I.; Ramkumar, T. Characterization of untreated and alkali treated new cellulosic fiber from an Areca palm leaf stalk as potential reinforcement in polymer composites. *Carbohydr. Polym.* **2018**, *195*, 566–575. [CrossRef]
26. Ganapathy, T.; Sathiskumar, R.; Senthamaraikannan, P.; Saravanakumar, S.; Khan, A. Characterization of raw and alkali treated new natural cellulosic fibres extracted from the aerial roots of banyan tree. *Int. J. Biol. Macromol.* **2019**, *138*, 573–581. [CrossRef]
27. Sghaier, A.E.O.B.; Chaabouni, Y.; Msahli, S.; Sakli, F. Morphological and crystalline characterization of NaOH and NaOCl treated Agave americana L. fiber. *Ind. Crop. Prod.* **2012**, *36*, 257–266. [CrossRef]
28. Cai, M.; Takagi, H.; Nakagaito, A.N.; Katoh, M.; Ueki, T.; Waterhouse, G.I.; Li, Y. Influence of alkali treatment on internal microstructure and tensile properties of abaca fibers. *Ind. Crop. Prod.* **2014**, *65*, 27–35. [CrossRef]
29. Vijay, R.; Singaravelu, D.L.; Vinod, A.; Sanjay, M.; Siengchin, S.; Jawaid, M.; Khan, A.; Parameswaranpillai, J. Characterization of raw and alkali treated new natural cellulosic fibers from Tridax procumbens. *Int. J. Biol. Macromol.* **2018**, *125*, 99–108. [CrossRef]
30. Arnata, I.W.; Suprihatin, S.; Fahma, F.; Richana, N.; Candra Sunarti, T. Cellulose production from sago frond with alkaline delignification and bleaching on various types of bleach agents. *Orient. J. Chem.* **2019**, *35*, 8–19. [CrossRef]
31. Darmanto, S.; Rochardjo, H.S.B.; Jamasri; Widyorini, R. Effect of sonication treatment on fibrilating snake fruit (Sallaca) frond fiber. *AIP Conf. Proc.* **2018**, *1931*, 030064. [CrossRef]
32. Chen, H.; Jiang, Z.; Qin, D.; Yu, Y.; Tian, G.; Lu, F.; Fei, B.; Wang, G. Contact angles of single bamboo fibers measured in different environments and compared with other plant fibers and bamboo strips. *BioResources* **2013**, *8*, 2827–2838. [CrossRef]
33. Yorseng, K.; Rangappa, S.M.; Parameswaranpillai, J.; Siengchin, S. Influence of accelerated weathering on the mechanical, fracture morphology, thermal stability, contact angle, and water absorption properties of natural fiber fabric-based epoxy hybrid composites. *Polymers* **2020**, *12*, 2254. [CrossRef]
34. Medina, J.D.C.; Woiciechowski, A.; Filho, Z.A.; Noseda, M.D.; Kaur, S.B.; Soccol, R.C. Lignin preparation from oil palm empty fruit bunches by sequential acid/alkaline treatment—A biorefinery approach. *Bioresour. Technol.* **2015**, *194*, 172–178. [CrossRef]
35. Wang, N.; Liu, W.; Huang, J.; Ma, K. The structure–mechanical relationship of palm vascular tissue. *J. Mech. Behav. Biomed. Mater.* **2014**, *36*, 1–11. [CrossRef] [PubMed]
36. Then, Y.Y.; Ibrahim, N.A.; Zainuddin, N.; Ariffin, H.; Yunus, W.M.Z.W.; Chieng, B.W. Static mechanical, interfacial, and water absorption behaviors of Alkali Treated Oil palm mesocarp fiber reinforced poly(butylene succinate) biocomposites. *BioResources* **2015**, *10*, 123–136. [CrossRef]
37. Krishnaiah, P.; Ratnam, C.T.; Manickam, S. Enhancements in crystallinity, thermal stability, tensile modulus and strength of sisal fibres and their PP composites induced by the synergistic effects of alkali and high intensity ultrasound (HIU) treatments. *Ultrason. Sonochem.* **2017**, *34*, 729–742. [CrossRef]
38. Moradbak, A.; Tahir, P.M.; Mohamed, A.Z.; Halis, R.B. Alkaline sulfite anthraquinone and methanol pulping of bamboo (*Gigantochloa scortechinii*). *BioResources* **2015**, *11*, 235–248. [CrossRef]
39. Munawar, S.S.; Umemura, K.; Kawai, S. Manufacture of oriented board using mild steam treatment of plant fiber bundles. *J. Wood Sci.* **2008**, *54*, 369–376. [CrossRef]
40. Baharin, A.; Fattah, N.A.; Abu Bakar, A.; Ariff, Z. Production of laminated natural fibre board from banana tree wastes. *Procedia Chem.* **2016**, *19*, 999–1006. [CrossRef]
41. Umemura, K.; Sugihara, O.; Kawai, S. Investigation of a new natural adhesive composed of citric acid and sucrose for particleboard. *J. Wood Sci.* **2013**, *59*, 203–208. [CrossRef]
42. Widyorini, R.; Umemura, K.; Septiano, A.; Soraya, D.K.; Dewi, G.K.; Nugroho, W.D. Manufacture and properties of citric acid-bonded composite board made from salacca frond: Effects of Maltodextrin Addition, Pressing Temperature, and Pressing Method. *BioResources* **2018**, *13*, 8662–8676. [CrossRef]
43. Khakpour, H.; Ayatollahi, M.R.; Akhavan-Safar, A.; da Silva, L.F.M. Mechanical properties of structural adhesives enhanced with natural date palm tree fibers: Effects of length, density and fiber type. *Compos. Struct.* **2020**, *237l*, 111950. [CrossRef]
44. Brígida, A.I.S.; Calado, V.M.A.; Gonçalves, L.R.B.; Coelho, M.A.Z. Effect of chemical treatments on properties of green coconut fiber. *Carbohydr. Polym.* **2010**, *79*, 832–838. [CrossRef]

45. Elanchezhian, C.; Ramnath, B.V.; Ramakrishnan, G.; Rajendrakumar, M.; Naveenkumar, V.; Saravanakumar, M.K. Review on mechanical properties of natural fiber composites. *Mater. Today Proc.* **2018**, *5*, 1785–1790. [CrossRef]
46. Sair, S.; Oushabi, A.; Kammouni, A.; Tanane, O.; Abboud, Y.; Oudrhiri Hassani, F.; Laachachi, A.; El Bouari, A. Effect of surface modification on morphological, mechanical and thermal conductivity of hemp fiber: Characterization of the interface of hemp -Polyurethane composite. *Case Stud. Therm. Eng.* **2017**, *10*, 550–559. [CrossRef]
47. Väisänen, T.; Batello, P.; Lappalainen, R.; Tomppo, L. Modification of hemp fibers (*Cannabis Sativa* L.) for composite applications. *Ind. Crop. Prod.* **2018**, *111*, 422–429. [CrossRef]
48. Santoso, M.; Widyorini, R.; Prayitno, T.A.; Sulistyo, J. Bonding performance of maltodextrin and citric acid for particleboard made from nipa fronds. *J. Korean Wood Sci. Technol.* **2017**, *45*, 432–443. [CrossRef]
49. Widyorini, R.; Umemura, K.; Isnan, R.; Putra, D.R.; Awaludin, A.; Prayitno, T.A. Manufacture and properties of citric acid-bonded particleboard made from bamboo materials. *Eur. J. Wood Prod.* **2016**, *74*, 7–65. [CrossRef]
50. Umemura, K.; Sugihara, O.; Kawai, S. Investigation of a new natural adhesive composed of citric acid and sucrose for particleboard II: Effects of board density and pressing temperature. *J. Wood Sci.* **2014**, *61*, 40–44. [CrossRef]
51. Zhao, Z.; Umemura, K.; Kanayama, K. Effects of the addition of citric acid on tannin-sucrose adhesive and physical properties of the particleboard. *BioResources* **2015**, *11*, 1333. [CrossRef]
52. Walther, T.; Kartal, S.N.; Hwang, W.J.; Umemura, K.; Kawai, S. Strength, decay and termite resistance of oriented kenaf fiberboards. *J. Wood Sci.* **2007**, *53*, 481–486. [CrossRef]
53. Hung, K.; Wu, T.; Chen, Y.; Wu, J. Assessing the effect of wood acetylation on mechanical properties and extended creep behavior of wood/recycled-polypropylene composites. *Constr. Build. Mater.* **2016**, *108*, 139–145. [CrossRef]
54. Hung, K.-C.; Wu, J.-H. Mechanical and interfacial properties of plastic composite panels made from esterified bamboo particles. *J. Wood Sci.* **2010**, *56*, 216–221. [CrossRef]
55. Kemalasari, D.; Widyorini, R. Karakteristik Papan Partikel dari Pelepah Salak Pondoh (*Saliacca* sp.) dengan Penambahan Asam Sitrat. In Proceedings of the Prosiding Seminar Nasional MAPEKI XVIII, Bandung, Indonesia, 4–5 November 2015; pp. 542–548.
56. Kusumah, S.S.; Astari, L.; Subyakto, W.; Zhao, Z. The utilization of citric acid as an environmentally friendly of chemical modification agent of the lignocellulosic materials: A review. *J. Lignocellul. Technol.* **2017**, *2*, 1–7.
57. Menezzi, C.; Del Amirou, S.; Pizzi, A.; Xi, X.; Delmotte, L. Reactions with wood carbohydrates and lignin of citric acid as a bond promoter of wood veneer panels. *Polymers* **2018**, *10*, 833. [CrossRef]

Article

Heat Transfer Modeling of Oriented Sorghum Fibers Reinforced High-Density Polyethylene Film Composites during Hot-Pressing

Chusheng Qi [1,*], Jinyue Wang [1] and Vikram Yadama [2]

[1] MOE Key Laboratory of Wood Material Science and Utilization, Beijing Forestry University, Beijing 100083, China; wjy123@bjfu.edu.cn
[2] Department of Civil & Environmental Engineering and Composite Materials and Engineering Center, Washington State University, Pullman, WA 99164, USA; vyadama@wsu.edu
* Correspondence: qichusehng@bjfu.edu.cn

Abstract: A one-dimensional heat transfer model was developed to simulate the heat transfer of oriented natural fiber reinforced thermoplastic composites during hot-pressing and provide guidance for determining appropriate hot-pressing parameters. The apparent heat capacity of thermoplastics due to the heat of fusion was included in the model, and the model was experimentally verified by monitoring the internal temperature during the hot-pressing process of oriented sorghum fiber reinforced high-density polyethylene (HDPE) film composites (OFPCs). The results showed that the apparent heat capacity of HDPE accurately described its heat fusion of melting and simplified the governing energy equations. The data predicted by the model were consistent with the experimental data. The thermal conduction efficiency increased with the mat density and HDPE content during hot-pressing, and a higher mat density resulted in a higher mat core temperature. The addition of HDPE delayed heat transfer, and the mat had a lower core temperature at a higher HDPE content after reaching the melting temperature of HDPE. Both the experimental and simulated data suggested that a higher temperature and/or a longer duration during the hot-pressing process should be used to fabricate OFPC as the HDPE content increases.

Keywords: wood–plastic composites; heat transfer modeling; sweet sorghum; high-density polyethylene; hot-pressing

Citation: Qi, C.; Wang, J.; Yadama, V. Heat Transfer Modeling of Oriented Sorghum Fibers Reinforced High-Density Polyethylene Film Composites during Hot-Pressing. *Polymers* **2021**, *13*, 3631. https://doi.org/10.3390/polym13213631

Academic Editor: Jingpeng Li

Received: 30 September 2021
Accepted: 18 October 2021
Published: 21 October 2021

Publisher's Note: MDPI stays neutral with regard to jurisdictional claims in published maps and institutional affiliations.

Copyright: © 2021 by the authors. Licensee MDPI, Basel, Switzerland. This article is an open access article distributed under the terms and conditions of the Creative Commons Attribution (CC BY) license (https://creativecommons.org/licenses/by/4.0/).

1. Introduction

Natural fibers reinforced thermoplastic composites have been commercially developed to take advantage of the low cost and high strength and stiffness of natural fibers to reinforce the matrix. In addition, these composites save petrochemical polymers and are almost climate-neutral as natural fibers absorb exactly the amount of harmful greenhouse gases from the atmosphere during their growth [1]. Hot-pressing or compression molding between two heated platens is commonly used to fabricate natural fiber composite panels because it can be used to produce final products with higher fiber loading and large size fibers/fiber bundles compared with extrusion and injection moulding. In our previously carried out studies [2–4], oriented sorghum fiber reinforced thermoplastic film composites (OFPCs) were manufactured with a modulus of rupture of 37.2 MPa, a modulus of elasticity of 4.7 GPa, and a thickness swelling of 2.8%. The OFPC can be used as indoor and outdoor building materials with excellent water resistance properties and without formaldehyde release. During the hot-pressing, a loosely-formed mat was compressed to its final thickness under high temperature and pressure. Heat is transported by conduction from the hot platens to the mat surface and then to the interior of the mat by conduction and convection. The hot press temperature and duration are two critical parameters for fabricating OFPCs, and modeling their heat transfer will help to understand the influence of changing these two parameters on the fabrication process and product quality.

Heat and mass transfer during hot-pressing of natural fiber composites with thermosetting resins have been investigated by many researchers [5–7]. However, thermoplastics have different heat characteristics than thermosetting resins. Thermoplastics will absorb heat and begin to soften below their melting temperature and flow under pressure above their melting point. Their heat capacity and conductivity change with temperature and show different trends before and after the melting point. In comparison, thermosetting resins or thermosets release heat and solidify during curing. Therefore, natural fiber and thermoplastic composites display different heat transfer characteristics than natural fiber composites bonded with thermosets. This primary difference prevents heat and mass transfer models for natural fiber composites with thermosets directly applied to natural fiber and thermoplastic composites.

Thermoplastic materials undergo a phase change, and the heat transfer is quite challenging to handle owing to the presence of solid–liquid surfaces [8]. Solving the energy equations for the liquid and solid phases separately [9] and solving it simultaneously using the enthalpy methods [10,11] are typically two approaches to tackle this challenge. Mantell and Springer [12] developed a model including three submodels, namely, the thermo-chemical model, consolidation and bonding model, and stress and strain model, to simulate the processing of thermoplastic matrix composites. Xiong et al. [13] recently developed a model to describe the consolidation behavior of thermoplastic composite prepregs during the thermoforming process based on a generalized Maxwell approach. These studies treated the melting and crystallization of thermoplastics as a complicated process, and heat absorption or generation rates were involved in their energy equation. Woo et al. [14] reported that the effective heat capacity of thermoplastics could be used to replace their heat absorption or generation during melting and crystallization. The use of effective heat capacity can simplify the energy equation to obtain its numerical solution. Still, such a method should be verified in the manufacture of natural fiber reinforced thermoplastic composites. Thermal conductivity is another critical parameter to simulate the heat transfer of natural fiber reinforced thermoplastic composites. This study will directly apply the results of our previous study about the thermal conductivity of OFPCs [3].

Many previous studies have shown that the fiber moisture content can significantly affect the heat and mass transfer of natural fiber-based composites that use thermosetting resins during hot-pressing [7,15], because water is vaporized and transfers heat from high-temperature to low-temperature regions through convection under pressure. For the fabrication of OFPC in this study, HDPE film layers inside the mat served as a barrier for vapor diffusion from the surface into the core [2]. Additionally, HDPE is hydrophobic and is incompatible with the hydrophilic sorghum fiber, and more moisture would interfere with mechanical interaction between these two materials. Therefore, oven-dried sorghum fiber was used, and the effect of moisture content on heat transfer of OFPC was ignored in this study.

As a thermoplastic and phase-change material, high-density polyethylene was assumed to gradually melt and flow into the gaps between sorghum fibers during hot-pressing. The convection caused by HDPE flow is limited, and ignored because most of the HDPE stayed in place according to previous research [4]. Our previous study showed that the vertical density profile of OFPC was not U-shaped or M-shaped, but displayed rather a zigzag fluctuation [2]. To simplify the model, a homogenous mat was assumed here as the HDPE layer was very thin.

This research aimed to develop a mathematical model capable of simulating the heat transfer of natural fiber reinforced thermoplastic composites using the apparent heat capacity of thermoplastics. The data obtained from the model were used to optimize the hot-press parameters of OFPC.

2. Materials and Methods

2.1. Materials

Extracted sweet sorghum bagasse (referred to as sorghum fiber) with a length of 20–100 mm was provided by ChloroFill, LLC in San Diego, CA, USA. Sorghum fiber was oven-dried at 103 °C to obtain a moisture content of 3% or conditioned at varied temperatures and humidity for at least one week to obtain a 6–12% moisture content. HDPE film without any additives and with a thickness of 0.1 mm and a melt flow index of 20 g/10 min at 190 °C/2.16 kg was purchased from Tee Group Films, Ladd, IL, USA.

2.2. Governing Equations

The description of various transport phenomena involves the solution of mass and energy conservation equations. The model equations contain four dependent variables and four governing equations. The four dependent variables were the fiber volume fraction (V_f), the plastic volume fraction (V_m), the void volume fraction (V_v), and mat thermal conductivity (k), W/(m.K). The four variables were a function of two independent variables, temperature T (K) and time t (s). The one-dimension energy balance equation is as follows [7]:

$$\frac{\partial}{\partial t}\left[V_f C_f \rho_f + V_m C_m \rho_m + V_v C_v \rho_v\right] T = \frac{\partial T}{\partial z}\left(k \frac{\partial T}{\partial z}\right) - G \quad (1)$$

where, C_f, C_m, and C_v are the specific heat capacity of the sorghum fiber cell wall, HDPE, and air in voids (J/(kg·K)), respectively. ρ_f, ρ_m, and ρ_v denote the density of the sorghum fiber cell wall, HDPE, and air (kg/m^3), respectively. G is the heat absorption rate of HDPE during a phase change. The energy balance equation indicates that the sum of energy consumption consumed to increase the mat temperature and the heat absorption due to the phase change of HDPE films should be equal to the heat conducted by contacting the hot-press plates. The HDPE phase change can also be represented as an equivalent internal energy change, and Equation (1) can be modified as follows [14]:

$$\frac{\partial}{\partial t}\left[V_f C_f \rho_f + V_m C_m \rho_m + V_m C_m^* \rho_m + V_v C_v \rho_v\right] T = \frac{\partial T}{\partial z}\left(k \frac{\partial T}{\partial z}\right) \quad (2)$$

where C_m^* is the apparent heat capacity of high-density polyethylene owing to the heat of fusion and was calculated later. V_f, V_m, and V_v can be obtained as follows:

$$V_f = m_f / (\rho_f * L * W * H) \quad (3)$$

$$V_m = m_m / (\rho_m * L * W * H) \quad (4)$$

$$V_v = 1 - V_f - V_m \quad (5)$$

where m_f is the weight of sorghum fibers (kg), m_m denotes the weight of HDPE (kg), L indicates the length of the mat (m), W is the width of the mat (m), and H is the thickness of the mat (m).

2.3. Testing and Simulation of HDPE Heat Capacity

The effective heat capacity of HDPE (C_m') was the sum of C_m and C_m^*, and was measured by differential scanning calorimetry (Mettler-Toledo DSC 822), according to ASTM E1269-11 [16]. An empty aluminum holder, standard sapphire pellet, and HDPE film (20 mg) were sequentially examined in the temperature range of −20~200 °C at a heating rate of 20 °C/min and N$_2$ gas flow of 50 mL/min. The sample was kept at −20 °C for 15 min in a liquid nitrogen environment to obtain a uniform temperature of the sample before testing. Two DSC scans were carried out for each material to ensure reproducibility. The C'_m was calculated according to the following:

$$C'_m = C_p(st) * D_s * W_{st} / (D_{st} * W_s) \quad (6)$$

where $C_p(st)$ is the specific heat capacity of the sapphire standard, and its value can be obtained from ASTM E1269-11. D_s is the vertical displacement (mW) between the empty holder and the HDPE sample heat flow curves at a given temperature, and D_{st} is the vertical displacement (mW) between the empty holder and the sapphire heat flow curves at a given temperature. W_{st} and W_s (mg) are the masses of the sapphire and HDPE samples, respectively.

The HDPE samples start to absorb heat during the melting stage, and the DSC heat flow outside of the melting temperate range reflects the specific heat capacity of HDPE. Therefore, the C'_m values obtained using Equation (6) in the temperature ranges of 0–100 °C and 160–200 °C were fitted by linear regression to obtain the C_m value. The difference between C'_m and C_m is C^*_m, and C^*_m were fitted by a Gauss and a Lorentz equation.

2.4. Material Properties

2.4.1. Heat Capacity of Sorghum Fiber and Air

The heat capacity of air (C_v) was considered to be constant (1000 J/(kg·K)). The heat capacity of sorghum fiber (C_f) in the oven-dried state was previously tested, and its prediction equation is as follows [16]:

$$C_f = 5.74T - 469.1 \quad (7)$$

2.4.2. Thermal Conductivity

Oriented sorghum reinforced HDPE composites consisted of sorghum fibers, HDPE, and air in the voids. Their thermal conductivity of OFPC was measured and simulated in Qi et al. [4]:

$$k = 0.2 \times 10^{-3} T + 0.21 \times 10^{-3} \rho + 0.19 V_m - 0.21 V_m^2 - 0.10 \quad (8)$$

where ρ is the mat density (kg/m³) and V_m is the mass ratio of HDPE, which ranged from 0 to 1.

2.5. Numerical Solution

Eight unknown variables (T, k, C_f, C_m, C^*_m, V_f, V_m, and V_v) and eight equations (Equations (2)–(5), (7), (8), (10) and (12)) were defined, allowing these equations to be solved. The energy equation (Equation (2)) is a nonlinear partial differential equation, and the initial and the boundary conditions must be defined to obtain its numerical solution. For this one-dimensional heat transfer model, one initial and two boundary conditions are required. The initial temperature (T_i) was ambient temperature of 25 °C. The Dirichlet boundary condition was employed to solve Equation (2) because the mat surface temperature rapidly increased to the target temperature. The boundary condition is as follows:

$$T(0,t) = T(H,t) = T_\infty \quad (9)$$

where T_∞ was the hot-press platen temperature (K). The known parameters are listed in Table 1. Equation (2) is a second-order nonlinear partial differential equation with varied heat capacity and thermal conductivity and no analytical solution [17]. Equation (2) was discretized over time and thickness variables; MATLAB® software was used to obtain its numerical solution using the central-difference approximation for the derivative. The detailed numerical solution method and MATLAB code could be found in our previous study [18].

Table 1. Parameters and values of the mathematical model.

Parameters	Symbols	Values	Unit
Specific heat capacity of air	C_v	1000	J/(kg·K)
Density of sorghum fiber cell wall	ρ_f	1500	kg/m³
Density of high-density polyethylene	ρ_m	940	kg/m³
Air density	ρ_v	1.225	kg/m³
Thermal conductivity of HDPE	k_m	0.44	W/(m·K)
Mat thickness	H	0.015	m

2.6. Experimental Evaluation of Heat Transfer

To better understand and simulate the effect of varying HDPE content, target mat density, and sorghum fiber moisture content on heat transfer through the mat during the hot-pressing process, an experimental design was devised, as shown in Table 2. The platen temperature was held constant in all cases at 160 °C. To study the influence of HDPE content in the mat on heat transfer, HDPE content values were changed while holding the target panel density and the moisture content of the panel constant at the values shown in Table 2. In order to precisely place the thermocouple (J type, EXTT-J-24-500, Omega, Norwalk, CT, USA) into the mat during hot-pressing, the OFPC fabrication method was modified from our previous study [1,2]. Sorghum fiber at a target moisture content level (as shown in Table 2) was evenly divided into four portions; each portion was formed by orienting the fibers with respect to the longitudinal axis, and pre-pressed at 35 MPa at room temperature for 4 min to obtain a layer of oriented sorghum fiber. HDPE films were divided into five portions (12.5%, 25%, 25%, 25%, and 12.5%). A portion of the HDPE film (12.5%) was first placed at the bottom, followed by placing a layer of sorghum fiber, and another portion of HDPE film (25%). The above steps were repeated till the second 12.5% HDPE film was placed on the top. All four layers of sorghum fibers were aligned in the longitudinal direction of the final mat, and each fiber layer was layered on both sides with 12.5% HDPE films. Double-sided silicone release papers were placed on the top and bottom of the mat in order to prevent HDPE from sticking to the metal caul plates during the hot-pressing process. Thermocouples were placed on the surfaces of the mat and in the middle of each layer of HDPE films (Figure 1). In this study, no adhesive or coupling agent was added in order to prevent heat release.

Table 2. Experimental design for the effects of HDPE content, mat density, and sorghum fiber moisture content on heat transfer of the mat during hot-pressing.

Variables	Values	Fixed Parameters	Platen Temperature
HDPE content	0, 10, 20, 30, 40%	Target mat density was 0.9 g/cm³, 3% moisture content of sorghum fiber	160 °C
Mat density	0.7, 0.8, 0.9, 1.0 g/cm³	10% HDPE, 3% moisture content of sorghum fiber	
Sorghum fiber moisture content	3, 6, 9, 12%	10% HDPE, target mat density was 0.9 g/cm³	

To fabricate OFPC, the mat of sorghum fibers and HDPE was hot-pressed between two platens heated to 160 °C for 10 min at a target thickness of 15 mm, followed by 30 min cold-pressing (cold water was piped into the hot-press plates). The mat temperature during hot-press was continuously recorded by a data acquisition system. Table 2 describes the experimental design to evaluate the influence of HDPE content, target density, and sorghum fiber moisture content on temperature distribution within the mat during the hot-pressing process. All the tests were repeated three times.

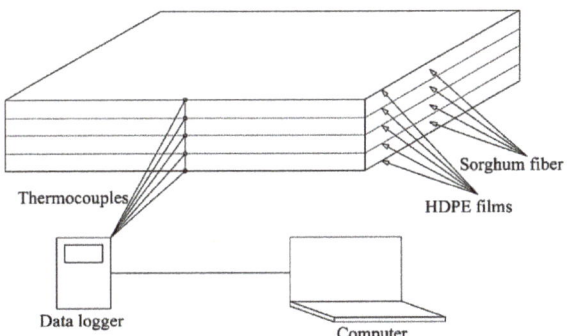

Figure 1. Schematic diagram of mat structure and thermocouple positions.

3. Results and Discussion

3.1. Heat Capacity and Heat Fusion of HDPE

The heat flow results for the empty aluminum holder, standard sapphire, and HDPE are shown in Figure 2a. The heat flows of the empty aluminum holder and standard sapphire changed linearly with temperature, while that of HDPE showed nonlinear changes. HDPE is a thermoplastic and exhibits endothermic behavior during melting. The onset temperature and ending temperature of HDPE melting were 121.2 °C and 151.3 °C, respectively, with a melting peak at 136.1 °C and a heat of fusion (ΔH_f) of 180.2 J/g. The ΔH_f value obtained in this study is close to the ΔH_f of 178.6 J/g previously reported by Sotomayor et al. [19].

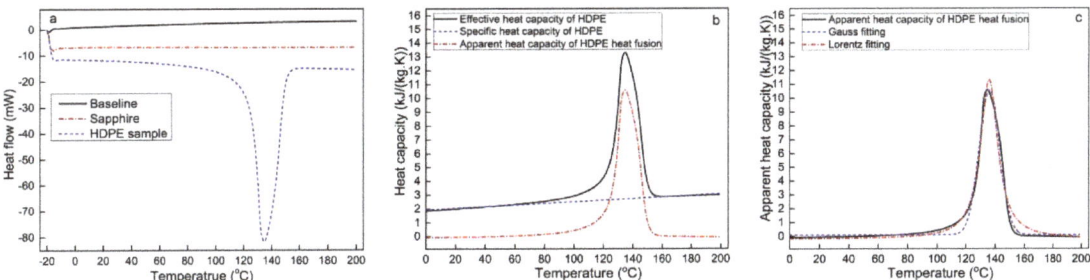

Figure 2. (a) Heat flow curves of heat capacity testing, (b) relationship between effective heat capacity and specific heat capacity of HDPE as well as the apparent heat capacity of the HDPE heat fusion, and (c) Gauss and Lorentz fits of the apparent heat capacity of the HDPE heat fusion.

Figure 2b shows the C'_m calculated according to Equation (6). C'_m in the temperature ranges of 0–100 °C and 160–200 °C outside the HDPE melting temperate was considered as the specific heat capacity of HDPE and was fitted using the following linear equation:

$$C_m = 5.7(T - 273.2) + 1930.1 \tag{10}$$

$$R^2 = 0.96$$

The R^2 value of Equation (10) was 0.96, and Figure 2b shows that Equation (10) predicts the baseline of C'_m well. The C_m of HDPE in this study was found to be 2077.1 J/(kg·K) at 20 °C and 2899.1 J/(kg·K) at 170 °C based on Equation (10). These values are very close to the value of approximately 2000 J/(kg·K) at 20 °C previously reported by Oh [20]. Li et al. [21] reported that HDPE has a specific heat capacity of 2480 J/(kg·K) at 170 °C, which is lower than the value obtained in our study, possibly owing to the different molecular weight of the HDPE used in their study.

The apparent heat capacity of HDPE due to the heat of fusion was obtained by subtracting the C_m from C'_m, as shown in Figure 2b. Fits to Gauss and Lorentz functional forms were applied to model C_m^*, obtaining the following functions:

Gauss fitting:

$$C_m^* = 101.3 + 10486.4 \exp\left(-0.01(T - 273.2 - 136.1)^2\right) \quad R^2 = 0.98 \tag{11}$$

Lorentz fitting:

$$C_m^* = -197.8 + 2.3 \times 10^6 / \left[4(T - 273.2 - 135.8)^2 + 198.8\right] \quad R^2 = 0.98 \tag{12}$$

The R^2 values of the Gauss and Lorentz fits are both 0.98, showing that both equations fit the C_m^* data very well. Figure 2c shows that the Gauss function describes the peak of C_m^* better than the Lorentz function. The Lorentz function with no exponential component has a simpler form for numerical calculations, enabling the easy numerical solution of the energy equation. Thus, the Lorentzian function was used in later modeling.

3.2. Heat Transfer Simulation of Pure HDPE

To verify whether Equation (2) could be used to reflect the phase change characteristics of HDPE, V_m was set to one when V_f and V_v were set to zero to simulate the heat transfer of pure HDPE. The initial temperature was 25 °C, T_∞ was set as 200 °C, and the HDPE thickness was 15 mm for the simulation. Figure 4 shows the temperature changes over time at different thickness locations within the mat. Both Figure 3a,b show that the core temperature lags behind between 120 and 150 °C, corresponding to the melting of HDPE. Figure 3a,b also show that the temperature lag increased from the surface to the core of the HDPE mat, as more energy was absorbed to melt the HDPE closer to the core. The temperature increased with time because more energy was input through the heated hot-press platens. The simulated heat transfer of pure HDPE in this study is consistent with Woo et al. 's research [14], and Equation (2) had a better prediction of the core temperature. Woo et al. [14] used the Gauss equation to describe the effective heat capacity of HDPE. Its prediction equation was divided into two segments with varied temperature factors to simulate HDPE heat transfer, which caused the predicted core temperature of HDPE during heating to abnormally increase above the melting peak temperature. However, the apparent heat capacity of HDPE predicted in Equation (11) is continuous and gradual, and it prevents an abrupt shift of the predicted temperature trend. These demonstrate that the C_m^* can be used to replace the heat absorption rate of the HDPE phase change (G), and the C_m^* of thermoplastics can be easily obtained from the DSC results. The use of C_m^* also simplified the numerical solution of Equation (2), as a differential factor was omitted.

3.3. Effects of Moisture Content on Heat Transfer

Figure 4 illustrates the effect of moisture content on the core temperature and one-quarter position temperature, and no obvious impact was observed. Owing to the closure of the hot press platens, the mat surface temperature rapidly increased to 150 °C in 20 s and can be described using the Dirichlet boundary condition. The surface temperature continued to increase to 160 °C during the next 300 s as the platens gradually reached the set target temperature. Additionally, the HDPE films at the surface absorbed energy during the melting process, decreasing the surface temperature. At the one-quarter position below the mats' surface, the temperature was much higher than that at the core, and this temperature difference decreased with time. The heat was transferred from the higher temperature region to the lower temperature one, and the heat gradually transferred from the surface to the core.

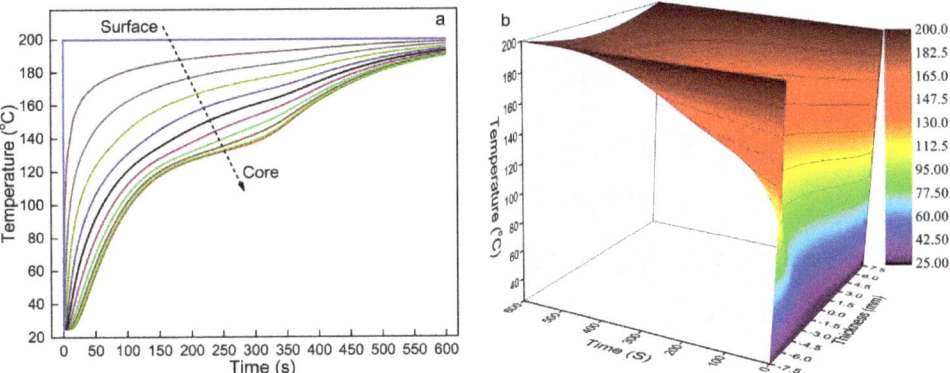

Figure 3. Heat transfer simulation results of pure HDPE at a hot-press temperature of 200 °C: (**a**) the change in temperature at different thicknesses and (**b**) 3D temperature distribution.

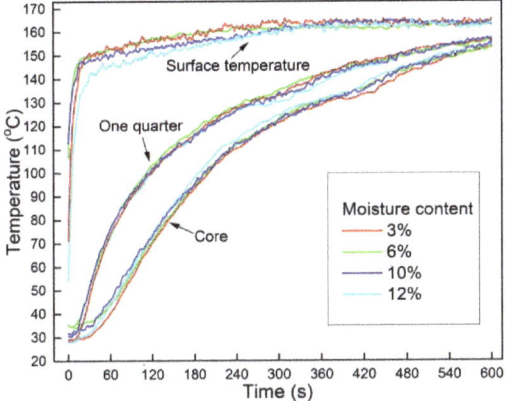

Figure 4. Effect of moisture content on heat transfer of OFPC during hot-pressing (the mat target density was 0.9 g/cm^3 and the HDPE content was 10%).

The temperature at the core and the one-quarter position did not obviously change when the moisture content of sorghum fiber increased from 3% to 12%, indicating that the fiber moisture content had little impact on the heat transfer of OFPC. Furthermore, unlike in a traditional natural fiber-based mat with liquid thermosetting resin, the HDPE films in the mat acted like barriers for water vapor flowing through the mat thickness. Therefore, it was reasonable to exclude the heat convection of vapor in the heat transfer model of the OFPC.

3.4. Effects of Mat Density on Heat Transfer

Figure 5 shows the effects of mat density on the mat core temperature during OFPC hot-pressing. The mat core temperature was higher at a higher mat density in both the experimental test (Figure 5a) and the mathematic model (Figure 5b). A close examination of Equation (8) shows that the thermal conductivity of the mat linearly increases with density, and a higher thermal conductivity results in higher efficiency of thermal conduction, supporting the experimental results that indicate an increase in the core temperature with increasing mat density. The mat contained more sorghum fiber and HDPE content per unit volume at a higher density. These materials, including more molecules in the mat, absorbed more energy to increase their internal energy. This had an inverse effect on the temperature increase of the mat, and Equation (2) also supports the inverse impact of mat density. The temperature increase due to energy absorption depends on the specific heat capacity. The mat core temperature increased with density under the combined effects of a higher heat transfer efficiency and higher heat absorption at a higher mat density.

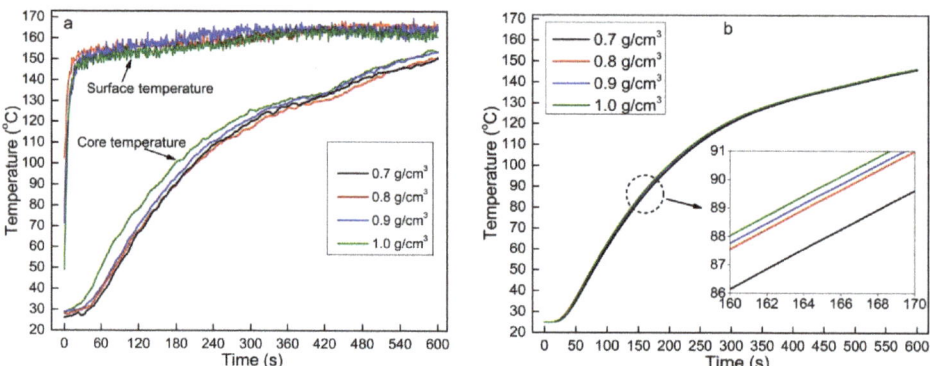

Figure 5. Effects of mat density on the mat core temperature during hot-pressing: (**a**) measured data and (**b**) modeling data (the sorghum fiber moisture content was 3% and HDPE content was 10%).

3.5. Effect of HDPE Content on Heat Transfer

Figure 6 illustrates the effect of HDPE content on the core temperature of the OFPC mat during hot-pressing. No temperature lag was observed in the experiment (Figure 6a) when HDPE was not added to the mat. However, the core temperature lagged between 300 and 480 s when 10% HDPE was incorporated into the mat. A higher HDPE content resulted in a lower core temperature after reaching the melting temperature of HDPE. HDPE is a thermoplastic and phase-change material that absorbs heat during melting; thus, more heat was absorbed with a higher HDPE content, which delayed the increase in the mat core temperature during melting. A similar trend was found in the mathematic modeling results (Figure 6b). The mat core temperature rapidly increased after all the HDPE melted (Figure 6a), as no additional heat was required to melt HDPE.

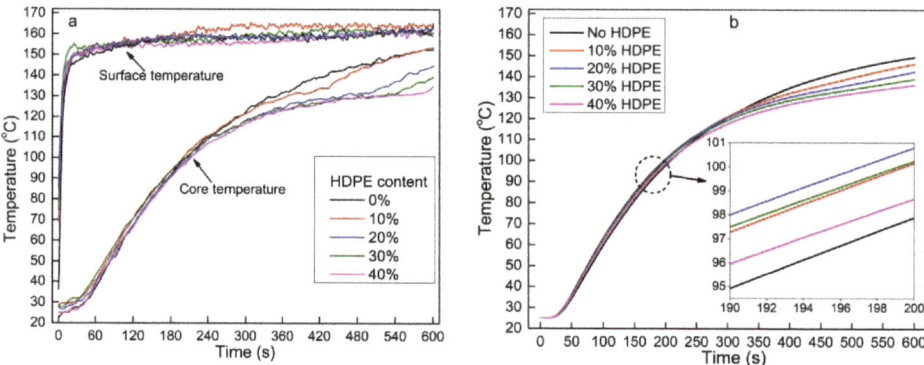

Figure 6. Effects of HDPE content on the mat core temperature during hot-pressing: (**a**) measured data and (**b**) modeling data (the sorghum fiber moisture content was 3% and mat target density was 0.9 g/cm^3).

During the initial stage of hot-pressing (below 100 °C), the core temperature of the mat with HDPE was higher than the mat without HDPE (Figure 6b), as the HDPE had a higher thermal conductivity (0.44 W/(m.K) at room temperature) and a higher thermal conduction efficiency than sorghum fiber (0.12–0.2 W/(m.K)). According to Equation (8), the thermal conductivity of the OFPC mat non-linearly increases with the HDPE content, and a higher thermal conductivity helps the thermal conduction from the surface to the core. In addition, the specific heat capacity of HDPE (2044.1 W/(m.K) at 20 °C) was much higher than that of the sorghum fiber (1213.9 W/(m.K) at 20 °C) [18]. Therefore, the mat with a higher HDPE content required more time to reach the same temperature at the same energy input. Owing to the combined effects of higher thermal conductivity and specific heat capacity of the mat at a higher HDPE content, the core temperature of the OFPC mat first increased with the HDPE content (0 to 20% HDPE), and then decreased at higher HDPE contents (30% and 40% HDPE) (Figure 6b). This trend was not obvious in the experimental results (Figure 6a), possibly owing to temperature fluctuations of the hot-press platens and an experimental error.

3.6. Temperature Distribution Prediction

Figure 7 shows the temperature prediction results of the OFPC mat during hot-pressing without HDPE (Figure 7a) and with 40% HDPE (Figure 7b). The temperature increased from the core to the surface and increased with time whether or not the mat contained HDPE. The surface temperature remained at 160 °C as the Dirichlet boundary condition was applied. After 10 min of hot-pressing, the core temperature of the mat without HDPE was 149.4 °C (Figure 7a), and it was 136.3 °C for that with 40% HDPE (Figure 7b). This 13.1 °C difference indicates that it is necessary to increase the temperature during the consolidation process and/or extend the hot-pressing duration to further increase the core temperature in the OFPC containing 40% HDPE, as a core temperature of 136.3 °C is not sufficient for HDPE to flow easily.

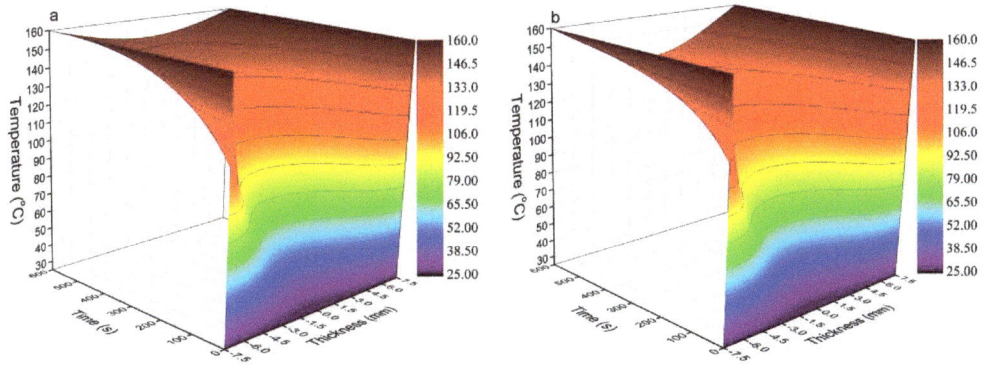

Figure 7. Heat transfer prediction results during the hot-pressing of sweet sorghum composites (**a**) without HDPE and (**b**) with 40% HDPE (the mat target density was 0.9 g/cm^3).

3.7. Comparison of Experimental Results with the Model Prediction

Figure 8 compares the measured temperatures at various locations within the OFPC mats with varying HDPE content during hot-pressing with those predicted by the model. In general, the predicted temperature at both the core and one-quarter positions was similar to that of the experimental values. The predicted temperature at the one-quarter thickness was slightly lower than that of the measured temperatures beyond 100 °C when no HDPE was added to the mat (Figure 8a). This could be because of convection heat transfer as there was no HDPE layer to act as a barrier preventing the transfer of water vapor from the surface to the core. However, in modeling this behavior, the energy equation (Equation (2)) did not consider the heat transfer due to convection resulting from the 3% moisture content of sorghum fiber.

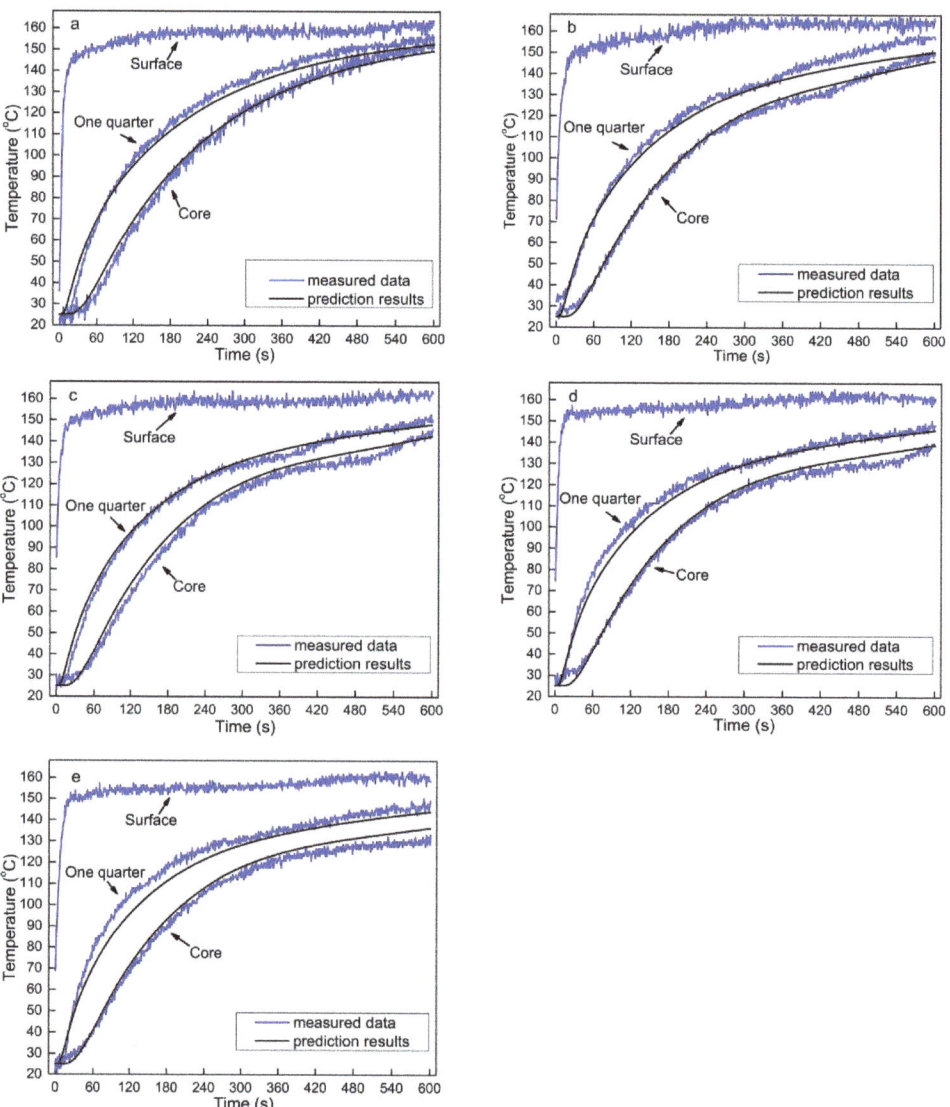

Figure 8. Comparison of heat transfer measured data of sweet sorghum fiber composites with predicted results: (**a**) without HDPE content, (**b**) 10%, (**c**) 20%, (**d**) 30%, and (**e**) 40% HDPE content (the sorghum fiber moisture content was 3% and mat target density was 0.9 g/cm^3).

A temperature lag was observed in the measured data between 120 and 140 °C when HDPE was incorporated into the OFPC (Figure 8b–e), but this phenomenon was not obvious in the model perdition, mainly because the HDPE in the OFPC was not evenly distributed and was present in layers. Additionally, controlling the temperature of the hot platens, thermocouple position, experimental test error, and model hypothesis also affected the difference between the measured data and model prediction data.

3.8. Optimization of Hot-Press Parameters

The most important function of mathematic modeling is to guide the manufacture of composite panels. As the onset and ending melting temperature of HDPE are 121.2 °C and 151.3 °C, respectively, the core temperature of the OFPC must reach at least 151.3 °C. The above analysis showed that a hot-press temperature of 160 °C and a duration of 10 min were not sufficient to attain the required core temperature to hot press a panel with 40% HDPE as the maximum core temperature was 136.3 °C. Figure 9 shows the predicted core temperature of OFPC containing 40% HDPE hot-pressed at 160 °C, 170 °C, 180 °C, 190 °C, and 200 °C for 20 min. A higher hot-pressing temperature and longer duration resulted in a higher core temperature. As a higher hot-pressing temperature consumes more energy and a lower hot-press duration results in higher productivity, the hot-pressing parameters should balance these two factors. Table 3 provides the predicted hot-pressing durations of OFPC at various HDPE contents when the core temperature reached 151.3 °C. A press temperature of 180 °C and a duration of 443 s were recommended for OFPC with 10% HDPE, and 180 °C and 574 s for OFPC with 40% HDPE.

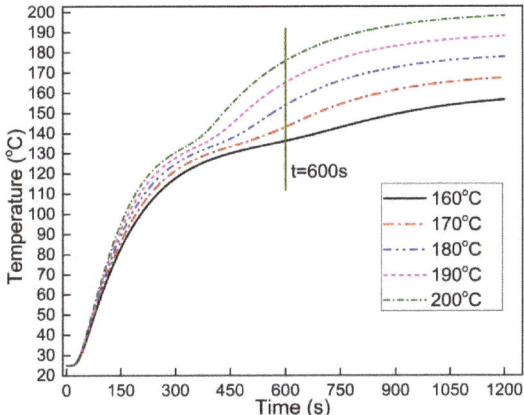

Figure 9. Core temperature simulation results of OFPC with 40% HDPE under different hot-press temperatures (the mat target density was 0.9 g/cm^3).

Table 3. The predicted hot-press duration of OFPC with varied HDPE content when the core temperature reached 151.3 °C.

Hot-Press Temperature (°C)	Hot-Pressing Duration (s)				
	No HDPE	10% HDPE	20% HDPE	30% HDPE	40% HDPE
160	645	710	781	861	954
170	490	533	582	637	702
180	410	443	480	523	574
190	360	385	415	451	494
200	323	344	370	400	437

4. Conclusions

A one-dimensional heat transfer model of natural fiber reinforced thermoplastic composites during hot-pressing was established. The novelty of this study is that the apparent heat capacity of thermoplastics was first simulated and then coupled with the heat transfer model to simulate the temperature distribution of natural fiber reinforced thermoplastics composites during hot-pressing. The heat transfer results predicted by the mathematical model were consistent with the experimental data. The moisture content of sorghum fiber had little effect on the heat transfer of OFPC. Both the experimental and model results showed that the mat core temperature slightly increased with the mat density. The addition of HDPE retarded the temperature increase after reaching its melting temperature, and a greater HDPE content resulted in a lower core temperature. A higher press temperature or a longer press time was required to fabricate an OFPC with a higher HDPE content. The thermal conductivity efficiency was higher at higher HDPE contents. A predictive tool was developed to assist manufacturers of OFPC panels in setting the processing parameters, namely press temperature and duration, depending on the HDPE content. The limitation of the heat transfer model in this study is that it does not consider the heat flow of molten thermoplastics.

Author Contributions: Conceptualization, methodology, validation, and writing—original draft preparation, C.Q.; investigation, J.W.; supervision and writing—review and editing, V.Y. All authors have read and agreed to the published version of the manuscript.

Funding: This research was funded by the Fundamental Research Funds for the Central Universities (No. 2021ZY30) and the National Natural Science Foundation of China (No. 31870536).

Institutional Review Board Statement: Not applicable.

Informed Consent Statement: Not applicable.

Data Availability Statement: The data presented in this research are available on request from the corresponding author.

Acknowledgments: The authors are grateful to the financial support of the Fundamental Research Funds for the Central Universities (NO. 2021ZY30) and the National Natural Science Foundation of China (No.31870536).

Conflicts of Interest: The authors declare no conflict of interest.

References

1. Friedrich, D. Thermoplastic Moulding of Wood-Polymer Composites (WPC): A Review on Physical and Mechanical Behaviour under Hot-Pressing Technique. *Compos. Struct.* **2021**, *262*, 113649. [CrossRef]
2. Qi, C.; Yadama, V.; Guo, K.; Wolcott, M.P. Preparation and Properties of Oriented Sorghum-Thermoplastic Composites Using Flat Hot-Pressing Technology. *J. Reinf. Plast. Compos.* **2015**, *34*, 1241–1252. [CrossRef]
3. Qi, C.; Yadama, V.; Guo, K.; Wolcott, M.P. Thermal Stability Evaluation of Sweet Sorghum Fiber and Degradation Simulation during Hot Pressing of Sweet Sorghum-Thermoplastic Composite Panels. *Ind. Crop. Prod.* **2015**, *69*, 335–343. [CrossRef]
4. Qi, C.; Yadama, V.; Guo, K.; Wolcott, M.P. Thermal Conductivity of Sorghum and Sorghum-Thermoplastic Composite Panels. *Ind. Crop. Prod.* **2013**, *45*, 455–460. [CrossRef]
5. Stanish, M.A.; Schajer, G.S.; Kayihan, F. A Mathematical Model of Drying for Hygroscopic Porous Media. *AIChE J.* **1986**, *32*, 1301–1311. [CrossRef]
6. Zombori, B.G.; Kamke, F.A.; Watson, L.T. Simulation of the Internal Conditions during the Hot-Pressing Process. *Wood Fiber Sci.* **2003**, *35*, 2–23.
7. Dai, C.; Yu, C. Heat and Mass Transfer in Wood Composite Panels during Hot-Pressing: Part I. A Physical-Mathematical Model. *Wood Fiber Sci.* **2004**, *36*, 585–597.
8. Erchiqui, F.; Kaddami, H.; Dituba-Ngoma, G.; Slaoui-Hasnaoui, F. Comparative Study of the Use of Infrared and Microwave Heating Modes for the Thermoforming of Wood-Plastic Composite Sheets. *Int. J. Heat Mass Transf.* **2020**, *158*, 119996. [CrossRef]
9. Kot, V.A. Solution of the Classical Stefan Problem: Neumann Condition. *J. Eng. Phys. Thermophys.* **2017**, *90*, 889–917. [CrossRef]
10. Khan, S.A.; Girard, P.; Bhuiyan, N.; Thomson, V. Improved Mathematical Modeling for the Sheet Reheat Phase during Thermoforming. *Polym. Eng. Sci.* **2012**, *52*, 625–636. [CrossRef]
11. Nedjar, B. An Enthalpy-Based Finite Element Method for Nonlinear Heat Problems Involving Phase Change. *Comput. Struct.* **2002**, *80*, 9–21. [CrossRef]
12. Mantell, S.C.; Springer, G.S. Manufacturing Process Models for Thermoplastic Composites. *J. Compos. Mater.* **1992**, *26*, 2348–2377. [CrossRef]
13. Xiong, H.; Hamila, N.; Boisse, P. Consolidation Modeling during Thermoforming of Thermoplastic Composite Prepregs. *Materials* **2019**, *12*, 2853. [CrossRef] [PubMed]
14. Woo, M.W.; Wong, P.; Tang, Y.; Triacca, V.; Gloor, P.E.; Hrymak, A.N.; Hamielec, A.E. Melting Behavior and Thermal Properties of High Density Polythylene. *Polym. Eng. Sci.* **1995**, *35*, 151–156. [CrossRef]
15. Thoemen, H.; Humphrey, P.E. Modeling the Continuous Pressing Process for Wood-Based Composites. *Wood Fiber Sci.* **2003**, *35*, 456–468.

16. ASTM E1269-11. *Standard Test Method for Determining Specific Heat Capacity by Differential Scanning Calorimetry*; ASTM International: West Conshohocken, PA, USA, 2018.
17. Ivanovic, M.; Svicevic, M.; Savovic, S. Numerical Solution of Stefan Problem with Variable Space Grid Method Based on Mixed Finite Element/Finite Difference Approach. *Int. J. Numer. Methods Heat Fluid Flow* **2017**, *27*, 2682–2695. [CrossRef]
18. Qi, C. Fabrication of Oriented Biomass-High Density Polyethylene Composites Using Hot Pressing Process and Its Molding Mechanism. Ph.D. Thesis, Northwest A & F University, Yangling, China, 2013.
19. Sotomayor, M.E.; Krupa, I.; Várez, A.; Levenfeld, B. Thermal and Mechanical Characterization of Injection Moulded High Density Polyethylene/Paraffin Wax Blends as Phase Change Materials. *Renew. Energy* **2014**, *68*, 140–145. [CrossRef]
20. Oh, D.W. Thermal Characterisation of High Density Polyethylene with Multi-Walled Carbon Nanotube. *Int. J. Nanotechnol.* **2018**, *15*, 747–752. [CrossRef]
21. Li, X.; Tabil, L.G.; Oguocha, I.N.; Panigrahi, S. Thermal Diffusivity, Thermal Conductivity, and Specific Heat of Flax Fiber-HDPE Biocomposites at Processing Temperatures. *Compos. Sci. Technol.* **2008**, *68*, 1753–1758. [CrossRef]

Article

Surfactant-Induced Reconfiguration of Urea-Formaldehyde Resins Enables Improved Surface Properties and Gluability of Bamboo

Lulu Liang, Yu Zheng, Yitian Wu, Jin Yang, Jiajie Wang, Yingjie Tao, Lanze Li, Chaoliang Ma, Yajun Pang *, Hao Chen, Hongwei Yu * and Zhehong Shen *

College of Chemistry and Materials Engineering, Zhejiang Provincial Collaborative Innovation Center for Bamboo Resources and High-Efficiency Utilization, National Engineering and Technology Research Center of Wood-Based Resources Comprehensive Utilization, Key Laboratory of Wood Science and Technology of Zhejiang Province, Zhejiang A&F University, Hangzhou 311300, China; m13070173339@163.com (L.L.); z541962382@icloud.com (Y.Z.); Y1tianWu@163.com (Y.W.); Yangjin_0127@163.com (J.Y.); laowangwjj0126@163.com (J.W.); taoyingjie123@126.com (Y.T.); lilanze112@163.com (L.L.); qq843543030@163.com (C.M.); haochen10@fudan.edu.cn (H.C.)
* Correspondence: yjpang@zafu.edu.cn (Y.P.); yhw416@sina.com (H.Y.); zhehongshen@zafu.edu.cn (Z.S.); Tel.: +86-0571-6374-1609 (Y.P.)

Citation: Liang, L.; Zheng, Y.; Wu, Y.; Yang, J.; Wang, J.; Tao, Y.; Li, L.; Ma, C.; Pang, Y.; Chen, H.; et al. Surfactant-Induced Reconfiguration of Urea-Formaldehyde Resins Enables Improved Surface Properties and Gluability of Bamboo. *Polymers* **2021**, *13*, 3542. https://doi.org/10.3390/polym13203542

Academic Editors: Jingpeng Li, Yun Lu and Huiqing Wang

Received: 1 October 2021
Accepted: 11 October 2021
Published: 14 October 2021

Publisher's Note: MDPI stays neutral with regard to jurisdictional claims in published maps and institutional affiliations.

Copyright: © 2021 by the authors. Licensee MDPI, Basel, Switzerland. This article is an open access article distributed under the terms and conditions of the Creative Commons Attribution (CC BY) license (https:// creativecommons.org/licenses/by/ 4.0/).

Abstract: The high-efficiency development and utilization of bamboo resources can greatly alleviate the current shortage of wood and promote the neutralization of CO_2. However, the wide application of bamboo-derived products is largely limited by their unideal surface properties with adhesive as well as poor gluability. Herein, a facile strategy using the surfactant-induced reconfiguration of urea-formaldehyde (UF) resins was proposed to enhance the interface with bamboo and significantly improve its gluability. Specifically, through the coupling of a variety of surfactants, the viscosity and surface tension of the UF resins were properly regulated. Therefore, the resultant surfactant reconfigured UF resin showed much-improved wettability and spreading performance to the surface of both bamboo green and bamboo yellow. Specifically, the contact angle (CA) values of the bamboo green and bamboo yellow decreased from 79.6° to 30.5° and from 57.5° to 28.2°, respectively, with the corresponding resin spreading area increasing from 0.2 mm^2 to 7.6 mm^2 and from 0.1 mm^2 to 5.6 mm^2. Moreover, our reconfigured UF resin can reduce the amount of glue spread applied to bond the laminated commercial bamboo veneer products to 60 g m^{-2}, while the products prepared by the initial UF resin are unable to meet the requirements of the test standard, suggesting that this facile method is an effective way to decrease the application of petroleum-based resins and production costs. More broadly, this surfactant reconfigured strategy can also be performed to regulate the wettability between UF resin and other materials (such as polypropylene board and tinplate), expanding the application fields of UF resin.

Keywords: surfactant; urea-formaldehyde resin; bamboo; laminated bamboo veneer; gluability

1. Introduction

As a natural organic material as well as a sustainable resource, wood has been widely relied on and developed for a long period all over the world due to its facile processing, light weight, good elasticity, impact resistance, low density, rich mesoporous structure, and other characteristics [1,2]. However, the supply of available wood is far from meeting the world's needs. Therefore, we urgently need to explore effective alternative materials [3]. Bamboo is a natural biomass composite with not only renewable, biodegradable, and carbon sequestration but also an impressive faster growth rate than wood, a high annual regeneration rate after harvesting, and excellent mechanical properties [4,5]. It is thus considered a promising ingredient substitute for wood.

Bamboo is hollow inside, and tubular parts are formed between the nodes, which can effectively resist the bending force [6]. However, the connection and bonding of circular cross-sections are difficult, and the pipes cannot be used in applications requiring flat surfaces [5]. In addition, the thickness of the stalk wall gradually becomes thinner from the base to the top of the stalk, and the bamboo fiber content is not uniformly distributed with the change of the wall thickness. These changes in geometric and mechanical properties also limit the structure of whole bamboo [7]. At present, in order to avoid the defects caused by the natural shape of bamboo, there have been developments of bamboo-based composite materials using industrialized production processes, including, for example, bamboo plywood, laminated bamboo lumber, and bamboo scrimber. They are widely used in furniture, flooring, construction, civil engineering, and other fields [8–10]. However, due to the absence of transverse ray cells in bamboo tissues and the unobvious flow, wetting, and penetration depth of the liquid adhesive on the surface of the bamboo, the crosslinking of the adhesive and the bamboo material is unable to achieve the mechanical coupling effect with the wood [11–13]. In addition, bamboo green and bamboo yellow contain a certain amount of hydrophobic substances, including wax, fat, SiO_2, bamboo film, and others, which will negatively affect the wettability and adhesion of bamboo [14]. As a result, the current bamboo production generally removes bamboo green and bamboo yellow, which leads to the waste of bamboo raw materials and a low utilization rate [15]. Therefore, improving the wettability and gluability properties of the adhesives in bamboo materials is of great significance to reduce the production costs of bamboo products and accelerate the development of the bamboo industry.

However, most of the research on wood and bamboo is focused on the surface engineering of the material, such as the introduction of coating and reasonable design [16,17]. Moreover, the research on enhancing the gluability properties of bamboo mostly focuses on the performance of the bamboo materials itself and rarely involves the regulation of the properties of the adhesives [18–23]. Currently, the main commonly used adhesives include amino, phenolic, and isocyanate, among which 95% of the adhesives are formaldehyde-based, with urea-formaldehyde (UF) as the most predominant [24]. Specifically, although there are some shortcomings, such as easy aging, poor water resistance, poor wettability, and the release of volatile organic compounds (VOCs) [25–27], the UF resin is a thermosetting resin formed by the condensation of urea and formaldehyde [28] that has been used in the manufacture of particleboard, medium-density fiberboard, and indoor plywood due to the characteristics of cheap and easy-to-obtain raw materials, a simple synthesis process, low curing temperature, short pressing time, wide curing conditions, colorless glue line, and good panel performance [29–32]. Therefore, in this present study, taking urea-formaldehyde (UF) resin as an example, the significant role of regulating the adhesive on the wettability and bonding performance is systematically studied. In a detailed manner, the surfactants that can reduce surface tension, increase solubility, and have the ability to improve wettability [33] are applied as modifiers to reconfigure UF resin, including three typical different types of surfactants (cetyltrimethylammonium bromide (CTAB) as an anionic surfactant, sodium dodecyl sulfate (SDS) as a cationic surfactant, and Span 80 as non-ionic surfactant). Impressively, the resultant surfactant reconfigured UF resin with regulated viscosity and surface tension can significantly improve its wettability and spreading performance to the surface of both bamboo green and bamboo yellow. When the optimized reconfigured UF resin is further applied for commercial bamboo veneer, the thus-formed laminated bamboo veneer product can also exhibit outstanding wettability, and the average peeling length is largely reduced during the dipping peel test. Moreover, our reconfigured UF resin can reduce the amount of glue spread to 60 g m^{-2}, while the products prepared by the initial UF resin are unable to meet the requirements of the test standard, suggesting that this surfactant-induced reconfiguration provides an effective way to decrease the application of petroleum-based resins and reduce production costs. In addition, this surfactant reconfigured strategy has also shown positive effects in improving the wettability between UF resin and other materials, such as polypropylene (PP) board and tinplate, supplementing a reference for expanding the application fields of UF resin.

2. Materials and Methods

2.1. Materials

Formaldehyde aqueous solution (HCHO, 37%) was obtained from Shanghai Zhanyun Chemical Co., Ltd., (Shanghai, China). Urea (CH_4N_2O, ≥99.0%) was obtained from Shanghai Lingfeng Chemical Reagent Co., Ltd., (Shanghai, China). Formic acid (HCOOH, 98%) was purchased from Guangdong Guanghua Sci-Tech Co., Ltd., (Guangdong, China). Sodium hydroxide (NaOH, 95%), cetyltrimethylammonium bromide (CTBA, AR), sodium dodecyl sulfate (SDS, AR), and Span 80 (SP) were purchased from Aladdin Industrial Corporation (Shanghai, China). All chemicals were used without further purification. Polypropylene (PP) boards were purchased from Guangzhou Chuangxin Rubber and Plastic Products Co., Ltd. (Guangzhou, China). Tinplates were purchased from Shenzhen Xinrong Stainless Steel Material Co., Ltd. (Shenzhen, China). Bamboo splits and bamboo veneers (produced by sawing the laminated bamboo lumbers into thin sheets) were kindly donated by Zhejiang Zhuangyi Furniture Co., Ltd. (Hangzhou, China). The molecular structures of all surfactants employed are listed in Figure 1.

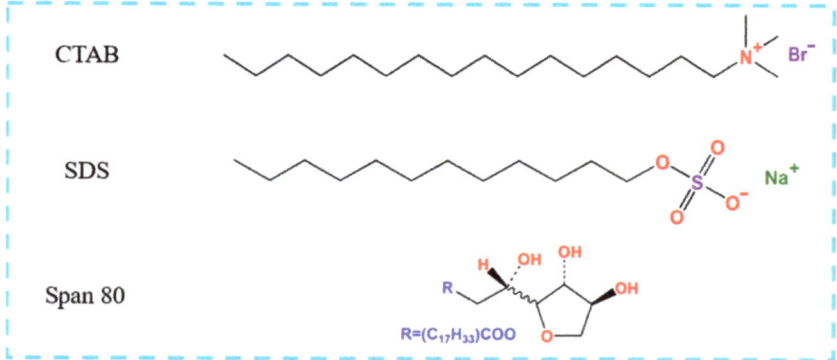

Figure 1. Molecular structure of surfactants.

2.2. Synthesis of Urea-Formaldehyde (UF) Resin and Reconfigured UF Resins

Urea-formaldehyde (UF) resin was prepared through the traditional "alkali-acid-alkali" three-step process. In detail, 264.6 g of formaldehyde solution was stirred in a four-neck reactor, and then the pH was adjusted to 8.0 with a 30% NaOH aqueous solution. Next, 95.14 g of urea was added into the formaldehyde solution for methylation reaction. The mixture was further heated to 90 °C and maintained at the temperature for 45 min. Next, under acidic conditions (pH = 5–5.5, adjusted with 30% HCOOH), the polycondensation reaction was carried out, and 18.98 g of urea was added until the reactant reached the target viscosity. When the temperature of the solution cooled to 70 °C, 6 g of urea was added to the reactor again. When the temperature of the solution in the reactor continued to cool to 40 °C, the pH of the solution was adjusted to alkaline conditions again (pH = 8.0, adjusted with 30% NaOH. As a result, the UF resin was thus obtained, which can be used after cooling to room temperature. The surfactant reconfigured UF resin was constructed by mixing UF resin with different amounts of CTAB at room temperature, including, for example, 0.5%, 1%, 1.5%, 2%, 2.5%, and 3%, which were based on the theoretical solid content of UF resin [34]. The resulting reconfigured UF resin was labeled x C-UF (x stands for CTAB content, C stands for CTAB). Similarly, S-UF resin and Sp-UF resin were SDS and Span 80 reconfigured UF resins, respectively.

2.3. Preparation of the Basal Substrates

The bamboo green and bamboo yellow used for contact angle measurement with the size of 50 mm × 3 mm × 10 mm (longitudinal × radial × tangential) were obtained from the bamboo splits without removing the bamboo green and bamboo yellow, respectively. The size of the bamboo veneer used for the dipping peel test was 75 mm × 75 mm × 3 mm. In addition, PP board, tinplate, and bamboo veneer were cut into a size of 40 mm × 40 mm × 10 mm for contact angle measurement.

2.4. Viscosity and Surface Tension Test

The viscosity of the resin was measured by NDJ-1 type rotational viscometer (Shanghai Lichen Bangxi Instrument Technology Co., Ltd., Shanghai, China) at a temperature of 25 ± 2 °C. The surface tension of the resin was tested with a contact angle measuring instrument (JC2000D1, Shanghai Zhongchen Digital Technology Instrument Co., Ltd., Shanghai, China) using the pendant drop method. Each sample was measured at 6 points, and the results were averaged.

2.5. Spreading Performance Test

A drop of different resin was dropped on the surface of the corresponding substrate, and then the recording of the contact angle (CA) images and optical photos of the resin and the substrate surface took place for 1 s and 60 s. After that, ImageJ software was used to measure the exact area of the resin before and after spreading on the substrate surface. By comparing the area value, the spreading performance of different resins on the surface of the corresponding substrate was studied. Each test was repeated 3 times.

2.6. Laminated Bamboo Veneers and Its Dipping Peel Test

Two bamboo veneers (75 mm × 75 mm × 3 mm) were bonded and formed laminated bamboo veneer using UF or 2.5% Sp-UF resin. The resin was manually applied to the surface of a bamboo veneer using the glue spread of 80 g m^{-2} and 60 g m^{-2}, respectively. Subsequently, the laminated bamboo veneers were hot-pressed for 10 min at 110 °C and with 3.8 MPa of pressing pressure. All the panels were stored under an ambient condition for 24 h prior to testing. The dipping peel analysis follows the Chinese National Standard (GB/T 20240-2017) in which the specimens were submerged in 63 ± 3 °C water bath for 3 h and then dried at 63 ± 3 °C for 10 h. Finally, the specimens were taken out to observe the degree of peeling of the glue lines. Each test has 6 duplicate test pieces.

2.7. Characterization

The CA of resin at substrate was analyzed at room temperature using a contact angle meter (JC2000D1, Shanghai Zhongchen Digital Technic Apparatus Co., Ltd., Shanghai, China), where the specific CA result was an average value recorded from tests performed at six locations on the same sample surface. The morphology was observed by scanning electron microscopy (SEM, TM3030, Hitachi, Tokyo, Japan). The surface composition of the product was identified by X-ray photoelectron spectroscopy (XPS, ESCALAB 250XI, Thermo Fisher Scientific, Waltham, America) in which the binding energy was calibrated using the C 1s reference peak at 284.8 eV. An optical microscopy image of the bamboo veneer surface was taken using microscopy (THMS600, Nikon, Tokyo, Japan).

3. Results

3.1. The Effect of the Surfactants on the Viscosity and Surface Tension of the UF Resins

Prior to applying the UF and reconfigured UF to the surface of the bamboo green and bamboo yellow, the viscosity and surface tension, these two significant characteristics of resin, were systematically investigated. It has been well-demonstrated that the viscosity of the resin will significantly affect its fluidity and permeability on the substrate [35]. As observed in Figure 2a, when CTAB and SDS are used to reconfigure the UF, the changes of their viscosity present a similar phenomenon; that is, the viscosity increases initially with the increase in the amount of surfactant and then reaches the maximum value, and, as the amount continues to increase, the viscosity will decrease instead. This is due to the fact that, for reconstructed UF resins containing such CTAB or SDS ionic-based surfactants, the increase in the ionic strength of the solution will increase the viscosity of the resin correspondingly, which will reach the maximum value when the micelles are formed in the solution [36]. However, when the proportion of surfactants continues to increase, there will be a large number of micelles in the solution, which repels the surfactant in the interfacial film, which will affect the accumulation mode and regularity of the surfactants and eventually lead to a corresponding weakening of the inhibitory force of particle movement, thus resulting in the decrease of viscosity [37]. To differentiate an ionic surfactant with a non-ionic surfactant, Span 80 has no electrostatic interaction with polymers. Therefore, the viscosity of the UF and Sp-UF shows negligible changes [38].

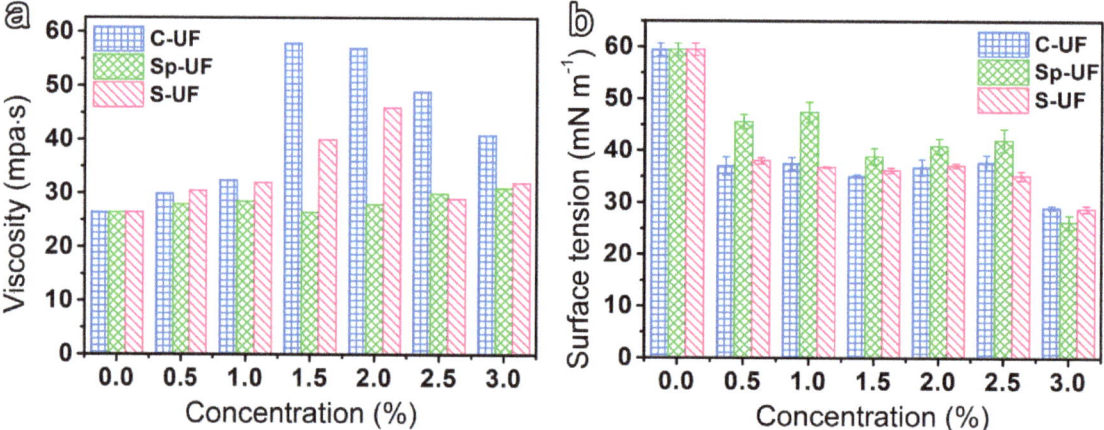

Figure 2. (a) Viscosity, and (b) surface tension of the UF and reconfigured UF resins.

Surface tension is an important factor in evaluating the wettability potential of resin on the substrate [39]. In general, the smaller the surface tension of the solution, the better its wetting performance. Figure 2b shows the surface tension of the different UF resins. Typically, the changing trend of the surface tension of the surfactants in the solution has three forms: (1) For most inorganic electrolytes, the surface tension of the solution increases slowly with the increase of the solute concentration. (2) The surface tension of the solution slowly decreases with the increase of the solute concentration, such as some low-molecular-weight polar organics with weak hydrophilicity. (3) For amphiphilic organic compounds containing more than eight carbon atoms, the surface tension of the solution can be significantly reduced at low concentrations, and, when the concentration increases to a certain value, the surface tension of the solution does not decrease or decreases slowly [40]. Thus, based on the molecular structure of the three surfactants, Span 80 belongs to the substance described by the second type, while CTAB and SDS belong to the substance described by the third type. As expected, it can be observed from Figure 2b that all the types conform to the corresponding rules.

3.2. The Wettability of UF and Reconfigured UF Resins on the Bamboo Green or Bamboo Yellow Surface

Considering the regulated viscosity and surface tension of the resin after the surfactant reconstruction, we further recorded the CA value of the resin on the surface of the bamboo green and bamboo yellow, respectively, to directly compare the wettability of the various UF resins on the bamboo substrate. The CA value of the UF resin on the bamboo green and bamboo yellow without any surfactant is as high as ~79.6° and 57.5°, while, impressively, all three of the reconfigured UF resins exhibit much lower CA values than that of the UF resin (Figure 3). This implies that a much better wetting behavior between the resin and the bamboo surface could be realized with the surfactant-induced reconfiguration. Among them, it is evident that the surface wettability of bamboo yellow is better than that of bamboo green, which can be attributed to more polar groups contained on the surface of bamboo yellow [41]. Furthermore, considering the wettability of various resins on the surface of the bamboo green and bamboo yellow, the 2.5% Sp-UF (CA value of 30.5°) and 2.5% C-UF (CA value of 28.2°) is the most suitable for coating on the surface of the bamboo green and bamboo yellow, respectively.

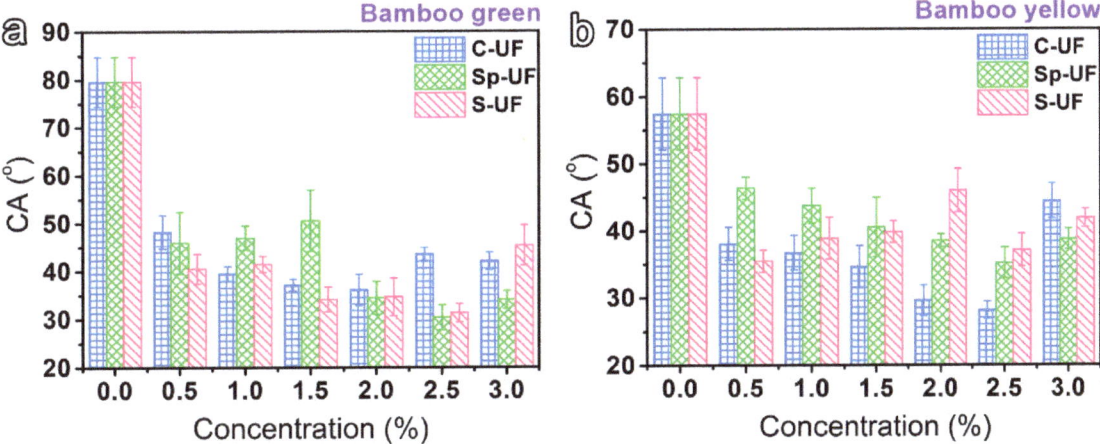

Figure 3. The CA curves of the UF and reconfigured UF resins on the surface of (**a**) bamboo green and (**b**) bamboo yellow.

3.3. Spreading Performance of the UF and Reconfigured UF Resins

As another factor affecting the adhesion performance, the diffusion ability of the UF, 2.5% Sp-UF, and 2.5% C-UF on the substrate was evaluated according to the results of the CA measurement. As shown in Figure 4, the recorded area changes over time when the UF and reconfigured UF resins are in contact with the same substrate. In detail, by comparing the optical photos and CA images of the UF and 2.5% Sp-UF resins dipped on the one bamboo green for 1 s, respectively, the area of a drop of 2.5% Sp-UF resin through software measurement can reach 31.5 mm^2, while the area of a drop of the UF resin was only 20.1 mm^2. More impressively, within 60 s, the spreading area of 2.5% Sp-UF resin on the bamboo green surface increased by 7.6 mm^2, while the spreading area of the UF resin exhibited no observable change. Similarly, the spreading area of 2.5% C-UF resin on the bamboo yellow increased from 32.1 to 37.7 mm^2 after waiting for 60 s; conversely, the spreading area of UF resin on bamboo yellow did not differ much. These results verify that the reconfigured UF resins possess a faster diffusion rate and stronger spreading ability on the bamboo substrates, which is consistent with the results in Figure 3. Therefore, the addition of surfactants can greatly enhance the spreading ability of UF resin on both bamboo green and bamboo yellow.

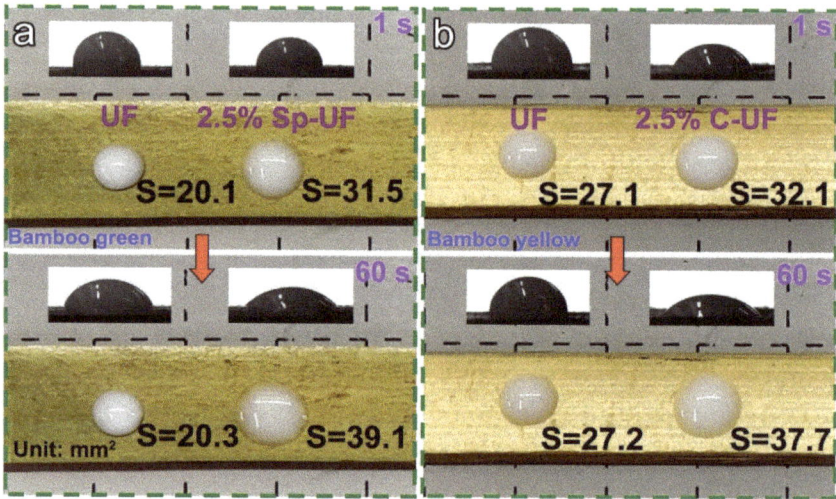

Figure 4. Optical photos and corresponding CA images of UF and reconfigured UF resins in contact with the surface of (**a**) bamboo green and (**b**) bamboo yellow for 1 s and 60 s.

3.4. Morphology and Compositions of UF and Reconfigured UF Resins

Considering the excellent wettability and spreading performance of our reconfigured UF resins, we further uniformly coated the corresponding three resins on the surface of the bamboo green and bamboo yellow, respectively, as illustrated in the schematic diagram of the production process (Figure 5a). The morphology of the UF and corresponding reconfigured UF resin cured on the surface of the bamboo green and bamboo yellow was first characterized by a scanning electron microscope (SEM), as shown in Figure 5b,c. Regarding the bamboo green substrate, the cured UF resin is granular on the surface, and the cured 2.5% Sp-UF resin is evenly distributed on the surface to form a thinner film, indicating that the coupling of Span 80 improves the spreading performance of the UF resin on the bamboo green surface. Figure 5d,e shows that the 2.5% C-UF resin exhibits the same effect for the surface of bamboo yellow. The UF resin solution not only cannot be uniformly distributed on the surface of bamboo yellow but is also difficult to penetrate the inside, resulting in the cured UF resin on the surface granular. However, the reconfigured resin solution can almost completely penetrate the inside of the bamboo substrate instead.

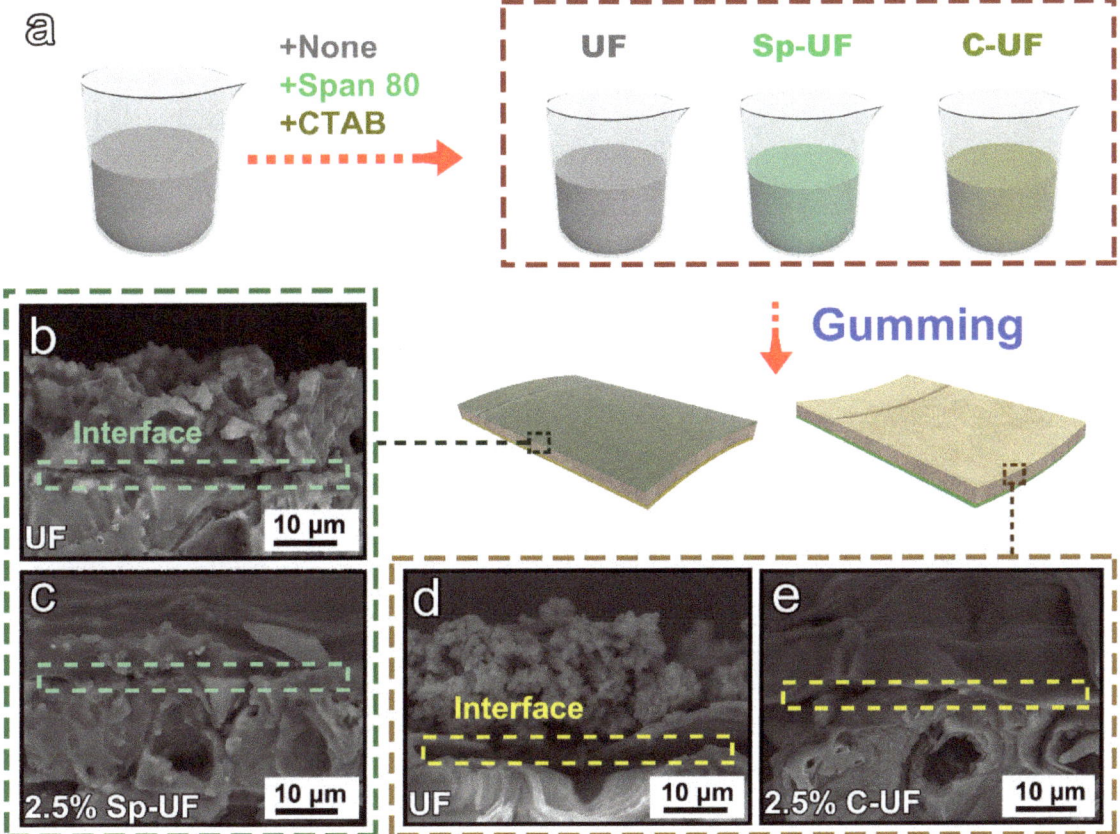

Figure 5. (**a**) Schematic procedure of the fabrication process. SEM images of (**b**) cured UF resin coated on the bamboo green, (**c**) cured 2.5% Sp-UF resin coated on the bamboo green, (**d**) cured UF resin coated on the bamboo yellow, and (**e**) 2.5% C-UF resin coated on the bamboo yellow.

The detailed compositions of the cured UF, 2.5% Sp-UF, and 2.5% C-UF resins were further investigated by X-ray photoelectron spectroscopy (XPS) spectrum. The full XPS survey spectra show the main components of the UF and two reconfigured UF resins are C, N, and O, which is also consistent with their structure (Figure 6a). Notably, unlike the others, an additional Br 3d is detected in the 2.5% C-UF resin due to its special molecular structure (see Figure 1), further suggesting that CTAB has been successfully coupled with the UF resin evenly. In addition, by comparing the C 1s of UF and 2.5% Sp-UF resin shown in Figure 6b, the high-resolution spectrum of the UF resin can be resolved into five peaks at 284.8, 286.7, 287.2, 287.8, and 288.8 eV, corresponding to the C–H, C–N, C–O, C=O, and –N–CO–N– components. In contrast, with the coupling of the Span 80 surfactant, there is a significant increase of the peak at 284.8 and a new peak at 288.6, which can be identified as the signals of the C–H and O–C=O groups of the Span 80, respectively [42–44]. This demonstrates that Span 80 is coupled with the UF resin. Moreover, regarding the high-resolution O 1s spectrum, the presence of the O–C=O group belonging to Sp 80 is also illustrated by the fitting result (Figure 6c) [45,46].

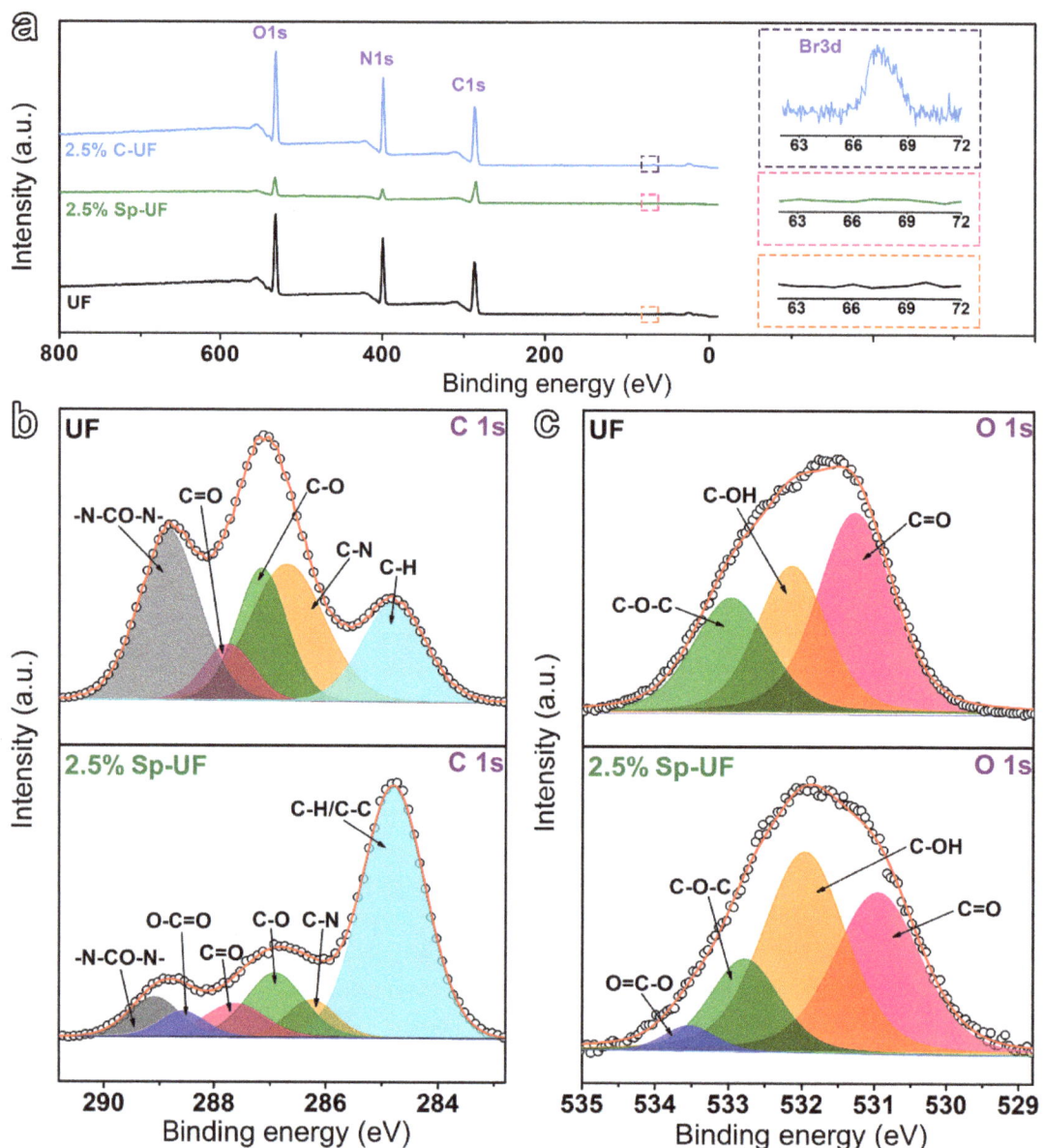

Figure 6. (**a**) XPS full spectra of the cured UF, 2.5% Sp-UF, and 2.5% C-UF resins. High-resolution (**b**) C 1s, and (**c**) O 1s XPS spectra of the UF and 2.5% Sp-UF resins.

The above results prove that the surfactant-coupled method we proposed can successfully realize the reconfiguration of the UF resin, thus regulating the viscosity and surface tension. The reconfigured UF resin overcame the problem of poor wettability and spreading performance on the surface of both the bamboo green and bamboo yellow, thereby obtaining a much-improved contact interface between the resin with the bamboo materials substrate. Therefore, it is believed that the reconfigured resins can be applied to improve the bonding performance between bamboo veneers, which are currently widely used as bamboo-based decoration and construction material in practical applications [47].

3.5. The Applications of Modified UF Resins on Laminated Bamboo Veneers

According to the comprehensive performance of various reconfigured UF resins, the UF and 2.5% Sp-UF resin were, therefore, selected to be applied to the laminated bamboo veneer. The resultant wettability and bonding performance are shown in Figure 7. Obviously, a lower CA value of 2.5% Sp-UF resin coated on the laminated bamboo veneer is realized as expected, implying enhanced wettability (Figure 7a). Figure 7b displays the optical photos of the bamboo veneer and its peeling of the glue line, where the four sides of the bamboo veneer are named L1, L2, L3, and L4, and the length of each side is 7.5 cm. It is worth pointing out that the above parameters are all selected according to the requirements of the test standard. We then performed the dipping peel test after the bamboo veneer was glued; a peeling length was thus obtained when the glue line on one side of the glue-laminated bamboo veneers fell off, and this was the basis for evaluating the adhesion performance of the resin. Specifically, we investigated the dipping peel test of glue-laminated bamboo veneers containing UF and 2.5% Sp-UF resin with glue spreads of 80 g m^{-2} and 60 g m^{-2}, respectively. The peeling length of each sample of six laminated bamboo veneers prepared with the UF or 2.5% Sp-UF resin under the glue spread of 80 g m^{-2} was recorded as presented in Figure 7c,d. Impressively, in comparison with the average peeling length of the UF resin, a peeling length of only ~0.52 cm of 2.5% Sp-UF was achieved, which is much lower than that of the UF resin without reconfiguration (Figure 7e). Moreover, a similar positive effect was also shown when the glue spread was decreased to 60 g m^{-2}, where the average peeling length was reduced from 6.38 cm to 0.9 cm (Figure 7f–h). In addition, it is worth pointing out that, according to the Chinese National Standards (GB/T 20240-2017), the total stripping length of at least five of the six specimens is less than 1/3 of the full length of the glue layer, so it can be determined that the products are in compliance with the standard requirement. Therefore, these results demonstrate our reconfigured UF resin can reduce the amount of glue spread to 60 g m^{-2}, while the products prepared by the initial UF resin are unable to meet the standard. In addition, by comparing the results with the previously reported literature, the amount of adhesive used in this study is much lower than the previous amounts (including 200 g m^{-2}, 220 g m^{-2}, and 250 g m^{-2}) [48–50], together with maintaining excellent gluability performance. These aspects prove that our method provides a way to reduce production costs.

To investigate the failure of the cured UF and 2.5% Sp-UF resins under the glue spread of 60 g m^{-2} in depth, optical microscopy images of the unglued bamboo veneer and glued bamboo veneers were observed (Figure 8). It can be seen that the UF resin is unevenly distributed on the surface of the bamboo veneer, resulting in a lack of adhesive in some places, while the 2.5% Sp-UF resin is evenly distributed and almost completely covers the surface. This phenomenon reveals that the interfacial properties of the resin and the bamboo surface, including the wettability and spreading properties, will greatly affect the distribution of the resin during the gluing process, thereby affecting the gluability performance. These results further prove that the surfactant-induced reconfigured resin can improve its wettability to the bamboo laminate, thus making it exhibit higher adhesion performance than the UF resin and hold great potential in promoting the comprehensive and efficient utilization of bamboo-based resources.

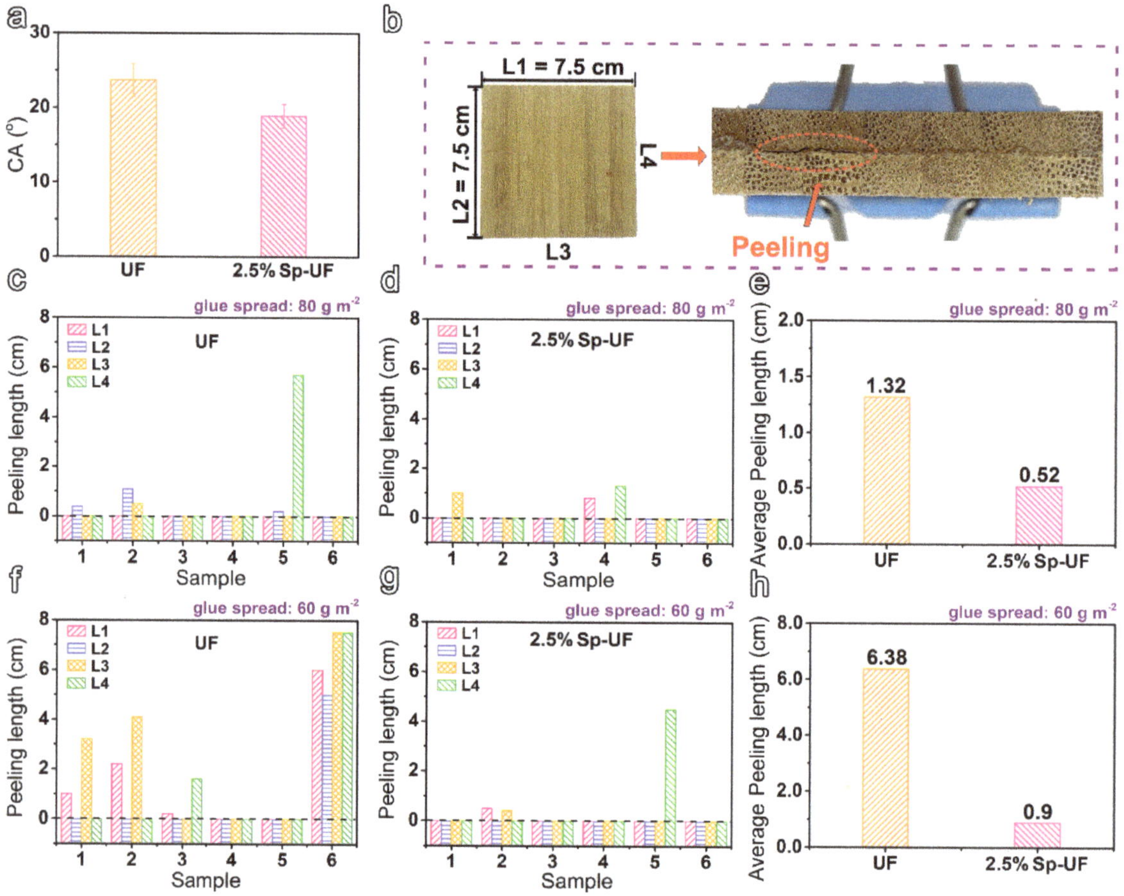

Figure 7. (**a**) The CA values curve of UF and 2.5% Sp-UF resins on the bamboo veneer surface, (**b**) Optical photos of the laminated bamboo veneer and its peeling of the glue line, (**c**,**d**) Peeling length graphs of UF and 2.5% Sp-UF resins and (**e**) the comparison graph of their average peeling length at glue spread of 80 g m^{-2}, (**f**,**g**) Peeling length graphs of UF and 2.5% Sp-UF resins and (**h**) the comparison graph of average peeling length at glue spread of 60 g m^{-2}.

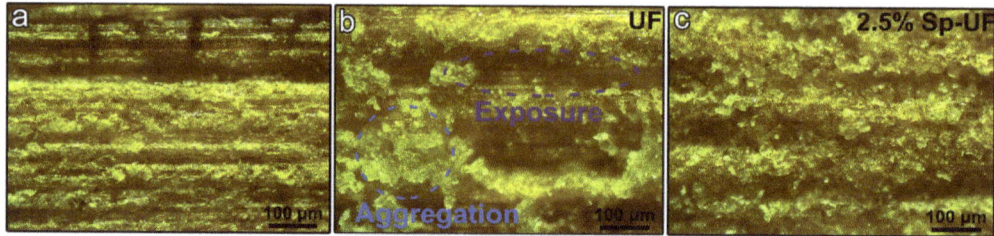

Figure 8. Optical microscopy images of (**a**) unglued bamboo veneer and glued bamboo veneers with cured (**b**) UF resin and (**c**) 2.5% Sp-UF resin at the glue spread of 60 g m^{-2}.

3.6. The Wettability Study of UF and Reconfigured UF Resins on the Other Substrates

In addition to the outstanding performance on the bamboo substrate, we also verified the application of our designed UF resin on other substrates, including PP board and tinplate. The corresponding CA results of various resins show that, although the regulation of wettability on the different substrates is inconsistent, for example, the S-UF possesses the smallest CA value on the PP board, this surfactant-induced reconfiguration can also indeed improve the wettability of UF resin on other substrates (Figure 9a,b). Additionally, we also compared the spreadability of the optimized reconfigured UF resins on the two substrates (Figure 9c,d). The resultant spreading area of 0.5% S-UF resin on the PP board and tinplate increased from 32.9 to 36.8 mm^2 and from 35.9 to 42.4%, respectively, which undoubtedly shows a faster spreading speed and stronger spreading ability than the UF resin. In short, the induction of the surfactants can also improve the wettability and spreadability of the UF resin to PP boards and tinplate, which provides a reference for expanding the application fields of UF resin.

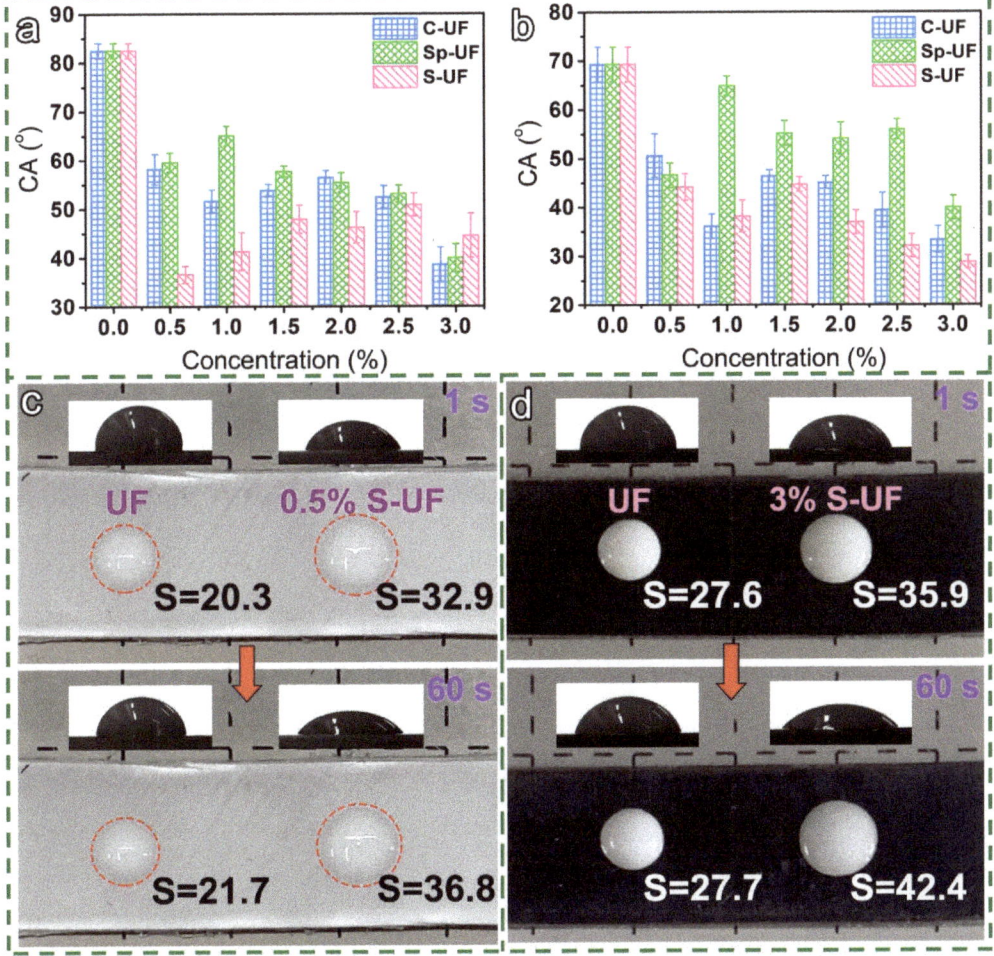

Figure 9. The CA curves of the UF and reconfigured UF resins on the surface of (**a**) PP board, and (**b**) tinplate. Optical photos and corresponding CA images of UF and reconfigured UF resins in contact with the surface of (**c**) PP board and (**d**) tinplate for 1 s and 60 s.

4. Conclusions

In summary, the wettability of urea-formaldehyde resin reconfigured with three surfactants (CTAB, Span 80, and SDS) on the series of bamboo substrates (bamboo green, bamboo yellow, and bamboo veneer) has been greatly improved. Specifically, the CA values of the reconfigured UF resins on the different substrates have been obviously reduced. Among them, the CA value of the optimized Span 80 reconfigured UF resin on the surface of the bamboo green decreased from 79.6° to 30.5°. Meanwhile, after adding the appropriate amount of CTAB, the CA value of the UF resin on the surface of the bamboo yellow decreased from 57.5° to 28.2°. Therefore, according to the comprehensive performance of the reconfigured UF resins, the 2.5% Sp-UF resin is selected to be applied to the laminated bamboo veneer. The final experimental results showed that the laminated bamboo veneers prepared with a lower glue spread (60 g m^{-2}) of 2.5% Sp-UF resin can still meet the requirement determined by the Chinese National Standard (GB/T 20240-2017), which proves that the research in this paper could provide an efficient way to cut down the production costs and reduce the application of the petrochemical-based resin. Furthermore, this facile strategy can also be extended to other materials, such as polypropylene plates and tinplate.

Author Contributions: Conceptualization, Y.P., H.Y. and Z.S.; methodology, L.L. (Lulu Liang), L.L. (Lanze Li), C.M. and Y.T.; software, Y.Z. and Y.W.; validation, L.L. (Lulu Liang) and Y.Z.; formal analysis, L.L. (Lulu Liang), Y.Z. and Y.W.; investigation, L.L. (Lulu Liang); resources, J.W.; data curation, L.L. (Lulu Liang) and J.Y.; writing—original draft preparation, L.L. (Lulu Liang); writing—review and editing, Y.P. and Z.S.; visualization, Y.Z.; supervision, Y.P., H.Y., H.C. and Z.S.; project administration, Z.S., H.C. and Y.P.; funding acquisition, Z.S., H.C., H.Y. and Y.P. All authors have read and agreed to the published version of the manuscript.

Funding: This research was funded by the Zhejiang Provincial Key Research and Development Project, grant number 2019C02037; the Fundamental Research Funds for the Provincial Universities of Zhejiang, grant number 2020YQ005; the Jiangsu North Special Project from Jiangsu Suqian, grant number BN2016176, and the Research Foundation of Talented Scholars of Zhejiang A&F University, grant number 2020FR069.

Institutional Review Board Statement: Not applicable.

Informed Consent Statement: Not applicable.

Data Availability Statement: Not applicable.

Acknowledgments: This work was supported by the Zhejiang Provincial Key Research and Development Project (2019C02037), the Fundamental Research Funds for the Provincial Universities of Zhejiang (2020YQ005), Jiangsu North Special Project from Jiangsu Suqian (BN2016176), and Research Foundation of Talented Scholars of Zhejiang A&F University (2020FR069).

Conflicts of Interest: The authors declare no conflict of interest.

References

1. Gao, L.; Xiao, S.; Gan, W.; Zhan, X.; Li, J. Durable superamphiphobic wood surfaces from Cu_2O film modified with fluorinated alkyl silane. *RSC Adv.* **2015**, *5*, 98203–98208. [CrossRef]
2. Zhu, M.; Song, J.; Li, T.; Gong, A.; Wang, Y.; Dai, J.; Yao, Y.; Luo, W.; Henderson, D.; Hu, L. Highly Anisotropic, Highly Transparent Wood Composites. *Adv. Mater.* **2016**, *28*, 5181–5187. [CrossRef]
3. Chaowana, P. Bamboo: An Alternative Raw Material for Wood and Wood-Based Composites. *J. Mater. Sci. Res.* **2013**, *2*, 90–102. [CrossRef]
4. Wang, Y.-Y.; Wang, X.-Q.; Li, Y.-Q.; Huang, P.; Yang, B.; Hu, N.; Fu, S.-Y. High-Performance Bamboo Steel Derived from Natural Bamboo. *ACS Appl. Mater. Interfaces* **2021**, *13*, 1431–1440. [CrossRef] [PubMed]
5. Huang, Y.; Ji, Y.; Yu, W. Development of bamboo scrimber: A literature review. *J. Wood. Sci.* **2019**, *65*, 25. [CrossRef]
6. Mahdavi, M.; Clouston, P.L.; Arwade, S.R. Development of Laminated Bamboo Lumber: Review of Processing, Performance, and Economical Considerations. *J. Mater. Civil. Eng.* **2011**, *23*, 1036–1042. [CrossRef]
7. Xiao, Y.; Yang, R.Z.; Shan, B. Production, environmental impact and mechanical properties of glubam. *Constr. Build. Mater.* **2013**, *44*, 765–773. [CrossRef]

8. Wu, Y.; Wang, Y.; Yang, F.; Wang, J.; Wang, X. Study on the Properties of Transparent Bamboo Prepared by Epoxy Resin Impregnation. *Polymers* **2020**, *12*, 863. [CrossRef]
9. Li, P.; Zhou, G.; Du, H.; Lu, D.; Mo, L.; Xu, X.; Shi, Y.; Zhou, Y. Current and potential carbon stocks in Moso bamboo forests in China. *J. Environ. Manag.* **2015**, *156*, 89–96. [CrossRef]
10. Meng, F.-D.; Yu, Y.-L.; Zhang, Y.-M.; Yu, W.-J.; Gao, J.-M. Surface chemical composition analysis of heat-treated bamboo. *Appl. Surf. Sci.* **2016**, *371*, 383–390. [CrossRef]
11. Yu, Y.; Zhu, R.; Wu, B.; Hu, Y.A.; Yu, W. Fabrication, material properties, and application of bamboo scrimber. *Wood Sci. Technol.* **2015**, *49*, 83–98. [CrossRef]
12. Trujillo, D.J.; López, L.F. 18-Bamboo material characterisation. In *Nonconventional and Vernacular Construction Materials*, 2nd ed.; Harries, K.A., Sharma, B., Eds.; Woodhead Publishing: Sawston, UK, 2020; pp. 491–520.
13. Guan, M.; Huang, Z.; Zeng, D. Shear Strength and Microscopic Characterization of a Bamboo Bonding Interface with Phenol Formaldehyde Resins Modified with Larch Thanaka and Urea. *BioResources* **2016**, *11*, 492–502. [CrossRef]
14. Deng, J.; Li, H.; Wang, G.; Chen, F.; Zhang, W. Effect of removing extent of bamboo green on physical and mechanical properties of laminated bamboo-bundle veneer lumber (BLVL). *Eur. J. Wood Wood Prod.* **2015**, *73*, 499–506. [CrossRef]
15. Wang, X.; Cheng, K.J. Effect of Glow-Discharge Plasma Treatment on Contact Angle and Micromorphology of Bamboo Green Surface. *Forests* **2020**, *11*, 1293. [CrossRef]
16. Yu, P.; Manalo, A.; Ferdous, W.; Abousnina, R.; Salih, C.; Heyer, T.; Schubel, P. Investigation on the physical, mechanical and microstructural properties of epoxy polymer matrix with crumb rubber and short fibres for composite railway sleepers. *Constr. Build. Mater.* **2021**, *295*, 123700. [CrossRef]
17. Khotbehsara, M.M.; Manalo, A.; Aravinthan, T.; Ferdous, W.; Benmokrane, B.; Nguyen, K.T.Q. Synergistic effects of hygrothermal conditions and solar ultraviolet radiation on the properties of structural particulate-filled epoxy polymer coatings. *Constr. Build. Mater.* **2021**, *277*, 122336. [CrossRef]
18. Rao, J.; Bao, L.; Wang, B.; Fan, M.; Feo, L. Plasma surface modification and bonding enhancement for bamboo composites. *Compos. Part B Eng.* **2018**, *138*, 157–167. [CrossRef]
19. Wu, J.; Yuan, H.; Wang, W.; Wu, Q.; Guan, X.; Lin, J.; Li, J. Development of laminated bamboo lumber with high bond strength for structural uses by O2 plasma. *Constr. Build. Mater.* **2021**, *269*, 121269. [CrossRef]
20. Chang, F.; Liu, Y.; Zhang, B.; Fu, W.; Jiang, P.; Zhou, J. Factors Affecting the Temperature Increasing Rate in Arc-shaped Bamboo Pieces during High-frequency Heating. *BioResources* **2020**, *15*, 2.
21. Zhang, H.; Pizzi, A.; Zhou, X.; Lu, X.; Wang, Z. The study of linear vibrational welding of moso bamboo. *J. Adhes. Sci. Technol.* **2018**, *32*, 1–10. [CrossRef]
22. He, Q.; Zhan, T.; Zhang, H.; Ju, Z.; Hong, L.; Brosse, N.; Lu, X. Robust and durable bonding performance of bamboo induced by high voltage electrostatic field treatment. *Ind. Crop. Prod.* **2019**, *137*, 149–156. [CrossRef]
23. Semple, K.E.; Vnučec, D.; Kutnar, A.; Kamke, F.A.; Mikuljan, M.; Smith, G.D. Bonding of THM modified Moso bamboo (Phyllostachys pubescens Mazel) using modified soybean protein isolate (SPI) based adhesives. *Eur. J. Wood Wood Prod.* **2015**, *73*, 781–792. [CrossRef]
24. Bekhta, P.; Noshchenko, G.; Réh, R.; Kristak, L.; Sedliačik, J.; Antov, P.; Mirski, R.; Savov, V. Properties of Eco-Friendly Particleboards Bonded with Lignosulfonate-Urea-Formaldehyde Adhesives and pMDI as a Crosslinker. *Materials* **2021**, *14*, 4875. [CrossRef]
25. Łebkowska, M.; Załęska-Radziwiłł, M.; Tabernacka, A. Adhesives based on formaldehyde–environmental problems. *BioTechnologia* **2017**, *98*, 53–65. [CrossRef]
26. Antov, P.; Savov, V.; Trichkov, N.; Krišťák, Ľ.; Réh, R.; Papadopoulos, A.N.; Taghiyari, H.R.; Pizzi, A.; Kunecová, D.; Pachikova, M. Properties of High-Density Fiberboard Bonded with Urea–Formaldehyde Resin and Ammonium Lignosulfonate as a Bio-Based Additive. *Polymers* **2021**, *13*, 2775. [CrossRef]
27. Kumar, R.; Pizzi, A.J.A.f.W.; Materials, L. *Urea-Formaldehyde Resins*; Wiley-Scrivener Publishing: Hoboken, NJ, USA, 2019; pp. 61–100.
28. Khanjanzadeh, H.; Behrooz, R.; Bahramifar, N.; Pinkl, S.; Gindl-Altmutter, W. Application of surface chemical functionalized cellulose nanocrystals to improve the performance of UF adhesives used in wood based composites-MDF type. *Carbohyd. Polym.* **2019**, *206*, 11–20. [CrossRef]
29. Li, X.; Gao, Q.; Xia, C.; Li, J.; Zhou, X. Urea Formaldehyde Resin Resultant Plywood with Rapid Formaldehyde Release Modified by Tunnel-Structured Sepiolite. *Polymers* **2019**, *11*, 1286. [CrossRef] [PubMed]
30. Gao, S.; Cheng, Z.; Zhou, X.; Liu, Y.; Chen, R.; Wang, J.; Wang, C.; Chu, F.; Xu, F.; Zhang, D. Unexpected role of amphiphilic lignosulfonate to improve the storage stability of urea formaldehyde resin and its application as adhesives. *Int. J. Biol. Macromol.* **2020**, *161*, 755–762. [CrossRef] [PubMed]
31. Liu, M.; Wang, Y.; Wu, Y.; He, Z.; Wan, H. "Greener" adhesives composed of urea-formaldehyde resin and cottonseed meal for wood-based composites. *J. Clean. Prod.* **2018**, *187*, 361–371. [CrossRef]
32. Dunky, M. Adhesives in the Wood Industry. In *Handbook of Adhesive Technology*, 2nd ed.; Springer: Berlin/Heidelberg, Germany, 2003; Volume 70. [CrossRef]
33. Sar, P.; Ghosh, A.; Scarso, A.; Saha, B. Surfactant for better tomorrow: Applied aspect of surfactant aggregates from laboratory to industry. *Res. Chem. Intermed.* **2019**, *45*, 6021–6041. [CrossRef]

34. Mansouri, H.R.; Thomas, R.R.; Garnier, S.; Pizzi, A. Fluorinated polyether additives to improve the performance of urea–formaldehyde adhesives for wood panels. *J. Appl. Polym. Sci.* **2007**, *106*, 1683–1688. [CrossRef]
35. Chen, L.; Wang, Y.; Ziaud, D.; Fei, P.; Jin, W.; Xiong, H.; Wang, Z. Enhancing the performance of starch-based wood adhesive by silane coupling agent(KH570). *Int. J. Biol. Macromol.* **2017**, *104*, 137–144. [CrossRef] [PubMed]
36. Chauhan, S.; Singh, R.; Sharma, K. Volumetric, compressibility, surface tension and viscometric studies of CTAB in aqueous solutions of polymers (PEG and PVP) at different temperatures. *J. Chem. Thermodyn.* **2016**, *103*, 381–394. [CrossRef]
37. Singh, R.; Chauhan, S.; Sharma, K. Surface Tension, Viscosity, and Refractive Index of Sodium Dodecyl Sulfate (SDS) in Aqueous Solution Containing Poly(ethylene glycol) (PEG), Poly(vinyl pyrrolidone) (PVP), and Their Blends. *J. Chem. Eng. Data* **2017**, *62*, 1955–1964. [CrossRef]
38. Wang, X.-L.; Yuan, X.-Z.; Huang, H.-J.; Leng, L.-J.; Li, H.; Peng, X.; Wang, H.; Liu, Y.; Zeng, G.-M. Study on the solubilization capacity of bio-oil in diesel by microemulsion technology with Span80 as surfactant. *Fuel Process. Technol.* **2014**, *118*, 141–147. [CrossRef]
39. Pan, Z.; Cheng, F.; Zhao, B. Bio-Inspired Polymeric Structures with Special Wettability and Their Applications: An Overview. *Polymers* **2017**, *9*, 725. [CrossRef] [PubMed]
40. Kumar, S.; Panigrahi, P.; Saw, R.K.; Mandal, A. Interfacial Interaction of Cationic Surfactants and Its Effect on Wettability Alteration of Oil-Wet Carbonate Rock. *Energ. Fuel.* **2016**, *30*, 2846–2857. [CrossRef]
41. Lu, K.-T.; Fan, S.-Y. Effects of ultraviolet irradiation treatment on the surface properties and adhesion of moso bamboo (Phyllostachys pubescens). *J. Appl. Polym. Sci.* **2008**, *108*, 2037–2044. [CrossRef]
42. Ju, Z.; Zhan, T.; Zhang, H.; He, Q.; Hong, L.; Yuan, M.; Cui, J.; Cheng, L.; Lu, X. Strong, Durable, and Aging-Resistant Bamboo Composites Fabricated by Silver In Situ Impregnation. *ACS Sustainable Chem. Eng.* **2020**, *8*, 16647–16658. [CrossRef]
43. Wang, Z.; Zhao, S.; Zhang, W.; Qi, C.; Zhang, S.; Li, J. Bio-inspired cellulose nanofiber-reinforced soy protein resin adhesives with dopamine-induced codeposition of "water-resistant" interphases. *Appl. Surf. Sci.* **2019**, *478*, 441–450. [CrossRef]
44. Yanhua, Z.; Jiyou, G.; Haiyan, T.; XiangKai, J.; Junyou, S.; Yingfeng, Z.; Xiangli, W. Fabrication, performances, and reaction mechanism of urea–formaldehyde resin adhesive with isocyanate. *J. Adhes. Sci. Technol.* **2013**, *27*, 2191–2203. [CrossRef]
45. Kwan, Y.C.G.; Ng, G.M.; Huan, C.H.A. Identification of functional groups and determination of carboxyl formation temperature in graphene oxide using the XPS O 1s spectrum. *Thin Solid Films* **2015**, *590*, 40–48. [CrossRef]
46. Kadiyala, A.K.; Sharma, M.; Bijwe, J. Exploration of thermoplastic polyimide as high temperature adhesive and understanding the interfacial chemistry using XPS, ToF-SIMS and Raman spectroscopy. *Mater. Des.* **2016**, *109*, 622–633. [CrossRef]
47. Sharma, B.; Gatóo, A.; Bock, M.; Ramage, M. Engineered bamboo for structural applications. *Constr. Build. Mater.* **2015**, *81*, 66–73. [CrossRef]
48. Li, T.; Cheng, D.-L.; Wålinder, M.E.P.; Zhou, D.-G. Wettability of oil heat-treated bamboo and bonding strength of laminated bamboo board. *Ind. Crop. Prod.* **2015**, *69*, 15–20. [CrossRef]
49. Chaowana, P.; Jindawong, K.; Sungkaew, S. Adhesion and Bonding Performance of Laminated Bamboo Lumber made from Dendrocalamus sericeus. In *Product Design and Technology, Proceedings of the 10th World Bamboo Congress, Damyang, Korea, 17–22 September 2015*; World Bamboo Organization: Antwerp, Belgium, 2015.
50. Anokye, R.; Bakar, E.S.; Ratnasingam, J.; Yong, A.C.C.; Bakar, N.N. The effects of nodes and resin on the mechanical properties of laminated bamboo timber produced from Gigantochloa scortechinii. *Constr. Build. Mater.* **2016**, *105*, 285–290. [CrossRef]

Article

Effect of Rosin Modification on the Visual Characteristics of Round Bamboo Culm

Na Su [1,2], Changhua Fang [1,2], Hui Zhou [1,2], Tong Tang [3], Shuqin Zhang [1,2], Xiaohuan Wang [4,*] and Benhua Fei [1,2,*]

1. Department of Biomaterials, International Center for Bamboo and Rattan, Beijing 100102, China; yuhesu122216@126.com (N.S.); cfang@icbr.ac.cn (C.F.); zhouhui@icbr.ac.cn (H.Z.); zhangshuqin@icbr.ac.cn (S.Z.)
2. SFA and Beijing Co-Built Key Laboratory of Bamboo and Rattan Science & Technology, State Forestry Administration, Beijing 100102, China
3. Environmental Design, Institute of Art & Design, Qilu University of Technology, Jinan 250353, China; Tangtong@icbr.ac.cn
4. Beijing Forestry Machinery Institute of National Forestry and Grassland Administration, Beijing 100013, China
* Correspondence: wxh811118@126.com (X.W.); feibenhua@icbr.ac.cn (B.F.)

Citation: Su, N.; Fang, C.; Zhou, H.; Tang, T.; Zhang, S.; Wang, X.; Fei, B. Effect of Rosin Modification on the Visual Characteristics of Round Bamboo Culm. *Polymers* **2021**, *13*, 3500. https://doi.org/10.3390/polym13203500

Academic Editors: Jingpeng Li, Yun Lu and Huiqing Wang

Received: 14 September 2021
Accepted: 7 October 2021
Published: 12 October 2021

Publisher's Note: MDPI stays neutral with regard to jurisdictional claims in published maps and institutional affiliations.

Copyright: © 2021 by the authors. Licensee MDPI, Basel, Switzerland. This article is an open access article distributed under the terms and conditions of the Creative Commons Attribution (CC BY) license (https://creativecommons.org/licenses/by/4.0/).

Abstract: Rosin was used to treat round bamboo culm using the impregnation method. The quantitative color and gloss measurements combined with a qualitative eye tracking experiment were used to evaluate the effect of rosin treatment under different temperatures on the visual characteristics of the bamboo surface. Surface morphology analysis was also used to explore the mechanism of modification. The results showed that proper heating of the modified system was conducive to the formation of a continuous rosin film, which increased the gloss value. The maximum gloss value of 19.6 achieved at 50 °C was 122.7% higher than the gloss value of the control group. Heating decreased the brightness of the bamboo culm and changed the color from the green and yellow tones to red and blue. Additionally, at temperatures higher than 60 °C, the bamboo epidermal layer was damaged or shed, and stripes formed on the culm surface. The density of these stripes increased with an increase in treatment temperature. Eye movement experiment and subjective evaluation showed that high gloss would produce dazzling feeling, such as at 50 °C, while low gloss will appear dim, such as at 80 °C, while the gloss at 40 °C and 60 °C were appropriate. Additionally, the solid color surface below 60 °C had a large audience of about 73%, and the striped surface above 60 °C was preferred by 27% of the subjects.

Keywords: rosin modification; round bamboo culm; visual characteristics; eye tracking

1. Introduction

Since ancient times, people have used bamboo and wood to decorate the interior environment and to make indoor furniture [1,2]. Round bamboo in particular complements various natural materials in construction and interior design where it is used to create a traditional oriental style [3]. Round bamboo displays an impressive range of attractive properties, including a high ratio of weight to strength, easy workability, natural aesthetics, and environmental sustainability [1,4,5]. Among these, the visual characteristics (such as color, gloss and texture) play one of the most visual roles [4,5]. Visual characteristics not only affect the aesthetic of furniture or architecture, but also the psychological and physiological well-being of users by influencing psychology, communication, psychosomatic effects, visual ergonomics, etc. [6]. For example, the impression of a color and the message it conveys is of utmost importance in creating the psychological mood or ambiance that supports the function of a space [7,8]. Each color gives a different impression and conveys a specific meaning. With round bamboo culm, fresh bamboo is mostly green, while dried bamboo is mostly yellow [2,9]. Green induces a balancing, natural, and calm state, conveying a message in the interior space is of simplicity, security, and balance in the interior space [10]. Pastel yellow gives the impression of a sunny, friendly, and soft environment, with a message of stimulation, brightness, and coziness. Gloss also influences warmth with a high gloss, producing a stiff and cold impression [11]. Lastly, stripes or scars on the surface of the bamboo will also offer a simpler, more natural decorative effect [12].

The color related chemical components in bamboo mainly arise from lignin and some extract with unsaturated structure while the gloss may be more related to microstructure [13,14]. A smooth, low porosity surface will obtain a greater gloss but when the surface cells are exposed or damaged, diffuse reflection or absorption of light will reduce the gloss [15]. Therefore, application of treatment or finishing can change the color, gloss and even texture of the bamboo surface [16,17]. Feng et al., 2020 [18] reported that heat treatment above 200 °C will turn the bamboo brown, and indicated that this is due to the change in lignin structure or the oxidation of phenol compounds to quinones. Transparent finishing can also improve the gloss of bamboo and enhance the contrast of texture. Despite this, the visual quality is ignored in many processing treatments, which limits the practical applications of bamboo.

Rosin is a natural resin derived from living pine trees [19]. It is a common polymer monomer due to the presence of reactive groups, including carboxyl groups and conjugated double bonds in its molecules as shown in Figure 1 [20–22]. Rosin can produce polymerized rosin, maleated rosin and disproportionated rosin through addition reaction and polymerization, and can further synthesize polymer materials with different properties, which are widely used in glass fiber reinforced plastics and coatings [23,24]. In addition, rosin is often used in modifier or coating to improve the water resistance and visual properties because of its good hydrophobicity and gloss [25,26]. Over recent years, rosin and its derivatives have been used to modify nanocomposites, wood materials, wood-based panels, and packaging [27–30]. Dahlen et al. (2008) [31] reported that only 3% rosin could significantly increase the water resistance of wood. Dong et al. (2016) [29] used rosin to treat fast-growing poplar wood under vacuum pressure impregnation and showed that the wood density increased from 0.34 to 0.44 g/cm^3 after the treatment, and the anti-swelling efficiency achieved 36.00%. Rosin treatment was used as a hydrophobic method for bamboo in our previous studies [25,32]. This treatment resulted in good hydrophobicity and improved the dimensional stability of round bamboo. Microtopography also showed that rosin formed a transparent film on the surface of the round bamboo culm. Since rosin exists as a light yellow transparent solid, as shown in Figure 1, rosin treatment of bamboo may change the visual characteristics of bamboo culm. However, this has not been studied so far.

Visual characteristics could be described by physical quantities and psychological quantities. The former can be described using physical parameters such as color, gloss and texture, while the latter consists of human visually induced psychological assessment. At present, measurements of color and gloss are conducted using colorimeters and glossmeters to obtain quantitative data, respectively. Visual quantitative psychological parameters can be obtained using eye tracking tests [33]. Eye movement technology is often used in human-computer interaction research, psychological exploration, on web pages and in graphic design evaluation [34–37]. It is based on the characteristics of directness, naturalness and bi-directionality of human line of sight [38]. The basic working principle of eye tracking is to use image processing technology to continuously record both the change in view with the infrared rays in the camera and the pupil reflex with a tracking camera which focuses on the eye [39]. The images are used to clarify the response behavior of the visual system to different objects by analyzing the trajectory of eye movement characteristics, such as saccades, gaze time and pupil switching. In recent years, this technology has been gradually used in the fields of furniture and architectural design [40,41].

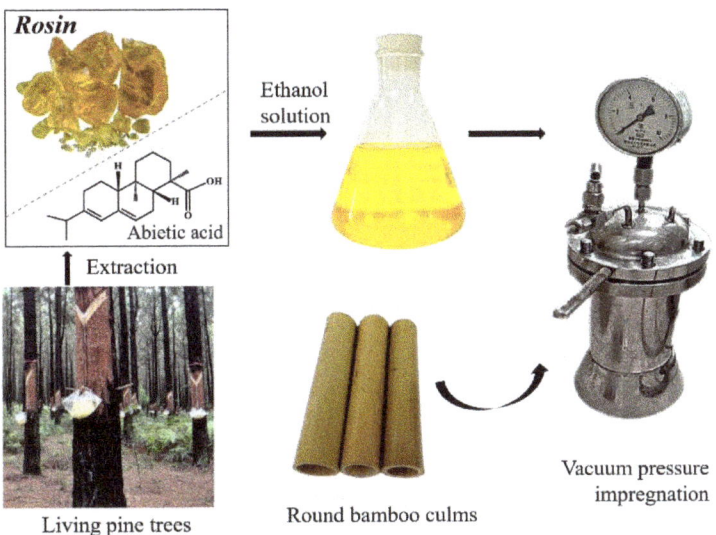

Figure 1. Schematic illustration of rosin impregnation of bamboo culms.

In this study, to investigate the effect of natural resin rosin modification on the surface visual effect of round bamboo culm, the color and gloss of bamboo culm before and after rosin modification were quantitatively determined. Meanwhile, the eye tracking technology was used to evaluate the users' preference for the visual characteristics of the bamboo culm surface. The combination of quantitative index and qualitative preference analysis to evaluate the visual effect of bamboo culm was used in this study to improve the scientificity and systematicness of the evaluation. In addition, to clarify the mechanism of color change, the surface microscopic morphology and chemical group of the bamboo culms were explored.

2. Materials and Methods

2.1. Impregnation of the Bamboo Culm with Rosin

The sampling of bamboo culm and preparation of rosin ethanol solution was consistent with our previous study [25,32]. Five-year-old Hong bamboo (*Phyllostachys iridencens* C.Y. Yao and S.Y. Chen), treated with steam at 130 °C for 1 h, were obtained from Hangzhou Suo Bamboo Industry Co., Ltd. (Suo Bamboo Industry Co., Ltd., Hangzhou, China). Culm internodes were selected for this study (as shown in Figure 1) from 1.5 to 4.0 m aboveground. The rosin was purchased from Jitian Chemical Co., Ltd. (Jitian Chemical Co., Ltd., Shenzhen, China). Rosin ethanol solutions with concentrations of 20 wt% were prepared (Figure 1). The rosin ethanol solutions were then used to impregnate bamboo culm with vacuum pressure impregnation, as outlined in our previous study [25,32]. However, different temperatures were used in the impregnation process to study the effect of temperature on the visual characteristics of bamboo culm. The temperature parameters are set to 25 °C (room temperature), 40 °C, 50 °C, 60 °C, 70 °C, and 80 °C. For all tests, eight replicates were performed.

2.2. Measurement of Color and Gloss

A surface gloss test was conducted according to ISO (2014) [42]. The gloss of bamboo culm samples was measured using a WGG 60 glossmeter (Shanghai Precision Scientific Instrument Co., Ltd., Shanghai, China) at six points on the bamboo surface before and after rosin treatment. The chosen geometry was an incidence angle of 60° and results were based on a specular gloss value of 100.

A color test was performed using color space CIE L*a*b* [43]. The color changes of bamboo culms before and after rosin treatment were determined using an SP60 spectrophotometer (CC-6834, BYK, Grazrid, Germany) with a D65 light (standard light source) and a 10° view angle. Color changes were assessed at six points on the bamboo surface before and after rosin treatment. The CIE L*a*b* system consists of three perpendicular axes to describe color. The L^* axis represents lightness, the a^* axis represents the red-green factor, and the b^* axis represents the yellow-blue factor. The overall color change (ΔE^*) was determined as follows:

$$\Delta E^* = [(\Delta a^*)^2 + (\Delta b^*)^2 + (\Delta L^*)^2]^{1/2} \qquad (1)$$

where Δa^*, Δb^*, and ΔL^* represent the changes in a^*, b^* and L^* between rosin-treated and control groups.

2.3. Micromorphology Characteristics

The surface micromorphology of bamboo culm before and after rosin treatment at different temperatures was characterized using a field emission scanning electron microscope (FE-SEM) (XL30, FEI, Hillsboro, OR, USA). To highlight the distribution of rosin on the bamboo surface, the multi-Otsu thresholding algorithm was used to obtain the thresholds of different surface components using FIJI/Image J software.

2.4. Micromorphology Characteristics

Fourier transform infrared (FTIR) spectroscopy of rosin and rosin-treated bamboo culm at different temperature was conducted with a standard FTIR spectroscope (Perkin Elmer Inc., Shelton, CT, USA). The spectra from 4000 to 500 cm^{-1} were recorded at a 4-cm^{-1} resolution across 64 scans.

2.5. Eye Tracking Test

In conjunction with the ClearView 2.7.0 software system, the Tobii X120 Eye-Tracker (Tobii X120, Tobii Technology AB, Danderyd, Sweden) was used to record the eye tracks of participants. This tracker consists of an observation host, computer and camera. Twenty-six volunteers (all students) were recruited from the Qilu University of Technology, including 25 males and 25 females. The health condition was tested, and all volunteers tested normal i.e., without any symptom of color blindness or weakness. The naked or corrected visual acuity for all volunteers was >1.0.

All bamboo culms (before and after rosin treatment) were placed about 600 mm away from the screen to follow the eye calibration procedure. The experimental aim was to explore the effect of different treatment temperatures on the visual effect of the bamboo culm surface. The total gaze time, gaze numbers, average gaze time and combined heat map (measured for each bamboo culm) were used as the eye-tracking indicators.

Meanwhile, subjective reviews were completed right after watching, including color comfort, gloss comfort, texture preference, and texture density comfort. Among these, the color comfort and texture density comfort levels were discomfort, general, and comfort; the gloss comfort level was dim, moderate and dazzling; texture preferences were solid color and texture. The proportion of each grade was then counted.

3. Results and Discussion

3.1. Color and Gloss

As shown in Figure 2, the bamboo color varied from yellow to dark yellow to light brown with some stripes presenting as the temperature of the rosin treatment increased. Specifically, when the temperature was between 25–50 °C, the bamboo surface presented a solid color, and the color change was minimal. At temperatures ranging between 60–80 °C, brown stripes appeared on the bamboo surface. The surface of the control group presented as yellow with a certain gloss. The 25 °C and 40 °C treatment hardly changed the color of bamboo culm. However, the treatment at room temperature (25 °C) decreased the gloss value, while treatment at 40 °C increased the gloss value. The gloss of the rosin-treated bamboo culm was mainly a result of the rosin film, because the rosin itself is light yellow and transparent. The 25 °C treatment did not promote the formation of a continuous film of rosin. With additional heating (40 °C and 50 °C), the gloss value, which indicated that heating could promote the formation of a continuous film of rosin. However, when the temperature was higher than 60 °C, brown stripes appeared on the surface of bamboo, and the gloss decreased gradually. Moreover, stripe formation increased with the increase of treatment temperature, and even patches were formed at 80 °C.

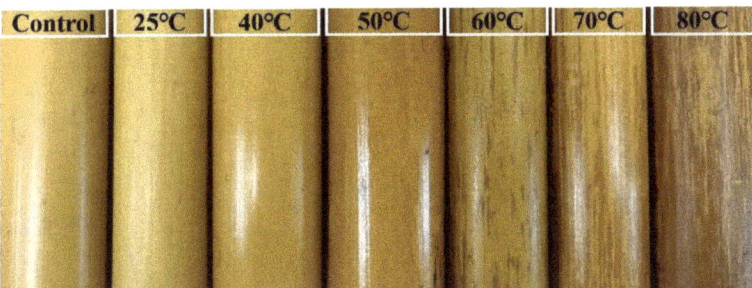

Figure 2. Effect of rosin treatment with different temperature on the color and gloss of bamboo culm.

The quantified chromaticity values are expressed in Table 1, while its trends are visually displayed in Figure 3. Firstly, the values of L^*, a^*, and b^* showed little change at 25 °C, which indicated that the rosin treatment at room temperature had no obvious effect on the color of bamboo culm. This is consistent with images displayed in Figure 2. The L^* value decreased with the increase in temperature, which indicated a decrease in brightness of the bamboo culm as treatment temperature increased. The a^* value initially decreased but then increased gradually with a rising temperature, whereas the b^* value decreased gradually with an increase in temperature (Table 1). These were indicative of color changes from the green and yellow zones to the red and blue zones. Feng, et al. found a similar phenomenon when investigating hygrothermal treatment on Moso bamboo [18]. Lastly, the change of ΔE^* shows that an increase in temperature will increase the color difference. This is also seen in Figure 1. The chemical components related to color in bamboo mainly come from lignin and extract, because the structure of lignin and extract contains a large number of aromatic compounds, chromogenic groups and color assisting groups [44]. In this study, the structure of lignin and extract may have remained unchanged because the treatment temperature was not high enough for such a change. The change in color may be relative to the addition reaction of the carboxyl group in rosin or the oxidative discoloration of rosin itself. Rosin contains conjugated double bonds and carboxyl groups, which are prone to addition reactions, especially oxidation and discoloration, at an elevated temperature [45].

Table 1. Change in color and gloss of bamboo culm after rosin treatment with different temperature.

Bamboo Culm Sample		L^*	a^*	b^*	Color ΔL^*	Δa^*	Δb^*	ΔE^*	Gloss
Control group		57.02	7.26	26.08	/	/	/	/	8.8
Rosin-treated group	25 °C	57.17	6.57	26.17	0.14	−0.69	0.09	0.71	5.6
	40 °C	56.52	7.50	26.74	−0.50	0.24	0.66	0.87	12.4
	50 °C	53.43	8.46	25.66	−3.60	0.20	−0.42	3.81	19.6
	60 °C	52.57	8.71	23.89	−4.45	1.45	−2.19	5.16	15.5
	70 °C	49.07	8.19	22.87	−7.95	0.92	−3.21	8.62	12.9
	80 °C	43.80	8.98	18.88	−13.22	0.72	−7.20	15.15	6.8

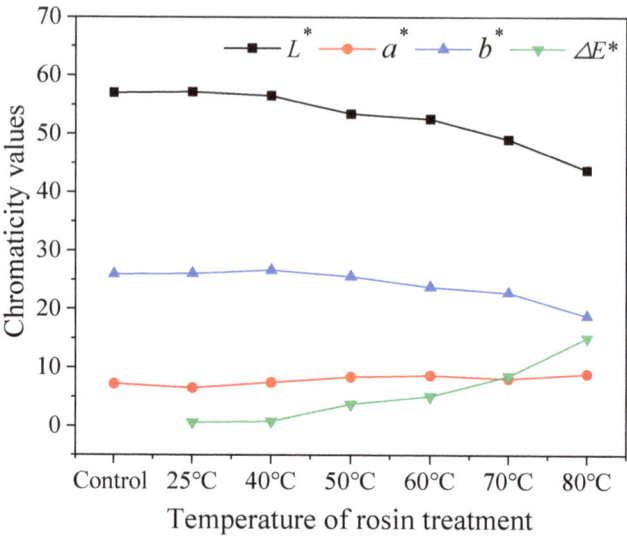

Figure 3. Chromaticity values of bamboo culm before and after the rosin treatment with different temperature.

Additionally, the gloss value in Table 1 shows that rosin treatment improved the gloss of bamboo culm at all temperatures, except at room temperature (25 °C) and 80 °C. In detail, the gloss decreased by 3.2 at 25 °C when compared to the control group (8.8). With additional heating, the gloss increased perceptibly. At 50 °C, the gloss reached the highest value of 19.6, which was 122.7% higher than that of the control group. With the continuous increase in temperature, the gloss showed a downward trend. The gloss value of the bamboo culm was just 6.8 at 80 °C, which was lower than the control group.

3.2. Surface Topography Characteristics

The Figure 4a shows the surface morphology of bamboo culm observed using a stereomicroscope to clearly examine the surface texture and the changes before and after rosin modification under different temperatures. Firstly, when the treatment temperature is lower than 60 °C, the bamboo surface is relatively smooth with a uniform color. In detail, the surface of bamboo treated at room temperature (25 °C) is rough with white granular bulges while the surfaces of rosin-treated bamboo at 40 and 50 °C are smooth and glossy. At 60 °C, 70 °C, and 80 °C, clear brown stripes appear on the bamboo surface. In Figure 4a it is observed that the brown stripes are concave, which indicates that stripes are formed by the shedding of the bamboo epidermis during rosin treatment. Comparing 60 °C, 70 °C and 80 °C, it is observed that more of the bamboo epidermis is shed with the increase in treatment temperature.

Figure 4b shows the surface micromorphology of bamboo culm using SEM. There is slight tearing in the epidermis of the control group. For bamboo culm treated at room temperature (25 °C), there are some granular bulges on the surface. Conversely, there are irregular depressions on the bamboo surface when treated at 40 °C. However, the surface of bamboo treated at 50 °C is very smooth. In order to more intuitively compare the surface morphology, images were threshold segmented using FIJI/Image J software, as shown in Figure 4c. In the thresholding images, the black represents the outline of the bamboo surface, while the white represents the outline of another substance, in this case rosin (Figure 4c). For the control group, the outline of the broken epidermis is significant. At 25 °C, 40 °C and 50 °C, the white dots are particles formed by rosin curing on the culm surface. It is evident that the particle size and the number both decrease with the increase in temperature as shown in Figure 4c. Overall, it is apparent that rosin forms the most continuous film at 50 °C, followed by 40 °C and 25 °C in succession. This may be related to the solubility of rosin at different temperatures. At room temperature, the particle size of the rosin ethanol system is larger than it is at 40 °C and 50 °C. Heating up the system can effectively reduce the particle size of rosin, which is conducive to the film-forming process. However, when the temperature reaches 60 °C, the bamboo epidermis is damaged to varying degrees. This may be a synergistic result of temperature, ethanol and rosin. In detail, there are a few but large depressions on the bamboo surface at 60 °C, while those depressions are more obvious and aggregated at 70 and 80 °C (Figure 4).

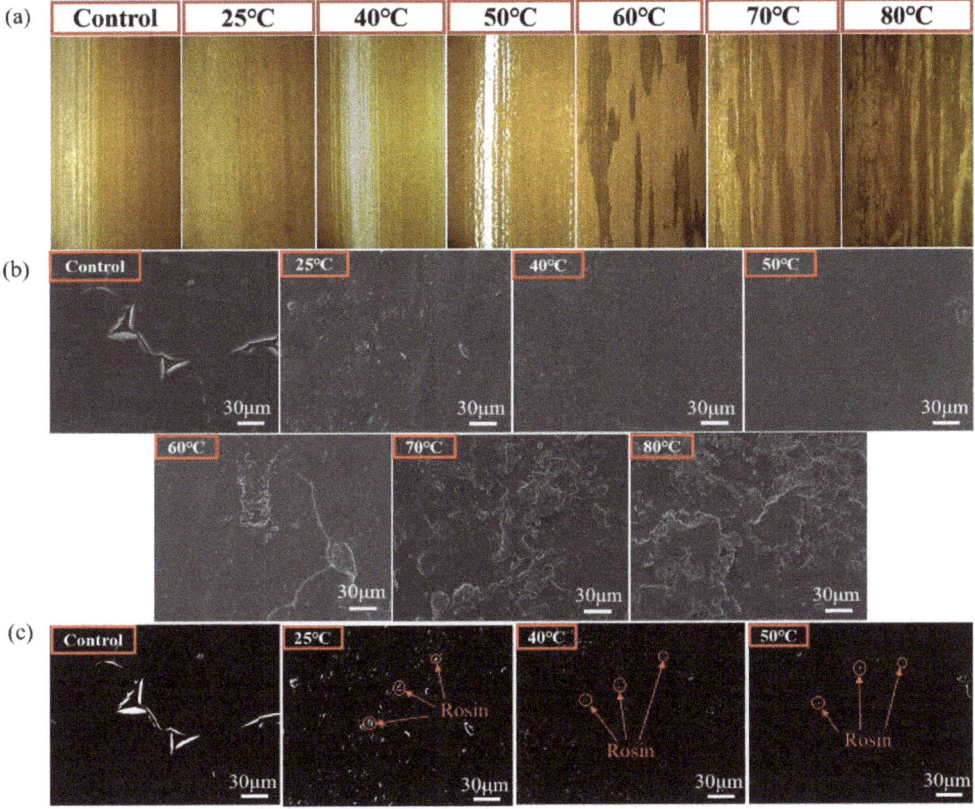

Figure 4. (**a**) Surface texture of bamboo culm observed using stereomicroscopy; (**b**) the surface micromorphology of bamboo culm observed using FE-SEM; (**c**) the surface micromorphology after thresholding by FIJI/Image J software.

The bamboo surface morphology (Figure 4) also explains the change in gloss value. Between 25–50 °C, the gloss is mainly determined by the distribution of rosin on the surface where the continuity of rosin film improves with the decrease in rosin particle size. The SEM results also show that the surface roughness of bamboo decreased at these temperatures. Since the gloss increases with the decrease of roughness, the gloss of bamboo culm increases from 25 °C to 50 °C as shown in Table 1. Between 60–80 °C, the gloss of the bamboo culm is related not only to the continuity of rosin film, but also to the destruction and abscission of the epidermal layer. The epidermal layer of bamboo is composed of long cells, embolic cells, siliceous cells and stomata [46]. Long cells account for most of the area and are arranged in parallel along the grain while embolic and siliceous cells are short and inserted between the ranks of long cells. The epidermal layer is closely arranged without gaps, interspersed only with stomata. However, the subcutaneous layer and cortical structure beneath the epidermal layer are loose. According to the basic principles of optics [47], the higher the degree of reflection of bamboo to incident light, the higher the gloss value. For a loose or rough surface, incident light will form multiple reflections on the bamboo surface, increasing the probability of absorption, and producing a low gloss value. The breaking or exfoliation of the bamboo epidermal layer after rosin treatment at 60–80 °C would decrease the gloss of the bamboo culm due to the exposure of the loosely structured subcutaneous layer. Similarly, for control group, the tearing of the epidermal layer also produces a low gloss.

3.3. FTIR Spectroscopy

The FTIR spectrum of surface materials of bamboo culm and rosin at different temperatures were obtained to determine the reasons of discoloration of bamboo surface. The FTIR spectrum results are shown in Figure 5. First, the FTIR spectrum of bamboo surface materials shows the typical transmittance peaks of bamboo, e.g., 3400, 2925, 1730, 1517 and 1458 cm^{-1} [48,49]. The peaks at 3400 cm^{-1} and 1730 cm^{-1} correspond to O–H and C=O stretching vibration, while the peaks at 2925, 1517 and 1458 cm^{-1} correspond to the C–H stretching vibration. However, compared to the control bamboo, these typical transmittance peaks hardly changed after rosin treatment at different temperatures. Therefore, the discoloration of bamboo surface may have little relationship with bamboo surface chemicals. Additionally, the Figure 5b shows the typical transmittance peaks of rosin. Among these, the peaks at 1694 cm^{-1} and 1276 cm^{-1} corresponds to C=O stretching vibration, which belong to carbonyl group in carboxyl group [22,32]. Carboxyl group is a chromogenic group, which is easy to be oxidized, resulting in discoloration [50]. Compared the rosin at different temperatures, it can be seen that the peak intensity at 1694 cm^{-1} and 1276 cm^{-1} decreased and broadened with the increase of temperature. This means that the stretching vibration of carbonyl group is weakened, which may be due to the oxidation of carboxyl group of rosin under the action of heat. In detail, when the temperature rises to 50 °C, both of peak's width at 1694 cm^{-1} and 1276 cm^{-1} gradually increase and the intensity gradually decrease. These are consistent with the discoloration of bamboo in Figure 2. The FTIR results also revealed that the discoloration of bamboo surface is caused by the oxidation discoloration of rosin under heating, rather than the material change of bamboo itself.

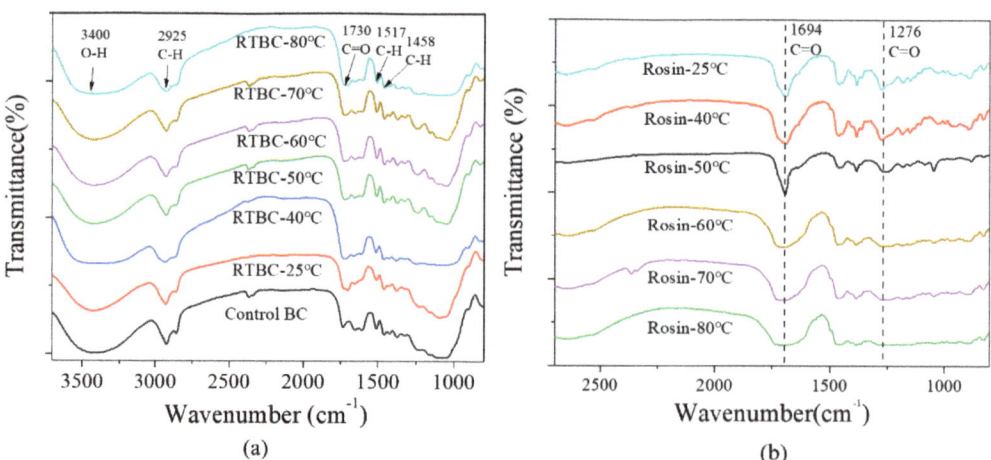

Figure 5. Fourier transform infrared spectroscopy spectra of (**a**) control bamboo culm and rosin-treated bamboo culm (RTBC) at different temperature; and (**b**) rosin at different temperature.

3.4. Eye Tracking Analysis

Figure 6 presents the total gaze time, gaze numbers, and average gaze time before and after rosin treatment at different temperatures. The total gaze time represents the total time each subject looked at a particular culm during the observation period while gaze numbers refer to the number of times subjects looked at a specific culm. The average gaze time is the average of the duration of the subject's line of sight on the bamboo culm each time. The gaze time and numbers reflect the volunteers' attention to the bamboo culms [38]. The average gaze time is used to measure the difficulty of information extraction [37,38]. The larger the value, the more complex the subject and its details, and the longer the interpretation time. First, it can be seen that the gaze time of rosin-treated bamboo culms are all higher than that of the control. This suggests that the rosin treatment has increased the attention given to the bamboo culms by participants. The gaze time of the 60 °C treatment was the highest. Combined with surface color and texture, the brown stripes may have been the key factor attracting the subject's attention. Research shows that certain knots on the wood surface will result in a certain decorative effect and give people a simple and natural feeling [51]. A similar phenomenon may have occurred with the bamboo surface. Additionally, the intensity of stripes may also affect feelings. Comparing the gaze time of bamboo culms treated at 60 °C, 70 °C and 80 °C, it is evident that with the increase of stripe density, the gaze time decreases, i.e., the degree of interest decreases. Furthermore, the lightness of the bamboo culm gradually decreases from 60 °C to 80 °C, suggesting that lightness may also influence people's interest. This is also reflected in the bamboo culm treated at 50 °C. This culm had the lowest gaze time (Figure 6) despite having the smoothest surface (Figure 4b,c) and highest gloss value (19.6—Table 1) among all groups. This may be related to the lower lightness of the bamboo. Moreover, a high gloss has been shown to induce dizziness [52]. This may also be the reason for the reduction of gaze time for 50 °C. The gaze numbers show a similar trend to the total gaze time, except at 80 °C. The gaze number for 80 °C is lower than that of the control group, which suggests that the bamboo culm treated at 80 °C has no better visual effect than the control culm. The gaze time of 80 °C is also only slightly higher than the control. Additionally, the average gaze times of the treated bamboo culm are all higher than the control group and are positively correlated with temperature. The culms treated at 60–80 °C have longer average gaze times (0.48 s, 0.51 s and 0.56 s) than those treated at 25–50 °C (0.44 s, 0.41 s, and 0.42 s) (Figure 6). This may be due to the brown stripes of the culms treated at higher temperatures which provide more complexity than the uniform surface of those treated at lower temperatures. The complex surface requires more time to observe and interpret. Furthermore, for bamboo culms with stripes, the average gaze time of bamboo treated at 80 °C is the highest (0.56 s), followed by 70 °C (0.51 s) and 60 °C (0.48 s) in succession. This could be due to the complexity of surface information increasing with an increase in stripe density correlated to temperature (Figure 4a).

The heat map showed the overall effect of all the gaze points on the samples [53]. The visual intensity mapped here reflects the degree of preference for each treatment at each temperature. The gaze frequency and time are represented using a color gradient between green-yellow-orange-red. Green represents a lower gaze frequency and shorter gaze time, while red represents a higher gaze frequency and longer gaze time. Figure 7 intuitively shows the gaze regions and visual presentation. From the heat map, the gaze point concentration area is distributed between the 60 °C and 40 °C treatments. The control group and bamboo treated with 70 °C and 80 °C have fewer gaze points. Considering the three factors of color, gloss and texture, the light color, appropriate gloss and sparse stripes may attract the most attention of all the bamboo culms. On the contrary, dark color, high gloss, and dense, uneven stripes attract less attention. The lower lightness and uneven texture at 80 °C received very little attention. Similarly, the bamboo culm treated at 50 °C with lower lightness and higher gloss obtained less attention than the culm treated at 40 °C.

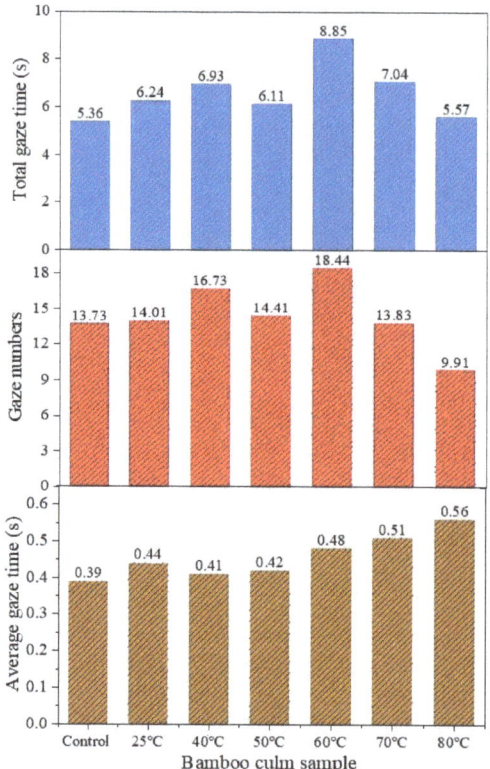

Figure 6. The total gaze time, gaze numbers, and average gaze time of bamboo culms.

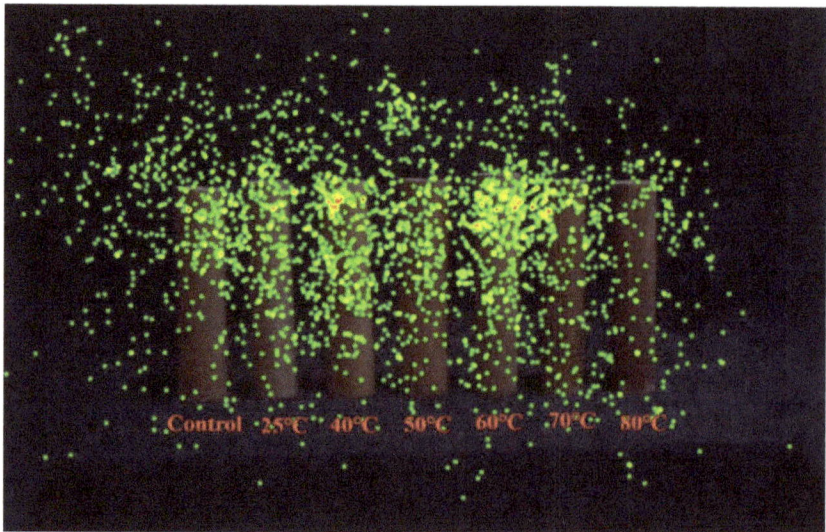

Figure 7. The heat map of bamboo culms in eye tracking test.

The subjective reviews respond to the "color", "gloss", and "texture" preference of testees. The results are shown in Figure 8. It shows that: (1) in terms of color comfort, the comfort preference of bamboo culm treated at 80 °C was obviously lower than that of the control group, and higher comfort recognition was obtained at 40 °C and 50 °C; (2) as regards the sense of gloss comfort, the bamboo culm treated at 25 °C, 50 °C, and 80 °C obtained the lower comfort recognition than control. Among these, the gloss at 25 °C and 80 °C were considered dull, while that at 50 °C was considered dazzling; these may be the reason why bamboo culms at 50 °C and 80 °C are less concerned than others; (3) by the texture preference, 73% of testees preferred a solid color bamboo surface, while 27% preferred texture one; this shows that the bamboo with solid color surface has a wider audience, and the textured surface is also loved by a small number of people; (4) additionally, as for texture density, the comfort recognition gradually decreased from 60 °C to 80 °C. It could be inferred that texture comfort weakened with the increase of texture density.

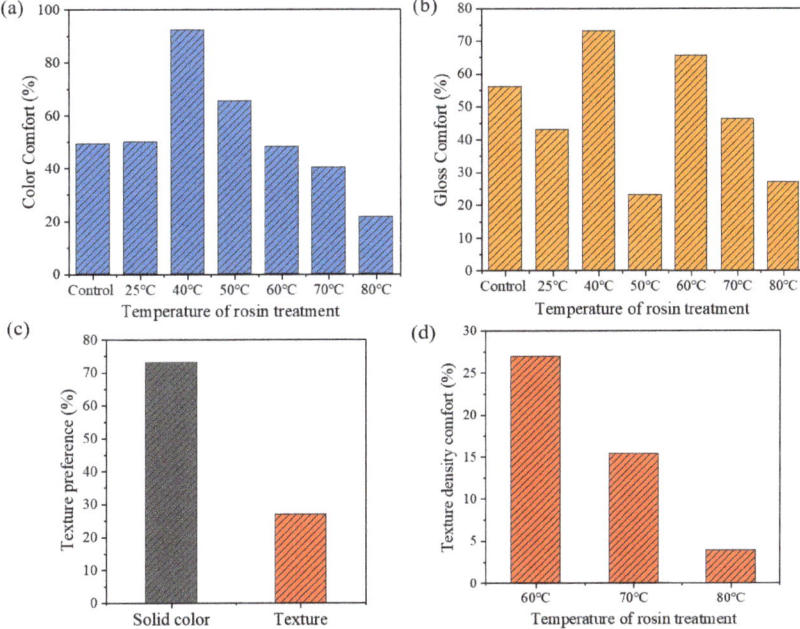

Figure 8. Subjective comfort evaluation results of color, gloss and texture preference of testees on bamboo culm surface. (**a**) color comfort; (**b**) gloss comfort; (**c**) texture preference; (**d**) texture density comfort.

From the results of subjective reviews, it could be inferred that the increased gaze time and number of bamboo culm treated at 60 °C and 70 °C may be results of its texture attracting more attention and requiring more time to observe. However, it could not infer that subjects prefer stripes from the more gaze time and numbers, because the proportion of testees who prefer stripes in subjective comfort evaluation is much less than that of pure color. Besides, the great color and gloss comfort recognition also contribute to higher attention.

4. Conclusions

In this study, natural resin rosin was used to modify bamboo culm at different temperatures. The visual characteristics were then analyzed quantitatively and qualitatively. At room temperature (25 °C), rosin was deposited in a granular form on the bamboo surface and increased the roughness of culm. This reduced the surface gloss but had little effect on the color of the bamboo. A proper heating treatment (40 °C and 50 °C) was conducive to the formation of a continuous rosin film, which then improved the gloss value. At these temperatures, the color of the bamboo culm also changed from green and yellow to red and blue. As temperature rose to 60 °C, the bamboo epidermis was shed. Some stripes were formed due to the shedding of the epidermal layer, and these became denser with the increase in treatment temperature. The results of the eye tracking experiment indicated that rosin modification at any temperature increased the observer's visual interest in the bamboo culm. From the rosin-treated culms, those with stripes formed by the shedding of the epidermis increased the gaze time of bamboo culm. However, compared with textured bamboo, solid-colored surfaces get more preference. Lastly, the appropriate improvement of brightness and gloss also increased the attractiveness of the bamboo culm. Quantitative color and gloss measurements, combined with qualitative eye tracking experiments, can more comprehensively analyze the effect of rosin on the visual characteristics of the bamboo surface. Natural rosin resin could effectively improve the visual characteristics of bamboo culm, and different visual effects on bamboo culm surfaces were obtained in different temperature ranges.

Author Contributions: Conceptualization, B.F.; methodology, X.W.; software, T.T.; investigation, C.F.; data curation, H.Z. and N.S.; writing—original draft preparation, N.S.; writing—review and editing, S.Z.; supervision, B.F.; project administration, B.F. All authors have read and agreed to the published version of the manuscript.

Funding: This work was financially supported by the Fundamental Research Funds for the International Centre for Bamboo and Rattan of the Ministry of Education of the People's Republic of China (Grant No. 1632020002). and National Promotion Project of Forestry Scientific and Technological Achievements of the Ministry of Finance of the People's Republic of China (Grant No. 2020133150).

Data Availability Statement: Not applicable.

Acknowledgments: The authors wish to thank the Zhucheng Antai Machinery Co., Ltd. for its impregnation equipment support.

Conflicts of Interest: The authors declare no conflict of interest.

References

1. Van der Lugt, P.; Van den Dobbelsteen, A.A.J.F.; Janssen, J.J.A. An environmental, economic and practical assessment of bamboo as a building material for supporting structures. *Constr. Build. Mater.* **2006**, *20*, 648–656. [CrossRef]
2. Liese, W.; Köhl, M. *Bamboo*; Springer: Cham, Switzerland, 2015.
3. Manandhar, R.; Kim, J.H.; Kim, J.T. Environmental, social and economic sustainability of bamboo and bamboo-based construction materials in buildings. *J. Asian Archit. Build. Eng.* **2019**, *18*, 49–59. [CrossRef]
4. Miclat, M.C. The visual poetry of Chinese bamboo: Some notes on traditional Chinese Xieyi painting. *Humanit. Diliman* **2000**, *1*, 22–35.
5. Wang, X.Y. Study on aesthetic application of the original bamboo in assembled bamboo buildings. *Adv. Educ. Res.* **2012**, *4*, 375–379.
6. Weibel, D.; Stricker, D.; Wissmath, B.; Mast, F.W. How Socially Relevant Visual Characteristics of Avatars Influence Impression Formation. *J. Media Psychol.* **2010**, *22*, 37–43. [CrossRef]
7. Küller, R.; Ballal, S.; Laike, T.; Mikellides, B.; Tonello, G. The impact of light and colour on psychological mood: A cross-cultural study of indoor work environments. *Ergonomics* **2006**, *49*, 1496–1507. [CrossRef]
8. Levy, B.I. Research into the psychological meaning of color. *Am. J. Art Ther.* **1984**, *23*, 58–62.
9. Jiang, Z. *Bamboo and Rattan in the World*; China Forestry Pub. House: Beijing, China, 2007.
10. Jacobs, K.W.; Blandino, S.E. Effects of color of paper on which the profile of mood states is printed on the psychological states it measures. *Percept. Mot. Ski.* **1992**, *75*, 267–271. [CrossRef]
11. Adams, W.J.; Kerrigan, I.S.; Graf, E.W. Touch influences perceived gloss. *Sci. Rep.* **2016**, *6*, 386855. [CrossRef]
12. Wastiels, L.; Schifferstein, H.N.; Heylighen, A.; Wouters, I. Relating material experience to technical parameters: A case study on visual and tactile warmth perception of indoor wall materials. *Build. Environ.* **2012**, *49*, 359–367. [CrossRef]
13. Rosu, D.; Teaca, C.-A.; Bodirlau, R.; Rosu, L. FTIR and color change of the modified wood as a result of artificial light irradiation. *J. Photochem. Photobiol. B Biol.* **2010**, *99*, 144–149. [CrossRef]
14. Temiz, A.; Yildiz, U.C.; Aydin, I.; Eikenes, M.; Alfredsen, G.; Çolakoglu, G. Surface roughness and color characteristics of wood treated with preservatives after accelerated weathering test. *Appl. Surf. Sci.* **2005**, *250*, 35–42. [CrossRef]

15. Kamdem, D.P.; Grelier, S. Surface Roughness and Color Change of Copper-Amine Treated Red Maple (*Acer rubrum*) Exposed to Artificial Ultraviolet Light. *Holzforschung* **2002**, *56*, 473–478. [CrossRef]
16. Chang, S.-T.; Wu, J.-H. Stabilizing Effect of Chromated Salt Treatment on the Green Color of Ma Bamboo (*Dendrocalamus latiflorus*). *Holzforschung* **2000**, *54*, 327–330. [CrossRef]
17. Zhou, T.; Shi, X.U.; Wang, Z.; Sun, D. Effect of heat treatment on texture and color changes in minimally processed water bamboo. *J. Wuxi Univ. Light Ind.* **2002**, *16*, 53–70.
18. Feng, Q.; Huang, Y.; Ye, C.; Fei, B.; Yang, S. Impact of hygrothermal treatment on the physical properties and chemical composition of Moso bamboo (*Phyllostachys edulis*). *Holzforschung* **2021**, *75*, 614–625. [CrossRef]
19. Karlberg, A.-T.; Magnusson, K. Rosin components identified in diapers. *Contact Dermat.* **1996**, *34*, 176–180. [CrossRef] [PubMed]
20. Shaw, D.N.; Sebrell, L.B. The Chemical Composition of Rosin. *Ind. Eng. Chem.* **1926**, *18*, 612–614. [CrossRef]
21. Su, Z.A.; Liang, Z.Q.; Qin, W.L.; Jiang, Z.R. Study on the principal chemical constituents of Chinese rosin and turpentine. *Sci. Silvae Sin.* **1980**, *16*, 214–220.
22. Wang, J.-F.; Lin, M.-T.; Wang, C.-P.; Chu, F.-X. Study on the synthesis, characterization, and kinetic of bulk polymerization of disproportionated rosin (β-acryloxyl ethyl) ester. *J. Appl. Polym. Sci.* **2009**, *113*, 3757–3765. [CrossRef]
23. Pathak, Y.V.; Dorle, A.K. Rosin and rosin derivatives as hydrophobic matrix materials for controlled release of drugs. *Drug Des. Deliv.* **1990**, *6*, 223–227.
24. Mayer, M.; Meuldijk, J.; Thoenes, D. Emulsion polymerization of styrene with disproportionated rosin acid soap as emulsifier. *J. Appl. Polym. Sci.* **2010**, *59*, 1047. [CrossRef]
25. Su, N.; Fang, C.; Yu, Z.; Zhou, H.; Wang, X.; Tang, T.; Zhang, S.; Fei, B. Effects of rosin treatment on hygroscopicity, dimensional stability, and pore structure of round bamboo culm. *Constr. Build. Mater.* **2021**, *287*, 123037. [CrossRef]
26. Satturwar, P.M.; Fulzele, S.V.; Dorle, A.K. Biodegradation and in vivo biocompatibility of rosin: A natural film-forming polymer. *AAPS PharmSciTech* **2003**, *4*, 434–439. [CrossRef] [PubMed]
27. Niu, X.; Liu, Y.; Song, Y.; Han, J.; Pan, H. Rosin modified cellulose nanofiber as a reinforcing and co-antimicrobial agents in polylactic acid/chitosan composite film for food packaging. *Carbohydr. Polym.* **2018**, *183*, 102–109. [CrossRef] [PubMed]
28. Moustafa, H.; El Kissi, N.; Abou-Kandil, A.I.; Abdel-Aziz, M.S.; Dufresne, A. PLA/PBAT Bionanocomposites with Antimicrobial Natural Rosin for Green Packaging. *ACS Appl. Mater. Interfaces* **2017**, *9*, 20132–20141. [CrossRef] [PubMed]
29. Dong, Y.; Yan, Y.; Wang, K.; Li, J.; Zhang, S.; Xia, C.; Shi, S.Q.; Cai, L. Improvement of water resistance, dimensional stability, and mechanical properties of poplar wood by rosin impregnation. *Eur. J. Wood Wood Prod.* **2016**, *74*, 177–184. [CrossRef]
30. Nguyen, T.T.H.; Li, S.; Li, J. The combined effects of copper sulfate and rosin sizing agent treatment on some physical and mechanical properties of poplar wood. *Constr. Build. Mater.* **2013**, *40*, 33–39. [CrossRef]
31. Dahlen, J.; Nicholas, D.D.; Schultz, T.P. Water Repellency and Dimensional Stability of Southern Pine Decking Treated with Waterborne Resin Acids. *J. Wood Chem. Technol.* **2008**, *28*, 47–54. [CrossRef]
32. Su, N.; Fang, C.; Zhou, H.; Tang, T.; Zhang, S.; Fei, B. Hydrophobic treatment of bamboo with rosin. *Constr. Build. Mater.* **2021**, *271*, 121507. [CrossRef]
33. Xu, J.F.; Zhang, H.N. Modern Furniture Color Image Based on Eye Tracking. *Appl. Mech. Mater.* **2012**, *157–158*, 410–414. [CrossRef]
34. Tuch, A.; Kreibig, S.; Roth, S.; Bargas-Avila, J.; Opwis, K.; Wilhelm, F. The Role of Visual Complexity in Affective Reactions to Webpages: Subjective, Eye Movement, and Cardiovascular Responses. *IEEE Trans. Affect. Comput.* **2011**, *2*, 230–236. [CrossRef]
35. Granka, L.A.; Joachims, T.; Gay, G. Eye-tracking analysis of user behavior in WWW search. In Proceedings of the International ACM SIGIR Conference on Research & Development in Information Retrieval, Sheffield, UK, 25–29 July 2004.
36. Jacob, R.J.K. Eye Tracking in Advanced Interface Design. *Virtual Environ. Adv. Interface Des.* **1995**, *258*, 288. [CrossRef]
37. Ozcelik, E.; Karakus, T.; Kursun, E.; Cagiltay, K. An eye-tracking study of how color coding affects multimedia learning. *Comput. Educ.* **2009**, *53*, 445–453. [CrossRef]
38. Holmqvist, K.; Nystrm, M.; Andersson, R.; Dewhurst, R.; Weijer, J. *Eye Tracking: A Comprehensive Guide to Methods and Measures*; Oxford University Press: Oxford, UK, 2011.
39. Kaufman, A.A.; Bandopadhay, A.; Piligian, G.J. Apparatus and Method for Eye Tracking Interface. U.S. Patent 5,360,971, 1 November 1994.
40. Wan, Q.; Wang, G.G.; Zhang, Y.C.; Song, S.S.; Fei, B.H.; Li, X.H. Cognitve processing torword traditional and new Chinese style furniture: Evidence from eye-tracking technology. *Wood Res.* **2018**, *63*, 727–740.
41. Suárez, L.A.D.L.F. Subjective experience and visual attention to a historic building: A real-world eye-tracking study. *Front. Arch. Res.* **2020**, *9*, 774–804. [CrossRef]
42. *ISO 2813:2014. Paints and Varnishes—Determination of Gloss Value at 20 Degrees, 60 Degrees and 85 Degrees*; CEN/TC 139—Paints and Varnishes; International Organization for Standardization: Geneva, Switzerland, 2014.
43. Brischke, C.; Welzbacher, C.R.; Brandt, K.; Rapp, A.O. Quality control of thermally modified timber: Interrelationship between heat treatment intensities and CIE $L*a*b*$ color data on homogenized wood samples. *Holzforschung* **2007**, *61*, 19–22. [CrossRef]
44. Habashi, F. Witt and the Theory of Dyeing. *Latest Trends Text. Fash. Des.* **2019**, *3*. [CrossRef]
45. Liu, J.L.; Liu, X.M.; Li, W.G.; Ma, L.; Shen, F. Kinetics of gum rosin oxidation under 365 nm ultraviolet irradiation. *Mon. Chem.* **2014**, *145*, 209–212. [CrossRef]
46. Liese, W. *The Anatomy of Bamboo Culms*; Inbar Technical Report; International Network for Bamboo and Rattan: Beijing, China, 1998.

47. Luxmoore, A.R. *Appendix—Basic Principles of Optics*; Springer: Dordrecht, The Netherlands, 1983.
48. Fahey, L.M.; Nieuwoudt, M.K.; Harris, P.J. Predicting the cell-wall compositions of *Pinus radiata* (radiata pine) wood using ATR and transmission FTIR spectroscopies. *Cellulose* **2017**, *24*, 5275–5293. [CrossRef]
49. Yin, Y.; Berglund, L.; Salmén, L. Effect of Steam Treatment on the Properties of Wood Cell Walls. *Biomacromolecules* **2011**, *12*, 194–202. [CrossRef]
50. Artaki, I.; Ray, U.; Gordon, H.; Gervasio, M. Thermal degradation of rosin during high temperature solder reflow. *Thermochim. Acta* **1992**, *198*, 7–20. [CrossRef]
51. Evans, P.D.; Thay, P.D.; Schmalzl, K.J. Degradation of wood surfaces during natural weathering. Effects on lignin and cellulose and on the adhesion of acrylic latex primers. *Wood Sci. Technol.* **1996**, *30*, 411–422. [CrossRef]
52. Ettwein, F.; Rohrer-Vanzo, V.; Langthaler, G.; Stern, T.; Moser, O.; Leitner, R.; Regenfelder, K.; Werner, A. Consumer's perception of high gloss furniture: Instrumental gloss measurement versus visual gloss evaluation. *Eur. J. Wood Wood Prod.* **2017**, *75*, 1009–1016. [CrossRef]
53. Tidén, J.; Klich, M. *Visualisering av Eye-Tracking Data—Heat Map Generator*; School of Computer Science & Communication: Stockholm, Sweden, 2011.

Article

Investigation of the Release Mechanism and Mould Resistance of Citral-Loaded Bamboo Strips

Rui Peng, Jingjing Zhang, Chungui Du *, Qi Li, Ailian Hu, Chunlin Liu, Shiqin Chen, Yingying Shan and Wenxiu Yin

College of Chemistry and Materials Engineering, Zhejiang A & F University, Hangzhou 311300, China; 18255276196@163.com (R.P.); jingjingzhang312@163.com (J.Z.); LQ950011@163.com (Q.L.); hal15857832323@163.com (A.H.); eustaceweaver7187@gmail.com (C.L.); 18768107239@163.com (S.C.); syy15968566686@163.com (Y.S.); yinwenxiu_110@163.com (W.Y.)
* Correspondence: chunguidu@163.com

Citation: Peng, R.; Zhang, J.; Du, C.; Li, Q.; Hu, A.; Liu, C.; Chen, S.; Shan, Y.; Yin, W. Investigation of the Release Mechanism and Mould Resistance of Citral-Loaded Bamboo Strips. *Polymers* **2021**, *13*, 3314. https://doi.org/10.3390/polym13193314

Academic Editors: Jingpeng Li, Yun Lu and Huiqing Wang

Received: 6 September 2021
Accepted: 24 September 2021
Published: 28 September 2021

Publisher's Note: MDPI stays neutral with regard to jurisdictional claims in published maps and institutional affiliations.

Copyright: © 2021 by the authors. Licensee MDPI, Basel, Switzerland. This article is an open access article distributed under the terms and conditions of the Creative Commons Attribution (CC BY) license (https://creativecommons.org/licenses/by/4.0/).

Abstract: In the present study, the sustained-release system loading citral was synthesised by using PNIPAm nanohydrogel as a carrier and analysed its drug-release kinetics and mechanism. Four release models, namely zero-order, first-order, Higuchi, and Peppas, were employed to fit the experimental data, and the underlying action mechanism was analysed. The optimised system was applied to treat a bamboo mould, followed by assessment of the mould-proof performance. Our experimental results revealed that the release kinetics equation of the system conformed to the first order; the higher the external temperature, the better the match was. In the release process, PNIPAm demonstrated a good protection and sustained-release effect on citral. Under the pressure of 0.5 MPa, immersion time of 120 min, and the system concentration ratio of 1, the optimal drug-loading parameters were obtained using the slow-release system with the best release parameters. Compared to the other conditions, bamboos treated with pressure impregnation demonstrated a better control effect on bamboo mould, while the control effect on *Penicillium citrinum*, *Trichoderma viride*, *Aspergillus niger*, and mixed mould was 100% after 28 days. Moreover, the structure and colour of bamboo remained unchanged during the entire process of mould control.

Keywords: nanohydrogels; citral; sustained-release; mechanism; bamboo; anti-mould

1. Introduction

Parallel to the excessive carbon dioxide emissions, the emission of greenhouse gases has increased considerably, making climate change a global challenge for mankind. Following the 'carbon neutral' goal of the Paris Agreement, the global market demand for low-carbon materials has been increasing [1,2]. As a natural low-carbon material, bamboo offers the advantages of sustainable logging and utilisation, high strength to weight ratio, high strength, and biodegradability [3–5]. It is not only commonly used in the construction, furniture, and interior decoration industries but also has significant application prospects in the aerospace domain [6–9]. Therefore, vigorous production and application of bamboo are crucial for promoting the transformation of social development mode from high carbon to green and low carbon. However, bamboo gets easily infected with mould under certain temperature and humidity conditions owing to its high sugar and protein contents, resulting in serious pollution and the loss of bamboo use value [10,11]. Thus, research on bamboo mould prevention can not only effectively improve the utilisation efficiency of bamboo but also play a positive role in realising the vision of a carbon-free future.

Nanomaterials are one of the most widely studied and applied materials owing to their unique nano-properties [12–14]. These materials have also become a research hotspot in the field of bamboo mould prevention to prepare a sustained-release system with nanomaterials as a carrier material. Slow release refers to the slowdown of the release rate of loaded fungicides that extends the associated release time under isolation and protection of the carrier material [15,16]. Notably, the release of mould agents is the key to determining the effect of mould. Based on the different properties of the carrier material, such as low critical solution temperature, permeability, viscosity, degradability, and concentration [17–20], the release parameters, including the release time, release amount, and release rate [21], can be adjusted to a certain extent, which in turn affects the performance of mould prevention in a slow-release system. Hydrophilic carriers are usually selected because they possess pore space to promote the release of mould inhibitors, especially for low-water-soluble drugs, which easily bind with the hydrophilic groups in bamboo. When the carrier comes into contact with the water medium, various phenomena such as degradation, dissolution, and swelling can occur [22]. Under the action of diffusion, convection, explosion, ion exchange, osmotic pressure, and other mechanisms, different release curves are produced [23]. In addition, the release parameters are related to the properties of fungicides, such as the solubility, particle size, and hydrophilicity [15]. The degradation rate of the carrier, the drug load, and the characteristics of the fungicide synergistically influence the release effect of the fungicide. Therefore, according to the choice of carrier and fungicide, different release mechanisms have been found conducive to the release of fungicide from the carrier material. However, depending on the proportion of participation, each mechanism can represent either a major release mechanism or a negligible one.

Although the sustained-release system with mould inhibitor has been applied and assessed in the field of mould control of bamboo, studies have mainly focused on the preparation process and performance, while the systematic studies on the mechanism of sustained release are rare [10,14]. Poly-n-isopropylacrylamide (PNIPAm) nanohydrogels are widely used in drug delivery systems as drug delivery carriers due to their good thermal sensitivity and biocompatibility, and their unique amphiphilic groups have a strong solubilizing effect on hydrophobic drugs. The natural antibacterial agent citral was encapsulated in PNIPAm nanohydrogel, which not only overcomes its own disadvantages of easy oxidation and volatile, but also facilitates the control of release, so that its hydrophilicity, chemical stability, antibacterial activity and utilization efficiency have been greatly improved. In this study, a PNIPAm/citral nanohydrogel (P/Cn) sustained-release system was prepared using PNIPAm nanohydrogel as the carrier material and a natural plant essential oil, citral, as the mould inhibitor. To explore the release mechanism of the sustained-release system, the release parameters of P/Cn were investigated through a release experiment, where the release kinetics equation was fitted. On this basis, the technology behind P/Cn for mould prevention treatment of bamboo was optimised, and the mould prevention performance was discussed. The schematic diagram of the experimental principle in this study is shown in Figure 1. The purpose of this study was to provide a theoretical and practical basis for obtaining the ideal release effect of P/Cn as well as to provide new ideas and methods for realising the efficient and lasting protection of bamboo material.

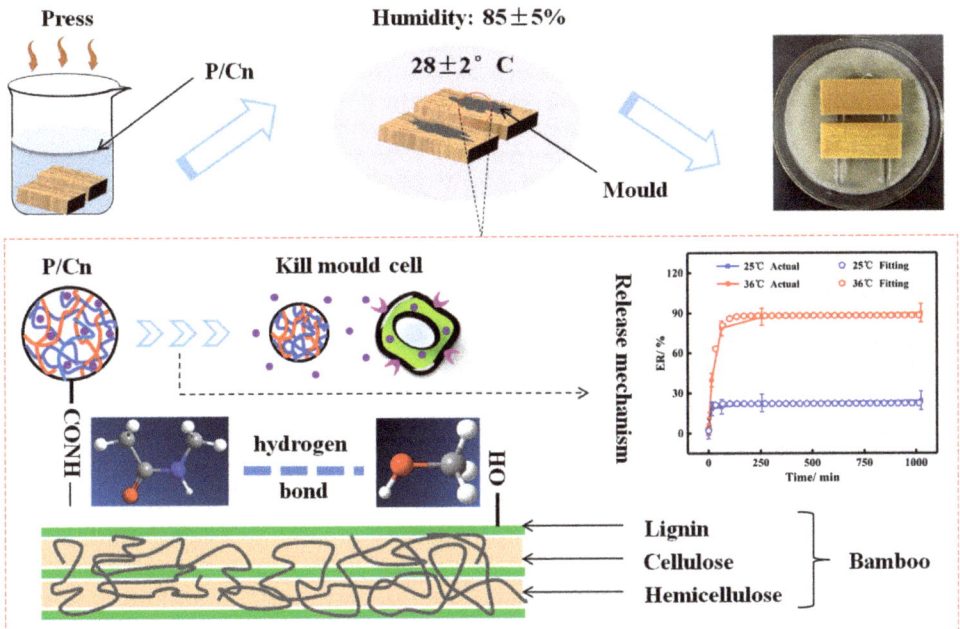

Figure 1. The schematic diagram of experimental principle.

2. Materials and Methods

2.1. Materials

P/Cns were synthesised in the laboratory. Details of the synthesis are presented in Supplementary Materials. Briefly, disodium hydrogen phosphate (AR) and sodium dihydrogen phosphate (AR) were both sourced from Sinophenol Chemical Reagent Co., Ltd. (Shanghai, China). Polyethylene glycol monostearate (PEGMS, n = 45, AR) was supplied by China Aladdin Reagent Co., Ltd. (Shanghai, China). Anhydrous ethanol (AR) was provided by Tianjin Yongda Chemical Reagent Co., Ltd. (Tianjin, China). Bamboo (moisture content of approximately 10%) was purchased from Zhejiang Yongyu Bamboo Industry Co., Ltd. (Hangzhou, China). and customised into a unified specification of 50 × 20 × 5 mm^3 strips in the laboratory. All reagents and solvents were used directly in this experiment without further purification.

2.2. Preparation of Phosphate-Buffered Saline

The preparation process of phosphate-buffered saline (PBS; 0.1 mol/L, pH = 6.8) was as follows: 35.822 g of disodium hydrogen phosphate and 15.603 g of disodium hydrogen phosphate were weighed and added to 2000 mL deionised water. Finally, 2.0 g of PEGMS was added and stirred until a clarified mixture was obtained, which was then stored at room temperature until further use.

2.3. The Standard Curve

First, the maximum absorption wavelength (λ_{Max}) of citral in PBS was measured by scanning the buffer at a full wavelength on an ultraviolet–visible (UV–Vis) spectrophotometer. Second, 0.01 g of citral was weighed accurately, and a small amount of PBS solution was added to the mixture. The concentrated storage solution of 100 μg/mL was finally obtained by transferring to achieve a constant volume in a 100-mL volumetric flask. Then, 2–8 μg/mL of standard citral solution was prepared in PBS as the solvent. Finally, the absorbance value of each citral standard solution was measured (measured thrice and then averaged) at λ_{Max}, and the absorbance was plotted against the concentration to draw a standard curve.

2.4. Release and Detection of Citral

Dialysis is the most common method to determine the extent of drug release in vitro in the nano-drug delivery systems [24,25]. Briefly, 1 g of the P/Cn solution was accurately weighed into a dialysis bag (MWCO: 8000–14,000) and placed in a beaker containing 100 mL of PBS. Then, the beaker was placed in a water bath and stirred at a rate of 100 rpm. Afterwards, 2 mL of the solution was removed from the beaker for detection within a specified time. The same volume of fresh PBS buffer was added to the beaker [26].

The absorbance of citral (at λ_{Max}) was determined through UV–Vis spectrophotometry after stirring the solution uniformly and adding fresh PBS solution to 20 mL. Then, the cumulative release L_2 of citral at different time points was calculated from the standard curve (measured thrice and then averaged), and the release rate of citral was calculated according to the following Formula (1):

$$\mathrm{ER}\ (\%) = L_2/L_1 \times 100\%, \tag{1}$$

where ER is the release rate of citral, %; L_1 is the mass of encapsulated citral (see the Supplementary Materials), mg; L_2 is the cumulative release of citral at different time points, mg.

2.5. Optimisation of Release Parameters

The addition of citral (mg), the reaction time (h), and the reaction temperature (°C) were selected as the investigation factors. The orthogonal test design of $L_9(3^4)$ was selected to determine the optimal release parameters of P/Cn under the release temperature of 36 °C.

2.6. Sustained-Release Kinetics

On the basis of the release experiment, the release curve of P/Cn was plotted. Four release models, including zero-order, first-order, Higuchi, and Peppas, were employed to fit the release curve [27], and the release mechanism was described in detail. The four dynamics equations applied are as follows:

$$Q = K\,t, \qquad \text{(Zero-order)}$$

$$Q = 1 - e^{(-k\,t)}, \qquad \text{(First-order)}$$

$$Q = K\,t^{1/2}, \qquad \text{(Higu-chi)}$$

$$Q = K\,t^n, \qquad \text{(Peppas)}$$

where Q and t represent the release rate and release time, respectively; K represents the release rate constant; and n represents the diffusion index.

2.7. Impregnation of Bamboo Strips

First, the bamboo strips with a moisture content of approximately 10% were weighed (m_1) and fully immersed in a beaker containing P/Cn. The strips were then placed in a pressurised tank for the dipping treatment. Then, the impregnated bamboo strips were weighed (m_2) and placed in a vacuum drying oven. The bamboo strips were dried to approximately 10% moisture content at 30 °C and then weighed (m_3). Finally, the samples were stored in a dryer for subsequent experiments.

2.8. Optimisation of the Impregnation Process

The value of pressure, impregnation time, and P/Cn concentration ratio (CR = P/Cn concentration of the treated bamboo strips divided by the P/Cn original concentration) were selected as experimental factors and used to investigate the influence of various factors on the drug-loading capacity of the bamboo strips. The wet drug load (R_1) and dry drug load (R_2) of the sample were calculated using the Formulas (2) and (3), respectively:

$$R_1 = \frac{(m_2 - m_1) \times C \times 10^6}{S}, \quad (2)$$

$$R_2 = \frac{(m_3 - m_1) \times C \times 10^6}{S}, \quad (3)$$

In the formula, R_1 and R_2 indicate wet and dry drug loadings, respectively, mg/m^2; m_1, m_2, and m_3 indicate the mass of bamboo strips before treatment, the wet weight after treatment, and the dry weight after drying, respectively, mg; C is the concentration (mass fraction) of citral in the system, %; and S is the sum of the surface area of the six surfaces of the bamboo strips, m^2.

2.9. Fourier-Transformed Infrared Spectroscopy

The bamboo strips were crushed with a micro-plant pulveriser and screened with a 300-purpose screen. The bamboo powder obtained was mixed with potassium bromide and pressed into thin slices according to the mass of 1:100 and finally scanned with IR Prestige-21 (Shimadzu, Shanghai, China). The resolution of Fourier-transform infrared spectroscopy (FT-IR) was kept at 4 cm^{-1}, and the wavelength range was 4000–400 cm^{-1}.

2.10. Anti-Mould Property

The bamboo strips were treated by P/Cn atmospheric pressure impregnation (OT) and pressure impregnation (PT) and treated by P/Cn pressure impregnation after the completion of the release experiment (RT). According to GB/T 18261-2013 [28], an indoor mould test was performed under the conditions of 28 ± 2 °C temperature and 85% ± 5% relative humidity. The control effects of the samples on *Penicillium citrinum* (PC), *Trichoderma viride* (TV), *Aspergillus niger* (AN), and mixed mould (mixed ratio of PC, TV, and AN was 1:1:1, PTA for short) were investigated, and the untreated bamboo strips (UT) were set as the blank control. During the mould prevention period, the degree of mould infection on the bamboo strips was recorded weekly (see the Supplementary Materials Table S1 for the evaluation criteria of the infection level), and the bamboo strips were photographed after 28 days of mould prevention to analyse the mould prevention results.

3. Results and Discussion

3.1. Draw the Standard Curve

The maximum absorption peak of citral was obtained at 238 nm (Figure 2A), while P/Cn showed a large absorption value at the same wavelength, which was similar to the maximum absorption wavelength of 240 nm and was not interfered with other substances. Accordingly, the standard solution of citral-PBS with a concentration of 2–8 μg/mL was prepared, and the absorbance value was measured at the wavelength of 238 nm. Figure 2B depicts the regression equation of the standard curve:

$$A = 0.08200\,C - 0.00571, \quad \left(R^2 = 0.999731\right)$$

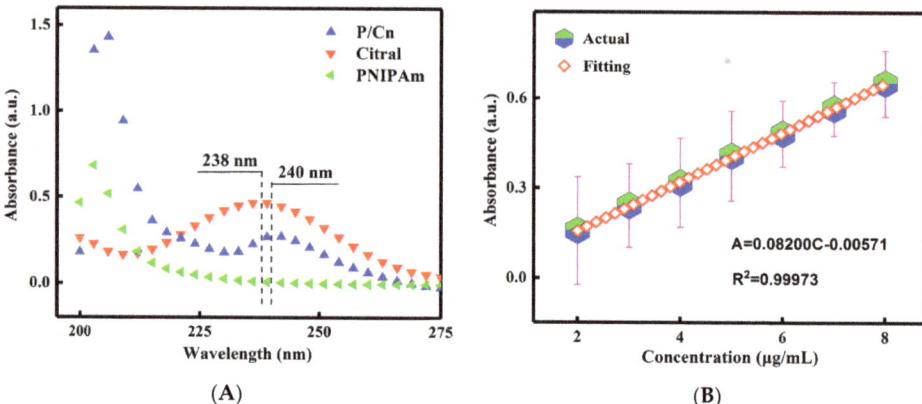

Figure 2. The Uv-vis spectra (**A**) and the standard curves (**B**) of P/Cn in PBS.

The fitting correlation coefficient R^2 of the standard curve of citral in PBS was ultra-close to 1, which indicated a favourable linear relationship between the concentration of citral and absorbance. The use of this standard curve can avoid experimental errors to the maximum extent and help meet the experimental requirements.

3.2. Orthogonal Test

To obtain the optimal release parameters of P/Cn (L_2 and ER), orthogonal tests were designed using the L_9 (3^4) orthogonal table (See the Supplementary Materials Table S2 for the factor level), and the results are shown in Table 1. According to the R-value, the influence of each factor on L_2 was in the order of A > C > B, with the optimal combination at $A_3B_2C_1$ (P/Cn8). In other words, under the citral amount of 50 mg, reaction time of 2 h, and reaction temperature of 25 °C, the cumulative release of citral was 33.469 mg, and the release rate was 89.986%. Conversely, the influence of each factor on ER was in the order of A > B > C, while the optimal combination was $A_3B_3C_2$ (P/Cn9). In other words, under the citral amount of 50 mg, reaction time of 3 h, and reaction temperature of 30 °C, the cumulative release amount of citral was 31.093 mg, and the release rate was 95.828%.

Table 1. The results of orthogonal test.

Test No.	Factor			L_2 (mg)	ER (%)
	A (mg)	B (h)	C (°C)		
1	1	1	1	5.099	81.919
2	1	2	2	5.628	85.675
3	1	3	3	4.942	81.919
4	2	1	2	19.820	82.893
5	2	2	3	19.748	82.197
6	2	3	1	22.515	90.542
7	3	1	3	27.132	85.953
8	3	2	1	33.469	89.986
9	3	3	2	31.093	95.828
k_1 (L_3)	5.223	17.35	20.361		
k_2 (L_3)	20.694	19.615	18.847		
k_3 (L_3)	30.565	19.517	17.274		
k_1 (ER)	83.171	83.588	87.482		
k_2 (ER)	85.211	85.953	88.132		
k_3 (ER)	90.589	89.43	83.356		
R (L_3)	25.342	2.265	3.087		
R (ER)	7.418	5.842	4.776		

Variance analysis was conducted for L_2 (Table 2). At the significance level of 1%, the amount of citral added showed a significant impact on the cumulative release of citral. As shown in Table 2, the amount of citral added also made a significant difference to the cumulative release of citral under the significance level of 10%. Hence, the amount of citral showed a significant effect on L_2 and ER. However, to maximise mould resistance, P/Cn8 with a higher L_2 value was considered as the optimal preparation result in this study, and its release mechanism was investigated.

Table 2. The results of ANOVA.

Source	DEVSQ		F		$F_{0.01}$		$F_{0.1}$		Significance	
	L_2	ER	L_2	ER	L_2	ER	L_2	ER	L_2	ER
A	978.986	88.113	610.721	9.245	99	9			*	*
B	9.831	51.801	6.133	5.435	99	9				
C	14.296	40.253	8.918	4.223	99	9				
Error	1.600	9.530								
Total	1004.713	189.697								

*: Represent the significant influence of parameter.

3.3. Release Kinetics Analysis

The release mechanism model of P/Cn can reveal the relationship between the release kinetics and the release architecture variables in essence [23]. Consequently, the release curves of P/Cn8 at 25 °C and 36 °C were plotted. Then, four kinetic models were used to fit the release curves. Figure 3 shows the fitting results of the release curves of P/Cn8 at different environment temperatures. Ultimately, the fitting effect was judged according to the fitting correlation coefficient R^2 (Table 3), followed by the kinetic analyses.

As depicted in Table 3, at 25 °C, the fitting correlation coefficients R^2 of the P/Cn8 release curve to the zero-order, first-order, Higuchi, and Peppas dynamics model were 0.11427, 0.96479, 0.38481, and 0.80165, respectively. At 36 °C, the corresponding fitting correlation coefficients R^2 of the P/Cn8 release curve to the four kinetic equations were 0.24557, 0.99905, 0.55642, and 0.88923, respectively. All P/Cn8 conformed to the first-order kinetic model at 25 °C and 36 °C, and the corresponding kinetic equations were $Q = 22.295 \times (1 - e^{-0.091t})$ and $Q = 88.143 \times (1 - e^{-0.036t})$, respectively. R^2 (36 °C) was significantly larger than R^2 (25 °C), indicating that the higher the release temperature was, the more consistent the release mechanism was with the first-order kinetic model. On one hand, the hydrogen bond formed between PNIPAm and water may be broken at high temperatures [29], and P/Cn8 releases citral under the action of hydrophobic groups. On the other hand, the diffusion rate of citral increases with an increase in temperature.

Figure 3 (Actual) depicts that 0–128 min was the burst-release stage of citral, during which citral was rapidly released. This could be partly because citral was in the free state or at the edge of PNIPAm hydrogel, and with the increase in the temperature, the diffusion rate of free citral increased, which increased the burst-release effect [30]. The moderate release process was noted at 128–512 min, and citral was released by free diffusion under the influence of concentration gradient [31]. With the passage of time, the concentration of citral in PNIPAm decreased and the release driving force decreased [15,27]; therefore, its release rate slowed down at 512–1024 min.

In addition, the cumulative release rate of P/Cn8 at 36 °C at 1024 min was 89.986%, and the embedded citral was almost completely released. After 1024 min, the cumulative release rate at 25 °C was only 24.348%. In conclusion, PNIPAm showed excellent protective and sustained-release effects on citral.

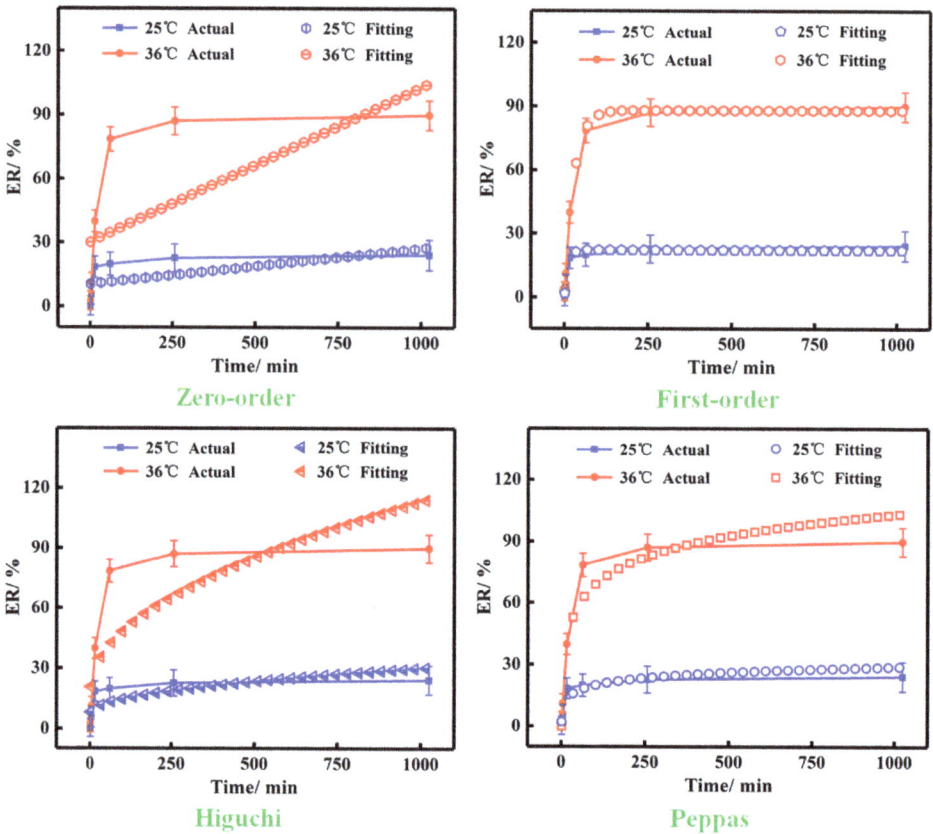

Figure 3. The release curves and fitting results of P/Cn8.

Table 3. Kinetic fitting results.

Equation	25 °C		36 °C	
	Fitted Equation	R^2	Fitted Equation	R^2
Zero-order	$Q = 0.017t + 10.586$	0.11427	$Q = 0.073t + 30.049$	0.24557
First-order	$Q = 22.295 \times (1 - e^{-0.091t})$	0.96479	$Q = 88.143 \times (1 - e^{-0.036t})$	0.99905
Higuchi	$Q = 0.719t^{1/2} + 7.528$	0.38481	$Q = 3.025t^{1/2} + 17.720$	0.55642
Peppas	$Q = 44{,}021.197t^{0.00088} - 44{,}018.933$	0.80165	$Q = 55{,}938.699t^{0.00027} - 55{,}938.810$	0.88923

3.4. Effect of Pressure

P/Cn8 was taken after dialysis when the impregnation time was 60 min and CR was 1. Bamboo strips were impregnated at different pressures; R_1 and R_2 were calculated, and then the two-drug loadings were plotted according to the pressure (Figure 4).

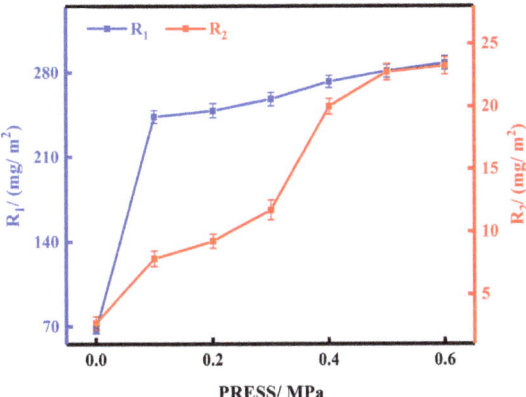

Figure 4. The effect of pressure on drug loading of bamboo strips.

As shown in the figure, both R_1 and R_2 increased with an increase in the pressure. When the pressure value was 0–0.5 MPa, the amount of the two types of drug loading increased, whereas when the pressure was 0.5 MPa, R_1 and R_2 were 281.420 mg/m^2 and 22.747 mg/m^2, respectively. However, with an increase in the pressure, the amount of increase in both the drug loads decreased. R_1 and R_2 of the bamboo strips only increased by 2.336% and 2.171% compared with 0.5 MPa, respectively, at 0.6 MPa. Under the premise of considering the actual production benefit, 0.5 MPa, therefore, was determined as the most appropriate pressure. In addition, after 0.5-MPa pressure impregnation, R_1 and R_2 of bamboo strips were 318.257% and 737.518% of that under normal pressure (the pressure value of 0 MPa), respectively. This observation indicates that pressure can significantly improve the drug loading of bamboo strips. Because under the action of pressure, the air inside the bamboo was discharged, the voids inside the bamboo were increased, so that the resistance of P/Cn diffusion into the bamboo was weakened, and the permeability and adsorption performance of the bamboo were greatly improved.

3.5. Effect of Impregnation Time

The P/Cn8 completed by dialysis was collected, and the bamboo strips were treated at different impregnation times under the pressure of 0.5 MPa and CR of 1. R_1 and R_2 were calculated, and the results of the two-drug loads were plotted according to the pressure (Figure 5).

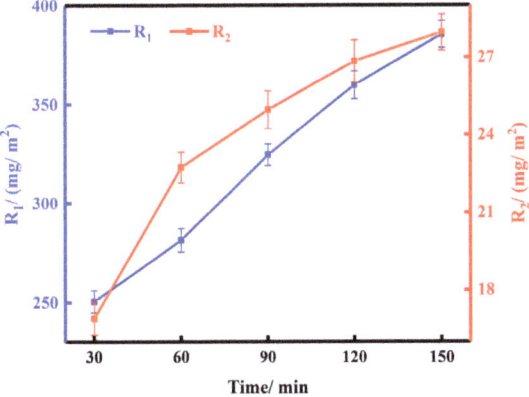

Figure 5. The effect of impregnation time on drug loading of bamboo strips.

As illustrated in Figure 5, R_1 and R_2 both increased with the extension of impregnation time. When the impregnation time was <120 min, the two-drug loading of bamboo strips increased rapidly with time, and the curve slope increased. At impregnation time of 120 min, R_1, and R_2 were 359.815 mg/m^2 and 26.852 mg/m^2, respectively. However, with the extension of impregnation time, the drug loading increased slowly. R_1 and R_2 only increased by 7.119% and 4.137% at the impregnation time of 150 min compared with that at 120 min. This result can be attributed to the fact that the surface of bamboo was saturated with P/Cn, and therefore, the increased impregnation time had only a slight effect on drug loading. Under the premise of considering the actual production efficiency, 120 min was determined as the optimum impregnation time.

3.6. Effect of CR

P/Cn8 after dialysis was collected and used to treat the bamboo strips under different CR, pressure of 0.5 MPa and impregnation time of 120 min. R_1 and R_2 were accordingly calculated, and the results of the two-drug loading were plotted according to the pressure (Figure 6).

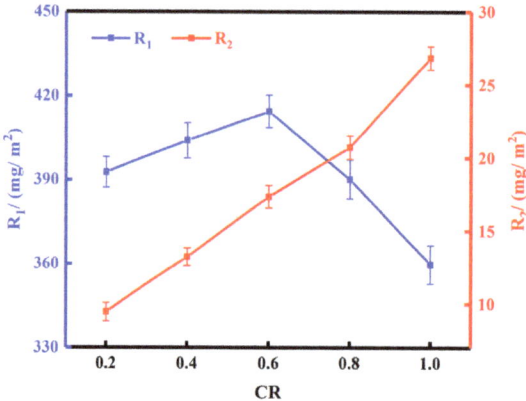

Figure 6. The effect of CR on drug loading of bamboo strips.

As observed in the Figure 6, R_1 and R_2 exhibited different trends with the change in CR. R_1 first showed an increasing trend and then a decreasing trend with increasing CR, while the maximum R_1 of bamboo was 414.506 mg/m^2 at CR = 0.6. This observation can be attributed to the fact that the greater the concentration of P/Cn, the greater the relative viscosity of its solution was, and the worse the fluidity and the obstruction of water infiltration into bamboo was. R_2 mainly depended on the concentration of P/Cn after evaporation of water in the bamboo strips. Therefore, an increase in the concentration of P/Cn could significantly improve R_2. However, considering that high concentration can easily cause gel precipitation that makes it difficult to impregnate bamboo [32,33], the value of 1.0 was eventually determined as the optimal CR. In other words, the original concentration of P/Cn8 was selected for bamboo strip processing. At this concentration, the R_2 of bamboo was as high as 26.852 mg/m^2.

In summary, the optimal process conditions for P/Cn impregnation of bamboo strips were as follows: 0.5 MPa pressure, 120 min impregnation time, and CR of 1.0. P/Cn8. The bamboo composites were prepared through this process for subsequent characterisation and mould-proof experiments.

3.7. FT-IR Analysis

Pure, untreated bamboo strips were used as blanks, and changes in the molecular structure of P/Cn8/bamboo were analysed through FT-IR. We found that, as per the spectrogram of blank bamboo strips shown in Figure 7, the absorption peak at wave number 1735 cm^{-1} was caused by the stretching vibration of the C=O hydroxyl groups on structures such as xylose and arabinose, which indicates the existence of hemicelluloses [34]. The absorption peaks related to the benzene ring structure at 1601, 1507, and 1244 cm^{-1} were the characteristic absorption peaks of lignin, the deformation vibration of CH$_2$ in lignin at 1460 cm^{-1}, and the stretching vibration of C-O in lignin at 1327 cm^{-1}. The absorption peaks at 1424, 1377, and 897 cm^{-1} represented the characteristic absorption peaks of cellulose [35]. However, all peaks of P/Cn8/bamboo remained unchanged in the spectrogram, which indicated that the chemical composition of the bamboo strips did not change after P/Cn8 dipping. In addition, the characteristic absorption peak of citral did not appear in this spectrum; hence, it was speculated that citral existed in bamboo in the form of a complex and that its molecular vibration was limited by PNIPAm. Therefore, the characteristic absorption peak of the aldehyde group of citral was not shown after encapsulation [36].

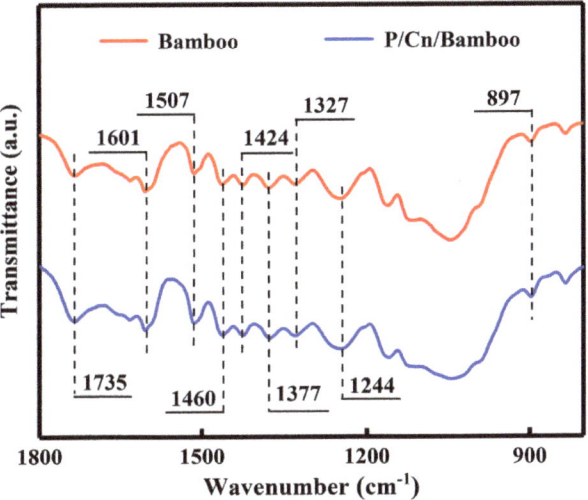

Figure 7. The FT−IR spectra of impregnated bamboo strips.

3.8. Analysis of Mould Resistance

Based on the previous experimental results, four groups, namely PT, OT, RT, and UT, were selected to conduct indoor mould prevention experiments. During the mould control period, the infection grades of bamboo slices were recorded weekly (Table 4), while the mould control results were photographed after 28 days (as shown in Figure 8) to analyse the influence of different treatment methods on the mould control performance of the bamboo strips.

As shown in Figure 8 and Table 4, the surface of UT was covered with clearly visible PC, TV, AN, and PTA, respectively, and the infection value reached 4.0, which fully met the requirements of the mould control experiment. In addition, all the bamboo pieces were discoloured, and the bamboo itself lost colour, which reduced the bamboo's use value. Therefore, it is necessary to treat bamboo with mould prevention. The surface of RT was also covered with clearly visible mould, and the infection grade on the surface of bamboo slices was 100%, indicating that P/Cn8 could not protect bamboo slices from mould despite sustained release.

Table 4. Infection grade of bamboo strips during mould control.

Moulds	Day 7	Day 14	Day 21	Day 28	Group
PC	0.1	0.6	0.9	1.2	OT
	0	0	0	0	PT
	1.2	2.2	3.5	4.0	UT
	1.0	2.0	3.0	4.0	RT
TV	0.7	1.3	1.9	2.1	OT
	0	0	0	0	PT
	1.5	2.5	3.6	4.0	UT
	1.1	2.3	3.2	4.0	RT
AN	1.3	2.1	2.6	3.2	OT
	0	0	0	0	PT
	2.0	3.5	4.0	4.0	UT
	1.8	3.1	3.9	4.0	RT
PTA	1.4	2.2	2.6	3.0	OT
	0	0	0	0	PT
	2.1	3.6	4	4	UT
	1.9	3.6	3.9	4	RT

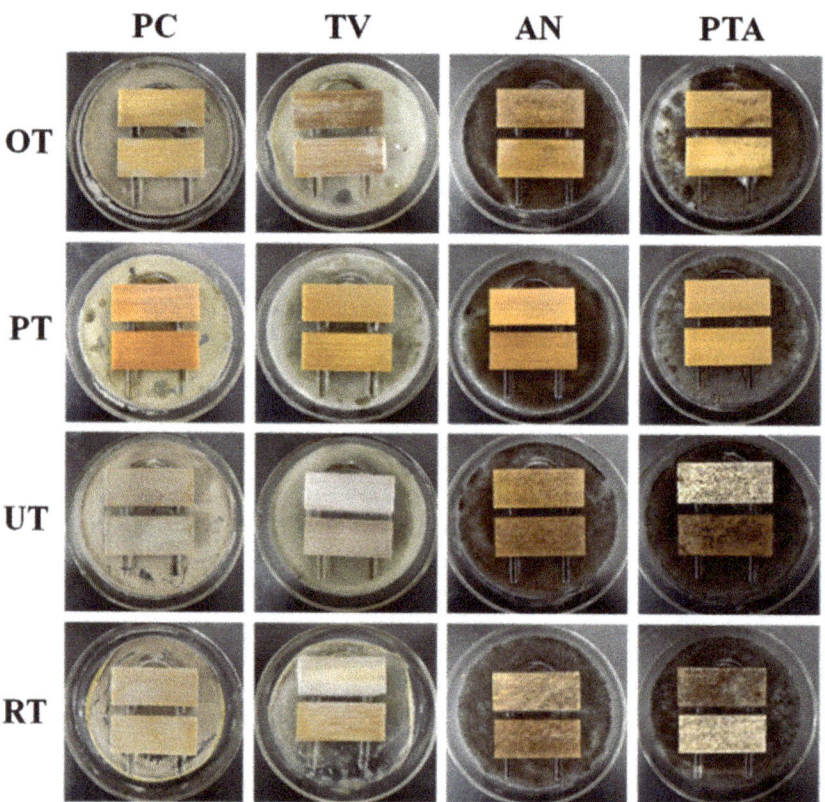

Figure 8. Mould control results of bamboo strips after 28 days.

However, the observed results of UT and RT infection with mould during the experiment showed that RT infection levels were lower than UT infection levels during the same time period (0–3 weeks). These results indicated that the P/Cn8 after sustained release of mould growth inhibition played a role, but, as a result of the reduced citral content, it could not prevent the fungus infection of bamboo. As shown in Figure 8 and Table 4, a small amount of mould remained on the surface of OT, and the corresponding infection values of PC, TV, AN, and PTA were 1.2, 2.1, 3.2, and 3.0, respectively. Correspondingly, PT was infected by all moulds at time 0 during the mould control period, and the bamboo strips remained completely natural and infused with a fresh lemon aroma. Therefore, the mould resistance of bamboo strips after P/Cn8-pressure impregnation treatment was obviously better than that of the atmospheric pressure impregnation treatment, which could be attributed to the promotion of P/Cn8 infiltration inside the bamboo tissue under pressure treatment, such that the bamboo strips could have a higher drug-loading capacity and hence a more lasting mould resistance effect.

4. Conclusions

In this study, a PNIPAm-supported citral sustained-release system was prepared through soap-free emulsion polymerisation, and the release mechanism, mould prevention process, and performance of the system were evaluated. The cumulative release parameters of the sustained-release system at different temperatures and time points were assessed, and the release mechanism was evaluated using a common drug-release model. The results show that the system conformed to the first-order kinetics model at 25 °C and 36 °C and that the higher the temperature, the better the first-order kinetics model match was. During the release process, PNIPAm exerted strong protective and sustained-release effects on citral. The effects of impregnating conditions on the drug-loading parameters of bamboo were investigated to reveal the following optimum conditions: pressure of 0.5 MPa, impregnation time of 120 min, and CR of 1.0. The laboratory mould control experiment results revealed that under the optimal conditions of release and impregnation time, the control efficiency of the bamboo treatment with pressure impregnation against the common bamboo moulds, such as *P. citrinum*, *T. viride*, *A. niger*, and mixed mould reached 100% after 28 days, and the original colour of the bamboo maintained during the mould control process. All in all, the relationship between the release mechanism and mildew prevention performance of the sustained-release system was fully revealed in this study, and it showed great application potential in bamboo mildew prevention. In addition, due to the good hydrophilicity of the system, and the mildew resistance of bamboo is more difficult than other substrates (bamboo has a higher density, which makes it difficult for the mildew inhibitor to penetrate), it has full potential to be applied to other hydrophilic biomass materials, such as wood and crop straw.

Supplementary Materials: The following are available online at https://www.mdpi.com/article/10.3390/polym13193314/s1, Table S1. Classification Standard of Surface Infection Levels of Samples. Table S2. Factor level.

Author Contributions: Conceptualization, C.D.; Data curation, R.P., S.C., Y.S. and W.Y.; Formal analysis, R.P., J.Z. and Q.L.; Funding acquisition, C.D.; Methodology, C.D.; Project administration, C.D.; Software, J.Z., A.H. and C.L.; Supervision, C.D.; Writing—original draft, R.P.; Writing—review & editing, C.D. All authors have read and agreed to the published version of the manuscript.

Funding: This research was funded by National Natural Science Foundation of China (31870541) and Zhejiang Provincial Key Research and Development Project (2019C02037).

Conflicts of Interest: The authors declare no conflict of interest.

References

1. Gielen, D.; Boshell, F.; Saygin, D. Climate and energy challenges for materials science. *Nat. Mater.* **2016**, *15*, 117–120. [CrossRef]
2. Brunet-Navarro, P.; Jochheim, H.; Cardellini, G.; Richter, K.; Muys, B. Climate mitigation by energy and material substitution of wood products has an expiry date. *J. Clean. Prod.* **2021**, *303*, 127026. [CrossRef]
3. Peng, R.; Yu, H.; Du, C.; Zhang, J.; Hu, A.; Li, Q.; Hua, Y.; Liu, H.; Chu, S. Preparation of uniformly dispersed N-isopropylacrylamide/Acrylic acid/nanosilver composite hydrogel and its anti-mold properties. *BioResources* **2020**, *16*, 441–454. [CrossRef]

4. Li, Z.; Chen, C.; Mi, R.; Gan, W.; Dai, J.; Jiao, M.; Xie, H.; Yao, Y.; Xiao, S.; Hu, L. A strong, tough, and scalable structural material from fast-growing bamboo. *Adv. Mater.* **2020**, *32*, 1906308. [CrossRef] [PubMed]
5. Chen, C.; Li, Z.; Mi, R.; Dai, J.; Xie, H.; Pei, Y.; Li, J.; Qiao, H.; Tang, H.; Yang, B.; et al. Rapid processing of whole bamboo with exposed, aligned nanofibrils toward a high-performance structural material. *ACS Nano* **2020**, *14*, 5194–5202. [CrossRef] [PubMed]
6. Li, J.; Wu, Z.; Bao, Y.; Chen, Y.; Huang, C.; Li, N.; He, S.; Chen, Z. Wet chemical synthesis of ZnO nanocoating on the surface of bamboo timber with improved mould-resistance. *J. Saudi Chem. Soc.* **2017**, *21*, 920–928. [CrossRef]
7. Chen, J.; Ma, Y.; Lin, H.; Zheng, Q.; Zhang, X.; Yang, W.; Li, R. Fabrication of hydrophobic ZnO/PMHS coatings on bamboo surfaces: The synergistic effect of ZnO and PMHS on anti-mildew properties. *Coatings* **2018**, *9*, 15. [CrossRef]
8. Ren, D.; Li, J.; Bao, Y.; Wu, Z.; He, S.; Wang, A.; Guo, F.; Chen, Y. Low-temperature synthesis of flower-like ZnO microstructures supported on TiO_2 thin films as efficient antifungal coatings for bamboo protection under dark conditions. *Colloids Surf. A Physicochem. Eng. Asp.* **2018**, *555*, 381–388. [CrossRef]
9. Sharma, B.; Gatoo, A.; Bock, M.; Ramage, M. Engineered bamboo for structural applications. *Constr. Build. Mater.* **2015**, *81*, 66–73. [CrossRef]
10. Li, J.; Yu, H.; Wu, Z.; Wang, J.; He, S.; Ji, J.; Li, N.; Bao, Y.; Huang, C.; Chen, Z.; et al. Room temperature synthesis of crystalline anatase TiO_2 on bamboo timber surface and their short-term antifungal capability under natural weather conditions. *Colloids Surf. A Physicochem. Eng. Asp.* **2016**, *508*, 117–123. [CrossRef]
11. Zhang, J.; Du, C.; Li, Q.; Hu, A.; Peng, R.; Sun, F.; Zhang, W. Inhibition mechanism and antibacterial activity of natural antibacterial agent citral on bamboo mould and its anti-mildew effect on bamboo. *R. Soc. Open Sci.* **2021**, *8*, 202244. [CrossRef] [PubMed]
12. Lee, J.; Ku, K.H.; Kim, M.; Shin, J.M.; Han, J.; Park, C.H.; Yi, G.-R.; Jang, S.G.; Kim, B.J. Nanostructured particles: Stimuli-responsive, shape-transforming nanostructured particles. *Adv. Mater.* **2017**, *29*, 1700608. [CrossRef]
13. Sultana, T.; Van Hai, H.; Park, M.; Lee, S.Y.; Lee, B.T. Controlled release of Mitomycin C from modified cellulose based thermo-gel prevents post-operative de novo peritoneal adhesion. *Carbohydr. Polym.* **2020**, *229*, 115552. [CrossRef] [PubMed]
14. Liu, G.; Lu, Z.; Zhu, X.; Du, X.; Hu, J.; Chang, S.; Li, X.; Liu, Y. Facile in-situ growth of Ag/TiO2 nanoparticles on polydopamine modified bamboo with excellent mildew-proofing. *Sci. Rep.* **2019**, *9*, 16496. [CrossRef] [PubMed]
15. Almeida, K.B.; Ramos, A.S.; Nunes, J.B.B.; Silva, B.O.; Ferraz, E.R.A.; Fernandes, A.S.; Felzenszwalb, I.; Amaral, A.C.F.; Roullin, V.G.; Falcão, D.Q. PLGA nanoparticles optimized by Box-Behnken for efficient encapsulation of therapeutic Cymbopogon citratus essential oil. *Colloids Surf. B Biointerfaces* **2019**, *181*, 935–942. [CrossRef]
16. Zhang, R.; Li, Y.; He, Y.; Qin, D. Preparation of iodopropynyl butycarbamate loaded halloysite and its anti-mildew activity. *J. Mater. Res. Technol.* **2020**, *9*, 10148–10156. [CrossRef]
17. Johnson, T.J.; Gupta, K.M.; Fabian, J.; Albright, T.H.; Kiser, P.F. Segmented polyurethane intravaginal rings for the sustained combined delivery of antiretroviral agents dapivirine and tenofovir. *Eur. J. Pharm. Sci.* **2010**, *39*, 203–212. [CrossRef] [PubMed]
18. Ma, C.; Shi, Y.; Pena, D.A.; Peng, L.; Yu, G. Thermally responsive hydrogel blends: A general drug carrier model for controlled drug release. *Angew. Chem.* **2015**, *127*, 7484–7488. [CrossRef]
19. Elashnikov, R.; Slepička, P.; Rimpelova, S.; Ulbrich, P.; Švorčík, V.; Lyutakov, O. Temperature-responsive PLLA/PNIPAM nanofibers for switchable release. *Mater. Sci. Eng. C* **2017**, *72*, 293–300. [CrossRef] [PubMed]
20. Wang, Z.; Wu, F.; Zhao, P.; Dai, N.; Zhai, Z.; Ai, T. Improving cracking resistance of cement mortar by thermo-sensitive poly N-isopropyl acrylamide (PNIPAM) gels. *J. Clean. Prod.* **2018**, *176*, 1292–1303. [CrossRef]
21. Rikhtegar, F.; Athanasiou, L.; Edelman, E. Endovascular drug-delivery and drug-elution systems. In *Biomechanics of Coronary Atherosclerotic Plaque*; Academic Press: Cambridge, MA, USA, 2021; pp. 595–631. [CrossRef]
22. Bode, C.; Kranz, H.; Fivez, A.; Siepmann, F.; Siepmann, J. Often neglected: PLGA/PLA swelling orchestrates drug release: HME implants. *J. Control. Release* **2019**, *306*, 97–107. [CrossRef]
23. Abbasnezhad, N.; Zirak, N.; Shirinbayan, M.; Kouidri, S.; Salahinejad, E.; Tcharkhtchi, A.; Bakir, F. Controlled release from polyurethane films: Drug release mechanisms. *J. Appl. Polym. Sci.* **2020**, *138*, 50083. [CrossRef]
24. Abdel-Mottaleb, M.M.A.; Lamprecht, A. Standardized in vitro drug release test for colloidal drug carriers using modified USP dissolution apparatus I. *Drug Dev. Ind. Pharm.* **2011**, *37*, 178–184. [CrossRef] [PubMed]
25. Zambito, Y.; Pedreschi, E.; Di Colo, G. Is dialysis a reliable method for studying drug release from nanoparticulate systems?—A case study. *Int. J. Pharm.* **2012**, *434*, 28–34. [CrossRef] [PubMed]
26. Aldawsari, H.M.; Badr-Eldin, S.M.; Labib, G.S.; El-Kamel, A.H. Design and formulation of a topical hydrogel integrating lemongrass-loaded nanosponges with an enhanced antifungal effect: In vitro/in vivo evaluation. *Int. J. Nanomed.* **2015**, *10*, 893–902. [CrossRef]
27. Almeida, K.B.; Araujo, J.L.; Cavalcanti, J.F.; Romanos, M.T.V.; Mourão, S.C.; Amaral, A.C.F.; Falcão, D.Q. In vitro release and anti-herpetic activity of Cymbopogon citratus volatile oil-loaded nanogel. *Rev. Bras. Farmacogn.* **2018**, *28*, 495–502. [CrossRef]
28. Test Method for Anti-Mildew Agents in Controlling Wood Mould and Stain Fungi. GB/T 18261-2013. 2013. Available online: http://www.nssi.org.cn/nssi/front/83564233.html (accessed on 10 September 2021).
29. Nisha, C.K.; Dhara, D.; Chatterji, P.R. Superabsorbency and volume phase transition in crosslinked poly[[3-(methacryloylamino)propyl]-trimethylammonium chloride] hydrogels. *J. Macromol. Sci. Part. A* **2000**, *37*, 1447–1460. [CrossRef]
30. Natrajan, D.; Srinivasan, S.; Sundar, K.; Ravindran, A. Formulation of essential oil-loaded chitosan–alginate nanocapsules. *J. Food Drug Anal.* **2015**, *23*, 560–568. [CrossRef]

31. Jovanović, J.; Krnjajić, S.; Ćirković, J.; Radojković, A.; Popović, T.; Branković, G.; Branković, Z. Effect of encapsulated lemongrass (*Cymbopogon citratus* L.) essential oil against potato tuber moth Phthorimaea operculella. *Crop. Prot.* **2020**, *132*, 105109. [CrossRef]
32. Jiang, Y.; Yan, R.; Pang, B.; Mi, J.; Zhang, Y.; Liu, H.; Xin, J.; Zhang, Y.; Li, N.; Zhao, Y.; et al. A novel temperature-dependent hydrogel emulsion with Sol/Gel reversible phase transition behavior based on Polystyrene-co-poly(N-isopropylacrylamide)/Poly(N-isopropylacrylamide) core–shell nanoparticle. *Macromol. Rapid Commun.* **2021**, *42*, 2000507. [CrossRef]
33. Cao-Luu, N.-H.; Pham, Q.-T.; Yao, Z.-H.; Wang, F.-M.; Chern, C.-S. Synthesis and characterization of poly(N-isopropylacrylamide-co-acrylamide) mesoglobule core–silica shell nanoparticles. *J. Colloid Interface Sci.* **2019**, *536*, 536–547. [CrossRef] [PubMed]
34. Rocky, B.; Thompson, A. Analyses of the chemical compositions and structures of four bamboo species and their natural fibers by infrared, laser, and x-ray spectroscopies. *Fibers Polym.* **2021**, *22*, 916–927. [CrossRef]
35. Sirisomboon, P.; Funke, A.; Posom, J. Improvement of proximate data and calorific value assessment of bamboo through near infrared wood chips acquisition. *Renew. Energy* **2020**, *147*, 1921–1931. [CrossRef]
36. Haiyee, Z.A.; Saim, N.; Said, M.; Illias, R.M.; Mustapha, W.A.W.; Hassan, O. Characterization of cyclodextrin complexes with turmeric oleoresin. *Food Chem.* **2009**, *114*, 459–465. [CrossRef]

Article

Outdoor Wood Mats-Based Engineering Composite: Influence of Process Parameters on Decay Resistance against Wood-Degrading Fungi *Trametes versicolor* and *Gloeophyllum trabeum*

Minzhen Bao [1,*], Neng Li [1], Yongjie Bao [1], Jingpeng Li [1], Hao Zhong [1], Yuhe Chen [1] and Yanglun Yu [2,*]

[1] Key Laboratory of High Efficient Processing of Bamboo of Zhejiang Province, China National Bamboo Research Center, Hangzhou 310012, China; lineng8657@sina.com (N.L.); bomithen@126.com (Y.B.); lijp@caf.ac.cn (J.L.); zhonghao@caf.ac.cn (H.Z.); yuhec@sina.com (Y.C.)

[2] Research Institute of Wood Industry, Chinese Academy of Forestry, Beijing 100091, China

* Correspondence: baominzhen@caf.ac.cn (M.B.); yuyanglun@caf.ac.cn (Y.Y.)

Citation: Bao, M.; Li, N.; Bao, Y.; Li, J.; Zhong, H.; Chen, Y.; Yu, Y. Outdoor Wood Mats-Based Engineering Composite: Influence of Process Parameters on Decay Resistance against Wood-Degrading Fungi *Trametes versicolor* and *Gloeophyllum trabeum*. *Polymers* **2021**, *13*, 3173. https://doi.org/10.3390/polym13183173

Academic Editor: Antonios N. Papadopoulos

Received: 30 August 2021
Accepted: 17 September 2021
Published: 18 September 2021

Publisher's Note: MDPI stays neutral with regard to jurisdictional claims in published maps and institutional affiliations.

Copyright: © 2021 by the authors. Licensee MDPI, Basel, Switzerland. This article is an open access article distributed under the terms and conditions of the Creative Commons Attribution (CC BY) license (https://creativecommons.org/licenses/by/4.0/).

Abstract: The process parameters significantly influence the preparation and final properties of outdoor wood mats-based engineering composite (OWMEC). During outdoor use, wood composites are susceptible to destruction by rot fungi. Herein, the role of process parameters such as density and resin content on OWMEC resistance to fungal decay was investigated. The poplar OWMEC samples were exposed to white-rot fungus *Trametes versicolor* and brown-rot fungus *Gloeophyllum trabeum* for a period of 12 weeks. The chemical composition, crystallinity, and morphology were evaluated to investigate the effect of process parameters on the chemical composition and microstructure of the decayed OWMEC. With an increase in the density and resin content, the mass loss of the decayed OWMEC decreased. The highest antifungal effect against *T. versicolor* (12.34% mass loss) and *G. trabeum* (19.43% mass loss) were observed at a density of 1.15 g/m^3 and resin content of 13%. As results of the chemical composition and microstructure measurements, the resistance of OWMEC against *T. versicolor* and *G. trabeum* fungi was improved remarkably by increasing the density and resin content. The results of this study will provide a technical basis to improve the decay resistance of OWMEC in outdoor environments.

Keywords: outdoor wood mats-based engineering composite; wood-degrading fungi; decay resistance; durability

1. Introduction

Wood, as a renewable material, is in high demand for construction and building applications. It can be converted into engineered wood composites with standardized dimensions, such as wood oriented strand board, glue laminated wood, and reconstituted wood lumber. Scrimber, a promising type of engineered wood composite, is marketed as being moisture resistant and suitable for outdoor structural use [1]. Wood mats-based engineering composite (WMEC), as a novel engineered scrimber composite, consists of mechanically defibered wood fiber mats soaked with a phenolic resin and compacted to up to three times the original density of the wood using hot or cold pressing [2]. The development of WMECs refers to the defibration technology, bonding technology, and forming process [3,4].

During outdoor use, wood and its composites are susceptible to discoloration by mold fungi, destruction by rot fungi, and attack by insects. Biological attack limits the utilization of wood materials because of physical, chemical, and biological changes occurring on the surface and inside the material. Discoloration by mold fungi can cause high-value loss of wood composites when exposed to humid or moist conditions [5]. Biodeterioration by white and brown rot fungi can alter the chemical composition of wood [6]. Witomski et al. [7] reported that the bending strength and compressive strength of Scots pine wood can decrease by up to 20% with only 7% mass loss. The outdoor service life of wood composites is closely related to their deterioration under ambient conditions.

The degradation rate of wood composites is related to internal parameters such as the moisture, density, and resin content. Kataoka et al. [8] reported that the spreading rate and extent of fungal degradation were dependent on the density of the composite. Furthermore, accelerated weathering is more likely to occur in low-density composites [9]. Previous studies have indicated that the dimensional stability and photostabilization of wood composites can be improved by increasing the concentration of phenol formaldehyde (PF) resin [10,11]. To improve the durability of wood composites, the common industrial practice is to adjust the resin content to over 13% and density up to 1.10 g/cm^3. The mass loss of bamboo scrimber after 12 weeks of fungal erosion was less than 5% [12,13]. However, the decay resistance of WMECs has not been adequately investigated, especially for different densities and resin contents.

In order to expand the applicability and extend service lifetime of outdoor WEMC (OWMEC), more attention should be paid to its decay resistance. Impregnation of PF resin and compression of wood have been shown to improve the durability and dimensional stability of wood [14,15]. Hence, the objective of this study was to investigate the resistance of OWMECs with different densities and resin contents exposed to *Trametes versicolor* and *Gloeophyllum trabeum* fungi for a period of 12 weeks. The wet chemical method, Fourier transform infrared spectroscopy (FTIR), X-ray diffraction (XRD), and scanning electron microscopy (SEM) were used to analyze the changes in the chemical and physical properties of the OWMECs before and after fungal erosion. These analyses provide technical support for the outdoor application of OWMECs.

2. Materials and Methods

2.1. Materials

Poplar wood (*Populus canadensis* Moench) with a diameter of approximately 400 mm and a basic density of 0.39 g/cm^3 was purchased from Langfang Senyuan Wood Co., Ltd., Hebei, China. PF resin was purchased from Dynea Chemical Industry Co., Guangdong, China. The pH, solid content, and viscosity of the PF resin were 10.5 ± 0.2, $46.8\% \pm 1.0\%$, and $41.2\% \pm 2.0\%$, respectively.

2.2. Preparation of Outdoor Wood Mats-Based Engineering Composites (OWMECs)

The wood was manufactured into OWMEC according to a previously published method [2]. Briefly, the wood logs were peeled and split into fiber mats with a thickness of 6 mm. After air-drying, the mats were impregnated with PF resin to achieve a target resin content based on the weight of dried mats. Thereafter, the resin-impregnated wood fiber mats were air dried again to obtain a moisture content of 12 wt.%. Finally, the fiber mats were laminated along the grain in a mold and hot-pressed at 140 °C for 30 min to obtain the OWMEC ($400 \times 150 \times 18$ mm^3). OWMEC samples with different resin contents (8, 13, and 18 wt.%) were obtained with a density of 0.95 g/cm^3. By changing the mat weight, OWMEC samples with different densities (0.85, 1.00, and 1.15 g/cm^3) were also obtained with a resin content of 13 wt.%. All the OWMECs were conditioned in a room at 65% relative humidity (RH) and 20 °C for 2 weeks prior to testing.

2.3. Decay Resistance

The resistance of the OWMEC specimens and control specimens (poplar wood) against decay by white rot *T. versicolor* fungi and brown rot *G. trabeum* fungi was assessed in accordance with the Chinese Standard GB/T 13942.1-2009 [16]. The dimensions of each specimen were $20 \times 20 \times 10$ mm^3. Six samples were tested in parallel for each group. White-rot *T. versicolor* fungi or brown-rot *G. trabeum* fungi were inoculated on malt agar medium and pre-incubated for 10 days. Once the mycelium had covered the entire surface of the malt-agar medium, the test specimens were introduced on the sterile glass sticks for 12 weeks at 28 °C and 80% RH. After incubation, the decayed samples were gently cleaned to remove the mycelium adhered to the surface of the samples and oven-dried at 103 °C to constant weight. The percentage mass loss (ML) of the samples was calculated after 12 weeks.

$$ML = \frac{W_1 - W_2}{W_1} \times 100\% \tag{1}$$

where W_1 is the weight of samples before decay and W_2 is the weight of the samples after 12 weeks of incubation.

2.4. Chemical Analysis

Changes in the chemical composition of the healthy and decayed OWMEC samples were evaluated following the Chinese standards. The specimens were oven dried, milled, and sieved through a mesh with holes of 0.4 mm. Then, the contents of acid-insoluble lignin (GB/T 2677.8-1994) [17], holocellulose (GB/T 2677.10-1995) [18], and α-cellulose (GB/T 744-1989) [19] were determined.

2.5. Fourier Transform Infrared (FTIR) Analysis

FTIR spectra of the OWMECs before and after 12 weeks of fungal exposure were obtained using an FTIR spectrometer (Vertex 70, Bruker, Japan). KBr disks containing 1% of the finely ground samples were employed. Each spectrum was recorded in absorbance units from 1800 to 800 cm^{-1} as an average of 16 scans at a spectral resolution of 4 cm^{-1}.

2.6. X-ray Diffraction (XRD) Analysis

The crystalline structures of the samples were identified using a Bruker D8 Advance X-ray diffractometer equipped with a Cu Kα X-ray source (λ = 1.5404 Å) operated at 40 kV and 40 mA. The X-ray patterns were plotted within the range of 10–80° at a rate of 2° min^{-1}. The degree of crystallinity (Cr) was determined using the following equation:

$$Cr = \frac{A_{\text{crystalline}}}{A_{\text{total}}} \times 100\% \tag{2}$$

where $A_{\text{crystalline}}$ is the sum of all areas of crystallographic reflections and A_{total} is the total area of both the crystalline and amorphous contributions.

2.7. Scanning Electron Microscopy (SEM) Analysis

Small blocks (3 × 5 × 5 mm^3) were cut from OWMECs before and after fungal exposure. Sections of 20 μm were sliced off from the cross-section of each block using a sliding microtome until a smooth, clear surface was obtained. Then, all block surfaces were gold-coated and examined using a scanning electron microscope (Hitachi-S4800) at an accelerating voltage of 10 kV.

2.8. Statistical Analysis

One-way analysis of variance (ANOVA) was conducted to study the effect of process parameters on decay resistance of the OWEMCs at the 0.05 significance level. Duncan's tests were employed to multiply compare the properties of OWEMCs with different resin contents and densities.

3. Results and Discussion

3.1. Mass Loss Analysis

The appearance of the specimens exposed to *T. versicolor* and *G. trabeum* fungi is shown in Figures 1 and 2. The original shape of the poplar wood was severely altered, leaving only a small piece of wood. In contrast, OWMEC specimens retained their shape. White decay marks were occasionally observed on the surface of the OWMEC specimens. Furthermore, there were many cracks and holes on the surface of the OWMEC samples with low resin content and density. This indicates that an increase in the resin content and density can improve the decay resistance of OWMEC.

Figure 1. Photographs of wood reference samples (control) and outdoor wood mats-based engineering composite (OWMEC) samples with different resin contents (8%, 13%, and 18%) after 12 weeks of incubation. (**a**) *T. versicolor* decay; (**b**) *G. trabeum* decay.

Figure 2. Photographs of wood reference samples (control) and OWMEC samples with different densities (0.85, 1.00, and 1.15 g/cm^3) after 12 weeks of incubation. (**a**) *T. versicolor* decay; (**b**) *G. trabeum* decay.

Mass loss analysis can predict the potential performance loss of materials. The mass losses of the specimens after 12 weeks of incubation with *T. versicolor* and *G. trabeum* fungi are shown in Figures 3 and 4. For the reference poplar wood samples, mass losses of approximately 92.62% and 93.41% were observed for white- and brown-rot fungi, respectively. This demonstrates that poplar wood is easily destroyed by these fungi. In comparison, the mass loss of the OWMEC samples was significantly reduced, indicating that the fungal resistance is greatly improved when poplar is made into poplar OWMEC. The mass loss of the OWMEC samples was dependent on the resin content and density. As the density and resin content increased, the mass loss decreased. The OWMEC with a resin content of 18% and density of 0.95 g/m^3 exhibited mass losses as low as 23.24% and 27.88% after fungal attack by *T. versicolor* and *G. trabeum*, respectively. These values increased to 29.05% and 31.27%, respectively, when the resin content and density were 13% and 0.95 g/m^3, respectively. These results confirm the results of a previous study, which reported that PF resin and densification had a certain inhibitory effect on fungal decay [20]. The OWMEC with a density and resin content of 1.15 g/m^3 and 13%, respectively, exhibited mass losses of just 12.34% and 19.43% after 12 weeks of incubation with *T. versicolor* and *G. trabeum* fungi, respectively. Therefore, poplar OWMEC with a high resin content and density shows excellent corrosion resistance, making it suitable for use outdoors.

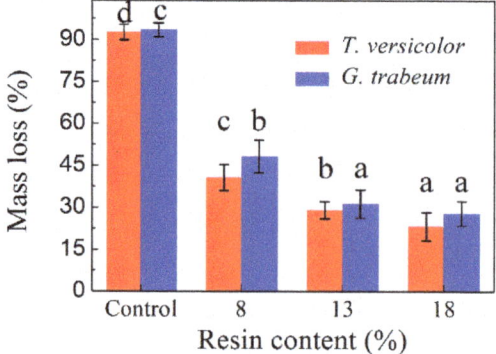

Figure 3. Mass loss of wood reference samples (control) and OWMEC samples with different resin contents after 12 weeks of incubation with *T. versicolor* and *G. trabeum* fungi. For each sample, value bars with the same letter (a, b, c, d) indicate no significant difference at the 0.05 level. Error bars represent the standard deviation.

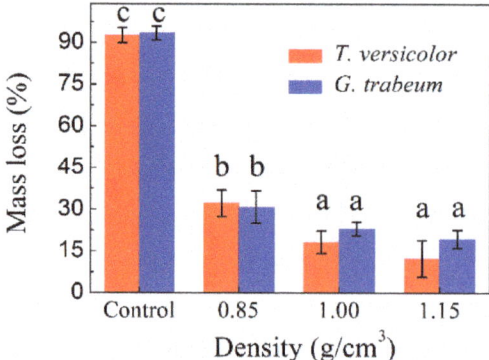

Figure 4. Mass loss of wood reference samples (control) and OWMEC samples with different densities after 12 weeks incubation with *T. versicolor* and *G. trabeum* fungi. For each sample, value bars with the same letter (a, b, c) indicate no significant difference at the 0.05 level. Error bars represent the standard deviation.

3.2. Chemical Analysis

Tables 1 and 2 show the holocellulose, α-cellulose, and acid-insoluble lignin content of the OWMEC samples before and after fungal decay. The percentage of holocellulose in the decayed OWMEC samples was lower than that in the healthy samples, whereas the percentage of acid-insoluble lignin was increased.

Table 1. Chemical compositions of OWMEC samples with different resin contents before and after 12 weeks of incubation with *T. versicolor* and *G. trabeum* fungi.

Resin Content (%)	Fungus	Holocellulose (%)	α-Cellulose (%)	Acid Insoluble Lignin (%)
8	-	68.47 (0.10) [c]	40.58 (0.17) [b]	25.09 (0.11) [a]
13	-	64.19 (0.21) [b]	38.61 (0.54) [a]	28.12 (0.16) [b]
18	-	62.64 (0.20) [a]	37.11 (0.13) [a]	31.26 (0.16) [c]
8	T. versicolor	63.13 (0.11) [c]	37.32 (0.08) [b]	28.86 (0.10) [a]
13	T. versicolor	61.53 (0.16) [b]	37.31 (0.14) [b]	33.41 (0.11) [b]
18	T. versicolor	60.46 (0.07) [a]	36.82 (0.10) [a]	34.44 (0.44) [c]
8	G. trabeum	60.25 (0.13) [a]	34.98 (0.20) [a]	31.16 (0.24) [a]
13	G. trabeum	60.39 (0.11) [a]	36.89 (0.23) [b]	33.75 (0.08) [b]
18	G. trabeum	60.11 (0.06) [a]	36.68 (0.10) [b]	35.48 (0.18) [c]

Values in parenthesis are standard deviations. For each parameter, average values with different letters ([a], [b], [c]) in each column indicate a significant difference at the 0.05 level (analysis of variance (ANOVA), followed by Duncan's multiple range test).

Table 2. Chemical compositions of OWMEC samples with different densities before and after 12 weeks of incubation with *T. versicolor* and *G. trabeum* fungi.

Density (g/cm^3)	Fungus	Holocellulose (%)	α-Cellulose (%)	Acid Insoluble Lignin (%)
0.85	-	63.78 (0.18) [a]	38.31 (0.13) [a]	29.63 (0.13) [c]
1.00	-	64.19 (0.21) [a]	38.61 (0.17) [a]	28.12 (0.16) [b]
1.15	-	65.32 (0.11) [b]	39.37 (0.08) [b]	27.31 (0.21) [a]
0.85	T. versicolor	60.29 (0.14) [a]	35.92 (0.18) [b]	32.48 (0.18) [a]
1.00	T. versicolor	63.55 (0.08) [c]	38.99 (0.16) [c]	32.01 (0.34) [a]
1.15	T. versicolor	62.18 (0.16) [b]	35.04 (0.11) [a]	32.64 (0.17) [a]
0.85	G. trabeum	59.32 (0.11) [a]	35.56 (0.13) [a]	32.52 (0.16) [a]
1.00	G. trabeum	61.24 (0.10) [c]	37.77 (0.13) [b]	32.47 (0.13) [a]
1.15	G. trabeum	60.90 (0.92) [b]	37.45 (0.06) [b]	33.18 (0.08) [b]

Values in parenthesis are standard deviations. For each parameter, average values with different letters ([a], [b], [c]) in each column indicate a significant difference at the 0.05 level (analysis of variance (ANOVA), followed by Duncan's multiple range test).

As shown in Table 1, the holocellulose and α-cellulose contents of the OWMEC samples decreased with increasing resin content, whereas the acid-insoluble lignin content increased. This is because the increase in resin content per unit volume and decrease in fiber content led to a decrease in the cellulose content. Regardless of the type of fungi, the holocellulose and α-cellulose contents of the decayed OWMEC decreased as the resin content increased, whereas the acid insoluble lignin increased. With an increase in the resin content, the holocellulose and α-cellulose contents in the *T. versicolor*-exposed OWMEC decreased, whereas the acid-insoluble lignin content increased. The holocellulose and α-cellulose contents in the *G. trabeum*-exposed OWMEC first increased and then decreased, whereas the acid-insoluble lignin gradually increased.

Table 2 shows that the holocellulose and α-cellulose contents of the OWMEC samples increased with increasing density, whereas the acid-insoluble lignin decreased. Regardless of the type of fungi, the holocellulose content of the decayed OWMEC decreased as the density increased, whereas the acid-insoluble lignin content increased.

Microorganisms change the chemical composition of wood during decay. Regardless of the type of fungi, the holocellulose content of the decayed OWMEC decreased as the resin content or density increased, whereas the acid insoluble lignin content increased. Both fungi can simultaneously decompose the major chemical components of wood cell walls [21–23]. However, the degradation rate of lignin was not as rapid as that of holocellulose, so the relative content of lignin showed an increasing trend. The amount of holocellulose decomposed by *G. trabeum* fungi was greater than that by *T. versicolor* fungi, indicating that *G. trabeum* has a higher ability to decompose holocellulose than *T. versicolor*. These observations are consistent with previous findings that brown-rot fungi primarily degrade holocellulose [24,25].

3.3. FTIR Analysis

Figures 5 and 6 show FTIR spectra of the OWMECs with different resin contents and densities before and after 12 weeks of fungal attack. The region from 1800 to 800 cm^{-1} is associated with various functional group characteristic of wood components (cellulose, hemicellulose, and lignin). The FTIR spectra of the healthy OWMEC specimen exhibited carbohydrate-associated bands at 1740, 1373, 1159, and 898 cm^{-1} and lignin-associated bands at 1600, 1510, 1462, 1425, 1333, and 1244 cm^{-1} [26,27]. Relative increases and decreases in the intensities of these characteristic absorption peaks indicate chemical changes in the OWMEC. After 12 weeks of decay, the intensity of each peak for each OWMEC sample decreased to some extent. However, the absorption intensity decreased slowly as the resin content or density increased. These results demonstrate that, not only were hemicellulose and lignin decomposed by fungi, but cellulose was degraded to different degrees. The results are similar to those in earlier studies, which showed that both white- and brown-rot fungi degrade carbohydrates and lignin in wood cell walls [28].

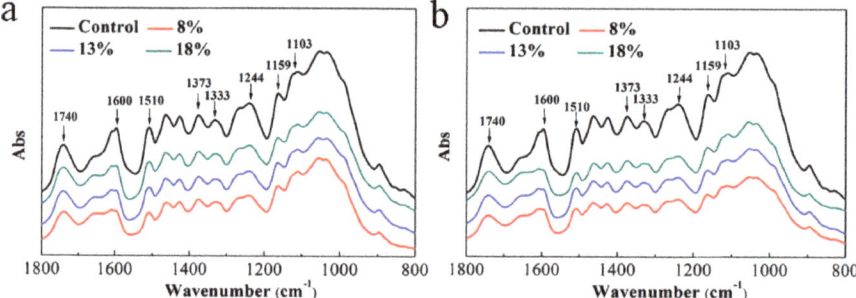

Figure 5. Fourier transform infrared (FTIR) spectra of healthy (control) and decayed OWMEC samples with different resin contents: (**a**) *T. versicolor* decay; (**b**) *G. trabeum* decay.

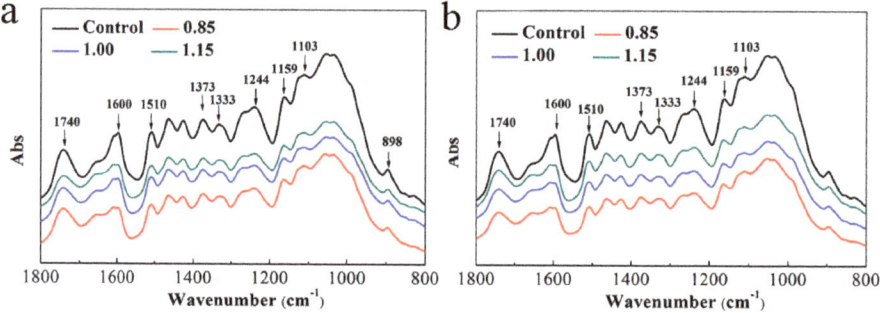

Figure 6. FTIR spectra of healthy (control) and decayed OWMEC samples with different densities: (**a**) *T. versicolor* decay; (**b**) *G. trabeum* decay.

In the FTIR spectra of the decayed OWMECs, the intensity of the peak near 1740 cm^{-1}, which corresponds to the C=O stretching vibration of the acetyl and carboxyl groups, was reduced. This indicates that the hemicellulose was degraded by both white- and brown-rot fungi. The intensities of the peaks located at 1600 and 1510 cm^{-1}, which represent the aromatic skeleton of lignin, decreased significantly, indicating that lignin was decomposed during the fungal decay tests. The intensities of the bands at 1373, 1159, and 898 cm^{-1}, which correspond mainly to polysaccharides (hemicellulose and cellulose), decreased after decay treatment. Moreover, the intensities of the peaks at 1333, 1244, and 1103 cm^{-1}, which are associated with lignin–carbohydrate complexes, also decreased. The change in the relative intensities of these peaks shows that the consumption of hemicellulose and lignin reached a new balance. The fungal decay-induced chemical changes to the lignin were greater in the OWMEC with a density of 0.85 g/cm^3 or resin content of 8% than in that with a density of 1.00 g/cm^3 or resin content of 13%. When the resin content was 18.0% or the density was 1.15 g/cm^3, the intensity of the aromatic ring structure of lignin in the *T. versicolor* or *G. trabeum*-decayed OWMEC did not change significantly.

3.4. XRD Analysis

The XRD patterns of the OWMECs presented in Figures 7 and 8 show sharp and strong diffraction peaks, indicating the crystalline nature of the OWMEC composites. The peaks at 16.3° and 22.5° were assigned to the (110) and (200) planes, respectively, revealing that the decayed OWMECs with different resin contents and densities possessed a typical wood phase. An increase in the signal of both the (110) and (200) peaks in decayed OWMECs can be observed in Figures 7 and 8. When the XRD intensity is normalized with the (200) peak, the valley at 2θ = 18° appears to be slightly lower for the decayed samples than that for the healthy (control) sample. The (200) diffraction peak of decayed OWMECs was narrower than that of the control OWMEC, indicating that the lattice structure of the cellulose crystal zone was destroyed during the decay process.

The *Cr* of the OWMECs increased with increasing resin content and density, regardless of the *T. versicolor* or *G. trabeum* decay treatment (see Tables 3 and 4). The *Cr* of the healthy (control) OWMEC was 16.34%. The *Cr* increased to 21.81% for the *T. versicolor*-decayed OWMEC with 18% resin content and 21.45% for the *G. trabeum*-decayed OWMEC with 18% resin content, indicating respective increases of 33.48% and 31.27% when compared to the control OWMEC. The *Cr* increased to 23.27% for the *T. versicolor*-decayed OWMEC with a density of 1.15 g/cm^3 and 23.79% for the *G. trabeum*-decayed OWMEC with a density of 1.15 g/cm^3, indicating respective increases of 42.41% and 45.59% when compared to the control OWMEC.

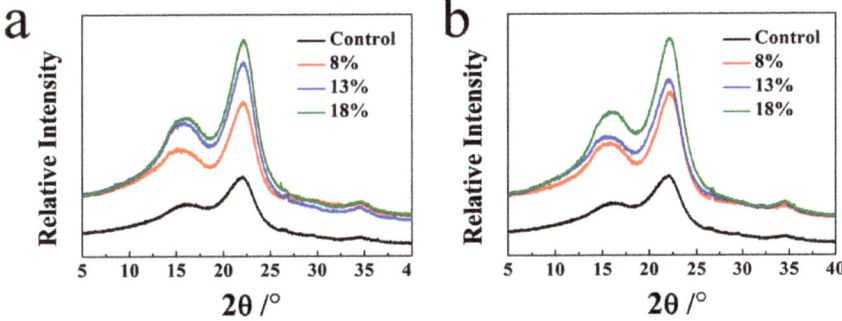

Figure 7. X-ray diffraction (XRD) patterns of healthy (control) and decayed OWMEC samples with different resin contents: (**a**) *T. versicolor* decay; (**b**) *G. trabeum* decay.

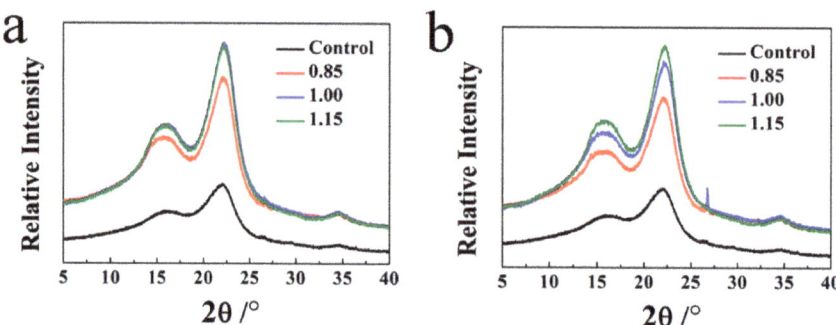

Figure 8. XRD patterns of healthy (control) and decayed OWMEC samples with different densities: (**a**) *T. versicolor* decay; (**b**) *G. trabeum* decay.

Table 3. Crystallinity (*Cr*) of healthy (control) and decayed OWMECs with different resin contents.

Resin Content (%)	Fungus	*Cr* (%)
Control	-	16.34
8.0		16.64
13.0	*T. versicolor*	21.10
18.0		21.81
8.0		18.65
13.0	*G. trabeum*	18.89
18.0		21.45

Table 4. Crystallinity (*Cr*) of healthy (control) and decayed OWMECs with different densities.

Density (g/cm^3)	Fungus	*Cr* (%)
Control	-	16.34
0.85		19.72
1.00	*T. versicolor*	22.81
1.15		23.27
0.85		19.21
1.00	*G. trabeum*	21.78
1.15		23.79

The XRD results showed that the *Cr* of the decayed OWMECs increased with increasing resin content and density. The greater *Cr* may be because the fungal degradation rate of hemicellulose is higher than that of cellulose. In contrast to hemicellulose, the decomposition of hydrogen bond-ordered cellulose is a complex procedure. Fackler et al. [29] found that amorphous polysaccharides are more susceptible than crystalline cellulose structures to fungal decay, which results in an increase in overall crystallinity. Furthermore, the higher the density, the higher the fiber content per unit volume.

3.5. SEM Analysis

Figure 9 shows the effect of the resin content on the microstructure of *T. versicolor*-decayed OWMEC. Exposure to *T. versicolor* fungi causes ruptures in the walls of all cell elements, such as vessels, fibers, and rays. The cell walls of the vessel of decayed OWMEC with 8% resin content formed rupture gaps and extended to the walls of other cells (Figure 9a). The wood ray was disintegrated, leaving only a small number of residual fragments (Figure 9b). The *T. versicolor* fungi attacked the fiber cells by thinning the cell walls and creating bore holes on the walls (Figure 9c). In contrast, the cell walls of the vessels of the decayed OWMEC with 18% resin content were not seriously damaged, and hyphae were occasionally found in the cells (Figure 9d). The fiber cells away from the vessel walls were found with intact walls, as shown in Figure 9d. A few cracks were observed in the ray cells (Figure 9e), and holes were observed between the fiber cells (Figure 9f). These results demonstrate that the vessels and rays were more vulnerable than the fibers, which were relatively resistant to fungal action. These relative differences are related to deviations in the cell wall thickness, in that the vessel and ray cell walls in poplar are much thinner than the fiber walls.

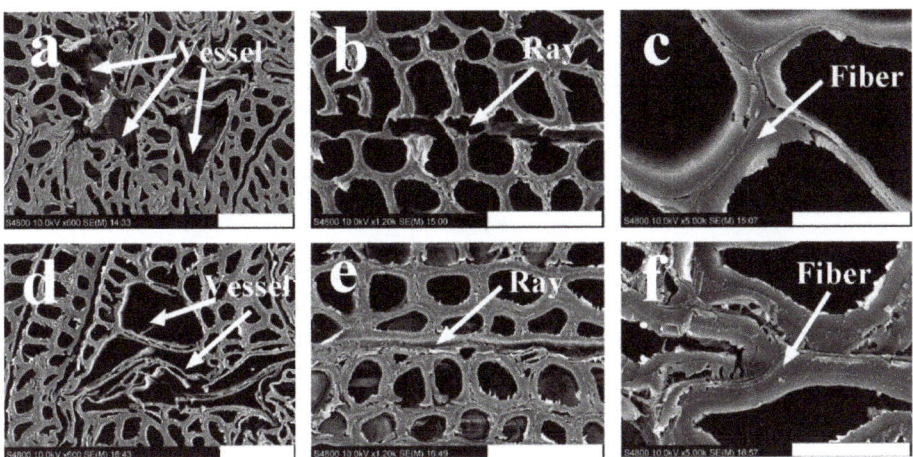

Figure 9. Scanning electron microscope (SEM) images of OWMEC with different resin contents after 12 weeks of incubation with *T. versicolor* fungi. (**a**–**c**) 8% resin content; (**d**–**f**) 18% resin content. Scale bars: (**a**,**d**) 50 μm; (**b**,**e**) 40 μm; (**c**,**f**) 10 μm.

Figure 10 shows the effect of density on the microstructure of *T. versicolor*-decayed OWMEC. Colonization of fungal hyphae within the vessel lumina and opening of the vessel walls were observed in the OWMEC with a density of 0.85 g/cm^3 (Figure 10a). The vessel cell walls contained several discontinuous gaps, indicating degradation by pit erosion. The rays and fibers were also deeply eroded by the *T. versicolor* fungi (Figure 10b,c). Erosion channels with a U-shaped incision appeared on the fiber cell walls, whereby two or more pores had fused together to form large pores on the fiber cell wall (Figure 10b). This illustrates the degradation of the cell walls. Figure 10c reveals that the thickness of the fiber wall decreased from the lumen side to the middle lamellae. Furthermore, erosion troughs formed in the fiber walls. Many large bore holes and loose fiber walls can also be observed in Figure 10c. Notably, in the OWMEC with a density of 1.15 g/cm^3, all of the cell types retained the compressed cell size and morphology of healthy OWMEC. The morphological changes to the OWMEC with a density of 1.15 g/cm^3 were not prominent, demonstrating that the extent of damage was greatly reduced.

The SEM images of the decayed OWMEC samples confirmed that the resin content and density were related to the decay resistance. In particular, a high resin content and density enhanced the decay resistance of the OWMEC. Phenolic resins containing various active groups react with the active groups of the cell walls of OWMEC to form stable cross-linking, which effectively improves the corrosion resistance [14,20,30]. The space between the cells was reduced in the higher-density OWMEC, indicating that the diffusion or penetration of hyphae and degrading enzymes was hindered. Moreover, the high density also increased the content of the fiber and phenolic resin, which effectively inhibited fungal corrosion.

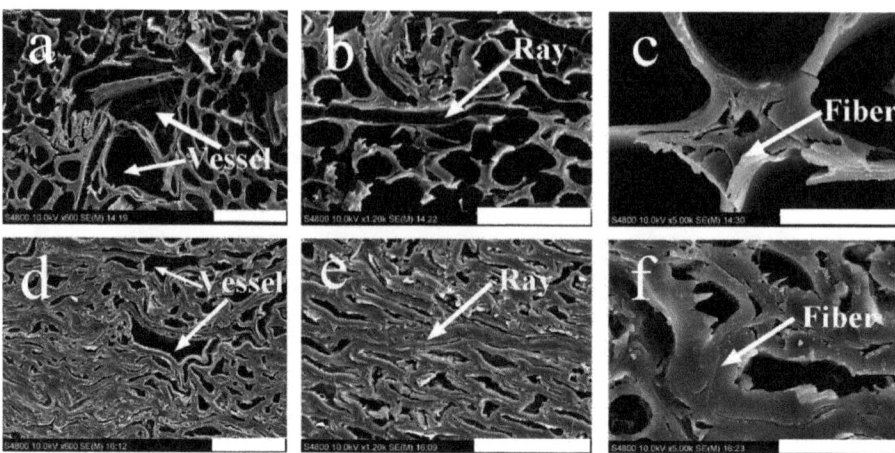

Figure 10. SEM images of OWMEC with different densities after 12 weeks of incubation with *T. versicolor* fungi. (**a–c**) 0.85 g/cm^3; (**d–f**) 1.15 g/cm^3. Scale bars: (**a,d**) 50 μm; (**b,e**) 40 μm; (**c,f**) 10 μm.

4. Conclusions

The effects of resin content and density on the resistance of OWMECs to fungal decay were investigated by fungal decay tests. Biological attack resulted in a loss of mass of OWMECs. Depending on the mass loss analysis, a decrease in mass from 32.20% to 12.34% for *T. versicolor* and 30.83% to 19.43% for *G. trabeum* was observed in the OWMECs (density from 0.85 g/cm^3 to 1.15 g/cm^3). A decrease in mass from 40.61% to 23.24% for *T. versicolor* and 48.19% to 27.88% for *G. trabeum* was also observed in the OWMECs (resin content from 8% to 18%). The decay resistance of the OWMEC could be enhanced by increasing the resin content or density. The chemical analysis and FTIR measurements showed that brown-rot fungus *G. trabeum* predominantly disintegrated the cellulose and hemicellulose, whereas white-rot fungus *T. versicolor* decayed both holocellulose and lignin; however, the chemical composition was less affected by fungal decal at a higher resin content and density. The Cr of the scrimbers increased with increasing resin content and density, regardless of the *T. versicolor* or *G. trabeum* decay treatment. The SEM results confirmed that the resin content and density were related to the decay resistance of OWMEC. In particular, a high resin content and density enhanced the decay resistance of the OWMEC. An appropriate process factor should be performed to improve the outdoor durability of the OWMEC.

Author Contributions: Investigation, M.B. and N.L.; writing-original draft preparation, M.B. and N.L.; writing–review and editing, Y.Y.; validation, Y.B., and J.L.; visualization, H.Z.; supervision, Y.C.; funding acquisition, M.B. All authors have read and agreed to the published version of the manuscript.

Funding: The authors appreciate the financial support from the Zhejiang Provincial Natural Science Foundation of China (LQ20C160001), Provincial Key Research Plan of Zhejiang Province (2021C02012) and the Fundamental Research Funds for the Non-profit Research Institution of Zhejiang Province (Study on Preparation and Application Technology of Bamboo Organic Anti-mildew Microcapsules).

Institutional Review Board Statement: Not applicable.

Informed Consent Statement: Not applicable.

Data Availability Statement: The data presented in this study are available on request from the corresponding author.

Conflicts of Interest: The authors declare no conflict of interest.

References

1. Zhang, Y.H.; Huang, Y.X.; Qi, Y.; Yu, W.J. Novel engineered scrimber with outstanding dimensional stability from finely fluffed poplar veneers. *Measurement* **2018**, *124*, 318–321. [CrossRef]
2. Bao, M.Z.; Li, N.; Huang, C.J.; Chen, Y.H.; Yu, W.J.; Yu, Y.L. Fabrication, physical–mechanical properties and morphological characterizations of novel scrimber composite. *Eur. J. Wood Wood Prod.* **2019**, *77*, 741–747. [CrossRef]
3. He, M.J.; Zhang, J.; Li, Z.; Li, M.L. Production and mechanical performance of scrimber composite manufactured from poplar wood for structural applications. *J. Wood Sci.* **2016**, *62*, 429–440. [CrossRef]
4. Zhang, Y.M.; Huang, X.A.; Zhang, Y.H.; Yu, Y.L.; Yu, W.J. Scrimber board (SB) manufacturing by a new method and characterization of SB's mechanical properties and dimensional stability. *Holzforschung* **2017**, *72*, 283–289. [CrossRef]
5. Kim, M.J.; Choi, Y.S.; Oh, J.J.; Kim, G.-H. Experimental investigation of the humidity effect on wood discoloration by selected mold and stain fungi for a proper conservation of wooden cultural heritages. *J. Wood Sci.* **2020**, *66*, 31. [CrossRef]
6. Karim, M.; Daryaei, M.G.; Torkaman, J.; Oladi, R.; Ghanbary, M.A.T.; Bari, E. In Vivo investigation of chemical alteration in oak wood decayed by *Pleurotus ostreatus*. *Int. Biodeterior. Biodegrad.* **2016**, *108*, 127–132. [CrossRef]
7. Witomski, P.; Olek, W.; Bonarski, J.T. Changes in strength of Scots pine wood (*Pinus silvestris* L.) decayed by brown rot (*Coniophora puteana*) and white rot (*Trametes versicolor*). *Constr. Build. Mater.* **2016**, *102*, 162–166. [CrossRef]
8. Kataoka, Y.; Kiguchi, M.; Fujiwara, T.; Evans, P.D. The effects of within-species and between-species variation in wood density on the photodegradation depth profiles of sugi (*Cryptomeria japonica*) and hinoki (*Chamaecyparis obtusa*). *J. Wood Sci.* **2005**, *51*, 531–536. [CrossRef]
9. Bao, M.Z.; Rao, F.; He, S.; Bao, Y.J.; Wu, Z.X.; Li, N.; Chen, Y.H. A note on the surface deterioration of scrimber composites exposed to artificial ageing. *Coatings* **2019**, *9*, 846. [CrossRef]
10. Zhang, Y.H.; Qi, Y.; Huang, Y.X.; Yu, Y.L.; Liang, Y.J.; Yu, W.J. Influence of veneer thickness, mat formation and resin content on some properties of novel poplar scrimbers. *Holzforschung* **2018**, *72*, 673–680. [CrossRef]
11. Evans, P.D.; Gibson, S.K.; Cullis, I.; Liu, C.; Sèbe, G. Photostabilization of wood using low molecular weight phenol formaldehyde resin and hindered amine light stabilizer. *Polym. Degrad. Stab.* **2013**, *98*, 158–168. [CrossRef]
12. Meng, F.; Liu, R.; Zhang, Y.; Huang, Y.; Yu, Y.; Yu, W. Improvement of the water repellency, dimensional stability, and biological resistance of bamboo-based fiber reinforced composites. *Polym. Compos.* **2019**, *40*, 506–513. [CrossRef]
13. Kumar, A.; Ryparovà, P.; Kasal, B.; Adamopoulos, S.; Hajek, P. Resistance of bamboo scrimber against white-rot and brown-rot fungi. *Wood Mater. Sci. Eng.* **2020**, *15*, 57–63. [CrossRef]
14. Bakar, E.S.; Hao, J.; Ashaari, Z.; Choo Cheng Yong, A. Durability of phenolic-resin-treated oil palm wood against subterranean termites a white-rot fungus. *Int. Biodeterior. Biodegrad.* **2013**, *85*, 126–130. [CrossRef]
15. Bao, M.Z.; Huang, X.A.; Zhang, Y.H.; Yu, W.J.; Yu, Y.L. Effect of density on the hygroscopicity and surface characteristics of hybrid poplar compreg. *J. Wood Sci.* **2016**, *62*, 441–454. [CrossRef]
16. *Durability of Wood—Part 1: Method for Laboratory Test of Natural Decay Resistance (GB/T13942.1-2009)*; Standard Administration of China: Beijing, China, 2009; pp. 1–5.
17. *Fibrous Raw Material—Determination of Acid-Insoluble Lignin (GB/T 2677.8-1994)*; Standard Administration of China: Beijing, China, 1994; pp. 213–215.
18. *Fibrous Raw Material—Determination of Holocellulose (GB/T 2677.10-1995)*; Standard Administration of China: Beijing, China, 1995; pp. 220–223.
19. *Pulps-Determination of A-Cellulose (GB/T 744-1989)*; Standard Administration of China: Beijing, China, 1989; pp. 156–158.
20. Freitag, C.; Kamke, F.A.; Morrell, J.J. Resistance of resin-impregnated VTC processed hybrid-poplar to fungal attack. *Int. Biodeterior. Biodegrad.* **2015**, *99*, 174–176. [CrossRef]
21. Bari, E.; Schmidt, O.; Oladi, R. A histological investigation of oriental beech wood decayed by *Pleurotus ostreatus* and *Trametes versicolor*. *For. Pathol.* **2015**, *45*, 349–357. [CrossRef]
22. Bari, E.; Nazarnezhad, N.; Kazemi, S.M.; Ghanbary, M.A.T.; Mohebby, B.; Schmidt, O.; Clausen, C.A. Comparison between degradation capabilities of the white rot fungi *Pleurotus ostreatus* and *Trametes versicolor* in beech wood. *Int. Biodeterior. Biodegrad.* **2015**, *104*, 231–237. [CrossRef]
23. Bari, E.; Taghiyari, H.R.; Naji, H.R.; Schmidt, O.; Ohno, K.M.; Clausen, C.A.; Bakar, E.S. Assessing the destructive behaviors of two white-rot fungi on beech wood. *Int. Biodeterior. Biodegrad.* **2016**, *114*, 129–140. [CrossRef]
24. Kirk, T.; Highley, T. Quantitative changes in structural components of conifer woods during decay by white-and brown-rot fungi. *Phytopathology* **1973**, *63*, 1338–1342. [CrossRef]
25. Irbe, I.; Andersons, B.; Chirkova, J.; Kallavus, U.; Andersone, I.; Faix, O. On the changes of pinewood (*Pinus sylvestris* L.) chemical composition and ultrastructure during the attack by brown-rot fungi *Postia placenta* and *Coniophora puteana*. *Int. Biodeterior. Biodegrad.* **2006**, *57*, 99–106. [CrossRef]
26. Pandey, K. A study of chemical structure of soft and hardwood and wood polymers by FTIR spectroscopy. *J. Appl. Polym. Sci.* **1999**, *71*, 1969–1975. [CrossRef]
27. Pandey, K.; Pitman, A. FTIR studies of the changes in wood chemistry following decay by brown-rot and white-rot fungi. *Int. Biodeterior. Biodegrad.* **2003**, *52*, 151–160. [CrossRef]
28. Schwarze, F.W.M.R. Wood decay under the microscope. *Fungal Biol. Rev.* **2007**, *21*, 133–170. [CrossRef]

29. Fackler, K.; Schwanninger, M. Polysaccharide degradation and lignin modification during brown rot of spruce wood: A polarised Fourier transform near infrared study. *J. Near Infrared Spectrosc.* **2010**, *18*, 403–416. [CrossRef]
30. Nabil, F.; Zaidon, A.; Anwar, U.; Bakar, E.; Lee, S.; Paridah, M. Impregnation of sesenduk (*Endospermum diadenum*) wood with phenol formaldehyde and nanoclay admixture: Effect on fungal decay and termites attack. *Sains Malays.* **2016**, *45*, 255–262.

MDPI
St. Alban-Anlage 66
4052 Basel
Switzerland
www.mdpi.com

Polymers Editorial Office
E-mail: polymers@mdpi.com
www.mdpi.com/journal/polymers

Disclaimer/Publisher's Note: The statements, opinions and data contained in all publications are solely those of the individual author(s) and contributor(s) and not of MDPI and/or the editor(s). MDPI and/or the editor(s) disclaim responsibility for any injury to people or property resulting from any ideas, methods, instructions or products referred to in the content.